Capacity Theory with Local Rationality

The Strong Fekete-Szegö Theorem on Curves

Mathematical
Surveys
and
Monographs

Volume 193

Capacity Theory with Local Rationality

The Strong Fekete-Szegö Theorem on Curves

Robert Rumely

American Mathematical Society
Providence, Rhode Island

2010 *Mathematics Subject Classification.* Primary 11G30, 14G40, 14G05;
Secondary 31C15.

This work was supported in part by NSF grants DMS 95-000892, DMS 00-70736, DMS 03-00784, and DMS 06-01037. Any opinions, findings and conclusions or recommendations expressed in this material are those of the author and do not necessarily reflect the views of the National Science Foundation.

For additional information and updates on this book, visit
www.ams.org/bookpages/surv-193

Library of Congress Cataloging-in-Publication Data

Rumely, Robert, 1952– author.
Capacity theory with local rationality : the strong Fekete-Szegö theorem on curves / Robert Rumely.
pages cm – (Mathematical surveys and monographs ; volume 193)
Includes bibliographical references and index.
ISBN 978-1-4704-0980-7 (alk. paper)
1. Curves, Algebraic. 2. Arithmetical algebraic geometry. I. Title.

QA565.R86 2014
512.7′4–dc23 2013034694

∞ The paper used in this book is acid-free and falls within the guidelines
established to ensure permanence and durability.
Visit the AMS home page at http://www.ams.org/

10 9 8 7 6 5 4 3 2 1 18 17 16 15 14 13

To Cherilyn, who makes me happy.

Contents

Introduction

The prototype for the Fekete-Szegö theorem with local rationality is Raphael Robinson's theorem on totally real algebraic integers in an interval:

Theorem (Robinson [**48**], 1964). *Let* $[a, b] \subset \mathbb{R}$. *If* $b - a > 4$, *then there are infinitely many totally real algebraic integers whose conjugates all belong to* $[a, b]$. *If* $b - a < 4$, *there are only finitely many.*

Robinson also gave a criterion for the existence of totally real units in $[a, b]$:

Theorem (Robinson [**49**], 1968). *Suppose* $0 < a < b \in \mathbb{R}$ *satisfy the conditions*

$$\log(\frac{b-a}{4}) > 0 , \tag{0.1}$$

$$\log(\frac{b-a}{4}) \cdot \log(\frac{b-a}{4ab}) - \Big(\log(\frac{\sqrt{b}+\sqrt{a}}{\sqrt{b}-\sqrt{a}})\Big)^2 > 0 . \tag{0.2}$$

Then there are infinitely many totally real units α *whose conjugates all belong to* $[a, b]$. *If either inequality is reversed, there are only finitely many.*

David Cantor's "Fekete-Szegö theorem with splitting conditions" on $\mathbb{P}^1$ ([**14**], Theorem 5.1.1, 1980) generalized Robinson's theorems, reformulated them adelically, and set them in a potential-theoretic framework.

In this work we prove a strong form of Cantor's result, valid for algebraic curves of arbitrary genus over global fields of any characteristic.

Let K be a global field, a number field or a finite extension of $\mathbb{F}_p(T)$ for some prime p. Let $\widetilde{K}$ be a fixed algebraic closure of K, and let $\widetilde{K}^{\mathrm{sep}} \subseteq \widetilde{K}$ be the separable closure of K. We will write $\mathrm{Aut}(\widetilde{K}/K)$ for the group of automorphisms $\mathrm{Aut}(\widetilde{K}/K) \cong \mathrm{Gal}(\widetilde{K}^{\mathrm{sep}}/K)$. Let $\mathcal{M}_K$ be the set of all places of K. For each $v \in \mathcal{M}_K$, let K_v be the completion of K at v, let $\widetilde{K}_v$ be an algebraic closure of K_v, and let $\mathbb{C}_v$ be the completion of $\widetilde{K}_v$. We will write $\mathrm{Aut}_c(\mathbb{C}_v/K_v)$ for the group of continuous automorphisms of $\mathbb{C}_v/K_v$; thus $\mathrm{Aut}_c(\mathbb{C}_v/K_v) \cong \mathrm{Aut}(\widetilde{K}_v/K_v) \cong \mathrm{Gal}(\widetilde{K}_v^{\mathrm{sep}}/K_v)$.

Let $\mathcal{C}/K$ be a smooth, geometrically integral, projective curve. If F is a field containing K, put $\mathcal{C}_F = \mathcal{C} \times_K \mathrm{Spec}(F)$ and let $\mathcal{C}(F) = \mathrm{Hom}_F(\mathrm{Spec}(F), \mathcal{C}_F)$ be the set of F-rational points; let $F(\mathcal{C})$ be the function field of $\mathcal{C}_F$. When $F = K_v$, we write $\mathcal{C}_v$ for $\mathcal{C}_{K_v}$. Let $\mathfrak{X} = \{x_1, \ldots, x_m\}$ be a finite, Galois-stable set points of $\mathcal{C}(\widetilde{K})$, and let $\mathbb{E} = \mathbb{E}_K = \prod_{v \in \mathcal{M}_K} E_v$ be a K-rational adelic set for $\mathcal{C}$, that is, a product of sets $E_v \subset \mathcal{C}_v(\mathbb{C}_v)$ such that each E_v is stable under $\mathrm{Aut}_c(\mathbb{C}_v/K_v)$. For each v, fix an embedding $\widetilde{K} \hookrightarrow \mathbb{C}_v$ over K, inducing an embedding $\mathcal{C}(\widetilde{K}) \hookrightarrow \mathcal{C}_v(\mathbb{C}_v)$. In this way $\mathfrak{X}$ can be regarded as a subset of $\mathcal{C}_v(\mathbb{C}_v)$: since $\mathfrak{X}$ is Galois-stable, its image is independent of the choice of embedding. The same is true for any Galois-stable set of points in $\mathcal{C}(\widetilde{K})$, such as the set of $\mathrm{Aut}(\widetilde{K}/K)$-conjugates of a point $\alpha \in \mathcal{C}(\widetilde{K})$.

We will call a set $E_v \subset \mathcal{C}_v(\mathbb{C}_v)$ an RL-domain ("Rational Lemniscate Domain") if there is a nonconstant rational function $f_v(z) \in \mathbb{C}_v(\mathcal{C}_v)$ such that $E_v = \{z \in \mathcal{C}_v(\mathbb{C}_v) : |f_v(z)|_v \leq 1\}$. This terminology is due to Cantor. By combining ([**26**], Satz 2.2) with ([**51**], Corollary 4.2.14), one sees that a set is an RL-domain if and only if it is a strict affinoid subdomain of $\mathcal{C}_v(\mathbb{C}_v)$, in the sense of rigid analysis.

Fix an embedding $\mathcal{C} \hookrightarrow \mathbb{P}^N_K = \mathbb{P}^N/\operatorname{Spec}(K)$ for an appropriate N, and equip $\mathbb{P}^N_K$ with a (K-rational) system of homogeneous coordinates. For each nonarchimedean v, this data determines a model $\mathfrak{C}_v/\operatorname{Spec}(\mathcal{O}_v)$. There is a natural metric $\|x, y\|_v$ on $\mathbb{P}^N_v(\mathbb{C}_v)$: the chordal distance associated to the Fubini-Study metric, if v is archimedean; the v-adic spherical metric, if v is nonarchimedean (see §3.4 below). The metric $\|x, y\|_v$ induces the v-topology on $\mathcal{C}_v(\mathbb{C}_v)$. Given $a \in \mathcal{C}_v(\mathbb{C}_v)$ and $r > 0$, we write $B(a, r)^- = \{z \in \mathcal{C}_v(\mathbb{C}_v) : \|z, a\|_v < r\}$ and $B(a, r) = \{z \in \mathcal{C}_v(\mathbb{C}_v) : \|z, a\|_v \leq r\}$ for the corresponding "open" and "closed" balls.

Sets in $\mathcal{C}_v(\mathbb{C}_v)$ that are well-behaved for capacity theory are called *algebraically capacitable* (see Definition 3.18 below). Finite unions of RL-domains and compact sets are algebraically capacitable (in the nonarchimedean case, this is follows from [**51**], Corollary 4.2.14 and Theorem 4.3.11). For the Fekete-Szegö theorem with local rationality, we need to restrict to a smaller class of sets:

DEFINITION 0.1. If $v \in \mathcal{M}_K$, and $E_v \subset \mathcal{C}_v(\mathbb{C}_v)$ is nonempty and stable under $\operatorname{Aut}_c(\mathbb{C}_v/K_v)$, we will say that E_v has a *finite K_v-primitive cover* if it can be written as a finite union $E_v = \bigcup_{\ell=1}^M E_{v,\ell}$, where

(A) If v is archimedean and $K_v \cong \mathbb{C}$, then each $E_{v,\ell}$ is compact, connected, and bounded by finitely many Jordan curves.

(B) If v is archimedean and $K_v \cong \mathbb{R}$, then each $E_{v,\ell}$ is either
- (1) compact, connected, and bounded by finitely many Jordan curves, or
- (2) is a closed subinterval of $\mathcal{C}_v(\mathbb{R})$ with nonempty interior.

(C) If v is nonarchimedean, then each $E_{v,\ell}$ is either
- (1) an RL-domain,
- (2) a ball $B(a_\ell, r_\ell)$ with radius r_ℓ in the value group of $\mathbb{C}_v^\times$, or
- (3) is compact, and has the form $\mathcal{C}_v(F_{w,\ell}) \cap D_v$ for some ball or RL-domain D_v, and some finite separable extension $F_{w,\ell}/K_v$.

Note that the sets $E_{v,\ell}$ can overlap, sets $E_{v,\ell}$ of more than one type can occur for a given v, and the extensions $F_{w,\ell}/K_v$ need not be Galois.

DEFINITION 0.2. If v is a nonarchimedean place of K, a set $E_v \subset \mathcal{C}_v(\mathbb{C}_v)$ will be called *$\mathfrak{X}$-trivial* if $\mathfrak{C}_v$ has good reduction at v, if the points of $\mathfrak{X}$ specialize to distinct points (mod v), and if $E_v = \mathcal{C}_v(\mathbb{C}_v) \backslash \bigcup_{i=1}^m B(x_i, 1)^-$.

If E_v is $\mathfrak{X}$-trivial, it consists of all points of $\mathcal{C}_v(\mathbb{C}_v)$ which are $\mathfrak{X}$-integral at v for the model $\mathfrak{C}_v$, i.e., which specialize to points complementary to $\mathfrak{X}$ (mod v). In particular, it is an RL-domain and is stable under $\operatorname{Aut}_c(\mathbb{C}_v/K_v)$.

DEFINITION 0.3. An adelic set $\mathbb{E} = \prod_{v \in \mathcal{M}_K} E_v \subset \prod_{v \in \mathcal{M}_K} \mathcal{C}_v(\mathbb{C}_v)$ will be called *K-rational* if each E_v is stable under $\operatorname{Aut}_c(\mathbb{C}_v/K_v)$. It will be called *compatible with $\mathfrak{X}$* if the following conditions hold:

(1) Each E_v is bounded away from $\mathfrak{X}$ in the v-topology.

(2) For all but finitely many v, E_v is $\mathfrak{X}$-trivial.

The properties of K-rationality and compatibility with $\mathfrak{X}$ are independent of the choice of projective embedding of $\mathcal{C}$.

When each E_v is algebraically capacitable, there is a potential-theoretic measure of size for the adelic set $\mathbb{E}$ relative to the set of global points $\mathfrak{X}$: the Cantor capacity $\gamma(\mathbb{E}, \mathfrak{X})$, defined in formula (0.10) below. Our main result is:

THEOREM 0.4 (The Fekete-Szegö Theorem with Local Rationality on Curves). *Let K be a global field, and let $\mathcal{C}/K$ be a smooth, geometrically integral, projective curve. Let $\mathfrak{X} = \{x_1, \ldots, x_m\} \subset \mathcal{C}(\widetilde{K})$ be a finite set of points stable under $\mathrm{Aut}(\widetilde{K}/K)$, and let $\mathbb{E} = \prod_v E_v \subset \prod_v \mathcal{C}_v(\mathbb{C}_v)$ be a K-rational adelic set compatible with $\mathfrak{X}$. (Thus, each E_v is bounded away from $\mathfrak{X}$ and stable under $\mathrm{Aut}_c(\mathbb{C}_v/K_v)$, and E_v is $\mathfrak{X}$-trivial for all but finitely many v.) Let $S \subset \mathcal{M}_K$ be a finite set of places containing all archimedean v and all nonarchimedean v such that E_v is not $\mathfrak{X}$-trivial. Assume that:*

(A) For each $v \in S$, E_v has a finite K_v-primitive cover.

(B) $\gamma(\mathbb{E}, \mathfrak{X}) > 1$.

Then there are infinitely many points $\alpha \in \mathcal{C}(\widetilde{K}^{\mathrm{sep}})$ such that for each $v \in \mathcal{M}_K$, the $\mathrm{Aut}(\widetilde{K}/K)$-conjugates of α all belong to E_v.

The primary content of the theorem is the local rationality assertion (the fact that the conjugates belong to E_v, for each v); the Fekete-Szegö theorem without local rationality, which constructs points α whose conjugates belong to arbitrarily small $\mathcal{C}_v(\mathbb{C}_v)$-neighborhoods of E_v, was proved in ([**51**], Theorem 6.3.2). In §2.4 we provide examples due to Daeshik Park, showing the need for the hypothesis of separability for the extensions $F_{w,\ell}/K_v$ in part (C) of the definition of a finite K_v-primitive cover.

Suppose that in the theorem, for each $v \in S$ we have $E_v \subset \mathcal{C}_v(K_v)$. Then for each $v \in S$, the conjugates of α belong to $\mathcal{C}_v(K_v)$, which means that v splits completely in $K(\alpha)$. In this case, following ([**52**]) and ([**53**]), we speak of "the Fekete-Szegö theorem with splitting conditions".

Sometimes it is the corollaries of a theorem, which are weaker but easier to apply, that are most useful. The following corollary of Theorem 0.4 strengthens Moret-Bailly's theorem for "Incomplete Skolem Problems on Affine Curves" ([**39**], Théorème 1.3, p.182), but does not require evaluating capacities.

Suppose $\mathcal{A}/K$ is an affine curve, embedded in $\mathbb{A}^N$ for some N. Let $z_1, \ldots, z_N$ be the coordinates on $\mathbb{A}^N$; given $v \in \mathcal{M}_K$ and a point $P \in \mathcal{A}(\mathbb{C}_v)$, write $\|P\|_{\mathcal{A},v} = \max(|z_1(P)|_v, \ldots, |z_N(P)|_v)$. We will say that a set $E_v \subset \mathcal{A}(\mathbb{C}_v)$ has a finite K_v-primitive cover relative to $\mathcal{A}$ if it is bounded under $\|\cdot\|_{\mathcal{A},v}$ and satisfies the conditions of Definition 0.1 with $\mathcal{C}$ replaced by $\mathcal{A}$, using balls for the metric $\|x-y\|_{\mathcal{A},v}$.

COROLLARY 0.5 (Fekete-Szegö for Skolem Problems on Affine Curves). *Let K be a global field, and let $\mathcal{A}/K$ be a geometrically integral (possibly singular) affine curve, embedded in $\mathbb{A}^N$. Fix a place v_0 of K, and let $S \subset \mathcal{M}_K \backslash \{v_0\}$ be a finite set of places containing all archimedean $v \neq v_0$. For each $v \in S$, let $E_v \subset \mathcal{A}_v(\mathbb{C}_v)$ be nonempty and stable under $\mathrm{Aut}_c(\mathbb{C}_v/K_v)$, with a finite K_v-primitive cover relative to $\mathcal{A}$. Assume that for each $v \in \mathcal{M}_K \backslash (S \cup \{v_0\})$ there is a point $P \in \mathcal{A}(\mathbb{C}_v)$ with $\|P\|_{\mathcal{A},v} \leq 1$. Then there is a bound $R = R(\mathcal{A}, \{E_v\}_{v \in S}, v_0) < \infty$ such that there are infinitely many points $\alpha \in \mathcal{A}(\widetilde{K}^{\mathrm{sep}})$ for which*

(1) for each $v \in S$, all the $\mathrm{Aut}(\widetilde{K}/K)$-conjugates $\sigma(\alpha)$ belong to E_v;

(2) for each $v \in \mathcal{M}_K \backslash (S \cup \{v_0\})$, all the conjugates satisfy $\|\sigma(\alpha)\|_{\mathcal{A},v} \leq 1$;

(3) for $v = v_0$, all the conjugates satisfy $\|\sigma(\alpha)\|_{\mathcal{A},v_0} \leq R$.

In Chapter 1 below, we will give several variants of Theorem 0.4, including one involving "quasi-neighborhoods" analogous to the classical theorem of Fekete and Szegö, one for more general sets $\mathbb{E}$ using the inner Cantor capacity $\overline{\gamma}(\mathbb{E}, \mathfrak{X})$, and two for sets on Berkovich curves. Theorem 0.4, Corollary 0.5, and the variants in Chapter 1 will be proved in Chapter 4.

Some History

The original theorem of Fekete and Szegö ([**25**], 1955) said that if $E \subset \mathbb{C}$ was compact and stable under complex conjugation, with logarithmic capacity $\gamma_\infty(E) > 1$, then every neighborhood U of E contained infinitely many conjugates sets of algebraic integers. (The neighborhood U was needed to 'fatten' sets like a circle $E = C(0, r)$ with transcendental radius r, which contain no algebraic numbers.)

A decade later Raphael Robinson gave the generalizations of the Fekete-Szegö theorem for totally real algebraic integers and totally real units stated above. Independently, Bertrandias gave an adelic generalization of the Fekete-Szegö theorem concerning algebraic integers with conjugates near sets E_p at a finite number of p-adic places as well as the archimedean place (see Amice [**3**], 1975).

In the 1970s David Cantor carried out an investigation of capacities on $\mathbb{P}^1$ dealing with all three themes: incorporating local rationality conditions, requiring integrality with respect to multiple poles, and formulating the theory adelically. In a series of papers culminating with ([**16**], 1980), he introduced the Cantor capacity $\gamma(\mathbb{E}, \mathfrak{X})$, which he called the extended transfinite diameter.

Cantor's capacity $\gamma(\mathbb{E}, \mathfrak{X})$ is defined by means of a minimax property which encodes a finite collection of linear inequalities; its definition is given in (0.10) below. The points in $\mathfrak{X}$ will be called the *poles* for the capacity. In the special case where $C = \mathbb{P}^1$ and $\mathfrak{X} = \{0, \infty\}$, Cantor's conditions are equivalent to those in Robinson's unit theorem. Among the applications Cantor gave in ([**16**]) were generalizations of the Pólya-Carlson theorem and Fekete's theorem, and the Fekete-Szegö theorem with splitting conditions. Unfortunately, as noted in ([**53**]), the part of the proof concerning the satisfiability of the splitting conditions had errors. However, many of Cantor's ideas are used in this work.

In the 1980's the author ([**51**]) extended Cantor's theory to curves of arbitrary genus, and proved the Fekete-Szegö theorem on curves, *without* splitting conditions. As an application he obtained a local-global principle for the existence of algebraic integer points on absolutely irreducible affine algebraic varieties ([**55**]), which had been conjectured by Cantor and Roquette ([**17**]).

Laurent Moret-Bailly and Lucien Szpiro recognized that the theory of capacities (which imposes conditions at all places) was stronger than was needed for the existence of integral points. They reformulated the local-global principle in scheme-theoretic language as an "Existence Theorem" for algebraic integer points, and gave a much simpler proof. Moret-Bailly subsequently gave far-reaching generalizations of the Existence Theorem ([**38**], [**39**], [**40**]), which allowed imposition of F_w-rationality conditions at a finite number of places, for a finite Galois extension F_w/K_v, and applied to algebraic stacks as well as schemes. However, the method required that there be at least one place v_0 where no conditions are imposed. Roquette, Green, and Pop ([**50**]) independently proved the Existence Theorem with F_w-rationality conditions, and Green, Matignon, and Pop ([**30**]) have given very general conditions on the base field K for such theorems to hold. The author ([**55**]), van den Dries ([**66**]), Prestel and Schmid ([**47**]), and others have given applications of these results to decision procedures in mathematical logic.

Recently Akio Tamagawa ([**63**]) proved an extension of the Existence Theorem in characteristic p, which produces points that are unramified outside v_0 and the places where the F_w-rationality conditions are imposed.

The Fekete-Szegö theorem with local rationality conditions constructs algebraic numbers satisfying conditions at *all* places. At its core it is analytic in character, while the Existence Theorem is algebraic. The proof of the Fekete-Szegö theorem involves a process called "patching", which takes an initial collection of local functions $f_v(z) \in K_v(\mathcal{C})$ with poles supported on $\mathfrak{X}$ and roots in E_v for each v, and constructs a global function $G(z) \in K(\mathcal{C})$ (of much higher degree) with poles supported on $\mathfrak{X}$, whose roots belong to E_v for all v. In his doctoral thesis, Pascal Autissier ([**6**]) gave a reformulation of the patching process in the context of Arakelov theory.

In ([**52**], [**53**]) the author proved the Fekete-Szegö theorem with splitting conditions for sets $\mathbb{E}$ in $\mathbb{P}^1$, when $\mathfrak{X} = \{\infty\}$. Those papers developed a method for carrying out the patching process in the p-adic compact case, and introduced a technique for patching together archimedean and nonarchimedean polynomials over number fields.

When $\mathcal{C} = \mathbb{P}^1/K$, with K a finite extension of $\mathbb{F}_p(T)$, the Fekete-Szegö theorem with splitting conditions was established in the doctoral thesis of Daeshik Park ([**45**]).

A Sketch of the Proof of the Fekete-Szegö Theorem

In outline, the proof of the classical Fekete-Szegö theorem ([**25**], 1955) is as follows. Let a compact set $E \subset \mathbb{C}$ and a complex neighborhood U of E be given. Assume E is stable under complex conjugation, and has logarithmic capacity $\gamma_\infty(E) > 1$. For simplicity, assume also that the boundary of E is piecewise smooth and the complement of E is connected.

Under these assumptions, there is a real-valued function $G(z,\infty;E)$, called the Green's function of E with respect to ∞, which is continuous on $\mathbb{C}$, 0 on E, harmonic and positive in $\mathbb{C}\backslash E$, and has the property that $G(z,\infty;E) - \log(|z|)$ is bounded as $z \to \infty$. (We write $\log(x)$ for $\ln(x)$.) The theorem on removable singularities for harmonic functions shows that the *Robin constant*, defined by

$$V_\infty(E) \;=\; \lim_{z\to\infty} G(z,\infty;E) - \log(|z|) \;,$$

exists. By definition $\gamma_\infty(E) = e^{-V_\infty(E)}$; our assumption that $\gamma_\infty(E) > 1$ means $V_\infty(E) < 0$. It can be shown that $V_\infty(E)$ is the minimum possible value of the "energy integral"

$$I_\infty(\nu) \;=\; \iint_{E\times E} -\log(|z-w|)\, d\nu(z) d\nu(w)$$

as ν ranges over all probability measures supported on E. There is a unique probability measure μ_∞ on E, called the equilibrium distribution of E with respect to ∞, for which

$$V_\infty(E) \;=\; \iint_{E\times E} -\log(|z-w|)\, d\mu_\infty(z) d\mu_\infty(w) \;.$$

The Green's function is related to the equilibrium distribution by

$$G(z, \infty; E) - V_\infty(E) = \int_E \log(|z-w|)\, d\mu_\infty(w) .$$

Because of its uniqueness, the measure μ_∞ is stable under complex conjugation. Taking a suitable discrete approximation $\mu_N = \frac{1}{N}\sum_{i=1}^N \delta_{x_i}(z)$ to μ_∞, stable under complex conjugation, one obtains a monic polynomial $f(z) = \prod_{i=1}^N (z - x_i) \in \mathbb{R}[z]$ such that $\frac{1}{N}\log(|f(z)|)$ approximates $G(z, \infty, E) - V_\infty(E)$ very well outside U. If the approximation is good enough, then since $V_\infty(E) < 0$, there will be an $\varepsilon > 0$ such that $\log(|f(z)|) > \varepsilon$ outside U.

One then uses the polynomial $f(z) \in \mathbb{R}[z]$ to construct a monic polynomial $G(z) \in \mathbb{Z}[z]$ of much higher degree, which has properties similar to those of $f(z)$. The construction is as follows. By adjusting the coefficients of $f(z)$ to be rational numbers and using continuity, one first obtains a polynomial $\phi(z) \in \mathbb{Q}[z]$ and an $R > 1$ such that $|\phi(z)| \ge R$ outside U. For suitably chosen n, the multinomial theorem implies that $\phi(z)^n$ will have a pre-designated number of high-order coefficients in $\mathbb{Z}$. By successively modifying the remaining coefficients of $G^{(0)}(z) := \phi(z)^n$ from highest to lowest order, writing $k = mN + r$ and adding $\delta_k \cdot z^r\phi(z)^m$ to change $a_k z^k$ with $a_k \in \mathbb{R}$ to $(a_k + \delta_k)z^k$ with $a_k + \delta_k \in \mathbb{Z}$ (the "patching" process), one obtains the desired polynomial $G(z) = G^{(n)}(z) \in \mathbb{Z}[z]$. One uses the polynomials $\delta_k z^r \phi(z)^m$ in patching, rather than simply the monomials $\delta_k z^k$, in order to control the sup-norms $\|z^r\phi(z)^m\|_E$. Each adjustment changes all the coefficients of order k and lower, but leaves the higher coefficients unchanged. Using a geometric series estimate to show that $|G(z)| > 1$ outside U, one concludes that $G(z)$ has all its roots in U. The algebraic integers produced by the classical Fekete-Szegö theorem are the roots of $G(z)^\ell - 1$ for $\ell = 1, 2, 3, \ldots$.

The proof of the Fekete-Szegö theorem with local rationality conditions on curves follows the same pattern, but with many complications. These arise from working on curves of arbitrary genus, from arranging that the zeros avoid the finite set $\mathfrak{X} = \{x_1, \ldots, x_m\}$ instead of a single point, from working adelically, and from imposing the local rationality conditions.

We will now sketch the proof in the situation where $E_v \subset \mathcal{C}_v(K_v)$ for each $v \in S$. The proof begins reducing the theorem to a setting where one is given a $\mathcal{C}_v(\mathbb{C}_v)$-neighborhood U_v of E_v for each v, with $U_v = E_v$ if $v \notin S$. One must then construct points $\alpha \in \mathcal{C}(\widetilde{K}^{\mathrm{sep}})$ whose conjugates belong to $U_v \cap \mathcal{C}_v(K_v)$ for each $v \in S$, and to U_v for each $v \notin S$. The strategy is to construct rational functions $G(z) \in K(\mathcal{C})$ with poles supported on $\mathfrak{X}$, whose zeros have the property above.

One first constructs an "initial approximating function" $f_v(z) \in K_v(\mathcal{C})$ for each $v \in S$. Each $f_v(z)$ has poles supported on $\mathfrak{X}$ and zeros in U_v, with the zeros in $\mathcal{C}_v(K_v)$ if $v \in S$. All the $f_v(z)$ have the same degree N, and they have the property that outside U_v the logarithms $\log_v(|f(z)|_v$ closely approximate a weighted sum of Green's functions $G(z, x_i; E_v)$. The weights are determined by $\mathbb{E}$ and $\mathfrak{X}$, through the definition of the Cantor capacity.

The construction of the initial approximating functions is one of the hardest parts of the proof. When working on curves of positive genus, one cannot simply take a discrete approximation to the equilibrium distribution, but must arrange that the divisor whose zeros come from that approximation and whose poles have the prespecified orders on the points in $\mathfrak{X}$, is principal. For places $v \in S$ there are additional constraints. When $K_v \cong \mathbb{R}$ and $E_v \subset \mathcal{C}_v(\mathbb{R})$, one must assure that $f_v(z)$ is real-valued and oscillates between large positive and negative values on E_v (a property like that of Chebyshev polynomials, first exploited by Robinson). In this work, we give a general potential-theoretic construction of oscillating functions. When K_v is nonarchimedean and $E_v \subset \mathcal{C}_v(K_v)$, one must arrange that the zeros of $f_v(z)$ belong to $U_v \cap \mathcal{C}_v(K_v)$ and are uniformly distributed with respect to a certain generalized equilibrium measure. Both cases are treated by constructing a nonprincipal divisor with the necessary properties, and then carefully moving some of its zeros to obtain a principal divisor. In this construction, the "canonical distance function" $[x, y]_\zeta$, introduced in ([**51**], §2.1), plays an essential role: given a divisor D of degree 0, the canonical distance tells what the v-adic absolute of a function with divisor D "would be", if such a function were to exist.

A further complication is that for archimedean v, one must arrange that the leading coefficients of the Laurent expansions of $f_v(z)$ at the points $x_i \in \mathfrak{X}$ have a property of "independent variability". When $K_v \cong \mathbb{C}$, this was established in ([**51**]) by using a convexity property of harmonic functions. When $K_v \cong \mathbb{R}$, we prove it by a continuity argument using the Brouwer Fixed Point theorem.

Once the initial approximating functions $f_v(z)$ have been constructed, we modify them to obtain "coherent approximating functions" $\phi_v(z)$ with specified leading coefficients, using global considerations. We then use the $\phi_v(z)$ to construct "initial patching functions" $G_v^{(0)}(z) \in K_v(\mathcal{C})$ of much higher degree which still have their zeros in U_v (and in $\mathcal{C}_v(K_v)$, for $v \in S$). The $G_v^{(0)}(z)$ are obtained by raising the $\phi_v(z)$ to high powers, or by composing them with Chebyshev polynomials or generalized Stirling polynomials if $v \in S$. (This idea goes back to Cantor [**16**].)

We next "patch" the functions $G_v^{(0)}(z)$, inductively constructing K_v-rational functions $(G_v^{(k)}(z))_{v \in S}$, $k = 1, 2, \ldots, n$, for which more and more of the high order Laurent coefficients (relative to the points in $\mathfrak{X}$) are K-rational and independent of v. In the patching process, we take care that the roots of $G_v^{(k)}(z)$ belong to U_v for all v, and belong to $\mathcal{C}_v(K_v)$ for each $v \in S$. In then end we obtain a global K-rational function $G^{(n)}(z) = G_v^{(n)}(z)$ independent of v, which "looks like" $G_v^{(0)}(z)$ at each $v \in S$.

The patching process has two aspects, global and local.

The global aspect concerns achieving K-rationality for $G(z)$, while assuring that its roots remain outside the balls $B_v(x_i, 1)^-$ for the infinitely many v where E_v is $\mathfrak{X}$-trivial. It is necessary to carry out the patching process in a Galois-invariant way. For this, we construct an $\mathrm{Aut}(\widetilde{K}/K)$-equivariant basis for the space of functions in $K(\mathbb{C})$ with poles supported on $\mathfrak{X}$, and arrange that when the functions $G_v^{(k)}(z)$ are expanded relative to this basis, their coefficients are equivariant under $\mathrm{Aut}_c(\mathbb{C}_v/K_v)$.

The most delicate step involves patching the leading coefficients: one must arrange that they be S-units (the analogue of monicity in the classical case). The argument can succeed only if the orders of the poles of the $f_v(z)$ at the x_i lie in a prescribed ratio to each other. The existence of such a ratio is intimately related to the fact that $\gamma(\mathbb{E}, \mathfrak{X}) > 1$, and is at the heart of the definition of the Cantor capacity, as will be explained below.

The remaining coefficients must be patched to be S-integers. As in the classical case, patching the high-order coefficients presents special difficulties. In general there are both archimedean and nonarchimedean places in S. It is no longer possible to use continuity and the multinomial theorem as in the classical case; instead, we use a phenomenon of "magnification" at the archimedean places, first applied in ([**53**]), together with a phenomenon of "contraction" at the nonarchimedean places. In the function field case, additional complications arise from inseparability issues. A different method is used to patch the high order coefficients than in the number field case: in the construction of initial patching functions, we arrange that the high order coefficients are all 0, and that the patching process for the leading coefficients preserves this property.

The local aspect of the patching process consists of giving "confinement arguments" showing how to keep the roots of the $G_v^{(k)}(z)$ in the sets E_v, while modifying the Laurent coefficients. Four confinement arguments are required, corresponding to the cases $K_v \cong \mathbb{C}$, $K_v \cong \mathbb{R}$ with $E_v \subset \mathcal{C}_v(\mathbb{R})$, K_v nonarchimedean with E_v being an RL-domain, and K_v nonarchimedean with $E_v \subset \mathcal{C}_v(K_v)$. The confinement arguments in the first and third cases are adapted from ([**51**]), and those in the second and fourth cases are generalizations of those in ([**53**]). The fourth case involves locally expanding the functions $G_v^{(k)}(z)$ as v-adic power series, and extending the Newton polygon construction in ([**53**]) from polynomials to power series. A crucial step involves moving apart roots which have come close to each other. This requires the theory of the universal function developed in Appendix C, and the local action of the Jacobian developed Appendix D.

The Definition of the Cantor Capacity

We next discuss the Cantor capacity $\gamma(\mathbb{E}, \mathfrak{X})$, which is treated more fully in ([**51**], §5.1). Our purpose here is to explain its meaning and its role in the proof of the Fekete-Szegö theorem. First, we will need some notation.

If v is archimedean, write $\log_v(x) = \ln(x)$. If v is nonarchimedean, let q_v be the order of the residue field of K_v, and write $\log_v(x)$ for the logarithm to the base q_v. Put $q_v = e$ if $K_v \cong \mathbb{R}$ and $q_v = e^2$ if $K_v \cong \mathbb{C}$.

Define normalized absolute values on the K_v by letting $|x|_v = |x|$ if v is archimedean, and taking $|x|_v$ to be the the modulus of additive Haar measure if v is nonarchimedean. For $0 \neq \kappa \in K$, the product formula reads

$$\sum_v \log_v(|\kappa|_v) \log(q_v) \;=\; 0 \ .$$

Each absolute value has a unique extension to $\mathbb{C}_v$, which we still denote by $|x|_v$.

For each $\zeta \in \mathcal{C}_v(\mathbb{C}_v)$, the *canonical distance* $[z,w]_\zeta$ on $\mathcal{C}_v(\mathbb{C}_v)\backslash\{\zeta\}$ (constructed in §2.1 of [**51**]) plays a role in the definition of $\gamma(\mathbb{E}, \mathfrak{X})$ similar to the role of the usual absolute value $|z-w|$ on $\mathbb{P}^1(\mathbb{C})\backslash\{\infty\}$ for the classical logarithmic capacity $\gamma(E)$. The canonical distance is a symmetric, real-valued, nonnegative function of $z, w \in \mathcal{C}_v(\mathbb{C}_v)$, with $[z,w]_\zeta = 0$ if and only if $z = w$. For each w, it has a "simple pole" as $z \to \zeta$. It is uniquely determined up to scaling by a constant. The constant can be specified by choosing a uniformizing parameter $g_\zeta(z) \in \mathbb{C}_v(\mathcal{C})$ at $z = \zeta$, and requiring that

$$\lim_{z \to \zeta} [z,w]_\zeta \cdot |g_\zeta(z)|_v \;=\; 1 \tag{0.3}$$

for each w. One definition of the canonical distance is that for each w,

$$[z,w]_\zeta = \lim_{n\to\infty} |f_n(z)|_v^{1/\deg(f_n)}$$

where the limit is taken over any sequence of functions $f_n(z) \in \mathbb{C}_v(\mathcal{C})$ having poles only at ζ whose zeros approach w, normalized so that

$$\lim_{z\to\zeta} |f_n(z) g_\zeta(z)^{\deg(f_n)}|_v = 1 .$$

A key property of $[z,w]_\zeta$ is that it can be used to factor the absolute value of a rational function in terms of its divisor: for each $f(z) \in \mathbb{C}_v(\mathcal{C})$, there is a constant $C(f)$ such that

$$|f(z)|_v = C(f) \cdot \prod_{x\neq\zeta} [z,x]_\zeta^{\mathrm{ord}_x(f)}$$

for all $z \neq \zeta$. For this reason, it is "right" kernel for use in arithmetic potential theory.

The Cantor capacity is defined in terms of Green's functions $G(z, x_i; E_v)$. We first introduce the Green's function for compact sets $H_v \subset \mathcal{C}_v(\mathbb{C}_v)$, where there is a potential-theoretic construction like the one in the classical case. Suppose $\zeta \notin H_v$. For each probability measure ν supported on H_v, consider the *energy integral*

$$I_\zeta(\nu) = \iint_{H_v\times H_v} -\log_v([z,w]_\zeta)\, d\nu(z) d\nu(w) .$$

Define the *Robin constant*

$$V_\zeta(H_v) = \inf_\nu I_\zeta(\nu) . \tag{0.4}$$

It can be shown that either $V_\zeta(H_v) < \infty$ for all $\zeta \notin E_v$, or $V_\zeta(H_v) = \infty$ for all $\zeta \notin E_v$ (see Lemma 3.15). In the first case we say that H_v has *positive inner capacity*, and the second case that it has *inner capacity* 0.

If H_v has positive inner capacity, there is a unique probability measure μ_ζ on H_v which achieves the infimum in (0.4). It is called the equilibrium distribution of H_v with respect to ζ. We define the Green's function by

$$G(z,\zeta;H_v) = V_\zeta(H_v) + \int_{H_v} \log_v([z,w]_\zeta)\, d\mu_\zeta(w) . \tag{0.5}$$

It is nonnegative and has a logarithmic pole as $z \to \zeta$. If H_v has inner capacity 0, we put $G(z,\zeta;H_v) = \infty$ for all z, ζ.

The Green's function is symmetric for $z, \zeta \notin H_v$, and is monotone decreasing in the set H_v: for compact sets $H_v \subset H_v'$ and $z, \zeta \notin E_v'$,

$$G(z,\zeta;H_v) \geq G(z,\zeta;H_v') . \tag{0.6}$$

If H_v has positive inner capacity, then for each neighborhood $U \supset H_v$, and each $\varepsilon > 0$, by taking a suitable discrete approximation to μ_ζ, one sees that there are an $N > 0$ and a function $f_v(z) \in \mathbb{C}_v(\mathcal{C})$ of degree N, with zeros in U and a pole of order N at ζ, such that

$$|G(z,\zeta;H_v) - \frac{1}{N}\log_v(|f_v(z)|_v)| < \varepsilon$$

for all $z \in \mathcal{C}_v(\mathbb{C}_v)\backslash(U \cup \{\zeta\})$.

In [**51**], Green's functions $G(z,\zeta;E_v)$ are defined for compact sets E_v in the archimedean case, and by a process of taking limits, for "algebraically capacitable" sets in the nonarchimedean case. Algebraically capacitable sets include all sets that are finite unions of compact sets and affinoid sets; see ([**51**], Theorem 4.3.11). In particular, the sets E_v in Theorem 0.4 are algebraically capacitable.

We next define local and global "Green's matrices". Let L/K be a finite normal extension containing $K(\mathfrak{X})$. For each place v of K and each w of L with $w|v$, after fixing an isomorphism $\mathbb{C}_w \cong \mathbb{C}_v$, we can pull back E_v to a set $E_w \subset \mathcal{C}_w(\mathbb{C}_w)$. The set E_w is independent of the isomorphism chosen, since E_v is stable under $\mathrm{Aut}_c(\mathbb{C}_v/K_v)$. If we identify $\mathcal{C}_v(\mathbb{C}_v)$ and $\mathcal{C}_w(\mathbb{C}_w)$, then for $z, \zeta \notin E_v$,

$$(0.7) \qquad G(z,\zeta;E_w)\log(q_w) \;=\; [L_w : K_v]\cdot G(z,\zeta;E_v)\log(q_v)\ .$$

For each $x_i \in \mathfrak{X}$, fix a global uniformizing parameter $g_{x_i}(x) \in L(\mathcal{C})$ and use it to define the upper Robin constants $V_{x_i}(E_w)$ for all w. For each w, let the "local upper Green's matrix" be

$$(0.8) \quad \Gamma(E_w,\mathfrak{X}) \;=\; \begin{pmatrix} V_{x_1}(E_w) & G(x_1,x_2;E_w) & \cdots & G(x_1,x_m;E_w) \\ G(x_2,x_1;E_w) & V_{x_2}(E_w) & \cdots & G(x_2,x_m;E_w) \\ \vdots & \vdots & \ddots & \vdots \\ G(x_m,x_1;E_w) & G(x_m,x_2;E_w) & \cdots & V_{x_m}(E_w) \end{pmatrix} .$$

Symmetrizing over the places of L, we define the "global Green's matrix" by

$$(0.9) \qquad \Gamma(\mathbb{E},\mathfrak{X}) \;=\; \frac{1}{[L:K]}\sum_{w\in\mathcal{M}_L}\Gamma(E_w,\mathfrak{X})\log(q_w)\ .$$

If $\mathbb{E}$ is compatible with $\mathfrak{X}$, the sum defining $\Gamma(\mathbb{E},\mathfrak{X})$ is finite. By the product formula, $\Gamma(\mathbb{E},\mathfrak{X})$ is independent of the choice of the $g_{x_i}(z)$. By (0.7) it is independent of the choice of L.

The global Green's matrix is symmetric and nonnegative off the diagonal. Its entries are finite if and only if each E_v has positive inner capacity.

Finally, for each K-rational $\mathbb{E}$ compatible with $\mathfrak{X}$, we define the *Cantor capacity* to be

$$(0.10) \qquad \gamma(\mathbb{E},\mathfrak{X}) \;=\; e^{-V(\mathbb{E},\mathfrak{X})}\ ,$$

where $V(\mathbb{E},\mathfrak{X}) = \mathrm{val}(\Gamma(\mathbb{E},\mathfrak{X}))$ is the value of $\Gamma(\mathbb{E},\mathfrak{X})$ as a matrix game. Here, for any $m \times m$ real-valued matrix Γ,

$$(0.11) \qquad \mathrm{val}(\Gamma) \;=\; \max_{\vec{s}\in\mathcal{P}^m}\ \min_{\vec{r}\in\mathcal{P}^m}\ {}^t\vec{s}\,\Gamma\vec{r}\ ,$$

where $\mathcal{P}^m = \{{}^t(s_1,\ldots,s_m) \in \mathbb{R}^m : s_1,\ldots,s_m \geq 0, \sum s_i = 1\}$ is the set of m-dimensional "probability vectors". Clearly $\gamma(\mathbb{E},\mathfrak{X}) > 0$ if and only if each E_v has positive inner capacity.

The hidden fact behind the definition is that $\mathrm{val}(\Gamma)$ is a function of matrices which, for symmetric real matrices Γ which are nonnegative off the diagonal, is negative if and only if Γ is negative definite; this is a consequence of Frobenius' Theorem (see ([**51**], p.328 and p.331) and ([**28**], p.53). Thus, $\gamma(\mathbb{E},\mathfrak{X}) > 1$ if and only if $\Gamma(\mathbb{E},\mathfrak{X})$ is negative definite.

If the matrix $\Gamma(\mathbb{E}, \mathfrak{X})$ is negative definite, there is a unique probability vector $\widehat{s} = {}^t(\widehat{s}_1, \ldots, \widehat{s}_m)$ such that

$$\Gamma(\mathbb{E}, \mathfrak{X})\,\widehat{s} \;=\; \begin{pmatrix} \widehat{V} \\ \vdots \\ \widehat{V} \end{pmatrix} \tag{0.12}$$

has all its coordinates equal. From the definition of val(Γ), it follows that $\widehat{V} = V(\mathbb{E}, \mathfrak{X}) < 0$. For simplicity, assume in what follows that $\widehat{s}$ has rational coordinates (in general, this fails; overcoming the failure is a major technical difficulty).

The probability vector $\widehat{s}$ determines the relative orders of the poles of the function $G(z)$ constructed in the Fekete-Szegö theorem. The idea is that the initial local approximating functions $f_v(z)$ should have polar divisor $\sum_{i=1}^m N\widehat{s}_i(x_i)$ for some N, and be such that for each v, outside the given neighborhood U_v of E_v,

$$\frac{1}{N}\log_v(|f_v(z)|_v) \;=\; \sum_{j=1}^m G(z, x_j; E_v)\widehat{s}_j \;.$$

(At archimedean places, this will only hold asymptotically as $z \to x_i$, for each x_i.) The fact that the coordinates of $\Gamma(\mathbb{E}, \mathfrak{X})\widehat{s}$ are equal means it is possible to scale the $f_v(z)$ so that in their Laurent expansions at x_i, the leading coefficients $c_{v,i}$ satisfy

$$\sum_v \log_v(|c_{v,i}|_v)\log(q_v) \;=\; 0,$$

compatible with the product formula, allowing the patching process to begin. Reversing this chain of ideas lead Cantor to his definition of the capacity.

For readers familiar with intersection theory, we remark that an Arakelov-like adelic intersection theory for curves was constructed in ([**56**]). The arithmetic divisors in that theory include all pairs $\mathcal{D} = (D, \{G(z, D; E_v)\}_{v \in \mathcal{M}_K})$ where $D = \sum_{i=1}^m s_i(x_i)$ is a K-rational divisor on $\mathcal{C}$ with real coefficients and $G(z, D; E_v) = \sum_{i=1} s_i G(z, x_i; E_v)$. If $\vec{s} = \widehat{s}$ is the probability vector constructed in (0.12), then relative to that intersection theory

$$V(\mathbb{E}, \mathfrak{X}) \;=\; {}^t\vec{s}\,\Gamma(\mathbb{E}, \mathfrak{X})\vec{s} \;=\; \mathcal{D}\cdot\mathcal{D} \;<\; 0 \;.$$

As noted by Moret-Bailly, this says that the Fekete-Szegö theorem with local rationality conditions can be viewed as a kind of arithmetic contractibility theorem.

Outline of the Book

In this section we outline the content and main ideas of this book.

The Introduction and Chapters 1 and 2 are expository, intended to give perspective on the Fekete-Szegö theorem. In Chapter 1 we state six variants of the theorem, which extend it in different directions. These include a version producing points in "quasi-neighborhoods" of $\mathbb{E}$, generalizing the classical Fekete-Szegö theorem; a version producing points in $\mathbb{E}$ under weaker conditions than those of Theorem 0.4; a version which imposes ramification conditions at finitely many primes outside S; a version for algebraically capacitable sets which expresses the Fekete/Fekete-Szegö dichotomy in terms of the global Green's matrix $\Gamma(\mathbb{E}, \mathfrak{X})$; and two versions for Berkovich curves.

In Chapter 2 we give numerical examples illustrating the theorem on $\mathbb{P}^1$, elliptic curves, Fermat curves, and modular curves. We begin by proving several formulas for capacities and Green's functions of archimedean and nonarchimedean sets, aiming to collect formulas useful for applications and going beyond those tabulated in ([**51**], Chapter 5). In the archimedean case, we give formulas for capacities and Green's functions of one, two, and arbitrarily many intervals in $\mathbb{R}$. The formulas for two intervals involve classical theta-functions, and those for multiple intervals (due to Harold Widom) involve hyperelliptic integrals. In the nonarchimedean case we give a general algorithm for computing capacities of compact sets. We determine the capacities and Green's functions of rings of integers, groups of units, and bounded tori in local fields. We also give the first known computation of a capacity of a nonarchimedean set where the Robin constant is not a rational number.

In the global case, we give numerical criteria for the existence/nonexistence of infinitely many algebraic integers and units satisfying various geometric conditions. The existence of such criteria, for which the prototypes are Robinson's theorems for totally real algebraic integers and units, is one of the attractive features of the subject. In applying a general theorem like the Fekete-Szegö theorem with local rationality conditions, it is often necessary to make clever reductions in order to obtain interesting results, and we have tried to give examples illustrating some of the reduction methods that can be used.

Our results for elliptic curves include a complete determination of the capacities (relative to the origin) of the integral points on Weierstrass models and Néron models. Our results for Fermat curves are based on McCallum's description of the special fiber for a regular model of the Fermat curve $\mathcal{F}_p$ over $\mathbb{Q}_p(\zeta_p)$. They show how the geometry of the model (in particular the number of "tame curves" in the special fibre) is reflected in the arithmetic of the curve. Our results for the modular curves $X_0(p)$ use the Deligne-Rapoport model. In combination, they illustrate a general principle that it is usually possible to compute nonarchimedean local capacities on a curve of higher genus, if a regular model of the curve is known.

Beginning with Chapter 3, we develop the theory systematically.

Chapter 3 covers notation, conventions, and foundational material about capacities and Green's functions used throughout the work. An important notion is the $(\mathfrak{X}, \vec{s})$-canonical distance $[z,w]_{\mathfrak{X},\vec{s}}$. Given a curve $\mathcal{C}/K$ and a place v of K, we will be interested in constructing rational functions $f \in \mathcal{C}_v(\mathbb{C}_v)$ whose poles are supported on a finite set $\mathfrak{X} = \{x_1, \ldots, x_m\}$ and whose polar divisor is proportional to $\sum_{i=1}^m s_i(x_i)$, where $\vec{s} = (s_1, \ldots, s_m)$ is a fixed probability vector. The $(\mathfrak{X}, \vec{s})$-canonical distance enables to treat $|f(z)|_v$ like the absolute value of a polynomial, factoring it in terms of the zero divisor of f as

$$|f(z)|_v \;=\; C(f) \cdot \prod_{\text{zeros } \alpha_i \text{ of } f} [z, \alpha_i]_{\mathfrak{X},\vec{s}} \; .$$

Furthermore, the product on the right, which we call an $(\mathfrak{X}, \vec{s})$-*pseudopolynomial*, is defined and continuous even for divisors which are not principal. This enables us to separate analytic and algebraic issues in the construction of f.

Put $L = K(\mathfrak{X}) = K(x_1, \ldots, x_m)$, and let L^{sep} be the separable closure of K in L. Other important technical tools from Chapter 3 are the L-rational and L^{sep}-rational bases. These are multiplicatively finitely generated sets of functions which can be used to expand rational functions with poles supported on $\mathfrak{X}$, much like the monomials $1, z, z^2, \ldots$ can be used to expand polynomials. As their names indicate, the functions in the L-rational basis are defined over L, and those in the L^{sep}-rational basis are defined over L^{sep}. The construction arranges that the transition matrix between the two bases is block diagonal, and has bounded norm at each place w of L.

In Chapter 4 we state a version of the Fekete-Szegö theorem with local rationality conditions for "K_v-simple sets" (Theorem 4.2), and we reduce Theorem 0.4, Corollary 0.5, and the variants stated in Chapter 1 to it. The rest of the book (Chapters 5–11 and Appendices A–D) is devoted to the proof of Theorem 4.2.

Chapters 5 and 6 contain the constructions of the initial approximating functions needed for Theorem 4.2. Four constructions are needed: for archimedean sets $E_v \subset \mathcal{C}_v(\mathbb{C})$ when the ground field is $\mathbb{C}$ and $\mathbb{R}$, and for nonarchimedean sets $E_v \subset \mathcal{C}_v(\mathbb{C}_v)$ which are RL-domains or are compact. The first and third were done in ([**51**]); the second and fourth are done here.

The probability vector $\vec{s}$ ultimately used in the construction is determined by $\mathbb{E}$ and $\mathfrak{X}$, through the global Green's matrix $\Gamma(\mathbb{E}, \mathfrak{X})$. This means that for each E_v, the local constructions must be carried out in a uniform way for all $\vec{s}$. In Appendix A we develop potential theory with respect to the kernel $[z, w]_{\mathfrak{X},\vec{s}}$. There are $(\mathfrak{X}, \vec{s})$-capacities, $(\mathfrak{X}, \vec{s})$-Green's functions, and $(\mathfrak{X}, \vec{s})$-equilibrium distributions with properties analogous to the corresponding objects in classical potential theory. The initial approximating functions are $(\mathfrak{X}, \vec{s})$-functions whose normalized logarithms $\deg(f)^{-1}\log_v(|f(z)|_v)$ closely approximate the $(\mathfrak{X}, \vec{s})$-Green's function outside a neighborhood of E_v, and whose zeros are roughly equidistributed like the $(\mathfrak{X}, \vec{s})$-equilibrium distribution.

Chapter 5 deals with the construction of initial approximating functions $f(z) \in \mathbb{R}(\mathcal{C}_v)$ when the ground field K_v is $\mathbb{R}$, for Galois-stable sets $E_v \subset \mathcal{C}_v(\mathbb{C})$ which are finite unions of intervals in $\mathcal{C}_v(\mathbb{R})$ and closed sets in $\mathcal{C}_v(\mathbb{C})$ with piecewise smooth boundaries. The desired functions must oscillate with large magnitude on the real intervals. The construction has two parts: a potential-theoretic part carried out in Appendix B, which constructs "$(\mathfrak{X}, \vec{s})$-pseudopolynomials" whose absolute value behaves like that of a Chebyshev polynomial, and an algebraic part which involves adjusting the divisor of the pseudopolynomial to make it principal. The first part of the argument requires subdividing the real intervals into "short" segments, where the notion of shortness depends only on the deviation of the canonical distance $[z, w]_{\mathfrak{X},\vec{s}}$ from $|z - w|$ in local coordinates, and is uniform over compact sets. The second part of the argument uses a variant of the Brouwer Fixed Point theorem. An added difficulty involves assuring that the "logarithmic leading coefficients" of f are independently variable over a range independent of $\vec{s}$, which is needed as an input to the global patching process in Chapter 7.

Chapter 6 deals with the construction of initial approximating functions $f \in K_v(\mathcal{C}_v)$ when the ground field K_v is a nonarchimedean local field, and the sets E_v are Galois-stable finite unions of balls in $\mathcal{C}_v(F_{w,\ell})$, for fields $F_{w,\ell}$ are which are finite separable extensions of K_v. Again the construction has two parts: an analytic part, which constructs an $(\mathfrak{X}, \vec{s})$-pseudopolynomial by transporting Stirling polynomials for the rings of integers of the $F_{w,\ell}$ to the balls, and an algebraic part, which involves moving some of the roots of the pseudopolynomial to make its divisor principal. When $\mathcal{C}_v$ has positive genus g, this uses an action of a neighborhood of the origin in $\mathrm{Jac}(\mathcal{C})(\mathbb{C}_v)$ on $\mathcal{C}_v(\mathbb{C}_v)^g$ constructed in Appendix D.

Chapter 7 contains the global patching argument for Theorem 4.2, which breaks into two cases: when $\mathrm{char}(K) = 0$, and when $\mathrm{char}(K) = p > 0$. The two cases involve different difficulties. When $\mathrm{char}(K) = 0$, the need to patch archimedean and nonarchimedean initial approximating functions together is the main constraint, and the most serious bottleneck involves patching the leading coefficients. The ability to independently adjust the logarithmic leading coefficients for the archimedean initial approximating functions allows us to accomplish this. When $\mathrm{char}(K) = p > 0$, the leading coefficients are not a problem, but separability/inseparability issues drive the argument. These are dealt with by simultaneously monitoring the patching process relative to the L-rational and L^{sep}-rational bases from Chapter 3.

Chapters 8–11 contain the local patching arguments needed for Theorem 4.2. Chapter 8 concerns the case when $K_v \cong \mathbb{C}$, Chapter 9 concerns the case when $K_v \cong \mathbb{R}$, Chapter 10 concerns the nonarchimedean case for RL-domains, and Chapter 11 concerns the nonarchimedean case for compact sets. Each provides geometrically increasing bounds for the amount the coefficients can be varied, while simultaneously confining the movement of the roots, as the patching proceeds from high order to low order coefficients.

Chapter 8 gives the local patching argument when $K_v \cong \mathbb{C}$. The aim of the construction is to confine the roots of the function to a prespecified neighborhood U_v of E_v, while providing the global patching construction with increasing freedom to modify the coefficents relative to the L-rational basis, as the degree of the basis functions goes down. For the purposes of the patching argument, the coefficients are grouped into "high-order", "middle" and "low-order". The construction begins by raising the initial approximating function to a high power n. A "magnification argument", similar to the ones in ([**52**]) and ([**53**]), is used to gain the freedom needed to patch the high-order coefficients.

Chapter 9 gives the local patching argument when $K_v \cong \mathbb{R}$. Here the construction must simultaneously confine the roots to a set U_v which is the union of $\mathbb{R}$-neighborhoods of the components of E_v in $\mathcal{C}_v(\mathbb{R})$, and $\mathbb{C}$-neighborhoods of the other components. We call such a set a "quasi-neighborhood" of E_v. The construction is similar to the one over $\mathbb{C}$, except that it begins by composing the initial approximating function with a Chebyshev polynomial of degree n. Chebyshev polynomials have the property that they oscillate with large magnitude on a real interval, and take a family of confocal ellipses in the complex plane to ellipses. Both properties are used in the confinement argument.

Chapter 10 gives the local patching construction when K_v is nonarchimedean and E_v is an RL-domain. The construction again begins by raising the initial approximating function to a power n. To facilitate patching the high-order coefficients, we require that n be divisible by a high power of the residue characteristic p. If K_v has characteristic 0, this makes the high order coefficients be p-adically small; if K_v has characteristic p, it makes them vanish (apart from the leading coefficients), so they do not need to be patched at all.

Chapter 11 gives the local patching construction when K_v is nonarchimedean and E_v is compact. This case is by far the most intricate, and begins by composing the initial approximating function with a Stirling polynomial. If K_v has characteristic 0, this makes the high order coefficients be p-adically small; if K_v has characteristic p, it makes them vanish. The confinement argument generalizes those in ([**52**], [**53**]), and the roots are controlled by tracking their positions within "ψ_v-regular sequences".

A ψ_v-regular sequence is a finite sequence of roots which are v-adically spaced like an initial segment of the integers, viewed as embedded in $\mathbb{Z}_p$ (see Definition 11.3). The local rationality of each root is preserved by an argument involving Newton polygons for power series. In the initial stages, confinement of the roots depends on the fact that the Stirling polynomial factors completely over K_v. Some roots may move quite close to others in early steps of the patching process, and the the middle part of argument involves an extra step of separating roots, first used in ([**52**]). This is accomplished by multiplying the partially patched function with a carefully chosen rational function whose zeros and poles are very close in pairs. This function is obtained by specializing the "universal function" constructed in Appendix C, which parametrizes all functions of given degree by means of their roots and poles and value at a normalizing point.

Appendix A develops potential theory with respect the kernel $[z, w]_{\mathfrak{X},\vec{s}}$, paralleling the classical development of potential theory over $\mathbb{C}$ given in ([**65**]). There are $(\mathfrak{X}, \vec{s})$-equilibrium distributions, potential functions, transfinite diameters, Chebyshev constants, and capacities with the same properties as in the classical theory. A key result is Proposition A.5, which asserts that "$(\mathfrak{X}, \vec{s})$-Green's functions", obtained by subtracting an "$(\mathfrak{X}, \vec{s})$-potential function" from an "$(\mathfrak{X}, \vec{s})$-Robin constant", are given by linear combinations of the Green's functions constructed in ([**51**]). Other important results are Lemmas A.6 and A.7, which provide uniform upper and lower bounds for the mass the $(\mathfrak{X}, \vec{s})$-equilibrium distribution can place on a subset, independent of $\vec{s}$; and Theorem A.13, which shows that nonarchimedean $(\mathfrak{X}, \vec{s})$-Green's functions and equilibrium distributions can be computed using linear algebra.

Appendix B constructs archimedean local oscillating functions for short intervals, and gives the potential-theoretic input for the construction of the initial approximating functions over $\mathbb{R}$ in Chapter 5. In classical potential theory, the equality of the transfinite diameter, Chebyshev constant, and logarithmic capacity of a compact set $E \subset \mathbb{C}$ is shown by means of a "rock-paper-scissors" argument proving in a cyclic fashion that each of the three quantities is greater than or equal to the next. Here, a rock-paper-scissors argument is used to prove Theorem B.13, which says that the probability measures associated to the roots of weighted Chebyshev polynomials for a set E_v converge to the $(\mathfrak{X}, \vec{s})$-equilibrium measure of E_v.

Appendix C studies the "universal function" of degree d on a curve, used in Chapter 11. We give two constructions for it, one by Robert Varley using Grauert's theorem, the other by the author using the theory of the Picard scheme. We then use local power series parametrizations, together with a compactness argument, to obtain uniform bounds for the change in the norm of a function outside a union of balls containing its divisor, if its zeros and poles are moved a distance at most δ (Theorem C.2).

Appendix D shows that in the nonarchimedean case, if the genus g of $\mathcal{C}$ is positive, then at generic points of $\mathcal{C}_v(\mathbb{C}_v)^g$ there is an action of a neighborhood of the origin of the Jacobian on $\mathcal{C}_v(\mathbb{C}_v)^g$, which makes $\mathcal{C}_v(\mathbb{C}_v)^g$ into a local principal homogeneous space. This is used in Chapters 6 and 11 in adjusting nonprincipal divisors to make them principal. The action is obtained by considering the canonical map $\mathcal{C}_v^g(\mathbb{C}_v) \to \mathrm{Jac}(\mathcal{C})(\mathbb{C}_v)$, which is locally an isomorphism outside a set of codimension 1, pulling back the formal group of the Jacobian, and using properties of power series in several variables. Theorem D.2 gives the most general form of the action.

Acknowledgments

The author thanks Pete Clark, Will Kazez, Dino Lorenzini, Ted Shifrin, and Robert Varley for help and useful conversations during the investigation. He also thanks the Institute Henri Poincaré in Paris, where the work was begun during the Special Trimestre on Diophantine Geometry in 1999, and the University of Georgia, where most of the research was carried out.

The author gratefully acknowledges the National Science Foundation's support of this project through grants DMS 95-000892, DMS 00-70736, DMS 03-00784, and DMS 06-01037. Any opinions, findings and conclusions or recommendations expressed in this material are those of the author and do not necessarily reflect the views of the National Science Foundation.

Symbol Table

Below are some symbols used. See §3.1, §3.2 for other conventions.

Symbol	Meaning	Defined
K	a global field	p. 61
$\mathcal{C}$	a smooth, projective, connected curve over K	p. 62
$g = g(\mathcal{C})$	the genus of $\mathcal{C}$	p. 65
$\widetilde{K}$	a fixed algebraic closure of K	p. 61
$\widetilde{K}^{\mathrm{sep}}$	the separable closure of K in $\widetilde{K}$	p. ix
K_v	the completion of K at a place v	p. 61
$\mathcal{O}_v$	the ring of integers of K_v	p. 61
$\widetilde{K}_v$	a fixed algebraic closure of K_v	p. 61
$\mathbb{C}_v$	the completion of $\widetilde{K}_v$	p. 61
$\mathrm{Aut}(\widetilde{K}/K)$	the group of continuous automorphisms of $\widetilde{K}/K$	p. 61
$\mathrm{Aut}_c(\mathbb{C}_v/K_v)$	the group of continuous automorphisms of $\mathbb{C}_v/K_v$	p. 61
$\mathfrak{X} = \{x_1, \ldots, x_m\}$	a finite, $\mathrm{Aut}_c(\widetilde{K}/K)$-stable set of points of $\mathcal{C}(\widetilde{K})$	p. 62
$\vec{s} = (s_1, \ldots, s_m)$	a probability vector weighting the points in $\mathfrak{X}$	p. xx
$L = K(\mathfrak{X})$	the field $K(x_1, \ldots, x_m)$	p. 62
L^{sep}	the separable closure of K in L	p. xx
$\mathcal{C}_v$	the curve $\mathcal{C} \times_K \mathrm{Spec}(K_v)$	p. ix
$\overline{\mathcal{C}}_v$	the curve $\mathcal{C}_v \times_{K_v} \mathrm{Spec}(\mathbb{C}_v)$	p. 179
$\|z, w\|_v$	the chordal distance or spherical metric on $\mathcal{C}_v(\mathbb{C}_v)$	p. 69ff
$\|f\|_{E_v}$	the sup norm $\sup_{z \in E_v} |f(z)|_v$	p. 62
$D(a, r)$	the "closed disc" $\{z \in \mathbb{C}_v : |z - a|_v \le r\}$	p. 70
$D(a, r)^-$	the "open disc" $\{z \in \mathbb{C}_v : |z - a|_v < r\}$	p. 70
$B(a, r)$	the "closed ball" $\{z \in \mathcal{C}_v(\mathbb{C}_v) : \|z, a\|_v \le r\}$	p. 70
$B(a, r)^-$	the "open ball" $\{z \in \mathcal{C}_v(\mathbb{C}_v) : \|z, a\|_v < r\}$	p. 70
q_v	the order of the residue field of K_v, if v is nonarchimedean	p. 61ff
w_v	the distinguished place of L over a place v of K	p. 62
$\mathrm{val}_v(x)$	the exponent of the largest power of q_v dividing $x \in \mathbb{N}$	p. 98
$\log_v(x)$	the logarithm to the base q_v, when v is nonarchimedean	p. 61
$\mathrm{ord}_v(z)$	the exponential valuation $-\log_v(|z|_v)$, for $z \in \mathbb{C}_v$	p. 61
$\log(x)$	the natural logarithm $\ln(x)$	p. 61
ζ	a point of $\mathcal{C}_v(\mathbb{C}_v)$	p. 71
$g_\zeta(z)$	a fixed uniformizing parameter at ζ	p. 71
$[z, w]_\zeta$	the canonical distance with respect to $\zeta \in \mathcal{C}_v(\mathbb{C}_v)$	p. 73
$[z, w]_{(\mathfrak{X}, \vec{s})}$	the $(\mathfrak{X}, \vec{s})$-canonical distance on $\mathcal{C}_v(\mathbb{C}_v)$	p. 76
E_v	a subset of $\mathcal{C}_v(\mathbb{C}_v)$	p. ix
$\mathrm{cl}(E_v)$	the topological closure of E_v	p. 3
$\gamma_\zeta(E_v)$	the capacity of E_v with respect to ζ and $g_\zeta(z)$	p. 78
$V_\zeta(E_v)$	the Robin constant of E_v with respect to ζ and $g_\zeta(z)$	p. 78
$G(z, \zeta; E_v)$	the Green's function of E_v	p. 81
$\mathrm{val}(\Gamma)$	the value of $\Gamma \in M_n(\mathbb{R})$ as a matrix game	p. xviii
$\mathbb{E} = \prod_v E_v$	an adelic set in $\prod_v \mathcal{C}_v(\mathbb{C}_v)$	p. x
$\Gamma(\mathbb{E}, \mathfrak{X})$	the global Green's matrix of $\mathbb{E}$ relative to $\mathfrak{X}$	p. xviii
$\gamma(\mathbb{E}, \mathfrak{X})$	the global Cantor capacity of $\mathbb{E}$ with respect to $\mathfrak{X}$	p. xviii
$\mathcal{C}_v^{\mathrm{an}}$	the Berkovich analytification of $\overline{\mathcal{C}}_v$	p. 5
$\mathbf{E}_v$	a subset of $\mathcal{C}_v^{\mathrm{an}}$	p. 5
$V_\zeta(\mathbf{E}_v)^{\mathrm{an}}$	the Robin constant of $\mathbf{E}_v$ with respect to ζ and $g_\zeta(z)$	p. 6
$G(z, \zeta; \mathbf{E}_v)^{\mathrm{an}}$	the Thuillier Green's function of $\mathbf{E}_v$	p. 6

Symbol	Meaning	Defined
$\gamma(\mathbb{E},\mathfrak{X})^{\mathrm{an}}$	the global capacity of a Berkovich set $\mathbb{E}=\prod_v \mathbf{E}_v$ relative to $\mathfrak{X}$	p. 6
$\mathcal{P}^m=\mathcal{P}^m(\mathbb{R})$	the set of probability vectors $\vec{s}=(s_1,\ldots,s_m)\in\mathbb{R}^m$	p. xviii
$\mathcal{P}^m(\mathbb{Q})$	the set of probability vectors with rational coefficients	p. 94
J	$J=2g+1$ if $\mathrm{char}(K)=0$, a power of p if $\mathrm{char}(K)=p>0$	p. 65ff
$\varphi_{ij}(z),\varphi_\lambda(z)$	functions in the L-rational basis	p. 65
$\widetilde{\varphi}_{ij}(z),\widetilde{\varphi}_\lambda(z)$	functions in the L^{sep}-rational basis	p. 66
Λ_0	number of low-order elements in the L and L^{sep}-rational bases	p. 65
Λ	number of basis elements deemed low-order in patching	p. 211
$f_v(z)$	an initial approximating function	p. 200
$c_{v,i}$	the leading coefficient of $f_v(z)$ at x_i	p. 225
$\Lambda_{x_i}(f_v,\vec{s})$	the logarithmic leading coefficient of $f_v(z)$ at x_i	p. 134
$\Lambda_{x_i}(E_v,\vec{s})$	the logarithmic leading coefficient of the Green's function of E_v	p. 134
$\phi_v(z)$	a coherent approximating function	p. 204
$\widetilde{c}_{v,i}$	the leading coefficient of $\phi_v(z)$ at x_i	p. 207
$\mathcal{I}$	the index set $\{(i,j)\in\mathbb{Z}^2 : 1\le i\le m, j\ge 0\}$	p. 213
$\prec_N$	the order on $\mathcal{I}$ determining how coefficients are patched	p. 213
$\mathrm{Band}_N(k)$	"Bands" of indices in $\mathcal{I}$ for the order $\prec_N$	p. 214
$\mathrm{Block}_N(i,j)$	the "Galois orbit" of the index $(i,j)\in\mathcal{I}$	p. 214
$G_v^{(k)}(z)$	the patching function at v in stage k of the patching process	p. 212
$A_{v,ij},A_{v,\lambda}$	the coefficients of $G_v^{(k)}(z)$ relative to the L-rational basis	p. 211
$\widetilde{A}_{v,ij},\widetilde{A}_{v,\lambda}$	the coefficients of $G_v^{(k)}(z)$ relative to the L^{sep}-rational basis	p. 234
$\Delta_{v,ij}^{(k)},\Delta_{v,\lambda}^{(n)}$	the changes in the coefficients of $G_v^{(k)}(z)$ in stage k of patching	p. 212
$\vartheta_{v,ij}^{(k)}(z),\vartheta_{v,\lambda}^{(n)}$	compensating functions for stage k of patching	p. 212
$\psi_v(k)$	the basic well-distributed sequence for the ring $\mathcal{O}_v$	p. 98
$S_{n,v}(z)$	the Stirling polynomial $\prod_{k=0}^{n-1}(z-\psi_v(k))$ for $\mathcal{O}_v$	p. 98
$E(a,b)$	the filled ellipse $\{z=x+iy\in\mathbb{C} : x^2/a^2+y^2/b^2\le 1\}$	p. 258
$T_n(z)$	the Chebyshev polynomial of degree n for $[-2,2]$	p. 258
$T_{n,R}(z)$	the Chebyshev polynomial of degree n for $[-2R,2R]$	p. 258
$\mathbb{F}_p[[t]]$	the ring of formal power series over $\mathbb{F}_p$	p. 38
$\mathbb{F}_p((t))$	the field of formal Laurent series over $\mathbb{F}_p$	p. 39
$\mathrm{Jac}(\mathcal{C}_v)$	the Jacobian of a curve $\mathcal{C}_v$ with genus $g>0$	p. 389
$J_{\mathrm{Ner}}(\mathcal{C}_v)$	the Néron model of $\mathcal{C}_v$	p. 408

CHAPTER 1

Variants

In this chapter we give six variants of Theorem 0.4, strengthening it in different ways. In Chapter 4, Theorem 0.4, Corollary 0.5 and the variants stated here will be reduced to yet another variant (Theorem 4.2) and we will spend most of the book proving the theorem in that form.

Our first variant is similar to the original theorem of Fekete and Szegö ([**25**]). In that theorem the sets $E_v \subset \mathbb{C}$ were compact, and the conjugates of the algebraic integers produced were required to lie in arbitrarily small open neighborhoods U_v of the E_v. In Theorem 1.2 below, we lift the assumption of compactness and replace the Cantor capacity with *inner Cantor Capacity* $\overline{\gamma}(\mathbb{E}, \mathfrak{X})$, which is defined for arbitrary adelic sets. We also replace the neighborhoods U_v with "quasi-neighborhoods", which are finite unions of open sets in $\mathcal{C}_v(\mathbb{C}_v)$ and open sets in $\mathbb{C}_v(F_w)$, for algebraic extensions F_w/K_v in $\mathbb{C}_v$.

The inner Cantor capacity $\overline{\gamma}(\mathbb{E}, \mathfrak{X})$ is similar to Cantor capacity except that it is defined in terms of upper Green's functions $\overline{G}(z, x_i; E_v)$. Here, we briefly recall the definitions of $\overline{G}(z, x_i; E_v)$ and $\overline{\gamma}(\mathbb{E}, \mathfrak{X})$ and some of their properties; they are studied in detail in §3.9 and §3.10 below.

Upper Green's functions are gotten by taking decreasing limits of Green's functions of compact sets. For an arbitrary $E_v \subset \mathcal{C}_v(\mathbb{C}_v)$, if $\zeta \notin E_v$ the upper Green's function is

$$\overline{G}(z, \zeta; E_v) = \inf_{\substack{H_v \subset E_v \\ H_v \text{ compact}}} G(z, \zeta; H_v) . \tag{1.1}$$

If ζ is not in the closure of E_v, the upper Robin constant $\overline{V}_\zeta(E_v)$ is finite and is defined by

$$\overline{V}_\zeta(E_v) = \lim_{z \to \zeta} \overline{G}(z, \zeta; E_v) + \log_v(|g_\zeta(z)|_v) , \tag{1.2}$$

where $g_\zeta(z)$ is the uniformizer from (0.3). By (0.6), if E_v is compact then by ([**51**], Theorem 4.4.4) $\overline{G}(z, \zeta; E_v) = G(z, \zeta; E_v)$ and $\overline{V}_\zeta(E_v) = V_\zeta(E_v)$. For nonarchimedean v, if E_v is assumed to be algebraically capacitable in the sense of ([**51**]), then $\overline{G}(z, \zeta; E_v) = G(z, \zeta; E_v)$ and $\overline{V}_\zeta(E_v) = V_\zeta(E_v)$. The upper Green's function is symmetric and nonnegative: for all $z, \zeta \notin E_v$, $\overline{G}(z, \zeta; E_v) = \overline{G}(\zeta, z; E_v) \geq 0$. It has functoriality properties under pullbacks and base change like those of $G(z, \zeta; E_v)$.

Now assume that each E_v is stable under $\mathrm{Aut}_c(\mathbb{C}_v/K_v)$, and that $\mathbb{E} = \prod_v E_v$ is compatible with $\mathfrak{X}$. Let L/K be a finite normal extension containing $K(\mathfrak{X})$. For each place v of K and each place w of L with $w|v$, after fixing an isomorphism $\mathbb{C}_w \cong \mathbb{C}_v$, we can pull back E_v to a set $E_w \subset \mathcal{C}_w(\mathbb{C}_w)$, which is independent of the isomorphism chosen. If we identify $\mathcal{C}_v(\mathbb{C}_v)$ with $\mathcal{C}_w(\mathbb{C}_w)$, then for $z, \zeta \notin E_v$

$$\overline{G}(z, \zeta; E_w) \log(q_w) = [L_w : K_v] \cdot \overline{G}(z, \zeta; E_v) \log(q_v) . \tag{1.3}$$

For each $x_i \in \mathfrak{X}$, fix a global uniformizing parameter $g_{x_i}(x) \in L(\mathcal{C})$ and use it to define the upper Robin constants $\overline{V}_{x_i}(E_w)$ for all places w of L. For each w, the "local upper Green's matrix" is

$$\overline{\Gamma}(E_w, \mathfrak{X}) = \begin{pmatrix} \overline{V}_{x_1}(E_w) & \overline{G}(x_1, x_2; E_w) & \cdots & \overline{G}(x_1, x_m; E_w) \\ \overline{G}(x_2, x_1; E_w) & \overline{V}_{x_2}(E_w) & \cdots & \overline{G}(x_2, x_m; E_w) \\ \vdots & \vdots & \ddots & \vdots \\ \overline{G}(x_m, x_1; E_w) & \overline{G}(x_m, x_2; E_w) & \cdots & \overline{V}_{x_m}(E_w) \end{pmatrix},$$

and the 'global upper Green's matrix' is

$$\overline{\Gamma}(\mathbb{E}, \mathfrak{X}) = \frac{1}{[L:K]} \sum_{w \in \mathcal{M}_L} \overline{\Gamma}(E_w, \mathfrak{X}) \log(q_w) .$$

Since $\mathbb{E}$ is compatible with $\mathfrak{X}$, all but finitely many of the $\overline{\Gamma}(E_w, \mathfrak{X})$ are 0. By the product formula, $\overline{\Gamma}(\mathbb{E}, \mathfrak{X})$ is independent of the choice of the $g_{x_i}(z)$. By (0.7) it is independent of the choice of L. It is symmetric and non-negative off the diagonal; its entries are finite if and only if each E_v has positive inner capacity.

For each K-rational $\mathbb{E}$ compatible with $\mathfrak{X}$, the *inner Cantor capacity* is

$$\overline{\gamma}(\mathbb{E}, \mathfrak{X}) = e^{-\overline{V}(\mathbb{E}, \mathfrak{X})} ,$$

where $\overline{V}(\mathbb{E}, \mathfrak{X}) = \mathrm{val}(\overline{\Gamma}(\mathbb{E}, \mathfrak{X}))$ is the value of $\overline{\Gamma}(\mathbb{E}, \mathfrak{X})$ as a matrix game. When the sets E_v are compact or algebraically capacitable, the inner Cantor capacity coincides with the Cantor capacity $\gamma(\mathbb{E}, \mathfrak{X})$ defined in ([**51**]). It reduces to the classical logarithmic capacity when $\mathcal{C} = \mathbb{P}^1/\mathbb{Q}$, $\mathfrak{X} = \infty$, and all the nonarchimedean E_v are trivial.

The reason the inner Cantor capacity is the appropriate capacity to use in the Fekete-Szegö theorem is that one of the initial reductions in the proof is to replace each E_v which is not $\mathfrak{X}$-trivial by a compact set $H_v \subset E_v$. Since the Green's function is a limit of Green's functions of compact sets, this can be done in such a way that $\Gamma(\mathbb{E}, \mathfrak{X})$ remains negative definite.

DEFINITION 1.1. Let v be a place of K. A set $U_v \subset \mathcal{C}_v(\mathbb{C}_v)$ will be called a *quasi-neighborhood* if there are open sets $U_{v,0}, U_{v,1}, \ldots, U_{v,M}$ in $\mathcal{C}_v(\mathbb{C}_v)$ and algebraic extensions $F_{w_1}/K_v, \ldots, F_{w_M}/K_v$ in $\mathbb{C}_v$ (of finite or infinite degree) such that

$$U_v = U_{v,0} \cup \bigcup_{\ell=1}^{M} \left(U_{v,\ell} \cap \mathcal{C}_v(F_{w_\ell})\right) .$$

We allow the possibility that one or more of the $U_{v,\ell}$ are empty. We will say that U_v is *K_v-symmetric* if it is stable under $\mathrm{Aut}_c(\mathbb{C}_v/K_v)$, and that it is *separable* if each F_{w_ℓ}/K_v is separable. If U_v contains a set E_v, we will say that U_v is a quasi-neighborhood of E_v.

Equivalently, a quasi-neighborhood $U_v \subset \mathcal{C}_v(\mathbb{C}_v)$ is the union of finitely many sets, each of which is either open in $\mathcal{C}_v(\mathbb{C}_v)$ or is open in $\mathcal{C}_v(F_{w_\ell})$ for some algebraic extension F_{w_ℓ}/K_v in $\mathbb{C}_v$. These sets need not be disjoint. For example, take $\mathcal{C} = \mathbb{P}^1$ and identify $\mathbb{P}^1(\mathbb{C}_v)$ with $\mathbb{C}_v \cup \infty$. Suppose v is nonarchimedean; let $F_{w_1}, \ldots, F_{w_M}$ be algebraic extensions of K_v contained in $\mathbb{C}_v$, and let $\mathcal{O}_{w_1}, \ldots, \mathcal{O}_{w_M}$ be their rings of integers. Then the set $U_v = \mathcal{O}_{v_1} \cup \cdots \cup \mathcal{O}_{w_M}$ is a quasi-neighborhood of the origin in $\mathbb{P}^1(\mathbb{C}_v)$.

If $\mathbb{E} = \prod_v E_v \subseteq \prod_v \mathcal{C}_v(\mathbb{C}_v)$ is an adelic set, we will call a set $\mathbb{U} = \prod_v U_v \subseteq \prod_v \mathcal{C}_v(\mathbb{C}_v)$ a *K-rational separable quasi-neighborhood of* $\mathbb{E}$ if each U_v is a separable quasi-neighborhood of E_v, stable under $\mathrm{Aut}_c(\mathbb{C}_v/K_v)$.

THEOREM 1.2 (FSZ with LR for Quasi-neighborhoods). *Let K be a global field, and let $\mathcal{C}/K$ be a smooth, connected, projective curve. Let $\mathfrak{X} = \{x_1, \ldots, x_m\} \subset \mathcal{C}(\widetilde{K})$ be a finite set of points stable under $\mathrm{Aut}(\widetilde{K}/K)$, and let $\mathbb{E} = \prod_v E_v \subset \prod_v \mathcal{C}_v(\mathbb{C}_v)$ be a K-rational adelic set compatible with $\mathfrak{X}$, so each E_v is stable under $\mathrm{Aut}_c(\mathbb{C}_v/K_v)$ and all but finitely many E_v are $\mathfrak{X}$-trivial.*

Suppose $\overline{\gamma}(\mathbb{E}, \mathfrak{X}) > 1$. Then for any K-rational separable quasi-neighborhood $\mathbb{U}$ of $\mathbb{E}$, there are infinitely many points $\alpha \in \mathcal{C}(\widetilde{K}^{\mathrm{sep}})$ such that for each $v \in \mathcal{M}_K$, the $\mathrm{Aut}(\widetilde{K}/K)$-conjugates of α all belong to U_v.

Our next variant is a stronger, but more technical, version of Theorem 0.4, which produces points with all their conjugates in $\mathbb{E}$. It uses the inner capacity, and weakens the conditions on the sets E_v.

Write $\mathrm{cl}(E_v)$ for the closure of E_v in $\mathcal{C}_v(\mathbb{C}_v)$. If v is an archimedean place of K, and a set $E_v \subset \mathcal{C}_v(\mathbb{C})$ and a subset $E_v' \subset E_v$ are given, we will say that a point $z_0 \in E_v$ is *analytically accessible* from E_v' if for some $r > 0$, there is a non-constant analytic map $f : D(0,r)^- \to \mathcal{C}_v(\mathbb{C})$ with $f(0) = z_0$, such that $f((0,r)) \subset E_v'$. (See Definition 3.29.)

THEOREM 1.3 (Strong FSZ with LR). *Let K be a global field, and let $\mathcal{C}/K$ be a smooth, geometrically integral projective curve. Let $\mathfrak{X} = \{x_1, \ldots, x_m\} \subset \mathcal{C}(\widetilde{K})$ be a finite set of points stable under $\mathrm{Aut}(\widetilde{K}/K)$, and let $\mathbb{E} = \prod_v E_v \subset \prod_v \mathcal{C}_v(\mathbb{C}_v)$ be a K-rational adelic set compatible with $\mathfrak{X}$, so each E_v is stable under $\mathrm{Aut}_c(\mathbb{C}_v/K_v)$ and all but finitely many E_v are $\mathfrak{X}$-trivial. Let $S \subset \mathcal{M}_K$ be a finite set of places v, containing all archimedean v, such that E_v is $\mathfrak{X}$-trivial for each $v \notin S$.*

Assume that $\overline{\gamma}(\mathbb{E}, \mathfrak{X}) > 1$. Assume also that for each $v \in S$, there is a (possibly empty) $\mathrm{Aut}_c(\mathbb{C}_v/K_v)$-stable Borel subset $e_v \subset \mathcal{C}_v(\mathbb{C}_v)$ of inner capacity 0 such that

(A) If v is archimedean and $K_v \cong \mathbb{C}$, then each point of $\mathrm{cl}(E_v)\backslash e_v$ is analytically accessible from the $\mathcal{C}_v(\mathbb{C})$-interior of E_v.

(B) If v is archimedean and $K_v \cong \mathbb{R}$, then each point of $\mathrm{cl}(E_v)\backslash e_v$ is

(1) analytically accessible from the $\mathcal{C}_v(\mathbb{C})$-interior of E_v, or

(2) is an endpoint of an open segment contained in $E_v \cap \mathcal{C}_v(\mathbb{R})$.

(C) If v is nonarchimedean, then E_v is the disjoint union of e_v and finitely many sets $E_{v,1}, \ldots, E_{v,M_v}$, where each $E_{v,\ell}$ is

(1) open in $\mathcal{C}_v(\mathbb{C}_v)$, or

(2) of the form $U_{v,\ell} \cap \mathcal{C}_v(F_{w,\ell})$, where $U_{v,\ell}$ is open in $\mathcal{C}_v(\mathbb{C}_v)$ and $F_{w,\ell}$ is a separable algebraic extension of K_v contained in $\mathbb{C}_v$ (possibly of infinite degree).

Then there are infinitely many points $\alpha \in \mathcal{C}(\widetilde{K}^{\mathrm{sep}})$ such that for each $v \in \mathcal{M}_K$, the $\mathrm{Aut}(\widetilde{K}/K)$-conjugates of α all belong to E_v.

Note that if v is archimedean, then the set e_v in Theorem 1.3 can be taken to belong to ∂E_v, since trivially each point of the $\mathcal{C}_v(\mathbb{C})$-interior of E_v or the $\mathcal{C}_v(\mathbb{R})$-interior of $E_v \cap \mathcal{C}_v(\mathbb{R})$ is analytically accessible. Any countable set has inner capacity 0, so the conditions in Theorem 0.4 imply those in Theorem 1.3.

If v is nonarchimedean, note that RL-domains and balls $B(a,r)^-$, $B(a,r)$, are both open and closed in the $\mathcal{C}_v(\mathbb{C}_v)$-topology. Thus if E_v is a finite union of sets which are RL-domains, open or closed balls, or their intersections with $\mathcal{C}_v(F_{w,\ell})$ for separable algebraic extensions $F_{w,\ell}/K_v$ in $\mathbb{C}_v$, the theorem applies with $e_v = \phi$.

For an example of an archimedean set satisfying the conditions of Theorem 1.3 but not Theorem 0.4, take $K = \mathbb{Q}$, $\mathcal{C} = \mathbb{P}^1$, and let v be the archimedean place of $\mathbb{Q}$. Identify $\mathbb{P}^1(\mathbb{C})$ with $\mathbb{C} \cup \infty$, and take $E_v = \{0\} \cup \big(\bigcup_{n=2}^{\infty} D(2/n, 1/n^2)\big)$. Then each point of $E_v \backslash \{0\}$ is analytically accessible from E_v^0. For an example where the conditions of Theorem 1.3 fail, let E_v be the union of a circle $C(0,r)$ and countably many pairwise disjoint discs $D(a_i, r_i)$ contained in $D(0,r)^-$ chosen in such a way that each point of $C(0,r)$ is a limit point of those discs.

Our third variant is a version of Theorem 0.4 which adds side conditions concerning ramification. It says that at a finite number of places outside S we can require that the algebraic numbers produced are ramified or unramified, "for free".

THEOREM 1.4 (FSZ with LR and Ramification Side Conditions). *Let K be a global field, and let $\mathcal{C}/K$ be a smooth, connected, projective curve. Let $\mathfrak{X} = \{x_1, \ldots, x_m\} \subset \mathcal{C}(\widetilde{K})$ be a finite, Galois-stable set of points, and let $\mathbb{E} = \prod_v E_v \subset \prod_v \mathcal{C}_v(\mathbb{C}_v)$ be a K-rational adelic set compatible with $\mathfrak{X}$, so each E_v is stable under $\mathrm{Aut}_c(\mathbb{C}_v/K_v)$ and all but finitely many E_v are $\mathfrak{X}$-trivial.*

Let $S, S', S'' \subset \mathcal{M}_K$ be finite (possibly empty) sets of places of K which are pairwise disjoint, such that the places in $S' \cup S''$ are nonarchimedean. Assume that $\overline{\gamma}(\mathbb{E}, \mathfrak{X}) > 1$, and that

(A) for each $v \in S$, E_v satisfies the conditions of Theorem 0.4 or Theorem 1.3.

(B) for each $v \in S'$, either E_v is $\mathfrak{X}$-trivial, or E_v is a finite union of closed isometrically parametrizable balls $B(a_i, r_i)$ whose radii belong to the value group of $K_v^\times$ and whose centers belong to an unramified extension of K_v;

(C) for each $v \in S''$, either E_v is $\mathfrak{X}$-trivial and $E_v \cap \mathcal{C}_v(K_v)$ is nonempty, or E_v is a finite union of closed and/or open isometrically parametrizable balls $B(a_i, r_i)$, $B(a_j, r_j)^-$ with centers in $\mathcal{C}_v(K_v)$.

Then there are infinitely many points $\alpha \in \mathcal{C}(\widetilde{K}^{\mathrm{sep}})$ such that

(1) for each $v \in \mathcal{M}_K$, the $\mathrm{Aut}(\widetilde{K}/K)$-conjugates of α all belong to E_v;

(2) for each $v \in S'$, each place of $K(\alpha)/K$ above v is unramified over v;

(3) for each $v \in S''$, each place of $K(\alpha)/K$ above v is totally ramified over v.

Our fourth variant involves a partial converse to the Fekete-Szegö theorem, known as Fekete's theorem, which asserts that if $\gamma(\mathbb{E}, \mathfrak{X}) < 1$ then for a sufficiently small neighborhood $\mathbb{U}$ of $\mathbb{E}$, there are only finitely many points $\alpha \in \mathcal{C}(\widetilde{K})$ whose conjugates all belong to $\mathbb{U}$. Fekete's theorem on curves is proved in ([**51**], Theorem 6.3.1). However, Fekete's theorem requires a different notion of capacity than we have been using here: it concerns the "outer capacity" $\underline{\gamma}(\mathbb{E}, \mathfrak{X})$, rather than the inner capacity $\overline{\gamma}(\mathbb{E}, \mathfrak{X})$.

Extending the definition of algebraic capacitability in ([**51**]) to both archimedean and nonarchimedean sets, we will say that E_v *algebraically capacitable* if it is closed in $\mathcal{C}_v(\mathbb{C}_v)$ and $\overline{\gamma}_\zeta(E_v) = \underline{\gamma}_\zeta(E_v)$ for each $\zeta \notin E_v$. If each E_v is algebraically capacitable, then $\overline{\gamma}(\mathbb{E}, \mathfrak{X})$ and $\underline{\gamma}(\mathbb{E}, \mathfrak{X})$ are equal, and coincide with the capacity $\gamma(\mathbb{E}, \mathfrak{X})$ in ([**51**]). Here

$$\overline{\gamma}_\zeta(E_v) = \sup_{\substack{H_v \subset E_v \\ H_v \text{ compact}}} \gamma(H_v) , \qquad \underline{\gamma}_\zeta(E_v) = \inf_{\substack{U_v \supset E_v \\ U_v \text{ a } PL_\zeta\text{-domain}}} \overline{\gamma}(U_v) .$$

A set U_v is a PL_ζ-domain if there is a nonconstant rational function $f(z) \in \mathbb{C}_v(\mathcal{C})$, whose only poles are at ζ, for which $U_v = \{z \in \mathcal{C}_v(\mathbb{C}_v) : |f(z)|_v \le 1\}$. In the nonarchimedean case, the compatibility of this definition with the one given in ([**51**], p.259) follows from ([**51**], Propositions 4.3.1 and 4.3.16). In ([**51**]), algebraic capacitability was not defined in the archimedean case, but all archimedean sets were required to be compact.

If v is archimedean, it follows from ([**51**], Proposition 3.3.3) that every compact set is algebraically capacitable. If v is nonarchimedean, it is shown in ([**51**], Theorem 4.3.13) that any set E_v which can be expressed as a finite combination of unions and intersections of compact sets and RL-domains, is algebraically capacitable.

Assuming algebraic capacitability for the sets E_v, the following result describes the dichotomy provided by Fekete's theorem and the Fekete-Szegö theorem in terms of the Green's matrix $\Gamma(\mathbb{E}, \mathfrak{X})$. Recall (see [**51**], §5.1) that $\gamma(\mathbb{E}, \mathfrak{X}) > 1$ if and only if $\Gamma(\mathbb{E}, \mathfrak{X})$ is negative definite, and that $\gamma(\mathbb{E}, \mathfrak{X}) < 1$ if and only if when the rows and columns of $\Gamma(\mathbb{E}, \mathfrak{X})$ are permuted to bring $\Gamma(\mathbb{E}, \mathfrak{X})$ into block diagonal form, then some eigenvalue of each block is positive.

THEOREM 1.5 (Fekete/FSZ with LR for Algebraically Capacitable Sets). *Let K be a global field and let $\mathcal{C}/K$ be a smooth, connected, projective curve. Let $\mathfrak{X} = \{x_1, \ldots, x_m\} \subset \mathcal{C}(\widetilde{K})$ be a finite, Galois-stable set of points, and let $\mathbb{E} = \prod_v E_v \subset \prod_v \mathcal{C}_v(\mathbb{C}_v)$ be a K-rational adelic set compatible with $\mathfrak{X}$, so each E_v is stable under* $\mathrm{Aut}_c(\mathbb{C}_v/K_v)$ *and all but finitely many E_v are $\mathfrak{X}$-trivial. Assume that each E_v is algebraically capacitable. Then*

(A) If all the eigenvalues of $\Gamma(\mathbb{E}, \mathfrak{X})$ are non-positive (that is, $\Gamma(\mathbb{E}, \mathfrak{X})$ is either negative definite or negative semi-definite), let $\mathbb{U} = \prod_v U_v$ be a separable K-rational quasi-neighborhood of $\mathbb{E}$ such that there is at least one place v_0 where E_{v_0} is compact and the quasi-neighborhood U_{v_0} properly contains E_{v_0}. If v_0 is archimedean, assume also that U_{v_0} meets each component of $\mathcal{C}_{v_0}(\mathbb{C}) \backslash E_{v_0}$ containing a point of $\mathfrak{X}$. Then there are infinitely many $\alpha \in \mathcal{C}(\widetilde{K}^{\mathrm{sep}})$ such that all the conjugates of α belong to $\mathbb{U}$.

(B) If some eigenvalue of $\Gamma(\mathbb{E}, \mathfrak{X})$ is positive (that is, $\Gamma(\mathbb{E}, \mathfrak{X})$ is either indefinite, nonzero and positive semi-definite, or positive definite), there is an adelic neighborhood $\mathbb{U}$ of $\mathbb{E}$ such that only finitely many points $\alpha \in \mathcal{C}(\widetilde{K})$ have all their conjugates in $\mathbb{U}$.

Finally, we formulate two Berkovich versions of the Fekete-Szegö Theorem with local rationality.

For each nonarchimedean place v of K, let $\mathcal{C}_v^{\mathrm{an}}$ be the Berkovich analytic space associated to $\mathcal{C}_v \times_{K_v} \mathrm{Spec}(\mathbb{C}_v)$ (see [**10**]). This is a locally ringed space whose underlying topological space is a compact, path connected Hausdorff space with $\mathcal{C}_v(\mathbb{C}_v)$ as a dense subset; it has the same sheaf of functions as the rigid analytic space associated to $\mathcal{C}_v \times_{K_v} \mathrm{Spec}(\mathbb{C}_v)$. In his doctoral thesis, Amaury Thuillier ([**64**]) constructed a potential theory on $\mathcal{C}_v^{\mathrm{an}}$ which includes a dd^c operator, harmonic functions, subharmonic functions, capacities, and Green's functions. When $\mathcal{C} \cong \mathbb{P}^1$, Baker and Rumely ([**7**]) constructed a similar theory in an elementary way.

Below, we assume familiarity with Berkovich analytic spaces and Thuillier's theory. For each compact, non-polar subset $\mathbf{E}_v \subset \mathcal{C}_v^{\mathrm{an}}$ and each $\zeta \in \mathcal{C}_v^{\mathrm{an}} \backslash \mathbf{E}_v$, Thuillier ([**64**], Théorème 3.6.15) has constructed a Green's function $g_{\zeta, \mathbf{E}_v}(z)$ which is non-negative, vanishes on $\mathbf{E}_v$ except possibly on a set of capacity 0, is subharmonic in $\mathcal{C}_v^{\mathrm{an}}$, harmonic in $\mathcal{C}_v^{\mathrm{an}} \backslash (\mathbf{E}_v \cup \{\zeta\})$, and satisfies the distributional equation

$dd^c g_{\zeta,\mathbf{E}_v} = \mu - \delta_\zeta$ where μ is a probability measure supported on K. We will write $G(z,\zeta;\mathbf{E}_v)^{\mathrm{an}}$ for $g_{\zeta,\mathbf{E}_v}(z)$, and regard it as a function of two variables. By Proposition 4.3 below, for all $z,\zeta \in \mathcal{C}_v^{\mathrm{an}} \backslash \mathbf{E}_v$ with $z \neq \zeta$,

$$G(z,\zeta;\mathbf{E}_v)^{\mathrm{an}} = G(\zeta,z;\mathbf{E}_v)^{\mathrm{an}} ,$$

and for each $\zeta \in \mathcal{C}_v(\mathbb{C}_v)\backslash \mathbf{E}_v$, the Robin constant

$$V_\zeta(\mathbf{E}_v)^{\mathrm{an}} = \lim_{\substack{z\to\zeta \\ z\in \mathcal{C}_v^{\mathrm{an}}}} G(z,\zeta;\mathbf{E}_v)^{\mathrm{an}} + \log(|g_\zeta(z)|_v)$$

exists. The group $\mathrm{Aut}_c(\mathbb{C}_v/K_v)$ acts on $\mathcal{C}_v^{\mathrm{an}}$ in a natural way, and for all $\sigma \in \mathrm{Aut}_c(\mathbb{C}_v/K_v)$

$$G(\sigma(z),\sigma(\zeta);\sigma(\mathbf{E}_v))^{\mathrm{an}} = G(z,\zeta;\mathbf{E}_v)^{\mathrm{an}} .$$

By Proposition 4.4 below, the Green's functions $G(z,\zeta;\mathbf{E}_v)^{\mathrm{an}}$ and the functions $G(z,\zeta;E_v)$ from this work are compatible up to a normalizing factor, in the sense that if $E_v \subset \mathcal{C}_v(\mathbb{C}_v)$ is algebraically capacitable (in particular, if E_v is a finite union of RL-domains and compact sets), and if $\mathbf{E}_v$ is the closure of E_v in $\mathcal{C}_v^{\mathrm{an}}$ for the Berkovich topology, then for all $z,\zeta \in \mathcal{C}_v(\mathbb{C}_v)\backslash E_v$,

$$G(z,\zeta;\mathbf{E}_v)^{\mathrm{an}} = G(z,\zeta;E_v)\log(q_v) .$$

If v is an archimedean place of K, we take $\mathcal{C}_v^{\mathrm{an}}$ to be the Riemann surface $\mathcal{C}_v(\mathbb{C})$, and for a set $\mathbf{E}_v = E_v \subset \mathcal{C}_v(\mathbb{C})$ we put $G(z,\zeta;\mathbf{E}_v)^{\mathrm{an}} = G(z,\zeta;E_v)$ and $V_\zeta(\mathbf{E}_v)^{\mathrm{an}} = V_\zeta(E_v)$.

Let $\mathfrak{X} = \{x_1,\ldots,x_m\} \subset \mathcal{C}(\widetilde{K})$ be a finite, Galois-stable set of points. We will now define the notion of a *compact Berkovich adelic set compatible with* $\mathfrak{X}$. For each place v of K, let $\mathbf{E}_v \subset \mathcal{C}_v^{\mathrm{an}}$ be a compact, nonpolar set disjoint from $\mathfrak{X}$. (A set $\mathbf{E}_v \subset \mathcal{C}_v^{\mathrm{an}}$ is nonpolar if and only if it has positive capacity: see ([**64**], §3.4.2 and Theorem 3.6.11).) We will say that $\mathbf{E}_v$ is $\mathfrak{X}$-trivial if v is nonarchimedean and the model $\mathfrak{C}_v/\operatorname{Spec}(\mathcal{O}_v)$ from Definition 0.2 has good reduction, the points of $\mathfrak{X}$ specialize to distinct points in the special fibre $r_v(\mathfrak{C}_v)$, and $\mathbf{E}_v$ consists of all points $z \in \mathcal{C}_v^{an}$ whose specialization $r_v(z) \in r_v(\mathfrak{C}_v)$ is distinct from $\{r_v(x_1),\ldots,r_v(x_m)\}$. Equivalently, $\mathbf{E}_v$ is $\mathfrak{X}$-trivial if it is the closure of the $\mathfrak{X}$-trivial set $E_v = \mathcal{C}_v(\mathbb{C}_v)\backslash(\bigcup_{i=1}^m B(x_i,1)^-)$ in $\mathcal{C}_v(\mathbb{C}_v)$. Then

$$\mathbb{E} := \prod_v \mathbf{E}_v \subset \prod_v \mathcal{C}_v^{\mathrm{an}}$$

is a *compact Berkovich adelic set compatible with* $\mathfrak{X}$ if each $\mathbf{E}_v$ satisfies the conditions above, and $\mathbf{E}_v$ is $\mathfrak{X}$-trivial for all but finitely many v.

If $\mathbb{E}$ is a compact Berkovich adelic set compatible with $\mathfrak{X}$, we define the local and global Green's matrices $\Gamma(\mathbf{E}_w,\mathfrak{X})^{\mathrm{an}}$ and $\Gamma(\mathbb{E},\mathfrak{X})^{\mathrm{an}}$ as in (0.8), (0.9), replacing $G(z,\zeta;E_v)$ by $G(z,\zeta;\mathbf{E}_v)^{\mathrm{an}}$ and $V_\zeta(E_v)$ by $V_\zeta(\mathbf{E}_v)^{\mathrm{an}}$, but omitting the weights $\log(q_v)$ at nonarchimedean places. We then define the global Robin constant $V(\mathbb{E},\mathfrak{X})^{\mathrm{an}}$ using the minimax formula (0.11) taking $\Gamma = \Gamma(\mathbb{E},\mathfrak{X})^{\mathrm{an}}$, and the global capacity by

$$\gamma(\mathbb{E},\mathfrak{X})^{\mathrm{an}} = e^{-V(\mathbb{E},\mathfrak{X})^{\mathrm{an}}} .$$

We will call a set

$$\mathbb{U} = \prod_v \mathbf{U}_v \subset \prod_v \mathcal{C}_v^{\mathrm{an}}$$

a *Berkovich adelic neighborhood* of $\mathbb{E}$ if $\mathbf{U}_v$ contains $\mathbf{E}_v$ for each v, and either $\mathbf{U}_v$ is an open set in $\mathcal{C}_v^{\mathrm{an}}$, or $\mathbf{E}_v$ is $\mathfrak{X}$-trivial and $\mathbf{U}_v = \mathbf{E}_v$. We will call $\mathbb{U}$ a *separable Berkovich quasi-neigbhorhood* of $\mathbb{E}$ if $\mathbf{U}_v$ contains $\mathbf{E}_v$ for each v, and either $\mathbf{U}_v$ is the union of a Berkovich open set and finitely many open sets in $\mathcal{C}_v(F_w)$ for finite separable extensions F_w/K_v, or $\mathbf{E}_v$ is $\mathfrak{X}$-trivial and $\mathbf{U}_v = \mathbf{E}_v$. We will say that $\mathbb{U}$ is *K-rational* if each U_v is stable under $\mathrm{Aut}_c(\mathbb{C}_v/K_v)$.

The following is the Berkovich analogue of Theorem 0.4:

THEOREM 1.6 (Berkovich FSZ with LR). *Let K be a global field, and let $\mathcal{C}/K$ be a smooth, geometrically integral, projective curve. Let $\mathfrak{X} = \{x_1, \ldots, x_m\} \subset \mathcal{C}(\widetilde{K})$ be a finite set of points stable under $\mathrm{Aut}(\widetilde{K}/K)$, and let $\mathbb{E} = \prod_v \mathbf{E}_v \subset \prod_v \mathcal{C}_v^{\mathrm{an}}$ be a compact Berkovic adelic set compatible with $\mathfrak{X}$, such that each $\mathbf{E}_v$ is stable under $\mathrm{Aut}_c(\mathbb{C}_v/K_v)$. Let $S \subset \mathcal{M}_K$ be a finite set of places, containing all archimedean v, such that $\mathbf{E}_v$ is $\mathfrak{X}$-trivial for each $v \notin S$.*

Assume that $\gamma(\mathbb{E}, \mathfrak{X}) > 1$. Assume also that for each $v \in S$,

(A) If v is archimedean and $K_v \cong \mathbb{C}$, then $\mathbf{E}_v$ is compact, and is a finite union of sets $E_{v,i}$, each of which is compact, connected, and bounded by finitely many Jordan curves.

(B) If v is archimedean and $K_v \cong \mathbb{R}$, then $\mathbf{E}_v$ is compact and is a finite union of sets $E_{v,\ell}$, where each $E_{v,\ell}$ is either

(1) compact, connected, and bounded by finitely many Jordan curves, or

(2) is a closed subinterval of $\mathcal{C}_v(\mathbb{R})$ with nonempty interior.

(C) If v is nonarchimedean, then $\mathbf{E}_v$ is compact and is a finite union of sets $E_{v,\ell}$, where each $E_{v,\ell}$ is either

(1) a strict closed Berkovich affinoid, or

(2) is a compact subset of $\mathcal{C}_v(\mathbb{C}_v)$ and has the form $\mathcal{C}_v(F_{w,\ell}) \cap B(a_\ell, r_\ell)$ for some finite separable extension $F_{w,\ell}/K_v$ in $\mathbb{C}_v$, and some ball $B(a_\ell, r_\ell)$.

Then there are infinitely many points $\alpha \in \mathcal{C}(\widetilde{K}^{\mathrm{sep}})$ such that for each $v \in \mathcal{M}_K$, the $\mathrm{Aut}(\widetilde{K}/K)$-conjugates of α all belong to $\mathbf{E}_v$.

Last, we give a Berkovich version of the Fekete-Szegö Theorem with local rationality for quasi-neighborhoods, generalizing Theorem 1.2 and ([**7**], Theorem 7.48):

THEOREM 1.7 (Berkovich Fekete/FSZ with LR for Quasi-neighborhoods). *Let K be a global field, and let $\mathcal{C}/K$ be a smooth, connected, projective curve. Let $\mathfrak{X} = \{x_1, \ldots, x_m\} \subset \mathcal{C}(\widetilde{K})$ be a finite set of points stable under $\mathrm{Aut}(\widetilde{K}/K)$, and let $\mathbb{E} = \prod_v \mathbf{E}_v \subset \prod_v \mathcal{C}_v^{\mathrm{an}}$ be a compact Berkovich adelic set compatible with $\mathfrak{X}$, such that each $\mathbf{E}_v$ is stable under $\mathrm{Aut}_c(\mathbb{C}_v/K_v)$.*

(A) If $\gamma(\mathbb{E}, \mathfrak{X})^{\mathrm{an}} < 1$, there is a K-rational Berkovich neighborhood $\mathbb{U} = \prod_v \mathbf{U}_v$ of $\mathbb{E}$ such that there are only finitely many points of $\mathcal{C}(\widetilde{K})$ whose $\mathrm{Aut}(\widetilde{K}/K)$-conjugates are all contained in $\mathbf{U}_v$, for each $v \in \mathcal{M}_K$.

(B) If $\gamma(\mathbb{E}, \mathfrak{X})^{\mathrm{an}} > 1$, then for any K-rational separable Berkovich quasi-neighborhood $\mathbb{U}$ of $\mathbb{E}$, there are infinitely many points $\alpha \in \mathcal{C}(\widetilde{K}^{\mathrm{sep}})$ such that for each $v \in \mathcal{M}_K$, the $\mathrm{Aut}(\widetilde{K}/K)$-conjugates of α all belong to $\mathbf{U}_v$.

CHAPTER 2

Examples and Applications

In this chapter we illustrate the Fekete-Szegö theorem with local rationality. We first apply it on $\mathbb{P}^1$, using it to construct algebraic integers and algebraic units satisfying various conditions. We then apply it on elliptic curves, Fermat curves, and modular curves.

1. Local Capacities and Green's Functions of Archimedean Sets

Suppose $K_v = \mathbb{R}$ or $K_v = \mathbb{C}$. In this section we give formulas for local capacities and Green's functions of sets in $\mathbb{P}^1(\mathbb{C})$ which arise naturally in arithmetic applications. Some involve closed formulas, others require numerical computations. Most of the formulas appear in the literature. Further examples, mainly concerning sets in $\mathbb{C}$ with geometric symmetry, are given in ([**51**], pp. 348-351).

For archimedean sets, the most effective way of determining capacities is by "guessing" the Green's function: given E and $\zeta \notin E$, if a function can be found which is continuous, 0 on E, and harmonic in the complement of E except for a positive logarithmic pole at ζ, then by the maximum modulus principle, it must be the Green's function. Then, given a uniformizing parameter $g_\zeta(z)$, the Robin constant and capacity of E with respect to ζ can be read off by

$$(2.1) \qquad V_\zeta(E) \;=\; \lim_{z \to \zeta} G(z, \zeta; E) + \log(|g_\zeta(z)|) \;, \quad \gamma_\zeta(E) = e^{-V_\zeta(E)} \;.$$

For the sets we are dealing with here, which are compact unions of continua, the upper Green's function $\overline{G}(z, \zeta; E)$ coincides with the usual Green's function $G(z, \zeta; E)$.

In the discussion below, we will identify $\mathbb{P}^1(\mathbb{C})$ with $\mathbb{C} \cup \{\infty\}$. When $\zeta = \infty$, we take $g_\zeta(z) = 1/z$; when $\zeta \in \mathbb{C}$, we take $g_\zeta(z) = z - \zeta$.

The Disc. The most basic example is when E is the disc $D(0, r) \subset \mathbb{C}$. Here

$$(2.2) \qquad G(z, \infty; E) \;=\; \log^+(|z/r|) \;=\; \begin{cases} \log(|z/r|) & \text{if } |z| > r \\ 0 & \text{if } |z| \le r \end{cases} \;.$$

Computing capacities relative to the parameter $g_\infty(z) = 1/z$, we find

$$(2.3) \qquad V_\infty(E) = \lim_{z \to \infty} G(z, \infty; E) - \log(|z|) = -\log(r) \;, \qquad \gamma_\infty(E) = e^{-V_\infty(E)} = r \;.$$

By applying a linear fractional transformation, one can find the Green's function of $D(0, r)$ with respect to an arbitrary point $\zeta \in \mathbb{C}$:

$$(2.4) \qquad G(z, \zeta; E) \;=\; \log^+\left(\left|\frac{r^2 - \overline{\zeta} z}{r(z - \zeta)}\right|\right) \;.$$

Computing capacities relative to $g_\zeta(z) = z - \zeta$, one has

$$(2.5) \qquad V_\zeta(E) = \lim_{z \to \zeta} G(z,\infty;E) + \log(|z-\zeta|) = \log(\frac{|\zeta|^2 - r^2}{r}) ,$$

$$(2.6) \qquad \gamma_\zeta(E) = e^{-V_\zeta(E)} = \frac{r}{|\zeta|^2 - r^2} .$$

The Segment. Another basic example is when E is a segment $[a,b] \subset \mathbb{R}$. Choosing the branch of $\sqrt{z}$ which is positive on the positive real axis and cut along the negative real axis, the map $z \mapsto w = \sqrt{(z-a)/(z-b)}$ takes $\mathbb{P}^1(\mathbb{C})\backslash[a,b]$ to the right half-plane and takes ∞ to 1; then $w \mapsto (w+1)/(w-1)$ takes the right half-plane to the exterior of the unit disc, and takes 1 to ∞. It follows that

$$(2.7) \qquad G(z,\infty;E) = \log^+\left(\left|\frac{\sqrt{(z-a)/(z-b)}+1}{\sqrt{(z-a)/(z-b)}-1}\right|\right) .$$

For an arbitrary $\zeta \in \mathbb{C}$, a similar computation (see [**16**], p.165) gives

$$(2.8) \qquad G(z,\zeta;E) = \log^+\left(\left|\frac{\sqrt{(z-a)/(z-b)} + \overline{\sqrt{(\zeta-a)/(\zeta-b)}}}{\sqrt{(z-a)/(z-b)} - \overline{\sqrt{(\zeta-a)/(\zeta-bB)}}}\right|\right) .$$

With $g_\infty(z) = 1/z$ and $g_\zeta(z) = z - \zeta$ for $\zeta \in \mathbb{C}\backslash E$, one finds

$$(2.9) \qquad V_\infty(E) = -\log((b-a)/4) , \qquad \gamma_\infty(E) = (b-a)/4 ,$$

$$(2.10) \qquad \gamma_\zeta(E) = e^{-V_\zeta(E)} = \frac{b-a}{4 \cdot \mathrm{Re}(\sqrt{(\zeta-a)|\zeta-a| \cdot (\overline{\zeta-b})|\zeta-b|})}$$

When $\zeta = \infty$ there is another expression for $G(z,\infty,E)$ which makes its geometric behavior clearer. For simplicity, assume $E = [-2r, 2r]$ where $0 < r \in \mathbb{R}$. It is well known, and easy to verify, that the Joukowski map

$$(2.11) \qquad z = J_r(w) = w + \frac{r^2}{w}$$

maps $\mathbb{C}\backslash D(0,r)$ conformally onto $\mathbb{C}\backslash[-2r,2r]$. For each $R > r$, it takes the circle $C(0,R)$ parametrized by $w = R\cos(\theta) + iR\sin(\theta)$ to the ellipse $\mathcal{E}(R + \frac{r^2}{R}, R - \frac{r^2}{R})$ parametrized by

$$(2.12) \quad z = x+iy = (R+\frac{r^2}{R})\cos(\theta)+i(R-\frac{r^2}{R})\sin(\theta) = J_r(R\cos(\theta)+iR\sin(\theta)) .$$

It maps the circle $C(0,R)$ in a $2-1$ manner to the interval $[-2r,2r]$, and takes ∞ to ∞.

The function $G_r(z) = \log(|J_r^{-1}(z)|/r)$ is harmonic on $\mathbb{C}\backslash E$, with a logarithmic pole at ∞; it has a continuous extension to $\mathbb{C}$ which takes the value 0 on E. By the characterization of Green's functions, $G(z,\infty;[-2r,2r]) = G_r(z)$. Thus, for each $R > r$,

$$(2.13) \qquad \{z \in \mathbb{C} : G(z,\infty;[-2r,2r]) = \log(R/r)\} = \mathcal{E}(R + \frac{r^2}{R}, R - \frac{r^2}{R}) .$$

Two Segments. When $E = [a,b] \cup [c,d] \subset \mathbb{R}$, there are closed formulas for the Green's function and capacity. When the segments have the same length, $G(z,\infty;E)$ and $\gamma_\infty(E)$ are given by elementary formulas. In general, they can be expressed in terms of theta-functions.

First suppose $E = [-b,-a] \cup [a,b] \subset \mathbb{R}$. Put $f(z) = z^2$; then $f^*((\infty)) = 2(\infty)$ and $E = f^{-1}([a^2,b^2])$. By the pullback formula for Green's functions (see (2.61) below),

$$G(z,\infty;E) \;=\; \frac{1}{2}\log^+\left(\left|\frac{\sqrt{(z^2-a^2)/(z^2-b^2)}+1}{\sqrt{(z^2-a^2)/(z^2-b^2)}-1}\right|\right) . \tag{2.14}$$

Using this, we find

$$V_\infty(E) = \frac{1}{2}\log(4/(b^2-a^2)) \;, \quad \gamma_\infty(E) = \frac{\sqrt{b^2-a^2}}{2} \;, \tag{2.15}$$

$$G(0,\infty;E) \;=\; G(\infty,0,E) \;=\; \frac{1}{2}\log(\frac{b+a}{b-a}) \;. \tag{2.16}$$

Similarly, when $\zeta = 0$, pulling back $[1/b^2, 1/a^2]$ by $f(z) = 1/z^2$, we get

$$G(z,0;E) \;=\; -\frac{1}{2}\log\left(\left|\frac{\sqrt{(z^2-b^2)/(z^2-a^2)}+b/a}{\sqrt{(z^2-b^2)/(z^2-a^2)}-b/a}\right|\right) , \tag{2.17}$$

$$V_0(E) = \frac{1}{2}\log(4a^2b^2/(b^2-a^2)) \;, \quad \gamma_0(E) = \frac{\sqrt{b^2-a^2}}{2ab} \;. \tag{2.18}$$

Before dealing with a general set $E = [a,b] \cup [c,d] \subset \mathbb{R}$, and arbitrary ζ, it will be useful to recall some of the properties of classical theta-functions (see Shimura, [**60**], [**67**]). For $u \in \mathbb{C}$, $\tau \in \mathfrak{H} = \{\mathrm{Im}(z) > 0\}$, and $r,s \in \mathbb{R}$, write $e(z) = e^{2\pi i z}$ and put

$$\theta(u,\tau;r,s) \;=\; \sum_{n\in\mathbb{Z}} e(\frac{1}{2}(n+r)^2\tau + (n+r)(u+s)) \;. \tag{2.19}$$

Because of the quadratic dependence on n in (2.19), the series defining $\theta(u,\tau;r,s)$ converges very rapidly. $\theta(u,\tau;r,s)$ is continuous in all four variables and is jointly holomophic in u and τ.

We will be particularly interested in $\theta(u,\tau;\frac{1}{2},\frac{1}{2})$. When $r,s \in \{0,1/2\}$, the functions $\theta(u,\tau;r,s)$ appear in the literature with several names. Our notation follows that of Krazer-Prym and Shimura; it differs from that of Riemann and Mumford (and of Whittaker-Watson [**67**]). The equivalences (KPS = RM = WW) are as follows:

$$\theta(u,\tau;\tfrac{1}{2},\tfrac{1}{2}) = \theta_{11}(u,\tau) = \vartheta_1(\pi u|\tau) \;, \qquad \theta(u,\tau;\tfrac{1}{2},0) = \theta_{10}(u,\tau) = \vartheta_2(\pi u|\tau) \;,$$
$$\theta(u,\tau;0,0) = \theta_{00}(u,\tau) = \vartheta_3(\pi u|\tau) \;, \qquad \theta(u,\tau;0,\tfrac{1}{2}) = \theta_{01}(u,\tau) = \vartheta_4(\pi u|\tau) \;.$$

Courant and Hilbert ([**13**]) use a fourth notation; the equivalence (KPS = CH) is:

$$\theta(u,\tau;0,\frac{1}{2}) = \theta_0(u) \qquad \text{and} \qquad \theta(u,\tau;\frac{1}{2},\frac{1}{2}) = \theta_1(u) \;.$$

Considering $\theta(u,\tau;r,s)$ as a function of u and using the definition, one sees that for all $a,b \in \mathbb{Z}$

$$\theta(u+za+b,\tau;r,s) = e(rb-as)\cdot e(-\frac{1}{2}a^2\tau - au)\cdot\theta(u,z;r,s) \;. \tag{2.20}$$

Applying the Argument Principle, it follows that $\theta(u,\tau,r,s)$ has a simple zero in each period parallelogram for the lattice $\langle 1,\tau\rangle \subset \mathbb{C}$; the zero occurs at $u \equiv (\frac{1}{2}-r)\tau + (\frac{1}{2}-s) \pmod{\langle 1,\tau\rangle}$ (see [**67**], p.465-466, and [**60**], formula (11), p.675).

Again using the definitions, one sees that $\theta(u,\tau;\frac{1}{2},\frac{1}{2})$ is an odd function of u, that $\theta(u+\frac{1}{2},\tau;\frac{1}{2},\frac{1}{2}) = -\theta(u,\tau;\frac{1}{2},0)$, and if τ is pure imaginary, then $\theta(\overline{u},\tau;\frac{1}{2},\frac{1}{2}) = \overline{\theta(u,\tau;\frac{1}{2},\frac{1}{2})}$. Similarly $\theta(u,\tau;\frac{1}{2},0)$ is an even function of u, and if τ is pure imaginary, then $\theta(\overline{u},\tau;\frac{1}{2},0) = \overline{\theta(u,\tau;\frac{1}{2},0)}$.

With these facts, one can check that if τ is pure imaginary, then for each $M \in \mathbb{C}$ with $\mathrm{Re}(M) \notin \frac{1}{2}\mathbb{Z}$, the function

$$\mathcal{G}(u) = \frac{\theta(u-M,\tau;\frac{1}{2},\frac{1}{2})}{\theta(u+\overline{M},\tau;\frac{1}{2},\frac{1}{2})} \tag{2.21}$$

satisfies $|\mathcal{G}(u+\tau)| = |\mathcal{G}(u+1)| = |\mathcal{G}(u)|$, and if $\mathrm{Re}(u) = 0$ or if $\mathrm{Re}(u) = \frac{1}{2}$ then $|\mathcal{G}(u)| = 1$. It has simple zeros at points $u \equiv M \pmod{\langle 1,\tau\rangle}$, simple poles at $u \equiv -\overline{M} \pmod{\langle 1,\tau\rangle}$, and no other zeros or poles.

Now consider a set $E = [a,b] \cup [c,d] \subset \mathbb{R}$, where $a < b < c < d$. We will give a (multivalued, periodic) conformal mapping of $\mathbb{C}\backslash E$ onto a vertical strip, which will enable us to express $G(z,\zeta;E)$ in terms of the function $\mathcal{G}(u)$ in (2.21). We follow Akhiezer ([**2**]) and Falliero and Sebbar ([**22**], [**23**]), but obtain a different expression for the capacity.

First, put

$$w = T(z) = \sqrt{\frac{(z-a)}{(z-b)}\frac{(d-b)}{(d-a)}}\,. \tag{2.22}$$

where $\sqrt{z}$ is positive for $z > 0$ and is slit along the negative real axis. $T(z)$ maps $\mathbb{C}\backslash E$ conformally onto the right half-plane with the segment $[1, 1/k]$ removed, where

$$k = \frac{1}{T(c)} = \sqrt{\frac{(c-b)}{(c-a)}\frac{(d-a)}{(d-b)}}\,. \tag{2.23}$$

$T(z)$ takes $a \mapsto 0$, $b \mapsto \infty$, $d \mapsto 1$, and $c \mapsto 1/k$. Since the linear fractional transformation $F(z) = (z-a)(d-b)/(z-b)(d-a)$ maps $\mathbb{R} \cup \infty$ to itself and preserves the cyclic order of a, b, c, d, ones sees that $T(c) > 1$ and $0 < k < 1$. Note that $1/k^2$ is the crossratio $(a, b; c, d)$.

Follow $T(z)$ with the elliptic integral

$$u = S(w) = \int_0^w \frac{dx}{\sqrt{(1-x^2)(1-k^2x^2)}}\,. \tag{2.24}$$

Here $S(w)$ is the Schwarz-Christoffel map which sends the upper half-plane to the rectangle with corners $\pm K, \pm K + iK'$, where

$$K = \int_0^1 \frac{dx}{\sqrt{(1-x^2)(1-k^2x^2)}} \tag{2.25}$$

$$iK' = \int_1^{1/k} \frac{dx}{\sqrt{(1-x^2)(1-k^2x^2)}} \tag{2.26}$$

and $K, K' > 0$. It takes the imaginary axis to itself, and sends $0 \mapsto 0$, $1 \mapsto K$, $1/k \mapsto K + iK'$, and $\infty \mapsto iK'$. By the Schwarz Reflection Principle, $S(w)$ extends to a multivalued holomorphic function taking $\{\mathrm{Re}(w) > 0\}\backslash[1, 1/k]$ to the vertical strip $0 < \mathrm{Re}(u) < K$, with period $2iK'$. The inverse function to $S(w)$ is the Jacobian elliptic function $w = \mathrm{sn}(u, k)$ (see [**67**], §22, and [**44**], §VI.3).

Now let $\tau = iK'/K$. Fix $\zeta \notin E$; put $u = S(T(z))$, $M = M(\zeta) = S(T(\zeta))$. Scaling $u \mapsto v = u/(2K)$ takes $0 < \mathrm{Re}(u) < K$ to the strip $0 < \mathrm{Re}(v) < 1/2$, with $2iK' \mapsto \tau$. We claim that

$$G(z, \zeta; E) \;=\; -\log\left(\left|\frac{\theta(\frac{u-M}{2K}, \tau; \frac{1}{2}, \frac{1}{2})}{\theta(\frac{u+\overline{M}}{2K}, \tau; \frac{1}{2}, \frac{1}{2})}\right|\right). \tag{2.27}$$

Indeed, by our discussion of theta-functions, the function on the right has the properties characterizing $G(z, \zeta; E)$: it is well-defined and continuous, vanishes on E, is harmonic on $\mathbb{P}^1(\mathbb{C})\backslash(E \cup \zeta)$, and has a positive logarithmic pole as $z \to \zeta$. This formula is one given by Falliero and Sebbar ([**22**]; [**23**], p.416).

Numerically, K and K' can be found using the hypergeometric function

$$F(a, b, c; z) \;=\; 1 + \frac{a \cdot b}{1 \cdot c} z + \frac{a(a+1) \cdot b(b+1)}{1 \cdot 2 \cdot c(c+1)} z^2 + \cdots$$

with $K = \frac{1}{2}\pi F(\frac{1}{2}, \frac{1}{2}, 1; k^2)$, $K' = \frac{1}{2}\pi F(\frac{1}{2}, \frac{1}{2}, 1; 1-k^2)$ (see [**67**], pp.499, 501); then $\tau = iK'/K$. Another way to determine τ, K and K' is by first solving for $q = e^{i\pi\tau}$ using the relation

$$\frac{(c-b)}{(c-a)}\frac{(d-a)}{(d-b)} = k^2 = \frac{\theta(0, \tau, \frac{1}{2}, 0)^4}{\theta(0, \tau, 0, 0)^4} = \frac{16(q^{1/4} + q^{9/4} + q^{25/4} + \cdots)^4}{(1 + 2q^4 + 2q^9 + \cdots)^4} \tag{2.28}$$

and then using the formulas

$$K = \frac{1}{2}\pi\theta(0, \tau, 0, 0)^2 \,, \qquad K' = -i\tau K \,. \tag{2.29}$$

Finally, M can be determined by solving

$$T(\zeta) \;=\; \mathrm{sn}(M, k) \;=\; \frac{1}{k}\frac{\theta(M/2K, \tau; \frac{1}{2}, \frac{1}{2})}{\theta(M/2K, \tau; 0, \frac{1}{2})} \,. \tag{2.30}$$

(See [**67**], pp.492, 501.)

We now determine the capacity of E. If $\zeta = \infty$, put $\widehat{z} = 1/z$; otherwise put $\widehat{z} = z - \zeta$. Then as $\widehat{z} \to 0$, we have $z \to \zeta$, $w \to T(\zeta)$, and $u \to M$. Using (2.27), it follows that

$$\begin{aligned} V_\zeta(E) &= \lim_{\widehat{z} \to 0} G(z, \zeta; E) + \log(|\widehat{z}|) \\ &= -\log\left(\left|\frac{\frac{d}{du}\theta(0, \tau; \frac{1}{2}, \frac{1}{2})}{\theta(\frac{M+\overline{M}}{2K}, \tau; \frac{1}{2}, \frac{1}{2})} \cdot \frac{1}{2K} \cdot \frac{dw}{du}(T(\zeta)) \cdot \frac{dT}{d\widehat{z}}(0)\right|\right) \end{aligned} \tag{2.31}$$

The last two terms can be computed in terms of ζ and a, b, c, d; the expression can then be simplified using the Jacobi identity

$$\frac{d}{du}\theta(0, \tau; \frac{1}{2}, \frac{1}{2}) = \pi\theta(0, \tau; 0, 0)\theta(0, \tau; \frac{1}{2}, 0)\theta(0, \tau; 0, \frac{1}{2}) \tag{2.32}$$

(see [**67**], p.470), together with (2.29) and (2.28). If $\zeta = \infty$ one obtains

$$\gamma_\infty(E) = e^{-V_\infty(E)} = \frac{\sqrt[4]{(c-a)(c-b)(d-a)(d-b)}}{2\left|\frac{\theta(\mathrm{Re}(M(\infty))/K, \tau; \frac{1}{2}, \frac{1}{2})}{\theta(0, \tau; 0, \frac{1}{2})}\right|} \;; \tag{2.33}$$

if $\zeta \in \mathbb{C}\backslash E$, then

$$\gamma_\zeta(E) = \frac{\sqrt[4]{(c-a)(c-b)(d-a)(d-b)}}{2\left|\frac{\theta(\mathrm{Re}(M(\zeta))/K, \tau; \frac{1}{2}, \frac{1}{2})}{\theta(0, \tau; 0, \frac{1}{2})}\right| \cdot |(\zeta-a)(\zeta-b)(\zeta-c)(\zeta-d)|^{1/2}} \,. \tag{2.34}$$

Numerical examples confirm the compatibility of (2.33) with (2.15). However the formula of Akhiezer reported in ([**23**], p.422) seems to be incorrect.

Three Segments. When $E = [a_1, b_1] \cup [a_2, b_2] \cup [a_3, b_3] \subset \mathbb{R}$ and $\zeta = \infty$, Thérèse Falliero has given formulas for the Green's function and capacity of E using theta-functions of genus 2; for these, we refer the reader to Falliero ([**22**]) and Falliero-Sebbar ([**23**]).

Multiple Segments. When $E = [a_1, b_1] \cup [a_2, b_2] \cup \cdots \cup [a_n, b_n] \subset \mathbb{R}$ with $a_1 < b_1 < a_2 < b_2 < \cdots < a_n < b_n$ and n arbitrary, Harold Widom ([**68**], pp.224ff) has given formulas for $G(z, \zeta; E)$ and $V_\zeta(E)$ which we recall below.

Let $q(z) = \prod_{j=1}^n (z - a_j)(z - b_j)$, and fix the branch of the square root

$$q(z)^{1/2} \;=\; \sqrt{\prod_{j=1}^{n} (z - a_j)(z - b_j)}$$

on $\mathbb{C}\backslash E$ which is positive as $z \to \infty$ along the real axis. This branch is well-defined throughout $\mathbb{C}\backslash E$ and positive on $\mathbb{R}\backslash E$. For each $x \in E$, the limiting values of $q(z)^{1/2}$ as $z \to x + i0^+$ and $z \to x + i0^-$ are pure imaginary, and are negatives of each other. For convenience, we will extend $q(z)^{1/2}$ to E by defining $q(x)^{1/2} = q(x + i0^-)^{1/2}$ when $x \in E$. Thus, $q(z)^{1/2}$ is pure imaginary on E.

First take $\zeta = \infty$. Fix a point $z_0 \in E$, and let $h(z) = h_0 + h_1 z + \cdots + h_{n-1} z^{n-1} \in \mathbb{R}(z)$ be a polynomial of degree $\le n-1$ with real coefficients. Consider the multiple-valued function

$$G_h(z) \;=\; \int_{z_0}^{z} h(w)/q(w)^{1/2}\, dw$$

on $\mathbb{C}$, where the integral is taken over any path from z_0 to z which is disjoint from E except for one or both of its endpoints. Since $G_h(z)$ has pure imaginary periods around ∞ and around each component $[a_j, b_j]$ of E, the function $\mathrm{Re}(G_h(z))$ is well-defined and continuous, and constant on each component of E. Since $G(z)$ has a holomorphic branch in a neighborhood of each point $w \in \mathbb{C}\backslash E$, $\mathrm{Re}(G_h(z))$ is harmonic in $\mathbb{C}\backslash E$.

Clearly $\mathrm{Re}(G_h(z)) \equiv 0$ on the component of E containing z_0. If

$$\int_{b_j}^{a_{j+1}} h(x)/q(x)^{1/2}\, dx \;=\; 0 \tag{2.35}$$

for each 'gap' (b_j, a_{j+1}), then $\mathrm{Re}(G_h(z)) \equiv 0$ on E. If in addition $h(z)$ is monic, there is a number $V \in \mathbb{R}$ such that $\mathrm{Re}(G_h(z))$ is asymptotic to $\log(|z|) + V$ as $z \to \infty$. In this setting, the characterization of Green's functions shows that

$$G(z, \infty; E) \;=\; \mathrm{Re}(G_h(z))\ . \tag{2.36}$$

We will now show that such an $h(z)$ exists.

Following Widom, put $A_{jk} = \int_{b_j}^{a_{j+1}} x^k/q(x)\, dx$ for $j = 1, \ldots, n-1$, $k = 0, \ldots, n-1$. Note that $1/q(z)^{1/2}$ has singularities at b_j and a_{j+1} of order $z^{-1/2}$, so each A_{jk} is finite and belongs to $\mathbb{R}$. We claim that there is a unique solution $h_0, h_1, \ldots, h_{n-1}$ to the system of linear equations

$$\begin{cases} \sum_{k=0}^{n-1} A_{jk} h_k \;=\; 0 & \text{for } j = 1, \ldots, n-1\ , \\ \qquad\quad h_{n-1} \;=\; 1\ . \end{cases} \tag{2.37}$$

If $h_0, \ldots, h_{n-1}$ satisfy (2.37), and $h(z)$ is the corresponding polynomial, then $h(z)$ is monic and the conditions (2.35) hold. Thus $G(z, \infty; E) = \mathrm{Re}(G_h(z))$. Solving the system (2.37) is called the 'Jacobi Inversion Problem'.

To see that (2.37) has a unique solution, it suffices to show that the $n \times n$ matrix associated to the system has rank n, or equivalently, that $h_0 = \cdots = h_{n-1} = 0$ is the only solution to the corresponding homogeneous system. Let $h_0, \ldots, h_{n-1} \in \mathbb{R}$ be any solution to the homogeneous system, and let $h(z)$ be the corresponding polynomial. Then $G_h(z)$ is harmonic on $\mathbb{C}\backslash E$, vanishes on E, and remains bounded as $z \to \infty$ since $h_{n-1} = 0$, so it extends to a function on $\mathbb{P}^1(\mathbb{C})\backslash E$ harmonic at ∞. By the maximum principle for harmonic functions, $G_h(z) \equiv 0$. Restricting $G_h(z)$ to $\mathbb{R}$, differentiating, and using the Fundamental Theorem of Calculus, we see that $h(z)/q(z)^{1/2} \equiv 0$ on $\mathbb{R}\backslash E$. Since $q(z)$ is nonzero except at the endpoints of E, it follows that $h(z) \equiv 0$, and hence that $h_0 = \cdots = h_{n-1} = 0$.

We remark that $h(z)$ has one zero in each gap (b_j, a_{j+1}), and no other zeros. Indeed $G(x, \infty; E)$ vanishes at b_j and a_{j+1}, and is real-valued and differentiable on (b_j, a_{j+1}), so by Rolle's Theorem there is a point $x_j^* \in (b_j, a_{j+1})$ where $G'(x_j^*, \infty; E) = 0$. The argument above shows that $h(x_j^*) = 0$. Since $h(z)$ has degree $n-1$, it has a unique zero in each gap, and these are its only zeros. Thus, $h(z) = \prod_{j=1}^{n-1}(z - x_j^*)$, and $h(z)$ has constant sign on each component of E.

Now consider the case when $\zeta \in \mathbb{C}\backslash E$. Again, fix $z_0 \in E$. We claim that for a suitable polynomial $h(z) = h_0 + h_1 z + \cdots h_{n-1}z^{n-1} \in \mathbb{C}[z]$, we have

$$G(z, \zeta; E) \;=\; \mathrm{Re}(G_h(z)) \;, \tag{2.38}$$

where now

$$G_h(z) \;=\; \int_{z_0}^{z} \frac{h(w)}{q(w)^{1/2}(w - \zeta)} \, dw \;. \tag{2.39}$$

Here $h(z)$ must be chosen so that the following properties are satisfied:

(1) The periods of $G_h(z)$ are pure imaginary, so $\mathrm{Re}(G_h(z))$ is well-defined.
(2) The value of $\mathrm{Re}(G_h(z))$ is 0 on each segment $[a_j, b_j]$.
(3) $\mathrm{Re}(G_h(z))$ has a singularity of type $-\log(|z - \zeta|)$ at ζ.

Let Γ_j be a loop about $[a_j, b_j]$, traversed counterclockwise. Using Cauchy's theorem, one sees that

$$\int_{\Gamma_j} \frac{h(w)}{q(w)^{1/2}(w - \zeta)} \, dw \;=\; 2\int_{a_j}^{b_j} \frac{h(x)}{q(x)^{1/2}(x - \zeta)} \, dx \;.$$

Thus for the periods of $G_h(z)$ about the intervals $[a_j, b_j]$ to be pure imaginary, we need

$$\mathrm{Re}\Big(\int_{a_j}^{b_j} \frac{h(x)}{q(x)^{1/2}(x - \zeta)} \, dx\Big) \;=\; 0 \qquad \text{for } j = 1, \ldots, n \;. \tag{2.40}$$

Let $\varepsilon > 0$ be small enough that the circle $C(\zeta, \varepsilon)$ is disjoint from E. Since the differential $h(w)\,dw/\big(q(w)^{1/2}(w - \zeta)\big)$ is holomorphic at ∞, applying Cauchy's theorem on the domain $\mathbb{P}^1(\mathbb{C})\backslash(E \cup \{\zeta\})$ we obtain

$$\sum_{j=1}^{n} \int_{\Gamma_j} \frac{h(w)}{q(w)^{1/2}(w - \zeta)} \, dw + \int_{C(\zeta,\varepsilon)} \frac{h(w)}{q(w)^{1/2}(w - \zeta)} \, dw \;=\; 0 \;.$$

When (2.40) holds, the period of $G_h(z)$ about ζ (namely $2\pi i h(\zeta)/q(\zeta)^{1/2}$) is pure imaginary as well.

If (2.40) holds, then $\mathrm{Re}(G_h(z))$ is well-defined, harmonic in $\mathbb{P}^1(\mathbb{C})\backslash(E \cup \{\zeta\})$, and constant on each segment $[a_j, b_j]$. Clearly its value on the segment containing z_0 is 0. For it to be identically 0 on E, we need

$$\mathrm{Re}\Big(\int_{b_j}^{a_{j+1}} \frac{h(x)}{q(x)^{1/2}(x-\zeta)}\,dx\Big) \;=\; 0 \qquad \text{for } j = 1, \ldots, n-1 \ . \tag{2.41}$$

Finally, for $\mathrm{Re}(G_h(z))$ to have a singularity of type $-\log(|z-\zeta|)$ at ζ, we need

$$-1 \;=\; \mathrm{Res}_{w=\zeta}\Big(\frac{h(w)}{q(w)^{1/2}(w-\zeta)}\Big) \;=\; \frac{h(\zeta)}{q(\zeta)^{1/2}} \ .$$

Since the period of $G_h(z)$ about ζ is imaginary we have $\mathrm{Im}(h(\zeta)/q(\zeta)^{1/2}) = 0$, and it is enough to require

$$-1 \;=\; \mathrm{Re}\Big(\frac{h(\zeta)}{q(\zeta)^{1/2}}\Big) \ . \tag{2.42}$$

Writing $h_k = c_k + d_k i$ for $k = 0, \ldots, n-1$, with $c_k, d_k \in \mathbb{R}$, the conditions (2.40), (2.41) and (2.42) represent a system of $2n$ linear equations with real coefficients in $2n$ real unknowns. To show that it has a unique solution, it is enough to show that the only solution to the corresponding homogeneous system is the trivial one.

Suppose that $h(z)$ arises from a solution to the homogeneous system. Then $\mathrm{Re}(G_h(z))$ is harmonic in $\mathbb{P}^1(\mathbb{C})\backslash(E \cup \{\zeta\})$ and extends to a function harmonic at ζ, with boundary values 0 on E. By the Maximum Principle, $\mathrm{Re}(G_h(z)) \equiv 0$. Differentiating, and using the Fundamental Theorem of Calculus on horizontal segments, we see that

$$\mathrm{Re}\Big(\frac{h(z)}{q(z)^{1/2}(z-\zeta)}\Big) \;=\; \frac{\partial}{\partial x}\big(\mathrm{Re}(G_h(z))\big) \;\equiv\; 0$$

on $\mathbb{C}\backslash(E \cup \{\zeta\})$. If the real part of an analytic function is identically 0, that function is constant, so we must have

$$\frac{h(z)}{q(z)^{1/2}(z-\zeta)} \;=\; C$$

for some purely imaginary constant C. However, $q(z)^{1/2}(z-\zeta)$ is not a polynomial, so this can hold only if $C = 0$. Thus $h(z) \equiv 0$, which means that $c_1 = d_1 = \cdots = c_n = d_n = 0$.

When $\zeta = \infty$, choosing $z_0 \neq 0$ and noting that $\log(|z|) = \mathrm{Re}\big(\int_{z_0}^{\infty} 1/w\,dw\big) - \log(|z_0|)$, Widom gives a formula for the Robin constant equivalent to

$$\begin{aligned} V_\infty(E) \;&=\; \lim_{z\to\infty}\big(G(z,\infty;E) - \log(|z|)\big) \\ &=\; \mathrm{Re}\Big(\int_{z_0}^{\infty} \frac{h(w)}{q(w)^{1/2}} - \frac{1}{w}\,dw\Big) \;+\; \log(|z_0|) \ . \end{aligned}$$

Similarly, when $\zeta \in \mathbb{C}\backslash E$,

$$\begin{aligned} V_\zeta(E) \;&=\; \lim_{z\to\zeta}\big(G(z,\zeta;E) + \log(|z-\zeta|)\big) \\ &=\; \mathrm{Re}\Big(\int_{z_0}^{\infty} \frac{h(w)}{q(w)^{1/2}(w-\zeta)} + \frac{1}{w-\zeta}\,dw\Big) \;-\; \log(|z_0-\zeta|) \ . \end{aligned}$$

When $\zeta = \infty$, there is a more illuminating formula for $V_\infty(E)$. Put $c = (a_1+b_n)/2$ and let $r = (b_n - a_1)/4$, so $E \subset [a_1, b_n] = [c-2r, c+2r]$. We claim that

$$V_\infty(E) \;=\; -\log(\frac{b_n - a_1}{4}) \;+\; \sum_{j=1}^{n-1} \int_{b_j}^{a_{j+1}} G(x,\infty;E)\,\frac{1}{\pi}\frac{dx}{\sqrt{4r^2-(x-c)^2}}\;. \tag{2.43}$$

Readers familiar with capacity theory will recognize $dx/(\pi\sqrt{4r^2-(x-c)^2})$ as the equilibrium distribution of $[a_1, b_n]$ relative to ∞.

To derive (2.43), assume for simplicity that $c = 0$, so $[a_1, b_n] = [-2r, 2r]$; this can always be arranged by a translation. Note that since $G(z,\infty;[-2r,2r]) \sim \log(|z|) + V_\infty([-2r,2r])$ as $z \to \infty$, and $V_\infty([-2r,2r]) = V_\infty([a_1,b_n]) = -\log((b_n - a_1)/4)$, we have

$$\begin{aligned} V_\infty(E) &:= \lim_{z\to\infty} G(z,\infty;E) - \log(|z|) \\ &= \lim_{z\to\infty} \big(G(z,\infty;E) - G(z,\infty;[-2r,2r])\big) - \log(\frac{b_n-a_1}{4})\;. \end{aligned} \tag{2.44}$$

The function $g(z) := G(z,\infty;E) - G(z,\infty;[-2r,2r])$ is harmonic in $\mathbb{C}\backslash[-2r,2r]$ and bounded as $z \to \infty$; hence it extends to a function harmonic in $\mathbb{P}^1(\mathbb{C})\backslash[-2r,2r]$, with

$$g(\infty) \;=\; \lim_{z\to\infty} \big(G(z,\infty;E) - G(z,\infty;[-2r,2r])\big)\;. \tag{2.45}$$

Let the Joukowski map $z = J_r(w) = w + r^2/w$ be as in (2.11). For each $R > r$, parametrize the ellipse $\mathcal{E}(R + r^2/R, R - r^2/R)$ by $z = J_r(R\cos(\theta) + iR\sin(\theta))$ as in (2.12). Let $\mathcal{D}_R = \mathbb{P}^1(\mathbb{C})\backslash D(0,R)$, and let $\mathcal{E}_R$ be the connected component of $\mathbb{P}^1(\mathbb{C})\backslash\mathcal{E}(R+r^2/R, R-r^2/R)$ containing ∞. The map $J_r(w)$ gives a conformal equivalence from $\mathcal{D}_R$ to $\mathcal{E}_R$, and takes ∞ to ∞. Thus $H(w) := g(J_r(w))$ is harmonic in $\mathcal{D}_R$, and $H(\infty) = g(\infty)$. By the mean value theorem for harmonic functions,

$$\begin{aligned} g(\infty) &= H(\infty) = \frac{1}{2\pi}\int_0^{2\pi} H(R\cos(\theta) + iR\sin(\theta))\,d\theta \\ &= \frac{1}{2\pi}\int_0^{2\pi} g((R+\frac{r^2}{R})\cos(\theta) + i(R - \frac{r^2}{R})\sin(\theta))\,d\theta\;. \end{aligned}$$

Since $\mathcal{E}(R+r^2/R, R-r^2/R)$ is the level curve $\log(R/r)$ for $G(z,\infty;[-2r,2r])$ (see (2.13)), it follows that

$$g(\infty) \;=\; \frac{1}{2\pi}\int_0^{2\pi} G((R+\frac{r^2}{R})\cos(\theta) + i(R-\frac{r^2}{R})\sin(\theta)),\infty;E)\,d\theta \;-\; \log(R/r)\;.$$

Since $G(z,\infty;E) = G_h(z)$ is continuous on $\mathbb{C}$, letting $R \to r$ we see that

$$g(\infty) \;=\; \frac{1}{2\pi}\int_0^{2\pi} G(2r\cos(\theta),\infty;E)\,d\theta\;.$$

Finally, making the change of variables $x = 2r\cos(\theta)$ yields

$$g(\infty) \;=\; \int_{a_1}^{b_n} G(x,\infty;E)\,\frac{1}{\pi}\frac{dx}{\sqrt{4r^2-x^2}} \;=\; \sum_{j=1}^{n-1}\int_{b_j}^{a_{j+1}} G(x,\infty;E)\,\frac{1}{\pi}\frac{dx}{\sqrt{4r^2-x^2}}\;. \tag{2.46}$$

Combining (2.44), (2.45) and (2.46) gives (2.43).

The Real Projective Line. If $E = \mathbb{P}^1(\mathbb{R})$, the components of its complement in $\mathbb{P}^1(\mathbb{C})$ are the upper and lower half-planes. Fix $\zeta \notin E$. Using the characterization of the Green's function, it is easy to check that if z and ζ belong to the same component of $\mathbb{P}^1(\mathbb{C})\backslash\mathbb{P}^1(\mathbb{R})$, then

$$G(z,\zeta;E) \;=\; -\log\left(\left|\frac{z-\zeta}{z-\overline{\zeta}}\right|\right). \tag{2.47}$$

If z and ζ are not in the same component, then $G(z,\zeta;E)=0$. Taking $g_\zeta(z)=z-\zeta$ and using (2.47) we obtain

$$V_\zeta(E) \;=\; \lim_{z\to\zeta} -\log\left(\left|\frac{z-\zeta}{z-\overline{\zeta}}\right|\right)+\log(|z-\zeta|) \;=\; \log(2|\operatorname{Im}(\zeta)|)\,, \tag{2.48}$$

$$\gamma_\zeta(E) \;=\; \frac{1}{2|\operatorname{Im}(\zeta)|}\,. \tag{2.49}$$

The Disc with Opposite Radial Arms. Take $L_1, L_2 \ge 0$, and let E be the union of $D(0,R)$ with the segment $[-L_1-R, R+L_2]$; thus E is a disc with opposite radial arms of length L_1, L_2. We claim that

$$\gamma_\infty(E) \;=\; \frac{1}{4}\left(2R+\frac{R^2+RL_1+L_1^2}{R+L_1}+\frac{R^2+RL_2+L_2^2}{R+L_2}\right). \tag{2.50}$$

To see this, first take $R=1$. Put $a_1=1+L_1$, $a_2=1+L_2$; then $E=D(0,1)\cup[-a_1,a_2]$. Let $w=\varphi(z)=(z-1)^2/z$. Then φ is the composite of the maps $z\to t=-1/(z+1)$, $t\to u=t+1/2$, $u\to v=u^2$, and $v\to w=-1/(v-1/4)$. Using standard properties of conformal maps one sees that $\varphi(z)$ maps $\mathbb{C}\backslash E$ conformally onto $\mathbb{C}\backslash[A,B]$, where

$$A=-\frac{(a_1+1)^2}{a_1}\,,\qquad B=\frac{(a_2-1)^2}{a_2}\,.$$

Clearly $\varphi(\infty)=\infty$. Since $\lim_{z\to\infty}\log(|w|/|z|)=0$, it follows that

$$\gamma_\infty(E) \;=\; \gamma_\infty([A,B]) \;=\; \frac{B-A}{4} \;=\; \frac{(a_1a_2+1)(a_1+a_2)}{4a_1a_2}\,. \tag{2.51}$$

In the general case, put $a_1=1+L_1/R$, $a_2=1+L_2/R$, and scale (2.51) by R; after simplification, one obtains (2.50). The expression (2.51) appears in ([**33**], p.82).

For the set $E=D(0,1)\cup[-a_1,a_2]$ discussed above, and for $z,\zeta\notin E$, one has

$$G(z,\zeta;E) \;=\; G(\varphi(z),\varphi(\zeta);[A,B])$$

where the Green's function of $[A,B]$ is given by (2.8). This can be used to find $\gamma_\zeta(E)$ for any $\zeta\in\mathbb{C}\backslash E$.

Two Concentric Circles. Fix $r>1$, and let E be the union of the circles $|z|=1/r$ and $|z|=r$. The complement of E has three components. If z and ζ belong to different components, then $G(z,\zeta;E)=0$. If they belong to the outer component, then $G(z,\zeta;E)=G(z,\zeta;D(0,r))$, while if they belong to the inner component then $G(z,\zeta;E)=G(1/z,1/\zeta;D(0,r))$.

$G(z,\zeta;E)$ is also known when z and ζ belong to the annular region between the circles. Courant and Hilbert ([**13**], pp. 386–388) derive a formula for it using the Schwarz Reflection Principle: define q by $q^{1/2}=1/r$, and suppose $1/r<|z|,|\zeta|<r$. Courant and Hilbert show that $G(z,\zeta;E)=-\log(|f_\zeta(z)|)$, where

$$f_\zeta(z)=|z|^{-\log(|\zeta|)/\log(q)}\cdot\frac{q^{1/4}(\sqrt{\frac{z}{\zeta}}-\sqrt{\frac{\zeta}{z}}\,)\prod_{n=1}^\infty(1-q^{2n}\frac{z}{\zeta})(1-q^{2n}\frac{\zeta}{z})}{\prod_{n=1}^\infty(1-q^{2n-1}\overline{\zeta}z)(1-q^{2n-1}\frac{1}{\overline{\zeta}z})}. \tag{2.52}$$

Recalling the product expansions of the theta functions, they note that the second term is a quotient of two theta functions, leading to the following expression: writing $\tau = 2i\log(r)/\pi$, $z = e^{2\pi i u}$ and $\zeta = e^{2\pi i\alpha}$, then for $1/r < |z|, |\zeta| < r$,

$$(2.53) \qquad G(z,\zeta;E) = -\frac{\log(|z|)\log(|\zeta|)}{2\log(r)} - \log\left(\left|\frac{\theta(u-\alpha,\tau;\frac{1}{2},\frac{1}{2})}{\theta(u-\overline{\alpha},\tau;0,\frac{1}{2})}\right|\right).$$

Here we have corrected a minor error in Courant-Hilbert, who state (2.52) for positive real ζ, and omit the conjugate on ζ in generalizing; this changes their $\theta(u+\alpha,\tau;0,\frac{1}{2})$ to $\theta(u-\overline{\alpha},\tau;0,\frac{1}{2})$. Using (2.53), we obtain

$$\begin{aligned}(2.54)\ V_\zeta(E) &= \lim_{z\to\zeta} G(z,\zeta;E) + \log(|z-\zeta|) \\ &= -\frac{(\log(|\zeta|))^2}{2\log(r)} + \log(|\theta(\alpha-\overline{\alpha},\tau;0,\frac{1}{2})|) \\ &\qquad - \log(|\frac{d}{du}\theta(0,\tau;\frac{1}{2},\frac{1}{2})| + \log(|\frac{dz}{du}(\alpha)|) \\ &= -\frac{(\log(|\zeta|))^2}{2\log(r)} + \log\left(\left|\frac{2\zeta\cdot\theta(\alpha-\overline{\alpha},\tau;0,\frac{1}{2})}{\theta(0,\tau;0,0)\theta(0,\tau;\frac{1}{2},0)\theta(0,\tau;0,\frac{1}{2})}\right|\right),\end{aligned}$$

where we have used Jacobi's identity (2.32) to simplify $\frac{d}{du}\theta(0,\tau;\frac{1}{2},\frac{1}{2})$. When $\zeta = 1$, this becomes

(2.55)
$$V_1(E) = \log\left(\frac{2}{|\theta(0,\tau;0,0)\theta(0,\tau;\frac{1}{2},0)|}\right), \qquad \gamma_1(E) = \frac{|\theta(0,\tau;0,0)\theta(0,\tau;\frac{1}{2},0)|}{2}.$$

We next consider some sets arising in Polynomial Dynamics:

Julia Sets. Let $\varphi(x) = a_0 + a_1x + \cdots + a_dx^d \in \mathbb{C}[x]$ be a polynomial of degree $d \geq 2$. By definition, the *filled Julia set* $\mathcal{K}_\varphi$ of $\varphi(x)$ is the set of all $z \in \mathbb{C}$ whose forward orbit $z, \varphi(z), \varphi(\varphi(z)), \ldots$ under φ remains bounded; the *Julia set* is its boundary $\mathcal{J}_\varphi = \partial\mathcal{K}_\varphi$. Let $\varphi^{(n)} = \varphi\circ\varphi\circ\cdots\circ\varphi$ be the n-fold iterate. For each sufficiently large R, we have $D(0,R) \supset \varphi^{-1}(D(0,R)) \supset (\varphi^{(2)})^{-1}(D(0,R)) \supset \cdots \supset \mathcal{K}_\varphi$, and

$$\mathcal{K}_\varphi = \bigcap_{n=1}^{\infty}(\varphi^{(n)})^{-1}(D(0,R)).$$

As in ([**62**], p.147), for each $z \in \mathbb{C}$ we have

$$(2.56) \qquad G(z,\infty;\mathcal{J}_\varphi) = G(z,\infty;\mathcal{K}_\varphi) = \lim_{n\to\infty}\frac{1}{d^n}\log^+(|\varphi^{(n)}(z)|)$$

(the "escape velocity" of z), and

$$(2.57)\quad V_\infty(\mathcal{J}_\varphi) = V_\infty(\mathcal{K}_\varphi) = \frac{\log(|a_d|)}{d-1}, \qquad \gamma_\infty(\mathcal{J}_\varphi) = \gamma_\infty(\mathcal{K}_\varphi) = |a_d|^{-1/(d-1)}.$$

The proofs of (2.56) and (2.57) are simple. It is easy to see that $\varphi^{(n)}(z)$ has degree d^n and leading coefficient $a_0^{d^{n-1}+d^{n-2}+\cdots+d+1}$. By the characterization of Green's functions it follows that $G(z,\infty;(\varphi^{(n)})^{-1}(D(0,R))) = d^{-n}\log^+(|\varphi^{(n)}(z)|)$, and that

$$V_\infty((\varphi^{(n)})^{-1}(D(0,R))) = \frac{d^n-1}{d^n(d-1)}\log(|a_0|) = \frac{1-1/d^n}{d-1}\log(|a_0|).$$

The functions $G(z,\infty;(\varphi^{(n)})^{-1}(D(0,R))$ decrease monotonically to $G(z,\infty;\mathcal{K}_\varphi)$, and the convergence is uniform outside any neighborhood of $\mathcal{K}_\varphi$, so (2.56) and (2.57) follow.

The Mandelbrot Set. Each quadratic polynomial $\varphi_c(x) = x^2+c \in \mathbb{C}[x]$ has 0 as a critical point. The *Mandelbrot set* $\mathcal{M}$ is the set of all $c \in \mathbb{C}$ for which the forward orbit $0, \varphi_c(0), \varphi_c^{(2)}(0), \ldots$ remains bounded; equivalently, $\mathcal{M}$ is the set of all $c \in \mathcal{C}$ for which 0 belongs to the filled Julia set of $\varphi_c(x)$. It is easy to see that $\varphi_c(0) = c$, $\varphi_c^{(2)}(0) = c^2-c$, and $\varphi_c^{(3)}(0) = c^4-2c^3+c^2-c$; in general $P_n(c) := \varphi_c^{(n+1)}(0)$ is a monic polynomial of degree 2^n in $\mathbb{Z}[c]$. It can be shown that $D(0,2) \supset P_1^{-1}(D(0,2)) \supset P_2^{-1}(D(0,2)) \supset \cdots \supset \mathcal{M}$, and that

$$\mathcal{M} = \bigcap_{n=1}^{\infty} P_n^{-1}(D(0,2)) \; ; \tag{2.58}$$

see ([**62**], p.158). By arguments like those for Julia sets, for each $c \in \mathbb{C}$ we have

$$G(c,\infty;\mathcal{M}) = \lim_{n\to\infty} \frac{1}{2^n}\log^+(|P_n(c)|) = \lim_{n\to\infty}\frac{1}{2^n}\log^+(\varphi_c^{(n+1)}(0)) \tag{2.59}$$

and

$$V_\infty(\mathcal{M}) = 0 \;, \qquad \gamma_\infty(\mathcal{M}) = 1 \;. \tag{2.60}$$

2. Local Capacities and Green's Functions of Nonarchimedean Sets

In this section, K_v will be a nonarchimedean local field. Identify $\mathbb{P}^1(\mathbb{C}_v)$ with $\mathbb{C}_v \cup \{\infty\}$. There are two methods of determining the Green's function for sets $E_v \subset \mathbb{P}^1(\mathbb{C}_v)$: using the pullback formula for Green's functions, for noncompact sets; or by guessing the equilibrium distribution based on symmetry, for compact sets. We are aided by the fact that the capacity is monotonic under containment of sets.

The pullback formula for Green's functions is as follows. Let $\mathcal{C}_1, \mathcal{C}_2/\mathbb{C}_v$ be smooth, complete curves, and let $f : \mathcal{C}_1 \to \mathcal{C}_2$ be a nonconstant rational map. Suppose $E_v \subset \mathcal{C}_2(\mathbb{C}_v)$ is an algebraically capacitable set of positive capacity. Fix $\zeta \in \mathcal{C}_2(\mathbb{C}_v)\backslash E_v$ and write $f^*((\zeta)) = \sum_{j=1}^m m_k(\xi_j)$ as a divisor. Then for each $z \in \mathcal{C}_1(\mathbb{C}_v)$,

$$G(f(z),\zeta;E_v) = \sum_{j=1}^m m_k G(z,\xi_j;f^{-1}(E_v)) \;. \tag{2.61}$$

This holds for both nonarchimedean and archimedean sets (see [**51**], Theorems 3.2.9, 4.4.19).

The Closed Disc. If $E_v = D(a,R) = \{z \in \mathbb{C}_v : |z-a|_v \le R\}$ then

$$G(z,\zeta;E_v) = \begin{cases} \log_v^+(|z-a|_v/R) & \text{if } \zeta = \infty \;, \\ \log_v^+\left(\frac{|\zeta-a|_v}{R}\cdot\left|\frac{z-a}{z-\zeta}\right|_v\right) & \text{if } \zeta \in \mathbb{C}_v\backslash D(a,R) \;. \end{cases} \tag{2.62}$$

The first formula is essentially the definition of the Green's function as given by Cantor ([**16**]); the second follows from the first, by applying the pullback formula (2.61) to the map $f(z) = (z-a)/(z-\zeta)$ which takes $D(a,R)$ to $D(0,R/|\zeta-a|_v)$ and takes ζ to ∞.

Taking $g_\infty(z) = 1/z$, and $g_\zeta(z) = z - \zeta$ if $\zeta \in \mathbb{C}_v \backslash D(a,R)$, we have

$$V_\infty(E_v) = -\log_v(R)\ , \qquad \gamma_\infty(E_v) = q_v^{-V_\infty(E_v)} = R\ , \tag{2.63}$$

$$V_\zeta(E_v) = -\log_v(R/|\zeta - a|_v^2)\ , \quad \gamma_\zeta(E_v) = R/|\zeta - a|_v^2\ . \tag{2.64}$$

The Open Disc. If $E_v = D(a,R)^- = \{z \in \mathbb{C}_v : |z|_v < R\}$, formulas (2.62), (2.63), and (2.64) for the Green's function, Robin constant and capacity remain valid.

If $\zeta = \infty$, this is because for $R_1 < R$ we have $D(a,R_1) \subset D(a,R)^- \subset D(a,R)$, and hence

$$G(z,\infty;D(a,R_1)) \le G(z,\infty;D(a,R)^-) \le G(z,\infty;D(a,R))\ ,$$
$$\gamma_\infty(D(a,R_1)) \le \gamma_\infty(D(a,R)^-) \le \gamma_\infty(D(a,R))\ .$$

In the limit as $R_1 \to R$, it follows from (2.62) and (2.63) that $G(z,\infty;D(a,R)^-) = G(z,\infty;D(a,R))$ and $\gamma_\infty(D(a,R)^-) = \gamma_\infty(D(a,R))$.

If $\zeta \in \mathbb{C}_v \backslash D(a,R)^-$, we can reduce to the case where $\zeta = \infty$ by applying the map $f(z) = (z-a)/(z-\zeta)$ and using the pullback formula (2.61). Thus (2.62), (2.63), and (2.64) hold when $E_v = D(a,R)^-$, for any $\zeta \notin E_v$.

The Punctured Disc. Suppose $E_v = D(a,R)\backslash(\bigcup_{i=1}^m D(a_i,R_i)^-)$, where $a_1,\ldots,a_m \in D(0,R)$ and $R_i \le R$ for each i. For each $\zeta \notin D(a,R)$, the Green's function and capacity are still given by (2.62), (2.63), and (2.64). Indeed, for any fixed $a_0 \in E_v$, we have $D(a_0,R)^- \subset E_v \subset D(a_0,R)$, so the result follows from the previous case.

If ζ belongs to one of the "holes" $D(a_i,R_i)^-$, then $D(a_i,R_i)^- = D(\zeta,R_i)^-$ and by applying $f(z) = 1/(z-\zeta)$ and using the pullback formula (2.61), we find that

$$G(z,\zeta;E_v) = G(\frac{1}{z-\zeta},\infty;D(0,\frac{1}{R_i})) = \log_v^+\left(\frac{R_i}{|z-\zeta|_v}\right)\ , \tag{2.65}$$

$$V_\zeta(E_v) = \log_v(R_i)\ , \quad \gamma_\zeta(E_v) = 1/R_i\ . \tag{2.66}$$

The Ring of Integers $\mathcal{O}_w$. We next determine the Green's function of the ring of integers of a finite extension F_w/K_v in $\mathbb{C}_v$.

PROPOSITION 2.1. *Let F_w/K_v be a finite extension in $\mathbb{C}_v$, with ramification index $e = e_{w/v}$ and residue degree $f = f_{w/v}$. Take $E_v = \mathcal{O}_w$, the ring of integers of F_w. Given $z \in \mathbb{C}_v$, put*

$$r = \|z,\mathcal{O}_w\|_v = \min_{x \in \mathcal{O}_w} |z-x|_v\ .$$

Let $M = \lfloor -e\log_v(r) \rfloor$ be the integer part of $-e\log_v(r)$, and let $\langle -e\log_v(r) \rangle$ be its fractional part. Then

$$G(z,\infty;\mathcal{O}_w) = \begin{cases} 0 & \text{if } z \in \mathcal{O}_w, \\ \frac{1}{e}\frac{1}{q_v^f-1}\frac{1}{q_v^{fM}} - \langle -e\log_v(r)\rangle \frac{1}{e}\frac{1}{q_v^{f(M+1)}} & \text{if } z \notin \mathcal{O}_w,\ |z|_v \le 1, \\ \frac{1}{e}\frac{1}{q_v^f-1} + \log_v(|z|_v) & \text{if } |z|_v > 1\ , \end{cases} \tag{2.67}$$

and if capacities are computed relative to the uniformizer $g_\infty(z) = 1/z$ then

$$V_\infty(\mathcal{O}_w) = \frac{1}{e}\frac{1}{q_v^f - 1}\ , \qquad \gamma_\infty(\mathcal{O}_w) = q_v^{-1/(e(q_v^f-1))}\ . \tag{2.68}$$

For any coset $a + b\mathcal{O}_w$ *where* $a \in \mathbb{C}_v$, $b \in \mathbb{C}_v^\times$,

$$G(z, \infty; a + b\mathcal{O}_w) \;=\; G((z-a)/b, \infty; \mathcal{O}_w) \;,$$

so that $V_\infty(a + b\mathcal{O}_w) = -\log_v(|b|_v) + V_\infty(\mathcal{O}_w)$ *and* $\gamma_\infty(a + b\mathcal{O}_w) = |b|_v \cdot \gamma_\infty(\mathcal{O}_w)$. *In particular, if* π_w *is a generator for the maximal ideal of* $\mathcal{O}_w$, *then*

(2.69)
$$V_\infty(a + \pi_w^m\mathcal{O}_w) \;=\; \frac{1}{e}\frac{1}{q_v^f - 1} + \frac{m}{e} \;, \qquad \gamma_\infty(a + \pi_w^m\mathcal{O}_w) \;=\; q_v^{-m/e - 1/(e(q_v^f - 1))} \;.$$

Proof. See ([**51**], Example 5.2.17). The equilibrium distribution of $\mathcal{O}_w$ is the additive Haar measure μ for F_w, normalized so that $\mu(\mathcal{O}_w) = 1$ (see [**51**], p.212). It follows that if we write $q_w = q_v^f$, and put $M = \lfloor -e\log_v(r)\rfloor$ if $|z|_v \le 1$, $M = -1$ if $|z|_v > 1$, then the potential function is given by

$$\begin{aligned} u_{\mathcal{O}_w}(z,\infty) &= \int_{\mathcal{O}_w} -\log_v(|z-x|_v)\,d\mu(x) \\ &= \sum_{k=0}^{M} \frac{k}{e}\cdot\frac{q_w - 1}{q_w^{k+1}} + \sum_{k=M+1}^{\infty} (-\log_v(r))\cdot\frac{q_w-1}{q_w^{k+1}} \\ &= \frac{1}{e}\frac{1}{q_w - 1}\cdot[1 - \frac{M+1}{q_w^M} + \frac{1}{q_w^{M+1}}] - \log_v(r)\cdot\frac{1}{q_w^{M+1}}. \end{aligned}$$

The potential function is invariant under translation by $\mathcal{O}_w$, so

$$V_\infty(\mathcal{O}_w) \;=\; u_{\mathcal{O}_w}(0,\infty) \;=\; \frac{1}{e(q_w - 1)} \;.$$

The expression (2.67) is obtained by simplifying

$$G(z,\infty;\mathcal{O}_w) \;=\; \frac{1}{e(q_w-1)} - u_{\mathcal{O}_w}(z,\infty) \;.$$

(Compare [**51**], Example 4.1.24, p.212). The assertions about cosets follow. □

We now recall a general procedure for computing capacities of finite disjoint unions of nonarchimedean sets (for more details, see Theorem A.13 and Corollary A.14 of Appendix A, or see [**51**], p.354).

Let $\mathcal{C}_v/K_v$ be a curve. Suppose $E_v = \bigcup_{i=1}^N E_{v,i} \subset \mathcal{C}_v(\mathbb{C}_v)$ is a finite disjoint union of compact sets $E_{v,i}$ with positive inner capacity, and that $\zeta \in \mathcal{C}_v(\mathbb{C}_v)$ is such that the canonical distance $[z,w]_\zeta$ (see §3.5) is constant on $E_{v,i} \times E_{v,j}$, for each $i \ne j$. For each i, let $\mu_{\zeta,i}$ be the equilibrium distribution of $E_{v,i}$ (see §3.8). Then each i, the potential function $u_{E_{v,i}}(z,\zeta) = \int_{E_{v,i}} -\log([z,w]_\zeta d\mu_{\zeta,i}(w)$ and Green's function $G(z,\zeta;E_{v,i}) = V_\zeta(E_v) - u_{E_{v,i}}(z,\zeta)$ are constant for $z \in E_{v,j}$, for each $j \ne i$.

We now show that we can compute $G(z,\zeta;E_v)$ and $V_\zeta(E_v)$ in terms of the potential functions $u_{E_{v,i}}(z,\zeta)$. Let capacities be defined in terms of the uniformizer $g_\zeta(z)$. For each $E_{v,i}$, put

$$W_{ii} \;=\; V_\zeta(E_{v,i})$$

and for each $i \neq j$ let W_{ij} be the value that $u_{E_{v,i}}(z,\zeta)$ assumes on $E_{v,j}$. Consider the system of $N+1$ linear equations in the variables V, $s_1, \ldots, s_N$:

$$\begin{aligned} 1 &= 0 \cdot V + 1 \cdot s_1 + 1 \cdot s_2 + \cdots + 1 \cdot s_N \ , \\ 0 &= V - W_{i1}s_1 - W_{i2}s_2 - \ldots - W_{iN}s_N, \\ & \qquad \text{for } i = 1, \ldots, N. \end{aligned} \tag{2.70}$$

We claim that this system of equations has a unique solution, given by

$$V_\zeta(E_v) = V \ , \tag{2.71}$$

$$G(z;\zeta;E_v) = \sum s_i G(z,\zeta;E_{v,i}) + \sum s_i W_{ii} - V, \tag{2.72}$$

and that in this solution we have $s_1, \ldots, s_N > 0$.

To see this, let μ be the equilibrium distribution of E_v with respect to ζ, and put $\widehat{s}_i = \mu(E_{v,i})$ for each i. Then $\widehat{s}_i > 0$: otherwise, μ would be supported on $E_v \backslash E_{v,i}$ and then $u_\zeta(z, E_v) = u_\zeta(z, E_v \backslash E_{v,i})$. By ([**51**], Corollary 4.1.12) we would have $u_\zeta(z, E_v \backslash E_{v,i}) < V_\zeta(E_v \backslash E_{v,i}) = V_\zeta(E_v)$ for all $z \in E_{v,i}$, contradicting that $u_\zeta(z; E_v)$ takes the value $V_\zeta(E_v)$ for all $z \in E_v$ except possibly on a set of inner capacity 0 ([**51**], Theorem 4.1.11). Consider the probability measure $\mu_i = \widehat{s_i}^{-1} \mu|_{E_{v,i}}$, and put $u_i(z,\zeta) = \int_{E_{v,i}} -\log_v([z,w]_\zeta)\, d\mu_i(w)$; by our hypothesis on the canonical distance, $u_i(z,\zeta)$ is constant on E_j, for each $j \neq i$. Then

$$\begin{aligned} u_{E_v}(z,\zeta) &= \int_{E_v} -\log_v([z,w]_\zeta)\, d\mu(z) \\ &= \sum_{i=1}^r \int_{E_{v,i}} -\log_v([z,w]_\zeta)\, d\mu(z) \ = \ \sum_{i=1}^r \widehat{s}_i u_i(z,\zeta) \ . \end{aligned}$$

For each i, since $u_{E_v}(z,\zeta)$ and the $u_j(z,\zeta)$ for $j \neq i$ are constant on $E_{v,i}$ except possibly on a set of inner capacity 0 , it follows that $u_i(z,\zeta)$ is constant on $E_{v,i}$ except possibly on a set of inner capacity 0. Since this property characterizes the equilibrium potential, it follows that μ_i must be the equilibrium distribution of $E_{v,i}$ with respect to ζ. Thus there are unique weights $\widehat{s}_1, \ldots, \widehat{s}_N > 0$ with $\sum_{i=1}^N \widehat{s}_i = 1$, for which

$$u_{E_v}(z,\zeta) = \sum_{i=1}^N \widehat{s}_i u_{E_{v,i}}(z,\zeta) \ . \tag{2.73}$$

Evaluating (2.73) at a generic point of each $E_{v,i}$, we see that $V = V_\zeta(E_v)$ and $\widehat{s}_1, \ldots, \widehat{s}_N$ are a solution to the system (2.70) with each $\widehat{s}_i > 0$. Conversely, any solution to (2.70) gives a system of weights for which $\mu = \sum s_i \mu_i$. The uniqueness of the equilibrium distribution ([**51**], Theorem 4.1.22) shows that $s_1, \ldots, s_N$, and in turn V, are unique. Thus $s_i = \widehat{s}_i$ for each i, and $V_\zeta(E_v) = V$. Since $G(z,\zeta;E_v) = V_\zeta(E_v) - u_{E_v}(z,\zeta)$, formula (2.71) follows.

The Group of Units $\mathcal{O}_w^\times$. Using the machinery above, we will now determine the Green's function and the capacity of the set $\mathcal{O}_w^\times$, relative to the point ∞.

PROPOSITION 2.2. *Let F_w/K_v be a finite extension, with ramification index $e = e_{w/v}$ and residue degree $f = f_{w/v}$. Let $\mathcal{O}_w^\times$ be the group of units of $\mathcal{O}_w$. For $z \in \mathbb{C}_v$, put $r_0 = \min_{x \in \mathcal{O}_w^\times} |z - x|_v$, $M_0 = \lfloor -e \log_v(r_0) \rfloor$; note that $r_0 = |z|_v$ if*

$|z|_v > 1$. *Then*

$$(2.74)\quad G(z,\infty;\mathcal{O}_w^\times) \;=\; \begin{cases} 0 & \text{if } z \in \mathcal{O}_w^\times, \\ \frac{q_v^f}{e(q_v^f-1)^2}\cdot\frac{1}{q_v^{fM_0}} - \langle -e\log_v(r_0)\rangle \frac{1}{e}\frac{1}{q_v^{fM_0}} & \text{if } 0 < r_0 \le 1, \\ \frac{q_v^f}{e(q_v^f-1)^2} + \log_v(|z|_v) & \text{if } |z|_v > 1. \end{cases}$$

If capacities are computed relative to the uniformizer $g_\infty(z) = 1/z$ then

$$(2.75)\qquad V_\infty(\mathcal{O}_w^\times) \;=\; \frac{1}{e}\frac{1}{q^f-1}\left(1+\frac{1}{q^f-1}\right) \;=\; \frac{q_v^f}{e(q_v^f-1)^2}\ .$$

PROOF. Put $N = q_v^f - 1$ and let $a_1,\ldots,a_N$ be coset representatives for the nonzero classes in $\mathcal{O}_w/\pi_w\mathcal{O}_w$. Then

$$\mathcal{O}_w^\times \;=\; \bigcup_{i=1}^N (a_i + \pi_w\mathcal{O}_w)$$

is a decomposition of the type needed to compute $G(z,\infty;\mathcal{O}_w^\times)$ in terms of the $G(z,\infty;a_i+\pi\mathcal{O}_w)$. Applying the last part of Proposition 2.1, solving the system (2.70) and simplifying (2.71), (2.72) gives the result. Here $W_{ij} = 0$ if $i \neq j$ and each $W_{ii} = q_v^f/(e(q_v^f-1))$, giving $V = q_v^f/(e(q_v^f-1)^2)$ and $s_i = 1/(q_v^f-1)$ for each i. □

COROLLARY 2.3. *Let K_v be nonarchimedean, and let π_v be a uniformizer for the maximal ideal of $\mathcal{O}_v$. Suppose $a_1,\ldots,a_N$ are representives for distinct cosets of $\mathcal{O}_v/\pi_v\mathcal{O}_v$, and put $E_v = \cup_{i=1}^N (a_i+\pi_v\mathcal{O}_v)$. Then*

$$V_\infty(E_v) \;=\; \frac{q_v}{N(q_v-1)}.$$

PROOF. The proof is similar to Proposition 2.2, with $s_i = 1/N$ for $i = 1,\ldots,N$. □

The Punctured $\mathcal{O}_v$-disc. Next we determine the capacity of a union of cosets of $\mathcal{O}_v^\times$, relative to the point $\zeta = \infty$. This computation has important consequences: it is used in the proof Proposition 3.30, which plays a key role in the reduction of Theorem 0.4 to Theorem 4.2.

PROPOSITION 2.4. *Put $E_{v,m} = \bigcup_{k=0}^m \pi_v^k\mathcal{O}_v^\times$, and take $\zeta = \infty$. Then*

$$(2.76)\qquad V_\infty(E_{v,m}) \;=\; \frac{1}{q_v-1} + \frac{1}{(q_v-1)^2(1+q_v^2+q_v^4+\cdots+q_v^{2m})},$$

$$(2.77)\qquad G(0,\infty;E_{v,m}) \;=\; \frac{q_v^{m+1}}{(q_v-1)^2(1+q_v^2+q_v^4+\cdots+q_v^{2m})}\ ,$$

and for each $k = 0,\ldots,m$ the mass of $\pi_v^k\mathcal{O}_v^\times$ under the equilibrium distribution μ_m of $E_{v,m}$ with respect to ∞ is

$$(2.78)\qquad \mu_m(\pi_v^k\mathcal{O}_v^\times) \;=\; \frac{q_v^k + q_v^{2m+1-k}}{1+q_v+q_v^2+q_v^3+\cdots+q_v^{2m+1}}\ .$$

PROOF. Write $V_m = V_\infty(E_{v,m})$. By Proposition 2.2, we have

$$(2.79)\qquad V_0 \;=\; \frac{q_v}{(q_v-1)^2} \;=\; \frac{1}{q_v-1} + \frac{1}{(q_v-1)^2}\ .$$

We will prove (2.76) by induction on m. Note that $E_{v,m} = \pi_v E_{v,m-1} \cup \mathcal{O}_v^\times$. For $z \in \pi_v E_{v,m-1}$ and $w \in \mathcal{O}_v^\times$, $-\log_v([z,w]_\infty) = -\log_v(|z-w|_v) = 0$, independent of z, w. Hence $u_{\pi_v E_{v,m-1}}(z,\infty) = 0$ if $z \in \mathcal{O}_w^\times$, and $u_{\mathcal{O}_w^\times}(z,\infty) = 0$ if $z \in \pi_v E_{v,m-1}$. By the scaling property of the capacity, $V_\infty(\pi_v E_{v,m}) = V_\infty(E_{v,m-1}) + 1 = V_{m-1}+1$. It follows from (2.70) that there are numbers $s_{1,m}, s_{2,m} > 0$ for which

$$(2.80) \qquad \left\{ \begin{array}{rcl} 1 & = & s_{1,m} + s_{2,m}, \\ V_m & = & (V_{m-1}+1)\cdot s_{1,m} + 0 \cdot s_{2,m}, \\ V_m & = & 0 \cdot s_{1,m} + V_0 \cdot s_{2,m} \end{array} \right. .$$

Solving (2.80) for V_m and inserting (2.79) leads to the recursion

$$V_m = \frac{q_v(1+V_{m-1})}{q_v + (q_v-1)^2 V_{m-1}}$$

whose solution is easily seen to be (2.76).

Once the V_m are known, one sees that

$$(2.81) \qquad s_{1,m} = \frac{q_v(1+q_v+\cdots+q_v^{2m-1})}{1+q_v+\cdots q_v^{2m+1}}, \qquad s_{2,m} = \frac{1+q_v^{2m+1}}{1+q_v+\cdots q_v^{2m+1}} .$$

To obtain (2.77), note that since $u_{\mathcal{O}_w^\times}(0,\infty) = 0$, we have $u_{E_{v,m}}(0,\infty) = s_{1,m}\cdot(1+u_{E_{v,m-1}}(0,\infty))$. Thus, recursively,

$$(2.82) \qquad u_{E_{v,m}}(0,\infty) = s_{1,m} + s_{1,m}s_{1,m-1} + \cdots + s_{1,m}s_{1,m-1}\cdots s_{1,1} .$$

One gets (2.77) by inserting (2.76), (2.81) and (2.82) in the formula

$$G(0,\infty;E_{v,m}) = V_m - u_{E_{v,m}}(0,\infty)$$

and simplifying. Finally, the weights of the cosets $\pi_v^k \mathcal{O}_w^\times$ under the equilibrium distribution μ_m can be found by using

$$\begin{aligned} \mu_m(\pi_v^k \mathcal{O}_w^\times) &= s_{1,m}\mu_{m-1}(\pi_v^{k-1}\mathcal{O}_w^\times) = \cdots \\ &= s_{1,m}s_{1,m-1}\cdots s_{1,m-k+1}\cdot \mu_{m-k}(\mathcal{O}_w^\times) \end{aligned}$$

where $\mu_{m-k}(\mathcal{O}_w^\times) = s_{2,m-k}$. Using (2.81), and simplifying, yields (2.78). Once the weights $\mu_m(\pi_v^k \mathcal{O}_v^\times)$ are known, $G(z,\infty;E_{v,m})$ can be found for any z. □

The Union of Two Rings of Integers. Let F_w be the unique unramified quadratic extension of K_v, and let F_u be a totally ramified quadratic extension. We will compute the capacity of the set $E_v = \mathcal{O}_w \cup \mathcal{O}_u$ with respect to ∞. This is the only nonarchimedean set known to the author whose Robin constant can be computed explicitly, and is not rational. The importance is not the result itself, but the method, which uses a partial self-similarity of E_v with itself, and can be applied to nondisjoint unions of much more general sets.

PROPOSITION 2.5. *Fix a nonarchimedean local field K_v. Let F_w/K_v be the unique unramified quadratic extension, and let F_u/K_v be a totally ramified quadratic extension. Put $E_v = \mathcal{O}_w \cup \mathcal{O}_u$ and let*

$$\begin{aligned} A &= 2q_v^4 + 2q_v^3 - 4q_v^2 + 2q_v - 2 , \\ B &= q_v^4 + 2q_v^3 - 2q_v^2 + 2q_v - 1 , \\ D &= q_v^8 + 4q_v^7 + 8q_v^6 + 12q_v^5 + 18q_v^4 + 12q_v^3 + 8q_v^2 + 4q_v + 1 . \end{aligned}$$

Then

$$(2.83) \qquad V_\infty(E_v) = \frac{-B+\sqrt{D}}{2A} .$$

Below are some numerical examples when $K_v = \mathbb{Q}_p$, for small primes p. We give the values of $V_\infty(\mathcal{O}_w)$ and $V_\infty(\mathcal{O}_u)$ for comparison.

	$q_v = 2$	$q_v = 3$	$q_v = 5$	$q_v = 7$	$q_v = 11$
$V_\infty(E_v)$	.2750820518	.1060035774	.0366954968	.0188065868	.0077456591
$V_\infty(\mathcal{O}_w)$	.3333333333	.1250000000	.0416666666	.0208333333	.0083333333
$V_\infty(\mathcal{O}_u)$	.5000000000	.2500000000	.1250000000	.0833333333	.0500000000

It can be shown that as $q_v \to \infty$, then $V_\infty(E_v) = 1/q_v^2 - 1/q_v^3 + O(1/q_v^4)$.

Proof of Proposition 2.5. Let $\pi = \pi_v$ be a generator for the maximal ideal of $\mathcal{O}_v$, and write $q = q_v$. Then $\#(\mathcal{O}_v/\pi\mathcal{O}_v) = q$; let $\gamma_1, \ldots, \gamma_q$ be coset representatives for $\mathcal{O}_v/\pi\mathcal{O}_v$. Put $E_{0,i} = \gamma_i + \pi E_v = \gamma_i + \pi(\mathcal{O}_w \cup \mathcal{O}_u)$, for $i = 1, \ldots, q$. There are $q^2 - q$ cosets of $\mathcal{O}_w/\pi\mathcal{O}_w$ which do not contain elements of $\mathcal{O}_v$; let these be $E_{1,j} = \alpha_j + \pi\mathcal{O}_w$, for $j = 1, \ldots, q^2 - q$. Similarly, there are $q^2 - q$ cosets of $\mathcal{O}_u/\pi\mathcal{O}_u$ which do not contain elements of $\mathcal{O}_v$; let these be $E_{2,k} = \beta_k + \pi\mathcal{O}_u$, for $k = 1, \ldots, q^2 - q$. Then the sets $E_{0,i}$, $E_{1,j}$ and $E_{2,k}$ are pairwise disjoint (in fact, they are contained in pairwise disjoint cosets $a + \pi\widehat{\mathcal{O}}_v$, where $\widehat{\mathcal{O}}_v = D(0,1)$ is the ring of integers of $\mathbb{C}_v$), and we can write

$$E_v = \Big(\bigcup_{i=1}^{q} E_{1,i}\Big) \cup \Big(\bigcup_{j=1}^{q^2-q} E_{2,j}\Big) \cup \Big(\bigcup_{k=1}^{q^2-q} E_{3,k}\Big) .$$

Let μ be the equilibrium distribution of E_v with respect to ∞, and put $w_{0,i} = \mu(E_{0,i})$, $w_{1,j} = \mu(E_{1,j})$, $w_{2,k} = \mu(E_{2,k})$ for all i, j, k. Then

(2.84)

$$u_\infty(z, E_v) = \sum_{i=1}^{q} w_{1,i} u_\infty(z, E_{0,i}) + \sum_{j=1}^{q^2-q} w_{0,j} u_\infty(z, E_{1,j}) + \sum_{k=1}^{q^2-q} w_{2,k} u_\infty(z, E_{2,k}) .$$

Let $V = V_\infty(E_v)$ be the (as yet unknown) Robin constant of $E_v = \mathcal{O}_w \cup \mathcal{O}_u$, and let $V_1 = V_\infty(\mathcal{O}_w)$, $V_2 = V_\infty(\mathcal{O}_u)$. Since $E_v \subset D(0,1)$, we must have $V \geq 0$. By Proposition 2.1,

$$V_1 = \frac{1}{q^2 - 1} , \qquad V_2 = \frac{1}{2(q-1)} . \tag{2.85}$$

In general, for any compact set $\widetilde{E} \subset \mathbb{C}_v$ of positive capacity, we have $V_\infty(a + \pi\widetilde{E}) = V_\infty(\widetilde{E}) + 1$ for each $a \in \mathbb{C}_v$. If $\widetilde{E} \subset D(a,r)$, then $u_\infty(z, \widetilde{E}) = -\log_v(|z-a|_v)$ for all $z \notin D(a,r)$. It follows that for each $E_{0,i}$, one has $u_\infty(z, E_{0,i}) = V + 1$ on $E_{0,i}$. On the $q - 1$ cosets $E_{2,k}$ contained in $\gamma_i + \sqrt{\pi}\widehat{\mathcal{O}}_v$, one has $u_\infty(z, E_{0,i}) = 1/2$. On the other $q^2 - 2q + 1$ cosets $E_{2,k}$ and the other $q - 1$ cosets $E_{0,i'}$, as well as all the cosets $E_{1,j}$, one has $u_\infty(z, E_{0,i}) = 0$. For each $E_{1,j}$, one has $u_\infty(z, E_{1,j}) = V_1 + 1$ on $E_{1,j}$, and $u_\infty(z, E_{1,j}) = 0$ on all the $E_{0,i}$, all the $E_{2,j}$ and all the $E_{1,j'}$ distinct from j. For each $E_{2,k}$, one has $u_\infty(z, E_{2,k}) = V_2 + 1$ on $E_{2,k}$. There are $q - 2$ other cosets $E_{2,k'}$ and one coset $E_{1,j}$ contained in $\beta_k + \sqrt{\pi}\widehat{\mathcal{O}}_v$. On those cosets we have $u_\infty(z, E_{2,k}) = 1/2$. On the remaining $q^2 - 2q + 1$ cosets $E_{2,k'}$ and on all the cosets $E_{1,j}$, one has $u_\infty(z, E_{2,k}) = 0$.

Evaluating $u_\infty(z, E_v)$ on each of the sets $E_{r,s}$ in turn yields a system of $2q^2 - q$ equations satisfied by V and the $w_{r,s}$. Since μ and $V = V_\infty(E_v)$ are unique, these equations uniquely determine the $w_{r,s}$. Hence for any permutation σ of the sets $E_{r,s}$

which takes sets of type $r = 0, 1, 2$ to sets of the same type, and which preserves distances between corresponding pairs of sets, we must have $w_{r,\sigma(s)} = w_{r,s}$ for all r, s. It is easy to see that there are enough permutations satisfying these conditions to assure that there are w_0, w_1, w_2 such that for all i, j, k

$$w_{0,i} = w_0 \ , \qquad w_{1,j} = w_1 \ , \qquad w_{2,k} = w_2 \ .$$

We can now determine V. From $\mu(E_v) = 1$, we obtain the mass equation

$$1 \ = \ (q) \cdot w_0 + (q^2 - q) \cdot w_1 + (q^2 - q) \cdot w_2 \ .$$

Evaluating $u_\infty(z, E_v)$ on the sets $E_{0,i}$, $E_{1,j}$ and $E_{2,k}$ gives the equations

$$\begin{aligned} V &= w_0 \cdot (V + 1) + w_2 \cdot (q-1) \cdot (1/2) \ , \\ V &= w_1 \cdot (V_1 + 1) \ , \\ V &= w_0 \cdot (1/2) + w_2 \cdot ((V_2 + 1) + (q-2) \cdot (1/2)) \ . \end{aligned}$$

Treating this as a linear system in w_0, w_1, w_2, solving it in terms of V, V_1, V_2, and inserting the resulting values in the mass equation leads to

$$1 \ = \ (q)\frac{V(\frac{1}{2} + V_2)}{(1+V)(V_2 + \frac{q}{2}) - \frac{q-1}{4}} + (q^2 - q)\frac{V}{1 + V_1} + (q^2 - q)\frac{V^2 + \frac{1}{2}V}{(1+V)(V_2 + \frac{q}{2}) - \frac{q-1}{4}} \ .$$

Clearing denominators and using the values for V_1, V_2 from (2.85) yields a quadratic in V. Its unique nonnegative root (simplified using Maple) is the one in (2.83). □

3. Global Examples on $\mathbb{P}^1$

As will be seen, capacity theory gives a "Calculus" for answering certain types of questions about algebraic integers and units. Note that $\alpha \in \widetilde{\mathbb{Q}}$ is an algebraic integer if and only if its conjugates all satisfy $|\sigma(\alpha)|_v \le 1$ for all nonarchimedean v, and it is a unit if and only if $|\sigma(\alpha)|_v = 1$ for all nonarchimedean v.

Algebraic Integers. The following is a trivial application of the Fekete-Szegö theorem with local rationality, but appears hard to prove without it.

EXAMPLE 2.6. Let $\mathcal{M}$ be the Mandelbrot set. Then

(A) There are infinitely many algebraic integers with all their conjugates in $\mathcal{M}$.

(B) For each $B > 0$ there are only finitely many algebraic integers α whose conjugates all belong to $\mathcal{M}$, and some prime $p \le B$ splits completely in $\mathbb{Q}(\alpha)$. Indeed, there is a neighborhood $U = U(B)$ of $\mathcal{M}$ with this property.

(C) On the other hand, for each neighborhood U of $\mathcal{M}$ in $\mathbb{C}$, there is a $C = C(U)$ such that for each prime $p > C$, there are infinitely many algebraic integers α such that all the conjugates of α belong to U, and p splits completely in $\mathbb{Q}(\alpha)$.

PROOF. Take $K = \mathbb{Q}$, $\mathcal{C} = \mathbb{P}^1$, and $\mathfrak{X} = \{\infty\}$.

Part (A) is well known. Indeed, put $\varphi_c(z) = z^2 + c$ and for each integer $n \ge 1$ put $P_n(c) = \varphi_c^{(n+1)}(0)$, as in the discussion preceding (2.58). Then $P_n(c)$ is a monic polynomial in $\mathbb{Z}[c]$ of degree 2^n. If α is a root of $P_n(c) = 0$, then $z = 0$ is periodic for $\varphi_\alpha(z)$ (with period dividing $n+1$) since $\varphi_\alpha^{(n+1)}(0) = 0$. The same is true for all the $\mathrm{Gal}(\widetilde{\mathbb{Q}}/\mathbb{Q})$-conjugates of α, so α is an algebraic integer whose conjugates all belong to $\mathcal{M}$.

There are many ways to see that as a collection, the $P_n(c)$ have infinitely many distinct roots. For example, note that $c = 0$ is the only number such that 0 is periodic for $\varphi_c(z)$ with period 1. Taking $n = p-1$ where p is prime, we obtain $2^n - 1$ values of c such that 0 is periodic for $\varphi_c(z)$ with exact period p; thus there are infinitely many algebraic integers whose conjugates all belong to $\mathcal{M}$.

For part (B), fix a prime p, let $E_\infty = \mathcal{M}$, $E_p = \mathbb{Z}_p$, and let $E_q = D(0,1) \subset \mathbb{C}_q$ for each prime $q \neq p$. Put $\mathbb{E} = \prod_v E_v$. Then $\mathbb{E}$ is algebraically capacitable, and

$$\gamma(\mathbb{E}, \mathfrak{X}) \;=\; \gamma_\infty(\mathcal{M}) \cdot \gamma_\infty(E_p) \;=\; p^{-1/(p-1)} \;<\; 1 \,.$$

By Theorem 1.5 there is an adelic neigbhorhood $\mathbb{U} = \mathbb{U}_p = \prod_v U_{p,v}$ of $\mathbb{E}$ such that there are only finitely many $\alpha \in \widetilde{\mathbb{Q}}$ which have all their conjugates in $\mathbb{U}_p$. Each algebraic integer α such that p splits completely in $\mathbb{Q}(\alpha)$ is such a number.

Given $B > 0$, put $U = U(B) = \cap_{p \le B} U_{p,\infty} \subset \mathbb{C}$. Then $U(B)$ has the desired properties.

For part (C), let $U \subset \mathbb{C}$ be any neighborhood of $\mathcal{M}$. By enlarging $\mathcal{M}$ within U (for example by choosing a point $a \in (U \cap \mathbb{R}) \backslash \mathcal{M}$ and adjoining a suitably small disc $D(a,r)$) we can obtain a set $\mathcal{M}_U \subset U$ which has $\gamma_\infty(\mathcal{M}_U) > 1$ and is stable under complex conjugation.

Fix a prime p, and take $E_\infty = \mathcal{M}_U$, $E_p = \mathbb{Z}_p$, and $E_q = D(0,1) \subset \mathbb{C}_p$ for each prime $q \neq p$. Put $\mathbb{E} = \prod_v E_v$. By (2.63) and (2.68), the capacity of $\mathbb{E}$ with respect to $\mathfrak{X}$ is

$$\gamma(\mathbb{E}, \mathfrak{X}) \;=\; \gamma_\infty(\mathcal{M}_U) \cdot p^{-1/(p-1)} \,.$$

It follows that if p is sufficiently large, then $\gamma(\mathbb{E}, \mathfrak{X}) > 1$. Put $U_\infty = U$, $U_p = \mathbb{Z}_p$, and $U_q = D(0,1)$ for each prime $q \neq p$. By Theorem 1.2, there are infinitely many $\alpha \in \widetilde{\mathbb{Q}}$ whose conjugates belong to U_v for each v. Each such α is an algebraic integer whose archimedean conjugates belong to U and whose conjugates in $\mathbb{C}_p$ belong to $\mathbb{Z}_p$, so p splits completely in $\mathbb{Q}(\alpha)$. □

We remark that Example 2.6 holds with $\mathcal{M}$ replaced by the Julia set of any monic polynomial $g(x) \in \mathbb{Z}[x]$ with degree $d > 1$ (see (2.57)). In this case, each repelling periodic point for $g(x)$ is an algebraic integer whose conjugates belong to the Julia set.

Our next result, originally formulated by Cantor ([**16**]) and proved in ([**52**]), generalizes Robinson's theorem on totally real algebraic integers ([**48**]) stated in the introduction.

EXAMPLE 2.7. Let $\mathcal{Q}$ be a finite set of primes of $\mathbb{Q}$, and let $[a,b] \subset \mathbb{R}$. If

$$b - a \;>\; 4 \cdot \prod_{q \in \mathcal{Q}} q^{1/(q-1)} \,,$$

there are infinitely many algebraic integers α whose conjugates all belong to $[a,b]$, such that the primes in $\mathcal{Q}$ split completely in $\mathbb{Q}(\alpha)$; if $b - a < 4 \cdot \prod_{q \in \mathcal{Q}} q^{1/(q-1)}$ there are only finitely many.

PROOF. Take $K = \mathbb{Q}$, $\mathcal{C} = \mathbb{P}^1$, and $\mathfrak{X} = \{\infty\}$. Put $E_\infty = [a,b]$, $E_q = \mathbb{Z}_q$ for $q \in \mathcal{Q}$, and $E_p = \widehat{\mathcal{O}}_p$ for finite $p \notin \mathcal{Q}$. Then for $\mathbb{E} = E_\infty \times \prod_p E_p$, using $g_\infty = 1/z$ to compute the local capacities, formulas (2.9) and (2.68) give

$$\gamma(\mathbb{E}, \mathfrak{X}) \;=\; \prod_{p,\infty} \gamma_\infty(E_p) \;=\; \frac{b-a}{4} \cdot \prod_{q \in \mathcal{Q}} q^{-1/(q-1)} \,.$$

Thus the result follows from Theorem 1.5. □

Over an arbitrary number field, we have the following generalization of Example 2.7, motivated by a result of Moret-Bailly ([**39**], Théorème 1.3, p.182).

EXAMPLE 2.8. Let K be a number field, with r_1 real places and r_2 complex places. Write $n = [K : \mathbb{Q}]$, and let $\mathcal{Q}$ be a finite set of nonarchimedean places of K. For each $v \in \mathcal{Q}$, let F_w/K_v be a finite Galois extension, with ramification index $e_{w/v}$ and residue degree $f_{w/v}$. If

$$R^n \ > \ 2^{r_1} \prod_{v \in \mathcal{Q}} q_v^{1/(e_{w/v}(q_v^{f_{w/v}}-1))} \tag{2.86}$$

then there are infinitely many algebraic integers α whose archimedean conjugates belong to $D(0,R)$ at each v where $K_v \cong \mathbb{C}$, to $[-R,R]$ at each v where $K_v \cong \mathbb{R}$, and are such that for each $v \in \mathcal{Q}$ all the conjugates in $\mathbb{C}_v$ belong to $\mathcal{O}_{F_w}$. If R^n is less than the bound in (2.86), there are only finitely many such algebraic integers.

PROOF. For each complex archimedean v, put $E_v = D(0,R)$; then $\gamma_\infty(E_v) = R$. For each real archimedean v, put $E_v = [-R,R] \subset \mathbb{R}$; then $\gamma_\infty(E_v) = R/2$. For each nonarchimedean $v \in \mathcal{Q}$, put $E_v = \mathcal{O}_w$, and write $e = e_{w/v}$, $f = f_{w/v}$; then $\gamma_\infty(E_v) = q_v^{-1/e(q_v^f-1)}$ by (2.68). For all other nonarchimedean v, put $E_v = D(0,1)$, and put $\mathbb{E} = \prod_v E_v$. By our convention about weights and absolute values in the complex archimedean case,

$$\begin{aligned}\gamma(\mathbb{E},\{\infty\}) &= \prod_{\text{real } v} \gamma_\infty(E_v) \cdot \prod_{\text{complex } v} \gamma_\infty(E_v)^2 \cdot \prod_{\text{finite } v} \gamma_\infty(E_v) \\ &= R^n \cdot 2^{-r_1} \cdot \prod_{v \in \mathcal{Q}} q_v^{-1/(q_v-1)} .\end{aligned}$$

Again the result follows from Theorem 1.5. □

In general, not much is known about the case when $R^n = 2^{r_1} \prod_{v \in S} q_v^{1/(q_v-1)}$. When K is totally real, and S is empty, we have $R = 2$, and the roots of Chebyshev polynomials belong to $[-2,2]$. When K is totally complex, $R = 1$ and the roots of unity belong to $D(0,1)$. Thus in these two cases there are infinitely many algebraic integers whose conjugates satisfy the required conditions. There are no known examples where there are only finitely many.

Algebraic numbers satisfing prescribed arithmetic conditions, with controlled archimedean conjugates, can often be constructed by imposing appropriate geometric conditions on the sets E_v. The following (admittedly contrived) example illustrates some of the possibilities.

EXAMPLE 2.9. For any $\varepsilon > 0$, there are infinitely many algebraic integers α such that

(1) each archimedean conjugate $\sigma(\alpha)$ is real and satisfies $0 < \sigma(\alpha) < 12\sqrt{5} + \varepsilon$;
(2) the primes above 2 in $\mathbb{Q}(\alpha)$ have residue degree 1, and $|\alpha|_v = 1$ at each $v|2$.
(3) 3 is unramified in $\mathbb{Q}(\alpha)$, and $\text{ord}_v(\alpha) = 1$ at all v above 3;
(4) 5 splits completely in $\mathbb{Q}(\alpha)$, and α is a quadratic nonresidue at each $v|5$;
(5) for all primes $\mathfrak{p}_v$ of $\mathbb{Q}(\alpha)$ above 7, we have $\alpha \equiv -1 \pmod{\mathfrak{p}_v}$.

PROOF. Take $K = \mathbb{Q}$ and let $L > 0$ be a parameter. Put $E_\infty = [0,L] \subset \mathbb{R}$, $E_2 = D(1,1)^-$, $E_3 = D(2/3,1/3)^-$, $E_5 = \mathbb{Z}_5 \cap (D(2,1)^- \cup D(3,1)^-)$, $E_7 = D(-1,1)^-$. Put $E_p = D(0,1)$ for all other primes p. Then $\gamma_\infty(E_\infty) = L/4$.

As seen in the discussion of capacities of nonarchimedean open discs, $\gamma_\infty(E_2) = \gamma_\infty(D(1,1)) = \gamma_\infty(D(0,1)) = 1$ and $\gamma_\infty(E_3) = \gamma_\infty(D(2/3,1/3)) = \gamma_\infty(D(0,1/3)) = 1/3$. Corollary 2.3 shows that $\gamma_\infty(E_5) = 5^{-2/4}$. At $p = 7$, we have $E_7 \subset D(0,1)$ so $\gamma_\infty(E_7) \le 1$; on the other hand for each totally ramified finite extension $F_w/\mathbb{Q}_7$ we have $-1 + \pi_w\mathcal{O}_w \subset E_7$, and Proposition 2.1 shows that $\gamma_\infty(-1+\pi_w\mathcal{O}_w) = 7^{-1/e_w} \cdot 7^{-1/6e_w}$. Letting $e_w \to \infty$, we see that $\gamma_\infty(E_7) = 1$. As noted in Theorem 1.4, the condition that the primes above 2 have residue degree 1, and the primes above 3 be unramified, can be imposed "for free". For Theorem 1.4 to be applicable, we need $L/4 \cdot 1/3 \cdot 5^{-1/2} > 1$. □

The following example illustrates a case in which some of the E_v are unions of "different types" of sets, with overlaps.

EXAMPLE 2.10. Take $K = \mathbb{Q}$; let $\mathfrak{X} = \{\infty\}$, put $E_\infty = D(0,1) \cup [1, 1+L]$ where $L \ge 0$, and put $E_3 = \mathcal{O}_{v_1} \cup \mathcal{O}_{v_2}$ where $\mathcal{O}_{v_1}$ is the ring of integers of the unramified extension $L_{v_1} = \mathbb{Q}_3(\sqrt{-1})$, and $\mathcal{O}_{v_2}$ is the ring of integers of the totally ramified extension $L_{v_2} = \mathbb{Q}_3(\sqrt{-3})$. For each finite prime $p \ne 2$, let $E_p = \widehat{\mathcal{O}}_v$ be the $\mathfrak{X}$-trivial set.

By formula (2.50) with $L_1 = 0$ and $L_2 = L$, and formula (2.83) with $q_v = 3$, we have

$$\begin{aligned} \gamma(\mathbb{E}, \{\infty\}) &= 3^{-(-61+\sqrt{6481})/184} \cdot \left(1 + \frac{L^2}{4(1+L)}\right) \\ &\cong 0.89000685 + 0.22251713 L^2/(1+L) \, . \end{aligned}$$

By Theorems 0.4 and 1.5, if $L > 0.99240793$ then there are infinitely many algebraic integers whose archimedean conjugates all lie in $D(0,1) \cup [1, 1+L]$ and whose $\mathbb{C}_3$-conjugates all lie in $\mathcal{O}_{v_1} \cup \mathcal{O}_{v_2}$, while if $L < 0.99240792$ there are only finitely many.

Our last result in this section is a continuation of an example of Cantor ([**16**], p.167). Suppose that instead of constructing algebraic integers, one has a rational function $f(x)$ and is interested in constructing numbers α for which $f(\alpha)$ is an algebraic integer.

For instance, let $f(x) = 1/(1+x^2)$. Using Fekete's Theorem, Cantor showed that there are only finitely many totally real α for which $f(\alpha)$ is an algebraic integer; indeed, $\alpha = 0$ and $\alpha = \infty$ are the only such points.

Suppose, however, that we were willing to accept numbers α, all of whose conjugates had a small imaginary part. How large would the imaginary parts have to be to guarantee the existence of infinitely many solutions?

EXAMPLE 2.11. Take $f(x) = 1/(1+x^2)$. Suppose $T > 3/4$. Then there are infinitely many $\alpha \in \widetilde{\mathbb{Q}}$, all of whose conjugates satisfy $|\operatorname{Im}(\sigma(\alpha))| < T$, for which $f(\alpha)$ is an algebraic integer. However, if $T < 3/4$, there are only finitely many.

PROOF. Take $K = \mathbb{Q}$, and let $\mathfrak{X} = \{i, -i\}$, the set of poles of $f(x)$. Put $g_i(z) = z - i$, $g_{-i}(z) = z + i$, and let $L = \mathbb{Q}(i)$. Fix $T > 0$. At the archimedean place of $\mathbb{Q}$, take

$$E_\infty = \{z \in \mathbb{C} : -T \le \operatorname{Im}(z) \le T\} \cup \{\infty\} \, .$$

At each finite prime p, put $E_p = f^{-1}(\widehat{\mathcal{O}}_p)$. One sees that $E_2 = \mathbb{P}^1(\mathbb{C}_2)\backslash B(1,1)^-$, while for each $p \geq 3$, $E_p = \mathbb{P}^1(\mathbb{C}_p)\backslash(B(i,1)^- \cup B(-i,1)^-)$ where the two balls are disjoint. Put $\mathbb{E} = E_\infty \times \prod_p E_p$.

To compute the Green's matrices we must make a base change to L. Recall that $\Gamma(\mathbb{E}, \mathfrak{X}) = [L:K]^{-1}\Gamma(\mathbb{E}_L, \mathfrak{X})$. There is one archimedean place of L, which we will denote w_∞. By (2.48), we have $V_i(E_{w_\infty}) = V_{-i}(E_{w_\infty}) = \log(2(1-T))$, while $G(-i,i;E_{w_\infty}) = G(i,-i;E_{w_\infty}) = 0$ since i and $-i$ belong to distinct components of $\mathbb{P}^1(\mathbb{C})\backslash E_\infty$. Since $L_{w_\infty} \cong \mathbb{C}$, we have $\log(q_{w_\infty}) = \log(e^2) = 2$. There is one place w_2 of L above 2; $L_{w_2}/\mathbb{Q}_2$ is totally ramified. Fixing an isomorphism $\mathbb{C}_{w_2} \cong \mathbb{C}_2$, identify E_{w_2} with E_2. Then $V_i(E_{w_2}) = V_{-i}(E_{w_2}) = 0$, while $G(-i,i;E_{w_2}) = G(i,-i;E_{w_2}) = -\log_2(|i-(-i)|_{w_2}) = 2$. We have $\log(q_{w_2}) = \log(2)$. For all other places v of L the Green's matrices are trivial. Thus

$$\Gamma(\mathbb{E}, \mathfrak{X}) = \begin{pmatrix} \log(2(1-T)) & \log(2) \\ \log(2) & \log(2(1-T)) \end{pmatrix} . \tag{2.87}$$

By definition $\gamma(\mathbb{E}, \mathfrak{X}) = \exp(-\operatorname{val}(\Gamma(\mathbb{E}, \mathfrak{X})))$, where

$$\operatorname{val}(\Gamma(\mathbb{E}, \mathfrak{X})) = \min_{\vec{r}\in\mathfrak{p}} \max_{\vec{s}\in\mathfrak{p}} {}^t\vec{r}\Gamma(\mathbb{E}, \mathfrak{X})\vec{s} = \max_{\vec{r}\in\mathfrak{p}} \min_{\vec{s}\in\mathfrak{p}} {}^t\vec{r}\Gamma(\mathbb{E}, \mathfrak{X})\vec{s}$$

is the value of $\Gamma(\mathbb{E}, \mathfrak{X})$ as a matrix game; here $\mathfrak{p}$ is the set of probability vectors in $\mathbb{R}^2$. If we take $\vec{s} = {}^t(\frac{1}{2}, \frac{1}{2})$ then both entries of $\Gamma(\mathbb{E}, \mathfrak{X})\vec{s}$ are equal to $\frac{1}{2}\log(4(1-T))$; combining the "mini-max" and "maxi-min" expressions for $\operatorname{val}(\Gamma(\mathbb{E}, \mathfrak{X}))$ shows that $\operatorname{val}(\Gamma(\mathbb{E}, \mathfrak{X})) = \frac{1}{2}\log(4(1-T))$.

Thus $\gamma(\mathbb{E}, \mathfrak{X}) > 1$ iff $T > 3/4$, while $\gamma(\mathbb{E}, \mathfrak{X}) < 1$ iff $T < 3/4$, and the result follows from the Fekete-Szegö Theorem 1.5. □

Algebraic Units. As in Example 2.9, the condition that an algebraic integer α be a unit at finitely many specified primes can be imposed "for free". However, if we want global algebraic units, the construction must assure that they avoid 0 (and ∞) at all nonarchimedean v. This can be accomplished by using the capacity relative to two points $\mathfrak{X} = \{0, \infty\}$.

Below is Robinson's unit theorem ([**49**]) cited in the Introduction, which was originally proved without using capacity theory. Cantor ([**16**]) was the first to recognize that Robinson's conditions arise naturally in the context of capacities.

Example 2.12 (**Robinson**). Suppose $0 < a < b \in \mathbb{R}$ satisfy the conditions

$$\log(\frac{b-a}{4}) > 0 , \tag{2.88}$$

$$\log(\frac{b-a}{4}) \cdot \log(\frac{b-a}{4ab}) - \Big(\log(\frac{\sqrt{b}+\sqrt{a}}{\sqrt{b}-\sqrt{a}})\Big)^2 > 0 . \tag{2.89}$$

Then there are infinitely many totally real units α whose conjugates belong to $[a, b]$.

Proof. We follow Cantor ([**16**], p.166). Take $K = \mathbb{Q}$, $\mathcal{C} = \mathbb{P}^1$, and $\mathfrak{X} = \{0, \infty\}$. Put $E_\infty = [a, b]$, and put $E_p = D(0,1)\backslash D(0,1)^-$ for each finite prime p. Each nonarchimedean E_p is the "$\mathfrak{X}$-trivial" set in $\mathbb{P}^1(\mathbb{C}_p)$, so we can take $\mathbb{E} = E_\infty \times \prod_p E_p$.

Let the uniformizing parameters used to compute capacities be $g_0(z) = z$, $g_\infty(z) = 1/z$. By formulas (2.9), (2.7), at the archimedean place $V_\infty([a,b]) = \log(4/(b-a))$ and $G(0,\infty;[a,b]) = \log((\sqrt{b}+\sqrt{a})/(\sqrt{b}-\sqrt{a}))$. Pulling back by $1/z$, we have $G(z,0;[a,b]) = G(1/z,\infty;[1/b,1/a])$. In view of our choices of the uniformizing parameters, this yields $V_0([a,b]) = V_\infty([1/b,1/a]) = \log(4ab/(b-a))$.

At each finite prime p, one sees easily that $V_0(E_p) = V_\infty(E_p) = G(0,\infty;E_p) = 0$. Thus

$$(2.90) \qquad \Gamma(\mathbb{E},\mathfrak{X}) \;=\; \Gamma(E_\infty,\mathfrak{X}) \;=\; \begin{pmatrix} \log(\frac{4ab}{b-a}) & \log(\frac{\sqrt{b}+\sqrt{a}}{\sqrt{b}-\sqrt{a}}) \\ \log(\frac{\sqrt{b}+\sqrt{a}}{\sqrt{b}-\sqrt{a}}) & \log(\frac{4}{b-a}) \end{pmatrix} .$$

The conditions (2.88), (2.89) in Robinson's theorem are simply determinant inequalities on the minors of $\Gamma(\mathbb{E},\mathfrak{X})$, necessary and sufficient for it to be negative definite. Hence the result follows from the Fekete-Szegö Theorem 1.5. □

In the next result, we bound the size of the units and their reciprocals, as well as imposing conditions at nonarchimedean places.

EXAMPLE 2.13. There are infinitely many totally real algebraic units α whose archimedean conjugates belong to $[-r,-1/r]\cup[1/r,r]$, if

$$r \;>\; 1+\sqrt{2}\ .$$

More generally, let $\mathcal{Q}$ be a finite set of primes, and put $A = \prod_{q\in\mathcal{Q}} q^{q/(q-1)^2}$. Then there are infinitely many totally real algebraic units α for which the primes $q\in\mathcal{Q}$ split completely in $\mathbb{Q}(\alpha)$, and whose archimedean conjugates belong to $[-r,-1/r]\cup[1/r,r]$ if

$$(2.91) \qquad r \;>\; A^2+\sqrt{A^4+1}\ .$$

If the opposite inequality holds, there are only finitely many.

PROOF. Take $K = \mathbb{Q}$, $\mathcal{C} = \mathbb{P}^1$, and $\mathfrak{X} = \{0,\infty\}$. Let the uniformizers be $g_0(z) = z$, $g_\infty(z) = 1/z$ as before. Take $r \geq 1$ and put $E_\infty = [-r,-1/r]\cup[1/r,r]\subset\mathbb{R}$. For each $q\in\mathcal{Q}$, put $E_q = \mathbb{Z}_q^\times$. For all other primes p, put $E_p = D(0,1)\backslash D(0,1)^-\subset\mathbb{C}_p$, then let $\mathbb{E} = E_\infty\times\prod_p E_p$.

By formulas (2.15), (2.16) and (2.18), we have

$$\Gamma(E_\infty,\mathfrak{X}) \;=\; \begin{pmatrix} \frac{1}{2}\log(\frac{4r^2}{r^4-1}) & \frac{1}{2}\log(\frac{r^2+1}{r^2-1}) \\ \frac{1}{2}\log(\frac{r^2+1}{r^2-1}) & \frac{1}{2}\log(\frac{4r^2}{r^4-1}) \end{pmatrix} .$$

For primes $q\in\mathcal{Q}$, formulas (2.74) and (2.75) give $V_\infty(E_q) = G(0,\infty;E_q) = q/(q-1)^2$. Pulling back by $1/z$ and using that $\mathbb{Z}_p^\times$ is stable under taking reciprocals, we have $G(z,0;E_q) = G(1/z,\infty;E_q)$ and hence $V_0(E_q) = G(\infty,0;E_q) = q/(q-1)^2$ as well. Thus

$$(2.92) \qquad \Gamma(E_q,\mathfrak{X}) \;=\; \begin{pmatrix} q/(q-1)^2 & q/(q-1)^2 \\ q/(q-1)^2 & q/(q-1)^2 \end{pmatrix} .$$

For all other p, $\Gamma(E_p,\mathfrak{X})$ is the 0 matrix. Hence

$$\begin{aligned} \Gamma(\mathbb{E},\mathfrak{X}) &= \Gamma(E_\infty,\mathfrak{X}) + \sum_p \Gamma(E_p,\mathfrak{X})\log(p) \\ &= \frac{1}{2}\begin{pmatrix} \log(\frac{4A^2r^2}{r^4-1}) & \log(A^2\frac{r^2+1}{r^2-1}) \\ \log(A^2\frac{r^2+1}{r^2-1}) & \log(\frac{4A^2r^2}{r^4-1}) \end{pmatrix} . \end{aligned}$$

Take $\vec{s} = {}^t(\frac{1}{2},\frac{1}{2})$. Then $\Gamma(\mathbb{E},\mathfrak{X})\vec{s}$ has equal entries

$$V \;=\; \frac{1}{2}\log\left(\frac{4A^4r^2}{(r^2-1)^2}\right) .$$

By the definition of the value of a matrix game, $V(\mathbb{E},\mathfrak{X}) := \mathrm{val}(\Gamma(\mathbb{E},\mathfrak{X})) = V$. Since $r \ge 1$, $\gamma(\mathbb{E},\mathfrak{X}) = e^{-V(\mathbb{E},\mathfrak{X})} = (r^2-1)/(2A^2r)$. It is easy to see that $\gamma(\mathbb{E},\mathfrak{X}) > 1$ if and only if condition (2.91) holds, and that $\gamma(\mathbb{E},\mathfrak{X}) < 1$ if and only if the opposite inequality holds. Hence the result follows from Theorem 1.5. □

If $r = 1+\sqrt{2}$ in the first part of Example 2.13, then there are infinitely many units whose conjugates lie in $[-r,-1/r]\cup[1/r,r]$. Note that this set is the pullback of $[-2,2]$ by $f(z) = z - 1/z$. For each $n \ge 1$, let $T_n(x)$ denote the Chebyshev polynomial of degree n. It is well known that $T_n(x)$ is a monic polynomial with integer coefficients, whose roots are simple and belong to the interval $[-2,2]$. Put $P_n(z) = z^n T_n(z-1/z)$. Then $P_n(z)$ is monic with integer coefficients, and has constant coefficient $(-1)^n$. Thus the roots of the $P_n(z)$ are the units we need.

Next we give an S-unit analogue of Example 2.13. By a trick, we are able to require that the S-units constructed be totally p-adic, while their archimedean conjugates all have absolute value 1:

EXAMPLE 2.14. Let $k = \mathbb{Q}$ and fix a (nonarchimedean) prime p. Let $\mathcal{Q}$ be a finite set of nonarchimedean primes of $\mathbb{Q}$, disjoint from $\{p\}$, and put $A = \prod_{q\in\mathcal{Q}} q^{q/(q-1)^2}$ as in Example 2.13. Suppose $0 < m \in \mathbb{Z}$ is such that

$$m\log(p) \;>\; 2\log(A) + \frac{p\,(p^{2m}+1)}{(p-1)(p^{2m+1}-1)}\log(p)\ . \tag{2.93}$$

Then there are infinitely many numbers $\alpha \in \widetilde{\mathbb{Q}}$ for which the primes $q \in \mathcal{Q}$ split completely in $\mathbb{Q}(\alpha)$, which are units at all nonarchimedean places v of $\mathbb{Q}(\alpha)$ not above p, whose archimedean conjugates all satisfy $|\sigma(\alpha)| = 1$, and whose conjugates in $\mathbb{C}_p$ all belong to $\mathbb{Q}_p$ and satisfy $|\operatorname{ord}_p(\sigma(\alpha))| \le m$.

If the opposite inequality to (2.93) holds, there are only finitely many.

PROOF. Take $K = \mathbb{Q}$, and let $\mathfrak{X} = \{0,\infty\}$. Choose the uniformizing parameters $g_0(z) = z$, $g_\infty(z) = 1/z$ as usual.

The proof makes use of two $\mathbb{Q}$-rational adelic sets, which we will denote $\mathbb{E}$ and $\mathbb{E}'$. To construct $\mathbb{E}$, let $E_\infty = C(0,1)$, the unit circle. For each $q \in \mathcal{Q}$, put $E_q = \mathbb{Z}_q^\times$, and put

$$E_p \;=\; \{x \in \mathbb{Q}_p : -m \le \operatorname{ord}_p(x) \le m\} \;=\; p^{-m}E_{p,2m}$$

where $E_{p,2m} = \bigcup_{k=0}^{2m} p^k\mathbb{Z}_p^\times$ is as in Proposition 2.4. For all other finite primes q take $E_q = \widehat{\mathcal{O}}_q^\times = D(0,1)\backslash D(0,1)^-$, the $\mathfrak{X}$-trivial set in $\mathbb{P}^1(\mathbb{C}_q)$. Set $\mathbb{E} = E_\infty \times \prod_{q\ne\infty} E_q$.

To construct $\mathbb{E}'$, first choose a square-free integer $d < 0$ which satisfies $d \equiv 1 \pmod 8$ and $d \equiv 1 \pmod q$ for each $q \in \mathcal{Q}\cup\{p\}$. Thus the primes in $\mathcal{Q}\cup\{p\}$ split completely in the quadratic imaginary field $F = \mathbb{Q}(\sqrt{d})$. Let

$$f(x) \;=\; \frac{x-\sqrt{d}}{x+\sqrt{d}}\ ,$$

and for each prime q (archimedean or nonarchimedean) put $E_q' = f^{-1}(E_q)$. Then $E_\infty' = \mathbb{P}^1(\mathbb{R})$, while for each $q \in \mathcal{Q}\cup\{p\}$ we have $E_q' \subset \mathbb{Q}_q$. For all other primes q, E_q' is the RL-domain in $\mathbb{P}^1(\mathbb{C}_q)$ gotten by omitting two open discs centered on $\pm\sqrt{d}$; for all but finitely many q these discs are disjoint and have radius 1. Note that for each q, the set E_q' is stable under $\mathrm{Aut}_c(\mathbb{C}_q/\mathbb{Q}_q)$. If $q \in \mathcal{Q}\cup\{p,\infty\}$ this is trivial; for all other q, note that for each $\sigma \in \mathrm{Aut}_c(\mathbb{C}_q/\mathbb{Q}_q)$, either $\sigma(f)(x) = f(x)$

or $\sigma(f)(x) = 1/f(x)$. Since $E_q = \widehat{\mathcal{O}}_q^\times$ is stable under inversion and $\mathrm{Aut}_c(\mathbb{C}_q/\mathbb{Q}_q)$, it follows that $x \in E_q'$ if and only if $\sigma(x) \in E_q'$.

Set $\mathbb{E}' = E_\infty' \times \prod_{q \neq \infty} E_q'$, and take $\mathfrak{X}' = \{\sqrt{d}, -\sqrt{d}\}$. We claim that

$$\Gamma(\mathbb{E}', \mathfrak{X}') \;=\; \Gamma(\mathbb{E}, \mathfrak{X}) \ .$$

This follows by pulling back using $f(x)$: let $\mathbb{E}_F$, $\mathbb{E}_F'$ be the F-rational adelic sets obtained from $\mathbb{E}$, $\mathbb{E}'$ by base change (see [**51**], §5.1). Then $\Gamma(\mathbb{E}_F, \mathfrak{X}) = [F : \mathbb{Q}] \cdot \Gamma(\mathbb{E}, \mathfrak{X})$ and $\Gamma(\mathbb{E}_F', \mathfrak{X}') = [F : \mathbb{Q}] \cdot \Gamma(\mathbb{E}', \mathfrak{X}')$ ([**51**], p.326, formula (9)). On the other hand $f(x)$ is rational over F, so by ([**51**], p.335, formula (16)) and the fact that $\deg(f) = 1$,

$$\Gamma(\mathbb{E}_F', \mathfrak{X}') \;=\; \Gamma(\mathbb{E}_F, \mathfrak{X}) \ .$$

This establishes the claim.

By the Fekete-Szegö Theorem 0.4, if $\Gamma(\mathbb{E}', \mathfrak{X}')$ is negative definite, there are infinitely many algebraic numbers whose archimdean conjugates belong to $E_\infty' = \mathbb{P}^1(\mathbb{R})$ and whose q-adic conjugates belong to E_q' for all nonarchimedean q. The images of these numbers under $f(x)$ will be the numbers α in the example.

Hence it suffices to show that $\Gamma(\mathbb{E}', \mathfrak{X}') = \Gamma(\mathbb{E}, \mathfrak{X})$ is negative definite under condition (2.93). We have

$$\Gamma(\mathbb{E}, \mathfrak{X}) \;=\; \Gamma(E_\infty, \mathfrak{X}) + \sum_{q \neq \infty} \Gamma(E_q, \mathfrak{X}) \log(q) \ .$$

Here $\Gamma(E_\infty, \mathfrak{X})$ is the 0 matrix. For each $q \in \mathcal{Q}$, $\Gamma(E_q, \mathfrak{X})$ is the same as in (2.92) in the proof of Example 2.13. To express $\Gamma(E_p, \mathfrak{X})$, write

$$\begin{aligned} B &= \frac{1}{p-1} + \frac{1}{(p-1)^2(1+p^2+p^4+\cdots+p^{4m})} \ , \\ C &= \frac{p^{2m+1}}{(p-1)^2(1+p^2+p^4+\cdots+p^{4m})} \ . \end{aligned}$$

By Proposition 2.4 and the scaling property of the capacity,

$$V_\infty(E_p) \;=\; V_\infty(p^{-m} E_{p,2m}) \;=\; -m + B \ .$$

The map $f(z) = 1/z$ stabilizes E_p but takes 0 to ∞, so our choice of uniformizing parameters gives $V_0(E_p) = V_\infty(E_p)$. Finally $f(z) = p^m z$ takes $0 \mapsto 0$, $\infty \mapsto \infty$, and E_p to $E_{p,2m}$, so by the pullback formula (2.61) and Proposition (2.4), $G(0, \infty; E_p) = G(0, \infty; E_{p,2m}) = C$. Thus

$$\Gamma(E_p, \mathfrak{X}) \;=\; \begin{pmatrix} -m+B & C \\ C & -m+B \end{pmatrix} \ .$$

For all other primes q, $\Gamma(E_q, \mathfrak{X})$ is the 0 matrix, so

$$\Gamma(\mathbb{E}, \mathfrak{X}) \;=\; \begin{pmatrix} \log(A) + (-m+B)\log(p) & \log(A) + C\log(p) \\ \log(A) + C\log(p) & \log(A) + (-m+B)\log(p) \end{pmatrix} \ .$$

Since $\Gamma(\mathbb{E}, \mathfrak{X})$ has equal row sums, as in the proof of Example 2.13 it follows that $\mathrm{val}(\Gamma(\mathbb{E}, \mathfrak{X})) = \frac{1}{2}\big(-m\log(p) + 2\log(A) + (B+C)\log(p)\big)$. Simplifying, we have $\mathrm{val}(\Gamma(\mathbb{E}, \mathfrak{X})) < 0$, and hence $\gamma(\mathbb{E}, \mathfrak{X}) > 1$, if and only if (2.93) holds; similarly, $\gamma(\mathbb{E}, \mathfrak{X}) < 1$ if and only if the opposite inequality holds. Thus the result follows from the Fekete-Szego Theorem 1.5. □

For instance, if $\mathcal{Q} = \{2\}$ in Example 2.14, then for $p = 3$ we need $m \geq 4$; for $5 \leq p \leq 17$ we can take $m = 2$, and for $p \geq 19$ we can take $m = 1$. If $\mathcal{Q} = \{2, 3\}$ in Example 2.14, then for $p = 5$ we need $m \geq 7$; for $7 \leq p \leq 11$ we can take $m = 5$; for $13 \leq p \leq 23$ we can take $m = 4$; for $29 \leq p \leq 109$ we can take $m = 3$; for $113 \leq p \leq 11673$ we can take $m = 2$; and for $p \geq 11677$ we can take $m = 1$.

Note that the S-units constructed in Example 2.14 are not roots of unity, because there are only finitely many roots of unity ζ_n for which p splits completely in $\mathbb{Q}(\zeta_n)$.

In the next example, a limit argument allows us deal with a situation where one of the points in $\mathfrak{X}$ belongs to E_∞:

EXAMPLE 2.15. Let $A > 0$. If $A \geq 4$, then there are infinitely many units whose conjugates all lie in $[0, 1] \cup [A, A + 1] \subset \mathbb{R}$. If $A < 4$, there are only finitely many.

PROOF. Write $E = [0, 1] \cup [A, A+1]$. If $A \leq 1$, then $E \subset [0, 2]$, so $\gamma_\infty(E) < 1/2$. Otherwise, E is a translate of $[-(A + 1)/2, -(A - 1)/2] \cup [(A - 1)/2, (A + 1)/2]$ and so $\gamma_\infty(E) = \sqrt{A}/2$ by formula (2.15). Thus if $A < 4$, we have $\gamma_\infty(E) < 1$, and Fekete's Theorem 1.5(B) shows there are only finitely many *algebraic integers*, and in particular finitely many units, whose conjugates lie in E.

Next suppose $A = 4$. We will explicitly construct infinitely many units whose conjugates lie in $[0, 1] \cup [4, 5]$. To do so, note that $[0, 1] \cup [4, 5]$ is the pullback of $[-2, 2]$ by $f(z) = z^2 - 5z + 2$. Let $T_n(x)$ denote the Chebyshev polynomial of degree n. As before, $T_n(x)$ is a monic polynomial with integer coefficients whose roots are simple and belong to $[-2, 2]$. It oscillates n times between ± 2 on $[-2, 2]$; in particular, $T_n(2) = 2$. Furthermore, $T_n(x)$ is an even function if n is even, and is an odd function if n is odd. Consider the polynomials $Q_n(z) = T_n(f(z))$. They are monic with integer coefficients, and have all their roots in $[0, 1] \cup [4, 5]$. Unfortunately, $Q_n(z)$ has constant coefficient $Q_n(0) = T_n(2) = 2$. However, if n is odd, then $Q_n(z)$ has $f(z)$ as a factor, so $P_n(z) := Q_n(z)/f(z)$ has constant coefficient 1. Thus the roots of the $P_n(z)$ for odd n are the required units.

Finally, suppose $A > 4$, and let $0 < \varepsilon < 1$. Consider the set $E_{\varepsilon,A} = [\varepsilon, 1] \cup [A, A + 1]$. Take $K = \mathbb{Q}$, $\mathcal{C} = \mathbb{P}^1$, $\mathfrak{X} = \{0, \infty\}$. Let $E_\infty = E_{\varepsilon,A}$, and for each finite prime p let $E_p = D(0, 1)\backslash D(0, 1)^- \subset \mathbb{C}_p$ be the $\mathfrak{X}$-trivial set. Put $\mathbb{E}_{\varepsilon,A} = E_{\varepsilon,A} \times \prod_p E_p$, and take $g_0(z) = z$, $g_\infty(z) = 1/z$ as before. Then

$$\Gamma(\mathbb{E}_{\varepsilon,A}, \mathfrak{X}) = \begin{pmatrix} V_0(E_{\varepsilon,A}) & G(\infty, 0; E_{\varepsilon,A}) \\ G(0, \infty; E_{\varepsilon,A}) & V_\infty(E_{\varepsilon,A}) \end{pmatrix} .$$

Formula (2.27) expresses the Green's function of two intervals in terms of a quotient of two theta functions. These theta-functions and their parameters vary continuously with ε and A, hence the Green's function varies continuously as well. Letting $\varepsilon \to 0$, formulas (2.15), (2.27), and (2.34) show that

$$\begin{aligned} \lim_{\varepsilon \to 0} V_\infty(E_{\varepsilon,A}) &= V_\infty(E) = -\frac{1}{2}\log(A/4) < 0 , \\ \lim_{\varepsilon \to 0} G(\infty, 0; E_{\varepsilon,A}) &= \lim_{\varepsilon \to 0} G(0, \infty; E_{\varepsilon,A}) = G(0, \infty; E) = 0 , \\ \lim_{\varepsilon \to 0} V_0(E_{\varepsilon,A}) &= -\infty . \end{aligned}$$

Thus for all sufficiently small $\varepsilon > 0$, $\Gamma(\mathbb{E}_{\varepsilon,A}, \mathfrak{X})$ is negative definite, and the Fekete-Szegö Theorems 0.4 and 1.5 yield the result. $\square$

As a whimsical side note, we remark that an argument similar to the one in Example 2.15 shows that $E = [0,1] \cup [A, A+.001]$ contains infinitely many conjugate sets of units if $A \geq 30.19249489$, but only finitely many if $0 < A < 30.19249488$. This is obtained by using Maple to evaluate $V_\infty(E)$ in formula (2.33).

It is also possible to use the Fekete-Szegö theorem to construct units whose conjugates *globally omit* residue classes. View $\mathbb{P}^1/\mathbb{Q}$ as the generic fibre of $\mathbb{P}^1_{\mathbb{Z}}/\operatorname{Spec}(\mathbb{Z})$. Given $\alpha, \beta \in \mathbb{P}^1(\widetilde{\mathbb{Q}})$, we will say that α is integral with respect to β if the Zariski closures of α and β in $\mathbb{P}^1_{\mathbb{Z}}$ do not meet. If $\beta_1, \ldots, \beta_N$ are the conjugates of β, this is equivalent to requiring that for every prime p, all the conjugates $\sigma(\alpha)$ in $\mathbb{P}^1(\mathbb{C}_p)$ belong to $\mathbb{P}^1(\mathbb{C}_p) \backslash (\bigcup_{i=1}^N B(\beta_i, 1)^-)$.

Recall that the (absolute, logarithmic) Weil height of a number $\alpha \in \widetilde{\mathbb{Q}}$ is

$$h(\alpha) = \frac{1}{[F:\mathbb{Q}]} \sum_{w \text{ of } F} \log_w^+(|\alpha|_w) \log(q_w) ,$$

where F is any finite extension $\mathbb{Q}(\alpha)$. The height is independent of the field used to compute it. By definition, $h(\infty) = 0$. The points of $\mathbb{P}^1(\widetilde{\mathbb{Q}})$ with $h(\alpha) = 0$ are precisely 0, ∞, and the roots of unity.

To put the following result in context, we remark that in ([**8**]) the authors show that if $h(\beta) \neq 0$, there are only finitely many roots of unity which are integral with respect to β.

EXAMPLE 2.16. Take $K = \mathbb{Q}$, and let $\beta \in \widetilde{\mathbb{Q}}$. Then there are infinitely many algebraic units $\eta \in \widetilde{\mathbb{Q}}$ which are integral with respect to β. For any $\varepsilon > 0$, these units can be required to have Weil height $h(\eta) < \varepsilon$.

PROOF. Put $L = \mathbb{Q}(\beta_1, \ldots, \beta_N)$, where $\beta_1, \ldots, \beta_N$ are the conjugates of β over $\mathbb{Q}$, and take $\mathfrak{X} = \{\infty, 0, \beta_1, \ldots, \beta_N\}$. Let $g_\infty(z) = 1/z$, $g_0(z) = z$, and $g_{\beta_i}(z) = z - \beta_i$ for each i. In the discussion below, we will assume that $\beta \neq 0$. If $\beta = 0$, then we are merely asking for units with height $h(\eta) < \varepsilon$, and the argument carries through in a simplified form.

Viewing the β_i as embedded in $\mathbb{C}$, let $r > 1$ be any number small enough that $r < \min_{|\beta_i|>1}(|\beta_i|)$, and then let $\rho > 0$ be any number small enough that the discs $D(\beta_i, \rho)$ for $i = 1, \ldots, N$ and $D(0, \delta)$ are pairwise disjoint and do not meet the circles $\{|z| = r\}$. (Note that r and ρ are independent of choice of the embedding of the β_i.). Eventually we will let $\rho \to 0$, and then let $r \to 1$. Put

$$E_\infty = \Big(D(0,r) \cup \big(\bigcup_{|\beta_i|>r} C(\beta_i, \rho)\big)\Big) \backslash \Big(D(0,\rho)^- \cup \bigcup_{|\beta_i|<r} D(\beta_i, \rho)^-\Big) .$$

Thus, E_∞ consists of a disc $D(0,r)$ with tiny holes deleted around 0 and the $\beta_i \in D(0,r)$, together with tiny circles adjoined around the $\beta_i \notin D(0,r)$. By construction E_∞ is stable under complex conjugation.

For each finite prime p, regarding the β_i as embedded in $\mathbb{C}_p$, put

$$E_p = \mathbb{P}^1(\mathbb{C}_p) \backslash \big(B(\infty,1)^- \cup B(0,1)^- \cup \bigcup_{i=1}^N B(\beta_i, 1)^-\big) .$$

Then E_p is an RL-domain, stable under $\text{Aut}_c(\mathbb{C}_p/\mathbb{Q}_p)$, and for all but finitely many p it is $\mathfrak{X}$-trivial.

Put $\mathbb{E} = E_\infty \times \prod_p E_p$. To compute $\Gamma(\mathbb{E},\mathfrak{X})$, we must first make a base change to the field L, over which the β_i are rational. By definition

$$(2.94)\qquad \Gamma(\mathbb{E},\mathfrak{X}) = \frac{1}{[L:K]}\Gamma(\mathbb{E}_L,\mathfrak{X}) = \frac{1}{[L:K]}\sum_{\text{places } w \text{ of } L}\Gamma(E_w,\mathfrak{X})\log(q_w)$$

where $\mathbb{E}_L = \prod_{w \text{ of } L} E_w$. Here, for each place w of L, if w lies over p, then E_w is gotten by choosing an embedding $\sigma : L \hookrightarrow \mathbb{C}_p$ which induces v, extending σ to an isomorphism $\overline{\sigma} : \mathbb{C}_w \to \mathbb{C}_p$, and setting $E_w = \overline{\sigma}^{-1}(E_p)$. Basically, E_w is the same as E_p, but the way the β_i are embedded depends w.

We now compute the matrices $\Gamma(\mathbb{E}_w,\mathfrak{X})$. First suppose $w|\infty$. By construction, each point of $\mathfrak{X}$ belongs to a different connected component of $\mathbb{P}^1(\mathbb{C}_w)\backslash E_w$, so $\Gamma(E_w,\mathfrak{X})$ is a diagonal matrix. We have $V_\infty(E_w) = -\log(r) - \delta(\rho)$ where $\delta(\rho) > 0$ and $\delta(\rho) \to 0$ as $\rho \to 0$, while $V_{\beta_i}(E_w) = -\log(\rho)$ for each i.

Next suppose w is nonarchimedean. Since E_w is obtained by deleting a finite number of open discs of radius 1 from $\mathbb{P}^1(\mathbb{C}_w)$, one of which is $B(\infty,1)^-$, we have $V_\infty(E_w) = V_\infty(D(0,1)) = 0$. The other entries of $\Gamma(E_w,\mathfrak{X})$ will not matter to us: $\Gamma(E_w,\mathfrak{X})$ is an $(N+2)\times(N+2)$ matrix whose $V_\infty(E_w)$ entry is 0. For all but finitely many w, E_w is $\mathfrak{X}$-trivial and $\Gamma(E_w,\mathfrak{X})$ is the 0 matrix.

By definition $\Gamma(\mathbb{E}_F,\mathfrak{X}) = \sum_w \Gamma(E_w,\mathfrak{X})\log(q_w)$; for archimedean w, $q_w = e$ if $L_w \cong \mathbb{R}$, while $q_v = e^2$ if $L_w \cong \mathbb{C}$. By (2.94)

$$\Gamma(\mathbb{E},\mathfrak{X}) = \begin{pmatrix} -\log(r)-\delta(\rho) & A_{12} & \cdots & A_{1,N+2} \\ A_{21} & -\log(\rho)+A_{22} & \cdots & A_{2,N+2} \\ \vdots & \vdots & \ddots & \vdots \\ A_{N+2,1} & A_{N+1,2} & \cdots & -\log(\rho)+A_{N+2,N+2} \end{pmatrix}$$

where the A_{ij} do not depend on r or ρ, and $A_{ij} = A_{ji}$ for all i,j. By the determinant criterion for negative definiteness from linear algebra (see for example [**51**], Proposition 5.1.8, p.331), for each fixed r if ρ is sufficiently small then $\Gamma(\mathbb{E},\mathfrak{X})$ is negative definite. Thus for any neighborhood U of E_∞ the Fekete-Szegö Theorem 1.5 produces infinitely many units η whose archimedean conjugates all lie in U, and whose nonarchimedean conjugates avoid the balls $B(\beta_i,1)^-$ at all places w of L.

To see why the numbers η can be assumed to have arbitrarily small height requires some understanding of the proof of the Fekete-Szegö theorem (either Theorem 6.3.2 of [**51**], or Theorem 4.2 in this work). We will now sketch the argument, assuming the reader is coarsely familiar with the proof.

Fix $r > 1$, and let ρ be small enough that $\Gamma(\mathbb{E},\mathfrak{X})$ is negative definite. Then there is a probability vector $\vec{s} = {}^t(s_1,\ldots,s_{N+2})$ for which $\Gamma(\mathbb{E},\mathfrak{X})\vec{s}$ has all its entries equal to $V(\mathbb{E},\mathfrak{X})$. These s_i are essentially the relative orders of the poles of the initial patching functions $G_v^{(0)}(z)$ at the points $x_i \in \mathfrak{X}$. As $\rho \to 0$, we will have $s_1 \to 1$ and $s_2,\ldots,s_{N+2} \to 0$ since the first row of $\Gamma(\mathbb{E},\mathfrak{X})$ (and hence $V(\mathbb{E},\mathfrak{X})$) remains bounded but the diagonal entries in the other rows approach ∞.

The archimedean initial local patching function $G_\infty^{(0)}(z)$ is chosen so that the discrete probability density of its zeros approximates $\sum_{i=1}^{N+2} s_i\mu_i$, where μ_i is the equilibrium distribution of E_v with respect to x_i. Here each μ_i is a probability measure supported on the boundary of the component of $\mathbb{P}^1(\mathbb{C}_v)\backslash E_v$ containing x_i. As $\rho \to 0$, the amount of mass which μ_1 (corresponding to $x_1 = \infty$) places on the circles $C(\beta_i, \rho)$ goes to 0. Thus the proportion of the zeros of $G_\infty^{(0)}(z)$ which lie near near $C(0, r)$ goes to 1. The remaining zeros all lie near the circles $C(\beta_i, \rho)$. If U is chosen small enough that each $C(\beta_i, \rho)$ outside $D(0, r)$ lies in a separate component of U, then the patching process preserves the number of zeros which lie in each component. Thus the final patched function $G^{(n)}(z)$, whose zeros are numbers constructed by the Fekete-Szegö theorem, has the same number of zeros in each component of U as the initial function $G_\infty^{(0)}(z)$.

Since the zeros of $G^{(n)}(z)$ (in our instance) are algebraic units, the only contribution to their height is from archimedean places. By the discussion above, that contribution approaches $\log(r)$ as $\rho \to 0$. So, if we first let $\rho \to 0$, and then let $r \to 1$, the Fekete-Szegö theorem produces numbers whose heights approach 0. □

Our final example constructs units which avoid the residue class of 1 at every prime, and whose archimedean conjugates all lie very close to the circle $C(0, r)$ or the circle $C(0, 1/r)$ (so $|\log(|\sigma(\alpha)|)| \approx \log(r)$), for suitable r.

EXAMPLE 2.17. Let r satisfy $1 < r < 2.96605206$. Then for any $\varepsilon > 0$, there are infinitely many units α whose conjugates all satisfy

$$||\sigma(\alpha)| - r| \ < \ \varepsilon \quad \text{or} \quad ||\sigma(\alpha)| - 1/r| \ < \ \varepsilon \ ,$$

and are such that $\sigma(\alpha) \not\equiv 1 \pmod{\mathfrak{p}}$ for each prime $\mathfrak{p}$ of $\mathcal{O}_{\mathbb{Q}(\sigma(\alpha))}$. If $r > 2.96605207$, there are only finitely many.

PROOF. Take $K = \mathbb{Q}$, $\mathcal{C} = \mathbb{P}^1$, and $\mathfrak{X} = \{0, 1, \infty\}$. Let $E_\infty = C(0, r) \cup C(0, 1/r)$, and for each finite prime let E_p be the $\mathfrak{X}$-trivial set

$$E_p \ = \ \mathbb{P}^1(\mathbb{C}_p) \backslash (B(0,1)^- \cup B(1,1)^- \cup B(\infty, 1)^-) \ .$$

Put $\mathbb{E} = E_\infty \times \prod_p E_p$, and take $g_0(z) = z$, $g_1(z) = z - 1$, $g_\infty(z) = 1/z$.

Note that 0, 1 and ∞ belong to different components of $\mathbb{P}^1(\mathbb{C}) \backslash E_\infty$. Then $V_\infty(E_\infty) = V_0(E_\infty) = -\log(r)$ by formula (2.3), while $V_1(E_\infty)$ is given by (2.55) with $\tau = 2i \log(r)/\pi$. At each nonarchimedean place, $\Gamma(E_p, \mathfrak{X})$ is the 0 matrix. Hence $\Gamma(\mathbb{E}, \mathfrak{X}) = \Gamma(E_\infty, \mathfrak{X})$ is the diagonal matrix

$$\Gamma(\mathbb{E}, \mathfrak{X}) \ = \ \begin{pmatrix} -\log(r) & 0 & 0 \\ 0 & -\log(\frac{|\theta(0,\tau;0,0)\theta(0,\tau;\frac{1}{2},0)|}{2}) & 0 \\ 0 & 0 & -\log(r) \end{pmatrix} . \tag{2.95}$$

Clearly $\Gamma(\mathbb{E}, \mathfrak{X})$ is negative definite if and only if the middle term is negative. A computation with Maple yields the result. □

4. Function Field Examples concerning Separability

In this section, we will take $K = \mathbb{F}_p(t)$ where p is a prime, $\mathbb{F}_p$ is the finite field with p elements, and t is transcendental over $\mathbb{F}_p$. We give three examples showing the need for the separability hypothesis in Theorem 0.4.C.2, and in Theorems 1.2, 1.3, and 1.5. This was discovered by Daeshik Park in his doctoral thesis ([**45**]).

Let v_0, v_1 and v_∞ be the valuations of $\mathbb{F}_p(t)$ for which $v_0(t) = 1$, $v_1(t-1) = 1$, and $v_\infty(1/t) = 1$, respectively. For each of the corresponding places, the residue field is $\mathbb{F}_p$, and we have $\mathcal{O}_{v_0} \cong \mathbb{F}_p[[t]]$, $\mathcal{O}_{v_1} \cong \mathbb{F}_p[[t-1]]$, and $\mathcal{O}_{v_\infty} \cong \mathbb{F}_p[[\frac{1}{t}]]$.

Our first example, which is due to Park, concerns a set where all the hypotheses of the Fekete-Szegö Theorem 0.4 are satisfied except for separability of the extension F_{w_0}/K_{v_0}, yet the conclusion of the theorem fails for r in a certain range.

EXAMPLE 2.18. Let $K = \mathbb{F}_p(t)$, and let $\mathcal{C} = \mathbb{P}^1/K$. Identify $\mathbb{P}^1(\mathbb{C}_v)$ with $\mathbb{C}_v \cup \{\infty\}$, and take $\mathfrak{X} = \{\infty\}$. Put $F_{w_0} = K_{v_0}(t^{1/p}) = \mathbb{F}_p((t^{1/p}))$, so that $\mathcal{O}_{w_0} = \mathbb{F}_p[[t^{1/p}]]$ and F_{w_0}/K_{v_0} is purely inseparable. Fix a place $v_2 \in \mathcal{M}_K$ distinct from v_0, v_1, v_∞ and define an adelic set $\mathbb{E} = \mathbb{E}(r) = \prod_{v \in \mathcal{M}_K} E_v$ by putting $E_{v_0} = \mathcal{O}_{w_0}$, taking $E_{v_1} = \mathcal{O}_{v_1}$, $E_{v_\infty} = \mathcal{O}_{v_\infty}$, $E_{v_2} = D(0,r)$, and letting $E_v = D(0,1)$ be $\mathfrak{X}$-trivial for all $v \neq v_0, v_1, v_2, v_\infty$. Then $\mathbb{E}$ is compatible with $\mathfrak{X}$ and satisfies all the hypotheses of Theorem 0.4 apart from the inseparability of F_{w_0}/K_{v_0}, and

$$\gamma(\mathbb{E}, \mathfrak{X}) = r \cdot p^{-\frac{2+1/p}{p-1}} .$$

However, if

$$p^{\frac{2+1/p}{p-1}} < r < p^{\frac{3}{p-1}}$$

then $\gamma(\mathbb{E}, \mathfrak{X}) > 1$, yet there are only finitely many numbers in $\widetilde{K}$ whose conjugates belong to E_v for each $v \in \mathcal{M}_K$.

PROOF. By Proposition 2.1 we have $\gamma_\infty(E_{v_1}) = \gamma_\infty(E_{v_\infty}) = p^{-1/(p-1)}$. The extension F_{w_0}/K_{v_0} is totally ramified, so by the same Proposition, $\gamma_\infty(E_{v_0}) = p^{-1/(p(p-1))}$. By (2.63) it follows that

$$\gamma(\mathbb{E}, \mathfrak{X}) = p^{-\frac{1}{p(p-1)}} \cdot p^{-\frac{1}{p-1}} \cdot p^{-\frac{1}{p-1}} \cdot r = r \cdot p^{-\frac{2+1/p}{p-1}} .$$

Suppose $\alpha \in \widetilde{K}$ has all its conjugates in E_v, for each $v \in \mathcal{M}_K$. Recall that $K_{v_0} = \mathbb{F}_p((t))$ and $K_{v_1} = \mathbb{F}_p((t-1))$ are separable over $K = F_p(t)$ (see Grothendieck, [**29**], EGA IV, 7.8.3ii, or Matsumura [**37**], Proposition 28.M, p.207). Since the conjugates of α in $\mathbb{C}_{v_1}$ all belong to $E_{v_1} = \mathcal{O}_{v_1}$, α must be separably algebraic over K. On the other hand each element of $F_{w_0} \backslash K_{v_0}$ is purely inseparable over K_{v_0}, so the only elements of $\mathcal{O}_{w_0}$ which can be separably algebraic over K are those in $\mathcal{O}_{v_0}$. Thus the conjugates of α in $\mathbb{C}_{v_0}$ (which à priori belong to $E_{v_0} = \mathcal{O}_{w_0}$) must actually belong to $\mathcal{O}_{v_0}$. It follows that the conjugates of α belong to

$$\mathbb{E}' = \mathcal{O}_{v_0} \times \mathcal{O}_{v_1} \times \mathcal{O}_{v_\infty} \times D(0,r) \times \prod_{v \neq v_0, v_1, v_2, v_\infty} E_v$$

whose capacity is $\gamma(\mathbb{E}', \mathfrak{X}) = r \cdot p^{-\frac{3}{p-1}}$.

Each local set occurring in $\mathbb{E}'$ is algebraically capacitable, so by Fekete's theorem ([**51**], Theorem 6.3.1, p.414), if $\gamma(\mathbb{E}', \mathfrak{X}) < 1$ there is an adelic neighborhood $\mathbb{U}$ of $\mathbb{E}'$ which contains only finitely many conjugate sets of numbers in $\widetilde{K}$. In particular, there are only finitely many $\alpha \in \widetilde{K}$ with all their conjugates in $\mathbb{E}'$.

When

$$p^{\frac{2+1/p}{p-1}} < r < p^{\frac{3}{p-1}}$$

we have $\gamma(\mathbb{E}, \mathfrak{X}) > 1$ but $\gamma(\mathbb{E}', \mathfrak{X}) < 1$, so all the hypotheses of Theorem 0.4 hold for $\mathbb{E}$ except for the inseparability of F_{w_0}/K_{v_0}, yet the conclusion of Theorem 0.4 fails. □

Our next example provides sets $\mathbb{E}$ of arbitrarily large capacity, where all the hypotheses of Theorem 0.4 are satisfied except for the separability condition, yet there are no $\alpha \in \widetilde{K}$ with all their conjugates in $\mathbb{E}$.

EXAMPLE 2.19. Let $K = \mathbb{F}_p(t)$, and let $\mathcal{C} = \mathbb{P}^1/K$. Identify $\mathbb{P}^1(\mathbb{C}_v)$ with $\mathbb{C}_v \cup \{\infty\}$, and take $\mathfrak{X} = \{\infty\}$. Again put $F_{w_0} = K_{v_0}(t^{1/p}) = \mathbb{F}_p((t^{1/p}))$, so that

F_{w_0}/K_{v_0} is purely inseparable. Fix a place $v_2 \in \mathcal{M}_K$ distinct from v_0 and v_1, and define $\mathbb{E} = \mathbb{E}(r) = \prod_{v \in \mathcal{M}_K} E_v$ by putting

$$E_{v_0} = t^{-1/p} + \mathcal{O}_{w_0} = B(t^{-1/p}, 1) \cap F_{w_0} ,$$

taking $E_{v_1} = \mathcal{O}_{v_1}$, $E_{v_2} = D(0, r)$, and letting $E_v = D(0, 1)$ be $\mathfrak{X}$-trivial for all $v \neq v_0, v_1, v_2$. Then $\mathbb{E}$ is compatible with $\mathfrak{X}$ and satisfies all the hypotheses of Theorem 0.4 apart from the inseparability of F_{w_0}/K_{v_0}. We have

$$\gamma(\mathbb{E}, \mathfrak{X}) = r \cdot p^{-\frac{1+1/p}{p-1}} .$$

If $r > p^{(1+1/p)/(p-1)}$ we have $\gamma(\mathbb{E}, \mathfrak{X}) > 1$, yet there are no $\alpha \in \widetilde{K}$ whose conjugates all belong to $\mathbb{E}$.

PROOF. The argument is the same as that in Example 2.18, except that in this case $E_{v_0} \subset F_{w_0} \backslash K_{v_0}$, so there are no $\alpha \in \widetilde{K}$ whose conjugates in $\mathbb{C}_{v_1}$ belong to E_{v_1} and whose conjugates in $\mathbb{C}_{v_0}$ belong to E_{v_0}. □

In the previous examples, the conclusion of Theorem 0.4 failed because of interactions between places of K where the extensions F_w/K_v were separable and inseparable. In our last example, the conclusion of Theorem 0.4 fails because of interaction between two places where F_v/K_v is inseparable, with different degrees of inseparability.

EXAMPLE 2.20. Let $K = \mathbb{F}_p(t)$, and let $\mathcal{C} = \mathbb{P}^1/K$. Identify $\mathbb{P}^1(\mathbb{C}_v)$ with $\mathbb{C}_v \cup \{\infty\}$, and take $\mathfrak{X} = \{\infty\}$. Put $F_{w_0} = K_{v_0}(t^{1/p}) = \mathbb{F}_p((t^{1/p}))$, so that F_{w_0}/K_{v_0} is purely inseparable of degree p, and put $F_{w_1} = K_{v_1}((t-1)^{1/p^2}) = \mathbb{F}_p(((t-1)^{1/p^2}))$, so that F_{w_1}/K_{v_1} is purely inseparable of degree p^2. Fix a place $v_2 \in \mathcal{M}_K$ distinct from v_0 and v_1, and define $\mathbb{E} = \mathbb{E}(r) = \prod_{v \in \mathcal{M}_K} E_v$ by putting

$$E_{v_0} = t^{-1/p} + \mathcal{O}_{w_0} = B(t^{-1/p}, 1) \cap F_{w_0} ,$$
$$E_{v_1} = (t-1)^{-1/p^2} + \mathcal{O}_{w_1} = B((t-1)^{-1/p^2}, 1) \cap F_{w_1} ,$$

taking $E_{v_2} = D(0, r)$, and letting $E_v = D(0, 1)$ be $\mathfrak{X}$-trivial for all $v \neq v_0, v_1, v_2$. Then $\mathbb{E}$ is compatible with $\mathfrak{X}$ and satisfies all the hypotheses of Theorem 0.4 apart from the inseparability of F_{w_0}/K_{v_0} and F_{w_1}/K_{v_1}, and

$$\gamma(\mathbb{E}, \mathfrak{X}) = r \cdot p^{-\frac{p+1}{p^2(p-1)}} .$$

If $r > p^{(p+1)/(p^2(p-1))}$ we have $\gamma(\mathbb{E}, \mathfrak{X}) > 1$, but there are no $\alpha \in \widetilde{K}$ whose conjugates all belong to $\mathbb{E}$.

PROOF. The argument is similar to that in Examples 2.18 and 2.19, except that here each element of $\widetilde{K} \cap E_{v_0}$ satisfies $[K(\alpha) : K]^{\text{insep}} = p$, while each element of $\widetilde{K} \cap E_{v_1}$ satisfies $[K(\alpha) : K]^{\text{insep}} = p^2$. □

5. Examples on Elliptic Curves

Capacities of Archimedean Sets on Elliptic Curves. In general it is hard to compute capacities of archimedean sets on curves of positive genus, but explicit formulas for the capacities of some sets can be obtained using pullbacks from $\mathbb{P}^1$.

Let K_v be $\mathbb{R}$ or $\mathbb{C}$, and suppose $\mathcal{E}_v/K_v$ is defined by a Weierstrass equation

$$y^2 + a_0xy + a_3y = x^3 + a_2x^2 + a_4x + a_6 . \tag{2.96}$$

We will compute capacities of sets relative to the origin $\bar{o} = \infty$, using $z = x/y$ as the uniformizing parameter.

Let $f \in K_v(\mathcal{E}_v)$ be a rational function of degree $d > 0$ whose only poles are at $\bar{o}$, and let $H \subset \mathbb{C}$ be a compact set of positive capacity. Take $E_v = f^{-1}(H) \subset \mathcal{E}_v(\mathbb{C})$. By the pullback formula (2.61),

$$G(p, \bar{o}; E_v) \ = \ \frac{1}{d} G(f(p), \infty; H) \ .$$

Assume that capacities of sets in $\mathbb{C}$ are computed using the standard uniformizing parameter at ∞, and that $\lim_{p \to \bar{o}} |f(p) \cdot z(p)^d| = A$. Then

$$(2.97) \qquad V_{\bar{o}}(E_v) \ = \ \lim_{p \to \bar{o}} G(p, \bar{o}; E_v) + \log(|z(p)|) \ = \frac{1}{d}(V_\infty(H) + \log(A)) \ .$$

In particular, taking $f(p) = x(p)$, then

$$(2.98) \qquad V_{\bar{o}}(E_v) \ = \ \frac{1}{2} V_\infty(H) \ .$$

If $f(p) = y(p)$, then

$$(2.99) \qquad V_{\bar{o}}(E_v) \ = \ \frac{1}{3} V_\infty(H) \ .$$

For example, if $E_v = \{p \in \mathcal{E}_v(\mathbb{C}) : |y(p)| \le R\} = y^{-1}(D(0, R))$, then $V_{\bar{o}}(E_v) = -\frac{1}{3}\log(R)$.

Now assume $K_v = \mathbb{R}$; we will compute some capacities of sets $E_v \subset \mathcal{E}_v(\mathbb{R})$. Completing the square on the left side of (2.96), we get

$$(2.100) \ (y + \frac{1}{2}a_0 x + \frac{1}{2}a_3)^2 \ = \ x^3 + (a_2 + \frac{1}{4}a_0^2)x^2 + (a_4 + \frac{1}{2}a_0 a_3)x + (a_6 + \frac{1}{4}a_3^2) \ .$$

Let $g(x)$ be the polynomial on the right side of (2.100). Then $g(x)$ has either one or three real roots.

If $g(x) = (x - a)(x - b)(x - c)$ with $a < b < c$, then $\mathcal{E}_v(\mathbb{R})$ has two components, the bounded loop $x^{-1}([a, b])$ and the unbounded loop $x^{-1}([c, \infty])$. If $E_v \subset \mathcal{E}_v(\mathbb{R})$ is the bounded loop, then by formula (2.9)

$$(2.101) \qquad V_{\bar{o}}(E_v) \ = \ -\frac{1}{2}\log(\frac{b-a}{4}) \ .$$

If $T > c$ and $E_v = \{p \in \mathcal{E}(\mathbb{R}) : x(p) \le T\}$, then $E_v = x^{-1}([a, b] \cup [c, T])$ and by formula (2.33)

$$(2.102) \qquad V_{\bar{o}}(E_v) \ = \ -\frac{1}{2}\log\left(\frac{\sqrt[4]{(c-a)(c-b)(T-a)(T-b)}}{2\left|\frac{\theta(\mathrm{Re}(M(\infty))/K, \tau; \frac{1}{2}, \frac{1}{2})}{\theta(0, \tau; 0, \frac{1}{2})}\right|}\right) \ .$$

If $g(x)$ has only one real root, $x = c$, then $\mathcal{E}_v(\mathbb{R}) = x^{-1}([c, \infty])$ has one component. If $T > c$ and we take $E_v = \{p \in \mathcal{E}(\mathbb{R}) : x(p) \le T\}$, then $E_v = x^{-1}([c, T])$ and by formula (2.9)

$$(2.103) \qquad V_{\bar{o}}(E_v) \ = \ -\frac{1}{2}\log(\frac{T-c}{4}) \ .$$

Capacities of Nonarchimedean Sets on Elliptic Curves. In this subsection we will compute the capacities of certain sets of integral points on Néron models and Weierstrass models.

THEOREM 2.21. *Suppose K_v is nonarchimedean. Let $\mathcal{E}/K_v$ be an elliptic curve, and let $\mathcal{E}_{\mathcal{N}}/\operatorname{Spec}(\mathcal{O}_v)$ be its Neron model. Let $\bar{o}$ be the origin of $\mathcal{E}$, and let $E_v \subset \mathcal{E}(K_v)$ be the set of K_v-rational points which do not specialize $\pmod v$ to the origin of the special fibre $\mathcal{E}_{\mathcal{N},v}$. Equivalently, if $\mathcal{E}_{\mathcal{W}}$ is the affine model of $\mathcal{E}$ defined by a minimal Weierstrass equation for $\mathcal{E}$, then $E_v = \mathcal{E}_{\mathcal{W}}(\mathcal{O}_v)$.*

Write k_v for the residue field of $\mathcal{O}_v$, and let q_v be its order. Let $g_{\bar{o}}(z) \in K_v(\mathcal{E})$ be a uniformizing parameter which specializes $\pmod v$ to a uniformizer at the origin of $\mathcal{E}_{\mathcal{N},v}$, so $g_{\bar{o}}(z)$ is a local coordinate function which defines the formal group at the origin of $\mathcal{E}$; for example, take $g_{\bar{o}}(z) = x/y$, in terms of the standard coordinates on a minimal Weierstrass model $\mathcal{E}_{\mathcal{W}}$.

Then the local Robin constant

$$V_{\bar{o}}(E_v) \ := \ \lim_{z \to \infty} G(z, \bar{o}; E_v) + \log_v(|g_{\bar{o}}(z)|_v)$$

is given by the following formulas, according to the reduction type of $\mathcal{E}$:

(A) Type I_0 : Good reduction. If $\#(\mathcal{E}_{\mathcal{N},v}(k_v)) = N$, then

$$V_{\bar{o}}(E_v) \ = \ \frac{q_v}{(N-1)(q_v-1)} \ . \tag{2.104}$$

(B) Type I_1 : Nodal reduction, one component,

$$V_{\bar{o}}(E_v) \ = \ \frac{q_v}{(q_v-2)(q_v-1)} \ ; \tag{2.105}$$

here we assume $q_v > 2$: if $q_v = 2$, then E_v is empty.

(C) Type I_n, $n \geq 2$: Multiplicative reduction, a loop of n lines.

(C1) Split multiplicative reduction. Let $\{P_k(x)\}_{k \geq 0}$ be the polynomials defined recursively by $P_0(x) = 1$, $P_1(x) = x$, $P_k(x) = xP_{k-1}(x) - P_{k-2}(x)$ for $k \geq 2$, so

$$P_k(x) \ = \ \frac{1}{2^{k+1}} \cdot \frac{(x+\sqrt{x^2+4})^{k+1} - (x-\sqrt{x^2+4})^{k+1}}{\sqrt{x^2+4}} \ . \tag{2.106}$$

Then

$$V_{\bar{o}}(E_v) \ = \ \frac{q_v P_{n-1}(q_v + \frac{1}{q_v})}{(q_v^2 - q_v + 2)P_{n-1}(q_v + \frac{1}{q_v}) - 2q_v P_{n-2}(q_v + \frac{1}{q_v}) - 2q_v} \ . \tag{2.107}$$

(C2) Nonsplit multiplicative reduction, n odd: one component with rational points,

$$V_{\bar{o}}(E_v) \ = \ \frac{1}{q_v - 1} \ . \tag{2.108}$$

(C3) Nonsplit multiplicative reduction, n even: two components with rational points,

$$V_{\bar{o}}(E_v) \ = \ \frac{nq_v^2 + 4q_v - n}{(q_v-1)(nq_v^2 + 8q_v - n + 4)} \ . \tag{2.109}$$

(D) Type II : Cuspidal reduction, one component,

$$V_{\bar{o}}(E_v) \ = \ \frac{q_v}{(q_v-1)^2} \ . \tag{2.110}$$

(E) Type III : Two lines tangent at a point,

$$V_{\bar{o}}(E_v) \ = \ \frac{q_v(q_v+1)}{(q_v-1)(q_v^2 + 2q_v - 1)} \ . \tag{2.111}$$

(*F*) *Type IV : Three lines meeting transversely at a point.*

(*F*1) *One k_v-rational component,*

$$V_{\bar{o}}(E_v) = \frac{q_v}{(q_v-1)^2}\,. \tag{2.112}$$

(*F*2) *All three components k_v-rational,*

$$V_{\bar{o}}(E_v) = \frac{q_v}{(q_v+1)(q_v-1)}\,. \tag{2.113}$$

(*G*) *Type I_0^* : Four lines of multiplicity* 1 *meeting a line of multiplicity* 2 *at distinct points.*

(*G*1) *One k_v-rational component of multiplicity* 1,

$$V_{\bar{o}}(E_v) = \frac{q_v}{(q_v-1)^2}\,. \tag{2.114}$$

(*G*2) *Two k_v-rational components of multiplicity* 1,

$$V_{\bar{o}}(E_v) = \frac{1}{q_v-1}\,. \tag{2.115}$$

(*G*3) *Four k_v-rational components of multiplicity* 1,

$$V_{\bar{o}}(E_v) = \frac{q_v(2q_v-1)}{(q_v-1)(2q_v^2+1)}\,. \tag{2.116}$$

(*H*) *Type I_n^*, $n \geq 1$: Two lines of multiplicity* 1 *at each end of a chain of $n+1$ lines of multiplicity* 2.

(*H*1) *Two k_v-rational components of multiplicity* 1 (*adjacent*),

$$V_{\bar{o}}(E_v) = \frac{1}{q_v-1}\,. \tag{2.117}$$

(*H*2) *Four k_v-rational components of multiplicity* 1,

$$V_{\bar{o}}(E_v) = \frac{q_v((n+2)q_v^2-(n-1)q_v-1)}{(q_v-1)((n+2)q_v^3-(n-2)q_v^2+q_v+1)}\,. \tag{2.118}$$

(*I*) *Type IV^* : Three lines of multiplicity* 1, *each meeting a line of multiplicity* 2, *which in turn meets a line of multiplicity* 3.

(*I*1) *One k_v-rational component of multiplicity* 1,

$$V_{\bar{o}}(E_v) = \frac{q_v}{(q_v-1)^2}\,. \tag{2.119}$$

(*I*2) *Three k_v-rational components of multiplicity* 1,

$$V_{\bar{o}}(E_v) = \frac{q_v(2q_v-1)}{(q_v-1)(2q_v^2-q_v+1)}\,. \tag{2.120}$$

(*J*) *Type III^* : A chain of lines with multiplicities $1-2-3-4-3-2-1$ with another line of multiplicity* 2 *meeting the component of multiplicity* 4,

$$V_{\bar{o}}(E_v) = \frac{q_v(3q_v-1)}{(q_v-1)(3q_v^2-2q_v+1)}\,. \tag{2.121}$$

(*K*) *Type II^* : A chain of lines with multiplicities $1-2-3-4-5-6-4-2$ with another line of multiplicity* 3 *meeting the component of multiplicity* 6,

$$V_{\bar{o}}(E_v) = \frac{q_v}{(q_v-1)^2}\,. \tag{2.122}$$

The proof requires a formula for the canonical distance in terms of intersection theory, derived for minimal models in ([**51**], §2.4) and for "well-adjusted" models in ([**19**]). In proving Theorem 2.21, we will need the intersection theory formula for an arbitrary model.

Let $\mathcal{C}_v/K_v$ be a smooth, connected, projective curve, and let $\mathfrak{C}_v/\operatorname{Spec}(\mathcal{O}_v)$ be any regular model of $\mathcal{C}_v$. Given a point $p \in \mathcal{C}_v(\mathbb{C}_v)$ (respectively, a divisor D on $\mathcal{C}_v$), write (p) (respectively $\mathrm{cl}(D)$) for its closure in $\mathfrak{C}_v$. Let $F_1, \ldots, F_m$ be the irreducible components of the special fibre of $\mathfrak{C}_v$. Recall that the $m \times m$ intersection matrix $(F_i \cdot F_j)$ is symmetric and negative semidefinite, with rank $m-1$. Its kernel consists of vectors which are multiples of the special fibre, and its image consists of all vectors orthogonal to the special fibre. In particular, any vector $\sum a_i F_i$ supported on components of multiplicity 1 in the special fibre, for which $\sum a_i = 0$, belongs to the image.

If $f \in K_v(\mathcal{C}_v)$ is a nonzero rational function, write $\operatorname{div}_{\mathcal{C}_v}(f)$ for its divisor on $\mathcal{C}_v$, and $\operatorname{div}_{\mathfrak{C}_v}(f)$ for its divisor on $\mathfrak{C}_v$; then there are integers $c_1, \ldots, c_m$ for which

$$\operatorname{div}_{\mathfrak{C}_v}(f) \;=\; \mathrm{cl}(\operatorname{div}_{\mathcal{C}_v}(f)) + \sum_{j=1}^{m} c_j F_j \ . \tag{2.123}$$

If $a \in \mathcal{C}_v(K_v)$, then

$$-\log_v(|f_v(a)|_v) \;=\; \operatorname{ord}_v(f(a)) \;=\; (a) \cdot \operatorname{div}_{\mathfrak{C}_v}(f) \ . \tag{2.124}$$

Suppose $\zeta \in \mathcal{C}_v(K_v)$. Then ζ specializes to a nonsingular closed point on the special fibre of $\mathfrak{C}_v$, and we can choose a uniformizer $g_\zeta(z) \in K_v(\mathcal{C})$ in such a way such that for all $t \in \mathcal{C}_v(K_v)$ which specialize to that same closed point,

$$-\log_v(|g_\zeta(t)|_v) \;=\; \operatorname{ord}_v(g_\zeta(t)) \;-\; (t) \cdot (\zeta)$$

Normalize the canonical distance $[x, y]_\zeta$ so that

$$\lim_{y \to \zeta} [x, y]_\zeta |g_\zeta(y)|_v \;=\; 1 \ . \tag{2.125}$$

Let $a \neq b$ be points of $\mathcal{C}_v(\mathbb{C}_v)\backslash\{\zeta\}$. By ([**51**], Proof of Uniqueness for Theorem 2.1.1, p.57), $[a, b]_\zeta$ is given by

$$-\log_v([a, b]_\zeta) \;=\; \lim_f \frac{1}{\deg(f)}(-\log_v(|f(a)|_v)) \tag{2.126}$$

where the limit is taken over any sequence of functions whose only poles are at ζ, whose zeros approach a, and which are normalized so that $\lim_{z\to\zeta} |f(z)|_v \cdot |g_\zeta(z)^{\deg(f}|_v = 1$. Consider $\operatorname{div}_{\mathfrak{C}_v}(f)$ for such an f. After relabeling the F_i, we can assume that (ζ) specializes to F_1. The normalization (2.125) determines the constant c_1 in (2.123): if the zeros of f are $b_1, \ldots, b_n$ then $c_1 = -\sum_{i=1}^n (\zeta)\cdot(b_i)$. The remaining c_j are determined by the equations $F_i \cdot \operatorname{div}_{\mathfrak{C}_v}(f) = 0$, for $i = 2, \ldots, m$. Put $\hat{c}_j = (c_j - c_1)/\deg(f)$.

Combining (2.124) and (2.126), passing to the limit in f, and using the asymptotic stability of the various terms in the intersection products, we obtain the intersection theory formula for the canonical distance:

PROPOSITION 2.22. *Let $\mathcal{C}_v/K_v$ be a smooth, connected, projective curve. Fix a regular model $\mathfrak{C}_v/\operatorname{Spec}(\mathcal{O}_v)$ of $\mathcal{C}_v$, and let $F_1, \ldots, F_m$ be the irreducible components of the special fibre of $\mathfrak{C}_v$. If $\zeta \in \mathcal{C}_v(K_v)$ and the canonical distance $[x, y]_\zeta$ is* *normalized as in* (2.125), *then for distinct* $a, b \in \mathcal{C}_v(K_v)\backslash\{\zeta\}$,

$$-\log_v([a,b]_\zeta) \;=\; (a)\cdot(b) - (a)\cdot(\zeta) - (b)\cdot(\zeta) + \sum_{j=1}^{m} \hat{c}_j\, F_j\cdot(a) \tag{2.127}$$

where $\hat{c}_1, \ldots, \hat{c}_m \in \mathbb{Q}$ *are uniquely determined by the equations*

$$\begin{cases} \sum_{j=1}^{m} \hat{c}_j\, F_i \cdot F_j \;=\; F_i\cdot(\zeta) - F_i\cdot(b) \quad \textit{for } i = 1,\ldots,m, \\ \hat{c}_1 = 0\ , \end{cases} \tag{2.128}$$

if ζ *specializes to* F_1.

In (2.128), the numbers $\hat{c}_j$ depend only on the components to which b and ζ specialize, and if a specializes to F_k, then in (2.127)

$$\sum_{j=1}^{m} \hat{c}_j\, F_j\cdot(a) \;=\; \hat{c}_k\ . \tag{2.129}$$

If b specializes to F_ℓ, we will write $j_\zeta(F_k, F_\ell)$ for $\hat{c}_k$. It is easily seen that $j_\zeta(F_k, F_\ell) \geq 0$, and that $j_\zeta(F_k, F_\ell) = j_\zeta(F_\ell, F_k)$. If the model $\mathfrak{C}_v$ is projective, and if $\|x, y\|_v$ is the spherical metric on $\mathcal{C}_v$ determined by the projective embedding of $\mathfrak{C}_v$, then

$$(a)\cdot(b) \;=\; -\log_v(\|a,b\|_v)\ . \tag{2.130}$$

Thus (2.127) can be rewritten

$$-\log_v([a,b]_\zeta) \;=\; -\log_v\left(\frac{\|a,b\|_v}{\|a,\zeta\|_v\|b,\zeta\|_v}\right) + j_\zeta(F_k, F_\ell)\ . \tag{2.131}$$

We now apply this to potential functions. Suppose $E_v \subset \mathcal{C}_v(K_v)$ is compact with positive capacity, and that all the points of E_v specialize to the same component F_ℓ. Suppose in addition that no point of E_v specializes to the same closed point of the special fibre as ζ. If μ is the equilibrium distribution of E_v with respect to ζ, and if $a \in \mathcal{C}_v(K_v)$ specializes to F_k, then by (2.131),

$$u_{E_v}(a,\zeta) \;=\; \int_{E_v} -\log_v(\|a,b\|_v)d\mu(b) + j_\zeta(F_k,F_\ell) - \log_v(\|a,\zeta\|_v)\ . \tag{2.132}$$

Note that if $a, b \in \mathcal{C}_v(K_v)$ specialize to different closed points of the special fibre, then $-\log_v(\|a,b\|_v) = 0$. If they specialize to the same closed point (which is necessarily nonsingular on the special fibre), and if we fix a K_v-rational isometric parametrization of the ball $B(b,1)^-$, then $-\log_v(\|a,b\|_v) = -\log_v(|a'-b'|_v)$ where $a', b' \in K_v$ correspond to a, b under the isometric parametrization. In particular, if $E_v = \mathcal{C}_v(K_v) \cap B(b_0,1)^-$, then the integral appearing in (2.132) is the same as the one studied in Proposition 2.1, and

$$V_\zeta(E_v) \;=\; 1 + \frac{1}{q_v - 1} + j_\zeta(F_\ell, F_\ell)\ . \tag{2.133}$$

For $a \in \mathcal{C}_v(K_v)\backslash(B(b_0,1)^- \cup B(\zeta,1)^-)$ specializing to F_k,

$$u_{E_v}(a,\zeta) \;=\; j_\zeta(F_k, F_\ell)\ . \tag{2.134}$$

If E_v consists of points belonging to several balls in a single component F_ℓ, the averaging procedure used in Corollary 2.3 applies. Thus, if $E_v = \mathcal{C}_v(K_v) \cap$

$(\bigcup_{i=1}^M B(b_i,1)^-)$ where $b_1,\ldots,b_M \in \mathcal{C}_v(K_v)$ specialize to distinct closed points of F_ℓ (and ζ does not specialize to any of those points), then

$$V_\zeta(E_v) \;=\; \frac{q_v}{M(q_v-1)} + j_\zeta(F_\ell,F_\ell)\ , \tag{2.135}$$

while for $a \in \mathcal{C}_v(K_v)\backslash(\bigcup_{i=1}^M B(b_i,1)^- \cup B(\zeta,1)^-)$ specializing to F_k,

$$u_{E_v}(a,\zeta) \;=\; j_\zeta(F_k,F_\ell)\ . \tag{2.136}$$

Finally, if E_v has points belonging to several components, we can find $V_\zeta(E_v)$ by solving the system of equations (2.70) for the potential functions of the sets $E_{v,\ell}$, where $E_{v,\ell} \subset E_v$ is the set of points specializing to F_ℓ.

PROOF OF THEOREM 2.21. As might be expected, the proof involves considering the various reduction types individually. One must solve the system of equations discussed above, in each case.

In cases (A), (B), (C2), (D), (F1), (G1), (I1) and (K), only the identity component of the special fibre has rational points. Since $E_v = \mathcal{C}_v(K_v)\backslash B(\overline{o},1)^-$, we can apply (2.135) with $M = \#\mathcal{E}_0(k_v)-1$ and $j_{\overline{o}}(F_1,F_1)=0$, where $F_1 = \mathcal{E}_0$ is the identity component. In case (B), $\#\mathcal{E}_0(k_v) = q_v - 1$; in case (C2), $\#\mathcal{E}_0(k_v) = q_v+1$; and in cases (D)–(K), $\#\mathcal{E}_0(k_v) = q_v$.

In cases (E), (F2), (G2), (G3), (I2) and (J), where the special fibre has a fixed number of components, the computations are similar except for details. For each, one first solves the system of equations (2.128) to determine the numbers $j_{\overline{o}}(F_k,F_\ell)$; then finds the potential functions $u_{E_{v,\ell}}(z,\zeta)$ corresponding to the various components F_ℓ, using (2.135) and (2.136) and taking into account the number of k_v-rational closed points on each component; and finally solves the system (2.70) to find $V_{\overline{o}}(E_v)$, using the potential functions $u_{E_{v,\ell}}(z,\zeta)$. The computations were carried out using Maple.

We will illustrate the method in case (F2), where $\mathcal{E}$ has Type IV additive reduction and the special fibre consists of three components meeting transversely at a point, each component being k_v-rational. Let these components be $F_1 = \mathcal{E}_0$, F_2, and F_3. Each has q_v k_v-rational closed points, so $E_{v,1}$ is formed from $q_v - 1$ balls, while $E_{v,2}$ and $E_{v,3}$ each have q_v balls. Trivially

$$j_{\overline{o}}(F_1,F_1) = j_{\overline{o}}(F_2,F_1) = j_{\overline{o}}(F_3,F_1) = 0\ .$$

To find the $j_{\overline{o}}(F_i,F_2)$, note that each $F_i^2 = -2$, while $F_i\cdot F_j = 1$ if $i \neq j$, and solve the system (2.128) which reads

$$\begin{cases} \hat{c}_1\cdot(-2)+\hat{c}_2\cdot 1+\hat{c}_3\cdot 1 = 1, \\ \hat{c}_1\cdot 1+\hat{c}_2\cdot(-2)+\hat{c}_3\cdot 1 = -1, \\ \hat{c}_1\cdot 1+\hat{c}_2\cdot 1+\hat{c}_3\cdot(-2) = 0, \\ \hat{c}_1 = 0, \end{cases}$$

giving $j_{\overline{o}}(F_1,F_2) = \hat{c}_1 = 0$, $j_{\overline{o}}(F_2,F_2) = \hat{c}_2 = 2/3$, $j_{\overline{o}}(F_3,F_2) = \hat{c}_3 = 1/3$. Similarly $j_{\overline{o}}(F_1,F_3) = 0$, $j_{\overline{o}}(F_2,F_3) = 1/3$, $j_{\overline{o}}(F_3,F_3) = 2/3$. The potential functions $u_{E_{v,i}}(z,\overline{o})$ are then given by (2.135) and (2.136), with $V_{\overline{o}}(E_{v,1}) = q_v/(q_v-1)^2$ and $V_{\overline{o}}(E_{v,2}) = V_{\overline{o}}(E_{v,3}) = 2/3 + 1/(q_v-1)$.

To find $V_{\bar{\mathfrak{o}}}(E_v)$, solve the system (2.70) which reads

$$\begin{cases} 1 &= 0V + s_1 + s_2 + s_3, \\ 0 &= V - \frac{q_v}{(q_v-1)^2} s_1 - 0s_2 - 0s_3, \\ 0 &= V - 0s_1 - (\frac{2}{3} + \frac{1}{q_v-1})s_2 - \frac{1}{3}s_3, \\ 0 &= V - 0s_1 - \frac{1}{3}s_2 - (\frac{2}{3} + \frac{1}{q_v-1})s_3, \end{cases}$$

giving $V = V_{\bar{\mathfrak{o}}}(E_v) = q_v/(q_v^2-1)$, and $s_1 = (q_v-1)/(q_v+1)$, $s_2 = s_3 = 1/(q_v+1)$. If necessary, the weights s_1, s_2, s_3 could be used to find $u_{E_v}(z,\zeta)$ for any $z \in \mathcal{E}(\mathbb{C}_v)$.

The remaining cases (C1), (C3) and (H), where the number of components depends on n, must be treated separately.

First consider case (C3), nonsplit multiplicative reduction with $n = 2N$. Among the n components $\mathcal{E}_0, \ldots, \mathcal{E}_{2N-1}$ (listed cyclically around the loop), only $\mathcal{E}_0$ and $\mathcal{E}_N$ have k_v-rational points (each with q_v+1), while the other components have none. By ([**51**], p.96), $j_{\bar{\mathfrak{o}}}(\mathcal{E}_\ell, \mathcal{E}_\ell) = \ell - \ell^2/n$, so $j_{\bar{\mathfrak{o}}}(\mathcal{E}_0, \mathcal{E}_0) = 0$ and $j_{\bar{\mathfrak{o}}}(\mathcal{E}_N, \mathcal{E}_N) = n/4$. Thus

$$V_{\bar{\mathfrak{o}}}(E_{v,0}) = \frac{1}{q_v-1}, \quad V_{\bar{\mathfrak{o}}}(E_{v,N}) = \frac{n}{4} + \frac{q_v}{q_v^2-1}.$$

The equations (2.70) read

$$\begin{cases} 1 &= 0V + s_1 + s_2, \\ 0 &= V - \frac{1}{q_v-1} s_1 - 0s_2, \\ 0 &= V - 0s_1 - (\frac{n}{4} + \frac{q_v}{q_v^2-1})s_2, \end{cases}$$

giving

$$V = V_{\bar{\mathfrak{o}}}(E_v) = \frac{nq_v^2 + 4q_v - n}{(q_v-1)(nq_v^2 + 8q_v - n + 4)},$$
$$s_1 = \frac{nq_v^2 + 4q_v - n}{nq_v^2 + 8q_v - n + 4}, \quad s_2 = \frac{4q_v + 4}{nq_v^2 + 8q_v - n + 4}.$$

Next, consider case (H): Type I_n^* additive reduction, $n \geq 1$. Let F_1, F_2, F_3, and F_4 be the four components of multiplicity 1, and let $G_1, \ldots, G_{n+1}$ be the components of multiplicity 2, listed sequentially along the chain; assume F_1 and F_2 meet G_1, and F_3 and F_4 meet G_{n+1}, with $F_1 = \mathcal{E}_0$ being the identity component.

We first determine the numbers $j_{\bar{\mathfrak{o}}}(F_k, F_\ell)$ and $j_{\bar{\mathfrak{o}}}(G_i, F_\ell)$.

Trivially $j_{\bar{\mathfrak{o}}}(F_k, F_1) = j_{\bar{\mathfrak{o}}}(G_i, F_1) = 0$ for all k and i. For F_2, F_3, and F_4 the equations (2.128) can be solved recursively. For F_2, one finds in turn $j_{\bar{\mathfrak{o}}}(F_1, F_2) = 0$, $j_{\bar{\mathfrak{o}}}(G_1, F_2) = 1$, $j_{\bar{\mathfrak{o}}}(F_2, F_2) = 1$, then $j_{\bar{\mathfrak{o}}}(G_i, F_2) = 1$ for $i = 2, \ldots, n+1$, and finally $j_{\bar{\mathfrak{o}}}(F_3, F_2) = j_{\bar{\mathfrak{o}}}(F_4, F_2) = 1/2$. For F_3, one finds $j_{\bar{\mathfrak{o}}}(F_1, F_3) = 0$, $j_{\bar{\mathfrak{o}}}(G_1, F_3) = 1$, $j_{\bar{\mathfrak{o}}}(F_2, F_3) = 1/2$, then $j_{\bar{\mathfrak{o}}}(G_i, F_2) = (i+1)/2$ for $i = 2, \ldots, n+1$, and finally $j_{\bar{\mathfrak{o}}}(F_3, F_3) = 1 + n/4$, $j_{\bar{\mathfrak{o}}}(F_4, F_3) = 1/2 + n/4$. For F_4, the values are the same as for F_3, except that $j_{\bar{\mathfrak{o}}}(F_3, F_4) = 1/2 + n/4$ and $j_{\bar{\mathfrak{o}}}(F_4, F_4) = 1 + n/4$.

In subcase (H1), F_1 and F_2 are k_v-rational but F_3 and F_4 are not. The computation is identical to the one in case G2), and one gets $V_{\bar{\mathfrak{o}}}(E_v) = 1/(q_v-1)$.

In subcase (H2), all of F_1, F_2, F_3, F_4 are k_v-rational. Each has q_v k_v-rational closed points. The equations (2.70) read

$$\begin{cases} 1 &= 0V + s_1 + s_2 + s_3 + s_4, \\ 0 &= V - \frac{q_v}{(q_v-1)^2} s_1 - 0s_2 - 0s_3 - 0s_4, \\ 0 &= V - 0s_1 - (1 + \frac{1}{q_v-1})s_2 - \frac{1}{2}s_3 - \frac{1}{2}s_4, \\ 0 &= V - 0s_1 - \frac{1}{2}s_2 - (1 + \frac{n}{4} + \frac{1}{q_v-1})s_3 - (\frac{1}{2} + \frac{n}{4})s_4, \\ 0 &= V - 0s_1 - \frac{1}{2}s_2 - (\frac{1}{2} + \frac{n}{4})s_3 - (1 + \frac{n}{4} + \frac{1}{q_v-1})s_4, \end{cases}$$

and Maple gives

$$V_{\overline{o}}(E_v) = V = \frac{q_v[(n+2)q_v^2 - (n-1)q_v - 1]}{(q_v-1)[(n+2)q_v^3 - (n-2)q_v^2 + q_v + 1]} .$$

Case (C1), split multiplicative reduction with $n \geq 2$ components, is the most difficult. Let the components (listed cyclically around the loop) be $\mathcal{E}_0, \ldots, \mathcal{E}_{n-1}$, where $\mathcal{E}_0$ is the identity component, and let $\bigcup_{i=0}^{n-1} E_{v,i}$ be the corresponding decomposition of E_v. There are $q_v - 1$ k_v-rational points on each $\mathcal{E}_i$, so $E_{v,0}$ consists of $q_v - 2$ balls and all the other $E_{v,i}$ consist of $q_v - 1$ balls. Put

$$\widehat{E}_v = \bigcup_{i=1}^{n-1} E_{v,i} ,$$

so $E_v = E_{v,0} \cup \widehat{E}_v$. We will first find $V_{\overline{o}}(\widehat{E}_v)$, and then use it to find $V_{\overline{o}}(E_v)$.

For this, we will need a lemma.

LEMMA 2.23 (**Cantor's Lemma**). *Suppose $A \in M_k(\mathbb{R})$ is symmetric and negative definite. Let $\vec{\mathbf{1}}$ be the row vector $(1, 1, \ldots, 1) \in \mathbb{R}^k$, and consider the matrix $B \in M_{k+1}(\mathbb{R})$ given in block form by*

$$B = \begin{pmatrix} 0 & \vec{\mathbf{1}} \\ {}^t\vec{\mathbf{1}} & A \end{pmatrix} .$$

Then B is invertible, and if $\vec{b} = -\vec{\mathbf{1}}A^{-1}$ and $\alpha^{-1} = -\vec{\mathbf{1}}(A^{-1})\,{}^t\vec{\mathbf{1}}$, then

$$B^{-1} = \begin{pmatrix} \alpha & \alpha\vec{b} \\ \alpha\,{}^t\vec{b} & A^{-1} + \alpha\,{}^t\vec{b} \cdot \vec{b} \end{pmatrix} . \tag{2.137}$$

PROOF. See ([**16**], Lemma 3.2.3) or ([**53**], p.406). The proof is a block by block verification that the matrix C in (2.137) satisfies $CB = I$. □

To find $V_{\overline{o}}(\widehat{E}_v)$, let $A \in M_{n-1}(\mathbb{R})$ be the matrix $(-j_{\overline{o}}(\mathcal{E}_k, \mathcal{E}_\ell))_{1 \leq k,\ell \leq n-1}$. By the equations (2.128) defining the $j_{\overline{o}}(\mathcal{E}_k, \mathcal{E}_\ell)$ and the fact that $j_{\overline{o}}(\mathcal{E}_0, \mathcal{E}_\ell) = 0$ for each ℓ, it follows that A is inverse to the tridiagonal matrix

$$\Delta = \begin{pmatrix} -2 & 1 & 0 & 0 & \cdots & 0 \\ 1 & -2 & 1 & 0 & \cdots & 0 \\ 0 & 1 & -2 & 1 & \cdots & 0 \\ \vdots & \vdots & \ddots & \ddots & \ddots & \vdots \\ 0 & 0 & \cdots & 1 & -2 & 1 \\ 0 & 0 & \cdots & 0 & 1 & -2 \end{pmatrix} .$$

It is well known (and easy to check) that Δ is negative definite, so A is also negative definite.

Let B be as in Lemma 2.23. Then $\alpha = 1/2$ and $\vec{b} = (1, 0, \ldots, 0, 1)$ in the formula for B^{-1} in that Lemma. Put $Q = q_v/(q_v - 1)^2$ and let

$$B_Q = \begin{pmatrix} 0 & \vec{\mathbf{1}} \\ {}^t\vec{\mathbf{1}} & A - QI_{n-1} \end{pmatrix}$$

where I_{n-1} is the $(n-1) \times (n-1)$ identity matrix. Then the system of equations (2.70) determining $V_{\bar{o}}(\widehat{E}_v)$ reads

$$B_Q \begin{pmatrix} V \\ s_1 \\ \vdots \\ s_{n-1} \end{pmatrix} = \begin{pmatrix} 1 \\ 0 \\ \vdots \\ 0 \end{pmatrix}.$$

Left-multiplying by B^{-1} yields the simpler system

$$\begin{pmatrix} 1 & -\frac{1}{2}Q & 0 & 0 & \cdots & 0 & -\frac{1}{2}Q \\ 0 & 1+\frac{3}{2}Q & -Q & 0 & \cdots & 0 & -\frac{1}{2}Q \\ 0 & -Q & 1+2Q & -Q & \cdots & 0 & 0 \\ \vdots & \vdots & \ddots & \ddots & \ddots & \vdots & \vdots \\ 0 & 0 & \cdots & 0 & -Q & 1+2Q & -Q \\ 0 & -\frac{1}{2}Q & 0 & \cdots & 0 & -Q & 1+\frac{3}{2}Q \end{pmatrix} \begin{pmatrix} V \\ s_1 \\ s_2 \\ \vdots \\ s_{n-2} \\ s_{n-1} \end{pmatrix} = \begin{pmatrix} 1/2 \\ 1/2 \\ 0 \\ \vdots \\ 0 \\ 1/2 \end{pmatrix}.$$

We now solve for V using Cramer's rule. It will be useful to write

$$P_k(x) = \det \begin{pmatrix} x & -1 & 0 & \cdots & 0 \\ -1 & x & -1 & \cdots & 0 \\ \vdots & \ddots & \ddots & \ddots & \vdots \\ 0 & \cdots & -1 & x & -1 \\ 0 & \cdots & 0 & -1 & x \end{pmatrix},$$

so $P_0(x) = 1$, $P_1(x) = x$, and in general $P_k(x) = xP_{k-1}(x) - P_{k-2}(x)$. Solving the linear recurrence yields the formula for $P_k(x)$ in Theorem 2.21. Cramer's rule then gives

$$V_{\bar{o}}(\widehat{E}_v) = V = \frac{\frac{1}{2}P_{n-1}(2+\frac{1}{Q})}{P_{n-1}(2+\frac{1}{Q}) - P_{n-2}(2+\frac{1}{Q}) - 1}. \tag{2.138}$$

With $V_{\bar{o}}(\widehat{E}_v)$ in hand, we can use the decomposition $E_v = E_{v,0} \cup \widehat{E}_v$ to find $V_{\bar{o}}(E_v)$. The equations (2.70) read

$$\begin{cases} 1 = 0V + s_0 + \hat{s}_1, \\ 0 = V - \frac{q_v}{(q_v-2)(q_v-1)} s_0 - 0\hat{s}_1, \\ 0 = V - 0s_0 - V_{\bar{o}}(\widehat{E}_v)\hat{s}_1, \end{cases}$$

giving

$$V_{\bar{o}}(E_v) = \frac{q_v P_{n-1}(2+\frac{1}{Q})}{(q_v^2 - q_v + 2)P_{n-1}(2+\frac{1}{Q}) - 2q_v P_{n-2}(2+\frac{1}{Q}) - 2q_v}. \tag{2.139}$$

If $q_v = 2$, then $E_{v,0}$ is empty and $V_{\bar{o}}(E_v) = V_{\bar{o}}(\widehat{E}_v)$. A quick check shows that (2.139) remains valid even in this case. Since $2 + 1/Q = q_v + 1/q_v$, this completes the proof of Theorem 2.21. □

The next proposition gives the capacity of the set $E_v = \{P \in \mathcal{E}_v(K_v) : |x(P)|_v \le q_v^k\}$ for an elliptic curve $\mathcal{E}_v/K_v$ in Weierstrass normal form,

$$y^2 + a_1xy + a_3y = x^3 + a_2x^2 + a_4x + a_6 \ , \tag{2.140}$$

whose coefficients belong to $\mathcal{O}_v$. Let Δ be its discriminant, and let Δ_0 be the discriminant of a minimal Weierstrass equation. Let π_v be a generator for the maximal ideal of $\mathcal{O}_v$. Then there is an integer $m \ge 0$ for which

$$\Delta \;=\; \pi_v^{12m}\Delta_0 \ .$$

This is the number of times that Step 11 in Tate's algorithm (replacing x by π_v^2x' and y by π_v^3y'; see [**61**], pp.364-368) is executed in computing the Néron model and a minimal Weierstrass equation for $\mathcal{E}_v$.

PROPOSITION 2.24. *Let v be a nonarchimedean place of K, and let $\mathcal{E}_v/K_v$ be the elliptic curve defined by the Weierstrass equation with integral coefficients* (2.140).

Let q_v be the order of the residue field of K_v, and let $Q_v = V_{\bar{o}}(E_v)$ be the number associated to the Nèron model of $\mathcal{E}_v$ in Theorem 2.21. *For each integer $\ell \ge 0$ put*

$$V_\ell(q_v, Q_v) = \frac{q_v[Q_v(q_v-1)(q_v^{2\ell-1}+1) + (q_v^{2\ell}-1)]}{(q_v-1)[Q_v(q_v-1)(q_v^{2\ell}-1) + (q_v^{2\ell+1}+1)]} \ . \tag{2.141}$$

Suppose $m \ge 0$ is the number of times Step 11 *of Tate's algorithm is executed in computing the Néron model of $\mathcal{E}_v$. Let $k \ge -m$ be an integer, and put $E_{v,k} = \{p \in \mathcal{E}_v(K_v) : |x(p)|_v \le q_v^{2k}\}$. Let $z = x/y$ be the standard uniforming parameter at the origin $\bar{o} = \infty$ of $\mathcal{E}_v$. Then, computing capacities relative to z,*

$$V_{\bar{o}}(E_{v,k}) \;=\; -k + V_{m+k}(q_v, Q_v) \ . \tag{2.142}$$

PROOF. Let

$$y_0^2 + a_{1,0}x_0y_0 + a_{3,0}y_0 = x_0^3 + a_{2,0}x_0^2 + a_{4,0}x_0 + a_{6,0} \tag{2.143}$$

be a minimal Weierstrass equation for $\mathcal{E}_v$. Then $z_0 = x_0/y_0$ can be used as the uniformizing parameter in Theorem 2.21. For each $\ell \ge 0$, put

$$E_v^{(\ell)} \;=\; \{P \in \mathcal{E}_v(K_v) : |x_0(P)|_v \le q_v^{2\ell}\} \ , \tag{2.144}$$

and put $z_\ell = \pi_v^{-\ell}z_0$. Then z_m is the uniformizing parameter $z = x/y$ in the proposition, and $E_{v,k} = E_v^{(m+k)}$.

Let $V_0 = Q_v$, and recursively define $V_1, V_2, \ldots$ by requiring that V_ℓ be determined by the system of equations

$$\begin{cases} 1 = 0V_\ell + s_1 + s_2, \\ 0 = V_\ell - \frac{q_v}{(q_v-1)^2}s_1 - 0s_2, \\ 0 = V_\ell - 0s_1 - (1+V_{\ell-1})s_2, \end{cases} \tag{2.145}$$

so that

$$V_\ell \;=\; \frac{q_v(1+V_{\ell-1})}{q_v + (q_v-1)^2(1+V_{\ell-1})} \quad \text{for } \ell \ge 1 \ . \tag{2.146}$$

An easy induction shows that $V_\ell = V_\ell(q_v, Q_v)$ is given by (2.141).

We claim that V_ℓ is the Robin constant of the set $E_v^{(\ell)}$ relative to the uniformizing parameter z_ℓ. To see this, note that $E_v^{(0)}$ coincides with the set E_v attached to the Néron model of $\mathcal{E}_v$ in Theorem 2.21, and for each $\ell \ge 1$,

$$E_v^{(\ell)} \;=\; E_v^{(\ell-1)} \bigcup \{P \in \mathcal{E}_v(K_v) : |x_0(P)| = q_v^{2\ell}\} \ .$$

By the intersection theory formula for the canonical distance, this decomposition satisfies the conditions needed to find $V_{\overline{o}}(E_v^{(\ell)})$ using a system (2.70). By Theorem 2.21, $V_{\overline{o}}(E_v^{(0)}) = Q_v$ when capacities are computed relative to z_0. Assume that $V_{\overline{o}}(E_v^{(\ell-1)}) = V_{\ell-1}$ when the capacity is computed relative to $z_{\ell-1}$. Relative to z_ℓ it is $1+V_{\ell-1}$. Hence the system of equations (2.70) for finding the capacity of $V_{\overline{o}}(E_v^{\ell})$ relative to z_ℓ is exactly (2.145), and our claim holds by induction.

If the Robin constant of $E_v^{(m+k)}$ relative to z_{m+k} is V_{m+k}, then relative to z_m it is $-k + V_{m+k}$. This yields the result. □

Global Examples on Elliptic Curves. In the following examples, N is the conductor of $\mathcal{E}$. We take $k = \mathbb{Q}$ and consider elliptic curves $\mathcal{E}/\mathbb{Q}$ defined by Weierstrass equations. If p is a prime, by $\mathcal{E}(\mathbb{Z}_p)$ or $\mathcal{E}(\widehat{\mathcal{O}}_p)$ we mean the set of integral points on the affine curve defined by the given equation.

EXAMPLE 2.25 (N = 50). Let $\mathcal{E}/\mathbb{Q}$ be the elliptic curve defined by the Weierstrass equation $y^2 + xy + y = x^3 - x - 2$, curve $50(A1)$ in Cremona's tables. Then for any $T \geq 41.898861528$ there are infinitely many points $\alpha \in \mathcal{E}(\widetilde{\mathbb{Q}})$ whose archimedean conjugates belong to $\mathcal{E}(\mathbb{R})$ and satisfy $x(\alpha) < T$, whose conjugates in $\mathcal{E}(\mathbb{C}_3)$ all belong to $\mathcal{E}(\mathbb{Z}_3)$, whose conjugates in $\mathcal{E}(\mathbb{C}_5)$ all belong to $\mathcal{E}(\mathbb{Z}_5)$, and whose conjugates in $\mathcal{E}(\mathbb{C}_p)$ belong to $\mathcal{E}(\widehat{\mathcal{O}}_p)$, for all primes $p \neq 3, 5$.

If $T \leq 41.898861527$, there are only finitely many $\alpha \in \mathcal{E}(\widetilde{\mathbb{Q}})$ satisfying these conditions.

PROOF. The given Weierstrass equation is minimal; after completing the square on the left side it becomes

$$(y + \frac{1}{2}x + \frac{1}{2})^2 \;=\; x^3 + \frac{1}{2}x^2 - \frac{1}{2}x - \frac{7}{4}\ . \tag{2.147}$$

At $p = 2$ it has reduction type I_1; at $p = 3$ it has good reduction, and at $p = 5$ it has reduction type IV with 3 rational components (see Cremona [**21**], p.93). We will compute capacities with respect to the uniformizing parameter $z = x/y$.

The real locus $\mathcal{E}(\mathbb{R})$ consists of one unbounded loop $x^{-1}([\alpha, \infty])$, where $\alpha \cong 1.256458778$ is the unique real root of the polynomial on the right side of (2.147). Take $E_\infty = x^{-1}([\alpha, T])$ where $T > \alpha$. By formula (2.103)

$$V_{\overline{o}}(E_\infty) \;=\; -\frac{1}{2}\ln(\frac{T-\alpha}{4})\ . \tag{2.148}$$

At the prime $p = 2$, the set $\mathcal{E}(\mathbb{Z}_2)$ is empty (see Case B of Theorem 2.21) so we cannot impose splitting and integrality conditions simultaneously; we require integrality by taking

$$E_2 \;=\; \mathcal{E}(\widehat{\mathcal{O}}_2) \;=\; \{P \in \mathcal{E}(\mathbb{C}_2) : |x(P)|_2 \leq 1\}\ .$$

so $V_{\overline{o}}(E_p) = 0$. At $p = 3$, where $\mathcal{E}$ has good reduction, we take $E_3 = \mathcal{E}(\mathbb{Z}_3)$. A simple check shows that $\mathcal{E}$ (mod 3) has $N = 3$ points rational over $\mathbb{F}_3$. By formula (2.104) $V_{\overline{o}}(E_3) = 3/4$. At $p = 5$, we take $E_5 = \mathcal{E}(\mathbb{Z}_5)$. By formula (2.113), $V_{\overline{o}}(E_5) = 5/24$. For $p > 5$, take $E_p = \mathcal{E}(\widehat{\mathcal{O}}_p)$. Since the given model of $\mathcal{E}$ and the parameter z have good reduction at p, $V_{\overline{o}}(E_p) = 0$.

Let $\mathbb{E} = \prod_{p,\infty} E_p$, and take $\mathfrak{X} = \{\overline{o}\}$. Then

$$V(\mathbb{E}, \mathfrak{X}) \;=\; -\frac{1}{2}\ln(\frac{T-\alpha}{4}) + \frac{3}{4}\ln(3) + \frac{5}{24}\ln(5)\ .$$

Maple shows that the value of T for which $V(\mathbb{E},\mathfrak{X}) = 0$ satisfies

$$41.898861527 < T < 41.898861528 ,$$

and the Fekete-Szegö Theorems 0.4 and 1.5 yield the result. □

EXAMPLE 2.26 (N = 32). Let $\mathcal{E}/\mathbb{Q}$ be the elliptic curve defined by the non-minimal Weierstrass equation $y^2 = x^3 - 256x$. There are infinitely many points $\alpha \in \mathcal{E}(\widetilde{\mathbb{Q}})$ whose archimedean conjugates all belong to the bounded real loop in $\mathcal{E}(\mathbb{R})$, whose conjugates in $\mathcal{E}(\mathbb{C}_2)$ all belong to $\mathcal{E}(\mathbb{Z}_2)$, and whose conjugates in $\mathcal{E}(\mathbb{C}_p)$ belong to $\mathcal{E}(\widehat{\mathcal{O}}_p)$, for each $p \geq 3$.

PROOF. The given Weierstrass equation is not minimal; the minimal equation is $y^2 = x^3 - x$ (curve $32(A2)$ in Cremona's tables [**21**]). We will use the parameter $z = x/y$ in computing capacities with respect to $\overline{o} = \infty$.

The bounded real loop is $E_\infty = x^{-1}([-16, 0])$, for which formula (2.101) gives

$$V_{\overline{o}}(E_\infty) = -\frac{1}{2}\ln(\frac{16}{4}) = -\ln(2) .$$

Take $E_2 = \mathcal{E}(\mathbb{Z}_2)$. The curve $y^2 = x^3 - x$ has Kodaira reduction type III at $p = 2$ (Cremona [**21**], p.91). In passing from $y^2 = x^3 - 256x$ to $y^2 = x^3 - x$ we have $m = 2$. By formulas (2.111) and (2.142), $V_{\overline{o}}(E_2) = 106/107$. For all other primes p, take $E_p = \mathcal{E}(\widehat{\mathcal{O}}_p)$, the trivial set with respect to $\overline{o}$. The model of $\mathcal{E}$ given by $y^2 = x^3 - 256x$ and the parameter z have good reduction outside 2, so $V_{\overline{o}}(E_p) = 0$.

Let $\mathbb{E} = \prod_{p,\infty} E_p$, and take $\mathfrak{X} = \{\overline{o}\}$. Then

$$V(\mathbb{E},\mathfrak{X}) = -\ln(2) + \frac{106}{107}\ln(2) < 0 ,$$

so the Fekete-Szegö Theorems 0.4 and 1.5 yield the result. □

EXAMPLE 2.27 (N = 48). Let $\mathcal{E}/\mathbb{Q}$ be the elliptic curve defined by the Weierstrass equation $y^2 = x^3 + x^2 - 24x + 36$, curve $48(A3)$ in Cremona's tables. Then for any $T \geq 28.890384202$ there are infinitely many points $\alpha \in \mathcal{E}(\widetilde{\mathbb{Q}})$ whose archimedean conjugates belong to $\mathcal{E}(\mathbb{R})$ and satisfy $x(\alpha) < T$, whose conjugates in $\mathcal{E}(\mathbb{C}_2)$ all belong to $\mathcal{E}(\mathbb{Z}_2)$, whose conjugates in $\mathcal{E}(\mathbb{C}_3)$ all belong to $\mathcal{E}(\mathbb{Z}_3)$, and whose conjugates in $\mathcal{E}(\mathbb{C}_p)$ belong to $\mathcal{E}(\widehat{\mathcal{O}}_p)$, for $p \geq 5$.

If $T \leq 28.890384201$, there are only finitely many $\alpha \in \mathcal{E}(\widetilde{\mathbb{Q}})$ satisfying the conditions above.

PROOF. The given Weierstrass equation is minimal; it factors as $y^2 = (x+6)(x-2)(x-3)$. At $p = 2$ it has reduction type I_2^*, with 4 components rational over k_2; at $p = 3$ it has split multiplicative reduction of type I_4 (see Cremona [**21**], p.93). We will compute capacities with respect to the uniformizing parameter $z = x/y$.

The real locus $\mathcal{E}(\mathbb{R})$ consists of the bounded loop $x^{-1}([-6, 2])$, whose Robin constant is $-\frac{1}{2}\ln(2)$, together with the unbound loop $x^{-1}([3, \infty])$. Take $E_\infty = x^{-1}([-6, 2] \cup [3, T])$ where $T \geq 3$. By formula (2.102)

$$V_{\overline{o}}(E_\infty) = f_\infty(T) := -\frac{1}{2}\ln\left(\frac{\sqrt[4]{9(T+6)(T-2)}}{2\left|\frac{\theta(\mathrm{Re}(M(\infty))/K,\tau;\frac{1}{2},\frac{1}{2})}{\theta(0,\tau;0,\frac{1}{2})}\right|}\right) . \tag{2.149}$$

Take $E_2 = \mathcal{E}(\mathbb{Z}_2)$. By formula (2.118), $V_{\bar{o}}(E_2) = 26/35$. Similarly, take $E_3 = \mathcal{E}(\mathbb{Z}_3)$. By formula (2.107) with $n = 4$, we have $V_{\bar{o}}(E_3) = 123/238$. For $p > 3$, take $E_p = \mathcal{E}(\widehat{\mathcal{O}}_p)$. Since the given model of $\mathcal{E}$ and the parameter z have good reduction at p, $V_{\bar{o}}(E_p) = 0$.

Let $\mathbb{E} = \prod_{p,\infty} E_p$, and take $\mathfrak{X} = \{\bar{o}\}$. Then

$$V(\mathbb{E}, \mathfrak{X}) \;=\; f_\infty(T) + \frac{26}{35}\ln(2) + \frac{123}{238}\ln(3) \ .$$

If we had taken E_∞ to be the real loop $x^{-1}([-6, 2])$, then by the Fekete-Szegö theorem there would be only finitely many $\alpha \in \mathcal{E}(\mathbb{Q})$ whose conjugates meet the given conditions. Maple shows that the value of T for which $V(\mathbb{E}, \mathfrak{X}) = 0$ satisfies

$$28.890384201 \;<\; T \;<\; 28.890384202 \ ,$$

and the Fekete-Szegö Theorems 0.4 and 1.5 yield the result. □

EXAMPLE 2.28 (N = 360). Let $\mathcal{E}/\mathbb{Q}$ be the elliptic curve defined by the Weierstrass equation $y^2 = x^3 + 117x + 918$, curve 360(*E*4) in Cremona's tables. Then for any $R \ge 142.388571238$ there are infinitely many $\alpha \in \mathcal{E}(\widetilde{\mathbb{Q}})$ whose archimedean conjugates satisfy $|y(\alpha)| \le R$, whose conjugates in $\mathcal{E}(\mathbb{C}_2)$ all belong to $\mathcal{E}(\mathbb{Z}_2)$, whose conjugates in $\mathcal{E}(\mathbb{C}_3)$ all belong to $\mathcal{E}(\mathbb{Z}_3)$, whose conjugates in $\mathcal{E}(\mathbb{C}_5)$ all belong to $\mathcal{E}(\mathbb{Z}_5)$, and whose conjugates in $\mathcal{E}(\mathbb{C}_p)$ belong to $\mathcal{E}(\widehat{\mathcal{O}}_p)$, for all primes $p > 5$.

If $R \le 142.388571237$, there are only finitely many such α.

PROOF. The given Weierstrass equation is minimal. At $p = 2$ it has reduction type III^*, with 2 rational components; at $p = 3$ it has reduction type I_0^*, with 2 rational components; and at $p = 5$ it has nonsplit multiplicative reduction ($n = 4$) with 2 rational components (see Cremona [**21**], p.133). We compute capacities with respect to the uniformizing parameter $z = x/y$.

Take $E_\infty = y^{-1}(D(0, R))$, where $R > 0$. By formula (2.99), $V_{\bar{o}}(E_\infty) = -\frac{1}{3}\ln(R)$. At $p = 2$, take $E_2 = \mathcal{E}(\mathbb{Z}_2)$. By formula (2.121), $V_{\bar{o}}(E_2) = 10/9$. At $p = 3$, take $E_3 = \mathcal{E}(\mathbb{Z}_3)$. By formula (2.104) $V_{\bar{o}}(E_3) = 1/2$. At $p = 5$, take $E_5 = \mathcal{E}(\mathbb{Z}_5)$. By formula (2.109), $V_{\bar{o}}(E_5) = 29/140$. For $p > 5$, take $E_p = \mathcal{E}(\widehat{\mathcal{O}}_p)$. Since the given model of $\mathcal{E}$ and the parameter z have good reduction at p, $V_{\bar{o}}(E_p) = 0$.

Let $\mathbb{E} = \prod_{p,\infty} E_p$, and take $\mathfrak{X} = \{\bar{o}\}$. Then

$$V(\mathbb{E}, \mathfrak{X}) \;=\; -\frac{1}{3}\ln(R) + \frac{10}{9}\ln(2) + \frac{1}{2}\ln(3) + \frac{29}{140}\ln(5) \ .$$

The value of R for which $V(\mathbb{E}, \mathfrak{X}) = 0$ satisfies

$$142.388571237 \;<\; R \;<\; 142.388571238 \ ,$$

and the Fekete-Szegö Theorems 0.4 and 1.5 yield the result. □

6. The Fermat Curve

Let $p \ge 3$ be an odd prime, and let $\zeta = e^{2\pi i/p}$. In this section we will apply the Fekete-Szegö theorem with local rationality to the Fermat curve

$$\mathcal{F}^p : X^p + Y^p \;=\; Z^p \ , \qquad (2.150)$$

taking the ground field to be $K = \mathbb{Q}$. Let $\mathfrak{F}^p/\operatorname{Spec}(\mathbb{Z})$ be the corresponding scheme. To obtain a nontrivial set $\mathbb{E}$, we make use of McCallum's description ([**42**]) of a regular model for $\mathfrak{F}^p_{v_p} := \mathfrak{F}^p \times \mathrm{Spec}(\mathcal{O}_{L,v_p})$, where $L = \mathbb{Q}(\zeta)$ and v_p is the unique place of L over p. The author thanks Dino Lorenzini for suggesting this example.

Writing $x = X/Z$, $y = Y/Z$, let the part of $\mathcal{F}^p$ in the coordinate patch $Z \neq 0$ be the affine curve

$$\mathcal{F}^{p,0} : \ x^p + y^p \ = \ 1 \ . \tag{2.151}$$

Let $\mathfrak{X} = \{\xi_1, \ldots, \xi_p\}$ be the set of points at infinity, where $\xi_k = (1 : -\zeta^k : 0)$, and take $\mathbb{E} = \prod_v E_v$ where the sets E_v are as follows: for the archimedean place, let

$$E_\infty \ = \ x^{-1}(D(0,R)) \ = \ \{z \in \mathcal{F}^p(\mathbb{C}) : |x(z)| \le R\} \ .$$

At the place p, take $E_p = \mathcal{F}^{p,0}(\mathcal{O}_{L,v_p})$, and for all the other nonarchimedean places q, take E_q to be the $\mathfrak{X}$-trivial set $E_q = \mathcal{F}^{p,0}(\widehat{\mathcal{O}}_q)$.

Let z vary over $\mathcal{F}^p(\mathbb{C})$. Writing (2.151) in the form

$$\prod_{k=1}^{p} (\frac{y}{x} + \zeta^k) \ = \ (\frac{1}{x})^p$$

we see that as $z \to \xi_k$, then $(y/x) + \zeta^k$ vanishes to order p; at each $\xi_k \in \mathfrak{X}$ we will take the local uniformizing parameter to be

$$g_{\xi_k}(z) = \frac{1}{x(z)} \ .$$

The Green's Matrix at the Archimedean Place. Since E_∞ and $|1/x(z)|$ are invariant under the automorphisms of $\mathcal{F}^p$ given by

$$(X : Y : Z) \ \mapsto \ (\zeta^k X : \zeta^\ell Y : Z) \ ,$$

while the ξ_k are permuted by those automorphisms, there are numbers A, B such that $G(\xi_k, \xi_\ell; E_\infty) = A$ and $V_{\xi_k}(E_\infty) = B$ for all $k \neq \ell$. Thus the archimedean local Green's matrix is

$$\Gamma(E_\infty, \mathfrak{X}) \ = \ \begin{pmatrix} B & A & \cdots & A \\ A & B & \cdots & A \\ \vdots & \vdots & \ddots & \vdots \\ A & A & \cdots & B \end{pmatrix} . \tag{2.152}$$

Although we are unable to determine the numbers A, B explicitly, we will see below that

$$(p-1)A + B \ = \ -\log(R) \ . \tag{2.153}$$

This relation will enable us to determine the capacities we need.

For the divisor (∞) on $\mathbb{P}^1$, we have $x^{-1}((\infty)) = (\xi_1) + \cdots + (\xi_p)$, so the pullback formula (2.61) shows that for each $z \in \mathcal{F}^p(\mathbb{C})$,

$$G(x(z), \infty; D(0,R)) \ = \ \sum_{\ell=1}^{p} G(z, \xi_\ell, E_\infty) \ .$$

Since $G(w, \infty; D(0,R)) = \log^+(|w/R|)$ in $\mathbb{P}^1$, for each ξ_k we have

$$\begin{aligned} -\log(R) &= \lim_{z \to \xi_k} G(x(z), \infty; D(0,R)) + \log(|1/x(z)|) \\ &= V_{\xi_k}(E_\infty) + \sum_{\ell \neq k} G(\xi_k, \xi_\ell; E_\infty) \ , \end{aligned} \tag{2.154}$$

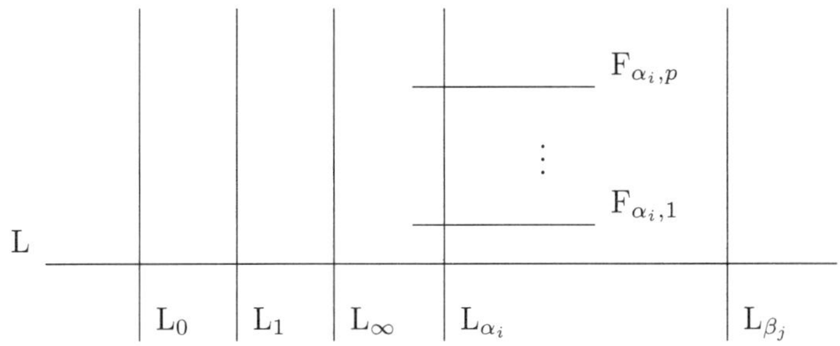

FIGURE 1. Fermat curve special fiber

and (2.153) follows.

The Green's Matrix at the Place p. Put $L = \mathbb{Q}(\zeta)$, and let v_p be the unique place of L above p; thus $\mathcal{O}_{L,v_p} \cong \mathbb{Z}_p[\zeta]$. Put $\pi_{v_p} = 1 - \zeta$. The residue field $k_v = \mathcal{O}_{L,v_p}/\pi_{v_p}\mathcal{O}_{L,v_p}$ is isomorphic to $\mathbb{F}_p$. Write $\mathcal{F}^p_{v_p} = \mathcal{F}^p \times_{\mathbb{Q}} \operatorname{Spec}(L_{v_p})$ and $\mathfrak{F}^p_{v_p} = \mathfrak{F}^p \times_{\mathbb{Z}} \operatorname{Spec}(\mathcal{O}_{L,v_p})$.

McCallum ([**42**], see Theorem 3, p.59; Diagram 3, p.69) has determined a regular model for $\mathcal{F}^p_{v_p}$. Put

$$\phi(x,y) = \frac{(x+y)^p - x^p - y^p}{p} .$$

Then $\phi(x,y)$ is a polynomial with integer coefficients, divisible by $xy(x+y)$. Let $\widetilde{\mathbb{F}}_p$ be the algebraic closure of $\mathbb{F}_p$; McCallum notes that $\phi(x,-y) \pmod{p}$ has a factorization over $\widetilde{\mathbb{F}}_p$ of the form

$$xy(x-y) \cdot \prod_i (x - \alpha_i y)^2 \cdot \prod_j (x - \beta_j y)$$

in which the $\alpha_i, \beta_j \in \widetilde{\mathbb{F}}_p$ are distinct, the α_i belong to $\mathbb{F}_p\backslash\{0,1\}$, and the β_j belong to $\widetilde{\mathbb{F}}_p\backslash\mathbb{F}_p$.

McCallum shows that there is a regular model $\mathfrak{G}^p_{v_p}/\operatorname{Spec}(\mathcal{O}_{L,v_p})$, gotten by blowing up $\mathfrak{F}^p_{v_p}$, whose geometric special fibre has the configuration shown in Figure 1. The components L_0, L_1, L_∞, L_{α_i} and L_{β_j} meeting L are indexed by the irreducible factors of $\phi(x,-y) \pmod{p}$, and for each α_i there are p components $\mathrm{F}_{\alpha_i,k}$ meeting L_{α_i}. All components are nonsingular and isomorphic to $\mathbb{P}^1/\widetilde{\mathbb{F}}_p$, and all intersections are transverse. The components L, L_0, L_1, L_∞, L_{α_i}, and $\mathrm{F}_{\alpha_i,j}$ are rational over $k_v = \mathbb{F}_p$; each L_{β_j} is rational over $\mathbb{F}_p(\beta_j)$. Furthermore L has multiplicity p and self-intersection -1; L_0, L_1, and L_∞ have multiplicity 1 and self-intersection $-p$; the L_{α_i} have multiplicity 2 and self-intersection $-p$; the L_{β_j} have multiplicity 1 and self-intersection $-p$; and the $\mathrm{F}_{\alpha_i,k}$ have multiplicity 1 and self-intersection -2.

The points of $\mathcal{F}^p_{v_p}(L_{v_p})$ specialize to the k_v-rational closed points of the k_v-rational multiplicity 1 components, which are not intersection points of components. There are p such points on each of L_0, L_1, and the components $\mathrm{F}_{\alpha_i,k}$. Each such point lifts to a subset of E_p isomorphic to $\pi_{v_p}\mathcal{O}_{L,v_p}$, and $E_p = \mathcal{F}^{p,0}(\mathcal{O}_{L,v_p})$ is the union of those subsets. On the other hand, $\xi_1, \ldots, \xi_p$ specialize to distinct k_v-rational closed points of L_∞.

Using Proposition 2.22 and the above description of E_p, we can determine $G(z, \xi_k; E_p)$ for each k. Since the computations are tedious, and the methods are the same as those in the proof of Theorem 2.21, we only give the final result: if n_p is the number of components L_{α_i}, and if

$$V = \frac{1}{p} + \frac{2p-1}{(2n_p+2)p - n_p}, \tag{2.155}$$

then for points $z \in \mathcal{F}^p(L_{v_p})$ specializing to L_∞,

$$G(z; \xi_k; E_p) = \frac{1}{p-1} \cdot \left(V + \log_{v_p}\left(((z)\cdot(\xi_k))_{\mathfrak{G}^p_{v_p}}\right)\right)$$

where $((z)\cdot(\xi_k))_{\mathfrak{G}^p_{v_p}}$ is the intersection number of the closures of z and ξ_k in the model $\mathfrak{G}^p_{v_p}$. The factor $1/(p-1)$ appears because the ramification index of $L_{v_p}/\mathbb{Q}_p$ is $p-1$. By analyzing the blowups in the construction of $\mathfrak{G}^p_{v_p}$, and writing $z \equiv_{\mathrm{L}_\infty} \xi_k$ if z and ξ_k specialize to the same closed point of L_∞, one further sees that

$$\log_{v_p}\left(((z)\cdot(\xi_k))_{\mathfrak{G}^p_{v_p}}\right) = \begin{cases} 0 & \text{if } z \not\equiv_{\mathrm{L}_\infty} \xi_k, \\ \log_{v_p}(|x(z)|_{v_p})) - 1 & \text{if } z \equiv_{\mathrm{L}_\infty} \xi_k. \end{cases}$$

Since $g_{\xi_k}(z) = 1/x(z)$ for each k, the local Green's matrix at p is

$$\Gamma(E_p, \mathfrak{X}) = \frac{1}{p-1} \begin{pmatrix} V-1 & V & \cdots & V \\ V & V-1 & \cdots & V \\ \vdots & \vdots & \ddots & \vdots \\ V & V & \cdots & V-1 \end{pmatrix}. \tag{2.156}$$

The Global Green's Matrix. For each prime q of $\mathbb{Q}$ with $q \neq p$, the model $\mathfrak{F}^p$ has good reduction at q, the points ξ_k specialize to distinct points of the special fibre, and the function $1/x(z)$ specializes to a nonconstant function $\pmod q$. Since E_q is $\mathfrak{X}$-trivial, $\Gamma(E_q, \mathfrak{X})$ is the zero matrix.

Thus the Global Green's matrix is

$$\Gamma(\mathbb{E}, \mathfrak{X}) = \Gamma(E_\infty, \mathfrak{X}) + \Gamma(E_p, \mathfrak{X}) \log(p).$$

When $\vec{s} = {}^t(\frac{1}{p}, \cdots, \frac{1}{p}) \in \mathcal{P}^p(\mathbb{R})$, entries of $\Gamma(\mathbb{E}, \mathfrak{X})\vec{s}$ are all equal, so using (2.153) we conclude that

$$\begin{aligned} V(\mathbb{E}, \mathfrak{X}) &= \frac{1}{p}\Big((B + (p-1)A) + \frac{1}{p-1}(pV-1)\Big) \\ &= \frac{1}{p}\big(-\log(R) + \frac{pV-1}{p-1}\big) \end{aligned}$$

Thus, by (2.155) and the Fekete-Szegö Theorems 0.4 and 1.5 we obtain:

THEOREM 2.29. *Let p be an odd prime. Then on the affine Fermat curve $x^p + y^p = 1$, if*

$$R > p^{\frac{p(2p-1)}{(p-1)^2((2n_p+2)p-n_p)}}, \tag{2.157}$$

there are infinitely many integral points α whose p-adic conjugates are all rational over L_{v_p} and whose archimedean conjugates satisfy $|x(\sigma(\alpha))| < R$.

If the inequality (2.157) is reversed, there are only finitely many.

For small primes, n_p can be computed using Maple. For $p = 2$ and $p = 5$, we have $n_p = 0$; for all primes with $5 < p < 75$ except $p = 59$, we have $n_p = 2$; for $p = 59$ we have $n_p = 13$. Below are some examples for the critical value of R:

p	n_p	critical R	
3	0	$3^{5/8}$	$\cong 1.987013346$
5	0	$5^{9/32}$	$\cong 1.572480664$
7	2	$7^{91/1440}$	$\cong 1.130851299$
53	2	$53^{5565/854464}$	$\cong 1.026195152$
59	13	$59^{6903/5513596}$	$\cong 1.005118113$
61	2	$61^{7381/1310400}$	$\cong 1.023425196$
73	2	$73^{10585/2260224}$	$\cong 1.020296147$

As McCallum remarks, n_p is the number of "tame curves" C_s for which $\mathrm{Jac}(C_s)$ is isogenous to a factor of $\mathrm{Jac}(\mathcal{F}_p)$. For $p = 59$ the abnormally large number of tame curves means the critical value of R is unusually small. It would be interesting to know if there are other phenomena related to this.

7. The Modular Curve $X_0(p)$

In this section we will give an example applying the Fekete-Szegö theorem with local rationality to the modular curve $X_0(p)/\mathbb{Q}$, where $p \geq 5$ is prime. The author thanks Pete Clark for suggesting this, and for help with properties of orders.

As is well known, $X_0(p)$ is the compactification of the moduli space for pairs (E, C) consisting of an elliptic curve and a cyclic subgroup of order p. As a Riemann surface, $X_0(p)(\mathbb{C})$ is gotten from $\Gamma_0(p)\backslash\mathfrak{H}$ by adjoining the "cusps" c_0 and c_∞; here $\mathfrak{H}$ is the complex upper half-plane and $\Gamma_0(p)$ is the congruence subgroup

$$\Gamma_0(p) = \left\{ \begin{pmatrix} a & b \\ c & d \end{pmatrix} \in SL_2(\mathbb{Z}) : c \equiv 0 \pmod{p} \right\} .$$

The function field of $X_0(p)/\mathbb{Q}$ is $\mathbb{Q}(j(z), j(pz))$ where $j(z)$ is the modular function

$$j(z) = \frac{1728 g_2^3}{g_2^3 - 27 g_3^2} = \frac{1}{q} + 744 + 196884q + \cdots .$$

Here $X = j(z)$ and $Y = j(pz)$ satisfy the "Modular Equation" $\Phi(X, Y) = 0$, where

$$\Phi(X,Y) = -(X^p - Y)(Y^p - X) + \sum_{\max(i,j) \leq p} a_{ij} X^i Y^j \in \mathbb{Z}[X,Y]$$

and each a_{ij} is divisible by p. The genus of $X_0(p)$ is

$$g_p = \begin{cases} (p-13)/12 & \text{if } p \equiv 1 \pmod{12}, \\ (p-5)/12 & \text{if } p \equiv 5 \pmod{12}, \\ (p-7)/12 & \text{if } p \equiv 7 \pmod{12}, \\ (p+1)/12 & \text{if } p \equiv 11 \pmod{12}, \end{cases}$$

Deligne and Rapoport determined a regular model $\mathfrak{M}_0(p)/\operatorname{Spec}(\mathbb{Z})$ for $X_0(p)$. It can be described as follows (see [**41**], Theorem 1.1, p.175). First, consider the projective normalization $M_0(p)$ of $\operatorname{Spec}(\mathbb{Z}[X,Y]/(\Phi(X,Y))$. It is smooth outside the points corresponding to supersingular elliptic curves in characteristic p with $j \neq 0, 1728$; its special fibre at p has two components, each isomorphic to $\mathbb{P}^1$, which meet transversely at the supersingular points. These components will be denoted

Z_0 and Z_∞; the reduction of j (that is, X) is a coordinate function on Z_∞. If $p \equiv 2 \pmod 3$ then $j = 0$ is supersingular in the fibre at p, and $M_0(p)$ has a singularity of type A_3 at the corresponding point; if $p \equiv 3 \pmod 4$ then $j = 1728$ is supersingular in the fibre at p and $M_0(p)$ has a singularity of type A_2 at the corresponding point.

The model $\mathfrak{M}_0(p)$ is obtained by resolving these singularities, introducing a chain of two components F_1, F_2 in the first case, and a single component G in the second. The special fibre of $\mathfrak{M}_0(p)$ is reduced, and all its components are rational over $\mathbb{F}_p$. There are $m = g_p + 1$ supersingular points, each of which is rational over $\mathbb{F}_{p^2}$. The components Z_0 and Z_∞ have self-intersection $-m$; the components F_1, F_2, and G (if present) have self-intersection -2. The cusps are rational over $\mathbb{Q}$; c_0 specializes to Z_0, and c_∞ specializes to Z_∞. Their images are not supersingular, and are the points "at infinity" on those components.

We will take $\mathfrak{X} = \{c_\infty, c_0\}$ to be the set of cusps, and we will take $\mathbb{E} = \prod_v E_v$, where

$$E_\infty \;=\; j^{-1}(D(0,R)) \;=\; \{z \in X_0(p)(\mathbb{C}) : |j(z)| \le R\}\ .$$

and where E_p is the set of points of $X_0(p)(\mathbb{Q}_p)$ specializing to the "ordinary" (i.e., nonsupersingular and noncuspidal) points of Z_∞. For all the other nonarchimedean places q, we will take E_q to be the $\mathfrak{X}$-trivial set

$$E_q \;= \mathfrak{M}_0(p)(\mathbb{C}_p)\backslash(B(c_0,1)^- \bigcup B(c_\infty,1)^-)\ .$$

We will take the uniformizing parameters to be $g_{c_\infty}(z) = 1/j(z)$, $g_{c_0}(z) = 1/j(pz)$.

This set $\mathbb{E}$ is chosen mainly because we can do explicit computations with it. However, it illustrates nicely how arithmetic and geometric information about a curve enter into capacities.

The Green's Matrix at the Archimedean Place. Let $\mathcal{D} = \{z \in \mathfrak{H} : -1/2 \le \mathrm{Re}(z) \le 1/2, |z| \ge 1\}$ be the standard closed fundamental domain for $SL_2(\mathbb{Z})$. As a function from $\mathfrak{H}$ to $\mathbb{C}$, $j(z)$ maps this region conformally onto $\mathbb{C}$, taking the ray from i to ∞ along the imaginary axis to the real interval $[1728,\infty)$, with $j(i) = 1728$; the circular arc at the bottom of $\mathcal{D}$ to the real interval $[0,1728]$ (covering it twice), with $j(e^{\pi i/3}) = j(e^{2\pi i/3}) = 0$; and the vertical sides of $\mathcal{D}$ to the real interval $[-\infty,0]$. It also takes the part of the imaginary axis from 0 to i to $[1728,\infty)$. A fundamental domain for $\Gamma_0(p)$ is given by

$$\mathcal{D}(p) \;=\; \mathcal{D} \cup \Big(\bigcup_{k=-(p-1)/2}^{(p-1)/2} f_k(\mathcal{D})\Big)$$

where $f_k(z) = -1/(z+k)$. Under the quotient $\Gamma_0(p)\backslash\mathfrak{H}$, the image of the circular arc at the bottom of $\mathcal{D}$ separates $X_0(p)(\mathbb{C})$ into two components, one containing c_∞ and the other containing c_0. On the other hand, the image of the imaginary axis joins the cusps c_0, c_∞.

By our choice of E_∞ and discussion above, it follows that when $j(z)$ is viewed as a map from $X_0(p)(\mathbb{C})$ to $\mathbb{P}^1(\mathbb{C})$, if $R \ge 1728$ then $X_0(p)(\mathbb{C})\backslash E_\infty$ has two connected components, while if $R < 1728$ it has one component.

As a divisor $j^{-1}((\infty)) = p(c_0) + (c_\infty)$, so the pullback formula (2.61) gives

$$G(j(z),\infty; D(0,R)) \;=\; pG(z,c_0;E_\infty) + G(z,c_\infty;E_\infty)\ .$$

Since $1/j(z)$ is the uniformizing parameter at c_∞, it follows that

$$-\log(R) \;=\; pG(c_\infty,c_0;E_\infty) + V_{c_\infty}(E_\infty)\ .$$

Similarly, since $\lim_{z\to c_0} j(z)^p/j(pz) = 1$, and since $1/j(pz)$ is the uniformizing parameter at c_0,

$$-\log(R) = G(c_0, c_\infty; E_\infty) + pV_{c_0}(E_\infty) .$$

Hence, writing $B(R) = G(c_\infty, c_0; E_\infty) = G(c_0, c_\infty; E_\infty)$, the archimedean local Green's matrix is

$$\Gamma(E_\infty, \mathfrak{X}) = \begin{pmatrix} -\log(R) & 0 \\ 0 & -\frac{1}{p}\log(R) \end{pmatrix} + B(R)\begin{pmatrix} -p & 1 \\ 1 & -1 \end{pmatrix} . \tag{2.158}$$

Here $B(R) = 0$ if $R \geq 1728$, while $B(R) > 0$ if $R < 1728$. It will turn out that $R > 1728$ in the situation of interest to us; however, note that in any case the second matrix in (2.158) is negative semi-definite.

The Green's Matrix at the Place p. Using Proposition 2.22 and the definition of E_p as the set of points of $X_0(\mathbb{Q}_p)$ specializing to ordinary points on the component Z_∞, we can determine $G(z, c_\infty; E_p)$ and $G(z, c_0; E_p)$.

Let $\mathcal{N}_p$ be the number of $\mathbb{F}_p$-rational ordinary points on Z_∞. For $z \in X_0(p)(\mathbb{C}_p)$, write $z \equiv_{Z_0} c_0$ if z specializes to same point of Z_0 as c_0, and write $z \equiv_{Z_\infty} c_\infty$ specializes to the same point of Z_∞ as c_∞. Using Proposition 2.22 and the methods in the proof of Theorem 2.21, we find that

$$G(z, c_\infty; E_p) = \begin{cases} \frac{1}{\mathcal{N}_p}\frac{p}{p-1} & \text{if } z \equiv_{Z_0} c_0, \\ \frac{1}{\mathcal{N}_p}\frac{p}{p-1} + \log_p(|j(z)|_p) & \text{if } z \equiv_{Z_\infty} c_\infty, \end{cases}$$

$$G(z, c_0; E_p) = \begin{cases} \frac{1}{\mathcal{N}_p}\frac{p}{p-1} & \text{if } z \equiv_{Z_\infty} c_\infty, \\ \frac{1}{\mathcal{N}_p}\frac{p}{p-1} + \frac{12}{p-1} + \log_p(|j(pz)|_p) & \text{if } z \equiv_{Z_0} c_0. \end{cases}$$

Here the number $12/(p-1)$ is actually the quantity $j_{c_0}(Z_\infty, Z_\infty)$ in the notation of (2.131), obtained by solving the equations (2.128) relating components. Since the special fibre of $\mathfrak{M}_0(p)$ at p has different configurations according as $p \equiv 1, 5, 7, 11 \pmod{12}$, it is somewhat surprising that the same value arises in all cases.

It follows that the local Green's matrix at p is

$$\Gamma(E_p, \mathfrak{X}) = \begin{pmatrix} \frac{1}{\mathcal{N}_p}\frac{p}{p-1} & \frac{1}{\mathcal{N}_p}\frac{p}{p-1} \\ \frac{1}{\mathcal{N}_p}\frac{p}{p-1} & \frac{1}{\mathcal{N}_p}\frac{p}{p-1} + \frac{12}{p-1} \end{pmatrix} . \tag{2.159}$$

The number $\mathcal{N}_p$ can be expressed in terms of the class number $h(-p)$ of the ring of integers of $\mathbb{Q}(\sqrt{-p})$. Indeed, $\mathcal{N}_p = p - n_{ss}(\mathbb{F}_p)$, where $n_{ss}(\mathbb{F}_p)$ is the number of $\mathbb{F}_p$-rational supersingular points on Z_∞. It is known (see for example [**20**], pp.75-76) that

$$n_{ss}(\mathbb{F}_p) = \frac{h'(-p) + h'(-4p)}{2}$$

where $h'(D)$ is the class number of the quadratic order of discriminant D if there is such an order, and is 0 otherwise. Using the formula relating class numbers of orders in quadratic fields to those of the maximal orders (see [**36**], Theorem 7, p.95), this simplifies to $n_{ss}(\mathbb{F}_p) = c_p h(-p)$, where

$$c_p = \begin{cases} 1/2 & \text{if } p \equiv 1 \pmod 8, \\ 2 & \text{if } p \equiv 3 \pmod 8, \\ 1/2 & \text{if } p \equiv 5 \pmod 8, \\ 1 & \text{if } p \equiv 7 \pmod 8). \end{cases} \tag{2.160}$$

Thus $\mathcal{N}_p = p - c_p h(-p)$. It is known that $\mathcal{N}_p$ is always positive, so E_p is nonempty.

The Global Green's Matrix. For each prime q of $\mathbb{Q}$ with $q \neq p$, the model $\mathfrak{M}_0(p)$ has good reduction at q, the cusps c_∞, c_0 specialize to distinct points of the special fibre, and the uniformizing parameters $g_{c_\infty}(z)$ and $g_{c_0}(z)$ specialize to nonconstant functions $\pmod q$. Since E_q is $\mathfrak{X}$-trivial, $\Gamma(E_q, \mathfrak{X})$ is the zero matrix.

Suppose for the moment that $R \geq 1728$; this assumption will be justified below. Then $B(R) = 0$, and the global Green's matrix $\Gamma(\mathbb{E}, \mathfrak{X})$ is

$$\begin{pmatrix} -\log(R) + \frac{1}{\mathcal{N}_p}\frac{p}{p-1}\log(p) & \frac{1}{\mathcal{N}_p}\frac{p}{p-1}\log(p) \\ \frac{1}{\mathcal{N}_p}\frac{p}{p-1}\log(p) & -\frac{1}{p}\log(R) + (\frac{1}{\mathcal{N}_p}\frac{p}{p-1} + \frac{12}{p-1})\log(p) \end{pmatrix} .$$

By the minimax definition of $V(\mathbb{E}, \mathfrak{X})$ (see formula (3.50) below)

$$V(\mathbb{E}, \mathfrak{X}) \;=\; \min_{\vec{s} \in \mathcal{P}^2(\mathbb{R})} \max_i \big(\Gamma(\mathbb{E}, \mathfrak{X})\vec{s}\big)_i \ .$$

Thus $V(\mathbb{E}, \mathfrak{X}) < 0$ if and only if for some $\vec{s} = {}^t(s_1, s_2) \in \mathcal{P}^2(\mathbb{R})$,

$$\begin{cases} s_1\big(-\log(R) + \frac{1}{\mathcal{N}_p}\frac{p}{p-1}\log(p)\big) + s_2\big(\frac{1}{\mathcal{N}_p}\frac{p}{p-1}\log(p)\big) \;<\; 0 \ , \\ s_1\big(\frac{1}{\mathcal{N}_p}\frac{p}{p-1}\log(p)\big) + s_2\big(-\frac{1}{p}\log(R) + \frac{1}{\mathcal{N}_p}\frac{p}{p-1} + \frac{12}{p-1}\log(p)\big) \;<\; 0 \ . \end{cases}$$

Equivalently, $V(\mathbb{E}, \mathfrak{X}) < 0$ if and only if for some $s \in \mathbb{R}$ with $0 < s < 1$,

$$\begin{cases} \frac{1}{s}\cdot\big(\frac{1}{\mathcal{N}_p}\frac{p}{p-1}\big) \;<\; \log_p(R) \ , \\ \frac{1}{1-s}\cdot\big(\frac{1}{\mathcal{N}_p}\frac{p^2}{p-1}\big) + \frac{12p}{p-1} \;<\; \log_p(R) \ . \end{cases} \tag{2.161}$$

The left side of the first inequality in (2.161) is decreasing with s, while that in the second inequality is increasing, so the extremal value of R is obtained when they are equal. Solving, and using the Fekete-Szegö Theorems 0.4 and 1.5, one obtains

THEOREM 2.30. *Let $p \geq 5$ be a prime, and consider the Deligne-Rapoport model $\mathfrak{M}_0(p)$ for modular curve $X_0(p)/\mathbb{Q}$. Put $\mathcal{N}_p = p - c_p h(-p)$, where c_p is as in* (2.160) *and $h(-p)$ is the class number of $\mathbb{Q}(\sqrt{-p})$. Then if*

$$R \;>\; p^{\frac{1+p+12\mathcal{N}_p + \sqrt{(1+p+12\mathcal{N}_p)^2 - 48\mathcal{N}_p}}{2\mathcal{N}_p}\cdot\frac{p}{p-1}} \ , \tag{2.162}$$

there are infinitely many $\alpha \in X_0(p)(\widetilde{\mathbb{Q}})$ with archimedean conjugates which satisfy $|j(\sigma(\alpha))| < R$, whose p-adic conjugates all belong to $X_0(p)(\mathbb{Q}_p)$ and specialize $\pmod p$ *to ordinary points in Z_∞, and whose conjugates in $X_0(\mathbb{C}_q)$ specialize* mod q *to noncuspidal points of $\mathfrak{M}_0(p)$, for all $q \neq p$.*

If the inequality (2.162) *is reversed, there are only finitely many.*

Note that the right side of (2.162) is greater than p^6, and for $p \geq 5$ this is at least 15625. By the second part of the theorem and the monotonicity of the sets $j^{-1}(D(0, R))$, the first part cannot hold for any $R < 15625$. This validates our assumption that $R > 1728$.

CHAPTER 3

Preliminaries

In this chapter we lay out notation, conventions, and foundational material used in the rest of the book.

This book can be regarded as a sequel to the author's monograph "Capacity theory on algebraic curves" ([**51**]), and we recall several results from that work. In particular, we consider spherical metrics on $\mathbb{P}^N$, the "canonical distance" $[z,w]_\zeta$ on an algebraic curve, sets of capacity 0, upper Green's functions, the inner Cantor capacity, and the L-rational basis for algebraic functions on a curve with poles supported on a finite set $\mathfrak{X}$.

1. Notation and Conventions

Throughout the book, we write $\log(x)$ for $\ln(x)$.

If K is a global field, $\widetilde{K}$ will be a fixed algebraic closure of K, and $\widetilde{K}^{\text{sep}}$ will be the separable closure of K in $\widetilde{K}$. We write $\text{Aut}(\widetilde{K}/K)$ for the group of automorphisms $\text{Aut}(\widetilde{K}/K) \cong \text{Gal}(\widetilde{K}^{\text{sep}}/K)$. Given a place v of K, let K_v be the completion of K at v, let $\widetilde{K}_v$ be an algebraic closure of K_v, let $\widetilde{K}_v^{\text{sep}}$ be the separable closure of K_v in $\widetilde{K}_v$, and let $\mathbb{C}_v$ be the completion of $\widetilde{K}_v$. If v is nonarchimedean, write $\widehat{\mathcal{O}}_v$ for the ring of integers of $\mathbb{C}_v$. Let $\text{Aut}_c(\mathbb{C}_v/K_v) \cong \text{Aut}(\widetilde{K}_v/K_v) \cong \text{Gal}(\widetilde{K}_v^{\text{sep}}/K_v)$ be the group of continuous automorphisms of $\mathbb{C}_v$ fixing K_v.

If v is archimedean, let $|z|_v = |z|$ be the usual absolute value on $\mathbb{R}$ or $\mathbb{C}$ for which the triangle inequality holds. For $0 < x \in \mathbb{R}$, write $\log_v(x) = \ln(x)$. If $K_v \cong \mathbb{R}$, put $q_v = e$; if $K_v \cong \mathbb{C}$, put $q_v = e^2$.

If v is nonarchimedean, let $|x|_v$ be the absolute value on K_v given by the modulus of additive Haar measure. Let $\mathcal{O}_v$ be the ring of integers of K_v, let π_v be a uniformizer for $\mathcal{O}_v$, and let $k_v = \mathcal{O}_v/\pi_v\mathcal{O}_v$ be the residue field. Put $q_v = \#(k_v)$; then $|\pi_v|_v = 1/q_v$. If $\text{char}(K) = p > 0$, then $q_v = p^{f_v}$ is a power of p. If $\text{char}(K) = 0$ and p is the rational prime under v, let e_v and f_v be the absolute ramification index and residue degree; then $q_v = p^{f_v}$ and $|p|_v = (1/p)^{-[K_v:\mathbb{Q}_p]} = 1/q_v^{e_v}$. This absolute value has a unique extension to $\mathbb{C}_v$, which we will continue to write as $|x|_v$. Let $\log_v(x)$ be the logarithm to the base q_v, and let $\text{ord}_v(x)$ be the additive valuation on $\mathbb{C}_v$ associated to $|x|_v$. Then $\text{ord}_v(x) = -\log_v(|x|_v) \in \mathbb{Q}$ for all $x \in \mathbb{C}_v^\times$.

Let $\mathcal{M}_K$ be the set of all places of K. For $0 \neq \kappa \in K$, the product formula reads

$$\sum_{v \in \mathcal{M}_K} \log_v(|\kappa|_v) \log(q_v) \ = \ 0 \ .$$

We will sometimes need to write the product formula multiplicatively. To do this, we introduce weights $D_v = \log_v(q_v)$, so that $D_v = 1$ unless $K_v \cong \mathbb{C}$, in which case

$D_v = 2$. Then

$$\prod_{v \in \mathcal{M}_K} |\kappa|_v^{D_v} = 1 .$$

We use this combination of normalized absolute values and weights for compatibility with the literature concerning capacities sets in $\mathbb{C}$.

If L/K is a finite extension, we use similar conventions in defining normalized absolute values $|x|_w$ for places w of L, as well as q_w, D_w and $\log_w(x)$. Thus for $0 \neq \lambda \in L$,

$$\sum_{w \in \mathcal{M}_L} \log_w(|\lambda|_w) \log(q_w) = 0 , \qquad \prod_{w \in \mathcal{M}_L} |\lambda|_w^{D_w} = 1 .$$

If v is a place of K and w is a place of L over K, then on $\mathbb{C}_w \cong \mathbb{C}_v$ we have the (extended) absolute values $|x|_v$ and $|x|_w$. For archimedean places, $|x|_w = |x|_v = |x|$ for all $x \in \mathbb{C}_w \cong \mathbb{C}_v \cong \mathbb{C}$. For nonarchimedean places, $|x|_w = |x|_v^{[L_w:K_v]}$ for $x \in \mathbb{C}_w \cong \mathbb{C}_v$. For all places, $|x|_w^{D_w} = |x|_v^{[L_w:K_v]D_v}$.

Let F be a field. By a 'variety' $\mathcal{V}/F$ we mean a separated scheme $\mathcal{V}$ of finite type over $\mathrm{Spec}(F)$. By a 'curve' $\mathcal{C}/F$, we mean a smooth, projective, geometrically integral variety of dimension 1 over $\mathrm{Spec}(F)$. If $\mathcal{V}/F$ is a variety and L is a field containing F, we write $\mathcal{V}_L = \mathcal{V} \times_F \mathrm{Spec}(L)$ and let $\mathcal{V}(L) = \mathrm{Hom}_F(\mathrm{Spec}(L), \mathcal{V}) \cong \mathrm{Hom}_L(\mathrm{Spec}(L), \mathcal{V}_L)$ be the set of L-rational points of $\mathcal{V}$. If $\mathcal{V}_L$ is irreducible, we write $L(\mathcal{V})$ for its function field.

If K is a global field, v is a place of K, and $\mathcal{V}/K$ is a variety, we abbreviate $\mathcal{V}_{K_v}$ by $\mathcal{V}_v$. Note that $\mathcal{V}_v(\mathbb{C}_v) \cong \mathcal{V}(\mathbb{C}_v)$ since K is embedded in K_v and $\mathbb{C}_v$. If $E_v \subset \mathcal{V}_v(\mathbb{C}_v)$ is a nonempty set, then for each $f \in \mathbb{C}_v(\mathcal{V})$, we write $\|f\|_{E_v} = \sup_{z \in E_v} |f(z)|_v$ for its sup norm.

2. Basic Assumptions

Throughout the paper we will assume that:

K is a global field and $\mathcal{C}/K$ is a curve (smooth, projective, and geometrically integral). Fix an embedding $\mathcal{C} \hookrightarrow \mathbb{P}_K^N = \mathbb{P}^N/\mathrm{Spec}(K)$ for an appropriate N, and equip $\mathbb{P}_K^N$ with a system of homogeneous coordinates. For each v, let $\|x, y\|_v$ be the v-adic metric associated to this embedding (see §3.4 below). For each nonarchimedean v, the choice of homogeneous coordinates yields gives an integral structure $\mathbb{P}^N/\mathrm{Spec}(\mathcal{O}_v)$. By taking the closure of $\mathcal{C}_v$ in $\mathbb{P}^N/\mathrm{Spec}(\mathcal{O}_v)$, we obtain a model $\mathfrak{C}_v/\mathrm{Spec}(\mathcal{O}_v)$.

$\mathfrak{X} = \{x_1, \ldots, x_n\} \subset \mathcal{C}(\widetilde{K})$ is a finite set of global algebraic points, stable under $\mathrm{Aut}(\widetilde{K}/K)$. We will call the points in $\mathfrak{X}$ *poles*.

$L = K(\mathfrak{X})$, so L/K is a finite normal extension. For each place v of K, fix an embedding $\iota_v : \widetilde{K} \hookrightarrow \mathbb{C}_v$ over K. This induces a distinguished place w_v of L lying over v, and an embedding $\mathfrak{X} \hookrightarrow \mathcal{C}_v(\mathbb{C}_v)$. In this way, we regard $\mathfrak{X}$ as a subset of $\mathcal{C}_v(\mathbb{C}_v)$, for each v.

$E_v \subset \mathcal{C}_v(\mathbb{C}_v)$ is a nonempty set, for each place v of K. A set E_v will be called $\mathfrak{X}$-*trivial* (for the model $\mathfrak{C}_v$) if v is nonarchimedean, $\mathfrak{C}_v$ has good reduction at v, the balls $B(x_i, 1)^- = \{z \in \mathcal{C}_v(\mathbb{C}_v) : \|z, x_i\|_v < 1\}$ are pairwise disjoint, and

$$E_v = \mathcal{C}_v(\mathbb{C}_v) \backslash \bigcup_{i=1}^m B(x_i, 1)^- .$$

Equivalently, E_v is $\mathfrak{X}$-trivial if $\mathfrak{C}_v$ has good reduction at v, the points in $\mathfrak{X}$ (identified with points of $\mathcal{C}_v(\mathbb{C}_v)$ and extended to sections of $\mathfrak{C}_v \otimes_{\mathcal{O}_v} \mathrm{Spec}(\widehat{\mathcal{O}}_v)$) specialize to distinct points $\pmod v$, and E_v is the set of points of $\mathcal{C}_v(\mathbb{C}_v)$ which specialize to points complementary to $\mathfrak{X} \pmod v$.

We will assume the sets E_v satisfy the following conditions:

(1) Each E_v is stable under $\mathrm{Aut}_c(\mathbb{C}_v/K_v)$.
(2) Each E_v is bounded away from $\mathfrak{X}$ in the v-topology.
(3) For all but finitely many v, E_v is $\mathfrak{X}$-trivial.

Here, property (3) is independent of the choice of the embedding $\mathcal{C} \hookrightarrow \mathbb{P}^N/K$ and the choice of homogeneous coordinates on $\mathbb{P}^N/K$. Note that E_v may be closed, open, or neither. However, an $\mathfrak{X}$-trivial set is both open and closed.

Put $\mathbb{E} = \prod_v E_v$. We will call $\mathbb{E}$ a *K-rational adelic set.* By our assumptions, $\mathbb{E}$ is compatible with $\mathfrak{X}$, in the terminology of the Introduction.

$U_v \subset \mathcal{C}_v(\mathbb{C}_v)$ is an open set containing E_v, for each place v of K. We do not assume that U_v is bounded away from $\mathfrak{X}$ or stable under $\mathrm{Aut}_c(\mathbb{C}_v/K_v)$, though we will reduce to that situation in the proof of the Fekete-Szegö theorem. The set $\mathbb{U} = \prod_v U_v$ will be called an *adelic neighborhood* of $\mathbb{E}$.

For each $x_i \in \mathfrak{X}$, fix a rational function $g_{x_i}(z) \in K(x_i)(\mathcal{C})$ with a simple zero at x_i. We require that the choices be made so that $g_{\sigma(x_i)}(z) = \sigma(g_{x_i})(z)$ for all $\sigma \in \mathrm{Aut}(\widetilde{K}/K)$. These uniformizing parameters will be used for normalizations throughout the paper.

We will often deal with objects on which $\mathrm{Aut}(\widetilde{K}/K)$ acts: for instance points in $\mathcal{C}(\widetilde{K})$ and functions in $\widetilde{K}(\mathcal{C})$. If L/K is Galois, the points of $\mathcal{C}(L)$ fixed by $\mathrm{Aut}(\widetilde{K}/K)$ are the K-rational points, and the functions in $L(\mathcal{C})$ fixed by $\mathrm{Aut}(\widetilde{K}/K)$ are the K-rational functions. Likewise, if v is a place of K, we will often deal with objects on which $\mathrm{Aut}_c(\mathbb{C}_v/K_v)$ acts. We will use the following terminology (due to Cantor [**16**]):

Definition 3.1 (K-symmetric, K_v-symmetric). Let K be a global field, and let Y be a collection of objects on which $\mathrm{Aut}(\widetilde{K}/K)$ acts. If an element $y_0 \in Y$ is fixed by that action, we will say that y_0 is K-symmetric. If a subset $Y_0 \subset Y$ is stable under $\mathrm{Aut}(\widetilde{K}/K)$, we will say that Y_0 is K-symmetric.

Similarly, let v be a place of K, and let Y be a set on which $\mathrm{Aut}_c(\mathbb{C}_v/K_v)$ acts. If an element $y_0 \in Y$ is fixed by that action, we will say that y_0 is K_v-symmetric. Likewise, if a subset $Y_0 \subset Y$ is stable under $\mathrm{Aut}_c(\mathbb{C}_v/K_v)$, we will say that Y_0 is K_v-symmetric.

Here is an important example.

Let $\mathfrak{X} = \{x_1, \ldots, x_m\} \subset \mathcal{C}(\widetilde{K})$ be as above. Define a permutation representation of $G = \mathrm{Aut}(\widetilde{K}/K)$ in S_m by $x_{\sigma(i)} = \sigma(x_i)$ for each $\sigma \in G$. There is an induced action of G on $\mathbb{R}^m$: given $\vec{s} = {}^t(s_1, \ldots, s_m) \in \mathbb{R}^m$, put $\sigma(\vec{s}) = {}^t(s_{\sigma(1)}, \ldots, s_{\sigma(m)})$. If $s_{\sigma(i)} = s_i$ for all $\sigma \in G$ and all i, we call $\vec{s}$ a *K-symmetric vector.* Similarly, G acts on $M_m(\mathbb{R})$ by simultaneously permuting the rows and columns. These two actions are compatible: for $\vec{s} \in \mathbb{R}^m$ and $\Gamma \in M_m(\mathbb{R})$, we have $\sigma(\Gamma\vec{s}) = \sigma(\Gamma)\sigma(\vec{s})$. If $\Gamma_{ij} = \Gamma_{\sigma(i),\sigma(j)}$ for all $\sigma \in G$ and all i, j, we call Γ a *K-symmetric matrix.*

For another example, let Y be the collection of all finite sets of functions in $\widetilde{K}(\mathcal{C})$. Then the set of uniformizing parameters $\{g_{x_i}(z)\}_{x_i \in \mathfrak{X}}$ chosen above is K-symmetric, since $\sigma(g_{x_i})(z) = g_{\sigma(x_i)}(z)$ for all i.

The following fact is well known, but we do not have a convenient reference.

PROPOSITION 3.2. *Let K be a global field, and let $\mathcal{C}/K$ be a smooth, projective, connected curve. Then $\mathcal{C}_v(\widetilde{K}_v^{\text{sep}})$ is dense in $\mathcal{C}_v(\mathbb{C}_v)$ under the v topology, for each place v of K.*

PROOF. We first show that $\widetilde{K}_v^{\text{sep}}$ is dense in $\mathbb{C}_v$. Since $\mathbb{C}_v$ is the completion of $\widetilde{K}_v$, it suffices to show that $\widetilde{K}_v^{\text{sep}}$ is dense in $\widetilde{K}_v$. If $\text{char}(K) = 0$ there is nothing to prove. Assume $\text{char}(K) = p > 0$, and take $\alpha \in \widetilde{K}_v \backslash \widetilde{K}_v^{\text{sep}}$. Let $P(x) \in K_v[x]$ be the minimal polynomial of α. Since α is inseparable over K_v, there is a polynomial $Q(x) \in K_v[x]$ such that $P(x) = Q(x^p)$. For each $0 \neq b \in K_v$, put $P_b(x) = P(x)+b\,x$. Then $P_b'(x) = b$, so $(P_b(x), P_b'(x)) = 1$, and $P_b(x)$ has distinct roots. In particular, its roots all belong to $\widetilde{K}_v^{\text{sep}}$. Fix $\varepsilon > 0$, and let $b \to 0$. By the continuity of roots of polynomials in $\mathbb{C}_v[x]$ under variation of the coefficients ([**34**], p.44), if $|b|_v$ is sufficiently small, then $P_b(x)$ has a root α_b with $|\alpha_b - \alpha|_v < \varepsilon$.

We next show that $\mathcal{C}_v(\widetilde{K}_v^{\text{sep}})$ is dense in $\mathcal{C}_v(\mathbb{C}_v)$. Since $\mathcal{C}/K$ is smooth, so is $\mathcal{C}_v/K_v$, hence the function field $K_v(\mathcal{C}_v)$ is separably generated over K_v (see [**51**], p.21). Since $K_v(\mathcal{C}_v)/K_v$ is finitely generated and has transcendence degree 1, there is an $f \in K_v(\mathcal{C})$ such that $K_v(\mathcal{C}_v)$ is finite and separable over $K_v(f)$. By the Primitive Element theorem, there is a $g \in K_v(\mathcal{C})$ such that $K_v(\mathcal{C}) = K_v(f, g)$. Let $F(x, y) \in K_v[x, y]$ be a nonzero polynomial of minimal degree for which $F(f, g) = 0$. Regarding $F(x, y)$ as a polynomial in y with coefficients in $K_v[x]$, write

$$F(x, y) \;=\; a_0(x)y^n + a_1(x)y^{n-1} + \cdots + a_n(x) \;.$$

Let $R(x)$ be the resultant of F and $\frac{\partial F}{\partial y}$. It is not the zero polynomial, since g is separable over $K_v(f)$. There are finitely many values of $x \in \mathbb{C}_v$ for which $R(x) = 0$; for all other x, the polynomial $F_x(y) = F(x, y)$ has n distinct roots.

Let $\mathcal{C}_{v,1}/K_v$ be the projective closure of the plane curve defined by $F(x, y) = 0$. There is a birational morphism $Q : \mathcal{C}_v \to \mathcal{C}_{v,1}$ defined over K_v. Let $S_1 \subset \mathcal{C}_{v,1}(\mathbb{C}_v)$ be the set consisting of all singular points, branch points of Q, points "at infinity", and points where $R(x) = 0$. Put $S = Q^{-1}(S_1) \subset \mathcal{C}_v(\mathbb{C}_v)$; both S_1 and S are finite. The map Q induces a topological isomorphism from $\mathcal{C}_v(\mathbb{C}_v)\backslash S$ onto $\mathcal{C}_{v,1}(\mathbb{C}_v)\backslash S_1$ which takes $\mathcal{C}_v(\widetilde{K}_v^{\text{sep}})\backslash S$ onto $\mathcal{C}_{v,1}(\widetilde{K}_v^{\text{sep}})\backslash S_1$. To show that $\mathcal{C}_v(\widetilde{K}_v^{\text{sep}})$ is dense in $\mathcal{C}_v(\mathbb{C}_v)$, it suffices to show that $\mathcal{C}_{v,1}(\widetilde{K}_v^{\text{sep}})\backslash S_1$ is dense in $\mathcal{C}_{v,1}(\mathbb{C}_v)\backslash S_1$.

Identify points of $\mathcal{C}_{v,1}(\mathbb{C}_v)\backslash S_1$ with solutions to $F(x, y) = 0$ in $\mathbb{C}_v^2$, and fix $P = (b, c) \in \mathcal{C}_{v,1}(\mathbb{C}_v)\backslash S_1$. Then c is a root of $F_b(y) = a_0(b)y^n + \cdots + a_0(b)$. Since $\widetilde{K}_v^{\text{sep}}$ is dense in $\mathbb{C}_v$, there is a sequence $b_1, b_2, \cdots \in \widetilde{K}_v^{\text{sep}}$ converging to b. We can assume that none of the b_i is the x-coordinate of a point in S_1. By the continuity of the roots of polynomials, there are $c_1, c_2, \cdots \in \widetilde{K}_v^{\text{sep}}$ such that each c_i is a root of $F_{b_i}(y)$, and the points $P_i = (a_i, b_i)$ converge to P. $\square$

3. The L-rational and L^{sep}-rational Bases

Let $L = K(\mathfrak{X})$, as above. Given finite extension F/K, let F^{sep} be the separable closure of K in F. In this section we will construct a K-symmetric (that is, $\text{Aut}(\widetilde{K}/K)$-equivariant) set of L-rational functions which we will use to expand functions in $\widetilde{K}(\mathcal{C})$ with poles supported on $\mathfrak{X}$. We will call this the L-rational basis. To deal with separability issues in the global patching construction, we also construct a related set of L^{sep}-rational functions, which we call the L^{sep}-rational basis.

When $\mathrm{char}(K) = 0$, the L-rational and L^{sep}-rational bases are the same, but when $\mathrm{char}(K) = p > 0$ they are different.

Given a $\widetilde{K}$-rational divisor D on $\mathcal{C}$, put

$$\widetilde{\Gamma}(D) = H^0(\mathcal{C}_{\widetilde{K}}, \mathcal{O}_{\mathcal{C}_{\widetilde{K}}}(D)) = \{f \in \widetilde{K}(\mathcal{C}) : \mathrm{div}(f) + D \geq 0\} .$$

A theorem of Weil (see for example Lang [**35**], Theorem 5, p.174) asserts that if D is rational over a finite extension F/K, then $\widetilde{\Gamma}(D)$ has a basis consisting of F-rational functions. In scheme-theoretic terms, this comes from the faithful flatness of $\mathrm{Spec}(\widetilde{K})/\mathrm{Spec}(F)$.

We first give the construction when $\mathrm{char}(K) = 0$. Let $g \geq 0$ be the genus of $\mathcal{C}$, and put $J = 2g + 1$.

Taking $D_0 = \sum_{i=1}^m J \cdot (x_i)$, put $\Lambda_0 = \dim_{\widetilde{K}}(\widetilde{\Gamma}((D_0))) \geq 1$. Noting that D_0 is rational over K, choose an arbitrary K-rational basis $\{\varphi_1, \ldots, \varphi_{\Lambda_0}\}$ for $\widetilde{\Gamma}(D_0)$. Next, choose a representative x_i from each $\mathrm{Aut}(L/K)$-orbit in $\mathfrak{X}$. For each $j = J, \ldots, 2J$, choose a $K(x_i)$-rational function $\varphi_{i,j}(z) \in \widetilde{\Gamma}(j \cdot (x_i))$ with a pole of exact order j at x_i, normalized in such a way that

$$\lim_{z \to x_i} \varphi_{i,j}(z) \cdot g_{x_i}(z)^j = 1 ,$$

and taking $\varphi_{i,2J}(z) = \varphi_{i,J}(z)^2$. Such functions exist by the Riemann-Roch Theorem. For each $\sigma(x_i)$ in the orbit of x_i and each $j = J, \ldots, 2J$, put $\varphi_{\sigma(i),j} = \sigma(\varphi_{i,j})$.

For each $j \geq 2J + 1$, we can uniquely write $j = \ell \cdot J + r$, where ℓ and r are integers with $\ell > 0$ and $J + 1 \leq r \leq 2J$. Define

$$\varphi_{i,j}(z) = (\varphi_{i,J}(z))^\ell \cdot \varphi_{i,r}(z). \tag{3.1}$$

Then $\varphi_{i,j}$ is rational over $K(x_i)$, has a simple pole of order j at x_i and no other poles, and is normalized so that

$$\lim_{z \to x_i} \varphi_{i,j}(z) \cdot g_{x_i}(z)^j = 1 . \tag{3.2}$$

For all $\sigma \in \mathrm{Aut}(\widetilde{K}/K)$ and all (i, j), we have $\varphi_{\sigma(i),j} = \sigma(\varphi_{i,j})$.

In this way we obtain a multiplicatively finitely generated, K-symmetric set of functions

$$\{\varphi_\lambda : \lambda = 1 \ldots, \Lambda_0\} \cup \{\varphi_{i,j} : i = 1, \ldots, m, j \geq J + 1\} \tag{3.3}$$

defined over L, which we will call the *L-rational basis* (note that the functions $\varphi_{i,J}$ do not belong to the L-rational basis, though they were used in constructing it). Each $f \in \widetilde{K}(\mathcal{C})$ with poles supported on $\mathfrak{X}$ can be uniquely expanded as a $\widetilde{K}$-linear combination of the φ_k and the $\varphi_{i,j}$, and if f has a pole of order n_i at each x_i, only the φ_λ and the $\varphi_{i,j}$ with $j \leq n_i$ are required in the expansion. Similarly, for each v, each $f \in \mathbb{C}_v(\mathcal{C})$ with poles supported on $\mathfrak{X}$ can be uniquely expanded as a $\mathbb{C}_v$-linear combination of the φ_k and the $\varphi_{i,j}$.

When $\mathrm{char}(K) = 0$, the L^{sep}-rational basis will be the same as the L-rational basis. However, we will write it as

$$\{\widetilde{\varphi}_\lambda : \lambda = 1 \ldots, \Lambda_0\} \cup \{\widetilde{\varphi}_{i,j} : i = 1, \ldots, m, j \geq J + 1\} , \tag{3.4}$$

with $\widetilde{\varphi}_\lambda = \varphi_\lambda$ and $\widetilde{\varphi}_{ij} = \varphi_{ij}$ for all λ, i, j.

Now let $\mathrm{char}(K) = p > 0$. For each x_i, let $[K(x_i) : K]^{\mathrm{insep}} = [K(x_i) : K(x_i)^{\mathrm{sep}}]$ be the inseparable degree of the extension $K(x_i)/K$. Let $J = p^A$ be the least power of p such that $p^A \geq \max\big(2g+1, \max_i([K(x_i):K]^{\text{insep}})\big)$. Put $D_0 = \sum_{i=1}^m J \cdot (x_i)$ and let $\Lambda_0 = \dim_{\widetilde{K}}(D_0) \geq 1$. By construction, D_0 is K-rational.

We first construct the L^{sep}-rational basis, then we use it to construct the L-rational basis. To assure L^{sep}-rationality, we must relax the condition that each $\widetilde{\varphi}_{i,j}$ has a pole of exact order j at x_i.

Fix a K-rational basis $\{\widetilde{\varphi}_1, \dots, \widetilde{\varphi}_{\Lambda_0}\}$ for $\widetilde{\Gamma}(J \cdot D)$. Then, for each $i = 1, \dots, m$, consider the divisors $D_{i,J} = J \cdot (x_i)$ and $D_{i,2J} = 2J \cdot (x_i)$. Both are rational over $K(x_i)^{\text{sep}}$, so $\widetilde{\Gamma}(D_{i,J})$ and $\widetilde{\Gamma}(D_{i,2J})$ have $K(x_i)^{\text{sep}}$-rational bases. By the Riemann-Roch theorem, there is a $K(x_i)^{\text{sep}}$-rational function $\widetilde{\varphi}_{i,J}$ with a pole of order precisely J at (x_i). Since J is divisible by $[K(x_i):K]^{\text{insep}}$, the function $g_{x_i}(z)^J$ is rational over $K(x_i)^{\text{sep}}$, and we can normalize $\widetilde{\varphi}_{i,J}$ so that

$$\lim_{z \to x_i} \widetilde{\varphi}_{i,J}(z) \cdot g_{x_i}(z)^J \;=\; 1\,.$$

Again by the Riemann-Roch theorem, $\dim_{\widetilde{K}}(\widetilde{\Gamma}(D_{i,2J})/\widetilde{\Gamma}(D_{i,J})) = J$. Choose $K(x_i)^{\text{sep}}$-rational functions $\widetilde{\varphi}_{i,J+1}, \dots, \widetilde{\varphi}_{i,2J} \in \widetilde{\Gamma}(D_{i,2J})$ in such a way that $\widetilde{\varphi}_{i,2J} = (\widetilde{\varphi}_{i,J})^2$ and the images of $\widetilde{\varphi}_{i,J+1}, \dots, \widetilde{\varphi}_{i,2J}$ in $\widetilde{\Gamma}(D_{i,2J})/\widetilde{\Gamma}(D_{i,J})$ form a basis for that space. For each $j > 2J$, we can uniquely write $j = \ell \cdot J + r$ with $\ell, r \in \mathbb{Z}$, $\ell \geq 1$, and $J+1 \leq r \leq 2J$; put

$$\widetilde{\varphi}_{i,j} \;=\; \widetilde{\varphi}_{i,J}^{\ell} \cdot \widetilde{\varphi}_{i,r}\,. \tag{3.5}$$

Thus, for each $j \geq J$, $\widetilde{\varphi}_{i,j}$ is rational over $K(x_i)^{\text{sep}}$. For an index $j > J$ not divisible by J, the function $\widetilde{\varphi}_{i,j}$ has a pole of order at most $J \cdot \lceil j/J \rceil$ at x_i, but its pole will not in general have exact order j.

We will require that the $\widetilde{\varphi}_{i,j}$ for different x_i be chosen in a $\mathrm{Gal}(L^{\text{sep}}/K)$-equivariant way, so $\widetilde{\varphi}_{\sigma(i),j} = \sigma(\widetilde{\varphi}_{i,j})$ for each $\sigma \in \mathrm{Aut}(\widetilde{K}/K)$. The collection of L^{sep}-rational functions

$$\{\widetilde{\varphi}_\lambda(z) : \lambda = 1 \dots, \Lambda_0\} \cup \{\widetilde{\varphi}_{i,j}(z) : i = 1, \dots, m, j \geq J+1\} \tag{3.6}$$

will be called the *L^{sep}-rational basis*. By construction, it is K-symmetric and multiplicatively finitely generated. Note that although $\widetilde{\varphi}_{i,J}$ was used in constructing the L^{sep}-rational basis, it is not an element of the basis.

Each $f \in \widetilde{K}(\mathcal{C})$ with poles supported on $\mathfrak{X}$ can be uniquely expanded in terms of the L^{sep}-rational basis, and if $f(z)$ has a pole of order n_i at each x_i, only the $\widetilde{\varphi}_\lambda$ and the $\widetilde{\varphi}_{i,j}$ with $j \leq J \cdot \lceil n_i/J \rceil$ are required in the expansion. Similarly, for each v, each $f \in \mathbb{C}_v(\mathcal{C})$ can uniquely be expanded as a $\mathbb{C}_v$-linear combination of the $\widetilde{\varphi}_k$ and the $\widetilde{\varphi}_{i,j}$.

We now use the L^{sep} rational basis to construct the L-rational basis.

Put $\varphi_\lambda = \widetilde{\varphi}_\lambda$ for $\lambda = 1, \dots, \Lambda_0$. For each x_i, put $\varphi_{i,J} = \widetilde{\varphi}_{i,J}$ and $\varphi_{i,2J} = \widetilde{\varphi}_{i,J}^2$. By the Riemann-Roch theorem, for each $j = J+1, \dots, 2J-1$ there is a $K(x_i)$-rational function $\varphi_{i,j}(z) \in \widetilde{\Gamma}(j \cdot (x_i))$ with a pole of exact order j at x_i. We will choose $\varphi_{i,j}(z)$ to be a $K(x_i)$-rational linear combination of $\widetilde{\varphi}_{i,J+1}, \dots, \widetilde{\varphi}_{i,2J}$, normalized so that

$$\lim_{z \to x_i} \varphi_{i,j}(z) \cdot g_{x_i}(z)^j \;=\; 1\,;$$

this is possible since $\widetilde{\varphi}_{i,J+1}, \dots, \widetilde{\varphi}_{i,2J}$ span $\widetilde{\Gamma}(2J \cdot (x_i))/\widetilde{\Gamma}(J \cdot (x_i))$. We will require that for distinct i, the $\varphi_{i,j}$ be chosen so that $\varphi_{\sigma(i),j} = \sigma(\varphi_{i.j})$ for all $\sigma \in \mathrm{Aut}(L/K)$.

For each $j > 2J$, we can uniquely write $j = \ell \cdot J + r$, where $0 \le \ell \in \mathbb{Z}$ and $J+1 \le r \le 2J$, and we define

$$\varphi_{i,j}(z) \;=\; (\varphi_{i,J}(z))^{\ell} \cdot \varphi_{i,r}(z). \tag{3.7}$$

Then $\varphi_{i,j}$ is rational over $K(x_i)$, has a simple pole of order j at x_i and no other poles, and is normalized so that

$$\lim_{z \to x_i} \varphi_{i,j}(z) \cdot g_{x_i}(z)^j \;=\; 1\,. \tag{3.8}$$

For all $\sigma \in \mathrm{Aut}(\widetilde{K}/K)$ and all (i,j), we have $\varphi_{\sigma(i),j} = \sigma(\varphi_{i,j})$.

The construction has a number of consequences, which we record for future use.

PROPOSITION 3.3 (Uniform Transition Coefficients). *Let* $\mathrm{char}(K)$ *be arbitrary. Then*

(A) $\widetilde{\varphi}_\lambda = \varphi_\lambda$ *for* $\lambda = 1, \ldots, \Lambda_0$*,*

(B) For each $i = 1, \ldots, m$ *and each* $\ell \ge 2$*,* $\varphi_{i,\ell J} = \widetilde{\varphi}_{i,\ell J} = \widetilde{\varphi}_{i,J}^{\ell}$ *is* $K(x_i)^{\text{sep}}$*-rational and belongs to both the* L*-rational and* L^{sep}*-rational bases; it has a pole of exact order* ℓJ *at* x_i*, and is normalized so that*

$$\lim_{z \to x_i} \varphi_{i,\ell J}(z) \cdot g_{x_i}(z)^{\ell J} \;=\; 1\,.$$

For each $j \ge J+1$ *we have* $\varphi_{i,\ell J}(z) \cdot \varphi_{ij}(z) = \varphi_{i,\ell J+j}(z)$ *and* $\widetilde{\varphi}_{i,\ell J}(z) \cdot \widetilde{\varphi}_{i,j}(z) = \widetilde{\varphi}_{i,\ell J+j}(z)$*.*

(C) For each $i = 1, \ldots, m$*, there is an invertible matrix* $\widetilde{\mathcal{B}}_i = (\widetilde{\mathcal{B}}_{i,jk})_{1 \le j,k \le J}$ *with coefficients in* $K(x_i)$ *such that for each* $\ell \ge 1$*, and each* $j = 1, \ldots, J$*,*

$$\varphi_{i,\ell J+j} \;=\; \sum_{k=1}^{J} \widetilde{\mathcal{B}}_{i,jk} \cdot \widetilde{\varphi}_{i,\ell J+k}\,. \tag{3.9}$$

Likewise, put $\mathcal{B}_i = \widetilde{\mathcal{B}}_i^{-1}$ *and write* $\mathcal{B}_i = (\mathcal{B}_{i,jk})_{1 \le j,k \le J}$*. Then the* $\mathcal{B}_{i,jk}$ *belong to* $K(x_i)$*, and for each* $\ell \ge 1$*, and each* $j = 1, \ldots, J$*,*

$$\widetilde{\varphi}_{i,\ell J+j} \;=\; \sum_{k=1}^{J} \mathcal{B}_{i,jk} \cdot \varphi_{i,\ell J+k}\,. \tag{3.10}$$

PROOF. When $\mathrm{char}(K) = 0$, the proposition is trivial, since the L-rational and L^{sep}-rational bases coincide.

When $\mathrm{char}(K) = p$, the low-order basis functions $\varphi_1, \ldots, \varphi_{\Lambda_0}$ coincide with $\widetilde{\varphi}_1, \ldots, \widetilde{\varphi}_{\Lambda_0}$. The high-order basis functions $\varphi_{i,j}$ and $\widetilde{\varphi}_{i,j}$ are closely related as well. For each i and each $\ell \ge 1$, the function $\varphi_{i,\ell J} = \widetilde{\varphi}_{i,\ell J} = \widetilde{\varphi}_{i,J}^{\ell}$ is defined over $K(x_i)^{\text{sep}}$; for $\ell \ge 2$, it belongs to both the L-rational and L^{sep}-rational bases. By construction, the functions $\varphi_{i,\ell J+1}, \ldots, \varphi_{i,(\ell+1)J}$ and $\widetilde{\varphi}_{i,\ell J+1}, \ldots, \widetilde{\varphi}_{i,(\ell+1)J}$, and $\widetilde{\varphi}_{i,\ell J+1}, \ldots, \widetilde{\varphi}_{i,(\ell+1)J}$ are $K(x_i)$-rational linear combinations of each other. Since $\varphi_{i,J} = \widetilde{\varphi}_{i,J}$, (3.5) and (3.7) show that for each block of J functions we have the same transition coefficients. □

For each place v of K, fix an embedding of $\widetilde{K}$ into $\mathbb{C}_v$, and use it to identify functions in $\widetilde{K}(\mathcal{C})$ with functions in $\mathbb{C}_v(\mathcal{C})$.

COROLLARY 3.4 (Uniform Comparison of Expansion Coefficients). *Let* char(K) *be arbitrary. For each place v of K, there are constants $B_v, \widetilde{B}_v > 0$ (with $B_v = \widetilde{B}_v = 1$ for all but finitely many v), such that for any $f \in \mathbb{C}_v(\mathcal{C})$ with poles supported on $\mathfrak{X}$, if we expand f using the L-rational and L^{sep}-rational bases as*

$$f = \sum_{i,j} A_{i,j}\varphi_{i,j} + \sum_{\lambda} A_\lambda \varphi_\lambda \ , \quad f = \sum_{i,j} \widetilde{A}_{i,j}\widetilde{\varphi}_{i,j} + \sum_{\lambda} \widetilde{A}_\lambda \widetilde{\varphi}_\lambda \ ,$$

then

$$\max_{i,j,\lambda}(|A_{i,j}|_v, |A_\lambda|_v) \ \leq \ \widetilde{B}_v \cdot \max_{i,j,\lambda}(|\widetilde{A}_{i,j}|_v, |\widetilde{A}_\lambda|_v)$$

and

$$\max_{i,j,\lambda}(|\widetilde{A}_{i,j}|_v, |\widetilde{A}_\lambda|_v) \ \leq \ B_v \cdot \max_{i,j,\lambda}(|A_{i,j}|_v, |A_\lambda|_v) \ .$$

COROLLARY 3.5 (Rationality Properties of Expansion Coefficients). *Let* char(K) *be arbitrary. Suppose $f \in K(\mathcal{C})$ is a K-rational function with poles supported on $\mathfrak{X}$. When f is expanded in terms of the L-rational and L-rational bases as*

$$(3.11)\quad f = \sum_{i=1}^{m}\sum_{j=J+1}^{N_i} A_{i,j}\varphi_{i,j} + \sum_{\lambda=1}^{\Lambda_0} A_\lambda \varphi_\lambda \ , \quad f = \sum_{i=1}^{m}\sum_{j=J+1}^{\widetilde{N}_i} \widetilde{A}_{i,j}\widetilde{\varphi}_{i,j} + \sum_{\lambda=1}^{\Lambda_0} \widetilde{A}_\lambda \widetilde{\varphi}_\lambda \ ,$$

then each A_{ij} belongs to $K(x_i)$, each A_λ belongs to K, each $\widetilde{A}_{ij}$ belongs to $K(x_i)^{\mathrm{sep}}$, each $\widetilde{A}_\lambda$ belongs to K, and the A_{ij}, A_λ, $\widetilde{A}_{ij}$, and $\widetilde{A}_\lambda$ are K-symmetric.

Similarly, for each place v of K, if $f \in K_v(\mathcal{C})$ is a K_v-rational function with poles supported on $\mathfrak{X}$, when f is expanded in terms of the L-rational and L-rational bases as in (3.11), *then each A_{ij} belongs to $K_v(x_i)$, each A_λ belongs to K_v, each $\widetilde{A}_{ij}$ belongs to $K_v(x_i)^{\mathrm{sep}}$, each $\widetilde{A}_\lambda$ belongs to K_v, and the A_{ij}, A_λ, $\widetilde{A}_{ij}$, and $\widetilde{A}_\lambda$ are K_v-symmetric.*

PROOF. We only give the proof in the global case, since the local case is similar. Let $f \in K(\mathcal{C})$ be a K-rational function with poles supported on $\mathfrak{X}$.

First consider the expansion of f in terms of the L^{sep}-rational basis. Since f, the $\widetilde{\varphi}_{ij}$, and the $\widetilde{\varphi}_\lambda$ are all defined over L^{sep}, the $\widetilde{A}_{ij}$ and $\widetilde{A}_\lambda$ belong to L^{sep}. Since the $\widetilde{\varphi}_{ij}$ are defined over $K(x_i)^{\mathrm{sep}}$ and are Galois-equivariant, and the $\widetilde{\varphi}_\lambda$ are defined over K, it follows from invariance of f under $\mathrm{Gal}(L^{\mathrm{sep}}/K)$ that each $\widetilde{A}_{ij}$ belongs to $K(x_i)^{\mathrm{sep}}$ and each $\widetilde{A}_\lambda$ belongs to K, and as a collection, the $\widetilde{A}_{ij}$ and $\widetilde{A}_\lambda$ are K-symmetric.

Next consider the expansion of f in terms of the L-rational basis. When the L^{sep}-rational basis is expressed in terms of the L-rational basis, for each (i,j), $\widetilde{\varphi}_{ij}$ is a $K(x_i)$-linear combination of the φ_{ik}, and for each λ, $\phi_\lambda = \widetilde{\varphi}_\lambda$. It follows that for each (i,j), A_{ij} is a $K(x_i)$-linear combination of the $\widetilde{A}_{ik}$, and for each λ, $A_\lambda = \widetilde{A}_\lambda$. Since each $\widetilde{A}_{ik}$ is $K(x_i)^{\mathrm{sep}}$-rational, it follows that each A_{ij} is $K(x_i)$-rational. Similarly, each A_λ is K_v-rational, and as a collection the A_{ij} and A_λ are K-symmetric. □

COROLLARY 3.6 (Good Reduction Almost Everywhere). *There is a finite set S of places of K, such that for each $v \notin S$, $\mathfrak{C}_v$ has good reduction at v and each of φ_λ, φ_{ij}, $\widetilde{\varphi}_\lambda$, and $\widetilde{\varphi}_{i,j}$ specializes to a well defined nonconstant function on $\mathfrak{C}_v$* (mod v).

PROOF. This follows from the fact that the L-rational and L^{sep}-rational bases are multiplicatively finitely generated. $\square$

For each place v of K, let U_v be the neighborhood of E_v chosen in §3.2. For any $\varphi \in \mathbb{C}_v(\mathcal{C})$ with poles supported on $\mathfrak{X}$, let $\|\varphi\|_{U_v} = \sup_{x \in U_v}(|\varphi(x)|_v)$ be the sup norm.

PROPOSITION 3.7 (Uniform Growth Bounds). *Suppose each U_v is bounded away from $\mathfrak{X}$ in the v-topology, and that U_v is $\mathfrak{X}$-trivial, for all but finitely many v. Then for each v, there is a constant $C_v > 0$ such that*

$$\begin{cases} \|\varphi_k\|_{U_v}, \ \|\widetilde{\varphi}_\lambda\|_{U_v} \le C_v & \text{for all } \lambda = 1, \ldots, \Lambda_0 \ , \\ \|\varphi_{i,j}\|_{U_v}, \ \|\widetilde{\varphi}_{i,j}\|_{U_v} \le C_v^j & \text{for all } i \text{ and all } j > J \ . \end{cases}$$

Moreover, for all but finitely many v, we can take $C_v = 1$.

PROOF. Since the L-rational and L^{sep}-rational bases are multiplicatively finitely generated, the proposition is immediate from the construction and our assumption that U_v is bounded away from $\mathfrak{X}$ and is $\mathfrak{X}$-trivial for all but finitely many v. $\square$

We now give an example comparing the L-rational and L^{sep}-rational bases when $\text{char}(K) = p > 0$. Let $K = \mathbb{F}_p(t)$ where t is transcendental over $\mathbb{F}_p$. Take $\mathcal{C} = \mathbb{P}^1/K$ and identify $\mathbb{P}^1$ with $\mathbb{A}^1 \cup \{\infty\}$, using z as the standard coordinate function on $\mathbb{A}^1$. Take $\mathfrak{X} = \{x_1\}$, where $x_1 = t^{1/p}$ in affine coordinates. Then $L = K(\mathfrak{X}) = K(t^{1/p})$ and $L^{\text{sep}} = K$. Choose $g_{x_1} = z - t^{1/p}$, noting that $(g_{x_i})^p = z^p - t$ is K-rational.

Then $J = p$, $D_0 = p \cdot (x_1)$, and $\Lambda_0 = \dim_{\widetilde{K}}(\widetilde{\Gamma}(D_0)) = p + 1$. We can take the low-order part of the L-rational and L^{sep} rational bases to be

$$\{\varphi_1, \ldots, \varphi_{p+1}\} = \{\widetilde{\varphi}_1, \ldots, \widetilde{\varphi}_{p+1}\} = \{1, \frac{1}{z^p - t}, \frac{z}{z^p - t}, \ldots, \frac{z^{p-1}}{z^p - t}\} .$$

For the high-order part of the L-rational basis we can take $\varphi_{1,j} = 1/(z - t^{1/p})^j$ for $j = p, \ldots, 2p - 1$, and for the the high-order part of the L^{sep}-rational basis we can take $\widetilde{\varphi}_{1,p} = 1/(z^p - t)$ and $\widetilde{\varphi}_{i,j} = z^{2p-j}/(z^p - t)^2$ for $j = p + 1, \ldots, 2p$. Thus in general for $j > p$, if we write $j = \ell \cdot p - s$ with $0 \le s < p$, then

$$\varphi_{1,j} = \frac{1}{(z - t^{1/p})^j} \quad \text{and} \quad \widetilde{\varphi}_{1,j} = \frac{z^s}{(z^p - t)^\ell} .$$

Observe that $\widetilde{\varphi}_{1,j}$ has a pole of order $p\lceil j/p \rceil$ at x_1. For each $\ell > 1$, $\varphi_{1,\ell p} = \widetilde{\varphi}_{1,\ell p} = 1/(z^p - t)^\ell$ is K-rational, with a pole of exact order ℓp at x_1.

4. The Spherical Metric and Isometric Parametrizability

Consider $\mathbb{P}^N/K$, equipped with a system of homogeneous coordinates $x_0, \ldots, x_N$. Write $\mathbb{A}_k^N$ for affine patch on which $x_k \neq 0$. There is a natural metric $\|z, w\|_v$ on $\mathbb{P}_v^N(\mathbb{C}_v)$ called the v-adic spherical metric (see [**51**], §1.1):

Write $z = (z_0 : \cdots : z_N)$, $w = (w_0 : \cdots : w_N)$. If v is archimedean, and we fix an isomorphism $\mathbb{C}_v \cong \mathbb{C}$, then $\|z, w\|_v$ is the chordal distance associated to the Fubini-Study metric: explicitly,

$$\|z, w\|_v = \frac{\sqrt{\sum_{0 \le i < j \le N} |z_i w_j - w_i z_j|^2}}{\sqrt{\sum_i |z_i|^2} \sqrt{\sum_j |w_j|^2}} .$$

It has the following geometric interpretation. Let φ be the length of the geodesic from z to w under the Fubini-Study metric on $\mathbb{P}_v^N(\mathbb{C})$, so $0 \le \varphi \le \pi$. Then $\|z,w\|_v = \sin(\varphi/2)$, the length of a chord subtending an central arc of measure φ in a circle of diameter 1 (see [**51**], p.26). When $N=1$, and if we write $\mathbb{P}^1(\mathbb{C}) = \mathbb{C} \cup \{\infty\}$, identifying the affine patch $\mathbb{A}_0^1(\mathbb{C})$ with $\mathbb{C}$, then for $z=(1:z_1)$ and $w=(1:w_1)$,

$$\|z,w\|_v \;=\; \frac{|z_1-w_1|}{\sqrt{1+|z_1|^2}\sqrt{1+|w_1|^2}}$$

is the usual chordal distance on $\mathbb{P}^1$.

If v is nonarchimedean, then $\|z,w\|_v$ is defined by

$$\|z,w\|_v \;=\; \frac{\max_{0i\le j\le N}|z_iw_j-w_iz_j|_v}{\max_i(|z_i|_v)\max_j(|w_j|_v)} \;.$$

If z, w belong to the affine patch $\mathbb{A}_0^N$ and are scaled so that $z_0=w_0=1$, then

$$\text{(3.12)} \qquad \|z,w\|_v \;=\; \frac{\max_{1\le i\le N}|z_i-w_i|_v}{(\max(1,\max_{1\le i\le N}|z_i|_v)\,(\max(1,\max_{1\le j\le N}|w_j|_v)} \;.$$

In particular, if $z,w\in\mathbb{A}_0^N(\widehat{\mathcal{O}}_v)$, then $\|z,w\|_v = \max_{1\le i\le N}(|z_i-w_i|_v)$.

If v is archimedean, then $\|z,w\|_v$ is invariant under the action of the unitary group $U(N+1,\mathbb{C})$ on $\mathbb{P}_v^N(\mathbb{C})$. If v is nonarchimedean, then $\|z,w\|_v$ is invariant under the action of $\mathrm{GL}(N+1,\widehat{\mathcal{O}}_v)$ on $\mathbb{P}_v^N(\mathbb{C}_v)$.

Clearly $0\le\|z,w\|_v\le 1$ for all $z,w\in\mathbb{P}_v^N(\mathbb{C}_v)$, with $\|z,w\|_v=0$ if and only if $z=w$. Furthermore, $\|z,w\|_v = \|w,z\|_v$. If v is archimedean, $\|z,w\|_v$ satisfies the triangle inequality; if v is nonarchimedean, it satisfies the ultrametric inequality ([**51**], p.26). In particular, $\|z,w\|_v$ is a metric on $\mathbb{P}_v^N(\mathbb{C}_v)$.

A deeper fact is that for each $\zeta\in\mathbb{P}_v^N(\mathbb{C}_v)$, the function

$$\text{(3.13)} \qquad [\![z,w]\!]_\zeta \;:=\; \frac{\|z,w\|_v}{\|z,\zeta\|_v\|w,\zeta\|_v}$$

is a metric on $\mathbb{P}_v^N(\mathbb{C}_v)\backslash\{\zeta\}$. If v is archimedean, $[\![z,w]\!]_\zeta$ satisfies the triangle inequality; if v is nonarchimedean, it satisfies the ultrametric inequality. For this, see ([**51**], Theorem 2.5.1, p.122).

If $\mathcal{C}/K$ is a smooth curve and $\iota:\mathcal{C}\hookrightarrow\mathbb{P}^N$ is a projective embedding, then we get an induced metric $\|z,w\|_v$ on $\mathcal{C}_v(\mathbb{C}_v)$, for each v.

It can be shown that $\|z,w\|_v$ is a Weil distribution for the diagonal divisor on $\mathcal{C}_v\times\mathcal{C}_v$ ([**51**], Theorem 1.1.1, p.27). From this, it follows that if $\iota_1:\mathcal{C}\hookrightarrow\mathbb{P}^{N_1}$ and $\iota_2:\mathcal{C}\hookrightarrow\mathbb{P}^{N_2}$ are two embeddings, and $\|z,w\|_{v,1}$, $\|z,w\|_{v,2}$ are the corresponding metrics on $\mathcal{C}_v(\mathbb{C}_v)$, they are equivalent: there are constants $C_1,C_2>0$ such that for all $z,w\in\mathcal{C}_v(\mathbb{C}_v)$,

$$\text{(3.14)} \qquad C_1\|z,w\|_{v,1} \;\le\; \|z,w\|_{v,2} \;\le\; C_2\|z,w\|_{v,1} \;.$$

We will call any such metric $\|z,w\|_v$ on $\mathcal{C}_v(\mathbb{C}_v)$ a spherical metric. We use the following notation for "discs" in $\mathbb{C}_v$, and "balls" in $\mathcal{C}_v(\mathbb{C}_v)$:

$$D(a,r)^- = \{z\in\mathbb{C}_v : |z-a|_v<r\}, \qquad D(a,r)=\{z\in\mathbb{C}_v : |z-a|_v\le r\},$$
$$B(a,r)^- = \{z\in\mathcal{C}_v(\mathbb{C}_v) : \|z,a\|_v<r\}, \quad B(a,r)=\{z\in\mathcal{C}_v(\mathbb{C}_v) : \|z,a\|_v\le r\}.$$

If $\zeta \in \mathcal{C}_v(\mathbb{C}_v)$, and $g_\zeta(z) \in \mathbb{C}_v(\mathcal{C})$ is a uniformizer at ζ, then there is a constant $C_\zeta > 0$ such that

$$\lim_{z \to \zeta} \frac{|g_\zeta(z)|_v}{\|z, \zeta\|_v} = C_\zeta \ . \tag{3.15}$$

This follows from the nonsingularity of the curve $\mathcal{C}$, and definition of $\|z, w\|_v$ in terms of local coordinate functions.

DEFINITION 3.8. Let $v \in \mathcal{M}_K$ be nonarchimedean. An open ball $B(a,r)^- \subset \mathcal{C}_v(\mathbb{C}_v)$ is *isometrically parametrizable* if it is contained in some affine patch $\mathbb{A}_k^N$, and there are power series $\lambda_1(z), \ldots, \lambda_N(z) \in \mathbb{C}_v[[z]]$ converging on the disc $D(0,r)^-$ such that the map $\Lambda : D(0,r)^- \to B(a,r)^-$ given in affine coordinates on $\mathbb{A}_k^N$ by $\Lambda(z) = (\lambda_1(z), \ldots, \lambda_N(z))$ is a surjective isometry: $\Lambda(D(0,r)^-) = B(a,r)^-$ and for all $x, y \in D(0,r)^-$,

$$\|\Lambda(x), \Lambda(y)\|_v = |x - y|_v \ .$$

We call the map Λ an *isometric parametrization.* If $F_u \subseteq \mathbb{C}_v$ is a field such that each $\lambda_i(z) \in F_u[[z]]$, we say that Λ is F_u-rational.

A closed ball $B(a,r)$ is *isometrically parametrizable* if it is contained in an isometrically parametrizable open ball $B(a,r_1)^-$ for some $r_1 > r$.

If v is nonarchimedean, and $\mathcal{C}$ is embedded in $\mathbb{P}^N$, then all sufficiently small balls with respect to the corresponding spherical metric $\|z, w\|_v$ are isometrically parametrizable:

THEOREM 3.9. *Let $v \in \mathcal{M}_K$ be nonarchimedean. Then there is a number $0 < R_v \le 1$, depending only on v and the embedding $\mathcal{C} \to \mathbb{P}^N$, such that each ball $B(a,r)^- \subset \mathcal{C}_v(\mathbb{C}_v)$ with $0 < r \le R_v$ is isometrically parametrizable. If $\mathcal{C}$ has good reduction at v for the given embedding, we can take $R_v = 1$.*

If $a \in \mathcal{C}_v(\mathbb{C}_v)$ and $0 < r \le R_v$, then for any point $a_0 \in B(a,r)^-$ and any complete field $F_u \subseteq \mathbb{C}_v$ such that $a_0 \in \mathcal{C}_v(F_u)$ and $K_v \subseteq F_u$, there is an F_u-rational isometric parametrization $\Lambda : D(0,r)^- \to B(a,r)^-$ with $\Lambda(0) = a_0$.

For any isometric parametrization $\widetilde{\Lambda} : D(0,r)^- \to B(a,r)^-$, there is an index i_0 such that for all $x, y \in D(0,r)^-$

$$\|\widetilde{\Lambda}(x), \widetilde{\Lambda}(y)\|_v = |\widetilde{\lambda}_{i_0}(x) - \widetilde{\lambda}_{i_0}(y)|_v = |x - y|_v \ . \tag{3.16}$$

Furthermore, if F_u is any field such that $\widetilde{\Lambda}$ is F_u-rational, and if $L_w \subset \mathbb{C}_v$ is a complete field containing F_u, then $\widetilde{\Lambda}(D(0,r)^- \cap L_w) = B(a,r)^- \cap \mathcal{C}_v(L_w)$.

PROOF. The existence of the number R_v and of isometric parametrizations with the specified properties relative to a_0 and F_u are proved in ([**51**], Theorem 1.2.3, p.31). The fact that we can take $R_v = 1$ when $\mathcal{C}$ has good reduction at v is proved in ([**51**], Corollary 1.2.4, p.39).

Now let $\widetilde{\Lambda} = (\widetilde{\lambda}_1, \ldots, \widetilde{\lambda}_N) : D(0,r)^- \to B(a,r)^-$ be an arbitrary isometric parametrization. We first show that there is an index i_0 for which (3.16) holds.

Put $a_0 = \widetilde{\Lambda}(0)$. After replacing a by a_0 and changing coordinates by a translation, we can assume that $a = a_0 = \vec{0}$. For each i, write

$$\widetilde{\lambda}_i(z) = \sum_{n=1}^{\infty} a_{i,n} z^n \ .$$

Since λ_i converges on $D(0,r)^-$, for each $0 < R < r$ we have $\lim_{n\to\infty} |a_{i,n}|R^n = 0$. Furthermore, by the Maximum Modulus Principle for power series, if $R \in |\mathbb{C}_v^\times|_v$ then

$$\|\widetilde{\lambda}_i\|_{D(0,R)} = \max_n |a_{i,n}|_v R^n \ . \tag{3.17}$$

Since $\widetilde{\Lambda}$ is an isometric parametrization with $\widetilde{\Lambda}(0) = \vec{0}$, for each i and each $0 < R < r$, if $x \in D(0,R)$ we have

$$|\widetilde{\lambda}_i(x)|_v = |\widetilde{\lambda}_i(x) - \widetilde{\lambda}_i(0)|_v \ \le \ \|\widetilde{\Lambda}(x), \widetilde{\Lambda}(0)\|_v \ = \ |x-0|_v \ \le \ R \ .$$

This means that $\|\widetilde{\lambda}_i\|_{D(0,R)} \le R$, for each i and R. Similarly, for all $x, y \in D(0,r)^-$

$$|\widetilde{\lambda}_i(x) - \widetilde{\lambda}_i(y)|_v \ \le \ \|\widetilde{\Lambda}(x), \widetilde{\Lambda}(y)\|_v \ = \ |x-y|_v \ . \tag{3.18}$$

On the other hand, for each R with $0 < R < r$, and each $x \in D(0,R)$, we have

$$\max_{1\le i\le N} |\widetilde{\lambda}_i(x)|_v \ = \ \|\widetilde{\Lambda}(x), \widetilde{\Lambda}(0)\|_v \ = \ |x-0|_v \ = \ |x|_v \ .$$

Letting $|x|_v$ approach R and using the Pigeon-hole Principle, we see for each R there is some i for which $\|\widetilde{\lambda}_i\|_{D(0,R)} = R$.

Take a sequence $0 < R_1 < R_2 < \cdots < r$ with $\lim_{\ell\to\infty} R_\ell = r$, such that each $R_\ell \in |\mathbb{C}_v^\times|_v$. By the Pigeon-hole Principle, there is an i_0 such that $\|\widetilde{\lambda}_{i_0}\|_{D(0,R_\ell)} = R_\ell$ for infinitely many ℓ. After replacing $\{R_\ell\}_{\ell\ge1}$ by a subsequence, we can assume this holds for all ℓ. For notational convenience, relabel the coordinates so that $i_0 = 1$.

For each ℓ, (3.17) shows that $|a_{1,n}|_v R_\ell^n \le \|\widetilde{\lambda}_1\|_{D(0,R_\ell)} = R_\ell$ for each n, with equality for some n. Let n_ℓ be the maximal index for which $|a_{1,n}|_v R_\ell^n = R_\ell$. We claim that $n_\ell = 1$ for each ℓ. Suppose to the contrary that $n_\ell \ge 2$ for some ℓ. Since the function $f_\ell(R) = |a_{1,n_\ell}|_v R^{n_\ell} - R$ is convex upward for $R > 0$, is negative for small positive R, and satisfies $f_\ell(R_\ell) = 0$, it must be positive for $R > R_\ell$. Hence for each $R \in |C_v^\times|_v$ with $R_\ell < R < r$ we would have

$$\|\widetilde{\lambda}_1\|_{D(0,R)} \ \ge \ |a_{i,n_\ell}|_v R^{n_\ell} \ > \ R \ ,$$

contradicting $\|\widetilde{\lambda}_1\|_{D(0,R)} \le R$. Thus $|a_{1,1}|_v R_\ell = R_\ell$ and $|a_{1,n}|_v R_\ell^n < R_\ell$ for each $n \ge 2$. Letting $\ell \to \infty$ and using the convexity of $|a_{1,n}|_v R^n$ for $n \ge 2$, we see that for $0 < R < r$

$$\begin{cases} |a_{1,1}|_v = 1 \ , \\ |a_{1,n}|_v R^{n-1} < 1 \quad \text{for } n \ge 2 \ . \end{cases} \tag{3.19}$$

Take $x, y \in D(0,r)^-$, and choose R with $\max(|x|_v, |y|_v) < R < r$. Then

$$\begin{aligned} |\widetilde{\lambda}_1(x) - \widetilde{\lambda}_1(y)|_v \ &= \ |\sum_{n=1}^\infty a_{1,n}(x^n - y^n)|_v \\ &= \ |x-y|_v \cdot |a_{1,1} + \sum_{n=2}^\infty a_{1,n}(\sum_{k=0}^{n-1} x^k y^{n-1-k})|_v \ = \ |x-y|_v \ , \end{aligned} \tag{3.20}$$

where the last step uses (3.19) and the ultrametric inequality. Combining (3.18) and (3.20) yields (3.16).

Now let F_u be any field over which $\widetilde{\Lambda}(z)$ is rational. Since $a_{1,1} \ne 0$, under composition of power series $\widetilde{\lambda}_1(z)$ has a formal inverse $\widetilde{\lambda}_1^{-1}(z) \in F_u[[z]]$. By (3.19) and a simple recursion, $\widetilde{\lambda}_1^{-1}(z)$ converges on $D(0,r)^-$: for each $x \in D(0,r)^-$,

$$\widetilde{\lambda}_1^{-1}(\widetilde{\lambda}_1(x)) \ = \ \widetilde{\lambda}_1(\widetilde{\lambda}_1^{-1}(x)) \ = \ x \ .$$

Thus $\widetilde{\lambda}_1$ and $\widetilde{\lambda}_1^{-1}$ induce inverse isometries from $D(0,r)^-$ onto itself.

If $F_u \subseteq L_w \subseteq \mathbb{C}_v$, then $\widetilde{\Lambda}$ is L_w-rational. Suppose L_w is complete. In this case $a_0 := \widetilde{\Lambda}(0) \in \mathcal{C}_v(L_w)$, and the initial reductions allowing us to assume $a_0 = \vec{0}$ do not affect the L_w-rationality of $\widetilde{\Lambda}$. Clearly $\widetilde{\Lambda}(D(0,r)^- \cap L_w) \subseteq B(a,r)^- \cap \mathcal{C}_v(L_w)$. For the opposite containment, take $b \in B(a,r)^- \cap \mathcal{C}_v(L_w)$. Write $b = (b_1, \ldots, b_N)$. Then $b_1 \in D(0,r)^- \cap L_w$; put $x_1 = \widetilde{\lambda}_1^{-1}(b_1)$. Since $\widetilde{\lambda}_1^{-1}$ is L_w-rational, it follows that $x_1 \in D(0,r)^- \cap L_w$. We claim that $\widetilde{\Lambda}(x_1) = b$. This is immediate, since there is some $x \in D(0,r)^-$ for which $\widetilde{\Lambda}(x) = b$; hence by (3.16),

$$\|b, \widetilde{\Lambda}(x_1)\|_v \ = \ \|\widetilde{\Lambda}(x), \widetilde{\Lambda}(x_1)\| \ = \ |b_1 - \widetilde{\lambda}_1(x_1)|_v \ = \ 0 \ . \qquad \square$$

5. The Canonical Distance and the $(\mathfrak{X}, \vec{s})$-Canonical Distance

Let v be a place of K. Consider the usual distance function $|z - w|_v$ on $\mathbb{C}_v = \mathbb{P}^1(\mathbb{C}_v)\backslash\{\infty\}$, which has the property that for each nonzero rational function $f \in \mathbb{C}_v(\mathbb{P}^1)$, if $\mathrm{div}(f) = \sum m_i(a_i)$, then there is a constant $C(f)$ such that for all $z \in \mathbb{C}_v$

$$(3.21) \qquad |f(z)|_v \ = \ C(f) \cdot \prod_{a_i \neq \infty} |z - a_i|_v^{m_i} \ .$$

Fix $\zeta \in \mathcal{C}_v(\mathbb{C}_v)$. In ([**51**], §2) a "canonical distance" $[z,w]_\zeta$ on $\mathcal{C}_v(\mathbb{C}_v)\backslash\{\zeta\}$ was introduced, generalizing $|z - w|_v$. The canonical distance $[z,w]_\zeta$ is a symmetric, nonnegative real-valued function of $z, w \in \mathcal{C}_v(\mathbb{C}_v)\backslash\{\zeta\}$, and is unique up to scaling by a constant. Its existence is shown in ([**51**], Theorem 2.1.1). It can be normalized by specifying a uniformizing parameter $g_\zeta(z)$, in which case it is characterized by the following three properties (see [**51**], Theorem 2.1.1, p.57, and Corollary 2.1.2, p.69):

(1) (Continuity): $[z,w]_\zeta$ is jointly continuous in z and w.
(2) (Factorization): Let $0 \neq f(z) \in \mathbb{C}_v(\mathcal{C})$ have divisor $\mathrm{div}(f) = \sum m_i(a_i)$. Then there is a constant $C(f)$ such that for all $z \in \mathcal{C}_v(\mathbb{C}_v)\backslash\{\zeta\}$,

$$(3.22) \qquad |f(z)|_v \ = \ C(f) \cdot \prod_{a_i \neq \zeta} [z, a_i]_\zeta^{m_i} \ .$$

(3) (Normalization): For each $w \in \mathcal{C}_v(\mathbb{C}_v)\backslash\{\zeta\}$,

$$\lim_{z \to \zeta} \ [z,w]_\zeta \cdot |g_\zeta(z)|_v \ = \ 1 \ .$$

Two other important properties of the canonical distance are as follows:

PROPOSITION 3.10. *For all* $z, w \in \mathcal{C}_v(\mathbb{C}_v)\backslash\{\zeta\}$
(4) (Symmetry): $[z,w]_\zeta = [w,z]_\zeta$,
(5) (Galois equivariance): *For each* $\sigma \in \mathrm{Aut}_c(\mathbb{C}_v/K_v)$,

$$[\sigma(z), \sigma(w)]_{\sigma(\zeta)} = [w,z]_\zeta$$

if $[x,y]_\zeta$ *and* $[x,y]_{\sigma(\zeta)}$ *are normalized compatibly* $\big($*e.g. if* $g_{\sigma(\zeta)}(z) = \sigma(g_\zeta)(z)\big)$.

PROOF. The canonical distance can be defined directly using rational functions. Fix $\zeta \in \mathcal{C}_v(\mathbb{C}_v)$ and fix a uniformizing parameter $g_\zeta(z)$. For each $w \neq \zeta$, there is a sequence of functions $f_n(z) \in \mathbb{C}_v(\mathcal{C})$ having poles only at ζ and whose zeros approach w in the v-topology. This follows from the "Jacobian Construction Principle" of ([**51**], Theorem 1.3.1, p.48), and depends on the fact that the residue field of $\mathbb{C}_v$ is the algebraic closure of the prime field $\mathbb{F}_p$ and its valuation group is the same as that of $\widetilde{K}_v$. Fixing a uniformizing parameter $g_\zeta(z)$, normalize the $f_n(z)$ so that

$$\lim_{z \to \zeta} |f_n(z) \cdot g_\zeta(z)^{\deg(f_n)}|_v = 1 .$$

Then one can define the canonical distance by

$$[z, w]_\zeta = \lim_{n \to \infty} |f_n(z)|_v^{1/\deg(f_n)} ;$$

see ([**51**], Theorem 2.1.1, pp.57-58). The limit is independent of the sequence $\{f_n\}$ and the convergence is uniform outside each ball $B(\zeta, r)^-$, with $r > 0$.

The fact that $[z, w]_\zeta$ is symmetric in z and w is proved in ([**51**], Theorem 2.1.1, p.57). Its Galois equivariance follows immediately from the Galois equivariance of functions. □

The fact that $[z, w]_\zeta$ can be approximated by absolute values of rational functions is the reason it is the kernel which appears in arithmetic potential theory, and is the key to the proof of the Fekete-Szegö theorem.

Several alternate constructions of the canonical distance are given in [**51**]. To clarify its relation with other objects in arithmetic geometry, we recall two of them:

First, the canonical distance is closely related to Néron's local height pairing. Recall that Néron's pairing $\langle \cdot, \cdot \rangle_v$ is a continuous, real-valued, $\mathrm{Aut}_c(\mathbb{C}_v/K_v)$-equivariant bilinear function defined for pairs of divisors on $\mathcal{C}_v(\mathbb{C}_v)$ of degree 0 with coprime support, having the property that for each $0 \neq f \in \mathbb{C}_v(\mathcal{C})$ and each $a, b \in \mathcal{C}_v(\mathbb{C}_v)$ disjoint from the support of $\mathrm{div}(f)$,

$$\langle \mathrm{div}(f), (a) - (b) \rangle_v = -\log_v(|f(a)/f(b)|_v) .$$

(Here we adopt the normalization of Néron's pairing used in [**51**], which differs from Néron's normalization by a factor $-1/\log(q_v)$.) Néron originally defined his pairing only for K_v-rational divisors: the fact that it can be extended to a Galois-equivariant pairing on $\mathbb{C}_v$-rational divisors follows from its continuity and invariance under base change; see ([**51**], pp.74-76).

The canonical distance, normalized as in (3.28), can be defined using Néron's pairing by coalescing the poles of $\langle (z) - (t), (w) - (\zeta) \rangle_v$; see ([**51**], §2.2):

$$-\log_v([z, w]_\zeta) = \lim_{t \to \zeta} \langle (z) - (t), (w) - (\zeta) \rangle_v + \log_v(|g_\zeta(t)|_v) .$$

Conversely, Néron's pairing can be recovered from the canonical distance. Suppose $D_1 = \sum m_i(a_i)$ and $D_2 = \sum n_j(b_j)$ are divisors of degree 0 with coprime support. Take ζ distinct from the a_i, b_j. Then

$$\langle D_1, D_2 \rangle_v = -\sum_{i,j} m_i n_j \log_v([a_i, b_j]_\zeta) .$$

Second, the canonical distance can be expressed in terms of Arakelov functions. If v is archimedean, identify $\mathbb{C}_v$ with $\mathbb{C}$ and let $((z, w))_v$ be an Arakelov function on

$\mathcal{C}_v(\mathbb{C}) \times \mathcal{C}_v(\mathbb{C})$. Then

$$[z,w]_\zeta \;=\; \frac{(\!(z,w)\!)_v}{(\!(z,\zeta)\!)_v (\!(w,\zeta)\!)_v} \tag{3.23}$$

is a canonical distance function: see ([**51**], §2.3). If v is nonarchimedean, functions $(\!(z,w)\!)_v$ on $\mathcal{C}_v(\mathbb{C}_v) \times \mathcal{C}_v(\mathbb{C}_v)$ for which (3.23) holds can be constructed using intersection theory and the semistable model theorem; see ([**51**], §2.4) and ([**19**], §2). They will also be called Arakelov functions.

For each v, the function $(\!(z,w)\!)_v$ is bounded, continuous, symmetric, and vanishes only on the diagonal. In ([**51**], §2.3, §2.4) it is shown there is a constant $C_v \geq 1$ (depending on the choice of the spherical metric $\|z,w\|_v$) such that for all $z, w \in \mathcal{C}_v(\mathbb{C}_v)$ we have

$$1/C_v \cdot \|z,w\|_v \;\leq\; (\!(z,w)\!)_v \;\leq\; C_v \|z,w\|_v. \tag{3.24}$$

From (3.23), one obtains the following 'change of pole' formula for the canonical distance: for any $\xi, \zeta \in \mathcal{C}_v(\mathbb{C}_v)$, there is a constant $C_{\xi,\zeta}$ such that for all $z, w \neq \xi, \zeta$,

$$[z,w]_\xi = C_{\xi,\zeta} \cdot \frac{[z,w]_\zeta}{[z,\xi]_\zeta [w,\xi]_\zeta} \; . \tag{3.25}$$

Using (3.23) or (3.25), one easily derives the following alternate form of (3.22): if $\deg(f) = N$ and we write $\mathrm{div}(f) = \sum_{i=1}^N (\alpha_i) - \sum_{i=1}^N (\xi_i)$, listing the zeros and poles of f with multiplicities, then there is a constant $\widetilde{C}(f)$ such that for all $z \in \mathcal{C}_v(\mathbb{C}_v) \backslash \{\xi_1, \ldots, \xi_N\}$,

$$|f(z)|_v \;=\; \widetilde{C}(f) \cdot \prod_{i=1}^N [z, \alpha_i]_{\xi_i} \; . \tag{3.26}$$

From (3.23) and the fact that

$$[\![z,w]\!]_\zeta \;=\; \frac{\|z,w\|_v}{\|z,\zeta\|_v \|w,\zeta\|_v}$$

is a metric on $\mathcal{C}_v(\mathbb{C}_v)$ (see (3.13)), it follows that $[z,w]_\zeta$ satisfies a weak triangle inequality: there is a constant B_v such that for each ζ and all $z, w, p \in \mathcal{C}_v(\mathbb{C}_v) \backslash \{\zeta\}$,

$$[z,w]_\zeta \;\leq\; B_v \cdot ([z,p]_\zeta + [p,w]_\zeta) \; .$$

This property justifies calling $[z,w]_\zeta$ a "distance". However, it seems not to be very important in practice, and examples show that one cannot always take $B_v = 1$ (see [**51**], p.128).

If v is nonarchimedean and $\mathcal{C}$ has good reduction at v for the projective embedding which induces $\|z,w\|_v$, then $\|z,w\|_v$ is an Arakelov function. Further, if $g_\zeta(z) \in K(\mathcal{C})$ is a uniformizing parameter at ζ, then for all but finitely many v

$$|g_\zeta(z)|_v \;=\; \|z,\zeta\|_v \tag{3.27}$$

on the ball $B(\zeta, 1)^-$. Hence, if the canonical distances are normalized as in (3.28), then for all but finitely many v,

$$[z,w]_\zeta \;=\; \frac{\|z,w\|_v}{\|z,\zeta\|_v \|w,\zeta\|_v} \; .$$

See [**51**], pp.90-92.

For most of this book, we will be interested in the case where ζ belongs to the K-symmetric set $\mathfrak{X} = \{x_1, \ldots, x_m\}$ in the Fekete-Szegö theorem. Let $g_{x_i}(z) \in K(\mathcal{C})$ be the fixed global uniformizing parameter chosen in §3.2. For each v we will normalize $[z,w]_{x_i}$ so that

$$\lim_{z \to x_i} [z,w]_{x_i} \cdot |g_{x_i}(z)|_v \ = \ 1 \ . \tag{3.28}$$

A mild generalization of the canonical distance, the "$(\mathfrak{X},\vec{s})$-canonical distance", will play an important role in this work. Given a probability vector $\vec{s} \in \mathcal{P}^m$, define

$$[z,w]_{\mathfrak{X},\vec{s}} \ = \ \prod_{i=1}^{m} ([z,w]_{x_i})^{s_i} \ , \tag{3.29}$$

where the $[z,w]_{x_i}$ are normalized as in (3.28). The case of interest is where $\mathfrak{X} \subset \mathcal{C}(\widetilde{K}) \subset \mathcal{C}_v(\mathbb{C}_v)$ is the set of global algebraic points in the Fekete-Szegö theorem, and the uniformizing parameters are the ones chosen in §3.2

If v is archimedean, and we identify $\mathbb{C}_v$ with $\mathbb{C}$, then $\mathcal{C}_v(\mathbb{C})$ is a Riemann surface. By a coordinate patch on $\mathcal{C}_v(\mathbb{C})$, we mean a simply connected open set $U \subset \mathcal{C}_v(\mathbb{C})$ for which there is a chart $\varphi : U \to \mathbb{C}$ giving an isomorphism of U with an open set $U' \subset \mathbb{C}$. Given $z, w \in U$, by abuse of notation we will write $|z-w|$ for $|\varphi(z)-\varphi(w)|$.

If v is nonarchimedean, recall from Theorem 3.9 that there is an $R_v > 0$ such that each ball $B(a,r)^- \subset \mathcal{C}_v(\mathbb{C}_v)$ with $r \le R_v$ is isometrically parametrizable by power series; if $\varphi : B(a,r)^- \to D(0,r)^-$ is the inverse map to an isometric parametrization, then for all $z, w \in B(a,r)^-$ we have $\|z,w\|_v = |\varphi(z)-\varphi(w)|_v$.

The following result, which is an immediate consequence of ([**51**], Proposition 2.1.3, p.69), asserts that $-\log_v([z,w]_{\mathfrak{X},\vec{s}})$ is "harmonic in z except for logarithmic singularities at w and the $x_i \in \mathfrak{X}$", and varies continuously with w. It will be used in developing potential theory for the kernel $-\log_v([z,w]_{\mathfrak{X},\vec{s}})$.

PROPOSITION 3.11. *Let $\mathcal{C}/K$ be a curve, and v a place of K. Fix $\mathfrak{X}$.*
(A) If v is archimedean:

(1) *If U and V are disjoint open sets not meeting $\mathfrak{X}$, then for each $\vec{s}$, the function $-\log([z,w]_{\mathfrak{X},\vec{s}})$ is continuous on $U \times V$ and is harmonic in each variable separately.*

(2) *On any coordinate patch $U \subset \mathcal{C}_v(\mathbb{C})$ not containing not meeting $\mathfrak{X}$, there are continuous, real-valued functions $\eta_{U,x_j}(z,w)$ on $U \times U$, harmonic in each variable separately, such that for all $z, w \in U$ and $\vec{s}$,*

$$-\log([z,w]_{\mathfrak{X},\vec{s}}) \ = \ -\log(|z-w|) + \sum_{j=1}^{m} s_j \eta_{U,x_j}(z,w) \ .$$

(3) *If U is a coordinate patch containing exactly one point $x_i \in \mathfrak{X}$, and V is a coordinate patch disjoint from U and $\mathfrak{X}$, then there are continuous real-valued functions $\eta_{U,V,x_j}(z,w)$ on $U \times V$, harmonic in each variable separately, such that for all $z \in U$ and $w \in V$ and all $\vec{s}$,*

$$-\log([z,w]_{\mathfrak{X},\vec{s}}) \ = \ s_i \log(|z-x_i|) + \sum_{j=1}^{m} s_j \eta_{U,V,x_j}(z,w) \ .$$

(B) If v is nonarchimedean:

(1) *If $U = B(a,r)^-$ and $V = B(b,s)^-$ are isometrically parametrizable balls disjoint from each other and from $\mathfrak{X}$, then $-\log_v([z,w]_{\mathfrak{X},\vec{s}})$ is constant on*

$U \times V$. More precisely, there are constants $\eta_{U,V,x_j} \in \mathbb{Q}$ such that for all $\vec{s}$ and all $z \in U$, $w \in V$,

$$-\log_v([z,w]_{\mathfrak{X},\vec{s}}) \;=\; \sum_{j=1}^{m} s_j \eta_{U,V,x_j} \;.$$

(2) *If $U = B(a,r)^-$ is an isometrically parametrizable ball not containing any points of $\mathfrak{X}$, then there are constants $\eta_{U,x_j} \in \mathbb{Q}$ such that for all $z, w \in U$ and all $\vec{s}$,*

$$-\log_v([z,w]_{\mathfrak{X},\vec{s}}) \;=\; -\log_v(\|z,w\|_v) + \sum_{j=1}^{m} s_j \eta_{U,x_j} \;.$$

(3) *If $U = B(a,r)^-$ is an isometrically parametrizable ball containing exactly one point $x_i \in \mathfrak{X}$, and $V = B(b,s)^-$ is an isometrically parametrizable ball disjoint from U and $\mathfrak{X}$, then there are constants $\eta_{U,V,x_j} \in \mathbb{Q}$ such that for all $z \in U$ and $w \in V$, and all $\vec{s}$,*

$$-\log_v([z,w]_{\mathfrak{X},\vec{s}}) \;=\; s_i \log_v(\|z,x_i\|_v) + \sum_{j=1}^{m} s_j \eta_{U,V,x_j} \;.$$

6. $(\mathfrak{X}, \vec{s})$-Functions and $(\mathfrak{X}, \vec{s})$-Pseudopolynomials

Fix a place v of K. Let $\mathfrak{X} = \{x_1, \ldots, x_m\} \subset \mathcal{C}(\widetilde{K})$ be the K-symmetric set from §3.2, and let the canonical distances $[z,w]_{x_i}$ be normalized as in (3.28), where the uniformizing parameters $g_{x_i}(z)$ are the ones from §3.2.

DEFINITION 3.12. Suppose $\vec{s} \in \mathcal{P}^m \cap \mathbb{Q}^m$. By an *$(\mathfrak{X}, \vec{s})$-function* we mean a rational function $f(z) \in \mathbb{C}_v(\mathcal{C})$ whose poles are supported on $\mathfrak{X}$, such that if $N = \deg(f)$, then $f(z)$ has a pole of exact order Ns_i at each $x_i \in \mathfrak{X}$.

DEFINITION 3.13. Let $\vec{s} \in \mathcal{P}^m$ be arbitrary. By an *$(\mathfrak{X}, \vec{s})$-pseudopolynomial* (or simply a *pseudopolynomial*) we mean a function $P : \mathcal{C}_v(\mathbb{C}_v) \to [0, \infty]$ of the form

$$P(z) \;=\; C \cdot \prod_{k=1}^{N} [z, \alpha_k]_{\mathfrak{X},\vec{s}} \;. \tag{3.30}$$

where $C > 0$ is a constant and $\alpha_1, \ldots, \alpha_N \in \mathcal{C}_v(\mathbb{C}_v) \backslash \mathfrak{X}$. We will call $\alpha_1, \ldots, \alpha_N$ the *roots* of $P(z)$. If $C = 1$ and we wish to emphasize that fact, we will say that $P(z)$ is *monic*. We call

$$\nu(z) \;=\; \frac{1}{N} \sum_{k=1}^{N} \delta_{\alpha_k}(z)$$

the probability measure associated to $P(z)$.

In the proof of the Fekete-Szegö theorem, $(\mathfrak{X}, \vec{s})$-functions occur naturally. The reason for introducing the $(\mathfrak{X}, \vec{s})$-canonical distance is that it allows us to view the absolute value of an $(\mathfrak{X}, \vec{s})$-function as an $(\mathfrak{X}, \vec{s})$-pseudopolynomial, factoring it in the form

$$|f(z)|_v \;=\; C(f) \cdot \prod_{k=1}^{N} [z, \alpha_k]_{\mathfrak{X},\vec{s}} \tag{3.31}$$

where $\alpha_1, \ldots, \alpha_N$ are the zeros of $f(z)$, listed with multiplicities. This follows by an easy symmetrization argument: suppose $f(z)$ is an $(\mathfrak{X}, \vec{s})$-function, and let $\xi_1, \ldots, \xi_N$ be the points $x_1, \ldots, x_m$ listed according to their multiplicities in $\mathrm{div}(f)$. Thus, each x_i occurs Ns_i times. For each permutation π of $\{1, \ldots, N\}$, by (3.26) there is a constant $C(f, \pi)$ such that $|f(z)|_v = C(f, \pi) \cdot \prod_{k=1}^N [z, \alpha_k]_{\xi_{\pi(k)}}$ for all $z \in \mathcal{C}_v(\mathbb{C}_v) \backslash \mathfrak{X}$. Taking the product over all π, and then extracting the $(N!)^{th}$ root, gives (3.31).

Note that a pseudopolynomial $P(z)$ makes sense even when its roots are not the zeros of an $(\mathfrak{X}, \vec{s})$-function $f(z)$, but it agrees with $|f(z)|_v$ (up to a multiplicative constant) when such a function exists. Furthermore, $P(z)$ varies continuously with its roots. This allows us to investigate absolute values of $(\mathfrak{X}, \vec{s})$-functions without worrying about questions of principality. This fact will play a key role in the construction of the initial local approximating functions in §5 and §6 below.

7. Capacities

In this section we define sets of capacity 0 and sets of positive capacity, and we introduce several numerical measures of capacity.

Fix a place v of K.

DEFINITION 3.14. If H is a compact subset of $\mathcal{C}_v(\mathbb{C}_v)$, we will say H has *positive capacity* if there is a positive measure ν supported on H for which

$$I(\nu) \ := \ \iint_{H \times H} -\log_v(\|z, w\|_v) \, d\nu(z) d\nu(w) \ < \ \infty \ .$$

If $I(\nu) = \infty$ for all positive measures ν on H, we say that H has *capacity* 0.

By (3.14) the property of having positive capacity or capacity 0 is independent of the choice of the spherical metric. Clearly it suffices to test $I(\nu)$ only for probability measures.

We next define the capacity of a compact set relative to a point.

Fix $\zeta \in \mathcal{C}_v(\mathbb{C}_v)$, and fix a uniformizing parameter $g_\zeta(z)$, giving a normalization of the canonical distance $[z, w]_\zeta$. Let $H \subset \mathcal{C}_v(\mathbb{C}_v) \backslash \{\zeta\}$ be compact. Given a probability measure ν supported on H, we define the *energy integral of ν with respect to ζ* by

$$I_\zeta(\nu) \ = \ \iint_{H \times H} -\log_v([z, w]_\zeta) \, d\nu(z) d\nu(w) \ .$$

If H is nonempty, the *Robin constant* of H with respect to ζ is

$$V_\zeta(H) \ = \ \inf_{\substack{\text{prob meas } \nu \\ \text{on } H}} I_\zeta(\nu) \ ,$$

where the infimum is taken over all probability measures supported on H. If H is empty, we put $V_\zeta(H) = \infty$. The *capacity* of H with respect to ζ is

$$\gamma_\zeta(H) \ = \ q_v^{-V_\zeta(H)} \ .$$

Thus, $\gamma_\zeta(H) > 0$ if and only if $I_\zeta(\nu) < \infty$ for some probability measure ν supported on H.

Likewise, given a probability vector $\vec{s} \in \mathcal{P}^m$, if $H \subset \mathcal{C}_v(\mathbb{C}_v)\backslash\mathfrak{X}$ is compact, then for any probability measure ν on H, we define the $(\mathfrak{X}, \vec{s})$-energy

$$I_{\mathfrak{X},\vec{s}}(\nu) \;=\; \iint_{H\times H} -\log_v([z,w]_{\mathfrak{X},\vec{s}})\,d\nu(z)d\nu(w)\ .$$

We put $V_{\mathfrak{X},\vec{s}}(H) = \inf_\nu I_{\mathfrak{X},\vec{s}}(\nu)$, and define the $(\mathfrak{X}, \vec{s})$-capacity

$$\gamma_{\mathfrak{X},\vec{s}}(H) \;=\; q_v^{-V_{\mathfrak{X},\vec{s}}(H)}\ .$$

The following lemma shows that for a given compact set H, either $\gamma_\zeta(H) > 0$ for all $\zeta \notin H$, or $\gamma_\zeta(H) = 0$ for all $\zeta \notin H$.

Lemma 3.15. *Let $H \subset \mathcal{C}_v(\mathbb{C}_v)$ be compact. Then the following are equivalent:*
(1) *H has capacity* 0.
(2) *For some $\zeta \in \mathcal{C}_v(\mathbb{C}_v)\backslash H$, $\gamma_\zeta(H) = 0$.*
(3) *For each $\zeta \in \mathcal{C}_v(\mathbb{C}_v)\backslash H$, $\gamma_\zeta(H) = 0$.*

If $H \subset \mathcal{C}_v(\mathbb{C}_v)\backslash\mathfrak{X}$, these are equivalent to:
(4) *For some $\vec{s} \in \mathcal{P}^m$, $\gamma_{\mathfrak{X},\vec{s}}(H) = 0$.*
(5) *For each $\vec{s} \in \mathcal{P}^m$, $\gamma_{\mathfrak{X},\vec{s}}(H) = 0$.*

Proof. If v is archimedean, the set $H = \mathcal{C}_v(\mathbb{C})$ is compact. In this case H has positive capacity, and the lemma is vacuously true. If v is nonarchimedean, then $\mathcal{C}_v(\mathbb{C})$ is not compact. Now suppose $H \neq \mathcal{C}_v(\mathbb{C})$. If $\zeta \notin H$, then $\|x, \zeta\|_v$ is uniformly bounded away from 0 for $x \in H$, because H is compact. Thus the lemma follows from (3.23) and (3.24). □

We next define the inner capacity and the outer capacity of a set, relative to a point.

Definition 3.16. For an arbitrary set $E_v \subset \mathcal{C}_v(\mathbb{C}_v)$, we say E_v has *positive inner capacity* if there is some compact set $H \subset E_v$ with positive capacity. If every compact set $H \subset E_v$ has capacity 0, we say that E_v has *inner capacity* 0.

For each $\zeta \in \mathcal{C}_v(\mathbb{C}_v)$, we define the inner capacity $\overline{\gamma}_\zeta(E_v)$ by

$$\overline{\gamma}_\zeta(E_v) \;=\; \sup_{\substack{H\subset E_v\backslash\{\zeta\}\\ H \text{ compact}}} \gamma_\zeta(H)\ . \tag{3.32}$$

Thus $0 \le \overline{\gamma}_\zeta(E_v) \le \infty$. When E_v is compact and $\zeta \notin E$, clearly $\overline{\gamma}_\zeta(E_v) = \gamma_\zeta(E_v)$.

Sets of inner capacity 0 are "negligible" for many purposes in potential theory. Each countable set E_v has inner capacity 0, because a probability measure supported on a compact subset of E_v necessarily consists of point masses. On the other hand, any set E_v which contains a nonempty open subset of $\mathcal{C}_v(\mathbb{C}_v)$, or a nonempty open subset of $\mathcal{C}_v(L_w)$ for some finite extension L_w/K_v, or a continuum (if v is archimedean), has positive inner capacity. This follows from ([**51**], Proposition 3.1.3, p.137) and ([**51**], Example 4.1.24, p.212).

To define the outer capacity, we will need the notion of a PL_ζ-domain.

Definition 3.17. If $\zeta \in \mathcal{C}_v(\mathbb{C}_v)$, a PL_ζ*-domain* is a set of the form

$$U = \{z \in \mathcal{C}_v(\mathbb{C}_v) : |f(z)|_v \le 1\}\ ,$$

where $f(z) \in \mathbb{C}_v(\mathcal{C})$ is a nonconstant function whose only poles are at ζ.

Fix $\zeta \in \mathcal{C}_v(\mathbb{C}_v)$, and fix a uniformizer $g_\zeta(z)$. If U is a PL_ζ-domain, let $f \in \mathbb{C}_v(\mathcal{C})$ be a function which defines it. Write $N = \deg(f)$ and define

$$V_\zeta(U) \;=\; \lim_{z\to\infty} \frac{1}{N} \log_v(f(z) \cdot g_\zeta(z)^N)$$

We then put

$$\gamma_\zeta(U) \;=\; q_v^{-V_\zeta(U)} \; . \tag{3.33}$$

Using ([**51**], Theorem 3.2.2 and Proposition 4.3.1) one sees that this definition is independent of the choice of f defining U. If v is archimdean, a PL_ζ-domain is compact, and ([**51**], Theorem 3.2.2) shows that the two definitions we have given for $\gamma_\zeta(U)$ coincide. If v is nonarchimedean, a PL_ζ-domain U is never compact; however, by ([**51**], Proposition 4.3.1) $\overline{\gamma}_\zeta(U) = \gamma_\zeta(U)$.

For an arbitrary set E_v, if $\zeta \notin E_v$, we define the outer capacity to be

$$\underline{\gamma}_\zeta(E_v) \;=\; \inf_{\substack{U \supset E_v \\ U \text{ a } \mathrm{PL}_\zeta\text{-domain}}} \gamma_\zeta(U) \; .$$

Trivially $\overline{\gamma}_\zeta(E_v) \le \underline{\gamma}_\zeta(E_v)$.

DEFINITION 3.18. Let $E_v \subset \mathcal{C}_v(\mathbb{C}_v)$ be arbitrary. If $\zeta \notin E_v$, and $\overline{\gamma}_\zeta(E_v) = \underline{\gamma}_\zeta(E_v)$, we say that E_v is *algebraically capacitable with respect to* ζ. If E_v is algebraically capacitable with respect to every $\zeta \notin E_v$, we simply say say that *algebraically capacitable.*

If E_v is algebraically capacitable with respect to ζ, we define its capacity $\gamma_\zeta(E_v)$ to be

$$\gamma_\zeta(E_v) \;=\; \overline{\gamma}_\zeta(E_v) \;=\; \underline{\gamma}_\zeta(E_v) \; . \tag{3.34}$$

In ([**51**]) algebraic capacitability was only defined for sets E_v at nonarchimedean places. If v is nonarchimedean, it is shown in ([**51**], Theorem 4.3.13) that compact sets, RL-domains, and finite unions of them are algebraically capacitable.

If v is archimedean, then each RL-domain is compact, and it follows from ([**51**], Propositions 3.1.17 and 3.3.3) that each compact archimedean set is algebraically capacitable.

Remark. The reason for introducing the notion of algebraic capacitability is that the inner capacity turns out to be the "right" notion of capacity for the Fekete-Szegö theorem, whereas the outer capacity is the right notion for Fekete's theorem (see Theorem 1.5). This is because the initial reductions in the proof of the Fekete-Szegö theorem involve replacing an arbitrary set E_v with a compact subset whose capacity is arbitrarily near $\overline{\gamma}_{x_i}(E_v)$, for each $x_i \in \mathfrak{X}$. Likewise, the initial reductions in the proof of Fekete's theorem involve replacing E_v with an algebraically defined neighborhood of itself.

Thus, algebraic capacitability is the hypothesis which makes a set permissible in both theorems. In this work, we are primarily interested in the Fekete-Szegö theorem, and in stating the most general versions of the theorem we use the inner capacity.

8. Green's Functions of Compact Sets

Next we define and study the Green's functions $G(z, \zeta; H_v)$ of compact sets.

Let $H_v \subset \mathcal{C}_v(\mathbb{C}_v)$ be compact. We first define $G(z, \zeta; H_v)$ when $\zeta \notin H_v$. Fix a uniformizing parameter $g_\zeta(z)$, which determines the normalization of $[z, w]_\zeta$. If H_v has positive inner capacity, so $V_\zeta(H) < \infty$, there is a unique probability measure μ_ζ supported on H_v for which $I_\zeta(\mu_\zeta) = V_\zeta(H_v)$. It is called the *equilibrium distribution* of H_v relative to ζ.

In the archimedean case, the existence of μ_ζ is shown in ([**51**], p.137), and its uniqueness in ([**51**], Theorem 3.1.12, p.145). In the nonarchimedean case, its existence is shown in ([**51**], p.190), its uniqueness in ([**51**], Theorem 4.1.22, p.211).

The *potential function* $u_{H_v}(z, \zeta)$ is defined by

$$u_{H_v}(z, \zeta) \quad = \quad \int_{H_v} - \log_v([z, w]_\zeta) \, d\mu_\zeta(w) \ . \tag{3.35}$$

Since $V_\zeta(H_v) = \int_{H_v \times H_v} - \log_v([z, w]_\zeta) \, d\mu_\zeta(z) d\mu_\zeta(w)$, clearly $V_\zeta(H_v) - u_{H_v}(z, \zeta)$ is independent of the normalization of $[z, w]_\zeta$.

DEFINITION 3.19. Let $H_v \subset \mathcal{C}_v(\mathbb{C}_v)$ be compact, and fix $\zeta \notin H_v$. If H_v has positive inner capacity, we define its Green's function with respect to ζ to be

$$G(z, \zeta; H_v) \quad = \quad V_\zeta(H_v) - u_{H_v}(z, \zeta) \tag{3.36}$$

for all $z \in \mathcal{C}_v(\mathbb{C}_v)$. If H_v has inner capacity 0, we put $G(z, \zeta; H_v) \equiv \infty$.

Remark. This definition of the Green's function for a compact set differs from the one in ([**51**]). In ([**51**], p.277) both 'upper' and 'lower' Green's functions $\overline{G}(z, \zeta; H_v)$ and $\underline{G}(z, \zeta; H_v)$ were defined. By ([**51**], Theorems 3.1.9 and 4.1.11), if H_v is algebraically capacitable (and in particular if H_v is compact) then $\overline{G}(z, \zeta; H_v)$ and $\underline{G}(z, \zeta; H_v)$ agree for all $z \notin H_v$; however $\underline{G}(z, \zeta; H_v) = 0$ for all $z \in H_v$ while $\overline{G}(z, \zeta; H_v)$ may be positive on a subset $e \subset H_v$ of inner capacity 0.

Our $G(z, \zeta; H_v)$ is the same as the upper Green's function $\overline{G}(z, \zeta; H_v)$, whereas in ([**51**]) $G(z, \zeta; H_v)$ was defined to be the lower Green's function $\underline{G}(z, \zeta; H_v)$.

We have made the change in order to simplify notation, and because of the author's conviction that the choice of $G(z, \zeta; H_v)$ made in ([**51**]) should have been reversed: $\overline{G}(z, \zeta; H)$ carries more information than $\underline{G}(z, \zeta; H)$, and is easier to work with.

The following proposition describes the main properties of the Green's function.

PROPOSITION 3.20. *Let $H_v \subset \mathcal{C}_v(\mathbb{C}_v)$ be a compact set of positive inner capacity, and fix $\zeta \notin H_v$. Then $G(z, \zeta; H_v)$ has the following properties: for each $\zeta \in \mathcal{C}_v(\mathbb{C}_v) \backslash H_v$,*

(1) *$G(z, \zeta; H_v) \geq 0$ for all $z \in \mathcal{C}_v(\mathbb{C}_v)$.*

(2) *If v is nonarchimedean, then $G(z, \zeta; H_v) > 0$ for all $z \notin H_v$. If v is archimedean, then $G(z, \zeta; H_v) > 0$ on the component of $\mathcal{C}_v(\mathbb{C}) \backslash H_v$ containing ζ, and is 0 on all other components. Furthermore $G(z, \zeta; H_v) = G(z, \zeta; \widehat{H}_v)$, where $\widehat{H}_v = \mathcal{C}_v(\mathbb{C}) \backslash D_\zeta$ and D_ζ is the component of $\mathcal{C}_v(\mathbb{C}) \backslash H_v$ containing ζ.*

(3) *$G(z, \zeta; H_v) = 0$ for $z \in H_v$, except on a (possibly empty) exceptional set $e_v \subset H_v$ of inner capacity 0, which is an F_σ set and in particular is Borel measurable. If v is archimedean, e_v is contained in the boundary ∂H_v and in fact is contained in the "outer boundary" ∂D_ζ. Each point of H_v which belongs to a*

continuum in H_v is nonexceptional. In particular, if H_v is a union of continua, the exceptional set is empty.

(4) *$G(z, \zeta; H_v)$ is continuous for each $z \notin H_v$, and at each $z_0 \in H_v$ where $G(z_0, \zeta; H_v) = 0$. If v is archimedean, then $G(z, \zeta; H_v)$ is harmonic on $\mathcal{C}_v(\mathbb{C})\backslash(H_v \cup \{\zeta\})$ and subharmonic on $\mathcal{C}_v(\mathbb{C})\backslash\{\zeta\}$.*

(5) *$G(z, \zeta; H_v)$ is upper semi-continuous everywhere; in fact, for each $z_0 \in \mathcal{C}_v(\mathbb{C}_v)$,*

$$\limsup_{z \to z_0} G(z, \zeta; H_v) = G(z_0, \zeta; H_v) ,$$

and if v is nonarchimedean and $z_0 \in H_v$ or if v is archimedean and $z_0 \in \partial D_\zeta$, then

$$\limsup_{\substack{z \to z_0 \\ z \notin H_v}} G(z, \zeta; H_v) = G(z_0, \zeta; H_v) .$$

(6) *If v is archimedean, then on any coordinate patch U with $\zeta \in U \subset D_\zeta$, there is a harmonic function $\eta_{H_v,\zeta}(z)$ such that $G(z, \zeta; H_v) = -\log_v(|z - \zeta|) + \eta_{H_v,\zeta}(z)$ on U. If v is nonarchimedean, then for any isometrically parametrizable ball $B(\zeta, r)^- \subset \mathcal{C}_v(\mathbb{C}_v)\backslash H_v$, there is a constant $\eta_{H_v,\zeta}$ such that $G(z, \zeta; H_v) = -\log_v(\|z, \zeta\|_v) + \eta_{H_v,\zeta}$ on $B(\zeta, r)^-$.*

PROOF. Parts (1)–(4) follow from ([**51**], Lemma 3.1.2, Theorem 3.1.7, Lemma 3.1.8, and Theorem 3.1.9) in the archimedean case, and ([**51**], Lemma 4.1.9, Theorem 4.1.11, and Corollary 4.1.12) in the nonarchimedean case. Part (5) is contained in ([**51**], Lemma 3.1.2) in the archimedean case, and follows from Proposition 3.11.B(2) and the definition of $u_{H_v}(z, \zeta)$ as an integral, in the nonarchimedean case. Part (6) is immediate from the definition. □

An important fact is that the Robin constant can be read off from the upper Green's function: if $g_\zeta(z)$ is the uniformizing parameter determining the normalization of $[z, w]_\zeta$, then

$$\lim_{z \to \zeta} G(z, \zeta; H_v) + \log_v(|g_\zeta(z)|_v) = V_\zeta(H_v) . \tag{3.37}$$

This follows trivially from the definition of $G(z, \zeta; E_v)$ in terms of the potential function. However, it emphasizes the fact that the Robin constant depends on the choice of the uniformizing parameter, while $G(z, \zeta; E_v)$ is absolute.

The Green's function is decreasing as a function of H_v:

LEMMA 3.21. *Let $H_v \subset H'_v$ be compact sets in $\mathcal{C}_v(\mathbb{C}_v)$, and suppose $\zeta \notin H'_v$. Then $G(z, \zeta; H_v) \geq G(z, \zeta; H'_v)$ for all z.*

PROOF. In the nonarchimedean case this is ([**51**], Proposition 4.1.21, p.209).

In the archimedean case (using our notation) it is shown in ([**51**], Lemma 3.2.5, p.157) that $\underline{G}(z, \zeta; H_v) \geq \underline{G}(z, \zeta; H'_v)$ for all z. By the discussion above, it follows that $G(z, \zeta; H_v) \geq G(z, \zeta; H'_v)$ except possibly on a set of capacity 0 contained in $\partial H'_v$. However, if $z_0 \in \partial H'_v$, then by Proposition 3.20(4),

$$G(z_0, \zeta; H'_v) = \limsup_{\substack{z \to z_0 \\ z \notin H'_v}} G(z, \zeta; H'_v) \leq \limsup_{\substack{z \to z_0 \\ z \notin H'_v}} G(z, \zeta; H_v) \leq G(z_0, \zeta; H_v) . \quad \square$$

The following result seems intrinsically obvious, but requires a surprising amount of work to prove. It will be used in the proof of Theorem 1.2.

PROPOSITION 3.22. *Fix v, and let $E_v \subset \mathcal{C}_v(\mathbb{C}_v)$ be a compact set of positive capacity. Let $H_{v,1} \subseteq H_{v,2} \subseteq \cdots \subseteq E_v$ be an exhaustion of E_v by an increasing sequence of compact sets. Then $\lim_{n\to\infty} V_\zeta(H_{v,n}) = V_\zeta(E_v)$ for each $\zeta \notin E_v$, and for each $z \neq \zeta$*

$$\lim_{n\to\infty} G(z,\zeta;H_{v,n}) \;=\; G(z,\zeta;E_v)\ .$$

PROOF. Fix $\zeta \in \mathcal{C}_v(\mathbb{C}_v)\backslash E_v$.

Without loss, we can assume that each $H_{v,n}$ has positive capacity. By the definition of the Robin constant, we have $V_\zeta(H_{v,1}) \geq V_\zeta(H_{v,2}) \geq \cdots \geq V_\zeta(E_v)$. Put

$$\widehat{V} \;=\; \lim_{n\to\infty} V_\zeta(H_{v,n})\ .$$

For each n, let μ_n be the equilibrium distribution of $H_{v,n}$ with respect to ζ, and let $u_n(z,\zeta) = u_{H_{v,n}}(z,\zeta)$ be the potential function. By definition

$$G(z,\zeta;H_{v,n}) \;=\; V_\zeta(H_{v,n}) - u_n(z,\zeta)\ .$$

By Proposition 3.20(3) there is an F_σ-set $e_n \subset H_{v,n}$ with inner capacity 0 such that $G(z,\zeta;H_{v,n}) = 0$, or equivalently $u_n(z,\zeta) = V_\zeta(H_{v,n})$, for all $z \in H_{v,n}\backslash e_n$. By Lemma 3.21, the functions $G(z,\zeta;H_{v,n})$ are nonnegative and decreasing with n. Put

$$G_\zeta(z) \;=\; \lim_{n\to\infty} G(z,\zeta;H_{v,n})\ .$$

Similarly, $G(z,\zeta;E_v) = V_\zeta(E_v) - u_{E_v}(z,\zeta)$, and there is an F_σ-set $e_0 \subset E_v$ of inner capacity 0 such that $G(z,\zeta;E_v) = 0$, or equivalently $u_{E_v}(z,\zeta) = V_\zeta(E_v)$, for all $z \in E_v\backslash e_0$. Let $e = \bigcup_{n=0}^{\infty} e_n$. By ([**51**], Propositions 3.1.15 and 4.1.14) the union of countably many Borel sets of inner capacity 0 itself has inner capacity 0, so e has inner capacity 0. For each $z \in E_v\backslash e$, we have

$$G_\zeta(z) \;=\; G(z,\zeta;E_v) \;=\; 0\ .$$

We will show, successively, that $\widehat{V} = V_\zeta(E_v)$, that the μ_n converge weakly to μ, and that $G_\zeta(z) = G(z,\zeta;E_v)$ for each $z \neq \zeta$.

By the discussion above, the potential function $u_{E_v}(z,\zeta)$ is identically equal to $V_\zeta(E_v)$ on $E_v\backslash e$. Since a Borel set of inner capacity 0 must have mass 0 for any positive Borel measure whose potential function is bounded above ([**51**], Lemmas 3.1.4 and 4.1.7), for each n the Fubini-Tonelli theorem gives

$$\begin{aligned}\int_{E_v} u_n(z,\zeta)\,d\mu(z) \;&=\; \iint_{E_v\times H_{v,n}} -\log_v([z,w]_\zeta)\,d\mu_n(w)d\mu(z)\\ &=\; \int_{H_{v,n}} u_{E_v}(w,\zeta)\,d\mu_n(w) \;=\; V_\zeta(E_v)\ . \qquad (3.38)\end{aligned}$$

On the other hand, pointwise for each $z_0 \in E_v\backslash e$, we have

$$\lim_{n\to\infty} u_n(z_0,\zeta) \;=\; \lim_{n\to\infty}\big(V_\zeta(H_{v,n}) - G(z,\zeta;H_{v,n})\big) \;=\; \widehat{V} - G_\zeta(z_0) = \widehat{V}\ .$$

Since E_v is bounded away from ζ, there is a $B_1 > -\infty$ such that $u_n(z,\zeta) \geq B_1$ on E_v, for all n. On the other hand, since $u_{H_{v,n}}(z,\zeta) \leq V_\zeta(H_{v,n})$ for all z and the $V_\zeta(H_{v,n})$ are finite and decreasing with n, there is a $B_2 < \infty$ such that $u_{H_{v,n}}(z,\zeta) \leq B_2$ on E_v, for all n. From (3.38), the fact that $\mu(e) = 0$, and the Dominated Convergence Theorem, it follows that

$$V_\zeta(E_v) \;=\; \lim_{n\to\infty}\int_{E_v\backslash e} u_n(z,\zeta)\,d\mu(z) \;=\; \int_{E_v\backslash e} \widehat{V}\,d\mu(z) \;=\; \widehat{V}\ . \qquad (3.39)$$

We next show that the μ_n converge weakly to μ. Let $\widehat{\mu}$ be any weak limit of a subsequence of $\{\mu_n\}_{n\geq 1}$. After passing to that subsequence, we can assume that the μ_n converge weakly to $\widehat{\mu}$. We will show that $\widehat{\mu} = \mu$.

For each $M \in \mathbb{R}$, write

$$-\log_v^{(M)}([z,w]_\zeta) \;=\; \min(M, -\log_v([z,w]_\zeta)) \ .$$

Since $-\log_v^{(M)}([z,w]_\zeta) \leq -\log_v([z,w]_\zeta)$, for each n and each M we have

$$\begin{aligned} &\iint -\log_v^{(M)}([z,w])_\zeta \, d\mu_n(z)d\mu_n(w) \\ &\leq \iint -\log_v([z,w])_\zeta \, d\mu_n(z)d\mu_n(w) \;=\; I_\zeta(\mu_n) \;=\; V_\zeta(H_{v,n}) \ . \end{aligned} \tag{3.40}$$

A standard argument shows that the measures $\mu_n \times \mu_n$ converge weakly to $\mu \times \mu$ on $E_v \times E_v$. Since the functions $-\log_v^{(M)}([z,w]_\zeta)$ are continuous on $E_v \times E_v$, for each M we have

$$\lim_{n\to\infty} \iint -\log_v^{(M)}([z,w])_\zeta \, d\mu_n(z)d\mu_n(w) \;=\; \iint -\log_v^{(M)}([z,w])_\zeta \, d\widehat{\mu}(z)d\widehat{\mu}(w) \ .$$

Combining this with (3.39) and (3.40) shows that for each M,

$$\iint -\log_v^{(M)}([z,w])_\zeta \, d\widehat{\mu}(z)d\widehat{\mu}(w) \;\leq\; \lim_{n\to\infty} V_\zeta(H_{v,n}) \;=\; \widehat{V} \;=\; V_\zeta(E_v) \ .$$

On the other hand, by the Monotone Convergence Theorem,

$$I_\zeta(\widehat{\mu}) \;=\; \lim_{M\to\infty} \iint -\log_v^{(M)}([z,w])_\zeta \, d\widehat{\mu}(z)d\widehat{\mu}(w) \ .$$

Thus $I_\zeta(\widehat{\mu}) \leq V_\zeta(E_v)$. However, by the definition of $V_\zeta(E_v)$ we must have $I_\zeta(\widehat{\mu}) \geq V_\zeta(E_v)$. Hence $I_\zeta(\widehat{\mu}) = V_\zeta(E_v)$, and the uniqueness of the equilibrium measure gives $\widehat{\mu} = \mu$.

Lastly, we show that $G_\zeta(z) = G(z,\zeta;E_v)$ for each $z \neq \zeta$. For each M, put

$$u_n^{(M)}(z,\zeta) = \int -\log_v^{(M)}([z,w]_\zeta)\, d\mu_n(w) \ , \quad u_{E_v}^{(M)}(z,\zeta) = \int -\log_v^{(M)}([z,w]_\zeta)\, d\mu_n(w) \ .$$

Since the kernels $-\log_v^{(M)}([z,w]_\zeta)$ are increasing with M, the Monotone Convergence Theorem shows that for each $z \neq \zeta$,

$$\lim_{M\to\infty} u_n^{(M)}(z,\zeta) \;=\; u_n(z,\zeta), \qquad \lim_{M\to\infty} u_{E_v}^{(M)}(z,\zeta) \;=\; u_{E_v}(z,\zeta) \ ,$$

where the limits are increasing.

Now fix $z_0 \neq \zeta$. For each M, since $-\log_v^{(M)}([z_0,w]_\zeta)$ is continuous on E_v as a function of w and the μ_n converge weakly to μ, we have

$$\lim_{n\to\infty} u_n^{(M)}(z_0,\zeta) \;=\; u_{E_v}^{(M)}(z_0,\zeta) \ .$$

Hence, for each $\varepsilon > 0$, there is an $N = N(M,\varepsilon)$ such that for all $n \geq N(M,\varepsilon)$, we have $u_n^{(M)}(z_0,\zeta) > u_{E_v}^{(M)}(z_0,\zeta) - \varepsilon$. By the monotonicity of $u_n^{(M)}(z_0,\zeta)$ in M, for each $n \geq N(M,\varepsilon)$ and each $M_1 > M$ we have

$$u_n^{(M_1)}(z_0,\zeta) \;\geq\; u_n^{(M)}(z_0,\zeta) \;>\; u_{E_v}^{(M)}(z_0,\zeta) - \varepsilon \ .$$

Letting $M_1 \to \infty$ and then letting $M \to \infty$, we see that for all sufficiently large n,

$$u_n(z_0,\zeta) \;\geq\; u_{E_v}(z_0,\zeta) - \varepsilon \ .$$

Consequently,

$$\begin{aligned} G_\zeta(z_0) &= \lim_{n\to\infty} G(z_0,\zeta;H_{v,n}) = \lim_{n\to\infty} V_\zeta(H_{v,n}) - u_n(z_0,\zeta) \\ &\le V_\zeta(E_v) - u_{E_v}(z_0,\zeta) + \varepsilon = G(z_0,\zeta;E_v) + \varepsilon \,. \end{aligned}$$

Letting $\varepsilon \to 0$, we see that $G_\zeta(z_0) \le G(z_0,\zeta;E_v)$. On the other hand, for each n the monotonicity of Green's functions shows that $G(z_0,\zeta;H_{v,n}) \ge G(z_0,\zeta;E_v)$, and hence trivially $G_\zeta(z_0) \ge G(z_0,\zeta;E_v)$. It follows that $G_\zeta(z_0) = G(z_0,\zeta;E_v)$.

This completes the proof. □

9. Upper Green's Functions

In this section we introduce the upper Green's function's $\overline{G}(z,\zeta;E_v)$ of a set.

DEFINITION 3.23. Given an arbitrary set $E_v \subset \mathcal{C}_v(\mathbb{C}_v)$, for each $z,\zeta \in \mathcal{C}_v(\mathbb{C}_v)$ we define the upper Green's function by

$$\overline{G}(z,\zeta;E_v) = \inf_{\substack{H_v \subset E_v \backslash \{\zeta\} \\ H_v \text{ compact}}} G(z,\zeta;H_v) \,, \tag{3.41}$$

and (fixing a uniformizing parameter $g_\zeta(z)$) we define the upper Robin constant by

$$\overline{V}_\zeta(E_v) = \inf_{\substack{H_v \subset E_v \backslash \{\zeta\} \\ H_v \text{ compact}}} V_\zeta(H_v) \,. \tag{3.42}$$

If E_v has inner capacity 0 then $\overline{G}(z,\zeta;E_v) \equiv \infty$. If E_v has positive inner capacity, then for each $z \ne \zeta$, $\overline{G}(z,\zeta;E_v)$ is finite and nonnegative, while $G(\zeta,\zeta;E_v) = \infty$. By Lemma 3.25, for a compact set H_v we have $\overline{G}(z,\zeta;H_v) = G(z,\zeta;H_v)$.

The following results will be needed later.

LEMMA 3.24. *Let $E_v \subset \mathcal{C}_v(\mathbb{C}_v)$ have positive inner capacity. For each $\zeta \in \mathcal{C}_v(\mathbb{C}_v)$, there is an increasing sequence of compact sets $H_{v,1} \subseteq H_{v,2} \subseteq \cdots \subset E_v\backslash\{\zeta\}$ such that $\overline{V}_\zeta(E_v) = \lim_{n\to\infty} V_\zeta(H_{v,n})$ and for each $z \in \mathcal{C}_v(\mathbb{C}_v)\backslash\{\zeta\}$,*

$$\overline{G}(z,\zeta;E_v) = \lim_{n\to\infty} G(z,\zeta;H_{v,n}) \,.$$

PROOF. Fix ζ, and let A be the collection of all compact sets $K \subset E_v\backslash\{\zeta\}$ with positive inner capacity. Consider the family $\{G(z,\zeta;K)\}_{K\in A}$. The set $\mathcal{C}_v(\mathbb{C})\backslash\{\zeta\}$ is a separable metric space, and by Proposition 3.20(5) each $G(z,\zeta;K)$ for $K \in A$ is upper semi-continuous in $\mathcal{C}_v(\mathbb{C})\backslash\{\zeta\}$. By a well-known property of upper semi-continuous functions (see [**32**], Lemma 2.3.2) there is a countable sequence $\{K_n\}_{n\ge1}$ of sets in A such that for each $z \in \mathcal{C}_v(\mathbb{C})\backslash\{\zeta\}$,

$$\overline{G}(z,\zeta;E_v) := \inf_{K\in A} G(z,\zeta;K) = \inf_{n\ge1} G(z,\zeta;K_n) \,.$$

Likewise, since $\overline{V}_\zeta(E_v) = \inf_{K\in A} V_\zeta(K)$, there is a sequence $\{K_n'\}_{n\ge1}$ of sets in A such that $\overline{V}_\zeta(E_v) = \lim_{n\to\infty} V_\zeta(K_n')$.

For each $n \ge 1$ put $H_{v,n} = (\bigcup_{i=1}^n K_i) \cup (\bigcup_{i=1}^n K_i')$. By the monotonicity of Green's functions of compact sets, for each n we have $G(z,\zeta;K_n) \ge G(z,\zeta;H_{v,n}) \ge G(z,\zeta;H_{v,n+1})$; similarly $V_\zeta(K_n') \ge V_\zeta(H_{v,n}) \ge V_\zeta(H_{v,n+1})$. It follows that

$$\overline{G}(z,\zeta;E_v) \le \lim_{n\to\infty} G(z,\zeta;H_{v,n}) \le \inf_{n\ge1} G(z,\zeta;K_n) = \overline{G}(z,\zeta;E_v)$$

and that $\overline{V}_\zeta(E_v) = \lim_{n\to\infty} V_\zeta(H_{v,n})$. □

LEMMA 3.25. *If E_v is compact, e_v has inner capacity* 0*, and $\zeta \notin E_v$, then*

$$\overline{G}(z,\zeta;E_v\backslash e_v) \ = \ G(z,\zeta;E_v)\ . \tag{3.43}$$

If $E_v \subset \mathcal{C}_v(\mathbb{C}_v)$ is arbitrary and e_v has inner capacity 0*, then for each $\zeta \in \mathcal{C}_v(\mathbb{C}_v)$,*

$$\overline{G}(z,\zeta;E_v\backslash e_v) \ = \ \overline{G}(z,\zeta;E_v)\ . \tag{3.44}$$

PROOF. We first prove (3.43). Suppose E_v is compact and $\zeta \notin E_v$. If E_v has inner capacity 0 the result holds trivially, so we can assume that E_v has positive inner capacity.

By ([**51**], Corollaries 3.1.16 and 4.1.15), we have $\overline{V}_\zeta(E_v\backslash e_v) = V_\zeta(E_v)$. (Note that in ([**51**]), our inner capacity $\overline{\gamma}(E_v)$ is denoted $\underline{\gamma}(E_v)$.) By Lemma 3.24 there is an increasing sequence of compact sets $H_{v,1} \subseteq H_{v,2} \subseteq \cdots \subseteq E_v\backslash e_v$ such that $\overline{V}_\zeta(E_v\backslash e_v) = \lim_{n\to\infty} V_\zeta(H_{v,n})$ and for each $z \in \mathcal{C}_v(\mathbb{C}_v)\backslash\{\zeta\}$,

$$\overline{G}(z,\zeta;E_v\backslash e_v) \ = \ \lim_{n\to\infty} G(z,\zeta;H_{v,n})\ .$$

Let μ_n be the equilibrium distribution of $H_{v,n}$ with respect to ζ, and let μ be the equilibrium distribution of E_v with respect to ζ. By the discussion above we have

$$\lim_{n\to\infty} V_\zeta(H_{v,n}) \ = \ V_\zeta(E_v)\ .$$

Now the same argument as in the part of the proof of Proposition 3.22 after formula (3.39) shows that the μ_n converge weakly to μ, and that $\overline{G}(z,\zeta;E_v\backslash e_v) = G(z,\zeta;E_v)$.

We can now deduce (3.44) formally. Let $E_v \subset \mathcal{C}_v(\mathbb{C}_v)$ and $\zeta \in \mathcal{C}_v(\mathbb{C}_v)$ be arbitrary. By definition,

$$\overline{G}(z,\zeta;E_v) = \inf_{\substack{H_v\subset E_v\backslash\{\zeta\}\\ H_v \text{ compact}}} G(z,\zeta;H_v)\ , \quad \overline{G}(z,\zeta;E_v\backslash e_v) = \inf_{\substack{H'_v\subset (E_v\backslash e_v)\backslash\{\zeta\}\\ H'_v \text{ compact}}} G(z,\zeta;H'_v)\ .$$

On the other hand, by what has been shown above, for each compact $H_v \subset E_v$ we have

$$G(z,\zeta;H_v) \ = \ \overline{G}(z,\zeta;H_v\backslash e_v) \ = \ \inf_{\substack{H'_v\subset (H_v\backslash e_v)\backslash\{\zeta\}\\ H'_v \text{ compact}}} G(z,\zeta;H'_v)\ .$$

It follows that $\overline{G}(z,\zeta;E_v) = \overline{G}(z,\zeta;E_v\backslash e_v)$. □

The upper Green's function has the following properties:

PROPOSITION 3.26. *Let $E_v \subset \mathcal{C}_v(\mathbb{C}_v)$. Then*

(1) (Finiteness): *$\overline{G}(z,\zeta;E_v)$ is valued in $[0,\infty]$. If E_v has inner capacity* 0*, then $\overline{G}(z,\zeta;E_v) = \infty$. If E_v has positive inner capacity, then $\overline{G}(z,\zeta;E_v)$ is finite and upper semi-continuous on $\mathcal{C}_v(\mathbb{C}_v)\backslash\{\zeta\}$. If v is archimedean and E_v has positive inner capacity, then $\overline{G}(z,\zeta;E_v)$ is subharmonic on $\mathcal{C}_v(\mathbb{C})\backslash\{\zeta\}$.*

(2) (Symmetry): *$\overline{G}(z,\zeta;E_v) = \overline{G}(\zeta,z;E_v)$ for all $z,\zeta \in \mathcal{C}_v(\mathbb{C}_v)$.*

(3) (Galois equivariance): *$\overline{G}(\sigma(z),\sigma(\zeta);\sigma(E_v)) = \overline{G}(z,\zeta;E_v)$ for all $z,\zeta \in \mathcal{C}_v(\mathbb{C}_v)$ and all $\sigma \in$* $\mathrm{Aut}_c(\mathbb{C}_v/K_v)$*. In particular, if E_v is stable under* $\mathrm{Aut}_c(\mathbb{C}_v/K_v)$*, then*

$$\overline{G}(\sigma(z),\sigma(\zeta);E_v) \ = \ \overline{G}(z,\zeta;E_v)\ .$$

(4) (Approximation): *Let $\mathfrak{X} = \{x_1, \ldots, x_m\} \subset \mathcal{C}_v(\mathbb{C}_v)$ be a finite set of points, none of which belongs to the closure of E_v, and fix $\varepsilon > 0$. Then there is a compact set $H_v \subset E_v$ such that for all $x_i, x_j \in \mathfrak{X}$,*

$$\begin{cases} \overline{G}(x_i, x_j; E_v) \le \overline{G}(x_i, x_j; H_v) \le \overline{G}(x_i, x_j; E_v) + \varepsilon & \text{if } i \ne j \ , \\ \overline{V}_{x_i}(E_v) \le \overline{V}_{x_i}(H_v) \le \overline{V}_{x_i}(E_v) + \varepsilon & \text{if } i = j \ . \end{cases}$$

(5) (Base Change): *Let L_w/K_v be a finite extension with ramification index $e_{w/v}$ (take $e_{w/v} = 1$ if v is archimedean). Fix an isomorphism $\iota_{w/v} : \mathbb{C}_w \to \mathbb{C}_v$ and put $E_w = \iota_{w/v}^{-1}(E_v) \subset \mathcal{C}_w(\mathbb{C}_w)$. Then*

$$\overline{G}(z, \zeta; E_w) = e_{w/v} \cdot \overline{G}(z, \zeta; E_v) \ .$$

(6) (Pullback): *Let $\mathcal{C}_1, \mathcal{C}_2/K_v$ be curves and let $f : \mathcal{C}_1 \to \mathcal{C}_2$ be a nonconstant rational map. Given $\zeta \in \mathcal{C}_2(\mathbb{C}_v)$, let its pullback divisor be $f^*((\zeta)) = \sum_i m_i(\xi_i)$. Then for each $E_v \subset \mathcal{C}_2(\mathbb{C}_v)$ and each $z \in \mathcal{C}_1(\mathbb{C}_v)$,*

$$\overline{G}(f(z), \zeta; E_v) = \sum_i m_i \overline{G}(z, \xi_i; f^{-1}(E_v)) \ .$$

PROOF. In [**51**] these properties are established for Green's functions of compact sets H_v, with $z, \zeta \notin H_v$. Using Definition 3.23, they carry over to arbitrary sets by taking limits:

(1) The finiteness properties in assertion (1) are immediate from the definition and Proposition 3.20. If E_v has positive inner capacity, then by Lemma 3.24 and Proposition 3.20, $G(z, \zeta; H_v)$ is a decreasing limit of upper semi-continuous functions on $\mathcal{C}_v(\mathbb{C}_v)\backslash\{\zeta\}$, hence is itself upper semi-continuous. If in addition v is archimedean, then $\overline{G}(z, \zeta; E_v)$ is the limit of a decreasing sequence of subharmonic functions which is bounded below, so it is subharmonic (see [**32**], Theorem 2.6.ii).

(2) By Lemma 3.25, since any finite set has inner capacity 0, for each fixed z_0, ζ_0 we have

$$\overline{G}(z_0, \zeta_0; E_v) = \overline{G}(z_0, \zeta; E_v\backslash\{z_0, \zeta\}) = \inf_{\substack{H_v \subset E_v\backslash\{z_0,\zeta_0\} \ , \\ H_v \text{ compact}}} G(z_0, \zeta_0; H_v) \ ,$$

and a similar formula holds for $\overline{G}(\zeta_0, z_0; E_v)$. By ([**51**], Theorem 3.2.7 and Theorem 4.4.14), for each compact $H_v \subset E_v\backslash\{z_0, \zeta_0\}$ we have $G(z_0, \zeta_0; H_v) = G(\zeta_0, z_0; H_v)$.

(3) For compact sets H_v we have $G(\sigma(z), \sigma(\zeta), \sigma(H_v)) = G(z, w; H_v)$ for $z, \zeta \notin H_v$, by the Galois-equivariance of $[z, w]_\zeta$ (Proposition 3.10 (B)) and formula (3.36). Taking limits, it follows that in general

$$\overline{G}(\sigma(z), \sigma(\zeta); \sigma(E_v)) = \overline{G}(z, \zeta; E_v) \ . \tag{3.45}$$

(4) The approximation property follows immediately from Definition 3.23 and Proposition 3.27.

(5) The base change property follows immediately from the normalizations of $|x|_v$,$|x|_w$, $\log_v(z)$ and $\log_w(z)$ (see §3.1)

(6) The pullback formula is established for compact sets in ([**51**], Theorem 3.2.9 and Theorem 4.4.19). Fix z_0 and ζ, and put $S = f^{-1}(\{z_0, \zeta\})$. As H_v runs over compact subsets of $E_v\backslash\{z_0, \zeta\}$, then $f^{-1}(H_v)$ runs over compact subsets of $f^{-1}(E_v)\backslash S$. These sets are cofinal in the compact subsets of $f^{-1}(E_v)\backslash S$, since for any compact $H_v' \subset f^{-1}(E_v)\backslash S$, its image $f(H_v') \subset E_v\backslash\{z_0, \zeta\}$ is compact, and $H_v' \subset f^{-1}(f(H_v'))$. Hence, the pullback formula follows in general by applying Lemma 3.25. □

We will now show that just as for compact sets, the Robin constant can be read off from the upper Green's function, using the chosen uniformizing parameter:

PROPOSITION 3.27. *If E_v has positive inner capacity and is bounded away from ζ, then the upper Robin constant is finite, and*

$$\overline{V}_\zeta(E_v) \;=\; \lim_{z\to\zeta} \overline{G}(z,\zeta;E_v) + \log_v(|g_\zeta(z)|_v)\ . \tag{3.46}$$

PROOF. In the archimedean case, fix a neighborhood U of ζ in $\mathcal{C}_v(\mathbb{C}_v)\backslash E_v$ such that the uniformizer $g_\zeta(z)$ has no zeros or poles in U except at ζ, and put $F = \mathcal{C}_v(\mathbb{C}_v)\backslash U$. Then F is a compact set of positive capacity and $E_v \subset F$, so $\overline{G}(z,\zeta,H_v) \ge \overline{G}(z,\zeta;F)$ for each $H_v \subset E_v$. For each compact $H_v \subset E_v$ of positive capacity, the function $\eta_{H_v}(z) = \overline{G}(z,\zeta;H_v) + \log_v(|g_\zeta(z)|_v)$ is harmonic on $U\backslash\{\zeta\}$ and is bounded in a neighborhood of ζ, so it extends to a harmonic function on U. Similarly $\eta_F(z) = \overline{G}(z,\zeta;F) + \log_v(|g_\zeta(z)|_v)$ extends to a harmonic function on U. By Lemma 3.24, there is an increasing sequence of compact sets $\{H_{v,n}\}$ contained in E_v such that $\overline{G}(z,\zeta;E_v) = \lim_{n\to\infty} G(z,\zeta;H_{v,n})$ and $\overline{V}_\zeta(E_v) = \lim_{n\to\infty} V_\zeta(H_{v,n})$. It follows that $\{\eta_{H_{v,n}}(z)\}$ is a decreasing sequence of functions harmonic in U, which is bounded below by $\eta_F(z)$. By Harnack's Principle, $\eta_{E_v}(z) := \lim_{n\to\infty} \eta_{H_{v,n}}(z)$ is harmonic in U and $\overline{G}(z,\zeta;E_v) = \eta_{E_v}(z) - \log_v(|g_\zeta(z)|_v)$ in $U\backslash\{\zeta\}$. Since $\eta_{H_{v,n}}(\zeta) = V_\zeta(H_{v,n})$ for each v, our assertion follows.

In the nonarchimedean case, let r_0 be as in Proposition 3.11(B), and take $0 < r < r_0$ small enough that $r < \|\zeta, E_v\|_v := \inf_{z\in E_v}(\|z,\zeta\|_v)$. By Proposition 3.11(B) and the definition (3.36), there is a constant C_ζ such that for each compact $H_v \subset E_v$ we have

$$\overline{G}(z,\zeta;H_v) \;=\; C_\zeta + V_\zeta(H_v) - \log_v(\|z,\zeta\|_v)$$

for all $z \in (B(\zeta,r)\backslash\{\zeta\})$. For each such z the monotonicity of the Green's functions of compact sets shows that the values $\overline{G}(z,\zeta;H_v)$ form a directed set, bounded below by 0, hence convergent. Thus for each $z \in (B(\zeta,r)\backslash\{\zeta\})$

$$\begin{aligned}\overline{G}(z,\zeta;E_v) \;&:=\; \inf_{H_v\subset E_v} \overline{G}(z,\zeta;H_v)\\ &=\; C_\zeta + \big(\inf_{H_v\subset E_v} V_\zeta(H_v) - \log_v(\|z,\zeta\|_v)\big)\end{aligned}$$

and it follows that

$$\overline{V}_\zeta(E_v) \;=\; \lim_{z\to\zeta} \overline{G}(z,\zeta;E_v) + \log_v(|g_\zeta(z)|_v) \;=\; \inf_{H_v\subset E_v} V_\zeta(H_v)\ . \qquad \square$$

An important fact is that for nonarchimedean v, the Green's functions and Robin constants of "nice" sets take on values in $\mathbb{Q}$:

PROPOSITION 3.28. *Let v be nonarchimedean. Let $E_v \subset \mathcal{C}_v(\mathbb{C}_v)$. Suppose one of the following conditions holds:*

(A) $E_v = \bigcup_{\ell=1}^D \big(B(a_\ell,r_\ell) \cap \mathcal{C}_v(F_{w_\ell})\big)$ where $B(a_1,r_1),\ldots,B(a_D,r_D) \subset \mathcal{C}_v(\mathbb{C}_v)$ are pairwise disjoint isometrically parametrizable balls and $F_{w_1},\ldots,F_{w_D}$ are finite extensions of K_v in $\mathbb{C}_v$, with $a_\ell \in \mathcal{C}_v(F_{w_\ell})$ and $r_\ell \in |F_{w_\ell}^\times|_v$ for each ℓ; or

(B) E_v is an RL*-domain, so $E_v = \{z \in \mathcal{C}_v(\mathbb{C}_v) : |f(z)|_v \le 1\}$ for some nonconstant $f(z) \in \mathbb{C}_v(\mathcal{C})$.*

Then for each $\zeta \in \mathcal{C}_v(\mathbb{C}_v)\backslash E_v$, we have $\overline{V}_\zeta(E_v) \in \mathbb{Q}$, and $\overline{G}(z,\zeta;E_v) \in \mathbb{Q}$ for all $z \neq \zeta$.

PROOF. In case (A), the assertion is proved in Corollary A.15 of Appendix A.

In case (B), the assertion follows from results proved in ([**51**]). Suppose E_v is an RL-domain. First note that by ([**51**], Theorem 4.3.3), each RL-domain is algebraically capacitable in the sense of ([**51**], Definition 4.3.2). By ([**51**], Theorem 4.4.4), this means that for $z \notin E_v$, the upper Green's function $\overline{G}(z,\zeta;E_v)$ coincides with the lower Green's function $\underline{G}(z,\zeta;E_v)$ defined in ([**51**], p.282). On the other hand, since each point of E_v has a neighborhood contained in E_v, trivially $\overline{G}(z,\zeta;E_v) = 0$ for $z \in E_v$. Since $\overline{G}(z,\zeta;E_v) \ge \underline{G}(z,\zeta;E_v) \ge 0$ for all z, we conclude that $\overline{G}(z,\zeta;E_v) = \underline{G}(z,\zeta;E_v)$ for all z.

Thus it suffices to work with the lower Green's function $\underline{G}(z,\zeta;E_v)$. By ([**51**], Theorem 4.2.15) E_v can be uniquely written in the form

$$E_v = \bigcap_{j=1}^{M} D_j$$

where $D_1,\ldots,D_M$ are "PL-domains" with pairwise disjoint complements. For each $\zeta \notin E_v$, there is a unique D_j with $\zeta \notin D_j$, and by ([**51**], Theorem 4.2.12) there is a function $h_\zeta(z) \in \mathbb{C}_v(\mathcal{C})$, whose only pole is at ζ, such that

$$D_j = \{z \in \mathcal{C}_v(\mathbb{C}_v) : |h_\zeta(z)|_v \le 1\}\,.$$

By ([**51**], Corollary 4.2.13), D_j is minimal among PL-domains containing E_v whose complement contains ζ.

By ([**51**], Proposition 4.4.1), if $\deg(h_\zeta) = N$, then

$$\underline{G}(z,\zeta;E_v) = \underline{G}(z,\zeta;D_j) = \begin{cases} \frac{1}{N}\log_v(|h_\zeta(z)|_v) & \text{if } z \notin D_j \quad , \\ 0 & \text{if } z \in D_j \quad . \end{cases}$$

It follows that $\overline{V}_\zeta(E_v) \in \mathbb{Q}$ and that $\overline{G}(z,\zeta;E_v) \in \mathbb{Q}$ for all $z \ne \zeta$. □

The next proposition plays a key role in the reduction of Theorem 0.4 to Theorem 4.2. We begin with a definition.

DEFINITION 3.29. Let a set $E_v \subset \mathcal{C}_v(\mathbb{C}_v)$ and a subset $E_v^0 \subset E_v$ be given. Let z_0 be a point in E_v.

If $v \in \mathcal{M}_K$ is archimedean, we will say that z_0 is *analytically accessible* from E_v^0 if for some $r > 0$, there is a nonconstant analytic map $f : D(0,r) \to \mathcal{C}_v(\mathbb{C})$ with $f(0) = z_0$, such that $f((0,r]) \subset E_v^0$.

If v is nonarchimedean, z_0 is *analytically accessible* from E_v^0 if there are an isometrically parametrizable ball $B(z_0,r)$ and a ($\mathbb{C}_v$-rational) isometric parametrization $f : D(0,r) \to B(z_0,r)$ with $f(0) = z_0$, such that $f((\mathcal{O}_v \cap D(0,r))\backslash\{0\}) \subset E_v^0$.

PROPOSITION 3.30. *Let $E_v \subset \mathcal{C}_v(\mathbb{C}_v)$ be a compact set of positive capacity, and let a subset $E_v^0 \subset E_v$ be given. Suppose there is a Borel subset $e \subset E_v$ of inner capacity* 0 *such that each point of $E_v\backslash e$ is analytically accessible from E_v^0. Then for each $\zeta \in \mathcal{C}_v(\mathbb{C}_v)\backslash E_v$, we have $\overline{V}_\zeta(E_v^0) = V_\zeta(E_v)$, and $\overline{G}(z,\zeta;E_v^0) = G(z,\zeta;E_v)$ for all $z \ne \zeta$.*

PROOF. Fix $\zeta \notin E_v$. We begin by showing that $\overline{G}(z,\zeta;E_v^0) = G(z,\zeta;E_v) = 0$ for each $z \in E_v\backslash e$.

First assume v is archimedean. Recall ([**32**], p.53) that a set $Z \subset \mathcal{C}_v(\mathbb{C})$ is said to be *thin* at a point z_0 if either z_0 is not a limit point of Z, or there are a neighborhood V of z_0 and a subharmonic function $u(z)$ on V such that

$$\limsup_{\substack{x \to z_0 \\ x \in Z \backslash \{z_0\}}} u(x) \ < \ u(z_0) \ .$$

Fix $z_0 \in E_v \backslash e$. Since z_0 is analytically accessible from E_v^0, there is a nonconstant analytic map $f : D(0,r) \to \mathcal{C}_v(\mathbb{C})$ with $f(0) = z_0$ such that $f((0,r]) \subseteq E_v^0$. Without loss we can assume r is small enough that $\zeta \notin f(D(0,r))$. By ([**32**], Corollary 4.8.5), $f([0,r])$ is not thin at z_0. On the other hand, for each $0 < \varepsilon < r$, the set $H_\varepsilon := f([\varepsilon, r])$ is a compact continuum contained in E_v^0. By Proposition 3.20.3, $G(z, \zeta; H_\varepsilon)$ is identically 0 on H_ε. Thus $\overline{G}(z, \zeta; E_v^0)$ is identically 0 on $f((0,r])$. By Proposition 3.20.5, $\overline{G}(z, \zeta; E_v^0)$ is subharmonic and nonnegative on $f(D(0,r))$, so

$$0 \ \leq \ G(z_0, \zeta; E_v) \ \leq \ \overline{G}(z_0, \zeta; E_v^0) \ \leq \ \limsup_{\substack{x \to z_0 \\ x \in f([0,r]) \backslash \{z_0\}}} \ \overline{G}(x, \zeta; E_v^0) \ = \ 0 \ .$$

Thus $G(z_0, \zeta; E_v) = \overline{G}(z_0, \zeta; E_v^0) = 0$.

Next let v be nonarchimedean. Given $z_0 \in E_v \backslash e$, let $f : D(0,r) \to B(z_0, r)$ be an isometric parametrization with $f(0) = z_0$, such that $f((D(0,r) \backslash \{0\}) \cap \mathcal{O}_v) \subseteq E_v^0$. Without loss we can assume r is small enough that $\zeta \notin B(z_0, r)$, and that $r \in |K_v^\times|$. For each $0 < \varepsilon < r$, put $K_\varepsilon = (D(0,r) \backslash D(0,\varepsilon)^-) \cap \mathcal{O}_v$ and put $H_\varepsilon = f(K_\varepsilon) \subset E_v^0$. Since $B(z_0, r)$ is an isometrically parametrizable ball disjoint from ζ, by Proposition 3.11(B.2), there is a constant C such that $[z, w]_\zeta = C \|z, w\|_v$ for all $z, w \in B(z_0, r)$. Pulling this back to $D(0,r)$, we see that $[f(x), f(y)]_\zeta = C|x - y|$ for all $x, y \in D(0,r)$,

Regarding K_ε as a subset of $\mathbb{C}_v = \mathbb{P}^1(\mathbb{C}_v) \backslash \{\infty\}$ and considering the definitions of $G(z, \infty; K_\varepsilon)$ and $G(z, \zeta; H_\varepsilon)$, it follows that $G(f(x), \zeta; H_\varepsilon) = G(x, \infty; K_\varepsilon)$ for each $x \in D(0,r)$. By the explicit computation in Proposition 2.4, together with a simple scaling argument, we have

$$\lim_{\varepsilon \to 0^+} G(z_0, \infty; H_\varepsilon) \ = \ \lim_{\varepsilon \to 0^+} G(0, \infty; K_\varepsilon) \ = \ 0 \ .$$

Since

$$0 \ \leq \ G(z_0, \zeta; E_v) \ \leq \ \overline{G}(z_0, \zeta; E_v^0) \ \leq \ \lim_{\varepsilon \to 0} G(z_0, \zeta; H_\varepsilon) \ = \ 0 \ ,$$

once again we see that $G(z_0, \zeta; E_v) = \overline{G}(z_0, \zeta; E_v^0) = 0$.

We next show that $\overline{V}_\zeta(E_v^0) = V_\zeta(E_v)$. The argument is similar to the one in Proposition 3.22, but since the context is somewhat different we give the details.

Write $\widehat{V} = \overline{V}_\zeta(E_v^0)$. By Lemma 3.24, there is an increasing sequence of compact sets $H_{v,1} \subseteq H_{v,2} \subseteq \cdots \subseteq E_v^0$ such that $\lim_{n \to \infty} G(z, \zeta; H_{v,n}) = \overline{G}(z, \zeta; E_v^0)$ for each $z \in \mathcal{C}_v(\mathbb{C}_v) \backslash \{\zeta\}$, and $\lim_{n \to \infty} V_\zeta(H_{v,n}) = \widehat{V}$; we can assume without loss that each $H_{v,n}$ has positive capacity. Write μ_n for the equilibrium distribution of $H_{v,n}$ with respect to ζ, and let μ be the equilibrium distribution of E_v with respect to ζ. After replacing $\{H_{v,n}\}_{n \geq 1}$ with a subsequence, if necessary, we can assume that the measures μ_n converge weakly to a probability measure $\widehat{\mu}$ on E_v.

By what has been shown above, the potential function $u_{E_v}(z, \zeta) = V_\zeta(E_v) - G(z, \zeta; E_v)$ is identically equal to $V_\zeta(E_v)$ on $E_v \backslash e$. Since a set of inner capacity 0 has mass 0 under any positive measure with a finite energy integral, for each n the

Fubini-Tonelli theorem gives

$$\begin{aligned}\int_{E_v} u_{H_{v,n}}(z,\zeta)\,d\mu(z) &= \iint_{E_v\times H_{v,n}} -\log_v([z,w]_\zeta)\,d\mu_n(w)d\mu(z)\\ &= \int_{H_{v,n}} u_{E_v}(w,\zeta)\,d\mu_n(w) \;=\; V_\zeta(E_v)\ . \end{aligned}\tag{3.47}$$

On the other hand, pointwise for each $z_0 \in E_v\backslash e$, we have

$$\begin{aligned}\lim_{n\to\infty} u_{H_{v,n}}(z_0,\zeta) &= \lim_{n\to\infty}\big(V_\zeta(H_{v,n}) - G(z_0,\zeta;H_{v,n})\big)\\ &= \widehat{V} - \overline{G}(z_0,\zeta;E_v^0) \;=\; \widehat{V}\ .\end{aligned}$$

Since E_v is bounded away from ζ, there is a $B_1 > -\infty$ such that $u_{H_{v,n}}(z,\zeta) \ge B_1$ on E_v, for all n. However, since $u_{H_{v,n}}(z,\zeta) \le V_\zeta(H_{v,n})$ for all z and the $V_\zeta(H_{v,n})$ are decreasing with n, there is a $B_2 < \infty$ such that $u_{H_{v,n}}(z,\zeta) \le B_2$ on E_v, for all n. By (3.47) and the Dominated Convergence Theorem, it follows that

$$V_\zeta(E_v) \;=\; \lim_{n\to\infty}\int_{E_v\backslash e} u_{H_{v,n}}(z,\zeta)\,d\mu(z) \;=\; \int_{E_v\backslash e}\widehat{V}\,d\mu(z) \;=\; \widehat{V}\ .$$

Hence $\overline{V}_\zeta(E_v^0) = \lim_{n\to\infty} V_\zeta(H_{v,n}) = \widehat{V} = V_\zeta(E_v)$.

The remainder of the proof is identical to the part of the proof of Proposition 3.22 after formula (3.39). Using the Monotone Convergence Theorem, one shows that $I_\zeta(\widehat{\mu}) = V_\zeta(E_v)$. The uniqueness of the equilibrium distribution implies $\widehat{\mu} = \mu$, and this in turn yields $\overline{G}(z,\zeta;E_v^0) = G(z,\zeta;E_v)$ for all $z \ne \zeta$. □

10. Green's Matrices and the Inner Cantor Capacity

Let notations and assumptions be as in §3.2. Thus, K is a global field and $\mathcal{C}/K$ is a curve. We are given a finite, Galois-stable set of points $\mathfrak{X} = \{x_1,\ldots,x_m\} \subset \mathcal{C}(\widetilde{K})$, and a K-rational adelic set $\mathbb{E} = \prod_v E_v$ compatible with $\mathfrak{X}$: each $E_v \subset \mathcal{C}_v(\mathbb{C}_v)$ is nonempty, stable under $\mathrm{Aut}_c(\mathbb{C}_v/K_v)$, and bounded away from $\mathfrak{X}$, with E_v being $\mathfrak{X}$-trivial for all but finitely many v. For each $x_i \in \mathfrak{X}$ we are given a uniformizing parameter $g_{x_i}(z) \in K(\mathcal{C})$, with $g_{\sigma(x_i)}(z) = \sigma(g_{x_i})(z)$ for each $\sigma \in \mathrm{Aut}(\widetilde{K}/K)$.

In this section we introduce the inner Cantor capacity, extending the definitions and results from ([**51**], §5.3) to arbitrary K-rational sets $\mathbb{E}$ compatible with $\mathfrak{X}$. When each E_v is algebraically capacitable (in particular, if each E_v is compact or $\mathfrak{X}$-trivial) then the upper Green's matrix $\overline{\Gamma}(\mathbb{E},\mathfrak{X})$ and the inner Cantor capacity $\overline{\gamma}(\mathbb{E},\mathfrak{X})$ defined here coincide with the Green's matrix $\Gamma(\mathbb{E},\mathfrak{X})$ and the Cantor capacity $\gamma(\mathbb{E},\mathfrak{X})$ from ([**51**], §5.3).

To define the inner Cantor capacity $\overline{\gamma}(\mathbb{E},\mathfrak{X})$, it is necessary to first make a base change to $L = K(\mathfrak{X})$. For each place v of K, and each place w of L over v, fix a continuous isomorphism $\iota_{w/v} : \mathbb{C}_w \to \mathbb{C}_v$, and put $E_w = \iota_{w/v}^{-1}(E_v) \subset \mathcal{C}_w(\mathbb{C}_w)$. Since E_v is stable under $\mathrm{Aut}_c(\mathbb{C}_v/K_v)$, the set E_w is independent of the choice of $\iota_{w/v}$. The Green's functions $\overline{G}(z,x_i;E_w)$ and Robin constants $\overline{V}_{x_i}(E_w)$ are defined the same way as the corresponding objects over K, but using the normalized absolute values $|x|_w$ and the normalized logarithms $\log_w(z)$, which means that

$$\overline{G}(x_i,x_j;E_w) = e_{w/v}\overline{G}(x_i,x_j;E_v)\ ,\quad \overline{V}_{x_i}(E_w) = e_{w/v}\overline{V}_{x_i}(E_v)$$

where $e_{w/v}$ is the ramification index (for archimedean places, our convention is that $e_{w/v} = 1$). Put $\mathbb{E}_L = \prod_w E_w$.

For each place w of L, we first define the *local upper Green's matrix* by

$$\overline{\Gamma}(E_w, \mathfrak{X}) = \begin{pmatrix} \overline{V}_{x_1}(E_w) & \overline{G}(x_1, x_2; E_w) & \dots & \overline{G}(x_1, x_m; E_w) \\ \overline{G}(x_2, x_1; E_w) & \overline{V}_{x_2}(E_w) & \dots & \overline{G}(x_2, x_m; E_w) \\ \vdots & \vdots & \ddots & \vdots \\ \overline{G}(x_m, x_1; E_w) & \overline{G}(x_m, x_2; E_w) & \dots & \overline{V}_{x_m}(E_w) \end{pmatrix} .$$

For all but finitely many w, each $g_{x_i}(z)$ has good reduction at w and E_w is $\mathfrak{X}$-trivial; for such w, $\overline{\Gamma}(E_w, \mathfrak{X})$ is the zero matrix (see [**51**], Proposition 5.1.2).

The global upper Green's matrix over L is defined by

$$\overline{\Gamma}(\mathbb{E}_L, \mathfrak{X}) = \sum_w \overline{\Gamma}(E_w, \mathfrak{X}) \log(q_w) . \tag{3.48}$$

By our remarks above, this is actually a finite sum.

The local and global upper Green's matrices over K are then defined by

$$\overline{\Gamma}(E_v, \mathfrak{X}) \log(q_v) = \frac{1}{[L:K]} \sum_{w|v} \overline{\Gamma}(E_w, \mathfrak{X}) \log(q_w) ,$$

$$\overline{\Gamma}(\mathbb{E}_K, \mathfrak{X}) = \frac{1}{[L:K]} \overline{\Gamma}(\mathbb{E}_L, \mathfrak{X}) ,$$

so that

$$\overline{\Gamma}(\mathbb{E}_K, \mathfrak{X}) = \sum_v \overline{\Gamma}(E_v, \mathfrak{X}) \log(q_v) .$$

The entries of $\overline{\Gamma}(\mathbb{E}_K, \mathfrak{X})$ are finite if and only if each E_v has positive innner capacity. Clearly $\overline{\Gamma}(\mathbb{E}_K, \mathfrak{X})$ and the $\overline{\Gamma}(E_v, \mathfrak{X})$ are symmetric and nonnegative off the diagonal; they are also K-symmetric in the sense of Definition 3.1, as shown by the following lemma:

LEMMA 3.31. *Let $\mathbb{E}$, $\mathfrak{X}$, L/K, and the uniformizers $g_{x_i}(z)$ be as above. Then for each $\sigma \in \mathrm{Aut}(L/K)$, each place w of L, and each $i \neq j$,*

$$\overline{G}(x_i, x_j; E_w) = \overline{G}(x_{\sigma(i)}, x_{\sigma(j)}, E_{\sigma(w)}) , \quad \overline{V}_{x_i}(E_w) = \overline{V}_{x_{\sigma(i)}}(E_{\sigma(w)}) .$$

PROOF. This is essentially a tautology.

We can view $\mathcal{C}(L)$ as embedded on the diagonal in $\oplus_{w|v} \mathcal{C}_w(L_w)$. Each $\sigma \in \mathrm{Aut}(L/K)$ acts on $L \otimes_K K_v$ through its action on L. Using the canonical isomorphism $L \otimes_K K_v \cong \oplus_{w|v} L_w$ (which holds when $\mathrm{char}(K) = p > 0$, as well as when $\mathrm{char}(K) = 0$; see ([**51**], p.321)) this action can also be described by a collection of isomorphisms $\tau_{\sigma,w} : L_w \to L_{\sigma(w)}$.

The action of σ on $\mathfrak{X} \subset \mathcal{C}(L)$ is described globally by the permutation representation $\sigma(x_i) = x_{\sigma(i)}$. When this is combined with the semilocal description of its action on $\oplus_{w|v} \mathcal{C}_w(L_w)$ (see §6.3), after identifying the points in $\mathfrak{X}$ with their images in $\mathcal{C}_w(L_w)$ and $\mathcal{C}_{\sigma(w)}(L_{\sigma(w)})$, we find that $\tau_{\sigma,w}$ takes $x_i \in \mathcal{C}_w(L_w)$ to $x_{\sigma(i)} = \sigma(x_i) \in \mathcal{C}_{\sigma(w)}(L_{\sigma(w)})$.

Extend each $\tau_{\sigma,w}$ to a continuous isomorphism $\overline{\tau}_{\sigma,w} : \mathbb{C}_w \to \mathbb{C}_{\sigma(w)}$. Under this isomorphism $E_w \subset \mathcal{C}_w(\mathbb{C}_w)$ is taken to $E_{\sigma(w)}$ since both are pullbacks of E_v, which is stable under $\mathrm{Aut}_c(\mathbb{C}_v/K_v)$. Viewing $\overline{\tau}_{\sigma,w}$ as an identification, we then have $\overline{G}(z, x_j, E_w) = \overline{G}(\overline{\tau}_{\sigma,w}(z), x_{\sigma(j)}; E_{\sigma(w)})$ for all $z \in \mathcal{C}_w(\mathbb{C}_w)$. Both assertions in the lemma now follow; the second uses Proposition 3.27 and the fact that the local uniformizers $g_{x_i}(z)$ are K-symmetric. □

We can now define the inner Cantor capacity. Let the set of m-dimensional real *probability vectors* be

$$\mathcal{P}^m = \mathcal{P}^m(\mathbb{R}) = \{\vec{s} \in \mathbb{R}^m : \sum s_i = 1,\ s_i \geq 0 \text{ for each } i\} ,$$

The global upper Robin constant $\overline{V}(\mathbb{E}_K, \mathfrak{X})$ is the value of $\overline{\Gamma}(\mathbb{E}_K, \mathfrak{X})$ as a matrix game:

$$\overline{V}(\mathbb{E}_K, \mathfrak{X}) = \mathrm{val}(\overline{\Gamma}(\mathbb{E}_K, \mathfrak{X})) := \max_{\vec{s} \in \mathcal{P}^m} \min_{\vec{r} \in \mathcal{P}^m} {}^t\vec{s}\,\overline{\Gamma}(\mathbb{E}_K, \mathfrak{X})\vec{r} \tag{3.49}$$

$$= \min_{\vec{s} \in \mathcal{P}^m} \max_i (\overline{\Gamma}(\mathbb{E}_K, \mathfrak{X})\vec{s})_i . \tag{3.50}$$

The equality of (3.49) and (3.50), along with many other similar expressions, follows from the Fundamental Theorem of Game Theory (see [**51**], p.327). The inner Cantor capacity is then defined by

$$\overline{\gamma}(\mathbb{E}_K, \mathfrak{X}) = e^{-\overline{V}(\mathbb{E}_K, \mathfrak{X})} .$$

Clearly $\overline{\gamma}(\mathbb{E}, \mathfrak{X}) > 0$ if and only if each E_v has positive inner capacity.

The inner Cantor capacity has properties like those of the classical Cantor capacity in ([**51**]). All of these are formal consequences of properties of matrices and Green's functions, so the proofs given in ([**51**]) carry over without change. The most important properties are as follows:

PROPOSITION 3.32. *$\overline{\gamma}(\mathbb{E}, \mathfrak{X}) > 1$ iff $\overline{\Gamma}(\mathbb{E}, \mathfrak{X})$ is negative definite.*

PROOF. See ([**51**], Proposition 5.1.8, p.331). The quantity $\mathrm{val}(\Gamma)$ is a statistic of symmetric, real-valued matrices such that $\mathrm{val}(\Gamma) < 0$ iff Γ is negative definite. □

PROPOSITION 3.33. *If $\overline{\gamma}(\mathbb{E}, \mathfrak{X}) > 1$, then there is a unique probability vector $\widehat{s} = {}^t(\widehat{s}_1, \ldots, \widehat{s}_m) \in \mathcal{P}^m(\mathbb{R})$ for which*

$$\overline{\Gamma}(\mathbb{E}, \mathfrak{X})\,\widehat{s} = \begin{pmatrix} \widehat{V} \\ \vdots \\ \widehat{V} \end{pmatrix}$$

has all of its entries equal. Furthermore, $\widehat{V} = \overline{V}(\mathbb{E}, \mathfrak{X}) < 0$, $\widehat{s}$ is K-symmetric, and $\widehat{s}_i > 0$ for each $i = 1, \ldots, m$.

PROOF. The proof is the same as ([**51**], Theorem 5.1.6, p.328), and goes back to ([**16**]). However, because this proposition plays a key role in the proof of the Fekete-Szegö theorem, we give the argument here.

For brevity, write $\Gamma = \overline{\Gamma}(\mathbb{E}, \mathfrak{X})$. By Proposition 3.32, Γ is negative definite, so all of its eigenvalues are negative. Choose $\alpha > 0$ large enough that $\Gamma + \alpha I$ is positive definite, where I is the $m \times m$ identity matrix. Then each entry of $\Gamma + \alpha I$ is nonnegative, and $0 < \lambda_i < \alpha$ for each eigenvalue λ_i of $\Gamma + \alpha I$. It follows that the series

$$-\Gamma^{-1} = (\alpha I - (\Gamma + \alpha I))^{-1} = \alpha^{-1}\left(I - \left(\frac{\Gamma + \alpha I}{\alpha}\right)\right)^{-1}$$

$$= \alpha^{-1} \sum_{k=0}^{\infty} \left(\frac{\Gamma + \alpha I}{\alpha}\right)^k$$

converges, so $-\Gamma^{-1}$ has only nonnegative entries.

Set

$$\vec{s}' = -\Gamma^{-1}\begin{pmatrix}1\\ \vdots\\ 1\end{pmatrix}.$$

By construction, each entry of $\vec{s}'$ is nonnegative; if some entry were 0, it would mean that every entry in the corresponding row of $-\Gamma^{-1}$ were 0, and hence $\Gamma^{-1}\cdot\Gamma = I$ would not be possible. Scale $\vec{s}'$ so as obtain a probability vector $\widehat{s}$. Then all the entries of $\widehat{s}$ are positive, and we have

$$\Gamma\,\widehat{s} = \begin{pmatrix}\widehat{V}\\ \vdots\\ \widehat{V}\end{pmatrix} \tag{3.51}$$

for some $\widehat{V}$. The minimax inequality (3.49) then shows $\widehat{V} = \overline{V}(\mathbb{E},\mathfrak{X}) < 0$.

The uniqueness of $\widehat{s}$ and $\widehat{V}$ satisfying (3.51) are clear. Finally, since $\overline{\Gamma}(\mathbb{E},\mathfrak{X})$ is K-symmetric, the uniqueness of $\widehat{s}$ means $\widehat{s}$ must be K-symmetric as well. □

We define the set of rational probability vectors to be

$$\mathcal{P}^m(\mathbb{Q}) = \mathcal{P}^m(\mathbb{R}) \cap \mathbb{Q}^m .$$

Unfortunately, $\widehat{s}$ need not be rational. This causes major technical difficulties in the proof of the Fekete-Szegö theorem. To overcome it, we show that in the number field case the "logarithmic leading coefficients" of the archimedean initial approximating functions can be "macroscopically independently varied", which allows us to replace $\widehat{s}$ with a suitable rational $\vec{s}$ closely approximating $\widehat{s}$. In the function field case, we show that if the sets E_v satisfy appropriate hypotheses, then all entries in $\overline{\Gamma}(\mathbb{E},\mathfrak{X})$ are rational multiples of $\log(p)$, and in that case $\widehat{s}$ belongs to $\mathcal{P}^m(\mathbb{Q})$.

We conclude this section by noting some properties of the inner Cantor capacity.

PROPOSITION 3.34.

(1) (Base Change): *If M/K is any finite extension, then after base change* $\overline{\Gamma}(\mathbb{E}_M,\mathfrak{X}) = [M:K]\,\overline{\Gamma}(\mathbb{E}_K,\mathfrak{X})$, $\overline{V}(\mathbb{E}_M,\mathfrak{X}) = [M:K]\,\overline{V}(\mathbb{E}_K,\mathfrak{X})$, *and*

$$\overline{\gamma}(\mathbb{E}_M,\mathfrak{X}) = \overline{\gamma}(\mathbb{E}_K,\mathfrak{X})^{[M:K]} .$$

(2) (Pullback): *Let $\mathcal{C}_1,\mathcal{C}_2/K$ be curves, and let $f:\mathcal{C}_1\to\mathcal{C}_2$ be a nonconstant rational map defined over K. If $\mathbb{E}$ is a K-rational adelic set on $\mathcal{C}_2$, compatible with $\mathfrak{X}\subset\mathcal{C}_2(\widetilde{K})$, then*

$$\overline{\gamma}(f^{-1}(\mathbb{E}),f^{-1}(\mathfrak{X})) = \overline{\gamma}(\mathbb{E},\mathfrak{X})^{1/\deg(f)} .$$

PROOF. These follow from the corresponding properties of upper Green's functions, by the arguments in the proofs of ([**51**], Theorems 5.1.13 and 5.1.14, p.333). □

11. Newton Polygons of Nonarchimedean Power Series

In this section we recall some some facts about Newton polygons of nonarchimedean power series needed for the proof of Theorem 6.3.

Given $r>0$, write $D(0,r) = \{z\in\mathbb{C}_v : |z|_v \le r\}$ for the "closed" disc of radius r in $\mathbb{C}_v$, and $D(0,r)^- = \{z\in\mathbb{C}_v : |z|_v < r\}$ for the "open" disc.

The Newton polygon of a power series $f(Z) = \sum_{k=0}^{\infty} c_k Z^k \in \mathbb{C}_v[[Z]]$ is the lower convex hull of the set of points $\{(k, \mathrm{ord}_v(c_k))\}$, with a vertical side above the point corresponding to the first nonzero coefficient. If $f(Z)$ is a polynomial of degree m, its Newton polygon is also considered to have a vertical side above $(m, \mathrm{ord}_v(c_m))$.

A power series of the form $h(Z) = 1 + \sum_{k=1}^{\infty} b_k Z^k \in \mathbb{C}_v[[Z]]$ will be called a *unit power series* for $D(0,r)$ if $h(Z)$ is a unit in the ring of power series converging on $D(0,r)$. This holds if and only if $|b_k|_v < 1/r^k$ for each $k \geq 1$, in which case $|h(z)|_v = 1$ for all $z \in D(0,r)$, and the series $h(Z)^{-1} = 1 + \sum_{k=1}^{\infty} b_k' Z^k$ with $h(Z) \cdot h(Z)^{-1} = 1$ also satisfies $|b_k'|_v < 1/r^k$ for each $k \geq 1$. If $h(Z) \in K_v[[Z]]$, then $h(Z)^{-1} \in K_v[[Z]]$.

LEMMA 3.35. *Suppose $f(Z) = \sum_{k=0}^{\infty} c_k Z^k \in \mathbb{C}_v[[Z]]$ converges on a disc $D(0,r)$, where $r > 0$ belongs to the value group of $\mathbb{C}_v^{\times}$. Then*

(A) $f(Z)$ has finitely many zeros in $D(0,r)$, and $f(Z)$ can be factored as $f(Z) = g_r(Z) \cdot h_r(Z)$ where $g_r(Z)$ is a polynomial having the same roots (with multiplicities) as $f(Z)$ in $D(0,r)$, and where $h_r(Z) = 1 + \sum_{k=1}^{\infty} b_k Z^k$ is a unit power series for $D(0,r)$. In particular, $|b_k|_v < 1/r^k$ for all $k \geq 1$. If $f(Z) \in F_w[[Z]]$ for some finite extension F_w/K_v, then $g_r(Z) \in F_w[Z]$ and $h_r(Z) \in F_w[[Z]]$.

(B) If $f(Z)$ has $m \geq 0$ roots in $D(0,r)$ (counted with multiplicities), then $(m, \mathrm{ord}_v(c_m))$ is a vertex of the Newton polygon of $f(Z)$, and the part of the Newton polygon of $f(Z)$ on and to the left of $(m, \mathrm{ord}_v(c_m))$ coincides with the Newton polygon of $g_r(Z)$.

PROOF. This is well known; we sketch the proof.

The fact that $f(Z)$ has finitely many zeros in $D(0,r)$, and the existence of the factorization $f(Z) = g_r(Z) \cdot h_r(Z)$, follow from the Weierstrass Preparation Theorem and a change of variables (see [**11**], p.201), or from Hensel's Lemma (as in [**4**], §2.5). The assertions about F_w-rationality hold because the Weierstrass Preparation Theorem and Hensel's Lemma are valid over any complete nonarchimedean field.

To establish the relation between the Newton polygons of $f(Z)$ and $g_r(Z)$, take $\beta \in \mathbb{C}_v$ with $|\beta|_v = r$. After replacing $f(Z)$ by $f(Z/\beta)$, which translates the Newton polygons of both $f(Z)$ and $g_r(Z)$ upwards by the line $y = x \log_v(r)$, we can assume that $r = 1$. Suppose $f(Z)$ has m roots in $D(0,1)$ and write $g(Z) = g_1(Z)$, $h(Z) = h_1(Z)$.

Consider the factorization $f(Z) = g(Z)h(Z)$. Write

$$f(Z) = \sum_{\ell=0}^{\infty} c_k Z^k, \quad g(Z) = \sum_{\ell=0}^{m} a_\ell Z^\ell ,$$

and expand

$$h(Z) \;=\; 1 + \sum_{j=1}^{\infty} b_j Z^j, \qquad h(Z)^{-1} \;=\; 1 + \sum_{j=1}^{\infty} b_j' Z^j .$$

Here $\mathrm{ord}_v(b_j), \mathrm{ord}_v(b_j') > 0$ for all $j \geq 1$. By the Weierstrass Factorization theorem, $|a_m|_v = |c_m|_v$. After dividing through by a_m, we can assume that $g(Z)$ is monic and that $\mathrm{ord}_v(c_k) \geq 0$ for all k.

Using that $g(Z) = f(Z)h(Z)^{-1}$, we see that if J is the smallest index for which $c_J \neq 0$, then $a_k = c_k = 0$ for $k < J$, while $a_J = c_J$. By hypothesis,

$\mathrm{ord}_v(a_m) = \mathrm{ord}_v(c_m) = 0$. For each ℓ with $J < \ell < m$, since $\mathrm{ord}_v(b'_j) > 0$ for all $j \geq 1$ and $\mathrm{ord}_v(c_k) \geq 0$ for all k,

$$\begin{aligned} \mathrm{ord}_v(a_\ell) &= \mathrm{ord}_v(c_\ell + c_{\ell-1}b'_1 + \ldots + c_0 b'_\ell) \\ &\geq \min(\mathrm{ord}_v(c_\ell), \mathrm{ord}_v(c_{\ell-1}), \ldots, \mathrm{ord}_v(c_0)) \ . \end{aligned} \tag{3.52}$$

If $(\ell, \mathrm{ord}_v(c_\ell))$ is a corner of the Newton polygon of $f(Z)$, then necessarily $\mathrm{ord}_v(c_\ell) < \mathrm{ord}_v(c_0), \ldots, \mathrm{ord}_v(c_{\ell-1})$ and it follows from (3.52) that $\mathrm{ord}_v(a_\ell) = \mathrm{ord}_v(c_\ell)$. Since $\mathrm{ord}_v(c_k) \geq 0$ for all k, the Newton polygon of $g(Z)$ lies on or below the initial part of the Newton polygon of $f(Z)$. Applying the same arguments to $f(Z) = g(Z)h(Z)$, we see that initial part of the Newton polygon of $f(Z)$ lies on or below the Newton polygon of $g(Z)$. □

For a polynomial, the absolute values of its roots and the slopes of the sides of its Newton polygon determine each other (see [**4**], §2.5): if the Newton polygon has a side with slope M, whose projection on the horizontal axis has length S, then $f(Z)$ has exactly S roots α for which $\log_v(|\alpha|_v) = M$. In this correspondence, the vertical ray above the vertex $(k, \mathrm{ord}_v(c_k))$ corresponding to the first nonzero coefficient is deemed to have projection length k.

The correspondence holds for power series as well. If the initial segment of the Newton polygon is a vertical ray above $(k, \mathrm{ord}_v(c_k))$, that ray is deemed to have slope $-\infty$ and projection length k. If $f(Z)$ is a polynomial, the vertical side above its rightmost vertex is deemed to have infinite projection length. There is also a special case when the radius of convergence r belongs to the value group of $\mathbb{C}_v^\times$ and $f(Z)$ converges in $D(0,r)$. In that situation the Newton polygon has a terminal ray of slope $\log_v(r)$ which can have at most a finite number of vertices on it, and the last such vertex is deemed the right endpoint of the rightmost side of finite length:

PROPOSITION 3.36. *Suppose $f(Z) = \sum_{k=0}^\infty c_k Z^k \in \mathbb{C}_v[[Z]]$ has radius of convergence $r > 0$. Then*

(A) The roots of $f(Z)$ correspond to the sides of the Newton polygon of $f(Z)$ of finite length in the same way as for a polynomial: for each finite length side with slope M and projection length S, $f(Z)$ has exactly S roots $\alpha_{M,1}, \ldots, \alpha_{M,S}$ (listed with multiplicity) for which $\log_v(|\alpha_{M,i}|_v) = M$.

(B) If $f(Z) \in F_w[[Z]]$, where F_w/K_v is a finite extension, then given a side of the Newton polygon with slope M and projection length S, the associated polynomial $\prod_{i=1}^S (Z - \alpha_{M,i})$ belongs to $F_w[Z]$. In particular, if $S = 1$, the unique root corresponding to that side is rational over F_w.

(C) The Newton polygon of $f(Z)$ is completely determined by the absolute value of the first nonzero coefficient of $f(Z)$, the absolute values of the roots of $f(Z)$, and the radius of convergence r.

PROOF. This too is well known. The domain of convergence of $f(Z)$ is either $D(0,r)$ or $D(0,r)^-$. Exhausting it by discs $D(0,r_1)$ where $0 < r_1 \leq r$ belongs to $|\mathbb{C}_v^\times|_v$, the roots of $f(Z)$ are accounted for by the roots of the polynomials $g_{r_1}(Z)$ in Lemma 3.35. By that Lemma and properties of Newton polygons of polynomials proved in ([**4**], §2.5), assertions (A) and (B) hold.

For (C), note that the absolute value of the first nonzero coefficient determines the location of the first corner of the Newton polygon. The absolute values of the nonzero roots determine the lengths and slopes of the sides of the Newton polygon of finite length. If there are infinitely many roots, the Newton polygon is completely determined; if there are only finitely many roots, the Newton polygon has a terminal ray with slope $\log_v(r)$. □

The following concept will play an important role in nonarchimedean constructions later:

DEFINITION 3.37. Let K_v be nonarchimedean. Let $B(a,\rho) \subset \mathcal{C}_v(\mathbb{C}_v)$ be an isometrically parametrizable ball and let $D(b,r) \subset \mathbb{C}_v$ be a disc. We will call a map $f : B(a,\rho) \to D(b,r)$ a *scaled isometry* if f is a $1-1$ correspondence satisfying $|f(z_1) - f(z_2)|_v = (r/\rho)\|z_1, z_2\|_v$ for all $z_1, z_2 \in B(a,\rho)$. Given two discs $D(a,\rho)$, $D(b,r)$, or two balls $B(a,\rho)$, $B(b,r)$, we define scaled a isometry $f : D(a,\rho) \to D(b,r)$ or $f : B(a,\rho) \to B(b,r)$ in a similar way.

We next give a criterion for a map defined by a power series to induce a scaled isometry.

PROPOSITION 3.38. *Suppose $f(Z) = \sum_{n=0}^{\infty} c_n Z^N \in \mathbb{C}_v[[Z]]$ converges on a disc $D(0,r)$, with $r > 0$. If f has a single zero in $D(0,r)$ (counted with multiplicity), then f induces a scaled isometry from $D(0,r)$ onto $D(0,R)$ where $R = |c_1|_v r$. If there is a finite extension F_w/K_v in $\mathbb{C}_v$ such that $f(Z) \in F_w[[Z]]$, then f maps $F_w \cap D(0,r)$ onto $F_w \cap D(0,R)$.*

If $H(Z) = \sum_{n=0}^{\infty} b_n Z^n$ is another power series which converges on $D(0,r)$, and H has no zeros in $D(0,r)$, then $|H(z)|_v$ takes the constant value $B = |b_0|$ for all $z \in D(0,r)$, and $f \cdot H$ induces a scaled isometry from $D(0,r)$ onto $D(0,BR)$. If there is a finite extension F_w/K_v in $\mathbb{C}_v$ such that $f(Z), H(Z) \in F_w[[Z]]$, then $f \cdot H$ maps $F_w \cap D(0,r)$ onto $F_w \cap D(0,BR)$.

PROOF. Since f converges on $D(0,r)$ and has a single zero there, the Newton polygon of $f(Z)$ lies on or above the line $y = (x-1)\log_v(r) + \mathrm{ord}_v(c_1)$, and the points $(n, \mathrm{ord}_v(c_n))$ for $n \geq 2$ lie strictly above that line. Thus $|c_0|_v \leq |c_1|_v r$ and $|c_n|_v < |c_1|_v / r^{n-1}$ for $n \geq 2$.

By the ultrametric inequality, for each $a \in D(0,r)$ we have $|f(a)|_v \leq |c_1|_v r$. On the other hand, for each $b \in D(0, |c_1|_v r)$ the Newton polygon of $f(Z) - b$ has the same properties as that of $f(Z)$, so there is a unique point $a \in D(0,r)$ such that $f(a) = b$. Thus f induces a $1-1$ correspondence from $D(0,r)$ onto $D(0, |c_1|_v r)$. In particular, $|c_1|_v = R/r$. To see that f is a scaled isometry, note that for all $z, w \in D(0,r)$

$$|f(z) - f(w)|_v = |z-w|_v \cdot |c_1 + \sum_{n=2}^{\infty} c_n \Big(\sum_{k=0}^{n-1} z^k w^{n-1-k} \Big)|_v = |c_1|_v \cdot |z-w|_v ,$$

where the last step follows from the ultrametric inequality using $|c_n|_v < |c_1|_v / r^{n-1}$.

If $f(Z) \in F_w[[Z]]$ for some finite extension F_w/K_v, clearly f maps $F_w \cap D(0,r)$ into $F_w \cap D(0, |c_1|_v r)$. On the other hand, if $b \in F_w \cap D(0, |c_1|_v r)$ then $f(Z) - b \in F_w[[Z]]$ and so the unique solution to $f(a) - b = 0$ belongs to $F_w \cap D(0,r)$ by Proposition 3.36(B). Thus f maps $F_w \cap D(0,r)$ onto $F_w \cap D(0, |c_1|_v r)$.

If $H(Z) = \sum_{n=0}^{\infty} b_n Z^n \in \mathbb{C}_v[[Z]]$ is another power series which converges on $D(0,r)$, but has no zeros there, then $H(Z) = b_0 \cdot h(Z)$ where $h(Z)$ is a unit power series. It follows that $|H(z)|_v = |b_0|_v$ for all $z \in D(0,R)$, and $|b_n|_v < B/r^n$ for all $n \geq 1$. If we write $f(Z)H(Z) = \sum_{n=0}^{\infty} a_n Z^n$, then $|a_1|_v = |b_0 c_1 + b_1 c_0|_v = BR/r$ since $|b_0 c_1|_v = BR/r$ while $|c_0 b_1|_v < BR/r$. We can now apply the previous discussion to $f \cdot H$. □

12. Stirling Polynomials and the Sequence $\psi_w(k)$

In this section we will consider v-adic Stirling polynomials, which play an important role in the construction of the nonarchimedean initial patching functions. They are also used in 'degree-raising' arguments in the global patching construction.

Let F_w/K_v be a finite extension in $\mathbb{C}_v$. Let $e_w = e_{w/v}$ be its ramification index and $f_w = f_{w/v}$ its residue degree, so $e_w f_w = [F_w : K_v]$.

Fix a prime element π_w for $\mathcal{O}_w$, and let $\mathrm{ord}_w(x)$ be the valuation on $\mathbb{C}_v$ for which $\mathrm{ord}_w(\pi_w) = 1$; thus $\mathrm{ord}_w(x) = e_w \,\mathrm{ord}_v(x)$. Put $q = q_w = q_v^{f_w}$, and write $\log_w(x)$ for the logarithm to the base q_w.

We first construct an explicit uniformly distributed sequence $\{\psi_w(k)\}_{k\geq 0}$ in $\mathcal{O}_w$. Let $\psi_w(0), \ldots, \psi_w(q-1)$ be the Teichmüller representatives for the cosets of $\mathcal{O}_w/\pi_w\mathcal{O}_w \cong \mathbb{F}_q$, with $\psi_w(0) = 0$. Thus when $\mathrm{char}(K) = 0$, the representatives are 0 and the $q-1^{st}$ roots of unity in F_w. When $\mathrm{char}(K) = p > 0$, so $F_w \cong \mathbb{F}_q((\pi_w))$, they are the elements of $\mathbb{F}_q$. This fact is important for the global patching argument (see Lemma 11.8 and Proposition 11.9).

Extend $\{\psi_w(k)\}_{0\leq k<q}$ to a sequence $\{\psi_w(k)\}_{0\leq k<\infty}$ as follows: for each $k \geq q$, write k in base q as

$$k \;=\; \sum_{i=0}^{N} d_i(k) q^i$$

where $N = \lfloor \log_v(k) \rfloor$ and $0 \leq d_i(k) \leq q-1$ for each i, then put

$$\psi_w(k) \;=\; \sum_{i=0}^{N} \psi_w(d_i(k))\pi_w^i \;\in\; \mathcal{O}_w \ .$$

For each m, the sequence $\{\psi_w(k)\}_{0\leq k<q^m}$ is a system of coset representatives for $\mathcal{O}_w/\pi_w^m\mathcal{O}_w$.

Define a function $\mathrm{val}_w(k)$ for integers $k \geq 0$ by letting $\mathrm{val}_w(k)$ be the smallest i for which $d_i(k) \neq 0$, if $k > 0$, and putting $\mathrm{val}_w(0) = \infty$. For each k it follows that

$$\mathrm{ord}_w(\psi_w(k)) \;=\; \mathrm{val}_w(k) \ . \tag{3.53}$$

Similarly, for all $k \neq \ell$,

$$\mathrm{ord}_w(\psi_w(k) - \psi_w(\ell)) \;=\; \mathrm{val}_w(|k-\ell|) \ . \tag{3.54}$$

This is because if $\mathrm{val}_w(|k-\ell|) = j$, then the digits $d_i(k)$ and $d_i(\ell)$ coincide for $i < j$, while $d_j(k) \neq d_j(\ell)$. It follows that for each $n > 0$, if $0 \leq k, \ell < n$ and $k \neq \ell$, then by (3.54)

$$\mathrm{ord}_w(\psi_w(k) - \psi_w(\ell)) \;<\; \log_w(n) \ , \tag{3.55}$$

since $\mathrm{val}_w(|k-\ell|) \leq \log_w(|k-\ell|) < \log_w(n)$.

DEFINITION 3.39. The *basic well-distributed sequence for* $\mathcal{O}_w$ is $\{\psi_w(k)\}_{0\leq k<\infty}$. The *Stirling polynomial of degree* n *for* $\mathcal{O}_w$ is

$$S_{n,w}(z) \;=\; \prod_{k=0}^{n-1} (z - \psi_w(k)) \ . \tag{3.56}$$

The polynomials $S_{n,w}(z)$ were first studied by Polya ([**46**]), and were used by Cantor in ([**16**]). The following proposition, which is similar to ([**53**], Lemma 8.7), summarizes their main properties:

PROPOSITION 3.40. *Let F_w/K_v be a finite, separable extension in $\mathbb{C}_v$. Let $S_{n,w}(z)$ be the Stirling polynomial of degree n for $\mathcal{O}_w$, and let $S'_{n,w}(z)$ be its derivative. Then*

(A) If $0 \le i, j < n$ and $i \ne j$ then $|\psi_v(i) - \psi_v(j)|_v > 1/n$.

(B) For each k, $0 \le k < n$, we have

$$\frac{n}{e_w(q_v^{f_w}-1)} - \frac{1}{e_w}\big(2\log_w(n)+3\big) \ < \ \mathrm{ord}_v\left(S'_{n,w}(\psi_w(k))\right) \ < \ \frac{n}{e_w(q_v^{f_w}-1)} \ . \tag{3.57}$$

(C) Fix $x \in \mathbb{C}_v$. If $0 \le J < n$ is such that $|x - \psi_w(J)|_v = \min_k(|x-\psi_w(k)|_v)$, then

$$\mathrm{ord}_v(S_{n,w}(x)) \ < \ \frac{n}{e_w(q_v^{f_w}-1)} + \mathrm{ord}_v(x - \psi_w(J)) \ . \tag{3.58}$$

If $x \notin D(0,1)$, then $\mathrm{ord}_v(S_{n,w}(x)) = n\,\mathrm{ord}_v(x)$.

PROOF. Assertion (A) is a reformulation of (3.55).

To prove assertion (B), it will be convenient to work over F_w rather than K_v. Fix $0 \le k < n$, and note that by (3.54), if $j \ne k$, then $\mathrm{ord}_w(\psi_w(k) - \psi_w(j)) = \mathrm{val}_w(|k-j|)$. This leads to a generalization of the well-known formula

$$\sum_{\ell=1}^{k} \mathrm{ord}_p(\ell) \ = \mathrm{ord}_p(k!) \ = \ \sum_{m \ge 1} \left\lfloor \frac{k}{p^m} \right\rfloor \ = \ \frac{k}{p-1} - \frac{1}{p-1}\sum_{i \ge 0} a_i \ ,$$

where the a_i are the base p digits of k. Writing $d_i = d_i(k)$, and $q = q_w$, for each $m \ge 1$ there are exactly $\lfloor k/q^m \rfloor$ integers $1 \le \ell \le k$ with $\mathrm{val}_w(\ell) \ge m$. Hence

$$\begin{aligned}
\sum_{\ell=1}^{k} \mathrm{val}_w(\ell) \ &= \ \sum_{m\ge 1} \left\lfloor \frac{k}{q^m} \right\rfloor \ = \ (d_1 + d_2 q + d_3 q^2 + \cdots) + (d_2 + d_3 q + \cdots) + \cdots \\
&= \ d_1 \cdot \frac{q-1}{q-1} + d_2 \cdot \frac{q^2-1}{q-1} + d_3 \cdot \frac{q^3-1}{q-1} + \cdots \\
&= \ \frac{d_0 + d_1 q + d_2 q^2 + \cdots}{q-1} - \frac{d_0 + d_1 + d_2 + \cdots}{q-1} \\
&= \ \frac{k}{q-1} - \frac{1}{q-1}\sum_{i \ge 0} d_i(k) \ . \qquad (3.59)
\end{aligned}$$

Consequently,

$$\begin{aligned}
\mathrm{ord}_w\Big(\prod_{\substack{\ell=0\\ \ell\ne k}}^{n-1}(\psi_w(k) - \psi_w(\ell))\Big) \ &= \ \sum_{\ell=1}^{k} \mathrm{val}_w(\ell) + \sum_{\ell=1}^{n-k-1} \mathrm{val}_w(\ell) \\
&= \ \frac{n}{q_w - 1} - \frac{\sum d_i(k) + \sum d_i(n-k-1) + 1}{q_w - 1} \ .
\end{aligned}$$

In particular,

$$0 < \ \sum d_i(k) + \sum d_i(n-k-1) + 1 \ < \ 2(q_w - 1)(\log_w(n) + 1) + 1$$

and hence

$$\frac{n}{q_w - 1} - 2\log_w(n) - 3 \ < \ \mathrm{ord}_w\left(S'_{n,w}(\psi_w(k))\right) \ < \ \frac{n}{q_w-1} \ . \tag{3.60}$$

Since $\mathrm{ord}_w(x) = e_w\,\mathrm{ord}_v(x)$, this is equivalent to (3.57).

For (C), fix x and let $0 \le J < n$ be an index for which $|x - \psi_w(J)|_v$ is minimal. For each $k \neq J$, we claim that $|x - \psi_w(k)|_v \ge |\psi_w(J) - \psi_w(k)|_v$. Suppose to the contrary that $|x - \psi_w(k)|_v < |\psi_w(J) - \psi_w(k)|_v$. By the ultrametric inequality,

$$\begin{aligned} |x - \psi_w(J)|_v &= \max(|x - \psi_w(k)|_v, |\psi_w(J) - \psi_w(k)|_v) \\ &= |\psi_w(J) - \psi_w(k)|_v \ > \ |x - \psi_w(k)|_v \ , \end{aligned}$$

contradicting our choice of J. Hence $|x - \psi_w(k)|_v \ge |\psi_w(J) - \psi_w(k)|_v$.

It follows that $\mathrm{ord}_v(S_{n,w}(x)) \le \mathrm{ord}_v(x - \psi_w(J)) + \mathrm{ord}_v(S'_{n,w}(\psi_w(J)))$. Thus the first assertion in (C) is a consequence of (B). The second is trivial, since if $x \notin D(0,1)$ then $|x - \psi_w(k)|_v = |x|_v$ for all k. □

COROLLARY 3.41. *Let F_w be a finite, separable extension of K_v in $\mathbb{C}_v$, and let $S_{n,w}(z)$ be the Stirling polynomial of degree n for $\mathcal{O}_w$. Given a radius R satisfying*

$$0 \ < \ R \ \le \ q_v^{-n/(e_w(q_v^{f_w}-1))} \cdot n^{-1/[F_w:K_v]} \ ,$$

put $\rho_k = R/|S'_{n,w}(\psi_w(k))|_v$ for $k = 0, \ldots, n-1$. Then

$$S_{n,w}^{-1}(D(0,R)) \ = \ \bigcup_{k=0}^{n-1} D(\psi_w(k), \rho_k) \ \subset \ D(0,1) \ ,$$

where the discs $D(\psi_w(k), \rho_k)$ are pairwise disjoint. For each k, $S_{n,w}(z)$ induces an F_w-rational scaled isometry from $D(\psi_w(k), \rho_k)$ onto $D(0,R)$. In particular, $S_{n,w}\big(F_w \cap D(\psi_w(k), \rho_k)\big) = F_w \cap D(0,R)$ and

$$S_{n,w}^{-1}\big(F_w \cap D(0,R)\big) \ = \ \bigcup_{k=0}^{n-1} \big(F_w \cap D(\psi_w(k), \rho_k)\big) \ . \tag{3.61}$$

Moreover, if $R \in |F_w^\times|_v$ then $\rho_k \in |F_w^\times|_v$ for each k.

PROOF. Note that $\log_v(n^{-1/[F_w:K_v]}) = -\log_v(n)/(e_w f_w) = -\log_w(n)/e_w$. By the definition of ρ_k, our assumption on R, and Proposition 3.40(A), for each k

$$\begin{aligned} -\log_v(\rho_k) &= -\log_v(R) - \mathrm{ord}_v(S'_{n,w}(\psi_w(k))) \\ &> \frac{1}{e_w}\Big(\frac{n}{q_w - 1} + \log_w(n)\Big) - \frac{1}{e_w}\frac{n}{q_w-1} \ = \ \frac{1}{e_w}\log_w(n) \ . \end{aligned} \tag{3.62}$$

On the other hand by (3.55), for all $\ell \neq k$

$$\mathrm{ord}_v(\psi_w(k) - \psi_w(\ell)) \ = \ \frac{1}{e_w}\mathrm{ord}_w(\psi_w(k) - \psi_w(\ell)) \ < \ \frac{1}{e_w}\log_w(n) \ .$$

Thus the discs $D(\psi_w(k), \rho_k)$ are pairwise disjoint and contained in $D(0,1)$.

Fix k and expand $S_{n,w}(x)$ about $\psi_w(k)$ as

$$S_{n,w}(z) \ = \ \sum_{\ell=1}^{N} b_\ell (z - \psi_w(k))^\ell$$

where $b_1 = S'_{n,w}(\psi_w(k))$.

The definition of ρ_k shows that $|b_1|_v \cdot \rho_k = R$. By Proposition 3.38, $S_{n,w}(z)$ induces an F_w-rational scaled isometry from $D(\psi_w(k), \rho_k)$ onto $D(0,R)$, which takes $F_w \cap D(\psi_w(k), \rho_k)$ onto $F_w \cap D(0,R)$.

Now let k vary. As noted above, the discs $D(\psi_w(k), \rho_k)$ are pairwise disjoint. Since $S_{n,w}(z)$ has degree n, for each $x \in D(0, R)$ the solutions to $S_{n,w}(z) = x$ in $\bigcup_{k=0}^{n-1} D(\psi_w(k), \rho_k)$ account for all the solutions in $\mathbb{C}_v$. Hence

$$S_{n,w}^{-1}\big(D(0,R)\big) \;=\; \bigcup_{k=0}^{n-1} D(\psi_w(k), \rho_k)\ .$$

Similar considerations show that

$$S_{n,w}^{-1}(F_w \cap D(0,R)) \;=\; \bigcup_{k=0}^{n-1} \big(F_w \cap D(\psi_w(k), \rho_k)\big)\ .$$

Finally, if $R \in |F_w^{\times}|_v$, then $\rho_k = R/|S'_{n,w}(\psi_w(k))|_v \in |F_w^{\times}|_v$ for each k. $\square$

CHAPTER 4

Reductions

In this chapter we will formulate a simplified version of Theorem 0.4, which is the form of the theorem we will spend most of this book proving. After stating this theorem (Theorem 4.2), we will use it to deduce Theorem 0.4, Corollary 0.5, and the variants given in Chapter 1.

DEFINITION 4.1. Let v be a place of K. A set $E_v \subset \mathcal{C}_v(\mathbb{C}_v)$ will be called K_v-*simple* if it is stable under $\mathrm{Aut}_c(\mathbb{C}_v/K_v)$ and is a union of finitely many pairwise disjoint, nonempty compact sets $E_{v,1}, \ldots, E_{v,D}$ such that:

(A) If $K_v \cong \mathbb{C}$, then each $E_{v,\ell}$ is simply connected, has a piecewise smooth boundary, and is the closure of its $\mathcal{C}_v(\mathbb{C})$-interior.

(B) If $K_v \cong \mathbb{R}$, then each $E_{v,\ell}$ is either

(1) disjoint from $\mathcal{C}_v(\mathbb{R})$, simply connected, has a piecewise smooth boundary, and is the closure of its $\mathcal{C}_v(\mathbb{C})$-interior, or

(2) is a closed subinterval of $\mathcal{C}_v(\mathbb{R})$ with nonempty interior.

(C) If K_v is nonarchimedean, then

(1) there are finite separable extensions $F_{w_1}, \ldots, F_{w_n}$ of K_v contained in $\mathbb{C}_v$, and pairwise disjoint isometrically parametrizable balls $B(a_1, r_1), \ldots, B(a_M, r_M)$, such that $E_{v,\ell} = \mathcal{C}_v(F_{w_\ell}) \cap B(a_\ell, r_\ell)$ for $\ell = 1, \ldots, M$; and

(2) the set of balls $\{B(a_1, r_1), \ldots, B(a_M, r_M)\}$ is stable under $\mathrm{Aut}_c(\mathbb{C}_v/K_v)$, and as σ ranges over $\mathrm{Aut}_c(\mathbb{C}_v/K_v)$, each ball $B(a_\ell, r_\ell)$ has $[F_{w_\ell} : K_v]$ distinct conjugates. For each σ, if $\sigma(B(a_\ell, r_\ell)) = B(a_j, r_j)$, then $\sigma(F_{w_\ell}) = F_{w_j}$ and $\sigma(E_{v,\ell}) = E_{v,j}$.

If E_v is K_v-simple, we will call a decomposition $E_v = \bigcup_{\ell=1}^{n} E_{v,\ell}$ of the type in Definition 4.1 a K_v-simple decomposition. If v is archimedean, a K_v-simple set has a unique K_v-simple decomposition. If v is nonarchimedean, a K_v-simple decomposition can always be refined to another K_v-simple decomposition with smaller balls and more sets.

THEOREM 4.2 (FSZ with LR for K_v-simple sets). *Let K be a global field, and let $\mathcal{C}/K$ be a smooth, geometrically integral, projective curve. Let $\mathfrak{X} = \{x_1, \ldots, x_m\} \subset \mathcal{C}(\widetilde{K})$ be a finite set of points stable under $\mathrm{Aut}(\widetilde{K}/K)$, and let $\mathbb{E} = \prod_v E_v \subset \prod_v \mathcal{C}_v(\mathbb{C}_v)$ be a K-rational adelic set compatible with $\mathfrak{X}$. Let $S \subset \mathcal{M}_K$ be a finite set of places containing all archimedean v and all nonarchimedean v such that E_v is not $\mathfrak{X}$-trivial.*

Assume that:

(A) For each $v \in S$, E_v is K_v-simple.

(B) $\gamma(\mathbb{E}, \mathfrak{X}) > 1$.

Then there are infinitely many points $\alpha \in \mathcal{C}(\widetilde{K}^{\mathrm{sep}})$ such that for each $v \in \mathcal{M}_K$, the $\mathrm{Aut}(\widetilde{K}/K)$-conjugates of α all belong to E_v.

We will now prove Theorems 0.4, 1.2, 1.3, 1.4, 1.5, 1.6, 1.7 and Corollary 0.5, assuming Theorem 4.2. For the convenience of the reader, we restate each theorem before proving it.

Theorem 0.4 (FSZ with LR for sets with a finite K_v-primitive cover)**.** *Let K be a global field, and let $\mathcal{C}/K$ be a smooth, geometrically integral, projective curve. Let $\mathfrak{X} = \{x_1, \ldots, x_m\} \subset \mathcal{C}(\widetilde{K})$ be a finite set of points stable under* $\mathrm{Aut}(\widetilde{K}/K)$*, and let $\mathbb{E} = \prod_v E_v \subset \prod_v \mathcal{C}_v(\mathbb{C}_v)$ be a K-rational adelic set compatible with $\mathfrak{X}$. (Thus, each E_v is bounded away from $\mathfrak{X}$ and stable under* $\mathrm{Aut}_c(\mathbb{C}_v/K_v)$*, and E_v is $\mathfrak{X}$-trivial for all but finitely many v.) Let $S \subset \mathcal{M}_K$ be a finite set of places containing all archimedean v and all nonarchimedean v such that E_v is not $\mathfrak{X}$-trivial. Assume that:*

(A) For each $v \in S$, E_v has a finite K_v-primitive cover.

(B) $\gamma(\mathbb{E}, \mathfrak{X}) > 1$.

Then there are infinitely many points $\alpha \in \mathcal{C}(\widetilde{K}^{\mathrm{sep}})$ such that for each $v \in \mathcal{M}_K$, the $\mathrm{Aut}(\widetilde{K}/K)$*-conjugates of α all belong to E_v.*

Proof of Theorem 0.4, assuming Theorem 4.2. The idea is to reduce Theorem 0.4 to the case where the E_v for $v \in S$ are K_v-simple. Note that for E_v to be K_v-simple is much stronger than for it to have a finite K_v-primitive cover (see Definition 0.1). If E_v is K_v-simple, the sets in the decomposition are compact and pairwise disjoint; when $K_v \cong \mathbb{R}$ they are either contained in $\mathcal{C}_v(\mathbb{R})$ or disjoint from it; and when K_v is nonarchimedean, the separable extensions $F_{w,\ell}/K_v$ are finite and the coset space $\mathrm{Aut}(\mathbb{C}_v/K_v)/\mathrm{Aut}(\mathbb{C}_v/F_{w,\ell})$ acts freely on the balls associated to $F_{w,\ell}$.

Assume Theorem 4.2 holds, and let $\mathbb{E} = \prod_v E_v \subset \prod_v \mathcal{C}_v(\mathbb{C}_v)$ be an adelic set compatible with $\mathfrak{X}$ which satisfies the hypotheses of Theorem 0.4. We will construct a new K-rational adelic set $\mathbb{E}' = \prod_v E'_v \subset \mathbb{E}$ compatible with $\mathfrak{X}$, which satisfies hypotheses of Theorem 4.2. We can then apply Theorem 4.2 to produce infinitely many points $\alpha \in \mathcal{C}(\widetilde{K}^{\mathrm{sep}})$ whose $\mathrm{Aut}(\widetilde{K}/K)$ conjugates belong to E'_v, hence E_v, for each $v \in \mathcal{M}_K$.

By hypothesis, $\gamma(\mathbb{E}, \mathfrak{X}) > 1$. Let Γ range over all symmetric matrices in $M_m(\mathbb{R})$. By (3.50) the value of Γ as a matrix game is a continuous function of its entries, so there is an $\varepsilon > 0$ such that for any Γ whose entries satisfy $|\Gamma(\mathbb{E}, \mathfrak{X})_{ij} - \Gamma_{ij}| < \varepsilon$ for all i, j, we have $\mathrm{val}(\Gamma) < 0$. Choose numbers $\varepsilon_v > 0$ for $v \in S$ such that $\sum_{v \in S} \varepsilon_v \log(q_v) < \varepsilon$. In constructing the sets E'_v for $v \in S$, in order to assure that $\gamma(\mathbb{E}', \mathfrak{X}) > 1$ it suffices to have

$$(4.1) \qquad \begin{cases} |G(x_i, x_j; E'_v) - G(x_i, x_j; E_v)| < \varepsilon_v & \text{for all } i \neq j\,, \\ |V_{x_i}(E'_v) - V_{x_i}(E_v)| < \varepsilon_v & \text{for each } i\,. \end{cases}$$

For each $v \notin S$, put $E'_v = E_v$. Now suppose $v \in S$; fix a finite K_v-primitive cover $E_v = \bigcup_{\ell=1}^{M_v} E_{v,\ell}$.

Case 1. If $K_v \cong \mathbb{C}$, then $\mathcal{C}_v(\mathbb{C})$ is a Riemann surface. Fix a triangulation $\mathcal{T}$ of $\mathcal{C}_v(\mathbb{C})$. Without loss we can assume that each edge of the triangulation is a smooth arc. For each $\delta > 0$, let $\mathcal{T}_\delta$ be a refinement of $\mathcal{T}$ such that each edge of $\mathcal{T}_\delta$ is a smooth arc and each triangle in $\mathcal{T}_\delta$ has diameter less than δ under the spherical distance $\|x, y\|_v$.

By assumption E_v is compact, connected, and bounded by finitely many Jordan curves. In particular, each point of ∂E_v is analytically accessible from the interior

E_v^0 of E_v. By Proposition 3.30, this means that $\overline{G}(z, x_i; E_v^0) = G(z, x_i; E_v)$ for each $\zeta \notin E_v$.

For each $\delta > 0$, let X_δ be the set of closed triangles in $\mathcal{T}_\delta$ which are contained in E_v^0, and let $E_{v,\delta}$ be the union of the triangles in X_δ. As $\delta \to 0$, the sets $E_{v,\delta}$ exhaust E_v^0. Hence there is some $\delta = \delta_0$ such that

$$\begin{cases} |G(x_i, x_j; E_{v,\delta_0}) - G(x_i, x_j; E_v)| < \varepsilon_v/2 & \text{for all } i \neq j \;, \\ |V_{x_i}(E_{v,\delta_0}) - V_{x_i}(E_v)| < \varepsilon_v/2 & \text{for each } i \,. \end{cases}$$

Let $U_v' \subset E_{v,\delta_0}$ be the union of the interiors of the triangles in X_{δ_0}.

Each point of E_{v,δ_0} is analytically accessible from U_v' in the sense of Definition 3.29. By exhausting the interior of each triangle in X_{δ_0} by an increasing sequence of closed subtriangles and applying Proposition 3.30 again, we can find a compact set $E_v' \subset U_v'$ which is a finite union of closed triangles, one contained in each connected component of U_v', such that

$$\begin{cases} |G(x_i, x_j; E_{v,\delta}) - G(x_i, x_j; E_v')| < \varepsilon_v/2 & \text{for all } i \neq j \;, \\ |V_{x_i}(E_{v,\delta}) - V_{x_i}(E_v')| < \varepsilon_v/2 & \text{for each } i \,. \end{cases}$$

Thus, $E_v' \subset E_v$, and E_v' satisfies (4.1). Moreover E_v' is compact and has finitely many connected components, each of which is simply connected, has a piecewise smooth boundary and is the closure of its interior. Thus it is $\mathbb{C}$-simple.

Case 2. If $K_v \cong \mathbb{R}$, again choose a triangulation $\mathcal{T}$ of $\mathcal{C}_v(\mathbb{C})$. After making adjustments to $\mathcal{T}$, if necessary, we can assume that $\mathcal{T}$ is stable under complex conjugation, that each edge of $\mathcal{T}$ is a smooth arc, and that $\mathcal{C}_v(\mathbb{R})$ is contained in the union of the edges of $\mathcal{T}$. For each $\delta > 0$, let $\mathcal{T}_\delta$ be a refinement of $\mathcal{T}$ with the properties in Case 1, such that each triangle in $\mathcal{T}_\delta$ has diameter less than δ under the spherical distance $\|x, y\|_v$.

By assumption, E_v is stable under complex conjugation and each $E_{v,\ell}$ is either compact, connected, and bounded by finitely many Jordan curves, or is a closed subinterval of $\mathcal{C}_v(\mathbb{R})$ with nonempty interior.

Let $E_{v,\mathbb{C}}$ be the union of the sets $E_{v,\ell}$ which are closures of their $\mathcal{C}_v(\mathbb{C})$-interiors, and let $E_{v,\mathbb{R}} = E_v \backslash E_{v,\mathbb{C}}$. Then $E_{v,\mathbb{C}}$ stable under complex conjugation and is the closure of its $\mathcal{C}_v(\mathbb{C})$-interior, and $E_{v,\mathbb{R}}$ is contained in $\mathcal{C}_v(\mathbb{R})$. Since $E_{v,\mathbb{C}}$ is closed, no component of $E_{v,\mathbb{R}}$ is reduced to a point.

We now apply a modification of the argument from Case 1. Let $E_{v,\mathbb{C}}^0$ be the $\mathcal{C}_v(\mathbb{C})$-interior of $E_{v,\mathbb{C}}$, and let $E_{v,\mathbb{C}}^{00} = E_{v,\mathbb{C}}^0 \backslash \mathcal{C}_v(\mathbb{R})$. Let $E_{v,\mathbb{R}}^0$ be the $\mathcal{C}_v(\mathbb{R})$-interior of $E_{v,\mathbb{R}}$, and put $E_v^1 = E_{v,\mathbb{C}}^{00} \cup E_{v,\mathbb{R}}^0$. Each point of $E_{v,\mathbb{C}}$ is analytically accessible from $E_{v,\mathbb{C}}^{00}$ and each point of $E_{v,\mathbb{R}}$ is analytically accessible from $E_{v,\mathbb{R}}^0$, so each point of E_v is analytically accessible from E_v^1. By Proposition 3.30, $\overline{G}(z, x_i; E_v^1) = G(z, x_i; E_v)$ for each $\zeta \notin E_v$.

For each $\delta > 0$, let $E_{v,\delta}$ be the union of the triangles in $\mathcal{T}_\delta$ contained in $E_{v,\mathbb{C}}^{00}$, together with the edges of $\mathcal{T}_\delta$ which are contained in $E_{v,\mathbb{R}}^0$. Each $E_{v,\delta}$ is compact and stable under complex conjugation. Furthermore, each compact subset of E_v^1 is contained in some $E_{v,\delta}$, so there is a $\delta_0 > 0$ such that

$$\begin{cases} |G(x_i, x_j; E_{v,\delta_0}) - G(x_i, x_j; E_v)| < \varepsilon_v/2 & \text{for all } i \neq j \;, \\ |V_{x_i}(E_{v,\delta_0}) - V_{x_i}(E_v)| < \varepsilon_v/2 & \text{for each } i \,. \end{cases}$$

Let U_v' be the union of the interiors of the triangles in $\mathcal{T}_{\delta_0}$ which are contained in E_{v,δ_0}, together the (real) interiors of the edges of $\mathcal{T}_{\delta_0}$ which are contained in $E_{v,\mathbb{R}}^0$.

Then U'_v is stable under complex conjugation, and is the disjoint union of finitely many open triangles in $\mathcal{C}_v(\mathbb{C})$ whose closures are disjoint from $\mathcal{C}_v(\mathbb{R})$, together with finitely many open segments in $\mathcal{C}_v(\mathbb{R})$.

Each point of E_{v,δ_0} is analytically accessible from U'_v. By exhausting the interior of each open triangle in U'_v by an increasing sequence of closed subtriangles and each open segment by an increasing sequence of closed subintervals, and applying Proposition 3.30 again, we can find a compact set $E'_v \subset U'_v$ which is a finite disjoint union of closed triangles and closed subintervals of $\mathcal{C}_v(\mathbb{R})$, one contained in each component of U'_v, such that

$$\begin{cases} |G(x_i, x_j; E_{v,\delta}) - G(x_i, x_j; E'_v)| < \varepsilon_v/2 & \text{for all } i \neq j \ , \\ |V_{x_i}(E_{v,\delta}) - V_{x_i}(E'_v)| < \varepsilon_v/2 & \text{for each } i\,. \end{cases}$$

By choosing the closed subtriangles appropriately, we can also arrange that E'_v is stable under complex conjugation.

Thus, E'_v is compact, stable under complex conjugation, $\mathbb{R}$-simple, contained in E_v, and satisfies (4.1).

Case 3. Suppose K_v is nonarchimedean. By assumption, $E_v = \bigcup_{\ell=1}^M E_{v,\ell}$ where each $E_{v,\ell}$ is an RL-domain, a ball $B(a_\ell, r_\ell)$ with $r_\ell \in |\mathbb{C}_v^\times|_v$, or a compact set of the form $B(a_\ell, r_\ell) \cap D_v$ where D_v is an RL-domain or a ball, and $F_{w,\ell}/K_v$ is finite and separable.

We begin by simplifying the form of the finite K_v-primitive cover. Each set of the form $E_{v,\ell} = B(a_\ell, r_\ell)$ is defined using the spherical metric $\|x, y\|_v$ attached to some projective embedding $\mathcal{C} \hookrightarrow \mathbb{P}^N$. Since $E_v \neq \mathcal{C}_v(\mathbb{C}_v)$ (note that $E_v \cap \mathfrak{X} = \phi$), such a ball must have radius $r_\ell < 1$. Thus it is contained in an affine patch $\mathbb{A}^N \subset \mathbb{P}^N$, and if $z_1, \ldots, z_N$ are the coordinates functions that patch, by (3.12) we have

$$B(a_\ell, r_\ell) = \{x \in \mathcal{C}_v(\mathbb{C}_v) : |z_i(x) - z_i(a_\ell)|_v \leq r_\ell\}.$$

The restriction of z_i to $\mathcal{C}_v(\mathbb{C}_v)$ is a nonconstant rational function $z_i(x)$, so $B(a_\ell, r_\ell)$ is the intersection of the RL-domains $\{x \in \mathcal{C}_v(\mathbb{C}_v) : |z_i(x)|_v \leq 1\}$. By ([**51**], Corollary 4.2.14) a finite intersection of RL-domains is again an RL-domain, so we can assume that the finite K_v-primitive cover of E_v contains only RL-domains and compact sets. Since a finite union of RL-domains is an RL-domain ([**51**], Theorem 4.2.15), we can assume that there is at most one RL domain in the decomposition.

By Theorem 3.9, each sufficiently small ball is isometrically parametrizable. Thus each $E_{v,\ell}$ of the form $\mathcal{C}_v(F_{w,\ell}) \cap D_v$, for an RL-domain or ball D_v, can be covered by finitely many isometrically parametrizable balls $B(a_{\ell,j}, r_{\ell,j})$ with radii in $|\mathbb{C}_v^\times|_v$. Hence without loss, we can assume that the compact sets in the finite K_v-primitive cover of E_v have the form $\mathcal{C}_v(F_{w,\ell}) \cap B(a_\ell, r_\ell)$, where $B(a_\ell, r_\ell)$ is isometrically parametrizable, and $r_\ell \in |\mathbb{C}_v^\times|_v$.

Our next goal is to reduce to the case where there are no RL-domains in the finite K_v-primitive cover of E_v. Suppose there were an RL-domain U_v. Let $E_v^{(1)}$ be the union of the compact sets $B(a_\ell, r_\ell) \cap \mathcal{C}_v(F_{w,\ell})$ in the decomposition. Both U_v and $E_v^{(1)}$ are stable under $\mathrm{Aut}_c(\mathbb{C}_v/K_v)$. In ([**51**], Theorem 4.3.11) it is shown that the union of an RL-domain and a compact set is algebraically capacitable. In fact, the proof of that theorem shows there is a compact set $E_v^{(2)} \subset U_v$, which itself is a finite union of compact sets of the form $B(a'_\ell, r'_\ell) \cap \mathcal{C}_v(F'_{w,\ell})$, with $B(a'_\ell, r'_\ell)$

isometrically parametrizable and $F'_{w,\ell}/K_v$ finite, such that

$$\begin{cases} |G(x_i, x_j; E_v^{(1)} \cup E_v^{(2)}) - G(x_i, x_j; E_v)| \ < \ \varepsilon_v/2 & \text{for all } i \neq j \ , \\ |V_{x_i}(E_v^{(1)} \cup E_v^{(2)}) - V_{x_i}(E_v)| \ < \ \varepsilon_v/2 & \text{for each } i \, . \end{cases}$$

The centers of the balls $B(a'_\ell, r'_\ell)$ can be required to belong to $\mathcal{C}_v(\widetilde{K}_v^{\text{sep}})$, since $\widetilde{K}_v^{\text{sep}}$ is dense in $\mathbb{C}_v$, and the extensions $F_{w,\ell}$ can be required to be separable extensions of K_v, since all that is needed for the proof of ([**51**], Theorem 4.3.11) is that the residue degree or the ramification index of $F_{w,\ell}/K_v$ can be taken arbitrarily large. The set $E_v^{(2)}$ need not be stable under $\text{Aut}_c(\mathbb{C}_v/K_v)$, but by its form it has only finitely many conjugates, and each of these is also contained in U_v. By replacing $E_v^{(2)}$ with the union of its conjugates, and using the monotonicity of the Green's functions (Lemma 3.21), we can arrange that $E_v^{(1)} \cup E_v^{(2)}$ is stable under $\text{Aut}_c(\mathbb{C}_v/K_v)$.

By replacing E_v with $E_v^{(1)} \cup E_v^{(2)}$ we can assume that there are no RL-domains in the decomposition of E_v. Write $E_v = \bigcup_{\ell=1}^M B(a_\ell, r_\ell) \cap \mathcal{C}_v(F_{w,\ell})$, and fix an $r \in |K_v^\times|_v$ with $0 < r \leq \min_\ell(r_\ell)$. After replacing each $B(a_\ell, r_\ell) \cap \mathcal{C}_v(F_{w,\ell})$ with finitely many sets $B(a_{\ell,j}, r) \cap \mathcal{C}_v(F_{w,\ell})$, we can assume that all the r_ℓ are equal to r, and that r belongs to the value group of $K_v^\times$.

Our next goal is to arrange that balls $B(a_\ell, r)$ are disjoint. Consider the sets $E_{v,\ell} = B(a_\ell, r) \cap \mathcal{C}_v(F_{w,\ell})$. Without loss, we can assume that none of the $E_{v,\ell}$ is properly contained in another. However, it is possible that the same ball $B(a_\ell, r)$ occurs with several different fields F_{w,ℓ_j}; given such a ball, let $\ell_1 = \ell, \ell_2, \ldots, \ell_t$ be the indices for which $B(a_{\ell_j}, r) = B(a_\ell, r)$. For each $j \geq 2$, the field $F_{w,\ell} \cap F_{w,\ell_j}$ is a proper subfield of $F_{w,\ell}$, and by a simple argument involving K_v-vector spaces, there is an element $u_\ell \in F_{w,\ell}$ which does not belong to $F_{w,\ell} \cap F_{w,\ell_j}$ for any $j \geq 2$. Let $0 \neq \pi_v \in K_v$ be such that $|\pi_v|_v < 1$. After replacing u_i by $1 + \pi_v^N u_\ell$ for sufficiently large N, we can assume that $u_\ell \in \mathcal{O}_{w_\ell}^\times$.

For each ℓ, put

$$X_\ell \ = \ E_{v,\ell} \backslash \Big(\bigcup_{\ell_j \neq \ell} E_{v,\ell_j} \Big) \ ,$$

and let $X = \bigcup_{\ell=1}^M X_\ell$. Note that X is stable under $\text{Aut}_c(\mathbb{C}_v/K_v)$.

We claim that each point of E_v is analytically accessible from X in the sense of Definition 3.29. Clearly each $b \in X$ is analytically accessible from X, since if $b \in X_\ell$ then there is a ball $B(b,s)$ for which $B(b,s) \cap \mathcal{C}_v(F_{w,\ell}) \subseteq X_\ell$. Suppose that $b \in E_{v,j} \cap E_{v,\ell}$ for some $j \neq \ell$. Then $b \in B(a_j, r) \cap \mathcal{C}_v(F_{w,j}) \cap \mathcal{C}_v(F_{w,\ell})$, so $b \in \mathcal{C}_v(F_{w,j} \cap F_{w,\ell})$. Let F_b be the field $K_v(b)$. By Theorem 3.9 there is an F_b-rational isometric parametrization $f_b : D(0,r) \to B(a_\ell, r)$ with $f_b(0) = b$. Consider the image of $D(0,r) \cap (\mathcal{O}_v \backslash \{0\})$ under the isometric parametrization $\widetilde{f}_b(z) := f_b(u_\ell z)$: we have

$$\widetilde{f}_b(D(0,r) \cap (\mathcal{O}_v \backslash \{0\})) \ = \ f_b(u_\ell \cdot (D(0,r) \cap (\mathcal{O}_v \backslash \{0\})) \, .$$

Since f_b is F_b-rational, for each complete field H with $F_b \subseteq H \subseteq \mathbb{C}_v$, f_b induces a $1-1$ correspondence between points of $D(0,r) \cap H$ and $B(a_i, r) \cap \mathcal{C}_v(H)$. By our choice of u_ℓ, we have $u_\ell \cdot (\mathcal{O}_v \backslash \{0\}) \subset \mathcal{O}_{w_\ell} \backslash (\bigcup_{j=1}^t F_{w,\ell_j})$. It follows that $\widetilde{f}_b(D(0,r) \cap (\mathcal{O}_v \backslash \{0\})) \subset X_\ell$. This establishes our claim.

By Proposition 3.30, $G(z,\zeta;E_v) = \overline{G}(z,\zeta;X)$ for each $\zeta \notin E_v$. Hence there is a compact set $Y \subset X$ such that

$$\begin{cases} |G(x_i,x_j;Y) - G(x_i,x_j;E_v)| \ < \ \varepsilon_v/2 & \text{for all } i \neq j \ , \\ |V_{x_i}(Y) - V_{x_i}(E_v)| \ < \ \varepsilon_v/2 & \text{for each } i \, . \end{cases}$$

The set Y thus constructed may not be stable under $\mathrm{Aut}_c(\mathbb{C}_v/K_v)$, but it only has finitely many conjugates and each of them is contained in X. By replacing Y with the union of its conjugates, and using the monotonicity of Green's functions, we can assume Y is stable under $\mathrm{Aut}_c(\mathbb{C}_v/K_v)$.

For each $\ell = 1,\ldots,M$, put $Y_\ell = Y \cap B(a_\ell,r) \cap \mathcal{C}_v(F_{w,\ell})$. Then Y_ℓ is compact and $Y_\ell \subset X_i$. Since the Y_ℓ are compact and pairwise disjoint, there is a number $0 < R \in |K_v^\times|_v$ such that

$$\min_{\ell \neq j} \ \min_{z \in Y_\ell, w \in Y_j} \|z,w\|_v \ > \ R \, .$$

Cover Y with finitely many balls $B(b_j,R)$, $j = 1,\ldots,N$, for points $b_j \in Y$.

We can now construct the K_v-simple set E_v'. For each $j = 1,\ldots,N$, if $b_j \in Y_\ell$, let $F_{w,\ell}' = F_{w,\ell}$, and put

$$E_v' \ = \ \bigcup_{j=1}^{N} B(b_j,R) \cap \mathcal{C}_v(F_{w,j}') \ .$$

Since $\mathrm{Aut}_c(\mathbb{C}_v/K_v)$ preserves the spherical distance, and stabilizes both X and Y, it follows that E_v' is stable under $\mathrm{Aut}_c(\mathbb{C}_v/K_v)$. By construction it is K_v-simple, stable under $\mathrm{Aut}_c(\mathbb{C}_v/K_v)$, contained in E_v, and satisfies (4.1). □

We next consider Corollary 0.5. Suppose $\mathcal{A}/K$ is an affine curve, embedded in $\mathbb{A}^N$. Recall that if $z_1,\ldots,z_N$ are coordinate functions on $\mathbb{A}^N$, then for each $v \in \mathcal{M}_K$ and each $P \in \mathcal{A}(\mathbb{C}_v)$, we write $\|P\|_{\mathcal{A},v} = \max(|z_1(P)|_v,\ldots,|z_N(P)|_v)$. A set $E_v \subset \mathcal{A}(\mathbb{C}_v)$ has a finite K_v-primitive cover relative to $\mathcal{A}$ if it is bounded and satisfies the conditions of Definition 0.1, using balls for the metric $\|x-y\|_{\mathcal{A},v}$.

Corollary 0.5 (FSZ for Skolem Problems on Affine Curves). *Let K be a global field, and let $\mathcal{A}/K$ be a geometrically integral (possibly singular) affine curve, embedded in $\mathbb{A}^N$. Fix a place v_0 of K, and let $S \subset \mathcal{M}_K \backslash \{v_0\}$ be a finite set of places containing all archimedean $v \neq v_0$. For each $v \in S$, let $E_v \subset \mathcal{A}_v(\mathbb{C}_v)$ be nonempty and stable under $\mathrm{Aut}_c(\mathbb{C}_v/K_v)$, with a finite K_v-primitive cover relative to $\mathcal{A}$. Assume that for each $v \in \mathcal{M}_K \backslash (S \cup \{v_0\})$ there is a point $P \in \mathcal{A}(\mathbb{C}_v)$ with $\|P\|_{\mathcal{A},v} \le 1$. Then there is a bound $R = R(\mathcal{A},\{E_v\}_{v\in S},v_0) < \infty$ such that there are infinitely many points $\alpha \in \mathcal{A}(K^{\mathrm{sep}})$ for which:*

(1) *for each $v \in S$, all the $\mathrm{Aut}(\widetilde{K}/K)$-conjugates $\sigma(\alpha)$ belong to E_v;*
(2) *for each $v \in \mathcal{M}_K \backslash (S \cup \{v_0\})$, all the conjugates satisfy $\|\sigma(\alpha)\|_{\mathcal{A},v} \le 1$;*
(3) *for $v = v_0$, all the conjugates satisfy $\|\sigma(\alpha)\|_{\mathcal{A},v_0} \le R$.*

Proof of Corollary 0.5, assuming Theorem 0.4. Let $\overline{\mathcal{A}}$ be the projective closure of $\mathcal{A}$, and let $\mathcal{C}/K$ be a desingularization of $\overline{\mathcal{A}}$. Then $\mathcal{C}$ is a smooth, geometrically integral, projective curve birational to $\overline{\mathcal{A}}$. Let $\pi : \mathcal{C} \to \overline{\mathcal{A}}$ be the natural morphism. Let $X = \overline{\mathcal{A}}(\widetilde{K}) \backslash \mathcal{A}(\widetilde{K})$ be the set of points at infinity for $\mathcal{A}$, and put $\mathfrak{X} = \pi^{-1}(X)$. Let Z be the set of singular points of $\overline{\mathcal{A}}$, and put $\mathcal{Z} = \pi^{-1}(Z)$. Then $\mathfrak{X}$, X, $\mathcal{Z}$ and Z are finite and stable under $\mathrm{Aut}(\widetilde{K}/K)$. For each $v \in \mathcal{M}_K$, π induces a surjective, continuous map $\pi_v : \mathcal{C}_v(\mathbb{C}_v) \to \overline{\mathcal{A}}_v(\mathbb{C}_v)$, and for each field

F_w with $K_v \subseteq F_w \subseteq \mathbb{C}_v$, it induces a homeomorphism $\mathcal{C}_v(F_w)\backslash\mathcal{Z} \cong \overline{\mathcal{A}}(F_w)\backslash Z$. For each $v \in \mathcal{M}_K$, and each $0 < r \in \mathbb{R}$, let $B_v(r) = \{x \in \mathcal{A}_v(\mathbb{C}_v) : \|x\|_{\mathcal{A},v} \le r\}$

We will construct a K-rational adelic set $\widetilde{\mathbb{E}} = \prod_v \widetilde{E}_v \subset \prod_v \mathcal{C}_v(\mathbb{C}_v)$ compatible with $\mathfrak{X}$, such that $\pi_v(\widetilde{E}_v) \subset E_v$ for each $v \in S$, $\pi_v(\widetilde{E}_v) = B_v(1)$ for each $v \notin S \cup \{v_0\}$, and such that there is a $0 < R \in \mathbb{R}$ with $\pi_{v_0}(\widetilde{E}_{v_0}) = B_{v_0}(R)$. The set $\widetilde{\mathbb{E}}$ will satisfy the conditions of Theorem 0.4, and the points in $\mathcal{A}(\widetilde{K}^{\mathrm{sep}})$ required by Corollary 0.5 will be the images $\pi(\alpha)$ of the $\alpha \in \mathcal{C}(\widetilde{K}^{\mathrm{sep}})$ produced by Theorem 0.4.

Before constructing $\widetilde{\mathbb{E}}$, we make some observations. Let $z_j(x)$, for $j = 1, \ldots, N$, be the restrictions of the coordinate functions on $\mathbb{A}^N$ to $\mathcal{A}$. Since the function fields $K(\mathcal{A})$ and $K(\mathcal{C})$ are canonically isomorphic, by abuse of notation we can regard the $z_j(x)$ as functions on $\mathcal{C}$. Their poles are supported on $\mathfrak{X}$, and each $x_i \in \mathfrak{X}$ is a pole of some $z_j(x)$. Similarly, for each v, the function fields $\mathbb{C}_v(\overline{\mathcal{A}})$ and $\mathbb{C}_v(\mathcal{C})$ are canonically isomorphic, and we will identify them. Given a nonconstant function $h(x) \in \mathbb{C}_v(\overline{\mathcal{A}})$ and an $r \in |\mathbb{C}_v^\times|_v$, put $D_{\overline{\mathcal{A}}}(h, r) = \{x \in \overline{\mathcal{A}}(\mathbb{C}_v) : |h(x)|_v \le r\}$ and $D_{\mathcal{C}}(h, r) = \{x \in \mathcal{C}(\mathbb{C}_v) : |h(x)|_v \le r\}$. Then $D_{\mathcal{C}}(h, r)$ is an RL-domain and

$$\pi_v^{-1}(D_{\overline{\mathcal{A}}}(h, r)) \;=\; D_{\mathcal{C}}(h, r)\ .$$

When v is nonarchimedean and $r \in |\mathbb{C}_v^\times|_v$, we claim that $\pi_v^{-1}(B_v(r))$ is an RL-domain. To see this, note that

$$B_v(r) \;=\; \{x \in \overline{\mathcal{A}}(\mathbb{C}_v) : |z_j(x)|_v \le r \text{ for } 1 \le j \le N\}\ .$$

It follows that $\pi_v^{-1}(B_v(r)) = \bigcap_{j=1}^N D_{\mathcal{C}}(z_j, r)$. By ([**51**], Corollary 4.2.14) $\pi_v^{-1}(B_v(r))$ is an RL-domain. Likewise, for each $P \in \mathcal{A}_v(\mathbb{C}_v)$, consider the ball

$$\begin{aligned} B_v(P, r) \;&=\; \{x \in \overline{\mathcal{A}}_v(\mathbb{C}_v) : \|x - P\|_{\mathcal{A},v} \le r\} \\ &=\; \{x \in \overline{\mathcal{A}}_v(\mathbb{C}_v) : |z_j(x) - z_j(P)|_v \le r \text{ for } 1 \le j \le N\}\ . \end{aligned}$$

Then when $r \in |\mathbb{C}_v^\times|_v$, $\pi_v^{-1}(B_v(P, r))$ is an RL-domain.

We now will construct the adelic set $\widetilde{\mathbb{E}} = \prod_v \widetilde{E}_v$.

When $v \in S$, E_v has a finite K_v-primitive cover in $\mathcal{A}$, say $E_v = \bigcup_{\ell=1}^M E_{v,\ell}$. In order to avoid problems with the singular set $Z \subset \overline{\mathcal{A}}_v(\mathbb{C}_v)$ in assuring that $\widetilde{E}_v$ has a finite K_v-primitive cover, we will lift a nonsingular subset of E_v to get $\widetilde{E}_v$.

If $K_v \cong \mathbb{C}$, then each $E_{v,\ell}$ is the closure of its $\mathcal{A}_v(\mathbb{C}_v)$-interior. Choose a ball $B_v(P, r)$ contained in $E_{v,1}\backslash Z$, and let $\widetilde{E}_v = \pi_v^{-1}(B_v(P, r))$; then $\widetilde{E}_v$ has finitely many components, each of which is bounded by finitely many analytic arcs. If $K_v \cong \mathbb{R}$, then each $E_{v,\ell}$ is the closure of its $\mathcal{A}_v(\mathbb{C})$-interior or is a closed interval in $\mathcal{A}_v(\mathbb{R})$ with nonempty interior. If there is an ℓ such that $E_{v,\ell}$ is a closed interval in $\mathcal{A}_v(\mathbb{R})$, choose a closed subinterval $[a, b]$ in $E_{v,\ell}\backslash Z$ with nonempty interior, and let $\widetilde{E}_v = \pi_v^{-1}([a, b])$. Otherwise, take a ball $B_v(P, r)$ in $E_{v,1}\backslash Z$, and let $\widetilde{E}_v$ be the union of $\pi_v^{-1}(B_v(P, r))$ and its complex conjugate. Finally, if v is nonarchimedean, then each $E_{v,\ell}$ is either a bounded domain $D_{\overline{\mathcal{A}}}(h_\ell, r)$ for some $h_\ell \in \mathcal{C}_v(\overline{\mathcal{A}})$ and $r_\ell \in |\mathbb{C}_v^\times|_v$, a ball $B_v(P_\ell, r_\ell)$ for some $P_\ell \in \mathcal{A}_v(\mathbb{C}_v)$ and some $r_\ell \in |\mathbb{C}_v^\times|_v$, or is the intersection of a bounded domain or ball $D_{v,\ell}$ with $\mathcal{A}_v(F_{w,\ell})$ for a finite separable extension $F_{w,\ell}/K_v$. If there is an $E_{v,\ell}$ which is a domain $D_{\overline{\mathcal{A}}}(h_\ell, r_\ell)$ or a ball $B_v(P_\ell, r_\ell)$, then $\pi_v^{-1}(E_{v,\ell})$ is an RL-domain. Any RL-domain in $\mathcal{C}_v(\mathbb{C}_v)$ can be defined by a function in $\widetilde{K}_v(\mathcal{C})$ (see the proof of Theorem 4.2.12 in [**51**]); let $\widetilde{E}_v$ be the union of $\pi_v^{-1}(E_{v,\ell})$ and its finitely many $\mathrm{Aut}_c(\mathbb{C}_v/K_v)$-conjugates. Otherwise, fix an ℓ and write $E_{v,\ell} = \mathcal{A}_v(F_{w,\ell}) \cap D_{v,\ell}$. Choose a point $Q_\ell \in E_{v,\ell} \backslash Z$ and a radius $r_{Q,\ell} \in |\mathbb{C}_v^\times|_v$ small enough that $B_v(Q_\ell, r_{Q,\ell}) \subset D_{v,\ell}$ and $B_v(Q_\ell, r_{Q,\ell}) \cap Z = \phi$, and let $\widetilde{E}_v$ be the union of $\pi_v^{-1}(B_v(Q_\ell, r_{Q,\ell})) \cap \mathcal{C}_v(F_{w,\ell})$ and its finitely many $\mathrm{Aut}_c(\mathbb{C}_v/K_v)$-conjugates. Note that in this case $\widetilde{E}_v$ is compact, so $\pi_v^{-1}(B_v(Q_\ell, r_{Q,\ell})) \cap \mathcal{C}_v(F_{w,\ell})$ and each of its conjugates can be covered by finitely many balls in $\mathcal{C}_v(\mathbb{C}_v)$.

When $v \notin (S \cup \{v_0\})$, put $\widetilde{E}_v = \pi_v^{-1}(B_v(1))$. When $v = v_0$, let $0 < R \in |C_{v_0}^\times|_{v_0}$ be a large number, to be specified shortly, and put

$$\widetilde{E}_{v_0} = \widetilde{E}_{v_0}(R) = \pi_{v_0}^{-1}(B_{v_0}(R)) .$$

By construction, each $\widetilde{E}_v$ is K_v-symmetric, bounded away from $\mathfrak{X}$, and has a finite K_v-primitive cover. We claim that $\widetilde{E}_v$ is $\mathfrak{X}$-trivial for all but finitely many v. Indeed, for all but finitely many nonarchimedean v, the model $\mathfrak{C}_v$ and the functions $z_i(x)$ have good reduction $\pmod v$, $\widetilde{E}_v = \pi_v^{-1}(B_v(1))$ is an RL-domain, and the points in $\mathfrak{X}$ specialize to distinct points $\pmod v$. For such v, is the $\mathfrak{X}$-trivial set.

Let

$$\widetilde{\mathbb{E}}(R) = \widetilde{E}_{v_0}(R) \times \prod_{v \neq v_0} \widetilde{E}_v .$$

Then $\widetilde{\mathbb{E}}(R)$ is an adelic set compatible with $\mathfrak{X}$. We claim that for all sufficiently large R, we have $\gamma(\widetilde{\mathbb{E}}(R), \mathfrak{X}) > 1$. Indeed, fix a spherical metric on $\mathcal{C}_{v_0}(\mathbb{C}_{v_0})$, and let $\varepsilon > 0$ be small enough that the balls $B_{v_0}(x_i, \varepsilon)$ for $x_i \in \mathfrak{X}$ are pairwise disjoint. If R is big enough then $\mathcal{C}_{v_0}(\mathbb{C}_{v_0}) \backslash E_{v_0}(R)$ will be contained in $\bigcup_{x_i \in \mathfrak{X}} B_{v_0}(x_i, \varepsilon)$, and in that case the local Green's matrix $\Gamma_{v_0}(E_{v_0}(R), \mathfrak{X})$ will be diagonal. By letting $R \to \infty$ we can make the diagonal entries arbitrarily large and negative. If R is sufficiently large the matrix $\Gamma(\widetilde{\mathbb{E}}(R), \mathfrak{X})$ will be negative definite. Fixing such an R and pushing forward the points produced by Theorem 0.4 yields the corollary. □

Theorem 1.2 (FSZ with LR for Quasi-neighborhoods). *Let K be a global field, and let $\mathcal{C}/K$ be a smooth, connected, projective curve. Let $\mathfrak{X} = \{x_1, \ldots, x_m\} \subset \mathcal{C}(\widetilde{K})$ be a finite set of points stable under $\mathrm{Aut}(\widetilde{K}/K)$, and let $\mathbb{E} = \prod_v E_v \subset \prod_v \mathcal{C}_v(\mathbb{C}_v)$ be a K-rational adelic set compatible with $\mathfrak{X}$, so each E_v is stable under $\mathrm{Aut}_c(\mathbb{C}_v/K_v)$ and all but finitely many E_v are $\mathfrak{X}$-trivial.*

Suppose $\overline{\gamma}(\mathbb{E}, \mathfrak{X}) > 1$. Then for any K-rational separable quasi-neighborhood $\mathbb{U}$ of $\mathbb{E}$, there are infinitely many points $\alpha \in \mathcal{C}(\widetilde{K}^{\mathrm{sep}})$ such that for each $v \in \mathcal{M}_K$, the $\mathrm{Aut}(\widetilde{K}/K)$-conjugates of α all belong to U_v.

Proof of Theorem 1.2, assuming Theorem 0.4. The idea is to adjust the sets E_v within their separable quasi-neighborhoods U_v (see Definition 1.1), and reduce Theorem 1.2 to the case where the E_v satisfy the conditions of Theorem 0.4. In particular, in the nonarchimedean case, we may need to cut down infinite algebraic extensions $F_{w,\ell}/K_v$ in the definition of the quasi-neighborhoods, to ones of finite degree.

Assume Theorem 0.4 holds. Let $\mathbb{E} = \prod_v E_v$ be a K-rational adelic set compatible with $\mathfrak{X}$, and let $\mathbb{U} = \prod_v U_v$ be a K-rational separable quasi-neighborhood of $\mathbb{E}$, for which the hypotheses of Theorem 1.2 apply. We will construct a new adelic set $\mathbb{E}' = \prod_v E_v' \subset \mathbb{U}$ such that the hypotheses of Theorem 0.4 hold for $\mathbb{E}'$. Let $S \subseteq \mathcal{M}_K$ be a finite set of places containing all archimedean places and all nonarchimedean places where E_v is not $\mathfrak{X}$-trivial. For each $v \notin S$, put $E_v' = E_v$.

By hypothesis, $\overline{\gamma}(\mathbb{E},\mathfrak{X}) > 1$. Let Γ range over all symmetric matrices in $M_m(\mathbb{R})$. By (3.50) there is an $\varepsilon > 0$ such that if the entries of Γ satisfy $\Gamma_{ij} < \Gamma(\mathbb{E},\mathfrak{X})_{ij} + \varepsilon$ for all i,j, we have $\mathrm{val}(\Gamma) < 0$. Choose numbers $\varepsilon_v > 0$ for $v \in S$ such that $\sum_{v\in S} \varepsilon_v \log(q_v) < \varepsilon$. In constructing the sets E_v' for $v \in S$, to assure $\gamma(\mathbb{E}',\mathfrak{X}) > 1$ it suffices to arrange that

$$(4.2) \qquad \begin{cases} G(x_i,x_j;E_v') \ < \ \overline{G}(x_i,x_j;E_v) + \varepsilon_v & \text{for all } i \neq j \ , \\ V_{x_i}(E_v') \ < \ \overline{V}_{x_i}(E_v) + \varepsilon_v & \text{for each } i \ . \end{cases}$$

We will now construct the sets E_v' for $v \in S$.

Case 1. If $K_v \cong \mathbb{C}$, then $U_v \subset \mathcal{C}_v(\mathbb{C})$ is open. By Proposition 3.26(4), there is a compact set $H_v \subset E_v$ such that

$$(4.3) \qquad \begin{cases} |G(x_i,x_j;H_v) - \overline{G}(x_i,x_j;E_v)| \ < \ \varepsilon_v & \text{for all } i \neq j \ , \\ |V_{x_i}(H) - \overline{V}_{x_i}(E_v)| \ < \ \varepsilon_v & \text{for each } i \ . \end{cases}$$

For each $a \in H_v$, there is an $r_a > 0$ such that the closed ball $B(a,r_a)$ is contained in U_v. The corresponding open balls $B(a,r_a)^-$ cover H_v. By compactness, finitely many of these balls, say $B(a_1,r_{a_1})^-,\ldots,B(a_n,r_{a_n})^-$ also cover H_v. Let

$$E_v' \ = \ B(a_1,r_{a_1}) \cup \cdots \cup B(a_n,r_{a_n}) \ \subset \ U_v \ .$$

Each $B(a_i,r_{a_i})$ is compact, has a smooth boundary, and is the closure of its interior, so E_v' satisfies the conditions of Theorem 0.4. Since $H_v \subset E_v'$, the monotonicity of Green's functions (Lemma 3.21) shows that 4.2 holds.

Case 2. If $K_v \cong \mathbb{R}$, then the quasi-neighborhood U_v of E_v is the union of a set $U_{v,0}$ open in $\mathcal{C}_v(\mathbb{C})$ and a set $U_{v,1}$ open in $\mathcal{C}_v(\mathbb{R})$. Since U_v is stable under complex conjugation, so is $U_{v,0}$. By Proposition 3.26(4), there is a compact set $H_v \subset E_v$ for which

$$(4.4) \qquad \begin{cases} |G(x_i,x_j;H_v) - \overline{G}(x_i,x_j;E_v)| \ < \ \varepsilon_v/2 & \text{for all } i \neq j \ , \\ |V_{x_i}(H_v) - \overline{V}_{x_i}(E_v)| \ < \ \varepsilon_v/2 & \text{for each } i \ . \end{cases}$$

We will construct the set E_v' in two steps. First, put $H_{v,1} = H_v \backslash U_{v,0}$; it is compact and contained in $U_{v,1}$. Next, for each $r > 0$, put $H_{v,0}(r) = \{x \in H_v : \|x,z\| \geq r \text{ for all } z \in H_{v,1}\}$, and let

$$H_v(r) \ = \ H_{v,0}(r) \cup H_{v,1} \ .$$

Then $H_{v,0}(r)$ is contained in $U_{v,0}$, and $H_{v,0}(r)$ and $H_v(r)$ are compact.

The sets $H_v(r)$ increase as $r \to 0$, and they form an exhaustion of H_v. Choose a sequence $r_1 > r_2 > \cdots > 0$ with $\lim_{m\to\infty} r_m = 0$. By Proposition 3.22, if we take M sufficiently large and put $X_v = H_v(r_M)$, then

$$(4.5) \qquad \begin{cases} |G(x_i,x_j;X_v) - G(x_i,x_j;H_v)| \ < \ \varepsilon_v/2 & \text{for all } i \neq j \ , \\ |V_{x_i}(X_v) - V_{x_i}(H_v)| \ < \ \varepsilon_v/2 & \text{for each } i \ . \end{cases}$$

For each $a \in H_{v,0}(r_M)$, there is an open ball $B(a,r)^-$ whose closure $B(a,r)$ is contained in $U_{v,0}$. Finitely many of these balls, say $B(a_1,r_1)^-,\ldots,B(a_{m_0},r_{m_0})^-$ cover $H_{v,0}(r_M)$. For each $a \in T_{v,1}$, there is an open interval $I_a \subset \mathcal{C}_v(\mathbb{R})$ containing x, whose closure $\overline{I}_a$ is contained in $U_{v,1}$. Finitely many of these intervals, say $I_{b_1},\ldots,I_{b_{m_1}}$, cover $H_{v,1}$. Put

$$E_v' \ = \ \Big(\bigcup_{i=1}^{m_0} B(a_i,r_i)\Big) \cup \Big(\bigcup_{i=1}^{m_0} B(\overline{a_i},r_i)\Big) \cup \big(\overline{I}_{b_1} \cup \ldots \cup \overline{I}_{b_{m_1}}\big) \ .$$

Then E'_v is compact, stable under complex conjugation, and contained in U_v. By construction, it satisfies the conditions of Theorem 0.4. Hence (4.4) and (4.5), together with the monotonicity of Green's functions, show that (4.2) holds.

Case 3. If K_v is nonarchimedean, then the quasi-neighborhood U_v has the form

$$U_v = U_{v,0} \cup \big(U_{v,1} \cap \mathcal{C}_v(F_{w,1})\big) \cup \cdots \cup \big(U_{v,M} \cap \mathcal{C}_v(F_{w,M})\big) \tag{4.6}$$

where $U_{v,0}, \ldots, U_{v,M}$ are open in $\mathcal{C}_v(\mathbb{C}_v)$ and each $F_{w,\ell}$ is a separable algebraic extension of K_v (possibly of infinite degree) contained in $\mathbb{C}_v$. By hypothesis, U_v is $\mathrm{Aut}_c(\mathbb{C}_v/K_v)$-stable and contains E_v.

In what follows, it will convenient to take $F_{w,0} = \mathbb{C}_v$, though we will be careful to make a distinction between $F_{w,0}$ and the $F_{w,\ell}$ for $\ell \geq 1$, which are separably algebraic over K_v. Define a *representation of* U_v to be a collection of pairs $(U_{v,\ell}, F_{w,\ell})$ such that (4.6) holds. We allow the possibility that some of the $U_{v,i} \cap \mathcal{C}_v(F_{w,i})$ may be empty. We will say that a set $U_{v,i}$ or a field $F_{w,i}$ *occurs*, if it is a component of one of the pairs.

There are many representation for U_v. We begin by adjusting the given representation to make it easier to work with. First, for each finite separable extension $F_{w,\ell}/K_v$ which occurs, there are finitely many intermediate fields $K_v \subseteq G_u \subseteq F_{w,\ell}$. Adjoin each of the pairs $(U_{v,\ell}, G_u)$ to the representation. Second, a given field may occur in several pairs. Replace those pairs with a single pair whose first component is the union of the sets in the original pairs. In this way, we can assume that the $F_{w,\ell}$ are distinct, and that whenever a finite separable extension $F_{w,\ell}/K_v$ occurs, so do all of its subextensions. In addition, if two finite separable extensions $F_{w,j} \subseteq F_{w,\ell}$ occur, then

$$U_{v,j} \cap \mathcal{C}_v(F_{w,j}) \subseteq U_{v,\ell} \cap \mathcal{C}_v(F_{w,\ell}) . \tag{4.7}$$

We will now construct a new representation $\{(W_{v,j}, F_{w,j})\}_{0 \leq j \leq n}$ with the same fields $F_{w,j}$, but giving a different decomposition of U_v. After reordering the pairs $(U_{v,\ell}, F_{w,\ell})$, we can assume that $F_{w,\ell}/K_v$ is of infinite degree for $\ell = 0, \ldots, M_0$ and that $F_{w,\ell}/K_v$ is finite for $\ell = M_0 + 1, \ldots, M$. After reordering them further, we can also assume that for each ℓ, $F_{w,\ell}$ is maximal among the $F_{w,j}$ with $j \geq \ell$ (under the partial order given by containment).

For each $j \leq M_0$, let $W_{v,j}$ be the union of all the isometrically parametrizable balls $B(a, r)$ such that $a \in U_{v,j} \cap F_{w,j}$ and $B(a, r) \subset U_{v,j}$. Since isometrically parametrizable balls are cofinal in the neighborhoods of a given point, we have $W_{v,j} \cap \mathcal{C}_v(F_{w,j}) = U_{v,j} \cap \mathcal{C}_v(F_{w,j})$. For each $j \geq M_0 + 1$, note that if $k > j$, then $\mathcal{C}_v(F_{w,k})$ is compact, hence closed in $\mathcal{C}_v(\mathbb{C}_v)$; put

$$W_{v,j} = U_{v,j} \backslash \Big(\bigcup_{k>j} \mathcal{C}_v(F_{w,k}) \Big) . \tag{4.8}$$

This means that the sets $W_{v,j} \cap \mathcal{C}_v(F_{v,j})$ corresponding to finite extensions $F_{w,j}/K_v$ are pairwise disjoint. However, by the construction of the ordering, for each $F_{w,k}$ with $K_v \subseteq F_{w,k} \subsetneq F_{w,j}$ we must have $k > j$. Hence by (4.7), we still have

$$U_v = \bigcup_{j=0}^{M} \big(W_{v,j} \cap \mathcal{C}_v(F_{w,j})\big) . \tag{4.9}$$

By Proposition 3.26(4), there is a compact set $H_v \subseteq E_v$ for which

$$(4.10)\qquad \begin{cases} |G(x_i, x_j; H_v) - \overline{G}(x_i, x_j; E_v)| \ < \ \varepsilon_v/3 & \text{for all } i \neq j \ , \\ |V_{x_i}(H_v) - \overline{V}_{x_i}(E_v)| \ < \ \varepsilon_v/3 & \text{for each } i \, . \end{cases}$$

We aim to pass from H_v (whose structure is unknown) to an $\mathrm{Aut}_c(\mathbb{C}_v/K_v)$-stable set E_v' contained in $U_v \cap \mathcal{C}_v(F_w)$ for some finite Galois extension F_w/K_v, which satisfies (4.2) and has the form required by Theorem 0.4. This will be done in four steps, first shrinking H_v to a disjoint union of compact sets respecting the decomposition (4.9), then enlarging those sets by means of a finite covering with balls, then shrinking them again to get a set contained in $\mathcal{C}_v(F_w)$ for a finite Galois extension F_w/K_v, and finally taking the union of its conjugates to get an $\mathrm{Aut}_c(\mathbb{C}_v/K_v)$-stable set.

For the first step, define $T_{v,-1} = H_v$. Put $H_{v,0} = H_v \cap W_{v,0}$ and put $T_{v,0} = H_v \backslash W_{v,0}$. Inductively, for $k = 1, \ldots, M$, put $H_{v,k} = T_{v,k-1} \cap W_{v,k}$ and $T_{v,k} = T_{v,k-1} \backslash W_{v,k}$. Then each $T_{v,k}$ is compact, and $T_{v,k-1} = H_{v,k} \cup T_{v,k}$. Since $H_v \subseteq W_{v,0} \cup \cdots \cup W_{v,n}$ it follows that $T_{v,M} = \phi$ and

$$H_v \ = \ H_{v,0} \cup H_{v,1} \cup \cdots \cup H_{v,M} \ .$$

By construction the $H_{v,k}$ are pairwise disjoint, but they are not in general compact.

If $M_0 + 1 \leq k \leq M$, we claim that $H_{v,k} \subseteq W_{v,k} \cap \mathcal{C}_v(F_{w,k})$. To see this, note that by definition $H_{v,k} \subseteq H_v \backslash (W_{v,0} \cup \cdots \cup W_{v,k-1})$. From (4.9) it follows that

$$H_{v,k} \ \subseteq \ \bigcup_{j=k}^{M} (W_{v,j} \cap \mathcal{C}_v(F_{w,j})) \ .$$

However, also $H_{v,k} \subseteq W_{v,k}$, and by (4.8) this means that $H_{v,k} \cap \mathcal{C}_v(F_{w,j}) = \phi$ for $j = k+1, \ldots, M$. Hence $H_{v,k} \subseteq W_{v,k} \cap \mathcal{C}_v(F_{w,k})$.

For each $r > 0$ and each $k = 0, \ldots, n$, put

$$H_{v,k}(r) \ = \ \{z \in T_{v,k-1} : \|z, a\|_v \geq r \text{ for each } a \in T_{v,k}\} \ \subseteq \ H_{v,k} \ ,$$

and then put

$$H_v(r) \ = \ \bigcup_{k=0}^{M} H_{v,k}(r) \ .$$

Since $T_{v,k-1}$ and $T_{v,k}$ are compact, each $H_{v,k}(r)$ is compact, and $H_v(r)$ is compact.

The sets $H_v(r)$ increase monotonically as r decreases. For each $z \in H_v$ there are an index k such that $z \in H_{v,k}$ and an $r > 0$ such that $\|z, a\|_v > r$ for all $a \in T_{v,k}$. Thus the $H_v(r)$ form an exhaustion of H_v. Choose a sequence $r_1 > r_2 > \cdots > 0$ with $\lim_{m \to \infty} r_m = 0$. By Proposition 3.22, if we take M sufficiently large and put $X_v = H_v(r_M)$, then

$$(4.11)\qquad \begin{cases} |G(x_i, x_j; X_v) - G(x_i, x_j; H_v)| \ < \ \varepsilon_v/3 & \text{for all } i \neq j \ , \\ |V_{x_i}(X_v) - V_{x_i}(H_v)| \ < \ \varepsilon_v/3 & \text{for each } i \, . \end{cases}$$

Combining (4.10) and (4.11) gives

$$(4.12)\qquad \begin{cases} G(x_i, x_j; X_v) \ < \ \overline{G}(x_i, x_j; E_v)| + 2\varepsilon_v/3 & \text{for all } i \neq j \ , \\ V_{x_i}(X_v) \ < \ \overline{V}_{x_i}(E_v) + 2\varepsilon_v/3 & \text{for each } i \, . \end{cases}$$

For each $k = 0, \ldots, M$ put $X_{v,k} = H_{v,k}(r_M) \subset W_{v,k}$. Then the $X_{v,k}$ are compact and pairwise disjoint, and

$$X_v \ = \ \bigcup_{k=0}^{M} X_{v,k} \ .$$

By the continuity of the spherical distance, there is an $R > 0$ such that for all $k \neq \ell$, and all $x \in X_{v,k}$, $y \in X_{v,\ell}$, we have $\|x, y\|_v > R$.

For the second step, we will enlarge X_v to a set Y_v which is the union of finitely many isometrically parametrizable balls and compact sets, as follows.

If $x \in X_{v,0}$, then $x \in W_{v,0}$, and by the definition of $W_{v,0}$ there is an isometrically parametrizable ball $B(a,r) \subseteq U_{v,0}$ which contains x. By Proposition 3.2, $\mathcal{C}_v(\widetilde{K}_v^{\text{sep}})$ is dense in $\mathcal{C}_v(\mathbb{C}_v)$, so we can assume that $a \in \mathcal{C}_v(\widetilde{K}_v^{\text{sep}})$. Since $X_{v,0}$ is compact, finitely many such balls $B(a_{0,1}, r_{0,1}), \ldots, B(a_{0,m_0}, r_{0,m_0})$ cover $X_{v,0}$. Without loss, we can assume that $r_{0,\ell} < R$ and $r_{0,\ell} \in |K_v^\times|_v$ for each ℓ; by construction each $a_{0,\ell} \in \mathcal{C}_v(\widetilde{K}_v^{\text{sep}})$.

If $1 \leq k \leq M_0$ and $x \in X_{v,k}$, then by the definition of $W_{v,k}$ there is an isometrically parametrizable ball $B(a,r)$, with $a \in W_{v,k} \cap \mathcal{C}_v(F_{w,k})$ and $B(a,r) \subset U_{v,k}$, such that $x \in B(a,r)$. Finitely many such balls $B(a_{k,1}, r_{k,1}), \ldots, B(a_{k,m_k}, r_{k,m_k})$ cover $X_{v,k}$, by compactness. Without loss, we can assume that $r_{k,\ell} < R$ and $r_{k,\ell} \in |K_v^\times|_v$ for each k, ℓ. By construction each $a_{k,\ell} \in \mathcal{C}_v(F_{w,k})$.

If $M_0 + 1 \leq k \leq M$, and $x \in X_{v,k}$, then $x \in H_{v,k}$ and by the discussion above we have $x \in W_{v,k} \cap \mathcal{C}_v(F_{w,k})$. As $W_{v,k}$ is open, there is an isometrically parametrizable ball centered at x for which $B(x,r) \subset W_{v,k}$. By the properties of an isometric parametrization, $B(x,r) \cap \mathcal{C}_v(F_{w,k})$ is open in $\mathcal{C}_v(F_{w,k})$. Since $X_{v,k}$ is compact and contained in $\mathcal{C}_v(F_{w,k})$, there are finitely many of these balls, say $B(a_{k,1}, r_{k,1}), \ldots, B(a_{k,m_k}, r_{k,m_k})$ for which

$$X_{v,k} \subseteq \bigcup_{\ell=1}^{m_k} \left(B(a_{k,\ell}, r_{k,\ell}) \cap \mathcal{C}_v(F_{w,k}) \right) . \tag{4.13}$$

Again we can assume that $r_{k,\ell} < R$ and $r_{k,\ell} \in |K_v^\times|_v$ for each ℓ; by construction each $a_{k,\ell} \in \mathcal{C}_v(F_{w,k})$. The right side of (4.13) is contained in $W_{v,k} \cap \mathcal{C}_v(F_{w,k})$, hence in U_v.

Put

$$Y_v = \Big(\bigcup_{k=0}^{M_0} \Big(\bigcup_{\ell=1}^{m_k} B(a_{k,\ell}, r_{k,\ell}) \Big) \Big) \cup \Big(\bigcup_{k=M_0+1}^{M} \Big(\bigcup_{\ell=1}^{m_k} B(a_{k,\ell}, r_{k,\ell}) \cap \mathcal{C}_v(F_{w,k}) \Big) \Big) .$$

Here, we have purposefully omitted the intersections with $\mathcal{C}_v(F_{w,k})$ for the balls $B(a_{k,\ell}, r_{k,\ell})$ with $k \leq M_0$. This means that Y_v need not be contained in U_v. However, Y_v contains X_v. By the monotonicity of Green's functions, together with (4.12), this gives

$$\begin{cases} \overline{G}(x_i, x_j; Y_v) < \overline{G}(x_i, x_j; E_v) + 2\varepsilon_v/3 & \text{for all } i \neq j \ , \\ \overline{V}_{x_i}(Y_v) < \overline{V}_{x_i}(E_v) + 2\varepsilon_v/3 & \text{for each } i \, . \end{cases} \tag{4.14}$$

For the third step, we will cut Y_v down to a set Z_v contained in $U_v \cap \mathcal{C}_v(F_w)$, where F_w is a suitable finite Galois extension of K_v. By construction, Y_v is the union of finitely many isometrically parametrizable balls whose radii belong to $|K_v^\times|_v$, and finitely many compact sets. By ([**51**], Theorems 4.2.16 and 4.3.11), it is algebraically capacitable. However, the proof of ([**51**], Theorem 4.3.11) gives more: the Green's function of Y_v is the limit of Green's functions of compact sets contained in finite extensions of K_v. An examination of the proof shows these sets can be chosen to lie in U_v.

Explicitly, this comes out as follows. Put $\widetilde{F}_{w,0} = \widetilde{K}_v^{\mathrm{sep}}$, and put $\widetilde{F}_{w,k} = F_{w,k}$ for $1 \le k \le M_0$. For each $k = 0, \ldots, M_0$, choose an exhaustion of $\widetilde{F}_{w,k}$ by an increasing sequence of finite separable extensions $\widetilde{F}_{w,k,j}/K_v$:

$$\widetilde{F}_{w,k,1} \subseteq \widetilde{F}_{w,k,2} \subseteq \widetilde{F}_{w,k,3} \subseteq \cdots \subseteq \widetilde{F}_{w,k} .$$

Without loss, we can assume $\widetilde{F}_{w,k,1}$ is large enough that each of $a_{k,1}, \ldots, a_{k,m_k}$ belongs to $\mathcal{C}_v(\widetilde{F}_{w,k,1})$. For each $j = 1, 2, 3, \ldots$, put

$$\begin{aligned} Z_{v,j} \quad = \quad & \Big(\bigcup_{k=0}^{M_0} \big(\bigcup_{\ell=1}^{m_k} B(a_{k,\ell}, r_{k,\ell}) \cap \mathcal{C}_v(\widetilde{F}_{w,k,j}) \big) \Big) \\ & \qquad \cup \Big(\bigcup_{k=M_0+1}^{M} \big(\bigcup_{\ell=1}^{m_k} B(a_{k,\ell}, r_{k,\ell}) \cap \mathcal{C}_v(F_{w,k}) \big) \Big) . \end{aligned}$$

The sets $Z_{v,j}$ play the same role as the sets H_j in ([**51**], p.270), though they are constructed somewhat differently. In ([**51**], p.269) the finite extensions $\widetilde{F}_{w,k,j}$ (denoted L_j there) were made to exhaust the algebraic closure $\widetilde{K}_v$, but all that is needed is the fact that for each $k = 0, \ldots, M_0$, as $j \to \infty$ either the ramification index $e_{w/v,k,j}$ or the residue degree $f_{w/v,k,j}$ of $\widetilde{F}_{w,k,j}/K_v$ grows arbitrarily large. This means that as $j \to \infty$,

$$\frac{1}{e_{w/v,k,j}} \cdot \frac{1}{q_v^{f_{w/v,k,j}} - 1} \longrightarrow 0 .$$

Hence the proof of ([**51**], Theorem 4.3.11) shows that if we take J large enough and put $Z_v = Z_{v,J}$ then

$$(4.15) \qquad \begin{cases} |G(x_i, x_j; Z_v) - \overline{G}(x_i, x_j; Y_v)| < \varepsilon_v/3 & \text{for all } i \ne j , \\ |V_{x_i}(Z_v) - \overline{V}_{x_i}(Y_v)| < \varepsilon_v/3 & \text{for each } i . \end{cases}$$

Let F_w be the Galois closure of the composite of the fields $\widetilde{F}_{w,0,J}, \ldots, \widetilde{F}_{w,M_0,J}$ and $F_{w,M_0+1}, \ldots, F_{w,n}$; then F_w is a finite Galois extension of K_v and $Z_v \subset U_v \cap \mathcal{C}_v(F_w)$.

For the last step, let E_v' be the union of the $\mathrm{Gal}(F_w/K_v)$-conjugates of Z_v. Then E_v' is compact. It is a finite union of sets of the kind in Theorem 0.4.C(2). Since U_v is stable under $\mathrm{Aut}_c(\mathbb{C}_v/K_v)$, E_v' is contained in U_v. From (4.14), (4.15), and the monotonicity of Green's functions, it follows that (4.2) holds; that is,

$$\begin{cases} G(x_i, x_j; E_v') < \overline{G}(x_i, x_j; E_v) + \varepsilon_v & \text{for all } i \ne j , \\ V_{x_i}(E_v') < \overline{V}_{x_i}(E_v) + \varepsilon_v & \text{for each } i . \end{cases}$$

This completes the proof. □

Theorem 1.3 (Strong FSZ with LR). *Let K be a global field, and let $\mathcal{C}/K$ be a smooth, geometrically integral projective curve. Let $\mathfrak{X} = \{x_1, \ldots, x_m\} \subset \mathcal{C}(\widetilde{K})$ be a finite set of points stable under* $\mathrm{Aut}(\widetilde{K}/K)$, *and let $\mathbb{E} = \prod_v E_v \subset \prod_v \mathcal{C}_v(\mathbb{C}_v)$ be a K-rational adelic set compatible with $\mathfrak{X}$, so each E_v is stable under* $\mathrm{Aut}_c(\mathbb{C}_v/K_v)$ *and all but finitely many E_v are $\mathfrak{X}$-trivial. Let $S \subset \mathcal{M}_K$ be a finite set of places v, containing all archimedean v, such that E_v is $\mathfrak{X}$-trivial for each $v \notin S$.*

Assume that $\overline{\gamma}(\mathbb{E}, \mathfrak{X}) > 1$. Assume also that for each $v \in S$, there is a (possibly empty) $\mathrm{Aut}_c(\mathbb{C}_v/K_v)$*-stable Borel subset $e_v \subset \mathcal{C}_v(\mathbb{C}_v)$ of inner capacity 0 such that:*

(A) If v is archimedean and $K_v \cong \mathbb{C}$, then each point of $\mathrm{cl}(E_v) \backslash e_v$ *is analytically accessible from the $\mathcal{C}_v(\mathbb{C})$-interior of E_v.*

(B) If v is archimedean and $K_v \cong \mathbb{R}$, then each point of $\mathrm{cl}(E_v)\backslash e_v$ *is*

(1) analytically accessible from the $\mathcal{C}_v(\mathbb{C})$-interior of E_v, or

(2) is an endpoint of an open segment contained in $E_v \cap \mathcal{C}_v(\mathbb{R})$.

(C) If v is nonarchimedean, then E_v is the disjoint union of e_v and finitely many sets $E_{v,1}, \ldots, E_{v,M_v}$, where each $E_{v,\ell}$ is

(1) open in $\mathcal{C}_v(\mathbb{C}_v)$, or

(2) of the form $U_{v,\ell} \cap \mathcal{C}_v(F_{w,\ell})$, where $U_{v,\ell}$ is open in $\mathcal{C}_v(\mathbb{C}_v)$ and $F_{w,\ell}$ is a separable algebraic extension of K_v contained in $\mathbb{C}_v$ (possibly of infinite degree).

Then there are infinitely many points $\alpha \in \mathcal{C}(\widetilde{K}^{\mathrm{sep}})$ such that for each $v \in \mathcal{M}_K$, the $\mathrm{Aut}(\widetilde{K}/K)$*-conjugates of α all belong to E_v.*

PROOF OF THEOREM 1.3, ASSUMING THEOREM 1.2. Assume Theorem 1.2, and let $\mathbb{E} = \prod_v E_v \subset \prod_v \mathcal{C}_v(\mathbb{C}_v)$ be an adelic set compatible with $\mathfrak{X}$ for which the hypotheses of Theorem 1.3 hold. In this case, apart from a set of inner capacity 0, each E_v is itself a separable quasi-neighborhood.

We will construct new adelic sets $\mathbb{E}' = \prod_v E_v'$ and $\mathbb{U}' = \prod_v U_v'$, with $\mathbb{E}' \subseteq \mathbb{U}' \subseteq \mathbb{E}$, such that the hypotheses of Theorem 1.2 hold for $\mathbb{E}'$ and $\mathbb{U}'$. In fact, we will have $\mathbb{E}' = \mathbb{U}'$ and $\overline{\gamma}(\mathbb{E}', \mathfrak{X}) = \overline{\gamma}(\mathbb{E}, \mathfrak{X})$. Let $S \subseteq \mathcal{M}_K$ be a finite set of places containing all archimedean places and all nonarchimedean places where E_v is not $\mathfrak{X}$-trivial. For each $v \notin S$, put $E_v' = U_v' = E_v$.

For each archimedean $v \in S$ such that $K_v \cong \mathbb{C}$, the set $\mathrm{cl}(E_v)$ is compact and there is a Borel subset $e_v \subset \mathcal{C}_v(\mathbb{C})$ of inner capacity 0 such that each point of $\mathrm{cl}(E_v)\backslash e_v$ is analytically accessible from the $\mathcal{C}_v(\mathbb{C})$-interior E_v^0 of E_v. Put $E_v' = U_v' = E_v^0$. Since U_v' is open, it is a quasi-neighborhood of E_v'. By Proposition 3.30, for each $\zeta \notin E_v$ we have $\overline{G}(z, \zeta; E_v') = G(z, \zeta; E_v)$.

For each archimedean $v \in S$ such that $K_v \cong \mathbb{R}$, the set $\mathrm{cl}(E_v)$ is compact and there is a Borel subset $e_v \subset \mathcal{C}_v(\mathbb{C})$ of inner capacity 0 such that each point of $\mathrm{cl}(E_v)\backslash e_v$ is analytically accessible from the $\mathcal{C}_v(\mathbb{C})$-interior E_v^0 of E_v or from the $\mathcal{C}_v(\mathbb{R})$-interior E_v^1 of $E_v \cap \mathcal{C}_v(\mathbb{R})$. Put $E_v' = U_v' := E_v^0 \cup E_v^1$. Since E_v is stable under complex conjugation, so are E_v^0 and E_v'. Again U_v' is a quasi-neighborhood of E_v', and by Proposition 3.30, for each $\zeta \notin E_v$ we have $\overline{G}(z, \zeta; E_v') = G(z, \zeta; E_v)$.

For each nonarchimedean $v \in S$, E_v is the disjoint union of a Borel subset e_v of inner capacity 0 and sets $E_{v,1}, \ldots, E_{v,n}$, each of which is either open in $\mathcal{C}_v(\mathbb{C}_v)$ or of the form $U_{v,i} \cap \mathcal{C}_v(F_{w,i})$ for some separable algebraic extension $F_{w,i}/K_v$ contained in $\mathcal{C}_v(\mathbb{C}_v)$. Put $E_v' = U_v' = E_{v,1} \cup \cdots \cup E_{v,n}$. By assumption, e_v is stable under $\mathrm{Aut}_c(\mathbb{C}_v/K_v)$, hence so is E_v'. By its construction, U_v' is a quasi-neighborhood of E_v'. Finally, by Lemma 3.25, removing a set of inner capacity 0 from a set does not change its Green's functions, so for each $\zeta \notin E_v$ we have $\overline{G}(z, \zeta; E_v') = \overline{G}(z, \zeta; E_v)$ □

Theorem 1.4 (FSZ with LR and Ramification Side Conditions). *Let K be a global field, and let $\mathcal{C}/K$ be a smooth, connected, projective curve. Let $\mathfrak{X} = \{x_1, \ldots, x_m\} \subset \mathcal{C}(\widetilde{K})$ be a finite, Galois-stable set of points, and let $\mathbb{E} = \prod_v E_v \subset \prod_v \mathcal{C}_v(\mathbb{C}_v)$ be a K-rational adelic set compatible with $\mathfrak{X}$, so each E_v is stable under* $\mathrm{Aut}_c(\mathbb{C}_v/K_v)$ *and all but finitely many E_v are $\mathfrak{X}$-trivial.*

Let $S, S', S'' \subset \mathcal{M}_K$ be finite (possibly empty) sets of places of K which are pairwise disjoint, such that the places in $S' \cup S''$ are nonarchimedean. Assume that $\overline{\gamma}(\mathbb{E}, \mathfrak{X}) > 1$, and that:

(A) For each $v \in S$, E_v satisfies the conditions of Theorem 0.4 *or Theorem* 1.3.

(B) For each $v \in S'$, either E_v is $\mathfrak{X}$-trivial, or E_v is a finite union of closed isometrically parametrizable balls $B(a_i, r_i)$ whose radii belong to the value group of $K_v^\times$ and whose centers belong to an unramified extension of K_v.

(C) For each $v \in S''$, either E_v is $\mathfrak{X}$-trivial and $E_v \cap \mathcal{C}_v(K_v)$ is nonempty, or E_v is a finite union of closed and/or open isometrically parametrizable balls $B(a_i, r_i)$, $B(a_j, r_j)^-$ with centers in $\mathcal{C}_v(K_v)$.

Then there are infinitely many points $\alpha \in \mathcal{C}(\widetilde{K}^{\text{sep}})$ such that

(1) for each $v \in \mathcal{M}_K$, the $\text{Aut}(\widetilde{K}/K)$-conjugates of α all belong to E_v;

(2) for each $v \in S'$, each place of $K(\alpha)/K$ above v is unramified over v;

(3) for each $v \in S''$, each place of $K(\alpha)/K$ above v is totally ramified over v.

Proof of Theorem 1.4, assuming Theorem 1.3. For each $v \in S' \cup S''$, the hypotheses in (B) and (C) will enable us to replace the given set E_v by sets of the form $\mathcal{C}_v(F_w) \cap E_v$ for suitably chosen finite Galois extensions F_w/K_v, which are unramified if $v \in S'$ and are totally ramified if $v \in S''$, in such a way that we still have $\overline{\gamma}(\mathbb{E}, \mathfrak{X}) > 1$.

More precisely, we claim that we can choose the extensions F_w/K_v so that the new Green's matrix is arbitrarily near the old one. Since $\overline{\gamma}(\mathbb{E}, \mathfrak{X}) > 1$ if and only if $\overline{\Gamma}(\mathbb{E}, \mathfrak{X})$ is negative definite, the new Green's matrix will be negative definite if it is sufficiently close to the old one. Thus, the theorem reduces to Theorem 1.3.

The claim is a consequence of explicit formulas for the Robin constants and Green's functions of the sets in question, derived in ([**51**], pp.353-359) and stated in (2.70) of this work. Fix a nonarchimedean place v, and fix a spherical metric on $\mathcal{C}_v(\mathbb{C}_v)$. Let $H_v = B(a, r) \subset \mathcal{C}_v(\mathbb{C}_v)$ be a closed isometrically parametrizable ball, and take $\zeta \in \mathcal{C}_v(\mathbb{C}_v) \backslash B(a, r)$. By ([**51**], Theorem 4.3.15, p.274) isometrically parametrizable balls are algebraically capacitable, and by the proof of ([**51**], Theorem 4.4.4) their Green's functions and upper Green's functions coincide. Fix a normalization for the canonical distance $[z, w]_\zeta$. Then there is an $R > 0$ such that $B(a, r) = \{z \in \mathcal{C}_v(\mathbb{C}_v) : [z, a]_\zeta \leq R\}$, and in terms of this R (see [**51**], p.357) we have

$$(4.16) \qquad \begin{aligned} V_\zeta(H_v) &= -\log_v(R) \ , \\ G(z, \zeta; H_v) &= \begin{cases} 0 & \text{if } z \in H_v, \\ \log_v([z, a]_\zeta / R) & \text{if } z \notin B(a, r). \end{cases} \end{aligned}$$

It will be useful to define $u_{H_v}(z, \zeta) = V_\zeta(H_v) - G(z, \zeta; H_v)$, so

$$(4.17) \qquad u_{H_v}(z, \zeta) = \begin{cases} -\log_v(R) & \text{if } z \in H_v, \\ -\log_v([z, a]_\zeta) & \text{if } z \notin B(a, r). \end{cases}$$

Furthermore, if F_w/K_v is a finite extension with ramification index e and residue degree f, if q_v is the order of the residue field of K_v, and if $a \in \mathcal{C}_v(F_w)$ and r belongs to the value group of $F_w^\times$, then for the set $\mathcal{C}_v(F_w) \cap H_v$ we have (see [**51**], p.358)

$$(4.18) \quad V_\zeta(\mathcal{C}_v(F_w) \cap H_v) = -\log_v(R) + \frac{1}{e(q_v^f - 1)} \ ,$$

$$(4.19) \quad u_{\mathcal{C}_v(F_w) \cap H_v}(z, \zeta) = \begin{cases} -\log_v(R) + \frac{1}{e(q_v^f - 1)} & \text{if } z \in \mathcal{C}_v(F_w) \cap H_v, \\ -\log_v([z, a]_\zeta) & \text{if } z \notin B(a, r). \end{cases}$$

More generally, if $H_v = \bigcup_{i=1}^N B(a_i, r_i)$ is a finite union of closed, pairwise disjoint isometrically parametrizable balls, write $H_{v,i} = B(a_i, r_i)$; then $V_\zeta(H_v)$ and $G(z, \zeta; H_v)$ can be determined by solving the following system of equations for $V, s_1, \ldots, s_N$ (see [**51**], p.359):

$$\text{(4.20)} \qquad \begin{cases} 1 = 0V + s_1 + s_2 + \cdots + s_N \ , \\ 0 = V - \sum_{i=1}^N s_i u_{H_{v,i}}(a_j, \zeta) \quad \text{for } j = 1, \ldots, N \end{cases}$$

By ([**51**], Proposition 4.2.7) the solution is unique, and has $s_1, \ldots, s_N > 0$; in terms of it,

$$\begin{aligned} V_\zeta(H_v) &= V \ , \\ G(z, \zeta; H_v) &= V - \sum_{i=1}^N s_i u_{H_{v,i}}(z, \zeta) \ . \end{aligned}$$

Similarly, if F_w/K_v is a finite extension with ramification index e and residue degree f, and if the a_i belong to $\mathcal{C}_v(F_w)$ and the r_i belong to the value group of $F_w^\times$, then we can determine $V_\zeta(\mathcal{C}_v(F_w) \cap H_v)$ and $G(z, \zeta; \mathcal{C}_v(F_w) \cap H_v)$ by solving for $V_w, s_{1,w}, \ldots, s_{N,w}$ in the following system of equations:

$$\text{(4.21)} \qquad \begin{cases} 1 = 0V_w + s_{1,w} + s_{2,w} + \cdots + s_{N,w} \ , \\ 0 = V_w - \sum_{i=1}^N s_{i,w} u_{\mathcal{C}_v(F_w) \cap H_{v,i}}(a_j, \zeta) \quad \text{for } j = 1, \ldots, N \, . \end{cases}$$

By the existence and uniqueness of the equilibrium measure for $\mathcal{C}_v(F_w) \cap H_v$ ([**51**], Theorem 3.1.12), again the solution is unique, with $s_{1,w}, \ldots, s_{N,w} > 0$; and

$$\begin{aligned} V_\zeta(\mathcal{C}_v(F_w) \cap H_v) &= V_w \ , \\ G(z, \zeta; \mathcal{C}_v(F_w) \cap H_v) &= V_w - \sum_{i=1}^N s_{i,w} u_{\mathcal{C}_v(F_w) \cap H_{v,i}}(z, \zeta) \ . \end{aligned}$$

Comparing the systems (4.20) and (4.21), as F_w passes through a sequence of extensions for which $1/(e(q_v^f - 1)) \to 0$, then the $V_\zeta(\mathcal{C}_v(F_w) \cap H_v)$ converge to $V_\zeta(H_v)$ and the $G(z, \zeta; \mathcal{C}_v(F_w) \cap H_v)$ converge (uniformly) to $G(z, \zeta; H_v)$.

We now apply this to the sets E_v for $v \in S' \cup S''$ in the theorem.

First suppose $E_v = \bigcup_{i=1}^N B(a_i, r_i)$ is a finite union of closed isometrically parametrizable balls. Without loss, we can assume the $B(a_i, r_i)$ are pairwise disjoint.

If $v \in S'$, there is a finite unramified extension F'_w/K_v with $a_1, \ldots, a_N \in \mathcal{C}_v(F'_w)$, and the r_i belong to the value group of $K_v^\times$. Letting F_w pass through all finite unramified extensions of K_v containing F'_w, for all $x_i \neq x_j \in \mathfrak{X}$ we can make $V_{x_i}(\mathcal{C}_v(F_w) \cap E_v)$ arbitrarily near $V_{x_i}(E_v)$, and we can make the $G(x_i, x_j; \mathcal{C}_v(F_w) \cap E_v)$ arbitrarily near the $G(x_i, x_j; E_v)$.

Similarly if $v \in S''$, then $a_1, \ldots, a_N \in \mathcal{C}_v(K_v)$. Letting F_w/K_v pass through a sequence of finite, Galois, totally ramified extensions for which $e_{w/v} \to \infty$ (for example, cyclotomic p-extensions, where p is the residue characteristic of K_v), we obtain the same conclusion as before. If $v \in S''$ and $E_v = (\bigcup_{i=1}^N B(a_i, r_i)) \cup (\bigcup_{j=N+1}^{N+M} B(a_j, r_j)^-)$, then by exhausting the open balls with closed balls $B(a_j, r'_j)$ and taking a limit as the $r'_j \to r_j$, we are reduced to the previous case. Note that any compact $H_v \subset E_v$ is contained in a set of the form $(\bigcup_{i=1}^N B(a_i, r_i)) \cup (\bigcup_{j=N+1}^{N+M} B(a_j, r'_j))$. Since the upper Green's function $\overline{G}(z, w; E_v)$ is by definition the (pointwise) limit of the upper Green's functions $\overline{G}(z,w;H_v)$ for compact $H_v \subset E_v$, and upper Green's functions are monotonic under containment, $\overline{G}(z,w;E_v)$ is the limit of the Green's functions for the unions of closed balls discussed above, and hence also of the Green's functions $G(z,w;\mathcal{C}_v(F_w)\cap E_v)$ as F_w passes through finite, Galois, totally ramified extensions of K_v.

Next, consider the case where E_v is $\mathfrak{X}$-trivial. The $\mathfrak{X}$-triviality implies that $\mathcal{C}_v$ has good reduction at v and the points of $\mathfrak{X}$ specialize to distinct points $\pmod v$. Furthermore, if $\|x,y\|_v$ is the spherical metric associated to the given embedding of $\mathcal{C}_v$, then the canonical distance (up to scaling by a constant) is given by $[x,y]_\zeta = \|x,y\|_v/(\|x,\zeta\|_v\|y,\zeta\|_v)$ (see [**51**], p.91). Relative to this normalization of the canonical distance, $\gamma_\zeta(E_v) = 1$ for each $\zeta \notin E_v$.

First suppose $v \in S'$. Let k_v be the residue field of the ring of integers of K_v, and let $\overline{k}_v$ be its algebraic closure. Then $\overline{k}_v$ is the residue field of K_v^{nr}, the maximal unramified algebraic extension of K_v. Write $\overline{\mathcal{C}}_v$ for the reduction of $\mathcal{C}_v$ $\pmod v$. By Hensel's lemma, each point of $\overline{\mathcal{C}}_v(\overline{k}_v)$ lifts to a point in $\mathcal{C}_v(K_v^{nr})$. Put $r = 1/q_v \in |K_v^\times|_v$. Then for arbitrarily large N we can find Galois-stable sets of the form $H_v(N) = \bigcup_{i=1}^N B(a_i,r) \subset E_v$, where each $a_i \in \mathcal{C}_v(K_v^{nr})$ and distinct a_i specialize to distinct points $\pmod v$. For each $\zeta \notin E_v$, and each ball $B(a_i,r)$, we have $B(a_i,r) = \{z \in \mathcal{C}_v(\mathbb{C}_v) : [z,a]_\zeta \le 1/q_v\}$, so by (4.17)

$$u_{B(a_i,r)}(z,\zeta) \;=\; \begin{cases} 1 & \text{if } z \in B(a_i,r), \\ -\log_v([z,a_i]_\zeta) & \text{if } z \notin B(a_i,r). \end{cases}$$

In particular, $u_{B(a_i,r)}(a_j,\zeta) = 0$ for each $j \neq i$. Inserting this in (4.20), we find that $V_\zeta(H_v(N)) = 1/N$, and that if $z \notin E_v$ then $G(z,\zeta;H_v(N)) = G(z,\zeta;E_v) + 1/N$. Thus by replacing E_v with $H_v(N)$ for a sufficiently large N, we are reduced to a previous case.

If $v \in S''$, then by hypothesis $\mathcal{C}_v(K_v) \cap E_v$ is nonempty; fix $a \in \mathcal{C}_v(K_v)\cap E_v$. Consider the open ball $B(a,1)^- \subset E_v$. Exhausting it by closed balls $B(a,r)$, and noting that $B(a,r) = \{z \in \mathcal{C}_v(\mathbb{C}_v) : [z,a]_\zeta \le r\}$ for each $\zeta \notin E_v$, it follows by (4.16) that $\overline{\gamma}_\zeta(B(a,1)^-) = 1$. By ([**51**], Lemma 4.4.7), $\overline{G}(z,\zeta;B(a,1)^-) = G(z,\zeta;E_v)$ for all $z,\zeta \notin E_v$. Thus, by replacing E_v with $B(a,1)^-$, again we are reduced to a case considered before. □

Theorem 1.5 (Fekete/Fekete-Szegö with LR for Algebraically Capacitable Sets). *Let K be a global field and let $\mathcal{C}/K$ be a smooth, connected, projective curve. Let $\mathfrak{X} = \{x_1,\ldots,x_m\} \subset \mathcal{C}(\widetilde{K})$ be a finite, Galois-stable set of points, and let $\mathbb{E} = \prod_v E_v \subset \prod_v \mathcal{C}_v(\mathbb{C}_v)$ be a K-rational adelic set compatible with $\mathfrak{X}$, so each E_v is stable under* $\mathrm{Aut}_c(\mathbb{C}_v/K_v)$ *and all but finitely many E_v are $\mathfrak{X}$-trivial. Assume that each E_v is algebraically capacitable. Then:*

(A) If all the eigenvalues of $\Gamma(\mathbb{E},\mathfrak{X})$ are nonpositive (that is, $\Gamma(\mathbb{E},\mathfrak{X})$ is either negative definite or negative semi-definite), let $\mathbb{U} = \prod_v U_v$ be a separable K-rational quasi-neighborhood of $\mathbb{E}$ such that there is at least one place v_0 where E_{v_0} is compact and the quasi-neighborhood U_{v_0} properly contains E_{v_0}. If v_0 is archimedean, assume also that U_{v_0} meets each component of $\mathcal{C}_{v_0}(\mathbb{C})\backslash E_{v_0}$ containing a point of $\mathfrak{X}$. Then there are infinitely many $\alpha \in \mathcal{C}(\widetilde{K}^{\mathrm{sep}})$ such that all the conjugates of α belong to $\mathbb{U}$.

(B) If some eigenvalue of $\Gamma(\mathbb{E},\mathfrak{X})$ is positive (that is, $\Gamma(\mathbb{E},\mathfrak{X})$ is either indefinite, nonzero and positive semi-definite, or positive definite), there is an adelic

neighborhood $\mathbb{U}$ *of* $\mathbb{E}$ *such that only finitely many points* $\alpha \in \mathcal{C}(\widetilde{K})$ *have all their conjugates in* $\mathbb{U}$.

PROOF OF THEOREM 1.5, ASSUMING THEOREM 1.2. Since each E_v is algebraically capacitable, we have $\overline{\gamma}(\mathbb{E}, \mathfrak{X}) = \gamma(\mathbb{E}, \mathfrak{X})$. Recall that a symmetric matrix $\Gamma \in M_k(\mathbb{R})$ with nonnegative off-diagonal entries is called *irreducible* ([**51**], p.328) if the graph on the set $\{1, \ldots, k\}$, for which there is an edge between i and j iff $\Gamma_{i,j} > 0$, is connected. By ([**51**], Lemma 5.1.7, p.328) if Γ is irreducible, then $\mathrm{val}(\Gamma)$ is positive, 0, or negative, depending on the largest eigenvalue of Γ.

By re-ordering the elements of $\mathfrak{X}$ if necessary, we can bring $\Gamma(\mathbb{E}, \mathfrak{X})$ to block-diagonal form $\mathrm{diag}(\Gamma_1, \ldots, \Gamma_r)$, where each Γ_i is irreducible. Note that $\Gamma(\mathbb{E}, \mathfrak{X})$ is negative definite if and only if each Γ_i is negative definite, and is negative semi-definite if and only if each Γ_i is negative definite or negative semi-definite.

If $\Gamma(\mathbb{E}, \mathfrak{X})$ is negative definite, then $\gamma(\mathbb{E}, \mathfrak{X}) > 1$, and the result follows from Theorem 1.2. If each Γ_i is negative semi-definite, then by enlarging the set E_0 within its quasi-neighborhood U_{v_0} (keeping E_{v_0} stable under $\mathrm{Aut}_c(\mathbb{C}_{v_0}/K_{v_0})$, we can decrease all the diagonal entries of $\Gamma(\mathbb{E}, \mathfrak{X})$, while either decreasing or leaving unchanged each off-diagonal entry. This makes $\Gamma(\mathbb{E}, \mathfrak{X})$ negative definite, and we can again apply Theorem 1.2.

If some Γ_i has a positive eigenvalue, let $\mathfrak{X}'$ be the subset of $\mathfrak{X}$ consisting of all $x_\ell \in \mathfrak{X}$ corresponding to blocks which have positive eigenvalues. Enlarge each E_v which is $\mathfrak{X}$-trivial to a set E'_v which is $\mathfrak{X}'$-trivial, and let $\mathbb{E}'$ be corresponding adelic set. Since the action of $\mathrm{Aut}(\widetilde{K}/K)$ on $\mathfrak{X}$ permutes the x_ℓ and hence the blocks Γ_i, the sets $\mathfrak{X}'$ and $\mathbb{E}'$ are Galois stable. By ([**51**], Lemma 5.1.7 and Theorem 5.1.6, p.328), $\mathrm{val}(\Gamma(\mathbb{E}', \mathfrak{X}')) > 0$ and hence $\gamma(\mathbb{E}', \mathfrak{X}') < 1$. The result now follows from Fekete's theorem applied to $\mathbb{E}'$ and $\mathfrak{X}'$ ([**51**], Theorem 6.3.1, p.414). □

We now prepare for the proofs of Theorems 1.6 and 1.7. In the following, we assume familiarity with Berkovich analytic spaces (see [**10**]) and Thuillier's potential theory on Berkovich curves ([**64**]). For nonarchimedean places v, Thuillier established the compatibility of capacities for sets in $\mathcal{C}_v^{\mathrm{an}}$, defined by him in ([**64**]), with capacities for sets in $\mathcal{C}_v(\mathbb{C}_v)$, as defined by Rumely in ([**51**]) and used in this work (see ([**64**], Appendix 5.1)). However, he did not explicitly state the compatibility of Green's functions. Before proving Theorems 1.6 and 1.7, we establish this.

Recall that for each compact, nonpolar subset $\mathbf{E}_v \subset \mathcal{C}_v^{\mathrm{an}}$ and each $\zeta \in \mathcal{C}_v^{\mathrm{an}} \backslash \mathbf{E}_v$, Thuillier ([**64**], Théorème 3.6.15) has constructed a Green's function $g_{\zeta, \mathbf{E}_v}(z)$ which is nonnegative, vanishes on $\mathbf{E}_v$ except possibly on a set of capacity 0, is subharmonic in $\mathcal{C}_v^{\mathrm{an}}$, harmonic in $\mathcal{C}_v^{\mathrm{an}} \backslash (\mathbf{E}_v \cup \{\zeta\})$, and satisfies the distributional equation $dd^c g_{\zeta, \mathbf{E}_v} = \mu - \delta_\zeta$ where μ is a probability measure supported on K. We write $G(z, \zeta; \mathbf{E}_v)^{\mathrm{an}}$ for $g_{\zeta, \mathbf{E}_v}(z)$, regarding it as a function of two variables.

By abuse of language, we write $\mathcal{C}_v^{\mathrm{an}}$ for the topological space underlying the ringed space $\mathcal{C}_v^{\mathrm{an}}$. Following Thuillier, let $I(\mathcal{C}_v^{\mathrm{an}}) := \mathcal{C}_v^{\mathrm{an}} \backslash \mathcal{C}_v(\mathbb{C}_v)$ be the set of non-classical points of $\mathcal{C}_v^{\mathrm{an}}$. Berkovich Green's functions have the following properties:

PROPOSITION 4.3. *Let* K *be a global field, and let* $\mathcal{C}/K$ *be a smooth, connected, projective curve. Let* v *be a nonarchimedean place of* K, *and let* $\mathcal{C}_v^{\mathrm{an}}$ *be the Berkovich analytification of* $\mathcal{C}_v \times_{K_v} \mathrm{Spec}(\mathbb{C}_v)$. *Let* $\mathbf{E}_v \subsetneq \mathcal{C}_v^{an}$ *be a proper compact, nonpolar subset of* $\mathcal{C}_v^{an}$. *Then:*

(A) For each $\zeta \in \mathcal{C}_v(\mathbb{C}_v)\backslash\mathbf{E}_v$*, if we fix a uniformizer* $g_\zeta(z)$*, the Robin constant*

$$V_\zeta(\mathbf{E}_v)^{\mathrm{an}} = \lim_{\substack{z\to\zeta \\ z\in\mathcal{C}_v^{\mathrm{an}}}} G(z,\zeta;\mathbf{E}_v)^{\mathrm{an}} + \log(|g_\zeta(z)|_v)$$

is well defined and finite.

(B) For all $x, y \in \mathcal{C}_v^{an}\backslash\mathbf{E}_v$ *with* $x \neq y$*,*

$$G(x,y;\mathbf{E}_v)^{\mathrm{an}} = G(y,x;\mathbf{E}_v)^{\mathrm{an}} .$$

(C) Let $\mathbf{E}_{v,1} \subseteq \mathbf{E}_{v,2}$ *be nonpolar, proper compact subsets of* $\mathcal{C}_v^{an}$*. Then for each* $y \in \mathcal{C}_v^{\mathrm{an}}\backslash\mathbf{E}_{v,2}$*, for all* $x \in \mathcal{C}_v^{\mathrm{an}}$ *with* $x \neq y$ *we have*

$$G(x,y;\mathbf{E}_{v,1})^{\mathrm{an}} \geq G(x,y;\mathbf{E}_{v,2})^{\mathrm{an}} .$$

This also holds when $y \in I(\mathcal{C}_v^{\mathrm{an}})\backslash\mathbf{E}_{v,2}$ *and* $x = y$*. For each* $\zeta \in \mathcal{C}_v(\mathbb{C}_v)\backslash\mathbf{E}_{v,2}$*,*

$$V_\zeta(\mathbf{E}_{v,1})^{\mathrm{an}} \geq V_\zeta(\mathbf{E}_{v,2})^{\mathrm{an}} .$$

(D) Let $\mathbf{K}_1 \supseteq \mathbf{K}_2 \supseteq \cdots \supseteq \mathbf{K}_n \cdots \supseteq \mathbf{E}_v$ *be a descending sequence of compact sets with* $\bigcap_{n=1}^\infty \mathbf{K}_n = \mathbf{E}_v$*. Then for all* $x, y \in \mathcal{C}_v^{\mathrm{an}}\backslash\mathbf{E}_v$ *such that* $x \neq y$*, or such that* $x = y \in I(\mathcal{C}_v^{\mathrm{an}})$*,*

$$\lim_{n\to\infty} G(x,y;\mathbf{K}_n)^{\mathrm{an}} = G(x,y;\mathbf{E}_v)^{\mathrm{an}} , \tag{4.22}$$

and for each $\zeta \in \mathcal{C}_v(\mathbb{C}_v)\backslash\mathbf{E}_v$*,*

$$\lim_{n\to\infty} V_\zeta(\mathbf{K}_n)^{\mathrm{an}} = V_\zeta(\mathbf{E}_v)^{\mathrm{an}} . \tag{4.23}$$

(E) Let $\mathbf{K}_1 \subseteq \mathbf{K}_2 \subseteq \cdots \subseteq \mathbf{K}_n \cdots \subseteq \mathbf{E}_v$ *be an ascending sequence of compact sets with* $\bigcup_{n=1}^\infty \mathbf{K}_n = \mathbf{E}_v$*. Then for all* $x, y \in \mathcal{C}_v^{\mathrm{an}}\backslash\mathbf{E}_v$ *such that* $x \neq y$*, or such that* $x = y \in I(\mathcal{C}_v^{\mathrm{an}})$*,*

$$\lim_{n\to\infty} G(x,y;\mathbf{K}_n)^{\mathrm{an}} = G(x,y;\mathbf{E}_v)^{\mathrm{an}} , \tag{4.24}$$

and for each $\zeta \in \mathcal{C}_v(\mathbb{C}_v)\backslash\mathbf{E}_v$*,*

$$\lim_{n\to\infty} V_\zeta(\mathbf{K}_n)^{\mathrm{an}} = V_\zeta(\mathbf{E}_v)^{\mathrm{an}} . \tag{4.25}$$

(F) For each $\sigma \in \mathrm{Aut}_c(\mathbb{C}_v/K_v)$*, and all* $x, y \in \mathcal{C}_v^{an}\backslash\mathbf{E}_v$ *with* $x \neq y$*,*

$$G(\sigma(x),\sigma(y);\sigma(\mathbf{E}_v))^{\mathrm{an}} = G(x,y;\mathbf{E}_v)^{\mathrm{an}} ,$$

and for each $\zeta \in \mathcal{C}_v(\mathbb{C}_v)\backslash\mathbf{E}_v$*, if the uniformizer* $g_{\sigma(\zeta)}(z)$ *is taken to be* $\sigma(g_\zeta)(z)$*, then*

$$V_{\sigma(\zeta)}(\sigma(\mathbf{E}_v))^{\mathrm{an}} = V_\zeta(\mathbf{E}_v)^{\mathrm{an}} .$$

PROOF. We first prove (A). When $\zeta \in \mathcal{C}_v(\mathbb{C}_v)\backslash\mathbf{E}_v$, the existence and finiteness of the limit defining $V_\zeta(\mathbf{E}_v)^{an}$ follows from the construction of $g_{\zeta,\mathbf{E}_v}$; see the proof of ([**64**], Théorème 3.6.15), noting that if V is the Berkovich closure of a suitably small isometrically parametrizable ball $B(\zeta, r)$ and y is its unique boundary point, the restriction of $\log(|g_\zeta|_v)$ to V satisfies $dd^c \log(|g_\zeta|_v) = \delta_\zeta - \delta_y$ and thus coincides, up to an additive constant, with the function $g_{\zeta,y}(z)$ from ([**64**], Lemma 3.4.14).

We next establish the diagonal case in (D) and (E). Let $\{\mathbf{K}_n\}_{n\geq 1}$ be a sequence of compact sets such that $\mathbf{K}_1 \subseteq \mathbf{K}_2 \subseteq \cdots \subseteq \mathbf{K}_n \cdots \subseteq \mathbf{E}_v$ and $\bigcup_{n=1}^\infty \mathbf{K}_n = \mathbf{E}_v$, or $\mathbf{K}_1 \supseteq \mathbf{K}_2 \supseteq \cdots \supseteq \mathbf{K}_n \cdots \supseteq \mathbf{E}_v$ and $\bigcap_{n=1}^\infty \mathbf{K}_n = \mathbf{E}_v$.

First suppose $x = y \in I(\mathcal{C}_v^{\mathrm{an}})\backslash\mathbf{E}_v$, and put $\Omega = \mathcal{C}_v^{\mathrm{an}}\backslash\{y\}$. For each compact nonpolar $\mathbf{K} \subset \Omega$, let $C(\mathbf{K},\Omega)$ be the capacity defined in ([**64**], §3.6). The construction in ([**64**], Théorème 3.6.15) shows that $g_{y,\mathbf{E}_v}(y) = C(\mathbf{E}_v,\Omega)^{-1}$ and

$g_{y,\mathbf{K}_n}(y) = C(\mathbf{K}_n, \Omega)^{-1}$ for all n. By ([**64**], Proposition 3.6.8, parts (ii) and (iv)), we have

$$C(\mathbf{E}_v, \Omega) = \lim_{n\to\infty} C(\mathbf{K}_n, \Omega) , \tag{4.26}$$

so (4.22) and (4.24) hold when $x = y \in I(\mathcal{C}_v^{\mathrm{an}})$.

A word is in order concerning the proof of ([**64**], Proposition 3.6.8). For an arbitrary subset $\mathbf{A} \subset \Omega$, Thuillier defines

$$C^*(\mathbf{A}, \Omega) = \sup_{\text{compact } \mathbf{K} \subseteq \mathbf{A}} C(\mathbf{K}, \Omega) ,$$

then shows that $C^*(\cdot, \Omega)$ is a Choquet capacity on subsets of Ω. Recall that this means $C^*(\cdot, \Omega)$ is an increasing set function, valued in $\mathbb{R}_{\geq 0}$ with $C(\Omega, \phi) = 0$, such that:

(1) For each increasing sequence $\mathbf{A}_1 \subseteq \mathbf{A}_2 \subseteq \cdots$ of arbitrary sets in Ω, if $A = \bigcup_{n=1}^{\infty} \mathbf{A}_n$ then $C^*(\Omega, A) = \lim_{n\to\infty} C^*(\Omega, A_n)$.
(2) For each decreasing sequence of *open* sets $\mathbf{U}_1 \supseteq \mathbf{U}_2 \supseteq \cdots$ in Ω, if $A = \bigcap_{n=1}^{\infty} \mathbf{U}_n$ then $C^*(\Omega, A) = \lim_{n\to\infty} C^*(\Omega, U_n)$.

In ([**64**], Proposition 3.6.8(ii)) Thuillier shows that for a compact subset $\mathbf{K} \subset \Omega$, one has $C(\mathbf{K}, \Omega) = C^*(\mathbf{K}, \Omega)$, and he deduces (4.26) for descending sequences of compact sets from this. His proof uses that $\mathbf{K} = \bigcap_{n=1}^{\infty} \mathbf{U}_n$ for a decreasing sequence of strict open affinoids $\mathbf{U}_1 \supseteq \mathbf{U}_2 \supseteq \cdots \supseteq \mathbf{K}$ (see the first line on ([**64**], p.112)). For a compact set on a Berkovich curve over an arbitrary complete valued field k this is not does not always hold (for example, if k has an uncountable residue field, take $\mathbf{K}$ to consist of a single type II point) but it does hold when $k = \mathbb{C}_v$. This is because $\mathbb{C}_v$ has a countable dense set, hence so do $\mathcal{C}_v(\mathbb{C}_v)$ and $\mathcal{C}_v^{\mathrm{an}}$. Consequently, if $\mathbf{K} \subseteq \mathcal{C}_v^{\mathrm{an}}$ is compact, then $\mathcal{C}_v^{\mathrm{an}} \backslash \mathbf{K}$ has countably many components. By ([**64**], Proposition 2.2.3), each component can be exhausted by an increasing sequence of strict closed affinoids. Using a diagonalization argument, one sees that $\mathbf{K}$ is the intersection of a decreasing sequence of strict open affinoids.

Next suppose $\zeta \in \mathcal{C}(\mathbb{C}_v) \backslash \mathbf{E}_v$. Let t be a tangent vector at ζ, and let $\mathrm{Cap}_{y,t}(\mathbf{K}) = \|t\|_{\mathbf{K}}^{c}$ be the function defined in ([**64**], Corollary 3.6.19). By ([**64**], Théorème 3.6.20), $\mathrm{Cap}_{y,t}(\mathbf{K})$ induces a Choquet capacity on subsets of $\mathcal{C}_v^{\mathrm{an}} \backslash \{y\}$. Thus if $\mathbf{K}_1 \subseteq \mathbf{K}_2 \subseteq \cdots \subseteq \mathbf{K}_n \cdots \subseteq \mathbf{E}_v$ and $\bigcup_{n=1}^{\infty} \mathbf{K}_n = \mathbf{E}_v$, or $\mathbf{K}_1 \supseteq \mathbf{K}_2 \supseteq \cdots \supseteq \mathbf{K}_n \cdots \supseteq \mathbf{E}_v$ and $\bigcap_{n=1}^{\infty} \mathbf{K}_n = \mathbf{E}_v$, then

$$\lim_{n\to\infty} \mathrm{Cap}_{y,t}(\mathbf{K}_n) = \mathrm{Cap}_{y,t}(\mathbf{E}_v) . \tag{4.27}$$

Now fix a uniformizing parameter $g_\zeta(z)$, and choose t so that $\langle t, g_\zeta \rangle = 1$. By the discussion on ([**64**], p.175), for each nonpolar compact $\mathbf{K} \subset \mathcal{C}_v^{\mathrm{an}} \backslash \mathbf{E}_v$,

$$V_\zeta(\mathbf{K})^{\mathrm{an}} = -\log(\mathrm{Cap}_{y,t}(\mathbf{K})) .$$

This yields (4.23) and (4.25).

We next prove a special case of (D). Suppose $\mathbf{K}_1 \supseteq \mathbf{K}_2 \supseteq \cdots \supseteq \mathbf{K}_n \cdots \supseteq \mathbf{E}_v$ is a descending sequence of compact sets with $\bigcap_{n=1}^{\infty} \mathbf{K}_n = \mathbf{E}_v$, and that in addition $\partial \mathbf{K}_n \subset I(\mathcal{C}_v^{\mathrm{an}})$ for each n. Fix $x, y \in \mathcal{C}_v^{\mathrm{an}} \backslash \mathbf{E}_v$ with $x \neq y$. After omitting finitely many $\mathbf{K}_n$ from the sequence, we can assume that $x, y \notin \mathbf{K}_1$.

If x and y belong to distinct components of $\mathcal{C}_v^{\mathrm{an}} \backslash \mathbf{E}_v$, they belong to distinct components of $\mathcal{C}_v^{\mathrm{an}} \backslash \mathbf{K}_n$ for all n, so $G(x, y; \mathbf{K}_n)^{\mathrm{an}} = G(x, y; \mathbf{E}_v) = 0$ for all n, and the result is trivial. Assume they belong to the same component U of $\mathcal{C}_v^{\mathrm{an}} \backslash \mathbf{E}_v$.

For all sufficiently large n, they belong to the same component U_n of $\mathcal{C}_v^{\mathrm{an}}\backslash\mathbf{K}_n$, and without loss we can assume they belong to U_n for all n. For each n, put

$$h_n(z) \;=\; G(z,y;\mathbf{E}_v)^{\mathrm{an}} - G(z,y;\mathbf{K}_n)^{\mathrm{an}} \;, \tag{4.28}$$

taking $h_n(y) = \lim_{z\to y} G(z,y;\mathbf{E}_v)^{\mathrm{an}} - G(z,y;\mathbf{K}_n)^{\mathrm{an}} = V_y(\mathbf{E}_v)^{\mathrm{an}} - V_y(\mathbf{K}_n)^{\mathrm{an}}$ if $y \in \mathcal{C}_v(\mathbb{C}_v)$. Then $h_n(z)$ is harmonic on U_n in the sense of ([**64**], §2.3). Note that $\partial U_n \subset \partial\mathbf{K}_n \subset I(\mathcal{C}_v^{\mathrm{an}})$. By ([**64**], Propositions 3.1.19 and 3.1.20), $G(z,y;\mathbf{K}_n)$ is continuous at each point of ∂U_n and vanishes on ∂U_n, so for each $p \in \partial U_n$

$$\liminf_{\substack{z\to p\\ z\in U_n}} h_n(z) \;\geq\; 0 \;.$$

Hence the maximum principle for harmonic functions ([**64**], Proposition 3.1.1) shows that $h_n(z) \geq 0$ for all $z \in U_n$, and in particular for all $z \in U_1$. Similarly, we see that $h_1(z) \geq h_2(z) \geq \cdots \geq 0$ for all $z \in U_1$.

To conclude the argument, we apply Harnack's Principle ([**64**], Proposition 3.1.2) to the functions $h_n(z)$ on U_1. By the diagonal case of (D) shown above, we have $\lim_{n\to\infty} h_n(y) = 0$. It follows from Harnack's Principle that the $h_n(z)$ converge uniformly to 0 on compact subsets of U_1 and, consequently,

$$\lim_{n\to\infty} G(x,y;\mathbf{K}_n)^{\mathrm{an}} \;=\; G(x,y;\mathbf{E}_v)^{\mathrm{an}} \;. \tag{4.29}$$

We can now prove (B), the symmetry of $G(x,y;\mathbf{E}_v)$. Fix $x,y \in \mathcal{C}_v^{\mathrm{an}}\backslash\mathbf{E}_v$ with $x \neq y$. If x and y belong to distinct components of $\mathcal{C}_v^{\mathrm{an}}\backslash\mathbf{E}_v$ then trivially $G(x,y;\mathbf{E}_v)^{\mathrm{an}} = 0 = G(y,x;\mathbf{E}_v)^{\mathrm{an}}$, so we can assume they belong to the same component U. Put $\mathbf{E}_v^U = \mathcal{C}_v^{\mathrm{an}}\backslash U$. The characterization of Green's functions in ([**64**], Théorème 3.6.15) shows that $G(z,y;\mathbf{E}_v^U)^{\mathrm{an}} = G(z,y;\mathbf{E}_v)^{\mathrm{an}}$ for all $z \in U$; in particular $G(x,y;\mathbf{E}_v^U)^{\mathrm{an}} = G(x,y;\mathbf{E}_v)^{\mathrm{an}}$.

By ([**64**], Proposition 2.2.23), there is an exhaustion of U by an increasing sequence of strict open affinoid domains $V_1 \subset V_2 \subset \cdots \subset U$ with $\partial V_n \subset V_{n+1}$ for each n; without loss, we can assume that $x, y \in V_1$. By the definition of an affinoid, $\partial V_n \subset I(\mathcal{C}_v^{\mathrm{an}})$ for each n. Let $g_y^{V_n}(z)$ be the Green's function of the domain V_n defined in ([**64**], Proposition 3.3.7(ii)). Then $g_y^{V_n}(z)$ is smooth, vanishes on ∂V_n, and satisfies the distributional equation $dd^c g_y^{V_n} = \mu_y^{V_n} - \delta_y$ where $\mu_y^{V_n}$ is a probability measure supported on ∂V_n. Put $\mathbf{K}_n = \mathcal{C}_v^{an}\backslash V_n$. Again by the characterization of Green's functions in ([**64**], Théorème 3.6.15), for all $z \in V_n$ we have $G(z,y;\mathbf{K}_n)^{an} = g_y^{V_n}(z)$.

Here $\mathbf{K}_1 \supset \mathbf{K}_2 \supset \cdots \supset \mathbf{E}_v^U$ is a descending sequence of compact sets with $\bigcap_{n=1}^\infty \mathbf{K}_n = \mathbf{E}_v^U$, and $\partial\mathbf{K}_n = \partial V_n \subset I(\mathcal{C}_v^{\mathrm{an}})$ for each n. By the special case of (E) shown above, we have

$$\lim_{n\to\infty} G(x,y;\mathbf{K}_n)^{\mathrm{an}} \;=\; G(x,y;\mathbf{E}_v^U)^{\mathrm{an}} \;=\; G(x,y;\mathbf{E}_v)^{\mathrm{an}} \;. \tag{4.30}$$

A similar formula holds with x and y interchanged. By ([**64**], Corollary 3.3.9(i)), for each n we have $g_x^{V_n}(y) = g_y^{V_n}(x)$. Combining these facts shows that $G(x,y;\mathbf{E}_v)^{\mathrm{an}} = G(y,x;\mathbf{E}_v)^{\mathrm{an}}$.

Part (C), the monotonicity of $G(x,y,\mathbf{E}_v)^{\mathrm{an}}$, follows by a related argument. Let $\mathbf{E}_{v,1} \subseteq \mathbf{E}_{v,2}$ be nonpolar, proper compact sets of $\mathcal{C}_v^{\mathrm{an}}$. Fix $x, y \in \mathcal{C}_v^{\mathrm{an}}$ with $x \neq y$ and $y \in \mathcal{C}_v^{\mathrm{an}}\backslash\mathbf{E}_{v,2}$, and let U be the component of $\mathcal{C}_v^{\mathrm{an}}\backslash\mathbf{E}_{v,2}$ containing y.

First suppose $x \in U$. Putting $\mathbf{E}_{v,2}^U = \mathcal{C}_v^{\mathrm{an}}\backslash\mathbf{E}_{v,2}$, let $\mathbf{K}_1 \supseteq \mathbf{K}_2 \supseteq \cdots \supseteq \mathbf{E}_{v,2}^U$ be a descending sequence of compact sets with $\bigcap_{n=1}^\infty \mathbf{K}_n = \mathbf{E}_{v,2}^U$ and $\partial\mathbf{K}_n = \partial V_n \subset$

$I(\mathcal{C}_v^{\mathrm{an}})$ for each n. By (4.30) we have

$$\lim_{n\to\infty} G(x,y;\mathbf{K}_n)^{\mathrm{an}} \ = \ G(x,y;\mathbf{E}_{v,2})^{\mathrm{an}} \ .$$

On the other hand, the same argument that gave (4.28) shows that

$$G(x,y;\mathbf{E}_{v,1}) \ \geq \ G(x,y;\mathbf{K}_n)^{\mathrm{an}}$$

for each n. Thus $G(x,y;\mathbf{E}_{v,1})^{\mathrm{an}} \ \geq \ G(x,y;\mathbf{E}_{v,2})^{\mathrm{an}}$.

Now suppose $x \notin U$. By the characterization of Green's functions in ([**64**], Théorème 3.6.15) we have $G(z,y;\mathbf{E}_v)^{\mathrm{an}} = G(z,y;\partial U)^{\mathrm{an}}$ for all z. First assume $x \notin \partial U$. Then x and y belong to distinct components of $\mathcal{C}_v^{\mathrm{an}} \backslash \partial U$, and trivially $G(x,y;\mathbf{E}_{v,1})^{\mathrm{an}} \geq 0 = G(x,y;\partial U) = G(x,y;\mathbf{E}_{v,2})^{\mathrm{an}}$. Last, assume $x \in \partial U$. Since $G(z,y;E_{v,1})$ and $G(z,y;E_{v,2})$ are subharmonic, necessarily they are upper semicontinuous (see [**64**], Définition 3.1.5). By what has been shown above,

$$G(x,y;\mathbf{E}_{v,2}) \ \geq \ \lim_{\substack{z\to x\\ z\in U}} G(z,y;\mathbf{E}_{v,2}) \ \geq \ \lim_{\substack{z\to x\\ z\in U}} G(z,y;\mathbf{E}_{v,1}) \ = \ G(x,y;\mathbf{E}_{v,1}) \ .$$

This establishes the desired inequality in all cases.

Finally, consider the Robin constants. If $\zeta \in \mathcal{C}_v^{\mathrm{an}} \backslash \mathbf{E}_{v,2}$, it follows that

$$\begin{aligned} V_\zeta(\mathbf{E}_{v,1})^{\mathrm{an}} \ &= \ \lim_{z\to\zeta} G(x,\zeta;\mathbf{E}_{v,1})^{\mathrm{an}} + \log(|g_\zeta(z)|_v) \\ &\geq \ \lim_{z\to\zeta} G(x,\zeta;\mathbf{E}_{v,2})^{\mathrm{an}} + \log(|g_\zeta(z)|_v) \ = \ V_\zeta(\mathbf{E}_{v,2})^{\mathrm{an}} \ . \end{aligned}$$

We can now prove (D) in full generality. Since we have already established the diagonal case, we only consider the nondiagonal case. Let $\{\mathbf{K}_n\}_{n\geq 1}$ be a sequence of compact sets with $\mathbf{K}_1 \supseteq \mathbf{K}_2 \supseteq \cdots \supseteq \mathbf{K}_n \cdots \supseteq \mathbf{E}_v$ such that $\bigcap_{n=1}^\infty \mathbf{K}_n = \mathbf{E}_v$, and fix $x,y \in \mathcal{C}_v^{\mathrm{an}} \backslash \mathbf{E}_v$ with $x \neq y$. For all sufficiently large n we have $x,y \notin \mathbf{K}_n$, so after omitting finitely many $\mathbf{K}_n$ we can assume without loss that $x,y \notin \mathbf{K}_1$.

Let U be the component of $\mathcal{C}_v^{\mathrm{an}} \backslash \mathbf{E}_v$ containing y. If $x \notin U$, then x and y belong to distinct components of K_n for all n, so $G(x,y;\mathbf{K}_n) = 0 = G(x,y;\mathbf{E}_v)$ for all n, and (4.22) is trivial. Suppose $x \in U$. After omitting finitely many K_n, we can assume that x and y belong to the same component U_1 of $\mathcal{C}_v^{\mathrm{an}} \backslash \mathbf{K}_1$. For each n, put $h_n(z) = G(z,y;E_v) - G(z,y;\mathbf{K}_n)$, taking $h_n(y) = V_y(\mathbf{K}_n) - V_y(\mathbf{E}_v)$ if $y \in \mathcal{C}_v(\mathbb{C}_v)$. Then $h_n(z)$ is harmonic in U_1, and by part (C), it is nonnegative. By the diagonal case of (D) shown above, we have $\lim_{n\to\infty} h_n(y) = 0$. Hence Harnack's Principle ([**64**], Proposition 3.1.2) gives that as $n \to \infty$, then $h_n(z) \to 0$ uniformly on compact subsets of U_1. In particular,

$$\lim_{n\to\infty} G(x,y;\mathbf{K}_n)^{\mathrm{an}} \ = \ G(x,y;\mathbf{E}_v)^{\mathrm{an}} \ .$$

The nondiagonal case of (E) follows by a similar argument, but uses the topology of $\mathcal{C}_v^{\mathrm{an}}$ in a stronger way. Let $\{\mathbf{K}_n\}_{n\geq 1}$ be a sequence of compact sets with $\mathbf{K}_1 \subseteq \mathbf{K}_2 \subseteq \cdots \subseteq \mathbf{K}_n \cdots \subseteq \mathbf{E}_v$ such that $\bigcup_{n=1}^\infty \mathbf{K}_n = \mathbf{E}_v$, and fix $x,y \in \mathcal{C}_v^{\mathrm{an}} \backslash \mathbf{E}_v$ with $x \neq y$.

Let $\Gamma_{x,y}$ be the union of all paths connecting x and y in $\mathcal{C}_v^{\mathrm{an}}$. Recall that there is a finite subgraph $\mathbf{S}$ of $\mathcal{C}_v^{\mathrm{an}}$, called its *skeleton*, such that there is retraction $\tau : \mathcal{C}_v^{\mathrm{an}} \to \mathbf{S}$ (see ([**64**], Théorème 2.210)). Each component of $\mathcal{C}_v^{\mathrm{an}} \backslash \mathbf{S}$ is a tree. It follows that $\Gamma_{x,y}$ is a graph with finitely many edges.

If x and y belong to distinct components of $\mathcal{C}_v^{\mathrm{an}} \backslash \mathbf{E}_v$, they belong to distinct components of $\Gamma_{x,y} \backslash \mathbf{E}_v$. Since $\Gamma_{x,y}$ has finite connectivity, there is a finite subset $P \subset \mathbf{E}_v \cap \Gamma_{x,y}$ which disconnects x from y. For all sufficiently large n we have

$P \subset \mathbf{K}_n$, and for such n it follows that $G(x,y;\mathbf{K}_n) = 0 = G(x,y;\mathbf{E}_v)$, so (4.24) is trivial.

Suppose x and y belong to the same component U of $\mathcal{C}_v^{\mathrm{an}} \backslash \mathbf{E}_v$. For each n, put $h_n(z) = G(z,y;K_n) - G(z,y;\mathbf{E}_v)$, taking $h_n(y) = V_y(\mathbf{E}_v) - V_y(\mathbf{K}_n)$ if $y \in \mathcal{C}_v(\mathbb{C}_v)$. Then $h_n(z)$ is harmonic in U, and by part (C), it is nonnegative. By the diagonal case of (E) shown above, we have $\lim_{n\to\infty} h_n(y) = 0$. Hence Harnack's Principle gives that as $n \to \infty$, then $h_n(z) \to 0$ uniformly on compact subsets of U, and again we conclude that

$$\lim_{n\to\infty} G(x,y;\mathbf{K}_n)^{\mathrm{an}} = G(x,y;\mathbf{E}_v)^{\mathrm{an}} .$$

Part (F), the functoriality of $G(x,y;\mathbf{E}_v)^{\mathrm{an}}$ and $V_\zeta(\mathbf{E}_v)^{\mathrm{an}}$ under $\mathrm{Aut}_c(\mathbb{C}_v/K_v)$, is immediate from the definition of the action of $\sigma \in \mathrm{Aut}_c(\mathbb{C}_v/K_v)$ on $\mathcal{C}_v^{\mathrm{an}}$ through its action on $\mathbb{C}_v$, and the characterization of $g_{y,\mathbf{E}_v}(z)$ in ([**64**], Théorème 3.6.15). $\square$

The following proposition establishes the compatibility of our Green's functions and Berkovich Green's functions. Note that in ([**64**], Théorème 5.1.2), Thuillier has already established the compatibility of our capacities with his. Throughout ([**64**]) Thuillier uses the natural logarithm $\log(x)$, while in v-adic constructions we use the logarithm $\log_v(x)$ to the base q_v in order to have $\log_v(|z|_v) \in \mathbb{Q}$ for $z \in \mathbb{C}_v^\times$. This gives rise to a factor of $\log(q_v)$ in comparing our Green's functions and his.

PROPOSITION 4.4 (Compatibility of Green's Functions). *Let K be a global field, and let $\mathcal{C}/K$ be a smooth, connected, projective curve. Let v be a nonarchimedean place of K, and let $\mathcal{C}_v^{\mathrm{an}}$ be the Berkovich analytification of $\mathcal{C}_v \times_{K_v} \mathrm{Spec}(\mathbb{C}_v)$.*

Suppose $E_v \subsetneq \mathcal{C}_v(\mathbb{C}_v)$ is an algebraically capacitable set with positive capacity, and $\mathbf{E}_v$ is its closure in $\mathcal{C}_v^{\mathrm{an}}$. Then $\mathbf{E}_v$ is a proper compact, nonpolar subset of $\mathcal{C}_v^{\mathrm{an}}$, and for all $z, \zeta \in \mathcal{C}_v(\mathbb{C}_v)\backslash E_v$,

$$G(z,\zeta;\mathbf{E}_v)^{\mathrm{an}} = G(z,\zeta;E_v)\log(q_v) , \qquad V_\zeta(\mathbf{E}_v)^{\mathrm{an}} = V_\zeta(E_v)\log(q_v) .$$

PROOF. We begin by considering two special cases: compact sets and PL_ζ-domains.

First, let $E_v \subset \mathcal{C}_v(\mathbb{C}_v)$ be a compact set with positive capacity. Since E_v is compact and the restriction of the topology on $\mathcal{C}_v^{\mathrm{an}}$ to $\mathcal{C}_v(\mathbb{C}_v)$ is the usual v-adic topology, E_v coincides with its Berkovich closure $\mathbf{E}_v$. Fix a point $\zeta \in \mathcal{C}_v(\mathbb{C}_v)\backslash E_v$ and a uniformizing parameter $g_\zeta(z)$, and let $[z,w]_\zeta$ be the canonical distance normalized so that $\lim_{z\to\zeta}[z,w]_\zeta \cdot |g_\zeta(z)|_v = 1$ for each $w \neq \zeta$ (see §3.5). By definition, our Robin constant is

$$V_\zeta(E_v) = \inf_\nu \iint -\log_v([x,y]_\zeta)\, d\nu(x)d\nu(y)$$

and our capacity is

$$\gamma_\zeta(E_v) = q_v^{-V_\zeta(E_v)} .$$

Let μ_ζ be the equilibrium distribution of E_v with respect to ζ: the unique probability measure on E_v which minimizes the energy integral

$$I_\zeta(\nu) = \inf_\nu \iint -\log_v([x,y]_\zeta)\, d\nu(x)d\nu(y)$$

(see §3.8). Then the potential function

$$u_{E_v}(z,\zeta) = \int -\log_v([z,w]_\zeta)\, d\mu_\zeta(w)$$

satisfies $u_{E_v}(z,\zeta) \le V_\zeta(E_v)$ for all $z \in \mathcal{C}_v(\mathbb{C}_v)\backslash\{\zeta\}$, takes the value $V_\zeta(E_v)$ on $E_v\backslash e_v$ where e_v is an F-sigma set of inner capacity 0, and has $u_{E_v}(z,\zeta) < V_\zeta(E_v)$ for all $z \in \mathcal{C}_v(\mathbb{C}_v)\backslash E_v$. In (3.36) we have defined

$$G(z,\zeta;E_v) \;=\; V_\zeta(z) - u_{E_v}(z,\zeta) \;=\; V_\zeta(E_v) + \int \log_v([z,w]_\zeta)\,d\mu_\zeta(w) \;.$$

In ([**64**], Théorème 5.1.5), Thuillier shows there is a unique extension of the canonical distance to a function on $(\mathcal{C}_v^{\mathrm{an}}\backslash\{\zeta\}) \times (\mathcal{C}_v^{\mathrm{an}}\backslash\{\zeta\})$, which we will denote $[z,w]_\zeta^{\mathrm{an}}$. The function $[z,w]_\zeta^{\mathrm{an}}$ is continuous, symmetric, and satisfies the Laplace equation $dd^c\log([z,w]_\zeta^{\mathrm{an}}) = \delta_w - \delta_\zeta$ for $w \ne \zeta$. Noting that $\log(x) = \log_v(x)\log(q_v)$, define

$$g_{\mu_\zeta}^{\mathrm{an}}(z) \;=\; V_\zeta(E_v)\log(q_v) + \int \log([z,w]_\zeta^{\mathrm{an}})\,d\mu_\zeta(w)$$

for $z \in \mathcal{C}_v^{\mathrm{an}}\backslash\{\zeta\}$. By arguments like those in ([**64**],Proposition 3.4.16), $g_{\mu_\zeta}^{\mathrm{an}}(z)$ is subharmonic on $\mathcal{C}_v^{\mathrm{an}}\backslash\{\zeta\}$, harmonic on $\mathcal{C}_v^{\mathrm{an}}\backslash(E_v \cup \{\zeta\})$, and satisfies $dd^c g_{\mu_\zeta}^{\mathrm{an}} = \mu_\zeta - \delta_\zeta$. Clearly $g_{\mu_\zeta}^{\mathrm{an}}(z) = G(z,\zeta;E_v)\log(q_v)$ for $z \in \mathcal{C}_v(\mathbb{C}_v)\backslash\{\zeta\}$. In particular, it vanishes on $\mathbf{E}_v = E_v$ except possibly on the set e_v. By ([**64**], Théorème 3.6.11 and Théorème 5.1.2), the set e_v is polar. The characterization of Green's functions in ([**64**], Théorème 3.6.15), shows that $G(z,\zeta;\mathbf{E}_v)^{an} = g_{\zeta,\mathbf{E}_v}(z) = g_{\mu_\zeta}^{\mathrm{an}}(z)$. Thus for all $z \in \mathcal{C}_v(\mathbb{C}_v)\backslash\{\zeta\}$,

$$G(z,\zeta;\mathbf{E}_v)^{\mathrm{an}} \;=\; G(z,\zeta;E_v)\log(q_v) \;.$$

It follows that $V_\zeta(\mathbf{E}_v)^{\mathrm{an}} \;=\; V_\zeta(E_v)\log(q_v)$.

Next, let $E_v \subsetneq \mathcal{C}_v(\mathbb{C}_v)$ be a PL$_\zeta$-domain in the sense of ([**51**], Definition 4.2.6): there is a nonconstant $f(z) \in \mathbb{C}_v(\mathcal{C})$ having poles only at ζ, such that $E_v = \{z \in \mathcal{C}_v(\mathbb{C}_v) : |f(z)|_v \le 1\}$. Given a function $f(z)$ defining E_v as a PL$_\zeta$-domain,

$$G(z,\zeta;E_v) \;=\; \begin{cases} \frac{1}{\deg}(f)\log_v(|f(z)|_v) & \text{if } z \in \mathcal{C}_v(\mathbb{C})\backslash E_v \;, \\ 0 & \text{if } z \in E_v \;, \end{cases}$$

where $\log_v(x)$ is the logarithm to the base q_v; by ([**51**], Proposition 4.4.1), $G(z,\zeta;E_v)$ is independent of the choice of f. The closure of E_v in $\mathcal{C}_v^{\mathrm{an}}$ is $\mathbf{E}_v = \{z \in \mathcal{C}_v^{\mathrm{an}} : |f(z)|_v \le 1\}$ and by the discussion on ([**64**], p.175)

$$G(z,\zeta;\mathbf{E}_v)^{\mathrm{an}} \;=\; \begin{cases} \frac{1}{\deg}(f)\log(|f(z)|_v) & \text{if } z \in \mathcal{C}_v^{\mathrm{an}}\backslash\mathbf{E}_v \;, \\ 0 & \text{if } z \in \mathbf{E}_v \;, \end{cases}$$

where $\log(x) = \ln(x)$. It follows that for all $z \in \mathcal{C}_v(\mathbb{C}_v)\backslash\{\zeta\}$

$$G(z,\zeta;\mathbf{E}_v)^{\mathrm{an}} \;=\; G(z,\zeta;E_v)\log(q_v) \tag{4.31}$$

and that

$$V_\zeta(\mathbf{E}_v)^{\mathrm{an}} \;=\; V_\zeta(E_v)\log(q_v) \;, \tag{4.32}$$

We can now deal with the general case. Let $E_v \subsetneq \mathcal{C}_v(\mathbb{C}_v)$ be an algebraically capacitable set with positive capacity, and let $\mathbf{E}_v$ be its closure in $\mathcal{C}_v^{\mathrm{an}}$. Note that E_v is closed in $\mathcal{C}_v(\mathbb{C}_v)$ by ([**51**], Proposition 4.3.15). Since the topology on $\mathcal{C}_v^{\mathrm{an}}$ restricts to v-adic topology on $\mathcal{C}_v(\mathbb{C}_v)$, this implies that $\mathbf{E}_v \cap \mathcal{C}_v(\mathbb{C}_v) = E_v$. In particular, $\mathbf{E}_v$ is a proper subset of $\mathcal{C}_v^{\mathrm{an}}$. It is clearly compact, and it is nonpolar since E_v contains compact subsets of $\mathcal{C}_v(\mathbb{C}_v)$ with positive capacity.

Fix $\zeta \in \mathcal{C}_v(\mathbb{C}_v)\backslash E_v$. By ([**51**], Definition 4.3.2) we have

$$\inf_{\text{compact } K \subseteq E_v} V_\zeta(K) = V_\zeta(E_v) , \qquad \sup_{\text{PL}_\zeta\text{-domains } U \supseteq E_v} V_\zeta(U) = V_\zeta(E_v) .$$

Since the union of finitely compact sets is compact, and the intersection of finitely many PL_ζ-domains is a PL_ζ-domain (see ([**51**], Corollary 4.2.13)), there are an ascending sequence of compact sets $K_1 \subseteq K_2 \subseteq \cdots \subseteq E_v$ with

$$\lim_{n\to\infty} V_\zeta(K_n) \;=\; V_\zeta(E_v) , \tag{4.33}$$

and a descending sequence of PL_ζ-domains $U_1 \supseteq U_2 \supseteq \cdots \supseteq E_v$ with

$$\lim_{n\to\infty} V_\zeta(U_n) \;=\; V_\zeta(E_v) . \tag{4.34}$$

By ([**51**], Lemma 4.4.7 and Definition 4.4.12), for each $z \in \mathcal{C}_v(\mathbb{C}_v)\backslash(E_v \cup \{\zeta\})$ we have

$$\lim_{n\to\infty} G(z,\zeta;K_n) = G(z,\zeta;E_v) , \qquad \lim_{n\to\infty} G(z,\zeta;U_n) = G(z,\zeta;E_v) . \tag{4.35}$$

By the compatibility of Green's functions for compact sets and PL_ζ-domains shown above, if $\mathbf{K}_n$ and $\mathbf{U}_n$ are the closures of K_n and U_n in $\mathcal{C}_v^{\text{an}}$ respectively, then for all $z \in \mathcal{C}_v(\mathbb{C}_v)\backslash\{\zeta\}$ and all n,

$$G(z,\zeta;\mathbf{K}_n)^{\text{an}} = G(z,\zeta;K_n)\log(q_v) , \qquad G(z,\zeta;\mathbf{U}_n)^{\text{an}} = G(z,\zeta;\mathbf{U}_n)\log(q_v) . \tag{4.36}$$

Clearly

$$\mathbf{K}_1 \;\subseteq\; \mathbf{K}_2 \;\subseteq\; \cdots \;\subseteq\; \mathbf{E}_v \;\subseteq \cdots \;\subseteq \mathbf{U}_2 \;\subseteq\; \mathbf{U}_1 ,$$

so by the monotonicity of Green's functions proved in Proposition 4.3(C), for all $z \in \mathcal{C}_v^{\text{an}}\backslash\{\zeta\}$

$$\begin{aligned} G(z,\zeta;\mathbf{K}_1)^{\text{an}} \;&\geq\; G(z,\zeta;\mathbf{K}_2)^{\text{an}} \;\geq\; \cdots \\ &\geq\; G(z,\zeta;\mathbf{E}_v)^{\text{an}} \;\geq \cdots \;\geq\; G(z,\zeta;\mathbf{U}_2)^{\text{an}} \;\geq\; G(z,\zeta;\mathbf{U}_1)^{\text{an}} . \end{aligned} \tag{4.37}$$

Combining (4.35), (4.36) and (4.37) shows that for each $z \in \mathcal{C}_v(\mathbb{C}_v)\backslash(E_v \cup \{\zeta\})$,

$$G(z,\zeta;\mathbf{E}_v)^{\text{an}} \;=\; G(z,\zeta;E_v)\log(q_v) .$$

In a similar way, from (4.33), (4.34), the compatibility of Robin constants for compact sets and PL_ζ domains, and the monotonicity of Robin constants proved in Proposition 4.3(C), we see that

$$V_\zeta(\mathbf{E}_v)^{\text{an}} \;=\; V_\zeta(E_v)\log(q_v) . \qquad \square$$

We can now prove Theorems 1.6 and 1.7.

Theorem 1.6 (Berkovich FSZ with LR). *Let K be a global field, and let $\mathcal{C}/K$ be a smooth, geometrically integral, projective curve. Let $\mathfrak{X} = \{x_1, \ldots, x_m\} \subset \mathcal{C}(\widetilde{K})$ be a finite set of points stable under $\text{Aut}(\widetilde{K}/K)$, and let $\mathbb{E} = \prod_v \mathbf{E}_v \subset \prod_v \mathcal{C}_v^{\text{an}}$ be a compact Berkovic adelic set compatible with $\mathfrak{X}$., uch that each $\mathbf{E}_v$ is stable under $\text{Aut}_c(\mathbb{C}_v/K_v)$. Let $S \subset \mathcal{M}_K$ be a finite set of places v, containing all archimedean v, such that $\mathbf{E}_v$ is $\mathfrak{X}$-trivial for each $v \notin S$.*

Assume that $\gamma(\mathbb{E},\mathfrak{X}) > 1$. Assume also that for each $v \in S$:

(A) If v is archimedean and $K_v \cong \mathbb{C}$, then $\mathbf{E}_v$ is compact, and is a finite union of sets $E_{v,i}$, each of which is compact, connected, and bounded by finitely many Jordan curves:

(B) If v is archimedean and $K_v \cong \mathbb{R}$, then $\mathbf{E}_v$ is compact and is a finite union of sets $E_{v,\ell}$, where each $E_{v,\ell}$ is either

(1) *compact, connected, and bounded by finitely many Jordan curves, or*

(2) *is a closed subinterval of* $\mathcal{C}_v(\mathbb{R})$ *with nonempty interior.*

(C) If v *is nonarchimedean, then* $\mathbf{E}_v$ *is compact and is a finite union of sets* $E_{v,\ell}$*, where each* $E_{v,\ell}$ *is either*

(1) *a strict closed Berkovich affinoid, or*

(2) *is a compact subset of* $\mathcal{C}_v(\mathbb{C}_v)$ *and has the form* $\mathcal{C}_v(F_{w,\ell}) \cap B(a_\ell, r_\ell)$ *for some finite separable extension* $F_{w,\ell}/K_v$ *in* $\mathbb{C}_v$*, and some ball* $B(a_\ell, r_\ell)$.

Then there are infinitely many points $\alpha \in \mathcal{C}(\widetilde{K}^{\mathrm{sep}})$ *such that for each* $v \in \mathcal{M}_K$*, the* $\mathrm{Aut}(\widetilde{K}/K)$*-conjugates of* α *all belong to* $\mathbf{E}_v$.

Proof of Theorem 1.6, assuming Theorem 0.4. For each $v \in \mathcal{M}_K$, put $E_v^0 = \mathbf{E}_v \cap \mathcal{C}_v(\mathbb{C}_v)$. By the hypotheses of Theorem 1.6, E_v^0 is algebraically capacitible and satisfies the hypotheses of Theorem 0.4. Those hypotheses in turn show that the Berkovich closure of E_v^0 is $\mathbf{E}_v$, so by Proposition 4.4 for all $x_i \neq x_j \in \mathfrak{X}$ we have

$$G(x_i, x_j; E_v^0) \;=\; G(x_i, x_j; \mathbf{E}_v)^{\mathrm{an}} \;, \qquad V_{x_i}(E_v^0) \;=\; V_{x_i}(\mathbf{E}_v)^{\mathrm{an}} \;.$$

Put $\mathbb{E}^0 = \prod_v E_v^0$. Then $\mathbb{E}^0$ is a K-rational adelic set compatible with $\mathfrak{X}$ and has $\gamma(\mathbb{E}^0, \mathfrak{X}) > 1$. By Theorem 1.6 there are infinitely many points $\alpha \in \mathcal{C}(\widetilde{K}^{\mathrm{sep}})$ such that for each $v \in \mathcal{M}_K$, the $\mathrm{Aut}(\widetilde{K}/K)$-conjugates of α all belong to E_v, hence to $\mathbf{E}_v$. □

Theorem 1.7 (Berkovich Fekete/FSZ with LR for Quasi-neighborhoods). *Let* K *be a global field, and let* $\mathcal{C}/K$ *be a smooth, connected, projective curve. Let* $\mathfrak{X} = \{x_1, \ldots, x_m\} \subset \mathcal{C}(\widetilde{K})$ *be a finite set of points stable under* $\mathrm{Aut}(\widetilde{K}/K)$*, and let* $\mathbb{E} = \prod_v \mathbf{E}_v \subset \prod_v \mathcal{C}_v^{\mathrm{an}}$ *be a compact Berkovich adelic set compatible with* $\mathfrak{X}$*, such that each* $\mathbf{E}_v$ *is stable under* $\mathrm{Aut}_c(\mathbb{C}_v/K_v)$.

(A) If $\gamma(\mathbb{E}, \mathfrak{X})^{\mathrm{an}} < 1$*, there is a* K*-rational Berkovich neighborhood* $\mathbb{U} = \prod_v \mathbf{U}_v$ *of* $\mathbb{E}$ *such that there are only finitely many points of* $\mathcal{C}(\widetilde{K})$ *whose* $\mathrm{Aut}(\widetilde{K}/K)$*-conjugates are all contained in* $\mathbf{U}_v$*, for each* $v \in \mathcal{M}_K$.

(B) If $\gamma(\mathbb{E}, \mathfrak{X})^{\mathrm{an}} > 1$*, then for any* K*-rational separable Berkovich quasi-neighborhood* $\mathbb{U}$ *of* $\mathbb{E}$*, there are infinitely many points* $\alpha \in \mathcal{C}(\widetilde{K}^{\mathrm{sep}})$ *such that for each* $v \in \mathcal{M}_K$*, the* $\mathrm{Aut}(\widetilde{K}/K)$*-conjugates of* α *all belong to* $\mathbf{U}_v$.

Proof of Theorem 1.7, assuming Theorem 1.2. We first prove (A). Suppose $\gamma(\mathbb{E}, \mathfrak{X})^{\mathrm{an}} < 1$. We begin by enlarging $\mathbb{E} = \prod_v \mathbf{E}_v$ to a set $\mathbb{F} = \prod_v \mathbf{F}_v$ with $\gamma(\mathbb{F}, \mathfrak{X}) < 1$, such that $\mathbf{F}_v$ is a strict closed affinoid for each nonarchimedean v. Let $\varepsilon > 0$ be small enough that if $\Gamma \in M_n(\mathbb{R})$ is a symmetric $n \times n$ matrix whose entries differ from those of $\Gamma(\mathbb{E}, \mathfrak{X})^{\mathrm{an}}$ by at most ε, then $\mathrm{val}(\Gamma) > 1$. Fix a nonempty finite set of places S of K containing all archimedean places and all nonarchimedean places where $\mathbf{E}_v$ is not $\mathfrak{X}$-trivial, and choose a set of numbers $\{\varepsilon_v\}_{v \in S}$ with $\varepsilon_v > 0$ for each v and $\sum_{v \in S} \varepsilon_v = \varepsilon$.

If v is archimedean, put $\mathbf{F}_v = \mathbf{E}_v$; likewise if $v \notin S$, so $\mathbf{E}_v$ is $\mathfrak{X}$-trivial, put $\mathbf{F}_v = \mathbf{E}_v$. Suppose $v \in S$ is nonarchimedean. By hypothesis $\mathbf{E}_v$ is compact, nonpolar, and stable under $\mathrm{Aut}_c(\mathbb{C}_v/K_v)$. As noted in the proof of Proposition 4.3, the fact that $\mathbb{C}_v$ has a countable dense set means there is a descending sequence of strict closed affinoids $\mathbf{K}_1 \supseteq \mathbf{K}_2 \supseteq \cdots \supseteq \mathbf{E}_v$ with $\bigcap_{n=1}^{\infty} \mathbf{K}_n = \mathbf{E}_v$. By Proposition 4.3(D), if n is large enough, then for all $x_i, x_j \in \mathfrak{X}$ with $i \neq j$ we have

$$|G(x_i, x_j; \mathbf{E}_v)^{\mathrm{an}} - G(x_i, x_j; \mathbf{K}_n)^{\mathrm{an}}| \;<\; \varepsilon_v \;,$$

and for each $x_i \in \mathfrak{X}$,

$$|V_{x_i}(\mathbf{E}_v)^{\mathrm{an}} - V_{x_i}(\mathbf{K}_n)^{\mathrm{an}}| \ < \ \varepsilon_v \ .$$

Fix such an n. Since $\widetilde{K}_v$ is dense in $\mathbb{C}_v$, the strict closed affinoid $\mathbf{K}_n$ can be defined by equations in $\widetilde{K}_v$ and has only finitely many distinct conjugates under $\mathrm{Aut}_c(\mathbb{C}_v/K_v)$. Put

$$\mathbf{F}_v \ = \ \bigcap_{\sigma \in \mathrm{Aut}_c(\mathbb{C}_v/K_v)} \sigma(\mathbf{K}_n) \ .$$

Since the intersection of finitely many strict closed affinoids is again a strict closed affinoid, $\mathbf{F}_v$ is a strict closed affinoid with $\mathbf{E}_v \subseteq \mathbf{F}_v \subseteq \mathbf{K}_n$. By construction it is stable under $\mathrm{Aut}_c(\mathbb{C}_v/K_v)$. The monotonicity of Green's functions in Proposition 4.3(C) shows that

$$|G(x_i, x_j; \mathbf{E}_v)^{\mathrm{an}} - G(x_i, x_j; \mathbf{F}_v)^{\mathrm{an}}| \ < \ \varepsilon_v \ ,$$

and for each $x_i \in \mathfrak{X}$,

$$|V_{x_i}(\mathbf{E}_v)^{\mathrm{an}} - V_{x_i}(\mathbf{F}_v)^{\mathrm{an}}| \ < \ \varepsilon_v \ .$$

We now reduce to the classical case. For each v, put $F_v^0 = \mathbf{F}_v \cap \mathcal{C}_v(\mathbb{C}_v)$. Thus, if $v \in S$ is archimedean, then $F_v^0 = \mathbf{F}_v$; if $v \in S$ is nonarchimedean, then F_v^0 is an RL-domain whose closure in $\mathcal{C}_v^{\mathrm{an}}$ is $\mathbf{F}_v$, and if $v \notin S$ then F_v^0 is $\mathfrak{X}$-trivial and again its closure in $\mathcal{C}_v^{\mathrm{an}}$ is $\mathbf{F}_v$. In particular, each F_v^0 is algebraically capacitable and stable under $\mathrm{Aut}^c(\mathbb{C}_v/K_v)$. Set $\mathbb{F}^0 = \prod_v F_v^0$. By Proposition 4.4, the Green's matrices $\Gamma(\mathbb{F}^0, \mathfrak{X})$ and $\Gamma(\mathbb{F}, \mathfrak{X})^{\mathrm{an}}$ coincide. Our choice of ε and the ε_v shows that the entries of $\Gamma(\mathbb{F}^0, \mathfrak{X})$ differ from those of $\Gamma(\mathbb{F}, \mathfrak{X})^{\mathrm{an}}$ by at most ε. Hence $\mathrm{val}(\mathbb{F}^0, \mathfrak{X}) > 1$, and $\gamma(\mathbb{F}^0, \mathfrak{X}) < 1$.

By ([**51**], Theorem 6.2.1), there is a function $f(z) \in K(\mathcal{C})$ with poles supported on $\mathfrak{X}$, such that for each $v \in S$

$$F_v^0 \ \subset \ \{z \in \mathcal{C}_v(\mathbb{C}_v) : |f(z)|_v < 1\} \ ,$$

and for each $v \notin S$

$$F_v^0 \ \subseteq \ \{z \in \mathcal{C}_v(\mathbb{C}_v) : |f(z)|_v \le 1\} \ .$$

For each $v \in S$, put $\mathbf{U}_v = \{z \in \mathcal{C}_v^{\mathrm{an}} : |f(z)|_v < 1\}$, and for each $v \notin S$, let $\mathbf{U}_v = \mathbf{F}_v \subseteq \{z \in \mathcal{C}_v^{\mathrm{an}} : |f(z)|_v \le 1\}$ be the $\mathfrak{X}$-trivial set. Then $\mathbb{U} = \prod_v \mathbf{U}_v$ is a K-rational Berkovich adelic neighborhood of $\mathbb{E}$.

We claim that there are only finitely many points of $\mathcal{C}(\widetilde{K})$ whose $\mathrm{Aut}(\widetilde{K}/K)$-conjugates belong to $\mathbf{U}_v$ for each v. If α is such a point, then $|N_{K(\alpha)/K}(f(\alpha))|_v < 1$ for each $v \in S$, and $|N_{K(\alpha)/K}(f(\alpha))|_v \le 1$ for all v, so

$$\prod_v |N_{K(\alpha)/K}(f(\alpha))|_v \ < \ 1 \ .$$

By the Product Formula, we must have $f(\alpha) = 0$. Since f has only finitely many zeros, the conclusion follows.

We now turn to the proof of (B). We are given a compact Berkovich adelic set $\mathbb{E} = \prod_v \mathbf{E}_v$ with $\gamma(\mathbb{E}, \mathfrak{X})^{\mathrm{an}} > 1$ and a K-rational separable Berkovich quasi-neighborhood $\mathbb{U} = \prod_v \mathbf{U}_v$ of $\mathbb{E}$. In this case, we will reduce the result to Theorem 1.2 by first shinking $\mathbb{E}$, then enlarging it within $\mathbb{U}$, and finally cutting back to classical points, obtaining a classical set $\mathbb{F}^0 = \prod_v F_v^0 \subset \prod_v \mathcal{C}_v(\mathbb{C}_v)$ with a K-rational separable quasi-neighborhood $U_v^0 = \prod_v U_v^0 \subset \prod_v \mathcal{C}_v(\mathbb{C}_v)$ satisfying the conditions of Theorem 1.2.

Since $\gamma(\mathbb{E},\mathfrak{X})^{\mathrm{an}} > 1$, we have $\mathrm{val}(\Gamma(\mathbb{E},\mathfrak{X})^{\mathrm{an}}) < 0$. Let $\varepsilon > 0$ be small enough that if $\Gamma \in M_n(\mathbb{R})$ is a symmetric $n \times n$ matrix whose entries differ from those of $\Gamma(\mathbb{E},\mathfrak{X})^{\mathrm{an}}$ by at most ε, then $\mathrm{val}(\Gamma) < 0$. Again fix a nonempty finite set of places S of K containing all archimedean places and all nonarchimedean places where $\mathbf{E}_v$ is not $\mathfrak{X}$-trivial, and choose a set of numbers $\{\varepsilon_v\}_{v\in S}$ with $\varepsilon_v > 0$ for each v and $\sum_{v\in S}\varepsilon_v = \varepsilon$.

If v is archimedean, put $F_v^0 = \mathbf{E}_v$, and let $U_v^0 = \mathbf{U}_v$; if $v \notin S$, so $\mathbf{E}_v$ is $\mathfrak{X}$-trivial, put $F_v^0 = U_v^0 = \mathbf{E}_v \cap \mathcal{C}_v(\mathbb{C}_v)$, so F_v^0 and U_v^0 are the classical $\mathfrak{X}$-trivial sets. For each nonarchimedean $v \in S$, the separable Berkovich quasi-neighborhood $\mathbf{U}_v$ of $\mathbf{E}_v$ can be written as

$$\mathbf{U}_v \;=\; \mathbf{U}_{v,0} \cup \big(U_{v,1} \cap \mathcal{C}_v(F_{w,1})\big) \cup \cdots \cup \big(U_{v,N} \cap \cap(F_{w,N})\big)\ ,$$

where $\mathbf{U}_{v,0} \subset \mathcal{C}_v^{\mathrm{an}}$ is a Berkovich open set, $U_{v,1},\ldots,U_{v,N} \subset \mathcal{C}_v(\mathbb{C}_v)$ are classical open sets, and $F_{w,1},\ldots,F_{w,N}$ are separable algebraic extensions of K_v. By hypothesis, $\mathbf{U}_v$ is stable under $\mathrm{Aut}_c(\mathbb{C}_v/K_v)$, which means that $\mathbf{U}_{v,0}$ is stable under $\mathrm{Aut}_c(\mathbb{C}_v/K_v)$ as well. Put

$$Y_v \;=\; \mathbf{E}_v \backslash \mathbf{U}_{v,0} \;\subset\; \mathcal{C}_v(\mathbb{C}_v)\ .$$

Since Y_v is compact as a subset of $\mathcal{C}_v^{\mathrm{an}}$, it is compact as a subset of $\mathcal{C}_v(\mathbb{C}_v)$.

For each $n = 1,2,3,\cdots$ we can choose a finite open cover of Y_v by balls $B(x_1,1/n)^-,\ldots,B(x_{k_n},1/n)^-$. Let $\mathbf{B}(x_1,1/n)^-,\ldots,\mathbf{B}(x_{k_n},1/n)^-$ be the corresponding Berkovich open sets. Put

$$\mathbf{K}_n \;=\; \Big(\mathbf{E}_v\backslash(\mathbf{B}(x_1,1/n)^- \cup \ldots \cup \mathbf{B}(x_{k_n},1/n)^-\Big) \cup Y_v\ .$$

(If Y_v is empty, take $\mathbf{K}_n = \mathbf{E}_v$ for each n.) Then $\mathbf{K}_1 \subseteq \mathbf{K}_2 \subseteq \cdots \subseteq \mathbf{K}_n \cdots \subseteq \mathbf{E}_v$ is an ascending sequence of compact sets with $\bigcup_{n=1}^{\infty}\mathbf{K}_n = \mathbf{E}_v$, so by Proposition 4.3(E) there is an n such that for all $x_i \neq x_j \in \mathfrak{X}$ with $i \neq j$ we have

$$(4.38)\quad |G(x_i,x_j;\mathbf{E}_v)^{\mathrm{an}} - G(x_i,x_j;\mathbf{K}_n)^{\mathrm{an}}| \;<\; \varepsilon_v\ ,\qquad |V_{x_i}(\mathbf{E}_v)^{\mathrm{an}} - V_{x_i}(\mathbf{K}_n)^{\mathrm{an}}| \;<\; \varepsilon_v\ .$$

Fix such an n and put

$$\mathbf{X}_v \;=\; \mathbf{K}_n \backslash \Big(\mathbf{E}_v \backslash \big(\mathbf{B}(x_1,1/n)^- \cup \ldots \cup \mathbf{B}(x_{k_n},1/n)^-\big)\Big)\ .$$

Then $\mathbf{X}_v$ is compact, $\mathbf{X}_v \subset \mathbf{U}_{v,0}$, and $\mathbf{K}_n = \mathbf{X}_v \cup Y_v$.

Since strict closed affinoids are cofinal in the closed neighborhoods of a compact Berkovich set, which in turn are cofinal in the open neighborhoods of the set, there is a strict closed Berkovich affinoid $\mathbf{A}_v$ with $\mathbf{X}_v \subseteq \mathbf{A}_v \subset \mathbf{U}_{v,0}$. As noted above, each strict closed Berkovich affinoid has finitely many conjugates under $\mathrm{Aut}_c(\mathbb{C}_v/K_v)$. Since the union of finitely many strict closed Berkovich affinoids is either a strict closed Berkovich affinoid or is all of $\mathcal{C}_v^{\mathrm{an}}$ (see [**64**], Corollaire 2.1.17), after replacing $\mathbf{A}_v$ with the union of its conjugates (which are contained in $\mathbf{U}_{v,0}$), we can assume that $\mathbf{A}_v$ is stable under $\mathrm{Aut}_c(\mathbb{C}_v/K_v)$. The intersection $A_v = \mathbf{A}_v \cap \mathcal{C}_v(\mathbb{C}_v)$ is a K_v-rational closed affinoid in the sense of rigid analysis. Since each rigid analytic strict closed affinoid is an RL-domain (see ([**26**], Satz 2.2) and ([**51**], Corollary 4.2.14), or Corollary C.5 of Appendix C below), there is a function $f(z) \in \mathbb{C}_v(\mathcal{C}_v)$ such that

$$A_v \;=\; \{z \in \mathcal{C}_v(\mathbb{C}_v) : |f(z)|_v \le 1\}\ ,\quad \mathbf{A}_v \;=\; \{z \in \mathcal{C}_v^{\mathrm{an}} : |f(z)|_v \le 1\}\ .$$

Since $\widetilde{K}_v$ is dense in $\mathbb{C}_v$, we can assume that $f(z) \in \widetilde{K}_v(\mathcal{C}_v)$, and after replacing it with its norm to K_v, that $f(z) \in K_v(\mathcal{C})$. Put $\mathbf{F}_v = \mathbf{A}_v \cup X_v$. Then $\mathbf{F}_v \subset \mathbf{U}_v$. Since

$\mathbf{K}_n \subseteq \mathbf{F}_v$, the monotonicity of Green's functions in Proposition 4.3(C) shows that for all $x_i \neq x_j \in \mathfrak{X}$ we have

$$(4.39) \qquad G(x_i, x_j; \mathbf{K}_n)^{\mathrm{an}} \ \geq \ G(x_i, x_j; \mathbf{F}_v)^{\mathrm{an}} \ , \qquad V_{x_i}(\mathbf{K}_n)^{\mathrm{an}} \ \geq \ V_{x_i}(\mathbf{F}_v)^{\mathrm{an}} \ .$$

Finally, put

$$F_v^0 \ = \ A_v \cup X_v \ = \ \mathbf{F}_v \cap \mathcal{C}_v(\mathbb{C}_v) \ , \qquad U_v^0 \ = \ \mathbf{U}_v \cap \mathcal{C}_v(\mathbb{C}_v) \ .$$

Then F_v^0 and U_v^0 are stable under $\mathrm{Aut}_c(\mathbb{C}_v/K_v)$, and $F_v^0 \subset U_v^0$. Furthermore, $U_v^0 = \big(\mathbf{U}_{v,0} \cap \mathcal{C}_v(\mathbb{C}_v)\big) \cup \big(U_{v,1} \cap \mathcal{C}_v(F_{w,1})\big) \cup \cdots \cup \big(U_{v,N} \cap \mathcal{C}_v(F_{w,N})\big)$ so U_v^0 is a K-rational separable quasi-neighborhood of F_v^0. Since F_v^0 is the union of an RL-domain and a compact set, its closure in $\mathcal{C}_v^{\mathrm{an}}$ is $\mathbf{F}_v$. By ([**51**], Theorem 4.3.11) it is algebraically capacitable. By Proposition 4.4, for all $x_i \neq x_j \in \mathfrak{X}$ we have

$$(4.40) \qquad G(x_i, x_j; F_v^0) \ = \ G(x_i, x_j; \mathbf{F}_v)^{\mathrm{an}} \ , \qquad V_{x_i}(F_v^0) \ = \ V_{x_i}(\mathbf{F}_v)^{\mathrm{an}} \ .$$

Globalizing, take $\mathbb{F}^0 = \prod_v F_v^0$ and $\mathbb{U}^0 = \prod_v U_v^0$. Then $\mathbb{F}^0$ is a K-rational adelic set in $\prod_v \mathcal{C}_v(\mathbb{C}_v)$, and $\mathbb{U}^0$ is a K-rational separable quasi-neighborhood of $\mathbb{F}^0$. Since $\gamma(\mathbb{F}, \mathfrak{X})^{\mathrm{an}} > 1$, by (4.38), (4.39) and (4.40) we have $\gamma(\mathbb{F}^0, \mathfrak{X}) > 1$ as well. By Theorem 1.2, there are infinitely many points $\alpha \in \mathcal{C}(\widetilde{K}^{\mathrm{sep}})$ whose $\mathrm{Aut}(\widetilde{K}/K)$-conjugates belong to U_v^0 (hence $\mathbf{U}_v$) for each v. □

CHAPTER 5

Initial Approximating Functions: Archimedean Case

Throughout this section v will be an archimedean place of K, so K is a number field and $K_v \cong \mathbb{R}$ or $K_v \cong \mathbb{C}$. Thus $\mathcal{C}_v(\mathbb{C})$ is a connected, compact Riemann surface. In this section we will construct the archimedean initial approximating functions needed for the proof of Theorem 4.2. When $K_v \cong \mathbb{R}$, the construction uses results about oscillating pseudopolynomials proved in Appendix B.

In Theorem 4.2, we are given a compact, K_v-simple set $E_v \subset \mathcal{C}_v(\mathbb{C})$, of positive inner capacity, which is disjoint from $\mathfrak{X}$. If $K_v \cong \mathbb{C}$, so E_v is $\mathbb{C}$-simple, this means that E_v is a finite union of pairwise disjoint compact sets $E_{v,i}$, each of which

(1) is simply connected, has a piecewise smooth boundary, and is the closure of its interior.

If $K_v \cong \mathbb{R}$, so E_v is $\mathbb{R}$-simple, then E_v is stable under complex conjugation and is a finite union of pairwise disjoint compact sets $E_{v,i}$, where each $E_{v,i}$ either

(1) is a closed subinterval of $\mathcal{C}_v(\mathbb{R})$ with positive length; or
(2) is disjoint from $\mathcal{C}_v(\mathbb{R})$ and is simply connected, has a piecewise smooth boundary, and is the closure of its interior.

Since E_v is compact the usual Green's functions and Robin constants, $G(z, x_i; E_v)$ and $V_{x_i}(E_v)$, coincide with the upper ones, $\overline{G}(z, x_i; E_v)$ and $\overline{V}_{x_i}(E_v)$.

If $K_v \cong \mathbb{C}$, let E_v^0 be the interior of E_v; if $K_v \cong \mathbb{R}$, let E_v^0 be the the union of the $\mathcal{C}_v(\mathbb{C})$-interiors of the components $E_{v,i}$ disjoint from $\mathcal{C}_v(\mathbb{R})$, together with the $\mathcal{C}_v(\mathbb{R})$-interiors of the components $E_{v,i}$ contained in $\mathcal{C}_v(\mathbb{R})$. We call E_v^0 the "quasi-interior" of E_v.

Fix $\varepsilon_v > 0$. In constructing the approximating functions, we first replace E_v with a K_v-simple set $\widetilde{E}_v \subset E_v^0$ such that for each $x_i \neq x_j \in \mathfrak{X}$,

$$|V_{x_i}(\widetilde{E}_v) - V_{x_i}(E_v)| < \varepsilon_v \ , \quad |G(x_i, x_j; \widetilde{E}_v) - G(x_i, x_j; E_v)| < \varepsilon_v \ .$$

Substituting $\widetilde{E}_v$ for E_v gives us "freedom of movement" in the constructions below. Next, let $U_v \subset \mathcal{C}_v(\mathbb{C})$ be an open set such that $U_v \cap E_v = E_v^0$. After shrinking U_v if necessary, we can assume it is bounded away from $\mathfrak{X}$, and that its connected components are in one-to-one correspondence with those of E_v. If $K_v \cong \mathbb{R}$, we can assume it is stable under complex conjugation as well.

Let $\vec{s} = (s_1, \ldots, s_m)$ be a K_v-symmetric probability vector with rational coefficients. By a K_v-rational $(\mathfrak{X}, \vec{s})$-function, we mean a function $f(z) \in K_v(\mathcal{C}_v)$, whose poles are supported on $\mathfrak{X}$, such that if N_i is the order of the pole of f at x_i and $N = \deg(f)$, then $\frac{1}{N}(N_1, \ldots, N_m) = \vec{s}$.

The initial approximating functions $f_v(z)$ will be K_v-rational $(\mathfrak{X}, \vec{s})$-functions having several properties:

First, $\frac{1}{N}\log_v(|f_v(z)|_v)$ will closely approximate $\sum_{i=1}^m s_i G(z, x_i; \widetilde{E}_v)$ outside U_v.

Second, $f_v(z)$ will have all its zeros in E_v^0, and satisfy $\{z \in \mathcal{C}_v(\mathbb{C}) : |f_v(z)|_v \leq 1\} \subset U_v$. If $K_v \cong \mathbb{R}$, we also require that $f_v(z)$ have a property like that of Chebyshev polynomials, oscillating between large positive and negative values on $E_v^0 \cap \mathcal{C}_v(\mathbb{R})$. This means that when $f(z)$ is perturbed slightly, its zeros continue to belong to E_v^0.

Third, for global aspects of the proof of the Fekete-Szegö theorem, we need to be able to independently vary the *logarithmic leading coefficients* of $f_v(z)$ at the points in $\mathfrak{X}$. We define the logarithmic leading coefficient of $f_v(z)$ at x_i to be

$$\Lambda_{x_i}(f_v, \vec{s}) = \lim_{z \to x_i} \left(\frac{1}{N} \log_v(|f_v(z)|_v) + s_i \log_v(|g_{x_i}(z)|_v) \right) .$$

Similarly, we define

$$\begin{aligned} \Lambda_{x_i}(\widetilde{E}_v, \vec{s}) &= \lim_{z \to x_i} \left((\sum_{j=1}^m s_j G(z, x_j; \widetilde{E}_v)) + s_i \log_v(|g_{x_i}(z)|_v) \right) \\ &= s_i V_{x_i}(\widetilde{E}_v) + \sum_{j \neq i} s_j G(x_i, x_j; \widetilde{E}_v) . \end{aligned}$$

We will require that for pre-specified numbers $\beta_1, \ldots, \beta_m$ belonging to an interval $[-\delta_v, \delta_v]$ depending only on $\widetilde{E}_v$ and U_v,

$$\Lambda_{x_i}(f_v, \vec{s}) = \Lambda_{x_i}(\widetilde{E}_v, \vec{s}) + \beta_i .$$

Here the β_i must be K_v-symmetric, but otherwise can be chosen arbitrarily. This "independent variability of the logarithmic leading coefficients" is needed to deal with the problem that the probability vector $\hat{s}$ for which $\Gamma(\mathbb{E}, \mathfrak{X})\hat{s}$ has equal entries (constructed in §3.10), may not have rational entries.

1. The Approximation Theorems

There are two cases to consider in constructing the initial approximating functions: when $K_v \cong \mathbb{C}$, and when $K_v \cong \mathbb{R}$. The case when $K_v \cong \mathbb{C}$ follows from results in [**51**]:

THEOREM 5.1. *Suppose $K_v \cong \mathbb{C}$. Let $E_v \subset \mathcal{C}_v(\mathbb{C})$ be a $\mathbb{C}$-simple set which is disjoint from $\mathfrak{X}$ and has positive capacity, and let $U_v = E_v^0$ be the $\mathcal{C}_v(\mathbb{C})$-interior of E_v. Fix $\varepsilon_v > 0$.*

Then there is a compact set $\widetilde{E}_v \subset U_v$ composed of a finite union of analytic arcs, with $\mathcal{C}_v(\mathbb{C}) \backslash \widetilde{E}_v$ connected, which has the following properties:

(A) For each $x_i \in \mathfrak{X}$,

$$|V_{x_i}(\widetilde{E}_v) - V_{x_i}(E_v)| < \varepsilon_v , \tag{5.1}$$

and for all $x_i, x_j \in \mathfrak{X}$ with $x_i \neq x_j$,

$$|G(x_i, x_j; \widetilde{E}_v) - G(x_i, x_j; E_v)| < \varepsilon_v . \tag{5.2}$$

(B) There is a $\delta_v > 0$ such that for any probability vector $\vec{s} = {}^t(s_1, \ldots, s_m)$ with rational entries, and any $\vec{\beta} = {}^t(\beta_1, \ldots, \beta_m) \in [-\delta_v, \delta_v]^m$, there is an integer $N_v \geq 1$ such that for each positive integer N divisible by N_v, there exists an $(\mathfrak{X}, \vec{s})$-function $f_v(z) \in K_v(\mathcal{C}_v)$ of degree N, satisfying

(1) *for each* $x_i \in \mathfrak{X}$,

$$\Lambda_{x_i}(f_v, \vec{s}) \;=\; \Lambda_{x_i}(\widetilde{E}_v, \vec{s}) + \beta_i \ . \tag{5.3}$$

(2) $\{z \in \mathcal{C}_v(\mathbb{C}) : |f_v(z)|_v \le 1\}$ *is contained in* E_v^0; *in particular, all the zeros of* $f_v(z)$ *belong to* E_v^0.

PROOF. We first construct the set $\widetilde{E}_v$. Since E_v is the closure of its interior, and its boundary is a finite union of smooth arcs, each point of E_v is analytically accessible from E_v^0. Hence Proposition 3.30 shows that $\overline{G}(z, x_i; E_v^0) = G(z, x_i, E_v)$ for each x_i. By the monotonicity of Green's functions, there is a compact set $E_v^* \subset E_v^0$ such that for all $x_i \ne x_j$

$$|V_{x_i}(E_v^*) - V_{x_i}(E_v)| \;<\; \varepsilon_v/2 \ , \quad |G(x_i, x_j; E_v^*) - G(x_i, x_j; E_v)| \;<\; \varepsilon_v/2 \ . \tag{5.4}$$

The remainder of the construction is a combination of results from [**51**]. By ([**51**], Proposition 3.3.2) there is a compact set $\widetilde{E}_v \subset U_v$ which is a finite union of analytic arcs, with $\mathcal{C}_v(\mathbb{C})\backslash\widetilde{E}_v$ connected, such that for each x_i,

$$|V_{x_i}(\widetilde{E}_v) - V_{x_i}(E_v^*)| \;<\; \varepsilon_v/2 \ , \quad |G(x_i, x_j; \widetilde{E}_v) - G(x_i, x_j; E_v^*)| \;<\; \varepsilon_v/2 \ . \tag{5.5}$$

The set $\widetilde{E}_v$ is obtained by first covering E_v^* with a finite collection of closed discs contained in U_v, then taking the union of the boundaries of those discs, and finally cutting short intervals out of each boundary arc to obtain a set such that $\mathcal{C}_v(\mathbb{C})\backslash\widetilde{E}_v$ connected. From (5.4) and (5.5) we obtain (5.1) and (5.2).

The existence of a number $\delta_v > 0$, and for each rational probability vector $\vec{s}$ and each $\vec{\beta} \in (-\delta_v, \delta_v)^m$, the existence of an integer $N_v \ge 1$ and an $(\mathfrak{X}, \vec{s})$-function $f_{v,0}(z) \in K_v(\mathcal{C}_v)$ of degree N_v, with the properties in the theorem, is proved in ([**51**], Theorem 3.3.7). After shrinking δ_v, one can replace the conditions $|\beta_i| < \delta_v$ in ([**51**], Theorem 3.3.7) with $|\beta_i| \le \delta_v$. We remark that the independent variability of the logarithmic leading coefficients is based on a convexity argument using that $-\log([z, w]_\zeta)$ is everywhere harmonic in z, apart from logarithmic singularities when $z = w$ or $z = \zeta$ (see [**51**], Lemma 3.3.9).

Note that properties (B1) and (B2) are preserved when $f_{v,0}(z)$ is raised to a power. Given an arbitrary multiple $N = kN_v$ we can obtain the approximating function of degree N by putting $f_v(z) = f_{v,0}(z)^k$. □

When $K_v \cong \mathbb{R}$, the approximation theorem we need is as follows. Note that if $f(z) \in \mathbb{R}(\mathcal{C}_v)$, then f is real-valued on $\mathcal{C}_v(\mathbb{R})$. Given a number $M > 0$, we say that f *oscillates* k *times between* $\pm M$ on an interval $I \subset \mathcal{C}_v(\mathbb{R})$ if it varies k times from $-M$ to M, or from M to $-M$, on I. In particular, it has at least k zeros in I. Conversely, if it has exactly k zeros in I and oscillates k times between $\pm M$, then each of those zeros is simple.

THEOREM 5.2. *Suppose* $K_v \cong \mathbb{R}$. *Let* E_v *be a compact* $\mathbb{R}$-*simple set which is disjoint from* $\mathfrak{X}$ *and has positive capacity, and let* E_v^0 *be the quasi-interior of* E_v. *Fix a* $\mathcal{C}_v(\mathbb{C})$-*open set* U_v *such that* $U_v \cap E_v = E_v^0$, *and which is stable under complex conjugation and bounded away from* $\mathfrak{X}$. *Take* $\varepsilon_v > 0$.

Then there is a $\mathbb{R}$-*simple compact set* $\widetilde{E}_v \subset E_v^0$ *such that* $\mathcal{C}_v(\mathbb{C})\backslash\widetilde{E}_v$ *is connected, which has the following properties:*

(*A*) *For each* $x_i \in \mathfrak{X}$,

$$|V_{x_i}(\widetilde{E}_v) - V_{x_i}(E_v)| \;<\; \varepsilon_v \ , \tag{5.6}$$

and for all $x_i, x_j \in \mathfrak{X}$ *with* $x_i \neq x_j$,

$$|G(x_i, x_j; \widetilde{E}_v) - G(x_i, x_j; E_v)| \ < \ \varepsilon_v \ . \tag{5.7}$$

(*B*) *Given* $0 < \mathcal{R}_v < 1$, *there is a* $\delta_v > 0$ *(depending on* $\widetilde{E}_v$, U_v, ε_v, *and* $\mathcal{R}_v$*) such that for each* K_v*-symmetric probability vector* $\vec{s} = {}^t(s_1, \ldots, s_m)$ *with rational entries, and for each* K_v*-symmetric* $\vec{\beta} = {}^t(\beta_1, \ldots, \beta_m) \in [-\delta_v, \delta_v]^m$, *there is an integer* $N_v \geq 1$ *such that for each positive integer* N *divisible by* N_v, *there is an* $(\mathfrak{X}, \vec{s})$*-function* $f_v(z) \in K_v(\mathcal{C}_v)$ *of degree* N *which satisfies:*

(1) *For each* $x_i \in \mathfrak{X}$,

$$\Lambda_{x_i}(f_v, \vec{s}) \ = \ \Lambda_{x_i}(\widetilde{E}_v, \vec{s}) + \beta_i \ . \tag{5.8}$$

(2) $\{z \in \mathcal{C}_v(\mathbb{C}) : |f_v(z)| \leq 1\} \ \subset \ U_v$.

(3) *All the zeros of* $f_v(z)$ *belong to* E_v^0, *and if* $E_{v,i}$ *is a component of* E_v *contained in* $\mathcal{C}_v(\mathbb{R})$ *and* $f_v(z)$ *has* N_i *zeros in* $E_{v,i}$, *then* $f_v(z)$ *oscillates* N_i *times between* $\pm\mathcal{R}_v^N$ *on* $E_{v,i}$.

The proof of Theorem 5.2 will occupy the rest of this chapter. For notational convenience we identify K_v with $\mathbb{R}$, and write $\log_v(x) = \log(x)$.

2. Outline of the Proof of Theorem 5.2

In this section we sketch the ideas behind the proof of Theorem 5.2. In §5.3 we establish an independence lemma, and in §5.4 we give the details of the proof.

If the $\mathbb{R}$-simple set $E_v = \bigcup_{i=1}^n E_{v,i}$ has no components in $\mathcal{C}_v(\mathbb{R})$, Theorem 5.2 follows from results in ([**51**]). For the remainder of the discussion below, assume that some $E_{v,i}$ is contained in $\mathcal{C}_v(\mathbb{R})$. By standard potential-theoretic arguments, we can construct a K_v-simple compact set $E_v^* \subset E_v^0$ such that for each $x_i \neq x_j$,

$$|V_{x_i}(E_v^*) - V_{x_i}(E_v)| < \varepsilon_v \ , \quad |G(x_i, x_j; E_v^*) - G(x_i, x_j; E_v)| < \varepsilon_v \ .$$

In doing so, we can arrange that each of the intervals making up $E_v^* \cap \mathcal{C}_v(\mathbb{R})$ is "short", in a sense to be made precise later.

To assist in constructing $(\mathfrak{X}, \vec{s})$-functions with prescribed logarithmic leading coefficients, we next adjoin a finite number of short intervals to E_v^*, which can be "wiggled" inside E_v^0. We will show that there are finitely many points $t_1, \ldots, t_d \subset (E_v^0 \cap \mathcal{C}_v(\mathbb{R}))\backslash E_v^*$ and a number $h > 0$, such that if $\widetilde{E}_{v,\ell} = [t_\ell - h, t_\ell + h]$ for $\ell = 1, \ldots, d$ (the intervals are defined using local coordinates at the points t_ℓ) then the set

$$\widetilde{E}_v \ := \ E_v^* \cup (\bigcup_{\ell=1}^d \widetilde{E}_{v,\ell})$$

meets the needs of the theorem, in particular, satisfying $|V_{x_i}(\widetilde{E}_v) - V_{x_i}(E_v)| < \varepsilon_v$ for each x_i, and $|G(x_i, x_j; \widetilde{E}_v) - G(x_i, x_j; E_v)| < \varepsilon_v$ for each $x_i \neq x_j$. The points $t_1, \ldots, t_d$ must be in "general position", in a sense to be described in §5.3, and h must be small enough that the intervals $\widetilde{E}_{v,\ell}$ are contained in $(E_v^0 \cap \mathcal{C}_v(\mathbb{R}))\backslash\mathfrak{X}$ and are disjoint from E_v^* and each other.

We next construct the functions $f_v(z)$. The first part of the construction is purely potential-theoretic, and is carried out in Appendices A and B.

Let each $[z,w]_{x_i}$ be normalized so $\lim_{z \to x_i} [z,w]_{x_i} \cdot |g_{x_i}(z)|_v = 1$. As in (3.29), given a probability vector $\vec{s} \in \mathcal{P}^m$, we define the $(\mathfrak{X}, \vec{s})$-*canonical distance* to be

$$[z,w]_{\mathfrak{X},\vec{s}} = \prod_{i=1}^{m} ([z,w]_{x_i})^{s_i} .$$

There is a potential theory for the $(\mathfrak{X}, \vec{s})$-canonical distance similar to the one for the usual canonical distance (see Section 1 of Appendix A):

Let $H_v \subset \mathcal{C}_v(\mathbb{C}) \backslash \mathfrak{X}$ be any compact set with positive capacity. For each probability measure ν supported on H_v, define the $(\mathfrak{X}, \vec{s})$-energy by

$$I_{\mathfrak{X},\vec{s}}(\nu) = \iint_{H_v \times H_v} -\log([z,w]_{\mathfrak{X},\vec{s}}) \, d\nu(z) d\nu(w)$$

and the $(\mathfrak{X}, \vec{s})$-Robin constant by

$$V_{\mathfrak{X},\vec{s}}(H_v) = \inf_{\nu} I_{\mathfrak{X},\vec{s}}(\nu) . \tag{5.9}$$

By Theorem A.2, there is a unique probability measure $\mu_{\mathfrak{X},\vec{s}}$ which achieves the infimum in (5.9); it will be called the $(\mathfrak{X}, \vec{s})$-equilibrium measure of H_v. By the same theorem, the $(\mathfrak{X}, \vec{s})$-potential function

$$u_{\mathfrak{X},\vec{s}}(z) = \int_{H_v} -\log([z,w]_{\mathfrak{X},\vec{s}}) \, d\mu_{\mathfrak{X},\vec{s}}(w)$$

satisfies $u_{\mathfrak{X},\vec{s}}(z) \le V_{\mathfrak{X},\vec{s}}(H_v)$ for all z, with $u_{\mathfrak{X},\vec{s}}(z) = V_{\mathfrak{X},\vec{s}}(H_v)$ on H_v if H_V has finitely many components and a piecewise smooth boundary.

By Proposition A.5, the $(\mathfrak{X}, \vec{s})$-Green's function $G_{\mathfrak{X},\vec{s}}(z; H_v) := V_{\mathfrak{X},\vec{s}}(H_v) - u_{\mathfrak{X},\vec{s}}(z)$ can be decomposed as

$$G_{\mathfrak{X},\vec{s}}(z; H_v) = \sum_{i=1}^{m} s_i G(z, x_i; H_v) \tag{5.10}$$

and the $(\mathfrak{X}, \vec{s})$-equilibrium measure of H_v is given by

$$\mu_{\mathfrak{X},\vec{s}} = \sum_{i=1}^{m} s_i \mu_{x_i}$$

where μ_{x_i} is the equilibrium measure of H_v with respect to x_i.

Recall that an $(\mathfrak{X}, \vec{s})$-function $f(z) \in K_v(\mathcal{C}_v)$ of degree N is a function with polar divisor $\sum_{i=1}^m N s_i (x_i)$. If the zeros of $f(z)$ are $\alpha_1, \ldots, \alpha_N$ (listed with multiplicities), then an easy symmetrization argument shows there is a constant C such that

$$|f(z)|_v = C \cdot \prod_{k=1}^{N} [z, \alpha_k]_{\mathfrak{X},\vec{s}} \tag{5.11}$$

for all $z \in \mathcal{C}_v(\mathbb{C}) \backslash \mathfrak{X}$: let $\xi_1, \ldots, \xi_N$ be the points $x_1, \ldots, x_m$ listed according to their multiplicities in $\mathrm{div}(f)$. For each permutation π of $\{1, \ldots, N\}$, by (3.26) there is a constant such that $|f(z)|_v = C(\pi) \cdot \prod_{k=1}^N [z, \alpha_k]_{\xi_{\pi(k)}}$. Taking the product over all π, and then extracting $(N!)^{th}$ roots, gives (5.11).

This motivates the definition of an $(\mathfrak{X}, \vec{s})$-*pseudopolynomial* (usually we will just say pseudopolynomial). Given a constant C and points $\alpha_1, \ldots, \alpha_N \in \mathcal{C}_v(\mathbb{C}) \backslash \mathfrak{X}$, the associated pseudopolynomial is the nonnegative real valued function

$$P(z) = P_{\vec{\alpha}}(z) = C \cdot \prod_{k=1}^{N} [z, \alpha_k]_{\mathfrak{X},\vec{s}} .$$

Write $N = \deg(P)$. We call $\mathrm{div}(P) := \sum_{i=1}^{N}(\alpha_i) - \sum_{i=1}^{m} N s_i(x_i)$ the *divisor* of P, we call the α_k the *roots* of P, and we call

$$\nu(z) = \nu_P(z) := \frac{1}{N}\sum_{k=1}^{N} \delta_{\alpha_k}(z)$$

the *probability measure associated to* P. Note that $P_{\vec{\alpha}}(z)$ makes sense even when the α_k are not the zeros of an $(\mathfrak{X},\vec{s})$-function $f(z)$, but it agrees with $|f(z)|_v$ (up to a multiplicative constant) when such a function exists. Furthermore, $P_{\vec{\alpha}}(z)$ varies continuously with its roots. This allows us to investigate absolute values of $(\mathfrak{X},\vec{s})$-functions with prescribed zeros, without worrying about principality of the divisors.

For each $x_i \in \mathfrak{X}$, we define the logarithmic leading coefficient of $P(z)$ at x_i as

$$\Lambda_{x_i}(P,\vec{s}) = \lim_{z \to x_i} \frac{1}{N}\log(P(z)) + s_i \log(|g_{x_i}(z)|_v) . \tag{5.12}$$

We now apply this to the $\mathbb{R}$-simple set $H_v = \widetilde{E}_v$. A detailed study of pseudopolynomials is carried out in Appendix B. There it is shown that if the components of $\widetilde{E}_v$ contained in $\mathcal{C}_v(\mathbb{R})$ are sufficiently short (the precise meaning of "short" is given in Definition B.15, in terms of the canonical distance functions $[z,w]_{x_i}$ relative to the $x_i \in \mathfrak{X}$), then by potential-theoretic methods one can show the existence of pseudopolynomials which behave like absolute values of classical Chebyshev polynomials, and have large oscillations on $\widetilde{E}_v$.

Let $D > d$ be the number of components of $\widetilde{E}_v$, and label those in E_v^* as $\widetilde{E}_{v,d+1},\ldots,\widetilde{E}_{v,D}$, so that $\widetilde{E}_v = \bigcup_{\ell=1}^{D}\widetilde{E}_{v,\ell}$. For each $\ell = 1,\ldots,D$, put $\sigma_\ell = \mu_{\mathfrak{X},\vec{s}}(\widetilde{E}_{v,\ell})$, and put $\vec{\sigma} = (\sigma_1,\ldots,\sigma_D)$. The following is a specialization of Theorem B.18 of Appendix B, formulated using the notation of this section. We write $\overline{z}$ for the complex conjugate of z.

THEOREM 5.3. *Suppose $K_v \cong \mathbb{R}$. Assume that $\mathfrak{X}$ is stable under complex conjugation, and that $\widetilde{E}_v \subset \mathcal{C}_v(\mathbb{C})\backslash\mathfrak{X}$ is K_v-simple, with components $\widetilde{E}_{v,1},\ldots,\widetilde{E}_{v,D}$. Assume also that each component $\widetilde{E}_{v,\ell}$ contained in $\mathcal{C}_v(\mathbb{R})$ is a "short interval" relative to $\mathfrak{X}$ in the sense of Definition* B.15.

Fix a K_v-symmetric probability vector $\vec{s} \in \mathcal{P}^m$. For each $\ell = 1,\ldots,D$ put $\sigma_\ell = \mu_{\mathfrak{X},\vec{s}}(\widetilde{E}_{v,\ell})$ and let $\vec{\sigma} = (\sigma_1,\ldots,\sigma_D)$. Given a K_v-symmetric vector $\vec{n} \in \mathbb{N}^D$ write $N = N_{\vec{n}} = \sum_\ell n_\ell$. Then there are a collection of $(\mathfrak{X},\vec{s})$-pseudopolynomials

$$\{Q_{\vec{n}}(z)\}_{\vec{n}\in\mathbb{N}^D} ,$$

and numbers $0 < R_{\vec{n}} \le 1$, with the following properties:

(A) For each $\vec{n}$, $Q_{\vec{n}}$ satisfies $\|Q_{\vec{n}}\|_{\widetilde{E}_v} = 1$, with $Q_{\vec{n}}(z) = Q_{\vec{n}}(\overline{z})$ for all $z \in \mathcal{C}_v(\mathbb{C})$. The roots of $Q_{\vec{n}}$ all belong to $\widetilde{E}_v$, with n_ℓ roots in each $\widetilde{E}_{v,\ell}$. For each $\widetilde{E}_{v,\ell}$ which is a short interval the roots of $Q_{\vec{n}}$ in $\widetilde{E}_{v,\ell}$ are distinct, and $Q_{\vec{n}}$ varies n_ℓ times from $R_{\vec{n}}^N$ to 0 to $R_{\vec{n}}^N$ on $\widetilde{E}_{v,\ell}$.

(B) Let $\{\vec{n}_k\}_{k\in\mathbb{N}}$ be a sequence with $N_{\vec{n}_k} \to \infty$ and $\vec{n}_k/N_{\vec{n}_k} \to \vec{\sigma}$. Then $\lim_{k\to\infty} R_{\vec{n}_k} = 1$, and the discrete measures $\omega_{\vec{n}_k}$ associated to the $Q_{\vec{n}_k}$ converge weakly to the equilibrium distribution $\mu_{\mathfrak{X},\vec{s}}$ of $\widetilde{E}_v$. For each neighborhood $\widetilde{U}_v$ of $\widetilde{E}_v$, the functions $\frac{1}{N_{\vec{n}_k}}\log(Q_{\vec{n}_k}(z))$ converge uniformly to $G_{\mathfrak{X},\vec{s}}(z,\widetilde{E}_v)$ on $\mathcal{C}_v(\mathbb{C}_v)\backslash(\widetilde{U}_v \cup \mathfrak{X})$, and for each $x_i \in \mathfrak{X}$,

$$\lim_{k\to\infty} \Lambda_{x_i}(Q_{\vec{n}_k},\vec{s}) = \Lambda_{x_i}(\widetilde{E}_v,\vec{s}) .$$

Theorem 5.3 is the potential-theoretic input to the construction. We will call the $(\mathfrak{X},\vec{s})$-pseudopolynomials $Q_{\vec{n}}(z)$ given by Theorem 5.3 *special pseudopolynomials* for $\widetilde{E}_v$.

However, we want to construct $(\mathfrak{X},\vec{s})$-functions, not just pseudopolynomials, with large oscillations on $\widetilde{E}_v$. The second part of the construction addresses this.

Let $\mathcal{R}_v$ be as in Theorem 5.2. Applying Theorem 5.3, for an appropriate $\vec{n}$ we obtain a pseudopolynomial $Q(z) = Q_{\vec{n}}(z)$ which varies n_ℓ times from $\mathcal{R}_v^N$ to 0 to $\mathcal{R}_v^N$ on each real component $\widetilde{E}_{v,\ell}$ of $\widetilde{E}_v$. If $\mathrm{div}(Q) := \sum_{k=1}^N(\alpha_k) - \sum_{i=1}^m Ns_i(x_i)$ were principal, then since it is K_v-symmetric there would be an $(\mathfrak{X},\vec{s})$-function $f(z) \in \mathbb{R}(\mathcal{C}_v)$ with $|f(z)| = Q(z)$ for all z. Moreover, by Theorem 5.3 the roots of $Q(z)$ in $\mathcal{C}_v(\mathbb{R})$ are simple, so each time $Q(z)$ varies from $\mathcal{R}_v^N$ to 0 to $\mathcal{R}_v^N$ on a real component of $\widetilde{E}_v$, then $f(z)$ oscillates from $\mathcal{R}_v^N$ to $-\mathcal{R}_v^N$, or from $-\mathcal{R}_v^N$ to $\mathcal{R}_v^N$.

Of course, it is unreasonable to expect $Q(z)$ to have a principal divisor. We must assume from the start that $\vec{s} = (s_1,\ldots,s_m) \in \mathbb{Q}^m$ and that $N\vec{s} \in \mathbb{Z}^m$, but still $\mathrm{div}(Q)$ will generally not be principal.

Our plan is to modify $Q(z)$ by scaling it and "sliding some of its roots along $\mathcal{C}_v(\mathbb{R})$" to make $\mathrm{div}(Q)$ principal. In this process, some of the roots may move outside $\widetilde{E}_v$, but they will remain inside E_v^0.

In the decomposition $\widetilde{E}_v = E_v^* \cup (\bigcup_{\ell=1}^d \widetilde{E}_{v,\ell})$, write $\widetilde{E}_{v,\ell} = [t_\ell - h, t_\ell + h]$ in suitable local coordinates. Suppose we are given a number $r > 0$ small enough that $[t_\ell - h - r, t_\ell + h + r]$ is contained in U_v and in the coordinate patch of $\widetilde{E}_{v,\ell}$, for each $\ell = 1,\ldots,d$. Suppose we are also given real numbers $\varepsilon_1,\ldots,\varepsilon_d$ with each $|\varepsilon_\ell| \le r$. Then for each $z \in \widetilde{E}_{v,\ell}$, it makes sense to speak of the point $z + \varepsilon_\ell$ in terms of the given local coordinates. Let ε_0 be another small real number, and put $\vec{\varepsilon} = (\varepsilon_0,\ldots,\varepsilon_d)$. If $Q(z) = C\cdot\prod_{k=1}^N[z,\alpha_k]_{\mathfrak{X},\vec{s}}$, define

$$Q^{\vec{\varepsilon}}(z) = \exp(\varepsilon_0)^N\cdot C\cdot \prod_{\alpha_k\in E_v^*}[z,\alpha_k]_{\mathfrak{X},\vec{s}}\cdot\prod_{\ell=1}^d\prod_{\alpha_k\in\widetilde{E}_{v,\ell}}[z,\alpha_k+\varepsilon_\ell]_{\mathfrak{X},\vec{s}} . \tag{5.13}$$

Using the continuity of $[z,w]_{\mathfrak{X},\vec{s}}$, one can choose r so that if each ε_ℓ is permitted to vary over the interval $[-r,r]$, then $Q^{\vec{\varepsilon}}(z)$ oscillates N times from $\mathcal{R}_v^N$ to 0 to $\mathcal{R}_v^N$ on U_v. (In the construction, r will be chosen before h, and h will be much smaller than r.)

We claim that for sufficiently large N, one can choose $\vec{\varepsilon}$ in such a way that $\mathrm{div}(Q^{\vec{\varepsilon}})$ becomes principal. If $\mathcal{C}_v$ has genus $g = 0$, there is nothing to prove. If $g > 0$, consider how $\mathrm{div}(Q)$ changes when Q is replaced by $Q^{\vec{\varepsilon}}$. Let $\mathrm{Jac}(\mathcal{C}_v) \cong \mathrm{Div}^0(\mathcal{C}_v)/P(\mathcal{C}_v)$ be the Jacobian of $\mathcal{C}_v$, where $\mathrm{Div}^0(\mathcal{C}_v)$ is the set of ($\mathbb{C}$-rational) divisors of degree 0, and $P(\mathcal{C}_v)$ is the subgroup of principal divisors. Fixing a base point $p_0 \in \mathcal{C}_v(\mathbb{R})$, there is a natural embedding $\varphi : \mathcal{C}_v(\mathbb{C}) \to \mathrm{Jac}(\mathcal{C}_v)(\mathbb{C})$, $\varphi(p) = \mathrm{cl}((p) - (p_0))$. Since $\mathcal{C}_v$ is defined over $\mathbb{R}$, the space of holomorphic differentials $H^1(\mathcal{C}_v, \mathbb{C})$ has a basis consisting of real differentials $\omega_1, \ldots, \omega_g \in H^1(\mathcal{C}_v, \mathbb{R})$. If $\mathcal{L} \subset \mathbb{C}^g$ is the corresponding period lattice, and we identify $\mathrm{Jac}(\mathcal{C}_v)(\mathbb{C})$ with $\mathbb{C}^g/\mathcal{L}$, then

$$\varphi(p) \;=\; (\int_{p_0}^{p} \omega_1, \ldots, \int_{p_0}^{p} \omega_g) \pmod{\mathcal{L}} .$$

The embedding φ induces a canonical surjective map

$$\Phi : \mathcal{C}_v(\mathbb{C})^g \to \mathrm{Jac}(\mathcal{C}_v)(\mathbb{C}) , \qquad \Phi((p_1, \ldots, p_g)) = \sum_{i=1}^{g} \varphi(p_i) .$$

This factors through the g-fold symmetric product $\mathrm{Sym}^{(g)}(\mathcal{C}_v)$ as

$$\Phi \;:\; \mathcal{C}_v(\mathbb{C})^g \;\to\; \mathrm{Sym}^{(g)}(\mathcal{C}_v) \;\to\; \mathrm{Jac}(\mathcal{C}_v)(\mathbb{C})$$

where the first map is finite of degree $g!$, and the second is a birational morphism. Hence the image $\Phi(\mathcal{C}_v(\mathbb{R})^g)$ contains a g-dimensional open subset of $\mathrm{Jac}(\mathcal{C}_v)(\mathbb{R})$. On the other hand, $\mathrm{Jac}(\mathcal{C}_v)(\mathbb{R})$ is a compact real Lie subgroup of $\mathrm{Jac}(\mathcal{C}_v)(\mathbb{C})$ of dimension at most g. Hence, $\mathrm{Jac}(\mathcal{C}_v)(\mathbb{R})$ must have dimension exactly g, and the identity component of $\mathrm{Jac}(\mathcal{C}_v)(\mathbb{R})$ must be isomorphic to a real torus $\mathbb{R}^g/\mathcal{L}_0$ for some lattice $\mathcal{L}_0 \subset \mathcal{L} \bigcap \mathbb{R}^g$, under our identification $\mathrm{Jac}(\mathcal{C}_v)(\mathbb{C}) \cong \mathbb{C}^g/\mathcal{L}$. The component group of $\mathrm{Jac}(\mathcal{C}_v)(\mathbb{R})$ is an elementary abelian 2-group, and if $N\vec{s} \in 2 \cdot \mathbb{N}^m$, one can arrange that $\mathrm{div}(Q)$ belongs to the identity component.

Note that since p_0 and the ω_i are real, the integrals $\int_{p_0}^{p} \omega_i(z)\,dz$ are real-valued on the component of $\mathcal{C}_v(\mathbb{R})$ containing p_0.

Now consider what happens when the roots $\alpha_k \in \widetilde{E}_{v,\ell} = [t_\ell - h, t_\ell + h]$ are translated by ε_ℓ. In terms of the local coordinate at t_ℓ, we can write $\omega_1 = h_1(z)\,dz, \ldots, \omega_g = h_g(z)\,dz$. The resulting change in $\varphi(\mathrm{div}(Q))$ is

$$\begin{aligned}(5.14)\quad \sum_{\ell=1}^{d} \sum_{\alpha_k \in \widetilde{E}_{v,\ell}} & (\int_{\alpha_k}^{\alpha_k+\varepsilon_\ell} \omega_1, \ldots, \int_{\alpha_k}^{\alpha_k+\varepsilon_\ell} \omega_g) \pmod{\mathcal{L}} \\ &\cong \sum_{\ell=1}^{d} N\sigma_\ell \cdot (\int_{t_\ell}^{t_\ell+\varepsilon_\ell} h_1(z)\,dz, \ldots, \int_{t_\ell}^{t_\ell+\varepsilon_\ell} h_g(z)\,dz) \pmod{\mathcal{L}} \\ &\cong \sum_{\ell=1}^{d} N\sigma_\ell\varepsilon_\ell \cdot (h_1(t_\ell), \ldots, h_g(t_\ell)) \pmod{\mathcal{L}} .\end{aligned}$$

However, it is more useful to consider the normalized lift $\widehat{\varphi} : \mathbb{R}^{d+1} \to \mathbb{R}^g$,

$$\begin{aligned}(5.15)\qquad \widehat{\varphi}(\vec{\varepsilon}) \;&:=\; \frac{1}{N} \sum_{\ell=1}^{d} \sum_{\alpha_k \in \widetilde{E}_{v,\ell}} (\int_{\alpha_k}^{\alpha_k+\varepsilon_\ell} h_1(z)\,dz, \ldots, \int_{\alpha_k}^{\alpha_k+\varepsilon_\ell} h_g(z)\,dz) \\ &\cong\; \sum_{\ell=1}^{d} \sigma_\ell\varepsilon_\ell \cdot (h_1(t_\ell), \ldots, h_g(t_\ell)) ,\end{aligned}$$

for which

$$(5.16)\qquad \varphi(\mathrm{div}(Q^{\vec{\varepsilon}})) - \varphi(\mathrm{div}(Q)) \;=\; N\widehat{\varphi}(\vec{\varepsilon}) \pmod{\mathcal{L}_0} .$$

We will show that for sufficiently large N, as $\vec{\varepsilon}$ varies over the ball $B(0,r) = \{\vec{x} \in \mathbb{R}^{d+1} : |\vec{x}| \leq r\}$, the image $\widehat{\varphi}(B(0,r))$ contains a fixed neighborhood of the origin in $\mathbb{R}^g$. Hence the nonnormalized change $N\widehat{\varphi}(\vec{\varepsilon})$ varies over a region containing a fundamental domain for $\mathcal{L}_0$, and $\vec{\varepsilon}$ can be chosen so $\mathrm{div}(Q^{\vec{\varepsilon}})$ is principal.

We also need to consider the change in the logarithmic leading coefficients produced by passing from $Q(z)$ to $Q^{\vec{\varepsilon}}(z)$. Because $\lim_{z \to x_i} [z,w]_{x_i} |g_{x_i}(z)|_v = 1$ for each x_i,

$$\begin{aligned} \Lambda_{x_i}(Q, \vec{s}) &= \lim_{z \to x_i} \left(\frac{1}{N} \log(Q(z)) + s_i \log(|g_i(z)|) \right) \\ &= V_{\mathfrak{X},\vec{s}}(\widetilde{E}_v) + \sum_{\substack{j=1 \\ j \neq i}}^{m} s_j \left(\sum_{k=1}^{N} \frac{1}{N} \log([x_i, \alpha_k]_{x_j}) \right) . \end{aligned} \tag{5.17}$$

For each $1 \leq \ell \leq d$, consider the contribution of the roots $\alpha_k \in \widetilde{E}_{v,\ell} = [t_\ell - h, t_\ell + h]$ to (5.17). Recall that $\sigma_\ell = \mu_{\mathfrak{X},\vec{s}}(\widetilde{E}_{v,\ell})$. If N is sufficiently large, $Q(z)$ has approximately $N\sigma_\ell$ roots in $\widetilde{E}_{v,\ell}$. If h is sufficiently small, the α_k belonging to $\widetilde{E}_{v,\ell}$ can be viewed as being essentially equal to t_ℓ, and will contribute approximately

$$\sum_{\substack{j=1 \\ j \neq i}}^{m} s_j \log([x_i, t_\ell]_{x_j})\, \sigma_\ell$$

to (5.17). If these roots are translated by ε_ℓ, and the contributions from all $\widetilde{E}_{v,\ell}$ are summed, along with the change due to scaling by $\exp(\varepsilon_0)^N$, we see that

$$\Lambda_{x_i}(Q^{\vec{\varepsilon}}, \vec{s}) - \Lambda_{x_i}(Q, \vec{s}) \cong \varepsilon_0 + \sum_{\ell=1}^{d} \sum_{\substack{j=1 \\ j \neq i}}^{m} s_j \left(\frac{d}{dt} \log([x_i, t_\ell]_{x_j}) \right) \Bigg|_{t_\ell} \cdot \sigma_\ell \varepsilon_\ell . \tag{5.18}$$

In the next section we will show that $t_1, \ldots, t_d$ can be chosen so that the quantities (5.15) and (5.18) are independent. Given this, a topological argument shows that there exist $\vec{\varepsilon}$ for which $\mathrm{div}(Q^{\vec{\varepsilon}})$ is principal and such that the logarithmic leading coefficients $\Lambda_{x_i}(Q^{\vec{\varepsilon}}, \vec{s})$ corresponding to distinct orbits $\mathfrak{X}_\ell$ can be specified arbitrarily, provided they are sufficently close to the $\Lambda_{x_\ell}(\widetilde{E}_v, \vec{s})$. It will also be seen that the construction can be carried out uniformly for all $\vec{s}$.

3. Independence

Let $\mathcal{C}_v/\mathbb{R}$ and $\mathfrak{X}$ be as in §5.2. If m_1 points in $\mathfrak{X}$ are real, and $2m_2$ points are in complex conjugate pairs, let $\mathfrak{X}_1, \ldots, \mathfrak{X}_{m_1+m_2}$ denote the corresponding orbits.

Fix a K_v-symmetric, rational probability vector $\vec{s}$, and consider formula (5.18) giving $\Lambda_{x_i}(Q^{\vec{\varepsilon}}, \vec{s}) - \Lambda_{x_i}(Q, \vec{s})$. Our first goal is to express the differential

$$\frac{d}{dt} \log([x_i, t]_{x_j})\, dt \tag{5.19}$$

in terms of meromorphic differentials on the Riemann surface $\mathcal{C}_v(\mathbb{C})$.

For each $x_i \neq x_j$, there is a multi-valued holomorphic function $\Omega_{i,j}(z)$ on $\mathcal{C}_v(\mathbb{C}) \backslash \{x_i, x_j\}$ whose real part coincides with $\log([x_i, z]_{x_j})$, and which has pure imaginary periods over all cycles of $\mathcal{C}_v(\mathbb{C})$ and loops around x_i, x_j. (See [**58**], or [**51**], pp. 64-65). In a given coordinate patch, $\Omega_{i,j}(z)$ is only defined up to a pure imaginary constant; however, the differential $d\Omega_{i,j}(z) = G_{i,j}(z)\,dz$ is a globally well-defined differential of the third kind on $\mathcal{C}_v(\mathbb{C})$, which is holomorphic except for simple poles with residue $+1$ at x_i and residue -1 at x_j.

Letting $\overline{z}$ be the complex conjugate of z, and writing τ for the permutation of $\{1,\ldots,m\}$ such that $\overline{x_i} = x_{\tau(i)}$, one sees that for a suitable normalization of $\Omega_{\tau(i),\tau(j)}(z)$,

$$\Omega_{\tau(i),\tau(j)}(\overline{z}) = \overline{\Omega_{i,j}(z)}\ .$$

With this normalization, for each $t \in \mathcal{C}_v(\mathbb{R})$,

$$\log([x_i,t]_{x_j}) = \frac{1}{2}\left(\Omega_{i,j}(t) + \Omega_{\tau(i),\tau(j)}(t)\right)\ .$$

Assuming t is the real part of the local coordinate function z, this means that at a point $t_\ell \in \mathcal{C}_v(\mathbb{R})$,

$$\left(\frac{d}{dt}\log([x_i,t]_{x_j})\right)\bigg|_{t_\ell} = \frac{1}{2}\left(G_{i,j}(t_\ell) + G_{\tau(i),\tau(j)}(t_\ell)\right)\ .$$

Note that if x_i is complex, and $x_j = x_{\tau(i)}$, then $[x_i,t]_{x_j}$ is constant for $t \in \mathcal{C}_v(\mathbb{R})$. This is because $[x_i,z]_{x_j} \cdot [x_j,z]_{x_i}$ is constant on $\mathcal{C}_v(\mathbb{C})$: its logarithm is harmonic everywhere except possibly at x_i, x_j; but the singularities at those points cancel so it is harmonic everywhere. On the other hand, applying τ to $[x_i,t]_{x_j}$, we have $[x_i,t]_{x_j} = [x_j,t]_{x_i}$. Combining these shows $[x_i,t]_{x_j}$ is constant. Thus the terms with $j = \tau(i)$ (those for which x_i and x_j belong to the same orbit $\mathfrak{X}_a$) can be omitted from (5.18).

Given an orbit $\mathfrak{X}_a$, fix $x_i \in \mathfrak{X}_a$ and define the differential

$$H_{\vec{s},a}(z)\,dz = \sum_{x_k \notin \mathfrak{X}_a} s_k \cdot \frac{1}{2}\left(G_{i,k}(z)\,dz + G_{\tau(i),\tau(k)}(z)\,dz\right)\ . \tag{5.20}$$

The same differential is obtained if x_i is replaced by $\tau(x_i)$. Clearly, $H_{\vec{s},a}(z)\,dz$ is holomorphic except at the points in $\mathfrak{X}$. If $\mathfrak{X}_a = \{x_i\}$ is real then $H_{\vec{s},a}(z)\,dz$ has a simple pole with residue $1 - s_i$ at x_i; if $\mathfrak{X}_a = \{x_i, x_{\tau(i)}\}$ is complex, it has simple poles with residue $\frac{1}{2} - s_i$ at x_i and $x_{\tau(i)}$. In both cases it has a simple pole with residue $-s_k$ at each $x_k \notin \mathfrak{X}_a$. Writing the $H_{\vec{s},a}(z)$ in appropriate local coordinates, then for $x_i \in \mathfrak{X}_a$, formula (5.18) becomes

$$\Lambda_{x_i}(Q^{\vec{\varepsilon}},\vec{s}) - \Lambda_{x_i}(Q,\vec{s}) \cong \varepsilon_0 + \sum_{\ell=1}^{d} H_{\vec{s},a}(t_\ell)\cdot\sigma_\ell\varepsilon_\ell\ . \tag{5.21}$$

We now consider what linear relations there might be between the meromorphic differentials $H_{\vec{s},a}(z)\,dz$ and the holomorphic differentials $\omega_j = h_j(z)\,dz$. Put $J = g + m_1 + m_2 - 1$, where g is the genus of $\mathcal{C}$.

Proposition 5.4. *For each probability vector $\vec{s} \in \mathcal{P}^m$, the meromorphic differentials*

$$H_{\vec{s},1}(z)\,dz\ ,\ \ldots\ ,\ H_{\vec{s},m_1+m_2}(z)\,dz$$

and the holomorphic differentials $\omega_1 = h_1(z)\,dz,\ldots,\omega_g = h_g(z)\,dz$ span a vector space of dimension $J = g + m_1 + m_2 - 1$. If the orbits $\mathfrak{X}_1,\ldots,\mathfrak{X}_{m_1}$ are real and $\mathfrak{X}_{m_1+1},\ldots,\mathfrak{X}_{m_1+m_2}$ are complex, and if x_i is a representative for $\mathfrak{X}_i$,

$i = 1, \ldots, m_1 + m_2$, *then every linear relation among the* $H_{\vec{s},i}(z)\,dz$ *and the* $\omega_j = h_j(z)\,dz$ *is a consequence of the relation*

$$\sum_{i=1}^{m_1} s_i H_{\vec{s},i}(z)\,dz + \sum_{i=m_1+1}^{m_1+m_2} 2s_i H_{\vec{s},i}(z)\,dz \;=\; 0\,. \tag{5.22}$$

PROOF. Note that $J = 0$ iff $g = 0$ and $\mathfrak{X}$ consists of a single point. In that case $m_1 = 1$, $m_2 = 0$, and $H_{\vec{s},1}(z)dz \equiv 0$, so (5.22) holds trivially. Suppose $J > 0$, and let

$$\sum_{i=1}^{m_1+m_2} c_i H_{\vec{s},i}(z)\,dz + \sum_{j=1}^{g} d_j h_j(z)\,dz \;=\; 0 \tag{5.23}$$

be an arbitrary relation. Considering the residues at the poles $x_1, \ldots, x_{m_1+m_2}$ and writing ${}^t\vec{c} = (c_1, \ldots, c_{m_1+m_2})$ we see that

$$\begin{pmatrix} 1-s_1 & \cdots & -s_1 & -s_1 & \cdots & -s_1 \\ \vdots & \ddots & \vdots & \vdots & & \vdots \\ -s_{m_1} & \cdots & 1-s_{m_1} & -s_{m_1} & \cdots & -s_{m_1} \\ -s_{m_1+1} & \cdots & -s_{m_1+1} & \frac{1}{2}-s_{m_1+1} & \cdots & -s_{m_1+1} \\ \vdots & & \vdots & \vdots & \ddots & \vdots \\ -s_{m_1+m_2} & \cdots & -s_{m_1+m_2} & -s_{m_1+m_2} & \cdots & \frac{1}{2}-s_{m_1+m_2} \end{pmatrix} \cdot \vec{c} \;=\; \vec{0}\,.$$

Put $C = c_1 + \ldots + c_{m_1+m_2}$. The equations above imply $c_i = s_i C$ if $\mathfrak{X}_i$ is real, and $c_i = 2s_i C$ if $\mathfrak{X}_i$ is complex. Conversely, for any C, we obtain a solution by taking $c_i = s_i C$ when $\mathfrak{X}_i$ is real and $c_i = 2s_i C$ when $\mathfrak{X}_i$ is complex, since $\sum_{i=1}^{m_1} s_i + \sum_{i=m_1+1}^{m_1+m_2} 2s_i = 1$.

Now consider the differential $\sum_{i=1}^{m_1+m_2} c_i H_{\vec{s},i}(z)\,dz$. It is meromorphic with pure imaginary periods and at worst simple poles at the points in $\mathfrak{X}$. We have just seen that it has residue 0 at each x_i, so it is everywhere holomorphic. However, the only holomorphic differential with pure imaginary periods is the 0 differential, so

$$\sum_{i=1}^{m_1+m_2} c_i H_{\vec{s},i}(z)\,dz \;=\; 0\,.$$

Inserting this into (5.23), we see that $\sum_{j=1}^{g} d_j \omega_j \;=\; 0$. Since the ω_j are linearly independent, the d_j must be 0. (If $g = 0$, there are no holomorphic differentials, and the result is vacuously true.) This yields the proposition. □

COROLLARY 5.5. *For any open subinterval* $I \subset C_v(\mathbb{R})\backslash\mathfrak{X}$, *there exist points* $t_1, \ldots, t_J \in I$ *such that the matrix*

$$\mathcal{G}_{\vec{s}}(t_1, \ldots, t_J) \;:=\; \begin{pmatrix} 1 & H_{\vec{s},1}(t_1) & \cdots & H_{\vec{s},1}(t_J) \\ \vdots & \vdots & & \vdots \\ 1 & H_{\vec{s},m_1+m_2}(t_1) & \cdots & H_{\vec{s},m_1+m_2}(t_J) \\ 0 & h_1(t_1) & \cdots & h_1(t_J) \\ \vdots & \vdots & & \vdots \\ 0 & h_g(t_1) & \cdots & h_g(t_J) \end{pmatrix} \tag{5.24}$$

is nonsingular.

PROOF. After relabeling the $\mathfrak{X}_i$ if necessary, we can assume that $s_1 > 0$. No nontrivial relation of the form

$$\sum_{i=2}^{m_1+m_2} c_i H_{\vec{s},i}(z)\,dz + \sum_{j=1}^{g} d_j h_j(z)\,dz \;=\; 0$$

can hold identically on I; otherwise, it would hold identically on $\mathcal{C}_v(\mathbb{C})$, contrary to Proposition 5.4. Hence we can find $t_1, \ldots, t_J \in I$ such that submatrix of (5.24) consisting of the last J rows and columns is nonsingular. But then, row reducing (5.24) by using the relation (5.22) and replacing the first row with the corresponding linear combination of the first $m_1 + m_2$ rows, that row becomes $(1, 0, \ldots, 0)$. Hence $\mathcal{G}_{\vec{s}}(t_1, \ldots, t_J)$ is nonsingular. □

4. Proof of Theorem 5.2

In this section we will prove Theorem 5.2. By assumption $K_v \cong \mathbb{R}$. We are given a K_v-simple set E_v which is bounded away from $\mathfrak{X}$ and has positive inner capacity, together with numbers $\varepsilon_v > 0$ and $0 < \mathcal{R}_v < 1$, and an open set $U_v \subset \mathcal{C}_v(\mathbb{C})$ such that $E_v \cap U_v = E_v^0$. After shrinking U_v we can assume it is bounded away from $\mathfrak{X}$, and is stable under complex conjugation. Note that for each component $E_{v,\ell}$ of E_v which not contained in $\mathcal{C}_v(\mathbb{R})$, the interior $E_{v,\ell}^0$ is one of the connected components of U_v.

The proof has several steps. We use the notation from §5.2 and §5.3.

Step 0. If E_v has no components contained in $\mathcal{C}_v(\mathbb{R})$, then Theorem 5.2 follows by the same argument as Theorem 5.1, using the assertions in ([**51**], Proposition 3.3.2) and ([**51**], Theorem 3.3.7) that deal with the case where $K_v \cong \mathbb{R}$, with E_v and $\mathfrak{X}$ stable under complex conjugation, and $\vec{\beta}$ and $\vec{s}$ being K_v-symmetric.

For the remainder of the proof, we will assume that at least one of the components of E_v is a closed interval in $\mathcal{C}_v(\mathbb{R})$.

Step 1. We first construct a K_v-simple set $E_v^* \subset E_v^0$, whose capacity is close to that of E_v, with each real interval in E_v^* "short" in the sense of Definition B.15.

Since each point of E_v is analytically accessible from E_v^0, by Proposition 3.30 we have $G(z, x_i; E_v) = \overline{G}(z, x_i, E_v^0)$ for each $x_i \in \mathfrak{X}$. Hence there is a compact subset $E_v^{**} \subset E_v^0$, such that for each $x_i, x_j \in \mathfrak{X}$ with $x_i \neq x_j$,

$$\begin{aligned} |V_{x_i}(E_v^{**}) - V_{x_i}(E_v)| \;&<\; \varepsilon_v/2 \quad \text{for each } i\ , \\ |G(x_i, x_j; E_v^{**}) - G(x_i, x_j; E_v)| \;&<\; \varepsilon_v/2 \quad \text{for all } i \neq j. \end{aligned} \tag{5.25}$$

Without loss we can assume that E_v^{**} is K_v-simple (and, in particular, stable under complex conjugation). Indeed, since E_v is K_v-simple, its quasi-interior E_v^0 has an exhaustion by K_v-simple sets.

To construct E_v^*, we will need a lemma. Fix a spherical metric $\|z, w\|_v$ on $\mathcal{C}_v(\mathbb{C})$. For $p \in \mathcal{C}_v(\mathbb{C})$ and $\delta > 0$, write $B(p, \delta)^- = \{z \subset \mathcal{C}_v(\mathbb{C}) : \|z, p\|_v < \delta\}$. For $p \in \mathcal{C}_v(\mathbb{R})$, let $I_p(\delta) = \{z \in \mathcal{C}_v(\mathbb{R}) : \|z, p\|_v < \delta\}$ and let $\overline{I}_p(\delta) = \{z \in \mathcal{C}_v(\mathbb{R}) : \|z, p\|_v \leq \delta\}$.

LEMMA 5.6. *Given a compact set $H \subset \mathcal{C}_v(\mathbb{C}) \backslash \mathfrak{X}$ and a number $\delta > 0$, let $H(\delta)$ be obtained from H in any of the following ways:*

(A) $H(\delta) = \{x \in \mathcal{C}_v(\mathbb{R}) : \|x, z\|_v \leq \delta$ *for some* $z \in H\}$.

(B) For some $p_1, \ldots, p_M \in H$, $\quad H(\delta) = H \backslash \left(\bigcup_{k=1}^M B(p_k, \delta)^-\right)$.

(C) For some $p_1, \ldots, p_M \in \mathcal{C}_v(\mathbb{R}) \backslash \mathfrak{X}$, $\quad H(\delta) = H \bigcup \left(\bigcup_{k=1}^M \overline{I}_{p_k}(\delta)\right)$.

Then for each $x_i \in \mathfrak{X}$,

$$\lim_{\delta \to 0} V_{x_i}(H(\delta)) = V_{x_i}(H) \tag{5.26}$$

and for each $x_i \neq x_j \in \mathfrak{X}$,

$$\lim_{\delta \to 0} G(x_i, x_j; H(\delta)) \ = \ G(x_i, x_j; H(\delta)) \ . \tag{5.27}$$

PROOF. This follows from ([**51**] , Corollary 3.1.16, p.149, Proposition 3.1.17, p.149, and Lemma 3.2.6, p.158). □

We next show that we can assume that each component of E_v^{**} contained in $\mathcal{C}_v(\mathbb{R})$ is a "short" interval. Let $C(E_v, \mathfrak{X})$ be the number gotten by taking $H = E_v$ in formula (B.101), and put

$$B \ = \ B(E_v, \mathfrak{X}) \ = \ \min(1/C(E_v, \mathfrak{X}), 1/\sqrt{2C(E_v, \mathfrak{X})}) \ .$$

Then any closed subinterval of $E_v^{**} \cap \mathcal{C}_v(\mathbb{R})$, with length at most B under $\|z, w\|_v$, is "short" in the sense of Definition B.15. Choose $p_1, \ldots, p_M \in E_v^{**} \cap \mathcal{C}_v(\mathbb{R})$ such that $(E_v^{**} \cap \mathcal{C}_v(\mathbb{R})) \backslash \{p_1, \ldots, p_M\}$ is composed of segments of length at most B. Lemma 5.6(B) then shows that by deleting small open balls about the p_k, we can find a K_v-simple compact set $E_v^* \subset E_v^{**}$ such that each real interval in E_v^* is "short", and

$$\begin{aligned} |V_{x_i}(E_v^*) - V_{x_j}(E_v)| \ &< \ \varepsilon_v \quad \text{for each } i \ , \\ |G(x_i, x_j; E_v^*) - G(x_i, x_j; E_v)| \ &< \ \varepsilon_v \quad \text{for all } i \neq j \ . \end{aligned} \tag{5.28}$$

The set E_v^* meets our needs.

Step 2. The choice of $t_1, \ldots, t_d$.

As in §5.3, put $J = g + m_1 + m_2 - 1$. Then $J \geq 0$, with $J = 0$ if and only if $g = 0$ and $\mathfrak{X} = \{x_1\}$. In that situation every divisor of degree 0 is principal, and the variation in the logarithmic leading coefficient at x_1 is accomplished by scaling alone (e.g. via ε_0 in (5.48) below). If $J = 0$, take $d = 0$ and ignore all constructions related to points t_ℓ in the rest of the proof.

Assume now that $J \geq 1$. Fix a closed interval $I \subset (E_v^0 \cap \mathcal{C}_v(\mathbb{R})) \backslash E_v^*$ with positive length. This interval will play an important role in the construction below; the points t_i, and the intervals we construct below, will belong to it.

Fix a local coordinate function z on a neighborhood of I, in such a way that z is real-valued on I. Write the differentials $H_{\vec{s},i}(z)\,dz$ and $\omega_j(z) = h_j(z)\,dz$ from §5.3 in terms of z. Translations of points, $z \mapsto z + \varepsilon$, will also be understood relative to this coordinate.

Let $\mathcal{P}_v^m \subset \mathcal{P}^m$ denote the set of K_v-*symmetric* real probability vectors. If $\vec{s}_0 \in \mathcal{P}_v^m$ and $\rho > 0$, let $B(\vec{s}_0, \rho) \subset \mathbb{R}^m$ denote the open Euclidean ball about $\vec{s}_0$ with radius ρ.

For a given $\vec{s}_0 \in \mathcal{P}_v^m$, Corollary 5.5 shows we can find points $t_1, \ldots, t_J$ in the interior of I such that the matrix $\mathcal{G}_{\vec{s}_0}(\vec{t})$ defined there is nonsingular. Fixing $\vec{t} = (t_1, \ldots, t_J)$, the function $w_{\vec{t}}(\vec{s}) := \det(\mathcal{G}_{\vec{s}})(\vec{t})$ is continuous for $\vec{s} \in \mathcal{P}_v^m$, so there is a $\rho = \rho(\vec{s}_0) > 0$ such that $\det(\mathcal{G}_{\vec{s}})(\vec{t})$ is nonsingular for all $\vec{s} \in \mathcal{P}_v^m \bigcap B(\vec{s}_0, \rho)$. Moreover, $|\det(\mathcal{G}_{\vec{s}}(\vec{t})|$ is uniformly bounded away from 0 if we restrict to $\vec{s} \in \mathcal{P}_v^m \bigcap B(\vec{s}_0, \rho/2)$. Since $\mathcal{P}_v^m$ is compact, we can cover it with a finite number of balls $B(\vec{s}_0, \rho(\vec{s}_0)/2)$. Let $\mathcal{T} = \{t_1, \ldots, t_d\}$ be the union of the sets $\{t_1, \ldots, t_J\}$ associated to these $\vec{s}_0$.

The set $\widetilde{E}_v$ will have the form

$$E_v^*(h) \ := \ E_v^* \cup \big(\bigcup_{\ell=1}^{d} [t_\ell - h, t_\ell + h]\big) \tag{5.29}$$

for a suitably small h which we will construct below, where the intervals $[t_\ell - h, t_\ell + h]$ (defined in terms of the local coordinate function z) are contained in the interior of I and are pairwise disjoint.

However, before proceeding further, we note two facts concerning $\mathcal{T}$:
First, given a square matrix $\mathcal{G}$ with real entries, let its L^2 operator norm be

$$\|\mathcal{G}\| \ = \ \max_{|\vec{x}|=1} |\mathcal{G}\vec{x}| \ .$$

By the construction of $\mathcal{T}$, there is a constant B_1 such that for each $\vec{s} \in \mathcal{P}_v^m$, we can find J points in $\mathcal{T}$ such that the corresponding matrix $\mathcal{G}_{\vec{s}}(\vec{t})$ from Corollary 5.5 is nonsingular, and

$$\|\mathcal{G}_{\vec{s}}(\vec{t})^{-1}\| \ \le \ B_1 \ . \tag{5.30}$$

Second, for all sufficiently small h, there are à priori bounds on the relative mass which the $(\mathfrak{X}, \vec{s})$-equilibrium distribution of $E_v^*(h)$ gives to each segment $e_\ell(h) := [t_\ell - h, t_\ell + h]$. This is a consequence of potential-theoretic results in Appendix A.2, as follows.

If $I = [a, b]$, fix a number $r_0 > 0$ small enough that $5r_0$ is less than the minimum of the distances $|t_k - t_\ell|$ for $t_k \ne t_\ell \in \mathcal{T}$, and such that $2r_0$ is less than the minimum of the distances to the endpoints, $|t_\ell - a|$, $|t_\ell - b|$, for each $t_\ell \in \mathcal{T}$. Thus the intervals $e_\ell(2r_0) \subset U_v$ are bounded away from each other, and are contained in the interior of I so they are bounded away from E_v^*. We will also require that r_0 be small enough that each $e_\ell(2r_0)$ is "short" in the sense of Definition B.15, permitting the construction of oscillating pseudopolynomials.

Consider $E_v^*(2r_0)$: it is K_v-simple (so in particular $\mathcal{C}_v(\mathbb{C}) \backslash E_v^*(2r_0)$ is connected), and each of the segments $e_\ell(2r_0)$ is a component of $E_v^*(2r_0)$. Put

$$B_2 \ = \ \min_{1 \le i \le m} \ \min_{1 \le \ell \le d} \ \min_{z \in e_\ell(2r_0)} \ G(z, x_i; E_v^*(2r_0) \backslash e_\ell(2r_0)) \ . \tag{5.31}$$

Then $B_2 > 0$, and by the monotonicity of Green's functions, for each $0 < h \le 2r_0$ each x_i and each ℓ, we have $G(z, x_i; E_v^*(h) \backslash e_\ell(h)) \ge B_2$ on $e_\ell(h)$.

By Proposition A.5, $V_{\mathfrak{X},\vec{s}}(E_v^*)$ is a continuous function of $\vec{s}$, so there is a finite upper bound B_3 for the values $V_{\mathfrak{X},\vec{s}}(E_v^*)$ as $\vec{s}$ ranges over the compact set $\mathcal{P}_v^m$. Trivially,

$$V_{\mathfrak{X},\vec{s}}(E_v^*(h)) \ \le \ V_{\mathfrak{X},\vec{s}}(E_v^*(h) \backslash e_\ell(h)) \ \le \ V_{\mathfrak{X},\vec{s}}(E_v^*) \ \le \ B_3$$

for all h, ℓ, and $\vec{s}$.

By Lemma A.8, there is a constant $A > 0$ such that for all $x_i \in \mathfrak{X}$, all $\ell = 1, \ldots, d$, and all sufficiently small $h > 0$,

$$-\log(h) - A \ < \ V_{x_i}(e_\ell(h)) \ < \ -\log(h) + A \ .$$

In particular, for all sufficiently small h, and all $x_i \in \mathfrak{X}$,

$$V_{x_i}(e_\ell(h)) \ > \ V_{x_i}(E_v^*) \ > \ V_{x_i}(E_v^*(h) \backslash e_\ell(h)) \ ,$$

validating the hypothesis of Lemma A.7.

For a given $\vec{s} \in \mathcal{P}_v^m$, let $\mu_{\mathfrak{X},\vec{s},h}$ be the $(\mathfrak{X},\vec{s})$-equilibrium distribution of $E_v^*(h)$. Then by Lemma A.6, there is a constant C such that for each sufficiently small h,

$$\mu_{\mathfrak{X},\vec{s},h}(e_\ell(h)) \leq \frac{V_{\mathfrak{X},\vec{s}}(E_v^*(h)) + C}{V_{\mathfrak{X},\vec{s}}(e_\ell(h)) + C} \leq \frac{B_3 + C}{-\log(h) - A + C} .$$

Likewise, by Lemma A.7, for sufficiently small h,

$$\begin{aligned} \mu_{\mathfrak{X},\vec{s},h}(e_\ell(h)) &\geq \frac{(B_2)^2}{2(V_{\mathfrak{X},\vec{s}}(E_v^*(h)\backslash e_\ell(h)) + C)(V_{\mathfrak{X},\vec{s}}(e_\ell(h)) + C + 2B_2)} \\ &\geq \frac{(B_2)^2}{2(B_3 + C)(-\log(h) + A + C + 2B_2)} . \end{aligned}$$

Put

$$B_4 = \frac{3d \cdot (B_3 + C)^2}{(B_2)^2} . \tag{5.32}$$

Then there is an $h_0 > 0$ such that for each $\vec{s} \in \mathcal{P}_v^m$, each $\ell = 1, \ldots, d$, and each $0 < h \leq h_0$,

$$\frac{\mu_{\mathfrak{X},\vec{s},h}(\bigcup_{k=1}^d e_k(h))}{\mu_{\mathfrak{X},\vec{s},h}(e_\ell(h))} \leq B_4 . \tag{5.33}$$

The parameter h will be chosen in Step 4 below. Given $0 < h \leq h_0$, we will put $\widetilde{E}_v = E_v^*(h) = \bigcup_{i=1}^D \widetilde{E}_{v,\ell}$, where $\widetilde{E}_{v,\ell} = [t_\ell - h, t_\ell + h]$ for $\ell = 1, \ldots, d$ and $\widetilde{E}_{v,d+1}, \ldots, \widetilde{E}_{v,D}$ are the components of E_v^*. We can apply Theorem 5.3 to $E_v^*(h)$, constructing $(\vec{X},\vec{s})$-pseudopolynomials $Q(z)$ with large oscillations on the real components of $E_v^*(h)$.

Step 3. The choice of r.

In later stages of the construction, we will need to move some of the roots of the pseudopolynomials $Q(z)$, in order to make their divisors principal and vary their logarithmic leading coefficients. We now define a number r which governs how far we can move the roots.

Let $0 < \mathcal{R}_v < 1$ be the oscillation bound required in the theorem. Fix a number $\widetilde{\mathcal{R}}_v$ with $\mathcal{R}_v < \widetilde{\mathcal{R}}_v < 1$, and put $\Delta_1 = \log(\widetilde{\mathcal{R}}_v) - \log(\mathcal{R}_v)$. Recall that U_v is the open set for which $E_v^0 = E_v \cap U_v$. Consider the set $E_v^* \cup I$, which is contained in U_v. The Green's functions $G(z, x_i; E_v^* \cup I)$ are continuous, and are positive in the complement of $E_v^* \cup I$. Since ∂U_v is compact and disjoint from $E_v^* \cup I$, there is a $\Delta_2 > 0$ such that for each $x_i \in \mathfrak{X}$, and $z \notin E_v^* \cup I$, we have $G(z, x_i; E_v^* \cup I) \geq \Delta_2$. For each $x_i \in \mathfrak{X}$, recall that there is a $\mathcal{C}^\infty$ function $\eta_i(z, w)$ such that $\log([z, w]_{x_\ell}) = \log(|z - w|) + \eta_i(z, w)$ on $I \times I$.

Let r be a number in the range

$$0 < r \leq r_0 , \tag{5.34}$$

(so in particular, the intervals $e_\ell(r)$ are pairwise disjoint, contained in I, and are "short" in the sense of Definition B.15), which satisfies

$$r < \min(\Delta_1, \Delta_2)/3 , \tag{5.35}$$

and which is small enough that for each $x_i \in \mathfrak{X}$, the following six conditions hold:

(1) for each $t_\ell \in \mathcal{T}$, and each $w \in E_v^*$,

$$\left| \max_{z \in e_\ell(r)} \log([z,w]_{x_i}) - \min_{z \in e_\ell(r)} \log([z,w]_{x_i}) \right| < \Delta_1/2 \; ; \tag{5.36}$$

(2) for each $t_\ell \in \mathcal{T}$, each $t_k \neq t_\ell \in \mathcal{T}$, and each $w \in e_k(2r_0)$,

$$\left| \max_{z \in e_\ell(r)} \log([z,w]_{x_i}) - \min_{z \in e_\ell(r)} \log([z,w]_{x_i}) \right| < \Delta_1/2 \; ; \tag{5.37}$$

(3) for each $t_\ell \in \mathcal{T}$,

$$\left| \max_{z,w \in e_\ell(2r)} \eta_i(z,w) - \min_{z,w \in e_\ell(2r)} \eta_i(z,w) \right| < \Delta_1/2 \; ; \tag{5.38}$$

(4) for each $x_i \neq x_j \in \mathfrak{X}$, and for each $t_\ell \in \mathcal{T}$ and each $z \in e_\ell(2r)$,

$$\left| \left(\log([x_i,z]_{x_j}) - \log([x_i,t_\ell]_{x_j}) \right) - (z-t_\ell) \cdot \frac{d}{dt} \log([x_i,t]_{x_j}) \Big|_{t=t_\ell} \right| < \frac{r}{24B_1B_4\sqrt{J+1}} \; ; \tag{5.39}$$

(5) for each $\omega_j = h_j(z)\,dz$, and for each $t_\ell \in \mathcal{T}$ and each $z \in e_\ell(2r)$,

$$|h_j(z) - h_j(t_\ell)| < 1/(24B_1B_4\sqrt{J+1}) \; ; \tag{5.40}$$

(6) for all $z \in \partial U_v$ and all $w_1, w_2 \in E_v^* \cup I$ with $\|w_1,w_2\|_v \leq r$,

$$|\log([z,w_1]_{x_i}) - \log([z,w_2]_{x_i})| < \Delta_2/3 \; . \tag{5.41}$$

Conditions (1), (2), and (3) hold for all sufficiently small r by the continuity of the functions $\log([z,w]_{x_i})$ and $\eta_i(z,w)$. Condition (4) holds for all sufficiently small r since the functions $\log([x_i,t]_{x_j})$ are $\mathcal{C}^\infty$ on the intervals $e_\ell(2r_0)$. Condition (5) holds for all sufficiently small r by the continuity of the $h_j(z)$. Condition (6) holds for all sufficiently small r since the functions $\log([z,w]_{x_i})$ are uniformly continuous for $(z,w) \in \partial U \times (E_v^* \cup I)$.

Fix such an r. By our assumptions on E_v^* and r, for each $0 < h \leq r$ we can apply Theorem 5.3 to the set $E_v^*(h)$, constructing $(\mathfrak{X},\vec{s})$-pseudopolynomials $Q(z)$. Recall if $\vec{s} \in \mathcal{P}_v^m$, then $G_{\mathfrak{X},\vec{s}}(z,E_v^*(h)) = \sum_{i=1}^m s_i G(z,x_i;E_v^*(h))$.

PROPOSITION 5.7. *Let r be as above. Given $0 < h \leq r$, put $\widetilde{E}_v = E_v^*(h) = E_v^* \cup \bigcup_{i=1}^d [t_\ell - h, t_\ell + h]$. Let $\vec{\varepsilon} = (\varepsilon_0, \varepsilon_1, \ldots, \varepsilon_d) \in \mathbb{R}^{d+1}$ be such that $|\varepsilon_\ell| \leq r$ for each ℓ, and put $\widetilde{E}_v(\vec{\varepsilon}) = E_v^* \cup \bigcup_{\ell=1}^d [t_\ell + \varepsilon_\ell - h, t_\ell + \varepsilon_\ell + h]$. Write $\widetilde{E}_{v,1}, \ldots, \widetilde{E}_{v,D}$ for the components of $\widetilde{E}_v$, where $\widetilde{E}_{v,\ell} = [t_\ell - h, m_\ell + h]$ for $\ell = 1, \ldots, d$, and $\widetilde{E}_{v,d+1}, \ldots, \widetilde{E}_{v,D}$ are the components of E_v^*. Write $\widetilde{E}_{v,1}(\vec{\varepsilon}), \ldots, \widetilde{E}_{v,D}(\vec{\varepsilon})$ for the corresponding components of $\widetilde{E}_v(\vec{\varepsilon})$.*

Fix $\vec{s} \in \mathcal{P}_v^m$. Given $\vec{n} = (n_1, \ldots, n_D) \in \mathbb{N}^D$, let $Q(z) = Q_{\vec{n}}(z) = C \cdot \prod_{k=1}^N [z,\alpha_k]_{\mathfrak{X},\vec{s}}$ be an $(\mathfrak{X},\vec{s})$-pseudopolynomial with n_ℓ roots in $\widetilde{E}_{v,\ell}$, for $\ell = 1, \ldots, D$. Suppose that $Q(z)$ oscillates n_ℓ times from $\widetilde{\mathcal{R}}^N$ to 0 to $\widetilde{\mathcal{R}}_v^N$ on each component $\widetilde{E}_{v,\ell}$ of $\widetilde{E}_v$ contained in $\mathcal{C}_v(\mathbb{R})$ and that $|\frac{1}{N}\log(Q(z)) - G_{\mathfrak{X},\vec{s}}(z,\widetilde{E}_v)| < r$ if $z \in \partial U_v$. Put

$$Q^{\vec{\varepsilon}}(z) := e^{N\varepsilon_0} \cdot C \cdot \prod_{x_i \in E_v^*} [z,\alpha_k]_{\mathfrak{X},\vec{s}} \cdot \prod_{\ell=1}^d \prod_{\alpha_k \in e_\ell(r)} [z,\alpha_k + \varepsilon_\ell]_{\mathfrak{X},\vec{s}} \; . \tag{5.42}$$

Then $Q^{\vec{\varepsilon}}(z)$ has n_ℓ roots in each $\widetilde{E}_{v,\ell}(\vec{\varepsilon})$, $Q^{\vec{\varepsilon}}(z)$ oscillates n_ℓ times from $\mathcal{R}^N$ to 0 to $\mathcal{R}_v^N$ on each component $\widetilde{E}_{v,\ell}(\vec{\varepsilon})$ in $\mathcal{C}_v(\mathbb{R})$, and $\{z \in \mathcal{C}_v(\mathbb{C}) : Q^{\vec{\varepsilon}}(z) \leq 1\} \subset U_v$.

PROOF. By its construction, $Q^{\vec{\varepsilon}}(z)$ has n_ℓ roots in $\widetilde{E}_{v,\ell}(\vec{\varepsilon})$, for each $\ell = 1, \ldots, D$. Note that since $|\varepsilon_\ell| \le r < r_0$, if $\alpha_k \in [t_\ell - h, t_\ell + h]$, then $\alpha_k + \varepsilon_\ell \in [t_\ell - 2r_0, t_\ell + 2r_0]$. We first show that $Q^{\vec{\varepsilon}}(z)$ oscillates n_ℓ times from $\mathcal{R}^N$ to 0 to $\mathcal{R}_v^N$ on each real component $\widetilde{E}_{v,\ell}(\vec{\varepsilon})$. There are two cases to consider.

First suppose that $1 \le \ell \le d$, so that $\widetilde{E}_{v,\ell} = [t_\ell - h, t_\ell + h]$ and $\widetilde{E}_{v,\ell}(\vec{\varepsilon}) = [t_\ell + \varepsilon_\ell - h, t_\ell + \varepsilon_\ell + h]$. Let $z_0 \in [t_\ell - h, t_\ell + h]$ be a point where $Q(z_0) = \widetilde{\mathcal{R}}_v^N$. By (5.37) and (5.38),

$$\begin{aligned}
&\Big|\frac{1}{N}\log(Q^{\vec{\varepsilon}}(z_0 + \varepsilon_\ell)) - \frac{1}{N}\log(Q(z_0))\Big| \\
&\le\ |\varepsilon_0| + \frac{1}{N}\sum_{\alpha_k \in E_v^*}\sum_{i=1}^m s_i \big|\log([z_0 + \varepsilon_\ell, \alpha_k]_{x_i}) - \log([z_0, \alpha_k]_{x_i})\big| \\
&\quad + \frac{1}{N}\sum_{\substack{j=1\\ j\ne\ell}}^d \sum_{\alpha_k \in [t_j - h, t_j + h]}\sum_{i=1}^m s_i \big|\log([z_0 + \varepsilon_\ell, \alpha_k + \varepsilon_j]_{x_i}) - \log([z_0, \alpha_k]_{x_i})\big| \\
&\quad + \frac{1}{N}\sum_{\alpha_k \in [t_\ell - h, t_\ell + h]}\sum_{i=1}^m s_i \big|\eta_i(z_0 + \varepsilon_\ell, \alpha_k + \varepsilon_\ell) - \eta_i(z_0, \alpha_k)\big| \\
&<\ r + \frac{1}{N}\cdot N \cdot \sum_{i=1}^m s_i \cdot r \ = \ 2r \ < \ \Delta_1 \ .
\end{aligned}$$

Since $\frac{1}{N}\log(Q(z_0)) = \log(\widetilde{\mathcal{R}}_v)$ and $\log(\widetilde{\mathcal{R}}_v) - \Delta_1 = \log(\mathcal{R}_v)$, it necessarily follows that $\frac{1}{N}\log(Q^{\vec{\varepsilon}}(z_0 + \varepsilon_\ell)) \ge \log(\mathcal{R}_v)$. If $Q(z)$ takes the values $\widetilde{\mathcal{R}}_v^N$, 0, $\widetilde{\mathcal{R}}_v^N$ at successive points z_0, α_k, and z_0' of $[t_\ell - h, t_\ell + h]$, then $Q^{\vec{\varepsilon}}(z)$ takes values $\ge \mathcal{R}_v^N$, 0, $\ge \mathcal{R}_v^N$ at the points $z_0 + \varepsilon_\ell$, $\alpha_k + \varepsilon_\ell$, and $z_0' + \varepsilon_\ell$ in $[t_\ell + \varepsilon_\ell - h, t_\ell + \varepsilon_\ell + h]$.

Next suppose $\ell \ge d + 1$, so $\widetilde{E}_{v,\ell}(\vec{\varepsilon}) = \widetilde{E}_{v,\ell}$ is a component of E_v^* contained in $\mathcal{C}_v(\mathbb{R})$. Let $z_0 \in \widetilde{E}_{v,\ell}$ be a point where $Q(z_0) = \widetilde{\mathcal{R}}_v^N$. By (5.36),

$$\begin{aligned}
&\Big|\frac{1}{N}\log(Q^{\vec{\varepsilon}}(z_0)) - \frac{1}{N}\log(Q(z_0))\Big| \\
&\le\ |\varepsilon_0| + \frac{1}{N}\sum_{\ell=1}^d \sum_{\alpha_k \in e_\ell(h)}\sum_{i=1}^m s_i \big|\log([z_0, \alpha_k]_{x_i}) - \log([z_0, \alpha_k + \varepsilon_\ell]_{x_i})\big| \\
&<\ r + \frac{1}{N}\cdot N \cdot \sum_{i=1}^m s_i \cdot r \ = \ 2r \ < \ \Delta_1 \ .
\end{aligned}$$

By the same argument as before, we conclude that $\frac{1}{N}\log(Q^{\vec{\varepsilon}}(z_0)) \ge \log(\widetilde{R}_v)$. Hence if $Q(z)$ takes the values $\widetilde{\mathcal{R}}_v^N$, 0, $\widetilde{\mathcal{R}}_v^N$ at successive points z_0, α_k, z_0' in $\widetilde{E}_{v,\ell}$, then $Q^{\vec{\varepsilon}}(z)$ takes values $\ge \mathcal{R}_v^N$, 0, $\ge \mathcal{R}_v^N$ at those points.

Finally, we show $\{z \in \mathcal{C}_v(\mathbb{C}) : Q^{\vec{\varepsilon}}(z) \le 1\} \subset U_v$. Note that $\frac{1}{N}\log(Q^{\vec{\varepsilon}}(z))$ is harmonic on $\mathcal{C}_v(\mathbb{C})\backslash(\widetilde{E}_v(\vec{\varepsilon}) \cup \mathfrak{X})$. For each $x_i \in \mathfrak{X}$, if $s_i > 0$ then $\lim_{z \to x_i} \frac{1}{N}\log(Q^{\vec{\varepsilon}}(z)) = \infty$, while if $s_i = 0$ then $\frac{1}{N}\log(Q^{\vec{\varepsilon}}(z))$ has a removable singularity at x_i. Hence if we show that $\frac{1}{N}\log(Q^{\vec{\varepsilon}}(z)) > 0$ on ∂U_v, the maximum principle for harmonic functions will give $\frac{1}{N}\log(Q^{\vec{\varepsilon}}(z)) > 0$ outside U_v.

Let z_0 be a point of ∂U_v. Since $\widetilde{E}_v = E_v^*(h) \subset E_v^* \cup I$, the monotonicity of Green's functions shows that $G(z_0, x_i; \widetilde{E}_v) \geq G(z_0, x_i; E_v^* \cup I) \geq \Delta_2$, and hence

$$G_{\mathfrak{X},\vec{s}}(z_0, \widetilde{E}_v) \;=\; \sum_{i=1}^m s_i G(z_0, x_i; \widetilde{E}_v) \;\geq\; \Delta_2 \;.$$

Since $|G_{\mathfrak{X},\vec{s}}(z_0, \widetilde{E}_v) - \frac{1}{N}\log(Q(z_0))| < r$ and $r < \Delta_2/3$, we have $\frac{1}{N}\log(Q(z_0)) > 2\Delta_2/3$. On the other hand, by (5.41)

$$\begin{aligned}
&\Big|\frac{1}{N}\log(Q^{\vec{\varepsilon}}(z_0)) - \frac{1}{N}\log(Q(z_0))\Big| \\
&\leq\; |\varepsilon_0| + \frac{1}{N}\sum_{\ell=1}^d \sum_{\alpha_k \in e_\ell(h)} \sum_{i=1}^m s_i \big|\log([z_0, \alpha_k]_{x_i}) - \log([z_0, \alpha_k + \varepsilon_\ell]_{x_i})\big| \\
&<\; r + \frac{1}{N}\cdot N \cdot \sum_{i=1}^m s_i \cdot r \;=\; 2r \;\leq\; 2\Delta_2/3 \;.
\end{aligned}$$

Hence $\frac{1}{N}\log(Q^{\vec{\varepsilon}}(z_0)) > 0$.

Since $\frac{1}{N}\log(Q^{\vec{\varepsilon}}(z)) > 0$ on ∂U_v, the maximum principle for harmonic functions (applied to $-\frac{1}{N}\log(Q^{\vec{\varepsilon}}(z))$ on $\mathcal{C}_v(\mathbb{C})\backslash(U_v \cup \mathfrak{X})$) shows that $\frac{1}{N}\log(Q^{\vec{\varepsilon}}(z)) > 0$ for all $z \notin U_v$. Thus $\{z \in \mathcal{C}_v(\mathbb{C}) : Q^{\vec{\varepsilon}}(z) \leq 1\} \subset U_v$. $\square$

Step 4. The choice of h and the construction of $\widetilde{E}_v$.

We will now choose the number h, and hence determine the set $\widetilde{E}_v$ in the theorem. Let $0 < h < r$ be small enough that the following three conditions are satisfied:

(1) For the set $E_v^*(h) = E_v^* \cup \big(\bigcup_{j=1}^d e_\ell(h)\big)$,

$$\left\{\begin{array}{ll} |V_{x_i}(E_v^*(h)) - V_{x_i}(E_v^*)| < \varepsilon_v/3 & \text{for each } i \;, \\ G(x_i, x_j; E_v^*(h)) - G(x_i, x_j; E_v^*)| < \varepsilon_v/3 & \text{for all } i \neq j \;. \end{array}\right. \tag{5.43}$$

(2) For all $i \neq j$, all $t_\ell \in \mathcal{T}$, all ε_ℓ with $|\varepsilon_\ell| \leq r$, and all $z \in [t_\ell - h, t_\ell + h]$,

$$\big|\log([x_i, z + \varepsilon_\ell]_{x_j}) - \log([x_i, t_\ell + \varepsilon_\ell]_{x_j})\big| \;<\; \frac{r}{24B_1B_4\sqrt{J+1}} \;. \tag{5.44}$$

(3) $h \leq h_0$, so the mass bounds (5.33) are valid.

Condition (1) holds for all small h by Lemma 5.6.C. Condition (2) holds for all small h by the continuity of the $\log([x_i, t]_{x_j})$. And, condition (3) clearly holds for all small h.

Fix h satisfying (1), (2) and (3), and write $\widetilde{E}_{v,\ell} = e_\ell(h) = [t_\ell - h, t_\ell + h]$, for $\ell = 1, \ldots, d$. Put

$$\widetilde{E}_v \;=\; E_v^* \cup \big(\bigcup_{\ell=1}^d [t_\ell - h, t_\ell + h]\big) \;. \tag{5.45}$$

Then $\widetilde{E}_v \subset E_v^0$, and $\widetilde{E}_v$ is K_v-simple; in particular, $\mathcal{C}_v(\mathbb{C})\backslash\widetilde{E}_v$ is connected. By (5.25), (5.28) and (5.43),

$$\left\{\begin{array}{ll} |V_{x_i}(\widetilde{E}_v) - V_{x_i}(E_v)| < \varepsilon_v & \text{for each } i, \\ G(x_i, x_j; \widetilde{E}_v) - G(x_i, x_j; E_v)| < \varepsilon_v & \text{for all } i \neq j. \end{array}\right. \tag{5.46}$$

as required by Theorem 5.2. Moreover, all the intervals making up $\widetilde{E}_v$ are "short" in the sense of Definition B.15, and Proposition 5.7 applies to $\widetilde{E}_v$.

Step 5. The total change map.

We now begin to address problem of constructing rational functions with large oscillations and specified logarithmic leading coefficients.

Fix $\vec{s} \in \mathcal{P}_v^m \bigcap \mathbb{Q}^m$. Consider a special $(\mathfrak{X}, \vec{s})$-pseudopolynomial for $\widetilde{E}_v$,

$$Q(z) \;=\; Q_{\vec{n}}(z) \;=\; C \cdot \prod_{k=1}^{N} [z, \alpha_k]_{\mathfrak{X},\vec{s}} \ , \tag{5.47}$$

as constructed in Theorem 5.3. In particular, $\|Q(z)\|_{\widetilde{E}_v} = 1$, and for each $\ell = 1, \ldots, D$, if $\vec{n} = (n_1, \ldots, n_D)$ then $Q(z)$ has n_ℓ roots in $\widetilde{E}_{v,\ell}$. Recalling that $J = g + m_1 + m_2 - 1$, fix a set of J points in $\mathcal{T}$ such that the matrix $\mathcal{G}_{\vec{s}}(\vec{t})$ in Corollary 5.5 is nonsingular and satisfies

$$\|\mathcal{G}_{\vec{s}}(\vec{t})^{-1}\| \;\le\; B_1 \ ,$$

as in (5.30); the construction of $\mathcal{T}$ shows that this can be done. Without loss, suppose these points are $t_1, \ldots, t_J$, and put $E_v^{00} = \widetilde{E}_v \backslash (\bigcup_{\ell=1}^J \widetilde{E}_{v,\ell})$. We will move the roots of $Q(z)$, constructing a pseudopolynomial $Q^{\vec{\varepsilon}}(z)$ as in (5.42); however, instead of moving the roots in all the intervals $\widetilde{E}_{v,\ell}$, $\ell = 1, \ldots, d$ we will only move the ones in $\widetilde{E}_{v,1}, \ldots, \widetilde{E}_{v,J}$. Equivalently, take $\varepsilon_\ell = 0$ for $\ell = J+1, \ldots, d$ in Proposition 5.7. Let $\vec{\varepsilon} = (\varepsilon_0, \varepsilon_1, \ldots, \varepsilon_J) \in \mathbb{R}^{J+1}$ be such that $|\varepsilon_\ell| \le r$ for each $\ell = 0, \ldots, J$, and consider

$$Q^{\vec{\varepsilon}}(z) \;=\; \exp(\varepsilon_0)^N \cdot C \cdot \prod_{\alpha_k \in E_v^{00}} [z, \alpha_k]_{\mathfrak{X},\vec{s}} \cdot \prod_{\ell=1}^{J} \prod_{\alpha_k \in [t_\ell - h, t_\ell + h]} [z, \alpha_k + \varepsilon_\ell]_{\mathfrak{X},\vec{s}} \ . \tag{5.48}$$

Let $\mathfrak{X}_1, \ldots, \mathfrak{X}_{m_1+m_2}$ be the orbits in $\mathfrak{X}$ under complex conjugation, as in §5.3. If $x_i \in \mathfrak{X}_a$, then passing from $Q(z)$ to $Q^{\vec{\varepsilon}}(z)$ produces the following change in the logarithmic leading coefficient at x_i:

$$\begin{aligned} \lambda_i(\vec{\varepsilon}) \;&:=\; \Lambda_{x_i}(Q^{\vec{\varepsilon}}, \vec{s}) - \Lambda_{x_i}(Q, \vec{s}) \\ &=\; \varepsilon_0 + \sum_{\ell=1}^{J} \sum_{\alpha_k \in [t_\ell - h, t_\ell + h]} \frac{1}{N} \sum_{\substack{j=1 \\ x_j \notin \mathfrak{X}_a}}^{m} s_k \left(\log([x_i, \alpha_k + \varepsilon_\ell]_{x_j}) - \log([x_i, \alpha_k]_{x_j}) \right) . \end{aligned}$$

This is immediate if $\mathfrak{X}_a = \{x_i\}$; if $\mathfrak{X}_a = \{x_i, x_j\}$ consists of two points, it follows from the fact that $[x_i, t]_{x_j}$ is constant on $\mathcal{C}_v(\mathbb{R})$, as shown in §5.3.

Recalling that $Q(z)$ has n_ℓ roots in $\widetilde{E}_{v,\ell}$, for each $a = 1, \ldots, m_1 + m_2$ fix a point $x_i \in \mathfrak{X}_a$ and put

$$\begin{aligned} L_a(\vec{\varepsilon}) \;&=\; \varepsilon_0 + \sum_{\ell=1}^{J} H_{\vec{s},a}(t_\ell) \cdot \frac{n_\ell}{N} \varepsilon_\ell \\ &=\; \varepsilon_0 + \sum_{\ell=1}^{J} \sum_{\substack{j=1 \\ x_j \notin \mathfrak{X}_a}}^{m} s_j \left(\frac{d}{dt} \log([x_i, t]_{x_j}) \right)\bigg|_{t=t_\ell} \cdot \frac{n_\ell}{N} \varepsilon_\ell \ . \end{aligned}$$

Then $L_a(\vec{\varepsilon})$ is a linear map which approximates $\lambda_i(\vec{\varepsilon})$, and

$$\begin{aligned}
&|\lambda_i(\vec{\varepsilon}) - L_a(\vec{\varepsilon})| \\
&\quad \leq \frac{1}{N}\sum_{\ell=1}^{J} \sum_{\alpha_k \in [t_\ell - h, t_\ell + h]} \sum_{\substack{j=1 \\ x_j \notin \mathfrak{X}_a}}^{m} s_j \left\{ \left|\log([x_i, \alpha_k + \varepsilon_\ell]_{x_j}) - \log([x_i, t_\ell + \varepsilon_\ell]_{x_j})\right| \right. \\
&\qquad\qquad + \left|\log([x_i, \alpha_k]_{x_j}) - \log([x_i, t_\ell]_{x_j})\right| \\
&\qquad \left. + \left|(\log([x_i, t_\ell + \varepsilon_\ell]_{x_j}) - \log([x_i, t_\ell]_{x_j})) - \left(\frac{d}{dt}\log([x_i, t]_{x_j})\right)\right|_{t=t_\ell} \varepsilon_\ell\right| \right\} .
\end{aligned}$$

By (5.39) and (5.44), the magnitude of each term in absolute values on the right side is at most $r/(24B_1B_4\sqrt{J+1})$. Hence

$$|\lambda_i(\vec{\varepsilon}) - L_a(\vec{\varepsilon})| \leq \frac{n_1 + \ldots + n_J}{N} \cdot \frac{3r}{24B_1B_4\sqrt{J+1}} . \tag{5.49}$$

Next, consider the "normalized divisor change" map $\widehat{\varphi}(\vec{\varepsilon})$ (see (5.15)) induced by passing from $Q(z)$ to $Q^{\vec{\varepsilon}}(z)$. If $g = 0$, every divisor of degree 0 is principal, and there are no nonzero holomorphic differentials; in that case ignore all constructions related to the $\omega_j = h_j(z)dz$ in the rest of the proof.

If $g \geq 1$, for each $j = 1, \ldots, g$, the jth coordinate of $\widehat{\varphi}(\vec{\varepsilon})$ is

$$\widehat{\varphi}_j(\vec{\varepsilon}) = \sum_{\ell=1}^{J} \sum_{\alpha_k \in [t_\ell - h, t_\ell + h]} \frac{1}{N} \int_{\alpha_k}^{\alpha_k + \varepsilon_\ell} h_j(z)\, dz .$$

By the Mean Value Theorem for integrals, for each α_k there is an $\alpha_k^* \in [\alpha_k, \alpha_k + \varepsilon_\ell]$ such that $\int_{\alpha_k}^{\alpha_k+\varepsilon_\ell} h_j(z)dz = h_j(\alpha_k^*) \cdot \varepsilon_\ell$. Consequently, if

$$f_j(\vec{\varepsilon}) := \sum_{\ell=1}^{J} h_j(t_\ell) \cdot \frac{n_\ell}{N} \varepsilon_\ell$$

is the linear approximation to $\widehat{\varphi}_j(\vec{\varepsilon})$ at the origin, then

$$|\widehat{\varphi}_j(\vec{\varepsilon}) - f_j(\vec{\varepsilon})| \leq \sum_{\ell=1}^{J} \sum_{\alpha_k \in [t_\ell - h, t_\ell + h]} \frac{1}{N} |h_j(\alpha_k^*) - h_j(t_\ell)|\, |\varepsilon_\ell| .$$

From (5.40) and the fact that $|\varepsilon_\ell| \leq r$ for each ℓ, it follows that

$$|\widehat{\varphi}_j(\vec{\varepsilon}) - f_j(\vec{\varepsilon})| \leq \frac{n_1 + \ldots + n_J}{N} \cdot \frac{r}{24B_1B_4\sqrt{J+1}} . \tag{5.50}$$

Let $B^{J+1}(0, r) = \{(\varepsilon_0, \ldots, \varepsilon_J) \in \mathbb{R}^{J+1} : |\vec{\varepsilon}| \leq r\}$. Assuming $x_1, \ldots, x_{m_1+m_2}$ form a set of representatives for the orbits $\mathfrak{X}_1, \ldots, \mathfrak{X}_{m_1+m_2}$, we define the "total change map" $F^Q : B^{J+1}(0,r) \to \mathbb{R}^{J+1}$ for $Q(z)$ by

$$(5.51) \qquad F^Q(\vec{\varepsilon}) = \begin{pmatrix} \lambda_1(\vec{\varepsilon}) \\ \vdots \\ \lambda_{m_1+m_2}(\vec{\varepsilon}) \\ \widehat{\varphi}_1(\vec{\varepsilon}) \\ \\ \vdots \\ \widehat{\varphi}_g(\vec{\varepsilon}) \end{pmatrix},$$

and note that its linear approximation

$$F_0^Q(\vec{\varepsilon}) := \begin{pmatrix} L_1(\vec{\varepsilon}) \\ \vdots \\ L_{m_1+m_2}(\vec{\varepsilon}) \\ f_1(\vec{\varepsilon}) \\ \vdots \\ f_g(\vec{\varepsilon}) \end{pmatrix} = \begin{pmatrix} \varepsilon_0 + \sum_{\ell=1}^J H_{\vec{s},1}(t_\ell) \cdot \frac{n_\ell}{N}\varepsilon_\ell \\ \vdots \\ \varepsilon_0 + \sum_{\ell=1}^J H_{\vec{s},m_1+m_2}(t_\ell) \cdot \frac{n_\ell}{N}\varepsilon_\ell \\ \sum_{\ell=1}^J h_1(t_\ell) \cdot \frac{n_\ell}{N}\varepsilon_\ell \\ \vdots \\ \sum_{\ell=1}^J h_g(t_\ell) \cdot \frac{n_\ell}{N}\varepsilon_\ell \end{pmatrix}$$

can be decomposed as

$$(5.52) \qquad F_0^Q(\vec{\varepsilon}) = \mathcal{G}_{\vec{s}}(\vec{t}) \cdot \mathcal{N} \cdot \vec{\varepsilon}$$

where $\mathcal{N}$ is the $(J+1) \times (J+1)$ diagonal matrix $\mathcal{N} = \operatorname{diag}(1, n_1/N, \ldots, n_J/N)$ and $\mathcal{G}_{\vec{s}}(\vec{t})$ is as in Corollary 5.5.

By (5.49), (5.50), (5.51) and (5.52), if

$$(5.53) \qquad d(\vec{\varepsilon}) = F^Q(\vec{\varepsilon}) - F_0^Q(\vec{\varepsilon})$$

for $\vec{\varepsilon} \in B^{J+1}(0,r)$, then each coordinate function $d_i(\vec{\varepsilon})$ of $d(\vec{\varepsilon})$ satisfies

$$(5.54) \qquad |d_i(\vec{\varepsilon})| \le \frac{r}{8B_1B_4\sqrt{J+1}} \cdot \sum_{\ell=1}^J \frac{n_\ell}{N} .$$

Next consider the renormalized map $\mathcal{F}^Q : B^{J+1}(0,r) \to \mathbb{R}^{J+1}$ given by

$$(5.55) \qquad \mathcal{F}^Q(\vec{\varepsilon}) = \mathcal{N}^{-1}\mathcal{G}_{\vec{s}}(\vec{t})^{-1} \cdot F^Q(\vec{\varepsilon}) ,$$

and the corresponding difference function

$$(5.56) \qquad D(\vec{\varepsilon}) = \mathcal{F}^Q(\vec{\varepsilon}) - \vec{\varepsilon} = \mathcal{N}^{-1}\mathcal{G}_{\vec{s}}(\vec{t})^{-1} \cdot d(\vec{\varepsilon}) .$$

By construction, the operator norm $\|\mathcal{G}_{\vec{s}}(\vec{t})^{-1}\|$ is bounded by B_1, and clearly

$$\|\mathcal{N}^{-1}\| = \frac{1}{\min_{1\le\ell\le J} n_\ell/N} .$$

It follows from (5.53) and (5.54) that for each $\vec{\varepsilon} \in B^{J+1}(0,r)$,

$$\begin{aligned} |D(\vec{\varepsilon})| &\le \frac{1}{\min_{1\le\ell\le J} n_\ell/N} \cdot B_1 \cdot \sqrt{J+1} \cdot \frac{r}{8B_1B_4\sqrt{J+1}} \cdot \sum_{\ell=1}^J n_\ell/N \\ (5.57) \qquad &= \frac{r}{8} \cdot \frac{1}{B_4} \cdot \frac{\sum_{\ell=1}^J n_\ell/N}{\min_{1\le\ell\le J} n_\ell/N} . \end{aligned}$$

We now cite a topological fact related to the Brouwer Fixed Point Theorem.[1]

PROPOSITION 5.8. *Fix* $0 < \eta < 1$ *and suppose* $g : B^{J+1}(0,r) \to \mathbb{R}^{J+1}$ *is a continuous map such that for each* $\vec{x}$ *in* $\partial B^{J+1}(0,r)$,

$$|g(\vec{x}) - \vec{x}| \ < \ \eta \cdot r \ .$$

Then $B^{J+1}(0,(1-\eta)r) \subset g(B^{J+1}(0,r))$.

PROOF. If $J = 0$ the result follows easily from the Intermediate Value theorem, so we can assume $J \ge 1$. Suppose there were some $\vec{x}_0 \in B(0,(1-\eta)r)$ which did not belong to $g(B^{J+1}(0,r))$. Then we could define a continuous map $g_0 : B^{J+1}(0,r) \to \partial B^{J+1}(0,r)$, such that $g_0(\vec{x})$ is the point where the ray from $\vec{x}_0$ through $g(\vec{x})$ meets $\partial B^{J+1}(0,r)$. For each $\vec{x} \in \partial B^{J+1}(0,r)$, $|g(\vec{x}) - \vec{x}| < \eta r$ while by hypothesis $|\vec{x}_0 - \vec{x}| > \eta r$. Hence there would be a homotopy from $g_0|_{\partial B^{J+1}(0,r)}$ to the identity, such that for each $t \in [0,1]$ and each $\vec{x} \in \partial B^{J+1}(0,r)$, $g_t(\vec{x})$ is the point where the ray from $\vec{x}_0$ through $tg(\vec{x}) + (1-t)\vec{x}$ meets $\partial B^{J+1}(0,r)$. Thus $g_0 : B^{J+1}(0,r) \to \partial B^{J+1}(0,r)$ would be a continuous map whose restriction to $\partial B^{J+1}(0,r)$ was homotopic to the identity. Now let $i : \partial B^{J+1}(0,r) \to B^{J+1}(0,r)$ be the inclusion, and consider the induced map $(g_0 \circ i)_* = (g_0)_* \circ i_*$ on homology:

$$H_J(\partial B^{J+1}(0,r)) \ \stackrel{i_*}{\longrightarrow} \ H_J(B^{J+1}(0,r)) \ \stackrel{(g_0)_*}{\longrightarrow} \ H_J(\partial B^{J+1}(0,r)) \ .$$

Here $H_J(B^{J+1}(0,r)) = 0$ so $(g_0)_* \circ i_*$ is the 0 map. However, $H_J(\partial B^{J+1}(0,r)) \cong \mathbb{Z}$ and $g_0 \circ i$ is homotopic to the identity, so $(g_0 \circ i)_*$ is the identity map. This is a contradiction, so $\vec{x}_0 \in g(B^{J+1}(0,r))$. $\square$

Let $\vec{\sigma}$ be the vector of weights of the components of $\widetilde{E}_v$ under $\mu_{\mathfrak{X},\vec{s}}$. For any special pseudopolynomial $Q_{\vec{n}}(z)$ for $\widetilde{E}_v$ with $N = \sum n_i$ sufficiently large, and with $\vec{n}/N$ close enough to $\vec{\sigma}$ that $|n_\ell/N - \mu_{\mathfrak{X},\vec{s}}(\widetilde{E}_{v,\ell})| < \frac{1}{2}\mu_{\mathfrak{X},\vec{s}}(\widetilde{E}_{v,\ell})$ for $\ell = 1, \ldots, J$ we will have

$$\frac{\sum_{\ell=1}^J n_\ell/N}{\min_{1\le\ell\le J} n_\ell/N} \ < \ \frac{3/2 \cdot \mu_{\mathfrak{X},\vec{s}}(\bigcup_{k=1}^J \widetilde{E}_{v,k})}{1/2 \cdot \min_{1\le\ell\le J} \mu_{\mathfrak{X},\vec{s}}(\widetilde{E}_{v,\ell})} \ = \ 3B_4. \tag{5.58}$$

If (5.58) holds, then in (5.57) we will have $|D(\vec{\varepsilon})| < r/2$. Applying Proposition 5.8 to $\mathcal{F}^Q(\vec{\varepsilon})$ with $\eta = 1/2$, it follows that

$$B^{J+1}(0,r/2) \ \subset \ \mathcal{F}^Q(B^{J+1}(0,r)) \ . \tag{5.59}$$

Step 6. The choice of δ_v.

Recalling that $\mu_{\mathfrak{X},\vec{s}} = \sum_{i=1}^m s_i\mu_i$ where μ_i is the equilibrium distribution of $\widetilde{E}_v$ with respect to x_i, put

$$B_5 \ = \ \min_{1\le i\le m} \min_{1\le\ell\le d} \mu_i(\widetilde{E}_{v,\ell}) \ > \ 0 \ . \tag{5.60}$$

Then for any $\vec{s} \in \mathcal{P}_v^m$, and any $\ell = 1, \ldots, d$,

$$\mu_{\mathfrak{X},\vec{s}}(\widetilde{E}_{v,\ell}) \ \ge \ B_5 \ . \tag{5.61}$$

Henceforth we will assume $\vec{n}$ and N are such that

$$\min_{1\le\ell\le d} n_\ell/N \ \ge \ B_5/2 \ . \tag{5.62}$$

[1] The author thanks Ted Shifrin for helpful discussions concerning this argument.

In the light of (5.59), (5.55), (5.62) and the construction of $\mathcal{N}$ and $\mathcal{G}_{\vec{s}}(\vec{t})$, for any such $\vec{n}$, if $Q(z) = Q_{\vec{n}}$ as in Theorem 5.3 then

$$B(0, \frac{rB_5}{4B_1}) \;=\; \frac{B_5}{2B_1} \cdot B(0, r/2) \;\subset\; F^Q(B^{J+1}(0,r)) \ . \tag{5.63}$$

This radius $\frac{rB_5}{4B_1}$ is uniform for all $\vec{s} \in \mathcal{P}_v^m$.

Let the number $\delta_v > 0$ in Theorem 5.2 be such that

$$\left(\prod_{i=1}^{m_1+m_2} [-2\delta_v, 2\delta_v] \right) \times B^g(0, \delta_v) \;\subset\; B(0, \frac{rB_5}{4B_1}) \ , \tag{5.64}$$

where $B^g(0,\delta_v) = \{\vec{y} \in \mathbb{R}^g : |\vec{y}| \le \delta_v\}$. Then if (5.58) and (5.62) hold,

$$[-2\delta_v, 2\delta_v]^{m_1+m_2} \times B^g(0, \delta_v) \;\subset\; F^Q(B^{J+1}(0,r)) \ . \tag{5.65}$$

Step 7. Achieving principality and varying the logarithmic leading coefficients.

As in §5.2, let $\mathrm{Jac}(\mathcal{C}_v)$ be the Jacobian of $\mathcal{C}_v$, and let $\omega_1, \ldots, \omega_g$ be a basis for the space of real holomorphic differentials of $\mathcal{C}_v$. Then $\mathrm{Jac}(\mathcal{C}_v)(\mathbb{C}) \cong \mathbb{C}^g/\mathcal{L}$ where $\mathcal{L}$ is the period lattice corresponding to $\omega_1, \ldots, \omega_g$. Let $\varphi : \mathcal{C}_v(\mathbb{C}) \to \mathrm{Jac}(\mathcal{C}_v)(\mathbb{C})$ be the canonical map associated to a base point $p_0 \in \mathcal{C}_v(\mathbb{R})$,

$$\varphi(p) = (\int_{p_0}^{p} \omega_1, \ldots, \int_{p_0}^{p} \omega_g) \pmod{\mathcal{L}} \tag{5.66}$$

and let $\Phi : \mathcal{C}_v(\mathbb{C})^g \to \mathrm{Jac}(\mathcal{C}_v)(\mathbb{C})$ be the corresponding summatory map.

Since $\mathrm{Jac}(\mathcal{C}_v)$ is nonsingular and defined over $\mathbb{R}$, and its origin is rational over $\mathbb{R}$, the implicit function theorem shows that $\mathrm{Jac}(\mathcal{C}_v)(\mathbb{R})$ is a g-dimensional real manifold. Since $\mathrm{Jac}(\mathcal{C}_v)$ is defined over $\mathbb{R}$, $\mathrm{Jac}(\mathcal{C}_v)(\mathbb{R})$ is a compact Lie subgroup of $\mathrm{Jac}(\mathcal{C}_v)(\mathbb{C})$. Thus its identity component must be isomorphic to a real torus $\mathbb{R}^g/\mathcal{L}_0$, where $\mathcal{L}_0 = \mathcal{L} \bigcap \mathbb{R}^g$ is a g-dimensional sublattice of $\mathcal{L}$.

Let $\mathrm{Jac}(\mathcal{C}_v)(\mathbb{R})_0$ be the identity component of $\mathrm{Jac}(\mathcal{C}_v)(\mathbb{R})$.

We claim that for each $p \in \mathcal{C}_v(\mathbb{R})$, the point $2\varphi(p)$ belongs to $\mathrm{Jac}(\mathcal{C}_v)(\mathbb{R})_0$. To see this, suppose that in (5.66), $z = \varphi(p) \in \mathbb{C}^g/\mathcal{L}$ is obtained by integrating over some path γ from p_0 to p. Applying complex conjugation to (5.66), we see that since p_0, p and the ω_i are fixed but γ is taken to its conjugate path, $z \pmod{\mathcal{L}}$ is changed to $\overline{z} \pmod{\mathcal{L}}$. Hence, lifting z to $\mathbb{C}^g$, it must be that

$$\overline{z} \;=\; z - \ell \tag{5.67}$$

for some $\ell \in \mathcal{L}$. We can find a basis $\ell_1, \ldots, \ell_{2g}$ for $\mathcal{L}$ such that $\ell_1, \ldots, \ell_g$ is a basis for $\mathcal{L}_0$. Writing

$$z \;=\; \sum_{i=1}^{g} a_i \ell_i + \sum_{i=1}^{g} b_i \ell_{g+i} \quad \text{where } a_i, b_i \in \mathbb{R} \ , \tag{5.68}$$

$$\ell \;=\; \sum_{i=1}^{g} c_i \ell_i + \sum_{i=1}^{g} d_i \ell_{g+i} \quad \text{where } c_i, d_i \in \mathbb{Z} \ , \tag{5.69}$$

it follows that

$$\overline{z} = \sum_{i=1}^{g} a_i \ell_i + \sum_{i=1}^{g} b_i \overline{\ell}_{g+i} \ . \tag{5.70}$$

But $\overline{\ell}_{g+i} \in \mathcal{L}$ for each i, and since $\overline{\ell}_{g+i} + \ell_{g+i} \in \mathbb{R}^g$, for each i,

$$\overline{\ell}_{g+i} = -\ell_{g+i} + \sum_{k=1}^{g} M_{ik}\ell_k \tag{5.71}$$

for certain integers M_{ik}. Inserting (5.68)–(5.71) in (5.67), we see that $2b_i = d_i \in \mathbb{Z}$ for each $i = 1, \ldots, g$. Thus, $2z \in \mathcal{L} + \mathbb{R}^g$, which shows that $2\varphi(p) \in \mathrm{Jac}(\mathcal{C}_v)(\mathbb{R})_0$.

Fix $\vec{s} \in \mathcal{P}_v^m \cap \mathbb{Q}^m$, and consider a special $(\mathfrak{X}, \vec{s})$-pseudopolynomial

$$Q(z) \;=\; Q_{\vec{n}}(z) \;=\; C_{\vec{n}} \cdot \prod_{k=1}^{N} [z, \alpha_k]_{\mathfrak{X},\vec{s}}$$

with roots in $\widetilde{E}_v$. Clearly roots belonging to the same segment in $\widetilde{E}_v$ map to the same connected component of $\mathrm{Jac}(\mathcal{C}_v)(\mathbb{R})$ under φ. Hence if we arrange that each coordinate n_ℓ of $\vec{n}$ is even, then $\sum_{i=1}^{N} \varphi(\alpha_k)$ will belong to $\mathrm{Jac}(\mathcal{C}_v)(\mathbb{R})_0$. Likewise, if $N = \sum n_\ell$ is such that $N\vec{s} \in 2\mathbb{Z}^m$, then $\varphi(\sum_{i=1}^{m} N s_i(x_i)) \in \mathrm{Jac}(\mathcal{C}_v)(\mathbb{R})_0$. If S is the least common denominator of the s_i, both of these conditions can be assured by requiring that

$$n_\ell \;\equiv\; 0 \pmod{2S} \quad \text{for all } \ell \;. \tag{5.72}$$

Suppose $\vec{n}/N$ is close enough to $\vec{\sigma}$ that (5.58) and (5.62) hold. Then

$$B^g(0, \delta_v) \;\subset\; \widehat{\varphi}(B^{J+1}(0, r)) \;.$$

If in addition N is sufficiently large, then $N \cdot B^g(0, \delta_v)$ will contain a fundamental domain for $\mathbb{R}^g/\mathcal{L}_0$.

Since $\varphi(\mathrm{div}(Q^{\vec{\varepsilon}}) - \varphi(\mathrm{div}(Q)) = N \cdot \widehat{\varphi}(\vec{\varepsilon})$, it follows that if (5.58), (5.62), and (5.72) are satisfied and N is sufficiently large, then $\vec{\varepsilon} \in B^{J+1}(0, r)$ can be chosen so that $Q^{\vec{\varepsilon}}(z)$ is principal. Furthermore, by (5.65) this can be done in such a way that for any K_v-symmetric $\vec{\beta}' = (\beta_1', \ldots, \beta_m') \in [-2\delta_v, 2\delta_v]^m$,

$$\Lambda_{x_i}(Q^{\vec{\varepsilon}}, \vec{s}) \;=\; \Lambda_{x_i}(Q, \vec{s}) + \beta_i' \quad \text{for } i = 1, \ldots, m \;. \tag{5.73}$$

Step 8. The choice of the number N_v, and conclusion of the proof.

Let $\widetilde{\mathcal{R}}_v$ be as in Proposition 5.7, let r be the number constructed in Step 3, and let δ_v be the number constructed in Step 7. Fix $\vec{s} = (s_1, \ldots, s_m) \in \mathcal{P}_v^m \cap \mathbb{Q}^m$, and let S be the least common denominator of the s_i.

Take a sequence of K_v-symmetric vectors $\vec{n}_k = (n_{k,1}, \ldots, n_{k,D}) \in \mathbb{N}^D$ satisfying the following three properties:

(1) for each k and ℓ, $n_{k,\ell} \equiv 0 \pmod{2S}$;
(2) for each k, $\sum_\ell n_{k,\ell} = k \cdot 2S$;
(3) as $k \to \infty$, writing $N_k = \sum_\ell n_{k,\ell}$, we have $\vec{n}_k/N_k \to \vec{\sigma}$.

Applying Theorem 5.3 to $\{\vec{n}_k\}_{k\in\mathbb{N}}$, we obtain a sequence of $(\mathfrak{X}, \vec{s})$-pseudopolynomials $Q_{\vec{n}_k}(z)$, whose roots belong to $\widetilde{E}_v$, with $Q_{\vec{n}_k}(z) = Q_{\vec{n}_k}(\overline{z})$ for all z and k, such that for all sufficiently large k,

(4) $|\frac{1}{N_k} \log(Q_{\vec{n}_k}(z)) - G_{\mathfrak{X},\vec{s}}(z, \widetilde{E}_v)| < r$ for all $z \in \mathcal{C}_v(\mathbb{C}) \backslash (U_v \cup \mathfrak{X})$;
(5) $|\Lambda_{x_i}(Q_{\vec{n}_k}, \vec{s}) - \Lambda_{x_i}(\widetilde{E}_v, \vec{s})| \le \delta_v$ for each $x_i \in \mathfrak{X}$;
(6) $Q_{\vec{n}_k}$ has $n_{k,\ell}$ roots in each component $\widetilde{E}_{v,\ell}$ of $\widetilde{E}_v$, and on each component contained in $\mathcal{C}_v(\mathbb{R})$, $Q_{\vec{n}_k}(z)$ varies $n_{k,\ell}$ times from $\widetilde{\mathcal{R}}_v^{N_k}$ to 0 to $\widetilde{\mathcal{R}}_v^{N_k}$.

Likewise, for all sufficiently large k, the set $N_k \cdot B^g(0, \delta_v)$ contains a fundamental domain for $\mathbb{R}^g/\mathcal{L}_0$, and (5.58) and (5.62) hold. Let k_0 be the least number such that these conditions hold for all $k \geq k_0$, and put $N_v = N_{k_0} = k_0 \cdot 2S$.

Let $N > 0$ be a multiple of N_v, say $N = t \cdot N_v$ for some integer $t \geq 1$. Put $k = t \cdot k_0$, let $\vec{n} = \vec{n}_k$, and write $Q(z) = Q_{\vec{n}_k}(z)$. Since N is a multiple of S, the divisor $\mathrm{div}(Q) = \sum_{k=1}^{N} \alpha_k - \sum_{i=1}^{m} Ns_i(x_i)$ is integral. Since $\mathrm{div}(Q)$ is K_v-symmetric and Ns_i is even for each i, the class of $\mathrm{div}(Q)$ in $\mathrm{Jac}(\mathcal{C}_v)(\mathbb{C})$ belongs to $\mathrm{Jac}(\mathcal{C}_v)(\mathbb{R})_0$. Let a K_v-symmetric vector $\vec{\beta} = (\beta_1, \ldots, \beta_m) \in \mathbb{R}^m$ with each $|\beta_i| \leq \delta_v$ be given. Putting $\beta_\ell' = \beta_i + \Lambda_{x_i}(\widetilde{E}_v, \vec{s}) - \Lambda_{x_i}(Q, \vec{s})$, we see that $|\beta_i'| \leq 2\delta_v$ for each i, and that $\vec{\beta}' = (\beta_1', \ldots, \beta_m')$ is K_v-symmetric. By (5.73) we can choose $\vec{\varepsilon} \in B^{J+1}(0, r)$ so that $\mathrm{div}(Q^{\vec{\varepsilon}})$ is principal and

$$\Lambda_{x_i}(Q^{\vec{\varepsilon}}, \vec{s}) \;=\; \Lambda_{x_i}(Q, \vec{s}) + \beta_i' \;=\; \Lambda_{x_i}(\widetilde{E}_v, \vec{s}) + \beta_i$$

for each i.

By Proposition 5.7, if $\widetilde{E}_{v,\ell}$ is a component of $\widetilde{E}_v$, then $Q^{\vec{\varepsilon}}$ has n_ℓ roots in the corresponding component $\widetilde{E}_{v,\ell}(\vec{\varepsilon})$ of $\widetilde{E}_v(\vec{\varepsilon})$. Furthermore, if $\widetilde{E}_{v,\ell}$ is contained in $\mathcal{C}_v(\mathbb{R})$, then $\widetilde{E}_{v,\ell}(\vec{\varepsilon})$ is contained in $\mathcal{C}_v(\mathbb{R})$, and $Q^{\vec{\varepsilon}}(z)$ varies n_ℓ times from $\mathcal{R}_v^N$ to 0 to $\mathcal{R}_v^N$ on $\widetilde{E}_{v,\ell}(\vec{\varepsilon})$.

Since $\mathrm{div}(Q^{\vec{\varepsilon}})$ is principal and stable under complex conjugation, there is a rational function $f_v(z) \in \mathbb{R}(\mathcal{C}_v)$ with $|f_v(z)| = Q^{\vec{\varepsilon}}(z)$ for all z. Thus $\Lambda_{x_i}(f_v, \vec{s}) = \Lambda_{x_i}(\widetilde{E}_v, \vec{s}) + \beta_i$ for each $x_i \in \mathfrak{X}$. Clearly $f_v(z)$ is real for all $z \in \mathcal{C}_v(\mathbb{R})$, and $\mathrm{div}(f_v) = \mathrm{div}(Q^{\vec{\varepsilon}})$. Since $\widetilde{E}_v(\vec{\varepsilon})$ is contained in E_v^0, all the roots of f_v belong to E_v^0. By Proposition 5.7,

$$\{z \in \mathcal{C}_v(\mathbb{C}) : |f_v(z)| \leq 1\} \;\subset\; U_v \;.$$

Let $E_{v,1}, \ldots, E_{v,k}$ be the components of the K_v-simple set E_v. For each $E_{v,i}$, put

$$N_i \;=\; \sum_{\widetilde{E}_{v,\ell} \subset E_{v,i}} n_\ell \;=\; \sum_{\widetilde{E}_{v,\ell}(\vec{\varepsilon}) \subset E_{v,i}} n_\ell \;.$$

Then f_v has N_i roots in $E_{v,i}$, counted with multiplicities. If $E_{v,i}$ is a component of E_v contained in $\mathcal{C}_v(\mathbb{R})$, then by construction $|f_v(z)| = Q^{\vec{\varepsilon}}(z)$ varies N_i times from $\mathcal{R}_v^N$ to 0 to $\mathcal{R}_v^N$ on $E_{v,i}$. Each of those oscillations accounts for at least one root of $f_v(z)$. Since f_v has exactly N_i roots in $E_{v,i}$, those roots must be simple. Since the only places $f_v(z)$ can change sign on $E_{v,i}$ are at the roots, each time $|f_v(z)|$ varies from $\mathcal{R}_v^N$ to 0 to $\mathcal{R}_v^N$ on $E_{v,i}$, the function $f_v(z)$ varies from $\mathcal{R}_v^N$ to $-\mathcal{R}_v^N$, or from $-\mathcal{R}_v^N$ to $\mathcal{R}_v^N$. Thus, the roots of f_v in $E_{v,i}$ are simple and $f_v(z)$ oscillates N_i times between $\pm\mathcal{R}_v^N$ on $E_{v,i}$. □

CHAPTER 6

Initial Approximating Functions: Nonarchimedean Case

In this section we will construct the nonarchimedean initial approximating functions needed for Theorem 4.2. Because of the ultrametric inequality, the nonarchimedean theory has a different flavor from the archimedean theory: the constructions are more rigid, but at the same time more explicit.

As in §3.2, let K be a global field, let $\mathcal{C}/K$ be a smooth, connected, projective curve, and let $\mathfrak{X} = \{x_1, \ldots, x_m\} \subset \mathcal{C}(\widetilde{K})$ be a finite, K-symmetric set of points. Put $L = K(\mathfrak{X})$, and let $\{g_{x_1}(z), \ldots, g_{x_m}(z)\}$ a system of uniformizing parameters, chosen in such a way that $g_{x_i}(z) \in K(x_i)(\mathcal{C})$ for each x_i, with $g_{x_{\sigma(i)}}(z) = \sigma(g_{x_i})(z)$ for each $\sigma \in \mathrm{Aut}(\widetilde{K}/K)$.

Let v be a nonarchimedean place of K. In the statement of Theorem 4.2, we are given a K_v-symmetric set $E_v \subset \mathcal{C}_v(\mathbb{C}_v)$ which is bounded away from $\mathfrak{X}$ and has positive capacity, and a finite set of places S of K. If $v \notin S$, then E_v is $\mathfrak{X}$-trivial; if $v \in S$, then E_v is K_v-simple (Definition 4.1). This means that E_v is compact, and there is a decomposition $E_v = \bigcup_{\ell=1}^{D} E_{v,\ell}$ with disjoint, nonempty compact sets $E_{v,\ell}$ such that:

(1) There are finite separable extensions $F_{w_1}, \ldots, F_{w_D}$ of K_v contained in $\mathbb{C}_v$, and disjoint isometrically parametrizable balls $B(a_1, r_1), \ldots, B(a_n, r_D)$, for which $E_{v,\ell} = \mathcal{C}_v(F_{w_\ell}) \cap B(a_\ell, r_\ell)$ for $\ell = 1, \ldots, D$.
(2) The set of balls $\{B(a_1, r_1), \ldots, B(a_D, r_D)\}$ is stable under $\mathrm{Aut}_c(\mathbb{C}_v/K_v)$, and as σ ranges over $\mathrm{Aut}_c(\mathbb{C}_v/K_v)$, each ball $B(a_\ell, r_\ell)$ has $[F_{w_\ell} : K_v]$ distinct conjugates. For each σ, if $\sigma(B(a_j, r_j)) = B(a_k, r_k)$, then $\sigma(F_{w_j}) = F_{w_k}$ and $\sigma(E_{v,j}) = E_{v,k}$.

Both $\mathfrak{X}$-trivial sets and compact sets are algebraically capacitable (see [**51**], Theorems 4.3.13, 4.3.15), so $\overline{G}(z, x_i; E_v) = G(z, x_i; E_v)$ and $\overline{V}_{x_i}(E_v) = V_{x_i}(E_v)$ for each x_i; throughout this section we will write $G(z, x_i; E_v)$ and $V_{x_i}(E_v)$ for the Green's functions and Robin constants.

Let $\vec{s} = (s_1, \ldots, s_m) \in \mathcal{P}^m$ be a K_v-symmetric probability vector. As in the archimedean case, the logarithmic leading coefficient of the Green's function of E_v at x_i is defined to be

$$\begin{aligned} \Lambda_{x_i}(E_v, \vec{s}) &= \lim_{z \to x_i} \Big(\sum_{j=1}^{m} s_j G(z, x_j; E_v) + s_i \log_v(|g_{x_i}(z)|_v) \Big) \\ &= s_i V_{x_i}(E_v) + \sum_{j \neq i} s_j G(x_i, x_j; E_v) \ . \end{aligned} \tag{6.1}$$

Likewise, if $\vec{s} \in \mathcal{P}^m(\mathbb{Q})$ and $f_v(z) \in K_v(\mathcal{C})$ is an $(\mathfrak{X}, \vec{s})$-function of degree N, the logarithmic leading coefficient of $f_v(z)$ at x_i is

$$(6.2) \qquad \Lambda_{x_i}(f_v, \vec{s}) = \lim_{z \to x_i} \left(\frac{1}{N} \log_v(|f_v(z)|_v) + s_i \log_v(|g_{x_i}(z)|_v) \right) .$$

Note that $|\mathbb{C}_v^\times|_v = q_v^{\mathbb{Q}}$ is dense in $\mathbb{R}_{>0}$ but is not equal to it. This means we cannot continuously vary the logarithmic leading coefficients $\Lambda_{x_i}(f_v, \vec{s})$ as in the archimedean case; this is one source of rigidity in the construction.

If E_v is $\mathfrak{X}$-trivial, then it is an RL-domain, and the initial local approximating function will be an $(\mathfrak{X}, \vec{s})$-function $f_v(z) \in K_v(z)$ which defines E_v as an RL-domain:

$$E_v = \{z \in \mathcal{C}_v(\mathbb{C}_v) : |f_v(z)|_v \le 1\} .$$

Here $\sum_{i=1}^m s_i G(z, x_i; E_v) = \frac{1}{N} \log_v(|f(z)|_v)$ for $z \notin E_v$, and $\Lambda_{x_i}(f_v, \vec{s}) = \Lambda_{x_i}(E_v, \vec{s})$ for each $x_i \in \mathfrak{X}$.

If E_v is K_v-simple, the initial local approximating function will be an $(\mathfrak{X}, \vec{s})$-function $f_v(z) \in K_v(z)$ whose zeros are distinct and belong to E_v, and are well-distributed relative to the $(\mathfrak{X}, \vec{s})$-equilibrium measure $\mu_{\mathfrak{X},\vec{s}}$ of a K_v-simple set $\widetilde{E}_v \subseteq E_v$ constructed below (see Appendix A regarding $\mu_{\mathfrak{X},\vec{s}}$). We will show that for any $\beta_v > 0$ in $\mathbb{Q}$, there are functions having these properties, such that for each $x_i \in \mathfrak{X}$,

$$\Lambda_{x_i}(f_v, \vec{s}) = \Lambda_{x_i}(\widetilde{E}_v, \vec{s}) + \beta_v .$$

In the K_v-simple case, the functions $f_v(z)$ will be constructed explicitly, by an argument generalizing the methods of ([**52**], [**53**]). The idea is to first define an infinite subsequence of E_v which is very uniformly distributed relative to $\mu_{\mathfrak{X},\vec{s}}$, using a lemma of Balinski and Young ([**9**]) originally proved for the purpose of apportioning seats in the US House of Representatives, and to then take an initial segment of that sequence and modify it to become the zeros of $f_v(z)$. The modification, which involves moving a finite number of points so as to obtain a principal divisor, is carried out by using an action of a small neighborhood of the origin in $\mathrm{Jac}(\mathcal{C}_v)(\mathbb{C}_v)$ on a suitably generic polydisc in $\mathcal{C}_v(\mathbb{C}_v)^g$ which makes the polydisc into a principal homogeneous space for the neighborhood. This action is studied in Appendix D.

1. The Approximation Theorems

There are two cases to consider in constructing the initial local approximating functions: the RL-domain case, and the compact case.

Recall that $\mathcal{C}_v(\mathbb{C}_v) \backslash E_v$ can be partitioned into equivalence classes called RL-components, which play the same role as the connected components of the complement of E_v in the archimedean case. The equivalence relation is defined by $z \equiv w$ iff $G(z, w; E_v) > 0$; see ([**51**], Theorems 4.2.11 and 4.4.17, and Definition 4.4.18).

For the RL-domains constructed in the initial reductions of the Fekete-Szegö theorem, each RL-component of $\mathcal{C}_v(\mathbb{C}_v) \backslash E_v$ automatically contains at least one point of $\mathfrak{X}$: If E_v is a finite union of isometrically parametrizable balls, then by ([**51**], Theorem 4.2.16 and Proposition 4.4.1(B)) the complement of E_v consists of a single RL-component. If E_v is $\mathfrak{X}$-trivial, then by definition $\mathcal{C}_v$ has good reduction at v, the points x_i specialize to distinct points $\pmod v$, and $E_v = \mathcal{C}_v(\mathbb{C}_v) \backslash \bigcup_{i=1}^m B(x_i, 1)^-$

relative to the spherical metric on $\mathcal{C}_v(\mathbb{C}_v)$. In this case the RL-components of $\mathcal{C}_v(\mathbb{C}_v)\backslash E_v$ are the balls $B(x_i, 1)^-$ for the $x_i \in \mathfrak{X}$.

The construction of the initial approximating functions when E_v is an RL-domain was already treated in ([**51**], §4):

THEOREM 6.1. *Suppose K_v is nonarchimedean, and let $E_v \subset \mathcal{C}_v(\mathbb{C}_v)\backslash\mathfrak{X}$ be an* RL-*domain such that each* RL-*component of $\mathcal{C}_v(\mathbb{C}_v)\backslash E_v$ contains at least one point of $\mathfrak{X}$. Let $\vec{s} \in \mathcal{P}^m(\mathbb{Q})$ be a K_v-symmetric rational probability vector, whose entries are all positive.*

Then there is a integer N_v with the following property. For each positive integer N divisible by N_v, there is an $(\mathfrak{X}, \vec{s})$-function $f_v(z) \in K_v(\mathcal{C})$ of degree N which defines E_v as an RL-*domain:*

$$E_v \;=\; \{z \in \mathcal{C}_v(\mathbb{C}_v) : |f_v(z)|_v \le 1\} \; . \tag{6.3}$$

For any such function $f_v(z)$ we have

$$\sum_{i=1}^{m} s_i G(z, x_i; E_v) \;=\; \begin{cases} \frac{1}{N}\log_v(|f_v(z)|_v) & \text{if } z \notin E_v\,, \\ 0 & \text{if } z \in E_v\,. \end{cases}$$

In particular, $\Lambda_{x_i}(f_v, \vec{s}) = \Lambda_{x_i}(E_v, \vec{s})$ for each $x_i \in \mathfrak{X}$.

Furthermore, if $\mathrm{char}(K_v) = p > 0$, we can require that for each x_i, the leading coefficient $c_{v,i} = \lim_{z\to x_i} f_v(z) \cdot g_{x_i}(z)^{Ns_i}$ belongs to $K_v(x_i)^{\mathrm{sep}}$.

PROOF. This is essentially ([**51**], Theorem 4.5.4, p.316). That theorem provides an integer N_0 and an $(\mathfrak{X}, \vec{s})$-function $f_0(z) \in K_v(\mathcal{C})$ of degree N_0 for which (6.3) holds. Put $N_v = N_0$. Given a multiple $N = kN_v$, we can take $f_v(z) = f_0(z)^k$.

If $\mathrm{char}(K_v) = p > 0$, put $p^B = \max_i([K_v(x_i) : K_v]^{\mathrm{insep}})$, and take $N_v = p^B N_0$ instead. Then if $N = kN_v$, we have $f_v(z) = (f_0(z)^{p^B})^k$. Since f_0 is K_v-rational, for each i its leading coefficient at x_i belongs to $K_v(x_i)$ (see Corollary 3.5), and the leading coefficient of $f_0(z)^{p^B}$ at x_i belongs to $K_v(x_i)^{\mathrm{sep}}$. □

Before stating the approximation theorem for the compact case, we will need a definition.

DEFINITION 6.2. Suppose K_v is nonarchimedean. We will say that K_v-simple sets E_v and $\widetilde{E}_v$ are *compatible*, or that $\widetilde{E}_v$ is *compatible with* E_v, if E_v and $\widetilde{E}_v$ are nonempty and have K_v-simple decompositions

$$E_v = \bigcup_{\ell=1}^{n} B(a_\ell, r_\ell) \cap \mathcal{C}_v(F_{w_\ell})\;, \quad \widetilde{E}_v = \bigcup_{k=1}^{\widetilde{n}} B(\widetilde{a}_k, \widetilde{r}_k) \cap \mathcal{C}_v(\widetilde{F}_{w_k})\;, \tag{6.4}$$

such that $\bigcup_{k=1}^{\widetilde{n}} B(\widetilde{a}_k, \widetilde{r}_k) \subseteq \bigcup_{\ell=1}^{n} B(a_\ell, r_\ell)$, and whenever $B(\widetilde{a}_k, \widetilde{r}_k) \subseteq B(a_\ell, r_\ell)$ we have $\widetilde{F}_{w_k} = F_{w_\ell}$. We will call a pair of K_v-simple decompositions (6.4) satisfying the conditions above *compatible decompositions.*

In the compact case, the theorem we need is the following. It may appear somewhat strange at first reading. The set $\widetilde{E}_v \subset E_v$ plays an auxiliary role in the construction: we will initially replace E_v by $\widetilde{E}_v$. This reserves an "unused" part $E_v\backslash\widetilde{E}_v$ of E_v, and allows us to move some zeros from $\widetilde{E}_v$ into E_v in constructing the approximating function $f_v(z)$. The final approximation set H_v is then defined in terms of $f_v(z)$; it is contained in E_v but not in general in $\widetilde{E}_v$.

THEOREM 6.3. *Suppose K_v is nonarchimedean. Let E_v be a compact K_v-simple set which is disjoint from $\mathfrak{X}$ and has positive capacity. Fix a K_v-simple decomposition*

$$E_v = \bigcup_{\ell=1}^{D} B(a_\ell, r_\ell) \cap \mathcal{C}_v(F_{w_\ell}) , \tag{6.5}$$

and fix $\varepsilon_v > 0$. Then there is a compact, K_v-simple set $\widetilde{E}_v \subseteq E_v$ compatible with E_v such that

(A) For each $x_i \in \mathfrak{X}$

$$|V_{x_i}(\widetilde{E}_v) - V_{x_i}(E_v)| < \varepsilon_v , \tag{6.6}$$

and for all $x_i, x_j \in \mathfrak{X}$ with $x_i \neq x_j$,

$$|G(x_i, x_j; \widetilde{E}_v) - G(x_i, x_j; E_v)| < \varepsilon_v . \tag{6.7}$$

(B) For each $0 < \beta_v \in \mathbb{Q}$ and each K_v-symmetric probability vector $\vec{s} = {}^t(s_1, \ldots, s_m)$ with rational entries, there is an integer $N_v \geq 1$ such that for each positive integer N divisible by N_v, there is an $(\mathfrak{X}, \vec{s})$-function $f_v \in K_v(\mathcal{C}_v)$ of degree N such that

(1) For each $x_i \in \mathfrak{X}$,

$$\Lambda_{x_i}(f_v, \vec{s}) = \Lambda_{x_i}(\widetilde{E}_v, \vec{s}) + \beta_v . \tag{6.8}$$

(2) The zeros $\theta_1, \ldots, \theta_N$ of f_v are distinct and belong to E_v;

(3) $f_v^{-1}(D(0,1)) \subseteq \bigcup_{\ell=1}^{D} B(a_\ell, r_\ell)$;

(4) There is a decomposition $f_v^{-1}(D(0,1)) = \bigcup_{h=1}^{N} B(\theta_h, \rho_h)$, where the balls $B(\theta_h, \rho_h)$ are pairwise disjoint and isometrically parametrizable. For each $h = 1, \ldots, N$, if $\ell = \ell(h)$ is such that $B(\theta_h, \rho_h) \subseteq B(a_\ell, r_\ell)$, put $F_{u_h} = F_{w_\ell}$. Then $\rho_h \in |F_{u_h}^\times|_v$ and f_v induces an F_{u_h}-rational scaled isometry from $B(\theta_h, \rho_h)$ to $D(0,1)$, with

$$f_v\big(B(\theta_h, \rho_h) \cap \mathcal{C}_v(F_{u_h})\big) = \mathcal{O}_{F_{u_h}} ,$$

such that $|f_v(z_1) - f_v(z_2)|_v = (1/\rho_h)\|z_1, z_2\|_v$ for all $z_1, z_2 \in B(\theta_h, \rho_h)$.

(5) The set $H_v := E_v \cap f_v^{-1}(D(0,1))$ is K_v-simple and compatible with E_v. Indeed,

$$H_v = \bigcup_{h=1}^{N} \big(B(\theta_h, \rho_h) \cap \mathcal{C}_v(F_{u_h})\big) \tag{6.9}$$

is a K_v-simple decomposition of H_v compatible with the K_v-simple decomposition (6.5).

(C) If $\mathrm{char}(K_v) = p > 0$, then for each x_i the leading coefficient $c_{v,i} = \lim_{z \to x_i} f_v(z) \cdot g_{x_i}(z)^{Ns_i}$ belongs to $K_v(x_i)^{\mathrm{sep}}$.

The proof of Theorem 6.3 will occupy the remainder of this chapter.

2. Reduction to a Set E_v in a Single Ball

In this section we will reduce proving Theorem 6.3 to proving it over a finite separable extension F_w/K_v, in the case where $E_v = \mathcal{C}_v(F_w) \cap B(a, r)$ for a single isometrically parametrizable ball.

To do this, we first recall some facts about nonarchimedean $(\mathfrak{X}, \vec{s})$-capacities established in Appendix A. Let $\vec{s} = (s_1, \ldots, s_m) \in \mathcal{P}^m$ be a probability vector. As in §3.3, define the $(\mathfrak{X}, \vec{s})$-canonical distance by

$$[z, w]_{\mathfrak{X},\vec{s}} = \prod_{i=1}^{m} ([z, w]_{x_i})^{s_i} ,$$

where the $[z, w]_{x_i}$ are normalized so that $\lim_{z \to x_i} [z, w]_{x_i} \cdot |g_{x_i}(z)|_v = 1$.

Given a compact set H_v disjoint from $\mathfrak{X}$, define its $(\mathfrak{X}, \vec{s})$-Robin constant by

$$V_{\mathfrak{X},\vec{s}}(H_v) = \inf_{\nu} \iint_{H_v \times H_v} -\log_v([z, w]_{\mathfrak{X},\vec{s}}) \, d\nu(z) d\nu(w) \tag{6.10}$$

where ν runs over all Borel probability measures supported on H_v, and its $(\mathfrak{X}, \vec{s})$-capacity by

$$\gamma_{\mathfrak{X},\vec{s}}(H_v) = q_v^{-V_{\mathfrak{X},\vec{s}}(H_v)} ,$$

where q_v is the order of the residue field of $\mathcal{O}_v$.

By Theorem A.2, if H_v has positive capacity, there is a unique probability measure $\mu_{\mathfrak{X},\vec{s}}$ supported on H_v for which the inf in (6.10) is achieved; this is called the $(\mathfrak{X}, \vec{s})$-equilibrium distribution. The $(\mathfrak{X}, \vec{s})$-potential function is defined by

$$u_{\mathfrak{X},\vec{s}}(z; H_v) = \int_{H_v} -\log_v([z, w]_{\mathfrak{X},\vec{s}}) \, d\mu_{\mathfrak{X},\vec{s}}(w) .$$

Here $u_{\mathfrak{X},\vec{s}}(z; H_v) = V_{\mathfrak{X},\vec{s}}(H_v)$ for all $z \in H_v$ except possibly a set of inner capacity 0, and $u_{\mathfrak{X},\vec{s}}(z; H_v) < V_{\mathfrak{X},\vec{s}}(H_v)$ for all $z \notin H_v$. By Proposition A.5, $V_{\mathfrak{X},\vec{s}}(H_v)$ is a continuous function of $\vec{s} \in \mathcal{P}^m$, and the $(\mathfrak{X}, \vec{s})$-equilibrium distribution and the $(\mathfrak{X}, \vec{s})$-Green's function $G_{\mathfrak{X},\vec{s}}(z, H_v) = V_{\mathfrak{X},\vec{s}}(H_v) - u_{\mathfrak{X},\vec{s}}(z; H_v)$ can be decomposed in terms of the corresponding objects for the individual points x_i:

$$\mu_{\mathfrak{X},\vec{s}} = \sum_{i=1}^{m} s_i \mu_i , \tag{6.11}$$

$$G_{\mathfrak{X},\vec{s}}(z; H_v) = \sum_{i=1}^{m} s_i G(z, x_i; H_v) , \tag{6.12}$$

where μ_i is the equilibrium distribution of H_v with respect to the point x_i, and $G(z, x_i; H_v) = V_{x_i}(H_v) - u_{x_i}(z; H_v)$.

If H_v is K_v-simple, with the K_v-simple decomposition

$$H_v = \bigcup_{\ell=1}^{D} \left(\mathcal{C}_v(F_{w_\ell}) \cap B(a_\ell, r_\ell) \right) , \tag{6.13}$$

then by Lemma A.9 and Proposition A.12 of Appendix A, the exceptional set of inner capacity 0, discussed above, is empty: $u_{\mathfrak{X},\vec{s}}(z; H_v) = V_{\mathfrak{X},\vec{s}}(H_v)$ for all $z \in H_v$. Moreover, the equilibrium distribution $\mu_{\mathfrak{X},\vec{s}}$ can be described as follows. For each ℓ, write $H_{v,\ell} = \mathcal{C}_v(F_{w_\ell}) \cap B(a_\ell, r_\ell)$ and let $\sigma_\ell : D(0, r_\ell) \to B(a_\ell, r_\ell)$ be an F_{w_ℓ}-rational isometric parametrization with $\sigma_\ell(0) = a_\ell$. Let μ_ℓ^* be the pushforward of additive Haar measure on $F_{w_\ell} \cap D(0, r_\ell)$ to $H_{v,\ell}$ by σ_ℓ, normalized to have mass 1. By Corollary A.14 of Appendix A, there are weights $w_\ell(\vec{s}) > 0$, satisfying

$\sum_{\ell=1}^{D} w_\ell(\vec{s}) = 1$, for which

$$\mu_{\mathfrak{X},\vec{s}} = \sum_{\ell=1}^{D} w_\ell(\vec{s})\mu_\ell^* \ . \tag{6.14}$$

The weights $w_\ell(\vec{s})$ are uniquely determined by the requirement that

$$u_{\mathfrak{X},\vec{s}}(z,H_v) = \sum_{\ell=1}^{D} w_\ell(\vec{s}) u_{\mathfrak{X},\vec{s}}(z,H_{v,\ell}) \tag{6.15}$$

takes the same value $V_{\mathfrak{X},\vec{s}}(H_v)$ on each $H_{v,\ell}$.

This description of $u_{\mathfrak{X},\vec{s}}(z,H_v)$ leads to the following system of linear equations. Writing $V = V_{\mathfrak{X},\vec{s}}(H_v)$ and $w_\ell = w_\ell(\vec{s})$, and evaluating $u_{\mathfrak{X},\vec{s}}(z,H_v)$ at a generic point of each $H_{v,\ell}$, we have (see Theorem A.13):

$$\begin{cases} 1 = 0 \cdot V + \sum_{\ell=1}^{D} w_\ell \ , \\ 0 = V + w_\ell \cdot \big(-V_{\mathfrak{X},\vec{s}}(H_{v,\ell})\big) + \sum_{\substack{j=1 \\ j\neq \ell}}^{D} w_j \cdot \log_v\big([a_\ell,a_j]_{\mathfrak{X},\vec{s}}\big) \\ \qquad \text{for } \ell = 1,\ldots,D. \end{cases} \tag{6.16}$$

By Theorem A.13 the system (6.16) has a unique solution. Since the $V_{\mathfrak{X},\vec{s}}(H_{v,\ell})$ and $[a_k,a_\ell]_{\mathfrak{X},\vec{s}}$ are continuous in $\vec{s}$ (see Proposition A.5 and Proposition 3.11(B1)), $V_{\mathfrak{X},\vec{s}}(H_v)$ and the $w_\ell(\vec{s})$ are continuous functions of $\vec{s}$. Since $w_\ell(\vec{s}) = \mu_{\mathfrak{X},\vec{s}}(H_{v,\ell}) = \sum_{i=1}^{m} s_i \mu_i(H_{v,\ell})$ and $\mu_i(H_{v,\ell}) > 0$ for each i and ℓ, there is a constant $W_0 = W_0(H_v,\mathfrak{X}) > 0$ such that $w_\ell(\vec{s}) \geq W_0$, uniformly for all ℓ and $\vec{s}$.

Since H_v is K_v-simple, if $\vec{s} \in \mathcal{P}^m(\mathbb{Q})$ the coefficients of the system (6.16) are rational, so the solutions $V_{\mathfrak{X},\vec{s}}(H_v)$ and $w_\ell(\vec{s})$ are rational as well. This fact is shown in Corollary A.14, but because it is crucial to our construction, and because the ideas motivate later parts of the construction, we repeat the reasoning here. It involves computations with explicit examples.

First suppose $\mathcal{C} = \mathbb{P}^1/K$. Identify $\mathbb{P}_v^1(\mathbb{C}_v)$ with $\mathbb{C}_v \bigcup \{\infty\}$, and take $\mathfrak{X} = \{\infty\}$. Let F_w/K_v be a finite, separable extension embedded in $\mathbb{C}_v$. We do not assume F_w/K_v is Galois. Consider the set $H_v^0 = \mathcal{O}_w$, where $\mathcal{O}_w$ is the ring of integers of F_w. The canonical distance $[x,y]_\infty$ (with respect to the uniformizing parameter $g_\infty(z) = 1/z$) is just $|x-y|_v$, which is unchanged when x and y are replaced by $x-a$ and $y-a$. It follows that the equilibrium measure $\mu = \mu_\infty$ is translation invariant under the additive group of $\mathcal{O}_w$, and hence must be additive Haar measure on $\mathcal{O}_w$. The Robin constant and potential function of $\mathcal{O}_w$ can be computed explicitly (see Lemma A.9, or [**51**], Example 4.1.24, p.212) and in particular, writing $q_w = q_v^{f_w}$ for the order of the residue field of $\mathcal{O}_w$, we have

$$u_\infty(z,\mathcal{O}_w) = \begin{cases} \frac{1}{e_w(q_w-1)} & \text{if } z \in \mathcal{O}_w \ , \\ -\log_v(|z|_v) & \text{if } |z|_v > 1 \ . \end{cases} \tag{6.17}$$

Next let $\mathcal{C}/K$ and $\mathfrak{X}$ be arbitrary, and consider the set $H_{v,\ell} = \mathcal{C}_v(F_{w_\ell}) \cap B(a_\ell,r_\ell)$ where $B(a_\ell,r_\ell)$ is isometrically parametrizable, F_{w_ℓ}/K_v is a finite separable extension in $\mathbb{C}_v$, $a_\ell \in \mathcal{C}_v(F_{w_\ell})$, and $r_\ell \in |F_{w_\ell}^\times|_v$. Put $q_{w_\ell} = q_v^{f_{w_\ell}}$. Fix an F_{w_ℓ}-rational isometric parametrization $\sigma_\ell : D(0,r_\ell) \to B(a_\ell,r_\ell)$ with $\sigma_\ell(0) = a_\ell$, and use it to identify $B(a_\ell,r_\ell)$ with $D(0,r_\ell)$. For each x_i there is a constant $A_{\ell,i} \in |\mathbb{C}_v^\times|_v$ such that $[y,z]_{x_i} = A_{\ell,i}|y-z|_v$ for all $z,w \in B(a_\ell,r_\ell)$. It follows that for $y,z \in B(a,r)$, we have $[y,z]_{\mathfrak{X},\vec{s}} = C_{\mathfrak{X},\vec{s}}\|z,w\|_v$ where $C_{\mathfrak{X},\vec{s},\ell} = \prod_i A_{\ell,i}^{s_i}$. By Proposition 3.11 the canonical distance is constant on pairwise disjoint isometrically parametrizable balls disjoint from $\mathfrak{X}$, so for $y \in B(a_\ell, r_\ell)$ and $z \notin B(a_\ell, r_\ell)$ we have $[z,y]_{\mathfrak{X},\vec{s}} = [z,a_\ell]_{\mathfrak{X},\vec{s}}$. The equilibrium distribution of $H_{v,\ell}$ is the pushforward of additive Haar measure on $F_{w_\ell} \cap D(0,r)$. Hence, by (6.17)

$$u_{\mathfrak{X},\vec{s}}(z, H_{v,\ell}) \ = \ \begin{cases} \frac{1}{e_{w_\ell}(q_{w_\ell}-1)} - \log_v(C_{\mathfrak{X},\vec{s},\ell}\cdot r_\ell) & \text{if } z \in H_{v,\ell}\ , \\ -\log_v([z,a_\ell]_{\mathfrak{X},\vec{s}}) & \text{if } z \notin B(a_\ell, r_\ell)\ . \end{cases} \tag{6.18}$$

In particular, $V_{\mathfrak{X},\vec{s}}(H_{v,\ell}) = 1/(e_{w_\ell}(q_{w_\ell}-1)) - \log_v(C_{\mathfrak{X},\vec{s},\ell}\cdot r_\ell)$. Note that $C_{\mathfrak{X},\vec{s},\ell}$ belongs to $|\mathbb{C}_v^\times|_v$ if all the s_i are rational. Thus, if $\vec{s} \in \mathcal{P}^m(\mathbb{Q})$, then $V_{\mathfrak{X},\vec{s}}(H_{v,\ell}) \in \mathbb{Q}$. Likewise, by Proposition 3.11(B1), $-\log_v([a_\ell,a_j]_{\mathfrak{X},\vec{s}}) \in \mathbb{Q}$ for all $\ell \neq j$.

Hence if $\vec{s} \in \mathcal{P}^m(\mathbb{Q})$, the coefficients of the system (6.16) are rational. It follows that the solution is rational as well.

We next apply the theory above to sets $\widetilde{E}_v \subset E_v \subset \mathcal{C}_v(\mathbb{C}_v)\backslash\mathfrak{X}$, where E_v and $\widetilde{E}_v$ are K_v-simple and $\widetilde{E}_v$ is compatible with E_v. Given a K_v-simple decomposition $E_v = \bigcup_{\ell=1}^D \mathcal{C}_v(F_{w_\ell}) \cap B(a_\ell, r_\ell)$, write $E_{v,\ell} = \mathcal{C}_v(F_{w_\ell}) \cap B(a_\ell, r_\ell)$, and put $\widetilde{E}_{v,\ell} = \widetilde{E}_v \cap B(a_\ell, r_\ell) \subset E_{v,\ell}$. Thus $E_v = \bigcup_{\ell=1}^D E_{v,\ell}$ and $\widetilde{E}_v = \bigcup_{\ell=1}^D \widetilde{E}_{v,\ell}$.

LEMMA 6.4. *Let $E_v \subset \mathcal{C}_v(\mathbb{C}_v)\backslash\mathfrak{X}$ be a K_v-simple set, with a K_v-simple decomposition $E_v = \bigcup_{\ell=1}^D \mathcal{C}_v(F_{w_\ell}) \cap B(a_\ell, r_\ell)$.*

Let $\varepsilon_v > 0$ be given. Then there is a $\delta_v > 0$ such that for any K_v-simple set $\widetilde{E}_v \subset E_v$, if $|V_{x_i}(\widetilde{E}_{v,\ell}) - V_{x_i}(E_{v,\ell})| < \delta_v$ for all $\ell = 1,\ldots,D$ and $i = 1,\ldots,m$, then for each $x_i \in \mathfrak{X}$ we have

$$|V_{x_i}(\widetilde{E}_v) - V_{x_i}(E_v)| \ < \ \varepsilon_v\ , \tag{6.19}$$

and for all $x_i, x_j \in \mathfrak{X}$ with $x_i \neq x_j$,

$$|G(x_i,x_j;\widetilde{E}_v) - G(x_i,x_j;E_v)| \ < \ \varepsilon_v\ . \tag{6.20}$$

PROOF. By Theorem A.13, there are systems of equations analogous to (6.16) for the sets E_v and $\widetilde{E}_v$: write $V = V_{\mathfrak{X},\vec{s}}(E_v)$ and $w_\ell = w_\ell(\vec{s})$ for the Robin constant and weights associated to E_v; let $\widetilde{V} = V_{\mathfrak{X},\vec{s}}(\widetilde{E}_v)$, $\widetilde{w}_\ell = \widetilde{w}_\ell(\vec{s})$ be the corresponding objects for $\widetilde{E}_v$. Then

(6.21)
$$\begin{cases} 1 = 0\cdot V + \sum_{\ell=1}^D w_\ell\ , \\ 0 = V + w_\ell \cdot \big(-V_{\mathfrak{X},\vec{s}}(E_{v,\ell})\big) + \sum_{\substack{j=1\\ j\neq\ell}}^D w_j \cdot \log_v\big([a_\ell,a_j]_{\mathfrak{X},\vec{s}}\big), \quad \text{for } \ell = 1,\ldots,D \end{cases}$$

and

(6.22)
$$\begin{cases} 1 = 0\cdot \widetilde{V} + \sum_{\ell=1}^D \widetilde{w}_\ell\ , \\ 0 = \widetilde{V} + \widetilde{w}_\ell \cdot \big(-V_{\mathfrak{X},\vec{s}}(\widetilde{E}_{v,\ell})\big) + \sum_{\substack{j=1\\ j\neq\ell}}^D \widetilde{w}_j \cdot \log_v\big([a_\ell,a_j]_{\mathfrak{X},\vec{s}}\big), \quad \text{for } \ell = 1,\ldots,D\,. \end{cases}$$

Fix $x_i \in \mathfrak{X}$ and specialize to the case where $[x,y]_{\mathfrak{X},\vec{s}} = [x,y]_{x_i}$; that is, take $\vec{s} = \vec{e}_i = (0,\ldots,1,\ldots,0)$. Note that the coefficients of the systems (6.21), (6.22) are the same except for their diagonal terms. Write $V_\ell = V_{\mathfrak{X},\vec{s}}(E_{v,\ell})$ and $\widetilde{V}_\ell = V_{\mathfrak{X},\vec{s}}(\widetilde{E}_{v,\ell})$, for $\ell = 1,\ldots,D$.

Henceforth we will regard $\widetilde{V}_1, \ldots, \widetilde{V}_D$ as variables, keeping $V_1, \ldots, V_D$ fixed. By Cramer's rule, the solution vectors $(V, w_1, \ldots, w_D)$ and $(\widetilde{V}, \widetilde{w}_1, \ldots, \widetilde{w}_D)$ are continuous functions of the coefficients in (6.21), (6.22). It follows that when $(\widetilde{V}_1, \ldots, \widetilde{V}_D)$ converges to $(V_1, \ldots, V_D)$, then $(\widetilde{V}, \widetilde{w}_1, \ldots, \widetilde{w}_D)$ converges to $(V, w_1, \ldots, w_D)$. In particular, $V_{x_i}(\widetilde{E}_v)$ converges to $V_{x_i}(E_v)$.

Furthermore, for each $z \neq x_i$ we have

$$(6.23) \qquad \begin{cases} G(z, x_i; E_v) &= V - u_{x_i}(x, E_v) &= V - \sum_{\ell=1}^D w_\ell \, u_{x_i}(z, E_{v,\ell}) \ , \\ G(z, x_i; \widetilde{E}_v) &= \widetilde{V} - u_{x_i}(x, \widetilde{E}_v) &= \widetilde{V} - \sum_{\ell=1}^D \widetilde{w}_\ell \, u_{x_i}(z, \widetilde{E}_{v,\ell}) \ . \end{cases}$$

Since $\mathfrak{X}$ is disjoint from $\bigcup_{\ell=1}^D B(a_\ell, r_\ell)$, it follows from (6.18) that for each $x_j \neq x_i$ we have $u_{x_i}(x_j, E_{v,\ell}) = u_{x_i}(x_j, \widetilde{E}_{v,\ell}) = -\log_v([x_j, a_\ell]_{x_i})$. Thus, when $(\widetilde{V}_1, \ldots, \widetilde{V}_D)$ converges to $(V_1, \ldots, V_D)$, then $G(x_j, x_i; \widetilde{E}_v)$ converges to $G(x_j, x_i; E_v)$ as well.

Since $\mathfrak{X}$ is finite, the lemma follows. $\square$

We now claim that in order to prove Theorem 6.3, it suffices to establish the following weaker version of the theorem for $E_{v,\ell}$ over F_{w_ℓ}, for each $\ell = 1, \ldots, D$:

PROPOSITION 6.5. *Let $E_{v,\ell} = \mathcal{C}_v(F_{w_\ell}) \cap B(a_\ell, r_\ell)$, where F_{w_ℓ} is a finite, separable extension of K_v in $\mathbb{C}_v$, $B(a_\ell, r_\ell) \subset \mathcal{C}_v(\mathbb{C}_v) \backslash \mathfrak{X}$ is an isometrically parametrizable ball, and $a_\ell \in \mathcal{C}_v(F_{w_\ell})$. Let $\varepsilon_{v,\ell} > 0$ be given. Then there is a compact set $\widetilde{E}_{v,\ell} \subseteq E_{v,\ell}$ for which*

(A) There are points $\alpha_{\ell,j} \in \mathcal{C}_v(F_{w_\ell}) \cap B(a_\ell, r_\ell)$ and pairwise disjoint isometrically parametrizable balls $B(\alpha_{\ell,j}, r_{\ell,j}) \subseteq B(a_\ell, r_\ell)$, for $j = 1, \ldots, d_\ell$, such that $\widetilde{E}_{v,\ell}$ has the form

$$(6.24) \qquad \widetilde{E}_{v,\ell} = \bigcup_{j=1}^{d_\ell} \left(\mathcal{C}_v(F_{w_\ell}) \cap B(\alpha_{\ell,j}, r_{\ell,j})\right)$$

and for each $x_i \in \mathfrak{X}$,

$$(6.25) \qquad |V_{x_i}(\widetilde{E}_{v,\ell}) - V_{x_i}(E_{v,\ell})| < \varepsilon_{v,\ell} \ .$$

(B) For each $0 < \beta_v \in \mathbb{Q}$ and each F_{w_ℓ}-symmetric probability vector $\vec{s} = {}^t(s_1, \ldots, s_m)$ with rational entries, there is an integer $N_{v,\ell} \geq 1$ such that for each positive integer N divisible by $N_{v,\ell}$, there is an $(\mathfrak{X}, \vec{s})$-function $f_{v,\ell} \in F_{w_\ell}(\mathcal{C}_v)$ of degree N such that:

(1) For all $z \in \mathcal{C}_v(\mathbb{C}_v) \backslash (B(a_\ell, r_\ell) \cup \mathfrak{X})$,

$$(6.26) \qquad \frac{1}{N} \log_v(|f_{v,\ell}(z)|_v) = G_{\mathfrak{X},\vec{s}}(z, \widetilde{E}_{v,\ell}) + \beta_v \ .$$

(2) The zeros $\theta_1, \ldots, \theta_N$ of $f_{v,\ell}$ are distinct and belong to $E_{v,\ell}$ (and $\mathcal{C}_v(F_{w_\ell})$).

(3) $f_{v,\ell}^{-1}(D(0,1)) = \bigcup_{h=1}^N B(\theta_h, \rho_h)$, where the balls $B(\theta_h, \rho_h)$ are pairwise disjoint and contained in $B(a_\ell, r_\ell)$.

Proposition 6.5 will be proved in Section 6.4.

PROOF OF THEOREM 6.3, ASSUMING PROPOSITION 6.5. Let $E_v \subset \mathcal{C}_v(\mathbb{C}_v)\backslash\mathfrak{X}$ be a K_v-simple set with the K_v-simple decomposition

$$E_v = \bigcup_{\ell=1}^{D} \left(\mathcal{C}_v(F_{w_\ell}) \cap B(a_\ell, r_\ell)\right) . \tag{6.27}$$

For each ℓ, write $E_{v,\ell} = \mathcal{C}_v(F_{w_\ell}) \cap B(a_\ell, r_\ell)$. By the definition of a K_v-simple decomposition, the collection of balls $\{B(a_\ell, r_\ell)\}_{1\le\ell\le D}$ is stable under $\mathrm{Aut}_c(\mathbb{C}_v/K_v)$ and for each ℓ the ball $B(a_\ell, r_\ell)$ has $[F_{w_\ell} : K_v]$ distinct conjugates. For each $\sigma \in \mathrm{Aut}_c(\mathbb{C}_v/K_v)$ and each $\ell = 1,\ldots,D$, let $\sigma(\ell)$ be the index such that $B(a_{\sigma(\ell)}, r_{\sigma(\ell)}) = B(a_\ell, r_\ell)$. Then $F_{\sigma(\ell)} = \sigma(F_\ell)$.

We first construct the set $\widetilde{E}_v$ in Theorem 6.3. Given $\varepsilon_v > 0$, let $\delta_v > 0$ be the number given by Lemma 6.4 for E_v and the K_v-simple decomposition (6.27). Suppose that under the action of $\mathrm{Aut}_c(\mathbb{C}_v/K_v)$ on $B(a_1,r_1),\ldots,B(a_D,r_D)$, there are T distinct orbits. We can assume without loss that $B(a_1,r_1),\ldots,B(a_T,r_T)$ are representatives for the orbits. For each $\ell = 1,\ldots,T$, take $\varepsilon_{v,\ell} = \delta_v$ and let $\widetilde{E}_{v,\ell} \subset E_{v,\ell}$ and $N_{v,\ell}$ be the F_{w_ℓ}-simple set and number given for $E_{v,\ell}$ and $\varepsilon_{v,\ell}$, by Proposition 6.5. Let $\widetilde{V}_\ell = V_{\mathfrak{X},\vec{s}}(\widetilde{E}_{v,\ell})$ be the $(\mathfrak{X},\vec{s})$-Robin constant of $\widetilde{E}_{v,\ell}$. We define the sets $\widetilde{E}_{v,\ell}$ for $T+1 \le \ell \le D$ by Galois conjugacy: given such an ℓ, there are a k with $1 \le k \le T$ and a $\sigma \in \mathrm{Aut}_c(\mathbb{C}_v/K_v)$ such that $\ell = \sigma(k)$. Put $\widetilde{E}_{v,\ell} = \sigma(\widetilde{E}_{v,k})$. It is easy to see that $\widetilde{E}_{v,\ell}$ is independent of the choice of σ, and that each $\widetilde{E}_{v,\ell}$ is F_{w_ℓ}-simple. With this definition, for each $1 \le \ell \le D$ and each $\sigma \in \mathrm{Aut}_c(\mathbb{C}_v/K_v)$ we have $\widetilde{E}_{v,\sigma(\ell)} = \sigma(\widetilde{E}_{v,\ell})$ and $V_{\mathfrak{X},\vec{s}}(\widetilde{E}_{v,\sigma(\ell)}) = V_{\mathfrak{X},\vec{s}}(\widetilde{E}_{v,\ell})$. Put

$$\widetilde{E}_v = \bigcup_{\ell=1}^{D} \widetilde{E}_{v,\ell} .$$

By construction $\widetilde{E}_v$ is K_v-simple and compatible with E_v. By Proposition 6.5, part (A) of Theorem 6.3 holds for $\widetilde{E}_v$ and E_v.

We next establish part (B). Fix $0 < \beta_v \in \mathbb{Q}$ and a K_v-symmetric probability vector $\vec{s} \in \mathcal{P}^m(\mathbb{Q})$. Let $\widetilde{V} = V_{\mathfrak{X},\vec{s}}(\widetilde{E}_v)$ and $\widetilde{w}_\ell = \widetilde{w}_\ell(\vec{s})$, for $\ell = 1,\ldots,D$, be the solutions to the system of equations (6.16) associated to $\widetilde{E}_v$. Since $\widetilde{E}_v$, $\mathfrak{X}$ and $\vec{s}$ are K_v-symmetric, we have $V_{\mathfrak{X},\vec{s}}(\widetilde{E}_{v,\sigma(\ell)}) = V_{\mathfrak{X},\vec{s}}(\widetilde{E}_{v,\ell})$ for all ℓ and σ, and each σ permutes the equations (6.16). Hence the weights $\widetilde{w}_\ell$ satisfy $\widetilde{w}_{\sigma(\ell)} = \widetilde{w}_\ell$ for all ℓ and all σ.

Since each $\widetilde{E}_{v,\ell}$ is F_{w_ℓ}-simple, Corollary A.14 (applied to an F_{w_ℓ}-simple decomposition of $\widetilde{E}_{v,\ell}$) shows that each $V_{\mathfrak{X},\vec{s}}(\widetilde{E}_{v,\ell})$ is rational. By Theorem A.13, $\widetilde{V} = V_{\mathfrak{X},\vec{s}}(\widetilde{E}_v)$ and the weights $\widetilde{w}_1,\ldots,\widetilde{w}_D$ are rational. Write $\widetilde{w}_\ell = P_\ell/Q_\ell$ with positive integers P_ℓ, Q_ℓ, and let Q be the least common multiple of $Q_1,\ldots,Q_T$. Write $V_{\mathfrak{X},\vec{s}}(\widetilde{E}_v) = X_0/Y_0$ and $V_{\mathfrak{X},\vec{s}}(\widetilde{E}_{v,\ell}) = X_\ell/Y_\ell$ with integers X_ℓ, Y_ℓ, and put $Y = \mathrm{LCM}(Y_0, Y_1, \ldots, Y_T)$.

Put $\widetilde{N}_v = \mathrm{LCM}(N_{v,1},\ldots,N_{v,T})$, and set $N_v = YQ\widetilde{N}_v$.

Suppose N is a multiple of N_v, say $N = kN_v$. For each $\ell = 1,\ldots,T$ put $n_\ell = \widetilde{w}_\ell N$, noting that $n_\ell \in \mathbb{N}$ and that $N_{v,\ell}|n_\ell$. Let $f_{v,\ell}(z) \in F_{w_\ell}(\mathcal{C})$ be the $(\mathfrak{X},\vec{s})$-function of degree n_ℓ given by Proposition 6.5 for $\widetilde{E}_{v,\ell}$, $E_{v,\ell}$, $\vec{s}$, and β_v. For the remaining sets $\widetilde{E}_{v,\ell}$ with $\ell = T+1,\ldots,D$, define $f_{v,\ell}$ by conjugacy, so that if $\ell = \sigma(k)$ with $1 \le k \le T$ and $\sigma \in \mathrm{Aut}_c(\mathbb{C}_v/K_v)$ then $f_{v,\ell} = (f_{v,k})^\sigma = \sigma \circ f_{v,k} \circ \sigma^{-1}$.

It follows that for each $\ell = 1, \ldots, D$, the function $f_{v,\ell}$ belongs to $F_{w_\ell}(\mathcal{C}_v)$, has degree $\widetilde{w}_\ell N$, and satisfies the conditions of Proposition 6.5 relative to $\widetilde{E}_{v,\ell}$, $\vec{s}$, and β_v. In particular, for all $z \in \mathcal{C}_v(\mathbb{C}_v)\backslash\big(B(a_\ell, r_\ell) \cup \mathfrak{X}\big)$, we have

$$\frac{1}{n_\ell} \log_v(|f_{v,\ell}(z)|_v) \ = \ G_{\mathfrak{X},\vec{s}}(z, \widetilde{E}_{v,\ell}) + \beta_v \ . \tag{6.28}$$

Clearly $f_{v,\sigma(\ell)} = (f_{v,\ell})^\sigma$ for all $\sigma \in \mathrm{Aut}_c(\mathbb{C}_v/K_v)$ and all ℓ.

Note that for each ℓ we have $u_{\mathfrak{X},\vec{s}}(z, \widetilde{E}_\ell) = V_{\mathfrak{X},\vec{s}}(\widetilde{E}_{v,\ell}) - u_{\mathfrak{X},\vec{s}}(z, \widetilde{E}_{v,\ell})$, and that

$$G_{\mathfrak{X},\vec{s}}(z, \widetilde{E}_v) \ = \ V_{\mathfrak{X},\vec{s}}(\widetilde{E}_v) - \sum_{\ell=1}^{D} \widetilde{w}_\ell \, u_{\mathfrak{X},\vec{s}}(z, \widetilde{E}_{v,\ell}) \ .$$

Thus if we put $C = V_{\mathfrak{X},\vec{s}}(\widetilde{E}_v) - \sum_{\ell=1}^{D} \widetilde{w}_\ell V_{\mathfrak{X},\vec{s}}(\widetilde{E}_{v,\ell})$, then

$$G_{\mathfrak{X},\vec{s}}(z, \widetilde{E}_v) \ = \ C + \sum_{\ell=1}^{D} \widetilde{w}_\ell \, G_{\mathfrak{X},\vec{s}}(z, \widetilde{E}_{v,\ell}) \ . \tag{6.29}$$

By our choice of N, we have $N \cdot C \in \mathbb{Z}$. Since $G_{\mathfrak{X},\vec{s}}(z, \widetilde{E}_v) = 0$ for all $z \in \widetilde{E}_v$, and $G_{\mathfrak{X},\vec{x}}(z, \widetilde{E}_{v,\ell}) \geq 0$ for all z and all ℓ, by evaluating both sides of (6.29)) at a point $z \in \widetilde{E}_v$ we see that $C \leq 0$.

Let π_v be a uniformizing element for the maximal ideal of $\mathcal{O}_v$, and define

$$f_v(z) \ = \ \pi_v^{-NC} \cdot \prod_{\ell=1}^{D} f_{v,\ell}(z) \ . \tag{6.30}$$

Since each $f_{v,\ell}$ is an $(\mathfrak{X}, \vec{s})$-function, so is f_v. By construction f_v is stable under $\mathrm{Aut}_c(\mathbb{C}_v/K_v)$. Since each F_{w_ℓ}/K_v is separable, f_v belongs to $K_v(\mathcal{C})$. It clearly has degree N.

We will now show that f_v satisfies the conditions of Theorem 6.3 relative to $\widetilde{E}_v$, $\vec{s}$, and β_v and the K_v-simple decomposition (6.27). First, by our hypotheses on the $f_{v,\ell}(z)$, for each $z \notin \bigcup_{\ell=1}^{D} B(a_\ell, r_\ell) \cup \mathfrak{X}$ we have

$$\begin{aligned}
\frac{1}{N} \log_v(|f_v(z)|_v) \ &= \ \frac{1}{N}\Big(NC + \sum_{\ell=1}^{D} \log_v(|f_{v,\ell}(z)|_v)\Big) \\
&= \ \frac{1}{N}\Big(NC + \sum_{\ell=1}^{D} \widetilde{w}_\ell \, N \cdot \big(G_{\mathfrak{X},\vec{s}}(z, \widetilde{E}_{v,\ell}) + \beta_v\big)\Big) \\
&= \ G_{\mathfrak{X},\vec{s}}(z, \widetilde{E}_v) + \beta_v \ ,
\end{aligned}$$

using (6.29) and the fact that $\sum_{\ell=1}^{D} \widetilde{w}_\ell = 1$. In particular, for each $x_i \in \mathfrak{X}$,

$$\Lambda_{x_i}(f_v, \vec{s}) \ = \ \lim_{z \to x_i} \big(G_{\mathfrak{X},\vec{s}}(z, \widetilde{E}_v) + s_i \log_v(|g_i(z)|_v)\big) + \beta_v \ = \ \Lambda_{x_i}(\widetilde{E}_v, \vec{s}) + \beta_v \ .$$

Next, we claim that the zeros of f_v are distinct and belong to E_v. Indeed, for each ℓ the zeros of $f_{v,\ell}$ are distinct and belong to $E_{v,\ell}$. This holds for $\ell = 1, \ldots, T$ by Proposition 6.5, and for the remaining ℓ by conjugacy. Since the sets $E_{v,\ell}$ are pairwise disjoint, our claim follows.

Fix ℓ. Recalling that $n_\ell = \deg(f_{v,\ell}) = \widetilde{w}_\ell N$, let $\theta_{\ell,1}, \ldots, \theta_{\ell,n_\ell}$ be the zeros of $f_{v,\ell}$. By Proposition 6.5 there pairwise disjoint balls $B(\theta_{\ell,1}, \rho_{\ell,1}), \ldots, B(a_{\ell,n_\ell}, \rho_{\ell,n_\ell})$

contained in $B(a_\ell, r_\ell)$ such that

$$f_{v,\ell}^{-1}(D(0,1)) \ = \ \bigcup_{j=1}^{n_\ell} B(\theta_{\ell,j}, \rho_{\ell,j}) \ .$$

Here, the balls $B(\theta_{\ell,j}, \rho_{\ell,j})$ are isometrically parametrizable since $B(a_\ell, r_\ell)$ is isometrically parametrizable. The $\theta_{\ell,j}$ belong to $\mathcal{C}_v(F_{w_\ell})$ since they belong to $E_{v,\ell}$. By choosing an F_{w_ℓ}-rational isometric parametrization $\varphi_{\ell,j} : D(0, r_\ell) \to B(a_\ell, r_\ell)$ with $\varphi_{\ell,j}(0) = \theta_{\ell,j}$, expanding $f_{v,\ell}$ as a power series $c_0 + c_1 Z + \cdots$, and applying Proposition 3.38, we see that $f_{v,\ell}$ induces a scaled isometry from $B(\theta_{\ell,j}, \rho_{\ell,j})$ to $D(0,1)$. Here $c_0, c_1, \cdots \in F_{w_\ell}$ since $f_{v,\ell} \in F_{w_\ell}(\mathcal{C}_v)$ and $\theta_{\ell,j} \in \mathcal{C}_v(F_{w_\ell})$. Proposition 3.38 gives $|c_1|_v \cdot \rho_{\ell,j} = 1$, so $\rho_{\ell,j} = 1/|c_1|_v \in |F_{w_\ell}^\times|_v$.

On the other hand, the function $H_\ell(z) = f_v(z)/f_{v,\ell}(z)$ is also F_{w_ℓ}-rational, and its zeros and poles are disjoint from $B(a_\ell, r_\ell)$. Hence there is a constant B_ℓ such that $|H_\ell(z)|_v = B_\ell$ for all $z \in B(a_\ell, r_\ell)$. Evaluating $H_\ell(z)$ at a point $z_\ell \in \widetilde{E}_{v,\ell}$, and using $G_{\mathfrak{X},\vec{s}}(z_\ell, \widetilde{E}_{v,\ell}) = 0$, (6.29), and $G_{\mathfrak{X},\vec{s}}(z_\ell, \widetilde{E}_v) = 0$, we see that

$$\begin{aligned}
\log_v(B_\ell) = \log_v\Big(|\pi_v^{-NC} \cdot \prod_{k\neq\ell} f_{v,k}(z_\ell)|_v\Big) &= NC + \sum_{k\neq\ell} \widetilde{w}_k N \cdot \big(G_{\mathfrak{X},\vec{s}}(z_\ell, \widetilde{E}_{v,k}) + \beta_v\big) \\
&= NC + \Big(\sum_{k=1}^{D} \widetilde{w}_k N \cdot G_{\mathfrak{X},\vec{s}}(z_\ell, \widetilde{E}_{v,k})\Big) + N \cdot (1 - \widetilde{w}_\ell)\beta_v \\
&= N \cdot G_{\mathfrak{X},\vec{s}}(z_0, \widetilde{E}_v) + N \cdot (1 - \widetilde{w}_\ell)\beta_v = N \cdot (1 - \widetilde{w}_\ell)\beta_v \geq 0 \, ,
\end{aligned}$$

so $B_\ell \geq 1$. Since $B_\ell = |\pi_v^{-NC} \cdot H_\ell(z_\ell)|_v$, where $H_\ell \in F_{w_\ell}(\mathcal{C}_v)$ and $z_\ell \in \mathcal{C}_v(F_{w_\ell})$, we have $B_\ell \in |F_{w_\ell}^\times|_v$.

Choose an F_{w_ℓ}-rational isometric parametrization of $B(\theta_{\ell,j}, \rho_{\ell,j})$, and expand $f_{v,\ell}$ and H_ℓ as power series. By Proposition 3.38, $f_v = f_{v,\ell} \cdot H_\ell$ induces an F_{w_ℓ}-rational scaled isometry from $B(\theta_{\ell,j}, B_\ell^{-1}\rho_{\ell,j})$ onto $D(0,1)$ which maps $\mathcal{C}_v(F_{w_\ell}) \cap B(\theta_{\ell,j}, B_\ell^{-1}\rho_{\ell,j})$ onto $\mathcal{O}_{w_\ell}$, for each j. Clearly $B_\ell^{-1}\rho_{\ell,j} \in |F_{w_\ell}^\times|_v$.

Now let ℓ vary. For each ℓ and each j we have

$$B(\theta_{\ell,j}, B_\ell^{-1}\rho_{\ell,j}) \ \subseteq \ B(\theta_{\ell,j}, \rho_{\ell,j}) \ \subseteq \ B(a_\ell, r_\ell) \ .$$

For a given ℓ the balls $B(\theta_{\ell,j}, \rho_{\ell,j})$ are pairwise disjoint, so the balls $B(\theta_{\ell,j}, B_\ell^{-1}\rho_{\ell,j})$ are pairwise disjoint. For different ℓ, the balls $B(a_\ell, r_\ell)$ are pairwise disjoint, so in fact the balls $B(\theta_{\ell,j}, B_\ell^{-1}\rho_{\ell,j})$ are pairwise disjoint for all ℓ and j. There are exactly $N = \sum_{\ell=1} n_\ell = \deg(f_v)$ such balls, so

$$f_v^{-1}(D(0,1)) \ = \ \bigcup_{\ell=1}^{D} \bigcup_{j=1}^{n_\ell} B(\theta_{\ell,j}, B_\ell^{-1}\rho_{\ell,j}) \ \subseteq \ \bigcup_{\ell=1}^{D} B(a_\ell, r_\ell) \ .$$

It follows that $H_v := E_v \cap f_v^{-1}\big(D(0,1)\big)$ is K_v-simple, and has the K_v-simple decomposition

$$H_v \ = \ \bigcup_{\ell=1}^{D} \bigcup_{j=1}^{n_\ell} \mathcal{C}_v(F_{w_\ell}) \cap B(\theta_{\ell,j}, B_\ell^{-1}\rho_{\ell,j})$$

which is compatible with the K_v-simple decomposition (6.27) of E_v. This completes the proof of part (B).

Finally, suppose $\mathrm{char}(K_v) = p > 0$. We will show that by modifying the construction above, we can arrange that the leading coefficients

$$c_{v,i} = \lim_{z \to x_i} f_v(z) \cdot g_{x_i}(z)^{N_i}$$

belong to $K_v(x_i)^{\mathrm{sep}}$, so that part (C) holds.

Fix a positive, K_v-symmetric probability vector $\vec{s} \in \mathbb{Q}^m$ and a number $0 < \beta_v \in \mathbb{Q}$ as before, and carry out the construction in part (B) for E_v, $\widetilde{E}_v$, and $\vec{s}$, but with β_v replaced by $\beta_v/2$. Let $N_{v,0} > 0$ be the integer given by part (B) for $\beta_v/2$; note that $N_{v,0} \cdot \beta_v/2 \in \mathbb{Z}$. Put $p^B = \max_{1 \le i \le m}([K_v(x_i) : K_v]^{\mathrm{insep}})$. After replacing $N_{v,0}$ by a multiple of itself if necessary, we can assume that

$$p^B N_{v,0} \cdot \frac{\beta_v}{2} \geq \frac{p^B}{q_v - 1} + \log_v(p^B) + 2 .$$

We will take $N_v = p^B N_{v,0}$.

Given a positive integer N divisible by N_v, put $N_0 = N/p^B$. Then N_0 is divisible by $N_{v,0}$; let $f_{v,0} \in K_v(\mathcal{C})$ be the $(\mathfrak{X}, \vec{s})$-function of degree N_0 constructed in part (B) for E_v and $\widetilde{E}_v$, relative to $\vec{s}$ and $\beta_v/2$. Noting that $N \cdot \beta_v/2 \in \mathbb{Z}$, compose $f_{v,0}$ with the Stirling polynomial $S_{p^B,\mathcal{O}_v}(z)$ and put

$$f_v(z) = \pi_v^{-N\beta_v/2} \cdot S_{p^B,\mathcal{O}_v}(f_{v,0}(z)) . \tag{6.31}$$

For each i, the leading coefficient $c_{v,i,0}$ of $f_{v,0}$ at x_i belongs to $K_v(x_i)$, since $f_{v,0} \in K_v(\mathcal{C})$ and g_{x_i} is rational over $K_v(x_i)$. This means that the leading coefficient of $f_v(z)$ at x_i is

$$c_{v,i} = \pi_v^{-N\beta_v/2} \cdot c_{v,i,0}^{p^B} ,$$

which belongs to $K_v(x_i)^{\mathrm{sep}}$. Thus part (C) of Theorem 6.3 holds for f_v.

We now show that $f_v(z)$ continues to satisfy properties (B.1)–(B.4) of of Theorem 6.3. First, note that since $\Lambda_{x_i}(f_{v,0}, \vec{s}) = \Lambda_{x_i}(\widetilde{E}_v, \vec{s}) + \beta_v/2$, we have

$$\begin{aligned}\Lambda_{x_i}(f_v, \vec{s}) &= \frac{1}{N}\log_v(|c_{v,i}|_v) = \frac{1}{N} \cdot N\beta_v/2 + \frac{1}{N} \cdot p^B \log_v(|c_{v,i,0}|_v) \\ &= \beta_v/2 + \Lambda_{x_i}(f_{v,0}, \vec{s}) = \Lambda_{x_i}(\widetilde{E}_v, \vec{s}) + \beta_v .\end{aligned}$$

This proves property (B.1).

For property (B.2), recall that the zeros of $S_{p^B,\mathcal{O}_v}(z)$ are distinct and belong to $\mathcal{O}_v$. For each $k = 1, \ldots, N_0$, the function $f_{v,0}$ induces a scaled isometry from $B(\theta_k, \rho_k)$ onto $D(0,1)$, which takes $B(\theta_k, \rho_k) \cap \mathcal{C}_v(F_{w_\ell(k)})$ onto $\mathcal{O}_{w_\ell(k)}$. The zeros of f_v in $B(\theta_k, \rho_k)$, which we will denote $\theta_{k,j}$ for $j = 0, \ldots, p^B - 1$, therefore belong to $B(\theta_k, \rho_k) \cap \mathcal{C}_v(F_{w_\ell(k)}) \subseteq E_v$. Letting k vary, we see that f_v has $N_0 \cdot p^B = N$ zeros in E_v. Since $\deg(f_v) = N$, these are all the zeros of f_v. Thus the zeros of f_v are distinct and belong to E_v.

For property (B.3), note that by Corollary 3.41, if $0 < R < q_v^{-p^B/(q_v-1)} \cdot (p^B)^{-1}$, then the inverse image of $D(0,R)$ under $S_{p^B,\mathcal{O}_v}(z)$ consists of p^B disjoint discs contained in $D(0,1)$, centered on the roots of $S_{p^B,\mathcal{O}_v}(z)$. In addition, if R belongs to the value group of $K_v^\times$, the radii of those discs belong to the value group of $K_v^\times$.

Take $R = q_v^{-\eta}$, where

$$\eta = \lceil \frac{p^B}{q_v - 1} + \log_v(p^B) \rceil + 1 < \frac{p^B}{q_v - 1} + \log_v(p^B) + 2 .$$

By our choice of N_v, we have $|\pi_v^{-N\beta_v/2}|_v \cdot R > 1$. Thus $D(0,1) \subset D(0, |\pi_v^{-N\beta_v/2}|_v R)$, and the inverse image of $D(0,1)$ under $\pi_v^{-N\beta_v/2} S_{p^B,\mathcal{O}_v}(z)$ consists of p^B disjoint discs in $D(0,1)$, centered on the roots of $S_{p^B,\mathcal{O}_v}(z)$ and having radii in $|K_v^\times|_v$. Since $f_{v,0}(z)$ induces an $F_{w_\ell(k)}$-rational scaled isometry from $B(\theta_k, \rho_k)$ onto $D(0,1)$ for each $k = 1, \ldots, N_0$, it follows that

$$f_v^{-1}(D(0,1)) \;=\; \bigcup_{k=1}^{N} \bigcup_{j=0}^{p^B-1} B(\theta_{k,j}, \rho_{k,j})$$

where the balls on the right are pairwise disjoint and isometrically parametrizable. Furthermore, $\rho_{k,j}$ belongs to the value group of $F_{w_\ell(k)}$, and $f_v(z)$ induces an $F_{w_\ell(k)}$-rational scaled isometry from $B(\theta_{k,j}, \rho_{k,j})$ onto $D(0,1)$, for all k, j. This establishes property (B.3).

It is clear from the discussion above that the set

$$(6.32) \qquad H_v \;=\; E_v \cap f_v^{-1}(D(0,1)) \;=\; \bigcup_{k=1}^{N} \bigcup_{j=0}^{p^B-1} \big(B(\theta_{k,j}, \rho_{k,j}) \cap \mathcal{C}_v(F_{w_\ell(k)})\big)$$

is K_v-simple, and that (6.32) is a K_v-simple decomposition compatible with the K_v-simple decomposition (6.5) of E_v. This yields property (B.4), and completes the proof. □

3. Generalized Stirling Polynomials

In this section we construct Stirling polynomials for sets of the form

$$H_w \;=\; \bigcup_{\ell=1}^{d} F_w \cap D(a_\ell, r_\ell) \;\subset\; \mathbb{C}_v \,.$$

These will play a key role in the proof of Theorem 6.3. The idea is that within any isometrically parametrizable ball $B(a,r)$ disjoint from $\mathfrak{X}$, the canonical distance $[x,y]_{\mathfrak{X},\vec{s}}$ is a multiple of the spherical distance $\|x,y\|_v$. Under an isometric parametrization, $\|x,y\|_v$ pulls back to a multiple of the usual distance $|X-Y|_v$ on the disc $D(0,r) \subset \mathbb{C}_v$. This means that potential-theoretic constructions on $B(a,r)$ relative to $[x,y]_{\mathfrak{X},\vec{s}}$ are essentially the same as potential-theoretic constructions on $\mathbb{C}_v$ relative to $[X,Y]_\infty = |X-Y|_v$.

First suppose $H_w = F_w \cap D(a,r)$ for a single disc, where $a \in F_w$ and $r \in |F_w^\times|_v$. Fix $b \in F_w$ with $|b|_v = r$. We can can obtain a well-distributed sequence in H_w by composing the basic well-distributed sequence $\{\psi_w(k)\}_{k \geq 0}$ for $\mathcal{O}_w$ with the affine map $a + bz$: for each integer $n \geq 1$, we define the Stirling polynomial $S_{n,H_w}(z)$ by

$$S_{n,H_w}(z) \;=\; \prod_{k=0}^{n-1} \big(z - (a + b\psi_w(k))\big) \;=\; b^n \cdot S_{n,\mathcal{O}_w}(\frac{z-a}{b}) \,.$$

Now consider the general case: suppose

$$H_w \;=\; \bigcup_{\ell=1}^{d} (F_w \cap D(a_\ell, r_\ell))$$

where $a_\ell \in F_w$, $r_\ell \in |F_w^\times|_v$ for each ℓ, and the discs $D(a_\ell, r_\ell)$ are pairwise disjoint. Put $H_{w_\ell} = F_w \cap D(a_\ell, r_\ell)$, so $H_w = \bigcup_{\ell=1}^d H_{w_\ell}$. By Corollary A.10 the potential function of H_{w_ℓ} is given by

$$(6.33) \qquad u_\infty(z, H_{w_\ell}) = \begin{cases} \frac{1}{e_w(q_w-1)} - \log_v(r_\ell) & \text{if } z \in H_{w_\ell} , \\ -\log_v(|z - a_\ell|_v) & \text{if } z \notin D(a_\ell, r_\ell) , \end{cases}$$

and in particular $V_\infty(H_{w_\ell}) = \frac{1}{e_w(q_w-1)} - \log_v(r_\ell)$.

Let μ_∞ be the equilibrium distribution of H_w relative to the point ∞, and let $V = V_\infty(H_w)$ be the Robin constant. For each ℓ, put $w_\ell = \mu_\infty(H_{w_\ell}) > 0$. As in §6.2, by Corollary A.14 the following system of linear equations uniquely determine V and the w_ℓ:

$$(6.34) \qquad \begin{cases} 1 = 0 \cdot V + \sum_{\ell=1}^d w_\ell , \\ 0 = V + w_\ell \cdot \big(\log_v(r_\ell) - \frac{1}{e_w(q_w-1)}\big) + \sum_{\substack{j=1 \\ j \neq \ell}}^d w_j \cdot \log_v(|a_\ell - a_j|_v) \\ \qquad \text{for } \ell = 1, \ldots, d. \end{cases}$$

Since the coefficients of this system are rational, V and the w_ℓ belong to $\mathbb{Q}$. Write $w_\ell = P_\ell/Q_\ell$ with positive integers P_ℓ, Q_ℓ, and put $Q = \mathrm{LCM}(Q_1, \ldots, Q_d)$.

For each ℓ, fix an element $b_\ell \in F_w$ with $|b_\ell|_v = r_\ell$. Then the affine map $\varphi_\ell(z) = a_\ell + b_\ell z$ takes $\mathcal{O}_w$ to H_{w_ℓ}.

Let n be a positive integer divisible by Q, and put $n_\ell = w_\ell \cdot n$ for $\ell = 1, \ldots, d$. Define the Stirling polynomial $S_{n,H_w}(z)$ by

$$(6.35) \qquad S_{n,H_w}(z) = \prod_{\ell=1}^d \prod_{k=0}^{n_\ell - 1} \big(z - (a_\ell + b_\ell \psi_w(k))\big) = \prod_{\ell=1}^d S_{n_\ell, H_{w_\ell}}(z) .$$

Note that Q and $S_{n,H_w}(z)$ depend on the decomposition $H_w = \bigcup_{\ell=1}^d H_{w_\ell}$ and the maps $\varphi_\ell(z) = a_\ell + b_\ell z$, not just H_w. For the rest of this section we will assume these are fixed.

The following proposition generalizes Proposition 3.40:

Proposition 6.6. *Let F_w/K_v be a finite, separable extension in $\mathbb{C}_v$. Suppose $H_w = \bigcup_{\ell=1}^d \big(F_w \cap D(a_\ell, r_\ell)\big)$, where the discs $D(a_\ell, r_\ell)$ are pairwise disjoint, and $a_\ell \in F_w$ and $r_\ell \in |F_w^\times|_v$ for each ℓ. Let $0 < w_1, \ldots, w_d \in \mathbb{Q}$ be the weights corresponding to the sets $H_{w_\ell} = F_w \cap D(a_\ell, r_\ell)$ by the system (6.34), and let Q be the least common multiple of their denominators.*

For each positive integer n divisible by Q, let $S_{n,H_w}(z) = \prod_{\ell=1}^d S_{n_\ell, H_{w_\ell}}(z)$ be the Stirling polynomial of degree n for H_w, and let $S_{n,H_w}'(z)$ be its derivative. Write $\vartheta_1, \ldots, \vartheta_n$ for its zeros, and for each ℓ let $\varphi_\ell(z) = a_\ell + b_\ell z : \mathcal{O}_w \to H_{w_\ell}$ be the affine map used in defining $S_{n,H_w}(z)$. Then

(A) $\vartheta_1, \ldots, \vartheta_n$ are distinct and belong to H_w. There is a constant $A > 0$, independent of n, such that for all $i \neq j$,

$$(6.36) \qquad |\vartheta_i - \vartheta_j|_v > A/n .$$

(B) For each k, $1 \le k \le n$, if $\vartheta_k \in D(a_\ell, r_\ell)$ then we have

$$(6.37) \qquad \mathrm{ord}_v\big(S_{n,H_w}'(\vartheta_k)\big) < n \cdot V_\infty(H_w) - \mathrm{ord}_v(b_\ell) .$$

(C) Given $x \in \mathbb{C}_v$, fix $1 \le J \le n$ with $|x-\vartheta_J|_v = \min_k(|x-\vartheta_k|_v)$. If $\vartheta_J \in H_{w_\ell}$, then

$$\operatorname{ord}_v(S_{n,H_w}(x)) \;<\; n \cdot V_\infty(H_w) + \operatorname{ord}_v\left(\frac{x-\vartheta_J}{b_\ell}\right) . \tag{6.38}$$

If $x \in \mathbb{C}_v \backslash \bigcup_{\ell=1}^d D(a_\ell, r_\ell)$, then

$$\operatorname{ord}_v(S_{n,H_w}(x)) \;=\; n \cdot \big(V_\infty(H_w) - G(x,\infty;H_w)\big) . \tag{6.39}$$

PROOF. We first prove the result when $H_w = F_w \cap D(a,r) = a + b\mathcal{O}_w$, with $a, b \in F_w$ and $|b|_v = r > 0$. For notational convenience, we relabel the zeros as $\vartheta_0, \ldots, \vartheta_{n-1}$, with $\vartheta_k = a + b\psi_w(k)$ for $k = 0, \ldots, n-1$.

In this case, part (A) holds with $A = r$, since

$$|\vartheta_i - \vartheta_j|_v \;=\; |b|_v|\psi_w(i) - \psi_w(j)|_v \;>\; r/n$$

by Proposition 3.40(A)

For part (B), note that $S_{n,H_w}(z) = b^n S_{n,\mathcal{O}_w}((z-a)/b)$. Hence $S'_{n,H_w}(\vartheta_k) = b^{n-1} S'_{n,\mathcal{O}_w}(\psi_w(k))$ for $k = 0, \ldots, n-1$, and since

$$V_\infty(H_w) \;=\; \frac{1}{e_w(q_w-1)} - \log_v(r) \;=\; V_\infty(\mathcal{O}_w) + \operatorname{ord}_v(b) ,$$

part (B) follows from Proposition 3.40(B).

For part (C), observe that if $x \in D(a,r)$ and $|x - \vartheta_J|_v = \min_{0 \le k < n}(|x - \vartheta_k|_v)$, then for $X := (x-a)/b \in D(0,1)$ we have $|(x-\vartheta_k)/b|_v = |X - \psi_w(J)|_v = \min_{0 \le k < n}(|X - \psi_w(k)|_v)$. Hence (6.38) follows from Proposition 3.40(C).

If $x \notin D(a,r)$, then by the ultrametric inequality we have $|x-\vartheta_k|_v = |x-a|_v$ for each k. Furthermore, $u_\infty(x, H_w) = -\log_v(|x-a|_v)$ by (6.33). Thus

$$\operatorname{ord}_v(S_{n,H_w}(x)) \;=\; n \cdot u_\infty(x, H_w) \;=\; n \cdot \big(V_\infty(H_w) - G(x,\infty;H_w)\big) .$$

This yields (6.39) since $G(x,\infty;H_w) > 0$.

Now consider the general case, where $H_w = \bigcup_{\ell=1}^d H_{w_\ell} = \bigcup_{\ell=1}^d \big(F_w \cap D(a_\ell, r_\ell)\big)$, with finitely many pairwise disjoint discs $D(a_\ell, r_\ell)$ such that $a_\ell \in F_w$ and $r_\ell \in |F_w^\times|_v$.

Fix $0 < n \in \mathbb{N}$ divisible by Q, put $n_\ell = n \cdot w_\ell$ for each ℓ, and consider $S_{n,H_w}(z) = \prod_{k=1}^n (z - \vartheta_k) = \prod_{\ell=1}^d S_{n_\ell, H_{w_\ell}}(z)$.

For part (A), put $A = \min_{1 \le \ell \le d}(r_\ell)$. Fix n divisible by Q, and let $\vartheta_i \ne \vartheta_j$ be distinct roots of $S_{n,H_w}(z)$. Necessarily $n \ge 2$. Since the balls $B(a_\ell, r_\ell)$ are pairwise disjoint, if ϑ_i and ϑ_j belong to distinct balls then $|\vartheta_i - \vartheta_j|_v > A > A/n$. On the other hand, if ϑ_i and ϑ_j belong to the same ball $B(a_\ell, r_\ell)$, then there are indices $k \ne h$ with $1 \le k, h \le n_\ell$ such that $\vartheta_i = a_\ell + b_\ell\psi_w(k)$ and $\vartheta_j = a_\ell + b_\ell\psi_w(h)$. In this case

$$|\vartheta_i - \vartheta_j|_v \;\ge\; |b_\ell|_v \cdot |\psi_w(k) - \psi_w(h)|_v \;>\; A/n_\ell \;\ge\; A/n .$$

For part (B), first note that if $x \in D(a_\ell, r_\ell)$, then for each $j \ne \ell$ and each $\vartheta_k \in D(a_j, r_j)$, we have $|x - \vartheta_k|_v = |a_\ell - a_j|_v$. Hence

$$|S_{n,H_w}(x)|_v \;=\; |S_{n_\ell, H_{w_\ell}}(x)|_v \cdot \prod_{\substack{j=1 \\ j \ne \ell}}^d |a_\ell - a_j|_v^{n_j} . \tag{6.40}$$

Similarly, if $\vartheta_h \in D(a_\ell, r_\ell)$ then

$$(6.41) \qquad |S'_{n,H_w}(\vartheta_h)|_v \;=\; |S'_{n_\ell,H_{w_\ell}}(\vartheta_h)|_v \cdot \prod_{\substack{j=1\\ j\neq \ell}}^{d} |a_\ell - a_j|_v^{n_j} \ .$$

On the other hand, since $n_j = n \cdot w_j$, for each ℓ the equations (6.34) give

$$(6.42) \qquad n \cdot V_\infty(H_w) \;=\; n_\ell V_\infty(H_{w_\ell}) + \sum_{\substack{j=1\\ j\neq \ell}}^{d} n_j \cdot \big(-\log_v(|a_\ell - a_j|_v)\big) \ .$$

One obtains part (B) by combining (6.37) for H_{w_ℓ} with (6.41) and (6.42), and using $\log_w(n_\ell) \le \log_w(n)$.

For part (C), if $\vartheta_J \in D(a_\ell, r_\ell)$ one obtains (6.38) by combining (6.38) for H_{w_ℓ} with (6.40) and (6.42). If $x \in \mathbb{C}_v \backslash \bigcup_{\ell=1}^d D(a_\ell, r_\ell)$, one obtains (6.39) by using (6.33) and noting that

$$\begin{aligned} -\log_v(|S_{n,H_w}(x)|_v) \;&=\; -\sum_{\ell=1}^{d} n_\ell \log_v(|x - a_\ell|_v) \;=\; n \cdot \Big(\sum_{\ell=1}^{d} w_\ell\, u_\infty(x, H_{w_\ell})\Big) \\ &=\; n \cdot u_\infty(x, H_w) \;=\; n \cdot \big(V_\infty(H_w) - G(x, \infty; H_w)\big) \ . \end{aligned}$$

□

Remark. In Proposition 6.6(B), one can show that if $x \in \mathbb{C}_v \backslash \bigcup_{\ell=1}^d D(a_\ell, r_\ell)$, then

$$\mathrm{ord}_v(S_{n,H_w}(x)) \;<\; n \cdot \Big(V_\infty(H_w) - \min_{1 \le \ell \le d}(w_\ell) \cdot \frac{1}{e_w(q_w - 1)}\Big) \ .$$

One can also prove the following generalization of Corollary 3.41: given an R satisfying

$$(6.43) \qquad 0 \;<\; R \;\le\; q_v^{-nV_\infty(H_w)} \cdot n^{-1/[F_w:K_v]} \ ,$$

put $\rho_k = R/|S'_{n,H_w}(\vartheta_k)|_v$ for $k = 1, \ldots, n$. Then the discs $D(\vartheta_k, \rho_k)$ are pairwise disjoint and

$$(6.44) \qquad S_{n,H_w}^{-1}\big(D(0,R)\big) \;=\; \bigcup_{k=1}^{n} D(\vartheta_k, \rho_k) \;\subseteq\; \bigcup_{\ell=1}^{d} D(a_\ell, r_\ell) \ .$$

4. Proof of Proposition 6.5

In this section we will prove Proposition 6.5, completing the proof of Theorem 6.3. The proof breaks into two cases, according as the genus $g(\mathcal{C}_v) = 0$ or $g(\mathcal{C}_v) > 0$.

Proposition 6.5 concerns each subset $E_{v,\ell} = \mathcal{C}_v(F_{w_\ell}) \cap B(a_\ell, r_\ell)$ individually. To simplify notation, we restate the Proposition, dropping the index ℓ and relabeling $E_{v,\ell}$ as E_w :

Proposition 6.5A. *Let F_w/K_v be a finite, separable extension in $\mathbb{C}_v$ and take $E_w = \mathcal{C}_v(F_w) \cap B(a, r)$, where $a \in \mathcal{C}_v(F_w)$ and $B(a, r) \subset \mathcal{C}_v(\mathbb{C}_v)\backslash\mathfrak{X}$ is an isometrically parametrizable ball. Let $\varepsilon_w > 0$ be given. Then there is a compact subset $\widetilde{E}_w \subseteq E_w$ for which:*

(A) There are points $\alpha_j \in \mathcal{C}_v(F_w) \cap B(a,r)$ and pairwise disjoint isometrically parametrizable balls $B(\alpha_1, r_1), \ldots, B(\alpha_d, r_d) \subseteq B(a,r)$, such that $\widetilde{E}_w$ has the form

$$\widetilde{E}_w \;=\; \bigcup_{j=1}^{d} \big(\mathcal{C}_v(F_w) \cap B(\alpha_j, r_j)\big) \tag{6.45}$$

and for each $x_i \in \mathfrak{X}$,

$$|V_{x_i}(\widetilde{E}_w) - V_{x_i}(E_w)| \;<\; \varepsilon_w \ . \tag{6.46}$$

(B) For each $0 < \beta_w \in \mathbb{Q}$ and each F_w-symmetric probability vector $\vec{s} = {}^t(s_1, \ldots, s_m)$ with rational entries, there is an integer $N_w \geq 1$ such that for each positive integer N divisible by N_w, there is an $(\mathfrak{X}, \vec{s})$-function $f_w \in F_w(\mathcal{C}_v)$ of degree N such that

(1) For all $z \in \mathcal{C}_v(\mathbb{C}_w) \backslash \big(B(a,r) \cup \mathfrak{X}\big)$,

$$\frac{1}{N} \log_v(|f_w(z)|_v) \;=\; G_{\mathfrak{X},\vec{s}}(z, \widetilde{E}_w) + \beta_w \ . \tag{6.47}$$

(2) The zeros $\theta_1, \ldots, \theta_N$ of f_w are distinct and belong to E_w (hence $\mathcal{C}_v(F_w)$).

(3) $f_w^{-1}(D(0,1)) = \bigcup_{h=1}^N B(\theta_h, \rho_h)$, where the balls $B(\theta_h, \rho_h)$ are pairwise disjoint and contained in $B(a,r)$.

The following lemma will be helpful in proving Proposition 6.5A:

LEMMA 6.7. *Let $E_w = \mathcal{C}_v(F_w) \cap B(a,r)$ be as Proposition 6.5A. Given $\varepsilon_w > 0$, suppose $\widetilde{E}_w \subseteq E_w$ is a compact subset satisfying part (A) of Proposition 6.5A.*

Then part (B) of Proposition 6.5A holds for E_w and $\widetilde{E}_w$ if for each F_w-symmetric $\vec{s} \in \mathcal{P}^m(\mathbb{Q})$, there are an integer $N_0 = N_0(\vec{s}, \widetilde{E}_w) > 0$, and constants $A = A(\vec{s}, \widetilde{E}_w) > 0$ and $B = B(\vec{s}, \widetilde{E}_w)$, such that $N_0\vec{s} \in \mathbb{N}^m$ and for each sufficiently large integer N divisible by N_0 there is a divisor D_N of the form

$$D_N \;=\; \sum_{h=1}^{N} (\theta_h) - \sum_{i=1}^{m} N s_i(x_i) \tag{6.48}$$

satisfying conditions (1), (2) and (3) below:

(1) D_N is principal.

(2) $\theta_1, \ldots, \theta_N$ are distinct and belong to E_w and, for all $i \neq j$,

$$\|\theta_i, \theta_j\|_v \;>\; A/N \ .$$

(3) The pseudopolynomial $Q_N(z) = \prod_{h=1}^N [z, \theta_h]_{\mathfrak{X},\vec{s}}$ with divisor D_N has the following property: given $z \in B(a,r)$, if θ_J is such that $\|z, \theta_J\|_v = \min_{1 \leq h \leq N} \|z, \theta_h\|_v$, then

$$-\log_v(Q_N(z)) \;\leq\; N \cdot V_{\mathfrak{X},\vec{s}}(\widetilde{E}_w) - \log_v(\|z, \theta_J\|_v) + B \ .$$

PROOF. Fix an F_w-symmetric probability vector $\vec{s} \in \mathcal{P}^m(\mathbb{Q})$ and a number $0 < \beta_w \in \mathbb{Q}$. Let $A > 0$, $B \geq 0$, and $0 < N_0 \in \mathbb{N}$ be the numbers given by Lemma 6.7 for E_w, $\widetilde{E}_w$, and $\vec{s}$. As in §6.2, there is a constant $C_{\mathfrak{X},\vec{s}} \in |\mathbb{C}_v^\times|_v$ such that $[x,y]_{\mathfrak{X},\vec{s}} = C_{\mathfrak{X},\vec{s}} \|x,y\|_v$ for all $x, y \in B(a,r)$. Let N_1 be the least positive integer such that $C_{\mathfrak{X},\vec{s}}^{N_1} \in |F_w^\times|_v$. By (6.45) and Corollary A.14, $V_{\mathfrak{X},\vec{s}}(\widetilde{E}_w) \in \mathbb{Q}$. By hypothesis, $\beta_w \in \mathbb{Q}$; let N_2 be the least common denominator for $V_{\mathfrak{X},\vec{s}}(\widetilde{E}_w)$ and β_w.

Let N_3 be the smallest natural number such that for each $N \geq N_3$ divisible by N_0, there is a divisor $D_N = \sum(\theta_h) - \sum N s_i(x_i)$ satisfying conditions (1), (2), and (3) of Lemma 6.7. Let N_4 be the smallest natural number such that for each $N \geq N_4$ we have

$$N\beta_w - B + \log_v(A) - \log_v(N) \ > \ 0 \ . \tag{6.49}$$

The number N_w in part (B) of Proposition 6.5A will be the least multiple of N_0, N_1, and N_2 which is greater than N_3 and N_4.

Given a positive integer N divisible by N_w, let D_N be the corresponding divisor. Since D_N is F_w-rational and principal, there is a function $f_N \in F_w(\mathcal{C}_v)$ with $\mathrm{div}(f_N) = D_N$. By the factorization property of the canonical distance, there is a constant C such that $|f_N(z)|_v = C\prod_{h=1}^N [z, \theta_h]_{\mathfrak{X},\vec{s}}$ for all $z \in \mathcal{C}_v(\mathbb{C}_v)$.

Fix a point $z_0 \in E_w \backslash \{\theta_1, \ldots, \theta_N\}$. Since f_N is F_w-rational and $z_0 \in \mathcal{C}_v(F_w)$, we have $f(z_0) \in F_w^\times$. Likewise, for each h we have $\|z_0, \theta_h\|_v \in |F_w^\times|_v$, so

$$\prod_{h=1}^N [z_0, \theta_h]_{\mathfrak{X},\vec{s}} \ = \ C_{\mathfrak{X},\vec{s}}^N \cdot \prod_{h=1}^N \|z_0, \theta_h\|_v \ \in \ |F_w^\times|_v \ .$$

Since $|f(z_0)|_v \ = \ C\prod_{h=1}^N [z_0, \theta_h]_{\mathfrak{X},\vec{s}}$ it follows that $C \in |F_w^\times|_v$. Thus, after scaling f_N by a suitable constant, we can assume that $C = 1$, and that $|f_N(z)|_v = \prod_{h=1}^N [z_0, \theta_h]_{\mathfrak{X},\vec{s}}$.

By construction, $N \cdot (V_{\mathfrak{X},\vec{s}}(\widetilde{E}_w) + \beta_w) \in \mathbb{Z}$. Let $C_N \in F_w^\times$ be such that

$$|C_N|_v \ = \ q_v^{NV_{\mathfrak{X},\vec{s}}(\widetilde{E}_w) + N\beta_w} \ , \tag{6.50}$$

and define the function f_w in part (B) of Proposition 6.5A to be $f_w = C_N \cdot f_N$. By assumption the zeros of f_w are distinct and belong to E_w. If A is the constant given Proposition 6.6(A) for H_w, then for all pairs of distinct roots $\theta_i \neq \theta_j$ we have

$$\|\theta_i, \theta_j\|_v \ = \ |\vartheta_i - \vartheta_j|_v \ > \ A/N \ .$$

Thus property (B2) in Proposition 6.5A holds.

We next show that property (B3) holds. Without loss, we can assume A is small enough that $B(\theta_h, A/N) \subseteq B(a, r)$ for each h. Thus $B(\theta_h, A/N)$ is isometrically parametrizable. Fix an F_w-rational isometric parametrization $\varphi_h : D(0, A/N) \to B(\theta_h, A/N)$ with $\varphi_h(0) = \theta_h$, and expand

$$f_w(\varphi_h(z)) \ = \ \sum_{k=1}^\infty c_{h,k} z^k \ \in \ F_w[[z]] \ .$$

The zeros of f_w are distinct, so $c_{h,1} \neq 0$. Hence if $|z|_v$ is sufficiently small, then

$$-\log_v(|f_w(\varphi_n(z))|_v) \ = \ -\log_v(|c_{h,1}|_v) - \log_v(|z|_v) \ .$$

On the other hand, by condition (3)

$$\begin{aligned} -\log_v(|f_w(\varphi_n(z))|_v) \ &\leq \ (B + NV_{\mathfrak{X},\vec{s}}(\widetilde{E}_w) - \log_v(|z|_v)) - (NV_{\mathfrak{X},\vec{s}}(\widetilde{E}_w) + N\beta_w) \\ &= \ B - N\beta_w - \log_v(|z|_v) \ , \end{aligned}$$

which means that $|c_{h,1}|_v \ \geq \ q_v^{-B+N\beta_w}$.

By condition (2) of Lemma 6.7, θ_h is the only zero of f_w in $B(\theta_h, A/N)$. Thus Proposition 3.38 shows that f_w induces an F_w-rational scaled isometry from $B(\theta_h, A/N)$ onto $D(0, |c_{h,1}|_v A/N)$. Since (6.49) holds, we have

$$|c_{h,1}|_v A/N \ \geq \ q_v^{-B+N\beta_w+\log_v(A)-\log_v(N)} \ > \ 1 \ .$$

Put $\rho_h = 1/|c_{h,1}|_v < A/N$; then f_w induces a scaled isometry from $B(\theta_h, \rho_h)$ onto $D(0,1)$, and $\rho_h \in |F_w^\times|_v$.

Again by condition (2), the balls $B(\theta_h, A/N)$ for $h = 1, \ldots, N$ are pairwise disjoint. Since $B(\theta_h, \rho_h) \subset B(\theta_h, A/N)$, the balls $B(\theta_h, \rho_h)$ are pairwise disjoint. Since f_w is a rational function of degree N, the N balls $B(\theta_h, \rho_h)$ account for all the solutions to $f(z) = x$ with $x \in D(0,1)$, and it follows that

$$f_w^{-1}(D(0,1)) \;=\; \bigcup_{h=1}^{N} B(\theta_h, \rho_h) \;\subset\; B(a,r)\ .$$

It remains to establish property (B1). Fix $z \in \mathcal{C}_v(\mathbb{C}_v)\backslash B(a,r)$. Since the canonical distance is constant on pairwise disjoint isometrically parametrizable balls, we have $[z, \theta_h]_{\mathfrak{X},\vec{s}} = [z,a]_{\mathfrak{X},\vec{s}}$ for each h, and so

$$\frac{1}{N}\log_v(|f_N(z)|_v) \;=\; \frac{1}{N}\sum_{h=1}^{N}\log_v([z,\theta_h]_{\mathfrak{X},\vec{s}}) \;=\; \log_v([z,a]_{\mathfrak{X},\vec{s}})\ .$$

However, by (6.18), $u_{\mathfrak{X},\vec{s}}(z, \widetilde{E}_w) \;=\; -\log_v([z,a]_{\mathfrak{X},\vec{s}})$. Since $f_w \;=\; C_N \cdot f_N$ and $G_{\mathfrak{X},\vec{s}}(z, \widetilde{E}_w) = V_{\mathfrak{X},\vec{s}}(\widetilde{E}_w) - u_{\mathfrak{X},\vec{s}}(z, \widetilde{E}_w)$, it follows from (6.50) that

$$\frac{1}{N}\log_v(|f_w(z)|_v) \;=\; (V_{\mathfrak{X},\vec{s}}(\widetilde{E}_w) + \beta_w) - u_{\mathfrak{X},\vec{s}}(z, \widetilde{E}_w) \;=\; G_{\mathfrak{X},\vec{s}}(z, \widetilde{E}_w) + \beta_w\ .$$

□

Proof of Proposition 6.5A when $g(\mathcal{C}_v) = 0$. In this case, the proof is relatively easy. We can take $\widetilde{E}_w = E_w = \mathcal{C}_v(F_w) \cap B(a,r)$, so (A) holds trivially.

For (B), let an F_w-symmetric probability vector $\vec{s} \in \mathcal{P}^m(\mathbb{Q})$ and a number $0 < \beta_w \in \mathbb{Q}$ be given. There is a constant $C_{\mathfrak{X},\vec{s}} \in |\mathbb{C}_v^\times|_v$ such that $[z,w]_{\mathfrak{X},\vec{s}} = C_{\mathfrak{X},\vec{s}}\|z,w\|_v$ for all $z, w \in B(a,r)$. Fix an F_w-rational isometric parametrization $\varphi : D(0,r) \to B(a,r)$ and let $H_w = F_w \cap D(0,r)$. Then $\varphi(H_w) = E_w = \widetilde{E}_w$.

Let $A > 0$ be the constant given in part (A) of Proposition 6.6 for H_w. The idea for constructing the functions f_w is to push forward the zeros of Stirling polynomials for H_w, and let them be the zeros of f_w. Let $0 < Q \in \mathbb{Z}$ be the number given for H_w for Proposition 6.6. Write $\beta_w = S/R$ with coprime integers S, R, where $R > 0$, and put $V = V_\infty(H_w)$. By Corollary A.14, $V \in \mathbb{Q}$; write $V = X/Y$ with coprime integers X, Y, where $Y > 0$. Finally, for each $i = 1, \ldots, m$ write $s_i = A_i/B_i$ with coprime integers A_i, B_i, where $B_i > 0$, and set $N_0 = QRY \cdot \mathrm{LCM}(B_1, \ldots, B_m)$.

Suppose $0 < N \in \mathbb{N}$ is a multiple of N_0. Let $S_{N,H_w}(z) = \prod_{k=1}^{N}(z - \vartheta_k)$ be the Stirling polynomial of degree N for H_w constructed in Proposition 6.6, and put $\theta_k = \varphi(\vartheta_k) \in E_w$ for $k = 1, \ldots, N$. Then

$$D_N \;:=\; \sum_{k=0}^{N-1}(\theta_k) - \sum_{i=1}^{m} N s_i(x_i)$$

is an F_w-rational divisor of degree 0. Since $\mathcal{C}_v(F_w)$ is nonempty, $\mathcal{C}_w = \mathcal{C}_v \times_{K_v} \mathrm{Spec}(F_w)$ is F_w-isomorphic to $\mathbb{P}^1/F_w$, and each divisor of degree 0 is principal. Thus condition (1) in Lemma 6.7 holds.

By construction $\theta_1, \ldots, \theta_N$ are distinct and belong to E_w. Since $\varphi : D(0,r) \to B(a,r)$ is an F_w-rational isometric parametrization, for all $i \neq j$ we have $\|\theta_i, \theta_j\|_v = |\vartheta_i - \vartheta_j|_v > A/N$. Thus condition (2) in Lemma 6.7 holds.

To show condition (3), let $Q_N(z) = \prod [z, \theta_k]_{\mathfrak{X},\vec{s}}$ be the $(\mathfrak{X}, \vec{s})$-pseudo-polynomial associated with D_N. Note that for $x, y \in D(0, r)$ we have $\|\varphi(x), \varphi(y)\|_v = |x - y|_v$, so $[\varphi(x), \varphi(y)]_{\mathfrak{X},\vec{s}} = C_{\mathfrak{X},\vec{s}}|x - y|_v$. Thus, for $x \in D(0, r)$,

$$Q_N(\varphi(x)) \;=\; \prod_{k=1}^{N} [\varphi(x), \theta_k]_{\mathfrak{X},\vec{s}} \;=\; C_{\mathfrak{X},\vec{s}}^N \prod_{k=1}^{N} |x - \vartheta_k|_v \;=\; C_{\mathfrak{X},\vec{s}}^N |S_{N,H_w}(x)|_v \;,$$

and so

$$-\log_v(Q_N(\varphi(x))) \;=\; -N \log_v(C_{\mathfrak{X},\vec{s}}) - \log_v(|S_{N,H_w}(x)|_v) \;. \tag{6.51}$$

Similarly

$$\begin{aligned} V_{\mathfrak{X},\vec{s}}(\widetilde{E}_w) \;&=\; \inf_{\text{prob meas } \nu \text{ on } \widetilde{E}_w} \iint_{\widetilde{E}_w \times \widetilde{E}_w} -\log_v([z, w]_{\mathfrak{X},\vec{s}})\, d\nu(z) d\nu(w) \\ &=\; \inf_{\text{prob meas } \nu \text{ on } H_w} \iint_{H_w \times H_w} -\log_v(C_{\mathfrak{X},\vec{s}}|x - y|_v)\, d\nu(x) d\nu(y) \\ &=\; V_\infty(H_w) - \log_v(C_{\mathfrak{X},\vec{s}}) \;. \qquad (6.52) \end{aligned}$$

Let $B = \log_v(r)$. Given $x \in D(0, r)$, let ϑ_J be the root of $S_N(z, H_w)$ for which $|x - \vartheta_J|_v$ is minimal. Using (6.51), (6.52) and Proposition 6.6(C), we obtain

$$\begin{aligned} -\log_v(Q_N(\varphi(x))) \;&=\; -N \log_v(C_{\mathfrak{X},\vec{s}}) - \log_v(|S_{N,H_w}(x)|_v) \\ &\le\; -N \log_v(C_{\mathfrak{X},\vec{s}}) + N \cdot V_\infty(H_w) - \log_v(|x - \vartheta_J|_v) + B \\ &=\; N \cdot V_{\mathfrak{X},\vec{s}}(\widetilde{E}_w) - \log_v(\|\varphi(x), \theta_J\|_v) + B \;. \end{aligned}$$

This yields condition (3) in Lemma 6.7.

Applying Lemma 6.7, we obtain part (B) of Proposition 6.5A. □

For the remainder of this section, we will assume that $g = g(\mathcal{C}_v) > 0$. To prove Theorem 6.3 when $g > 0$, we must first do some preparations. Given $\varepsilon_w > 0$, we will construct a subset $\widetilde{E}_w \subset E_w$ of the form

$$\widetilde{E}_w \;=\; E_w \backslash \Big(\bigcup_{j=1}^{g} B(\alpha_j, \rho_j) \Big) \cup \Big(\bigcup_{j=1}^{g} \mathcal{C}_v(F_w) \cap B(\alpha_j, \widetilde{\rho}_j) \Big) \;,$$

with $0 < \widetilde{\rho}_j < \rho_j$ for each j, such that $|V_{x_i}(E_w) - V_{x_i}(\widetilde{E}_w)| < \varepsilon_w$ for each $x_i \in \mathfrak{X}$. That is, we first remove finitely many discs from E_w, and replace them with the F_w-rational points in smaller discs having the same centers. The idea is that given a suitable divisor $D_N^* = \sum_{k=1}^N (\theta_k^*) - \sum_{i=1}^N N s_i(x_i)$ with $\theta_1^*, \ldots, \theta_N^* \in \widetilde{E}_w$, we will be able to create a principal divisor $D_N = \sum_{k=1}^N (\theta_k) - \sum_{i=1}^N N s_i(x_i)$ with $\theta_1, \ldots, \theta_N \in E_w$ by moving some of the θ_k^* into $E_w \backslash \widetilde{E}_w$.

The construction of $\widetilde{E}_w$ is based on two facts, proved in the Appendices. First, removing small balls from E_w does not significantly change its capacity:

PROPOSITION 6.8. *Let $E_w \subset \mathcal{C}_v(\mathbb{C}_v) \backslash \mathfrak{X}$ be as in Proposition 6.5A. Take a finite set of points $\alpha_1, \ldots, \alpha_t \in E_w$. Then for each $\varepsilon_w > 0$, there is an $R_1 > 0$ such that for any compact set $\widetilde{E}_w$ satisfying*

$$E_w \backslash \Big(\bigcup_{j=1}^{t} B(\alpha_j, R_1) \Big) \;\subseteq\; \widetilde{E}_w \;\subseteq\; E_w \;,$$

we have $|V_{x_i}(E_w) - V_{x_i}(\widetilde{E}_w)| < \varepsilon_w$ for each $x_i \in \mathfrak{X}$.

PROOF. This is a special case of Proposition A.16 of Appendix A. $\square$

Second, for each generic, sufficiently small polyball $\prod_{j=1}^{g} B(\alpha_j, \rho) \subset \mathcal{C}_v(\mathbb{C}_v)^g$, there is an action of a neighborhood of the origin in $\mathrm{Jac}(\mathcal{C}_v)(\mathbb{C}_v)$ which makes the polyball into a principal homogeneous space. This action will enable us to move the points θ_k^* and obtain the principal divisor D_N.

Let $J_{\mathrm{Ner}}(\mathcal{C}_v)/\operatorname{Spec}(\mathcal{O}_v)$ be the Néron model of $\mathrm{Jac}(\mathcal{C}_v)$. By ([**12**], Theorem 1, p.153), $J_{\mathrm{Ner}}(\mathcal{C}_v)$ is quasi-projective. We regard it as embedded in $\mathbb{P}^N/\operatorname{Spec}(\mathcal{O}_v)$, for some N, and identify $\mathrm{Jac}(\mathcal{C}_v)$ with generic fibre of $J_{\mathrm{Ner}}(\mathcal{C}_v)$ in $\mathbb{P}^N/\operatorname{Spec}(K_v)$.

Let $\|x, y\|_{J,v}$ be the corresponding spherical metric on $\mathrm{Jac}(\mathcal{C}_v)(\mathbb{C}_v)$, and let O be the origin in $\mathrm{Jac}(\mathcal{C}_v)(\mathbb{C}_v)$. Then the unit ball $B_J(O,1)^- := \{z \in \mathrm{Jac}(\mathcal{C}_v)(\mathbb{C}_v) : \|z, O\|_{J,v} < 1\}$ is a subgroup. Since O is nonsingular on the special fibre of the Néron model, there is a K_v-rational isometric parametrization

$$\Psi : D(\vec{0},1)^- \ \to\ B_J(O,1)^- \tag{6.53}$$

by power series converging on $D(\vec{0},1)^- \subset \mathbb{C}_v^g$, taking $\vec{0}$ to O (Theorem 3.9). Pulling the group action back to $D(\vec{0},1)^-$ using Ψ yields the formal group of $\mathrm{Jac}(\mathcal{C}_v)$.

Put $\overline{\mathcal{C}}_v = \mathcal{C}_v \times_{K_v} \operatorname{Spec}(\mathbb{C}_v)$, and let $\mathbf{Pic}_{\overline{\mathcal{C}}_v/\mathbb{C}_v}$ be its Picard scheme. Given a divisor D on $\mathcal{C}_v(\mathbb{C}_v)$, let $[D]$ be its class in $\mathbf{Pic}_{\overline{\mathcal{C}}_v/\mathbb{C}_v}(\mathbb{C}_v)$. The identity component $\mathbf{Pic}^0_{\overline{\mathcal{C}}_v/\mathbb{C}_v}$ is canonically isomorphic to $\mathrm{Jac}(\mathcal{C}_v) \times_{K_v} \operatorname{Spec}(\mathbb{C}_v)$, and we identify $\mathbf{Pic}^0_{\overline{\mathcal{C}}_v/\mathbb{C}_v}(\mathbb{C}_v)$ with $\mathrm{Jac}(\mathcal{C}_v)(\mathbb{C}_v)$. Given $\alpha \in \mathcal{C}_v(\mathbb{C}_v)$, the Abel map $\mathbf{j}_\alpha : \mathcal{C}_v(\mathbb{C}_v) \to \mathrm{Jac}(\mathcal{C}_v)(\mathbb{C}_v)$ is defined by

$$\mathbf{j}_\alpha(x) \ =\ [(x) - (\alpha)]\ .$$

The Abel map is continuous for the v-topology, and if $\alpha \in \mathcal{C}_v(F_w)$ for some finite extension F_w/K_v, it is F_w-rational. Given $\vec{\alpha} = (\alpha_1, \ldots, \alpha_g) \in \mathcal{C}_v(\mathbb{C}_v)^g$, let $\mathbf{J}_{\vec{\alpha}} : \mathcal{C}_v(\mathbb{C}_v)^g \to \mathrm{Jac}(\mathcal{C}_v)(\mathbb{C}_v)$ be the map

$$\mathbf{J}_{\vec{\alpha}}(\vec{x}) \ =\ [\sum_{j=1}^{g}(x_j) - \sum_{j=1}^{g}(\alpha_j)] \ =\ \sum_{j=1}^{g} \mathbf{j}_{\alpha_j}(x_j)\ .$$

If $\alpha_1, \ldots, \alpha_g \in \mathcal{C}_v(F_w)$, then $\mathbf{J}_{\vec{\alpha}}$ is F_w-rational. It is shown in Appendix D that $\mathbf{J}_{\vec{\alpha}}$ is nonsingular at $\vec{\alpha}$ for a dense set of $\vec{\alpha}$, and if $\mathbf{J}_{\vec{\alpha}}$ is nonsingular at $\vec{\alpha}$, then for each sufficiently small $\rho > 0$, the image of $\prod_{j=1}^{g} B(\alpha_j, \rho)$ under $\mathbf{J}_{\vec{\alpha}}$ is an open subgroup $W_{\vec{\alpha}}(\rho)$ of $\mathrm{Jac}(\mathcal{C}_v)(\mathbb{C}_v)$. Furthermore, if $\dot{+}$ is addition in $\mathbf{Pic}_{\overline{\mathcal{C}}_v/\mathbb{C}_v}(\mathbb{C}_v)$, then $\prod_{j=1}^{g} B(\alpha_j, \rho)$ is a principal homogeneous space for $W_{\vec{\alpha}}(\rho)$ under the action

$$w \ddot{+} \vec{x} \ =\ \mathbf{J}_{\vec{\alpha}}^{-1}(w \dot{+} \mathbf{J}_{\vec{\alpha}}(\vec{x}))\ ,$$

and if we write $[\vec{x}]$ for $[(x_1) + \cdots + (x_g)]$, then

$$[w \ddot{+} \vec{x}] \ =\ w \dot{+} [\vec{x}]\ .$$

Below are the properties of the action we will need; for a more general statement, see Theorem D.2 of Appendix D.

THEOREM 6.9. *Let K_v be a nonarchimedean local field, and let $\mathcal{C}_v/K_v$ be a smooth, projective, geometrically integral curve of genus $g > 0$. Then the points $\vec{\alpha} = (\vec{\alpha}_1, \ldots, \vec{\alpha}_g) \in \mathcal{C}_v(\mathbb{C}_v)^g$ such that $\mathbf{J}_{\vec{\alpha}} : \mathcal{C}_v(\mathbb{C}_v)^g \to \mathrm{Jac}(\mathcal{C}_v)(\mathbb{C}_v)$ is nonsingular at $\vec{\alpha}$ are dense in $\mathcal{C}_v(\mathbb{C}_v)^g$ for the v-topology. If F_w/K_v is a finite extension in $\mathbb{C}_v$ and $\mathcal{C}_v(F_w)$ is nonempty, they are dense in $\mathcal{C}_v(F_w)^g$.*

Fix such an $\vec{\alpha}$*; then* $\alpha_1, \ldots, \alpha_g$ *are distinct, and for each* $0 < \eta < 1$*, there is a number* $0 < R_2 < 1$ *(depending on* $\vec{\alpha}$ *and* η*) such that* $B(\alpha_1, R_2), \ldots, B(\alpha_g, R_2)$ *are pairwise disjoint and isometrically parametrizable, the map* $\mathbf{J}_{\vec{\alpha}} : \mathcal{C}_v(\mathbb{C}_v)^g \to \mathrm{Jac}(\mathcal{C}_v)(\mathbb{C}_v)$ *is injective on* $\prod_{j=1}^g B(\alpha_j, R_2)$*, and for each* $0 < \rho \le R_2$ *the following properties hold:*

(A) (Subgroup) *The set* $W_{\vec{\alpha}}(\rho) := \mathbf{J}_{\vec{\alpha}}\big(\prod_{j=1}^g B(\alpha_j, \rho)\big)$ *is an open subgroup of* $\mathrm{Jac}(\mathcal{C}_v)(\mathbb{C}_v)$.

(B) (Limited Distortion) *For each* $j = 1, \ldots g$*, let* $\varphi_j : D(0, \rho) \to B(\alpha_j, \rho)$ *be an isometric parametrization such that* $\varphi_j(0) = \alpha_j$*; consider the associated map* $\Phi_{\vec{\alpha}} = (\varphi_1, \ldots, \varphi_g) : D(0, \rho)^g \to \prod_{j=1}^g B(\alpha_j, \rho)$*. Let* $\Psi : D(\vec{0}, 1)^- \to B_J(O, 1)^-$ *be the isometric parametrization inducing the formal group, and let* $L_{\vec{\alpha}} : \mathbb{C}_v^g \to \mathbb{C}_v^g$ *be the linear map* $(\Psi^{-1} \circ \mathbf{J}_{\vec{\alpha}} \circ \Phi_{\vec{\alpha}})'(\vec{0})$.

Then $W_{\vec{\alpha}}(\rho) = \Psi(L_{\vec{\alpha}}(D(0, \rho)^g))$*. If we give* $D(0, \rho)^g$ *the structure of an additive subgroup of* $\mathbb{C}_v^g$*, the map* $\Psi \circ L_{\vec{\alpha}}$ *induces an isomorphism of groups*

$$D(0, \rho)^g / D(0, \eta\rho)^g \;\cong\; W_{\vec{\alpha}}(\rho) / W_{\vec{\alpha}}(\eta\rho) \tag{6.54}$$

with the property that for each $\vec{x} \in D(0, \rho)^g$*,*

$$\mathbf{J}_{\vec{\alpha}}(\Phi_{\vec{\alpha}}(\vec{x})) \;\equiv\; \Psi(L_{\vec{\alpha}}(\vec{x})) \quad (\mathrm{mod}\ W_{\vec{\alpha}}(\eta\rho)) \ . \tag{6.55}$$

(C) (Action) *There is an action* $(\omega, \vec{x}) \mapsto \omega \ddot{+} \vec{x}$ *of* $W_{\vec{\alpha}}(\rho)$ *on* $\prod_{j=1}^g B(\alpha_j, \rho)$ *which makes* $\prod_{j=1}^g B(\alpha_j, \rho)$ *into a principal homogeneous space for* $W_{\vec{\alpha}}(\rho)$*. If we restrict the domain of* $\mathbf{J}_{\vec{\alpha}}$ *to* $\prod_{j=1}^g B(\alpha_j, \rho)$*, then* $\omega \ddot{+} \vec{x} = \mathbf{J}_{\vec{\alpha}}^{-1}(\omega \dot{+} \mathbf{J}_{\vec{\alpha}}(\vec{x}))$*. For each* $\omega \in W_{\vec{\alpha}}(\rho)$ *and* $\vec{x} \in \prod_{j=1}^g B(\alpha_j, \rho)$*,*

$$[\omega \ddot{+} \vec{x}] \;=\; \omega \dot{+} [\vec{x}] \ . \tag{6.56}$$

(D) (Uniformity) *For each* $\vec{\beta} \in \prod_{i=1}^g B(\alpha_i, \rho)$*,*

$$W_{\vec{\alpha}}(\eta\rho) \ddot{+} \vec{\beta}' \;=\; \prod_{j=1}^g B(\beta_j, \eta\rho) \quad \text{and} \quad \mathbf{J}_{\vec{\beta}}\Big(\prod_{j=1}^g B(\beta_j, \eta\rho)\Big) \;=\; W_{\vec{\alpha}}(\eta\rho) \ . \tag{6.57}$$

(E) (Rationality) *If* F_w/K_v *is a finite extension in* $\mathbb{C}_v$*, and* $\vec{\alpha} \in \mathcal{C}_v(F_w)^g$*, then*

$$\mathbf{J}_{\vec{\alpha}}\Big(\prod_{j=1}^g (B(\alpha_j, \rho) \cap \mathcal{C}_v(F_w))\Big) \;=\; W_{\vec{\alpha}}(\rho) \cap \mathrm{Jac}(\mathcal{C}_v)(F_w) \ , \tag{6.58}$$

$$(W_{\vec{\alpha}}(\rho) \cap \mathrm{Jac}(\mathcal{C}_v)(F_w)) \ddot{+} \vec{\alpha} \;=\; \prod_{j=1}^g \big(B(\alpha_j, \rho) \cap \mathcal{C}_v(F_w)\big) \ . \tag{6.59}$$

(F) (Trace) *If* F_w/K_v *is finite and separable, there is a constant* $C = C(F_w, \vec{\alpha}) > 0$*, depending on* F_w *and* $\vec{\alpha}$ *but not on* ρ*, such that*

$$B_J(O, C\rho) \cap \mathrm{Jac}(\mathcal{C}_v)(K_v) \;\subseteq\; \mathrm{Tr}_{F_w/K_v}\big(W_{\vec{\alpha}}(\rho) \cap \mathrm{Jac}(\mathcal{C}_v)(F_w)\big) \ . \tag{6.60}$$

PROOF. This is a specialization of Theorem D.2 of Appendix D. In particular, R_2 is the number R from Theorem D.2. We restrict to the case where the vectors $\vec{r} = (r_1, \ldots, r_g)$ in that theorem have equal coordinates, so the condition $\eta \cdot \max(r_i) \le \min(r_i)$ is automatic. For $\vec{r} = (\rho, \ldots, \rho) \in \mathbb{R}^g$, the subgroups denoted $W_{\vec{\alpha}}(\vec{r})$ in Appendix D are written $W_{\vec{\alpha}}(\rho)$ here. The balls denoted $B_{\mathcal{C}_v}(a_j, r)$ in Appendix D are the balls $B(\alpha_j, \rho) \subset \mathcal{C}_v(\mathbb{C}_v)$ here. □

PROOF OF PROPOSITION 6.5A WHEN $g = g(\mathcal{C}_v) > 0$. Let F_w be a finite, separable extension of K_v in $\mathbb{C}_v$, and let $E_w = \mathcal{C}_v(F_w) \cap B(a,r)$, where $a \in \mathcal{C}_v(F_w)$ and $B(a,r) \subset \mathcal{C}_v(\mathbb{C}_v) \backslash \mathfrak{X}$ is isometrically parametrizable.

We first construct the set $\widetilde{E}_w$.

Let $\varepsilon_w > 0$ be given. Since E_w is open in $\mathcal{C}_v(F_w)$, by Theorem 6.9 we can choose $\alpha_1, \ldots, \alpha_g \in E_w$ such that the map $\mathbf{J}_{\vec{\alpha}} : \mathcal{C}_v(\mathbb{C}_v)^g \to \mathrm{Jac}(\mathcal{C}_v)(\mathbb{C})$ is injective at $\vec{\alpha} = (\alpha_1, \cdots, \alpha_g)$. Let R_1 be the number given by Proposition 6.8 for $\vec{\alpha}$ and ε_w. Fix a uniformizer π_w for the maximal ideal of $\mathcal{O}_w$. If the residue characteristic of F_w is odd, take $\eta = |\pi_w|_v$. If the residue characteristic of F_w is 2, take $\eta = |\pi_w^2|_v$. Let R_2 be the number given by Theorem 6.9 for $\vec{\alpha}$ and η. Without loss we can assume that R_1 and R_2 belong to $|F_w^\times|_v$. Set $\rho = \min(R_1, R_2, r)$, and put

$$\widetilde{E}_w \;=\; \Big(E_w \backslash \bigcup_{j=1}^{g} B(\alpha_j, \rho)\Big) \cup \Big(\bigcup_{j=1}^{g} (\mathcal{C}_v(F_w) \cap B(\alpha_j, \eta\rho))\Big) \;. \tag{6.61}$$

By Theorem 6.9(A), $W_{\vec{\alpha}}(\eta\rho)$ is an open subgroup of $\mathrm{Jac}(\mathcal{C}_v)(\mathbb{C}_v)$. For each $\tau \in \widetilde{E}_w$ the continuity of the Abel map $j_\tau : \mathcal{C}_v(\mathbb{C}_v) \to \mathrm{Jac}(\mathcal{C}_v)(\mathbb{C}_v)$ shows there is a radius $r_\tau > 0$ such that $\mathbf{j}_\tau(B(\tau, r_\tau)) \subseteq W_{\vec{\alpha}}(\eta\rho)$. Without loss we can assume that $r_\tau \le \eta\rho$ and that $r_\tau \in |F_w^\times|_v$. If $\tau \in B(\alpha_j, \eta\rho)$ for some j, then Theorem 6.9(D) shows that $\mathbf{j}_\tau(B(\tau, \eta\rho)) \subseteq W_{\vec{\alpha}}(\eta\rho)$, because if $\vec{\beta} = (\alpha_1, \ldots, \alpha_{j-1}, \tau, \alpha_{j+1}, \ldots, \alpha_g)$ and $\vec{x} = (\alpha_1, \ldots, \alpha_{j-1}, x, \alpha_{j+1}, \ldots, \alpha_g)$ then $\mathbf{j}_\tau(x) = \mathbf{J}_{\vec{\beta}}(\vec{x})$.

As $E_w \backslash \bigcup_{j=1}^g B(\alpha_j, \rho)$ is compact, there are $\alpha_{g+1}, \ldots, \alpha_d \in E_w \backslash \bigcup_{j=1}^g B(\alpha_j, \rho)$ such that the balls $B(\alpha_k, r_{\alpha_k})$ for $k = g+1, \ldots, d$ cover $E_w \backslash \bigcup_{j=1}^g B(\alpha_j, \rho)$. For notational simplicity, write r_k for r_{α_k}. Since any two balls are either pairwise disjoint or one is contained in the other, we can assume that $B(\alpha_{g+1}, r_{g+1}), \ldots, B(\alpha_d, r_d)$ are pairwise disjoint. They are also disjoint from $B(\alpha_1, \rho), \ldots, B(\alpha_g, \rho)$. For $k = 1, \ldots, g$, put $r_k = \eta\rho$. Then $B(\alpha_1, r_1), \ldots, B(\alpha_d, r_d)$ are pairwise disjoint and cover $\widetilde{E}_w$, and

$$\widetilde{E}_w \;=\; \bigcup_{j=1}^{d} \big(\mathcal{C}_v(F_w) \cap B(\alpha_j, r_j)\big) \tag{6.62}$$

is an F_w-simple decomposition of $\widetilde{E}_w$.

Since $\rho \le R_1$ and (6.61) holds, Proposition 6.8 shows $|V_{x_i}(E_w) - V_{x_i}(\widetilde{E}_w)| < \varepsilon_w$ for each $x_i \in \mathfrak{X}$. Thus $\widetilde{E}_w$ satisfies part (A) of Proposition 6.5A.

We next apply Lemma 6.7 to show that part (B) of Proposition 6.5A holds.

Fix an F_w-rational isometric parametrization $\varphi : D(0,r) \to B(a,r)$ with $\varphi(0) = a$. Let $\widetilde{\alpha}_1, \ldots, \widetilde{\alpha}_d \in F_w \cap D(0,r)$ be the points with $\varphi(\widetilde{\alpha}_j) = \alpha_j$, and put

$$\widetilde{H}_w \;=\; \bigcup_{j=1}^{d} \big(F_w \cap D(\widetilde{\alpha}_j, r_j)\big) \;.$$

Then $\varphi(\widetilde{H}_w) = \widetilde{E}_w$.

Let an F_w-symmetric probability vector $\vec{s} \in \mathcal{P}^m(\mathbb{Q})$ and a number $0 < \beta_w \in \mathbb{Q}$ be given. To construct the $(\mathfrak{X}, \vec{s})$-functions f_w in Proposition 6.5A, we begin with Stirling polynomials for $\widetilde{H}_w$, as in the proof when $g(\mathcal{C}_v) = 0$, but we then modify them.

Let J_w be the group $\mathrm{Jac}(\mathcal{C}_v)(F_w)$, and let

$$J_w(\rho) = W_{\vec{a}}(\rho) \cap \mathrm{Jac}(\mathcal{C}_v)(F_w) , \qquad J_w(\eta\rho) = W_{\vec{a}}(\eta\rho) \cap \mathrm{Jac}(\mathcal{C}_v)(F_w) .$$

By Theorem 6.9(A), $J_w(\eta\rho)$ is open in J_w. Since J_w is compact, the quotient group $J_w/J_w(\eta\rho)$ is finite. Let I be its order. Let $0 < Q \in \mathbb{Z}$ be the number given for $\widetilde{H}_w$ for Proposition 6.6, related to the construction of Stirling polynomials; in particular, if w_j is the weight of $\mathcal{C}_v(F_w) \cap B(\alpha_j, r_j)$ for the equilibrium distribution $\mu_{\mathfrak{X},\vec{s}}$ of $\widetilde{H}_w$, then $0 < Qw_j \in \mathbb{Z}$ for each j. Write $\beta_w = S/R$ with coprime integers S, R, where $R > 0$, and put $V = V_\infty(H_w)$. By Corollary A.14, $V \in \mathbb{Q}$; write $V = X/Y$ with coprime integers X, Y, where $Y > 0$. Finally, for each $i = 1, \ldots, m$ write $s_i = A_i/B_i$ with coprime integers A_i, B_i, where $B_i > 0$, and set $N_0 = 4IQRY \cdot \mathrm{LCM}(B_1, \ldots, B_m)$.

Suppose $0 < N \in \mathbb{N}$ is a multiple of N_0. Let $S_{N,\widetilde{H}_w}(z) = \prod_{k=1}^N (z - \vartheta_k)$ be the Stirling polynomial of degree N for $\widetilde{H}_w$ given by Proposition 6.6, and put $\theta_k^* = \varphi(\vartheta_k) \in \widetilde{E}_w$ for $k = 1, \ldots, N$. Then

$$D_N^* := \sum_{k=0}^{N-1} (\theta_k^*) - \sum_{i=1}^{m} N s_i(x_i)$$

is an F_w-rational $(\mathfrak{X}, \vec{s})$-divisor of degree 0 on $\mathcal{C}_v$. Its positive part is supported on $\widetilde{E}_w$ and its polar part is supported on $\mathfrak{X}$.

Condition (1) of Lemma 6.7 may fail for D_N^*, since it need not be principal, but we claim that conditions (2) and (3) hold for it.

For condition (2), note that by Proposition 6.6(A) applied to $\widetilde{H}_w$, there is a number $\widetilde{A} > 0$ such that $|\vartheta_i - \vartheta_j|_v > \widetilde{A}/N$ for all N and all $i \neq j$. Since $\varphi : D(0,r) \to B(a,r)$ is an F_w-rational isometric parametrization, we have $\|\theta_i^*, \theta_j^*\|_v = |\vartheta_i - \vartheta_j|_v > \widetilde{A}/N$ for all $i \neq j$.

For condition (3), let $Q_N^*(z) = \prod [z, \theta_k^*]_{\mathfrak{X},\vec{s}}$ be the $(\mathfrak{X}, \vec{s})$-pseudo-polynomial associated with D_N^*. By Proposition 3.11(B2), there is a number $C_{\mathfrak{X},\vec{s}} \in |\mathbb{C}_v^\times|_v$ such that $[z, w]_{\mathfrak{X},\vec{s}} = C_{\mathfrak{X},\vec{x}} \|z, w\|_v$ for all $z, w \in B(a, r)$. For $x, y \in D(0, r)$ we have $\|\varphi(x), \varphi(y)\|_v = |x - y|_v$, so $[\varphi(x), \varphi(y)]_{\mathfrak{X},\vec{s}} = C_{\mathfrak{X},\vec{s}}|x - y|_v$. Thus, for $x \in D(0, r)$,

$$Q_N^*(\varphi(x)) = \prod_{k=1}^{N} [\varphi(x), \theta_k^*]_{\mathfrak{X},\vec{s}} = C_{\mathfrak{X},\vec{s}}^N \prod_{k=1}^{N} |x - \vartheta_k|_v = C_{\mathfrak{X},\vec{s}}^N |S_{N,\widetilde{H}_w}(x)|_v ,$$

and so

$$-\log_v(Q_N^*(\varphi(x))) = -N \log_v(C_{\mathfrak{X},\vec{s}}) - \log_v(|S_{N,\widetilde{H}_w}(x)|_v) . \tag{6.63}$$

As in (6.52),

$$V_{\mathfrak{X},\vec{s}}(\widetilde{E}_w) = V_\infty(\widetilde{H}_w) - \log_v(C_{\mathfrak{X},\vec{s}}) . \tag{6.64}$$

Put $\widetilde{B} = \max_{1 \le j \le d}(\log_v(r_j))$. Given $z \in B(a, r)$, let $x \in D(0, r)$ be such that $\varphi(x) = z$, and let ϑ_J be the root of $S_N(z, \widetilde{H}_w)$ for which $|x - \vartheta_J|_v$ is minimal. Using (6.63), (6.64) and Proposition 6.6(C), we obtain

$$\begin{aligned} -\log_v(Q_N^*(z)) &= -N \log_v(C_{\mathfrak{X},\vec{s}}) - \log_v(|S_{N,\widetilde{H}_w}(x))|_v \\ &\le -N \log_v(C_{\mathfrak{X},\vec{s}}) + N \cdot V_\infty(\widetilde{H}_w) - \log_v(|x - \vartheta_J|_v) + \widetilde{B} \\ &= N \cdot V_{\mathfrak{X},\vec{s}}(\widetilde{E}_w) - \log_v(\|z, \theta_J^*\|_v) + \widetilde{B} . \end{aligned} \tag{6.65}$$

Thus condition (3) holds.

We will now modify D_N^* to obtain a principal divisor D_N which satisfies all the conditions of Lemma 6.7.

Put $\delta = [D_N^*]$. We claim that $\delta \in J_w(\eta\rho)$. To see this, for each $k = 1, \ldots, N$ let $j(k)$ denote the index $1 \le j \le d$ for which $\theta_k^* \in B(\alpha_j, r_j)$. For each $j = 1, \ldots, d$ put $P_j = N \cdot w_j$ and for each $i = 1, \ldots, m$ put $Q_i = N \cdot s_i$. The numbers P_j and Q_i belong to $\mathbb{N}$ since $N_0 | N$, and we can write

$$\delta \;=\; [D_N^*] \;=\; \sum_{k=1}^{N}[(\theta_k^*) - (\alpha_{j(k)})] + \sum_{j=1}^{d} P_j[(\alpha_j) - (a)] - \sum_{i=1}^{m} Q_i[(x_i) - (a)] \ .$$

By the construction of the balls $B(\alpha_j, r_j)$, for each k we have $[(\theta_k^*) - (\alpha_{j(k)})] \in J_w(\eta\rho)$. Since α_j and a belong to $\mathcal{C}_v(F_w)$, for each j we have $[(\alpha_j) - (a)] \in J_w$. By our choice of N_0, each P_j is divisible by I, so for each j we have $P_j[(\alpha_j) - (a)] \in J_w(\eta\rho)$. Similarly, each Q_i is divisible by I, and $\vec{s}$ and $\mathfrak{X}$ are F_w-symmetric, so $\sum_{i=1}^m Q_i[(x_i) - (a)] \in J_w(\eta\rho)$. Thus $\delta \in J_w(\eta\rho)$.

Let ℓ be such that $|\pi_w^\ell|_v = \rho$. If the residue characteristic of F_w is odd, then by parts (B) and (E) of Theorem 6.9, together with our choice of η, the group $J_w(\rho)/J_w(\eta\rho)$ is isomorphic to $(\pi_w^\ell \mathcal{O}_w / \pi_w^{\ell+1}\mathcal{O}_w)^g$. If the residue characteristic of F_w is 2, then $J_w(\rho)/J_w(\eta\rho)$ is isomorphic to $(\pi_w^\ell \mathcal{O}_w / \pi_w^{\ell+2}\mathcal{O}_w)^g$.

Let $\Psi : D(\vec{0}, 0)^- \to B_J(O,1)^-$ be the F_w-rational isometric parametrization in Theorem 6.9(B). Using the F_w-rational isometric parametrization $\varphi : D(0,r) \to B(a,r)$, we get F_w-rational isometric parametrizations $\varphi_j : D(0,\rho) \to B(\alpha_j, \rho)$, for $j = 1, \ldots, g$, by setting $\varphi_j(z) = \varphi(\widetilde{\alpha}_j + z)$ where $\varphi(\widetilde{\alpha}_j) = \alpha_j$. Define $\Phi : D(0,\rho)^g \to \prod_{j=1}^g B(\alpha_j, \rho)$ by $\Phi(\vec{z}) = (\varphi_1(z_1), \ldots, \varphi_g(z_g))$. Then $\Psi((\pi_w^\ell \mathcal{O}_w)^g) = J_w(\rho)$, and $\varphi_j(\pi_w^\ell \mathcal{O}_w) = \mathcal{C}_v(F_w) \cap B(\alpha_j, \rho)$ for each j. Let $L_{\vec{\alpha}} : \mathbb{C}_v^g \to \mathbb{C}_v^g$ be the F_w-rational linear map defined by $L_{\vec{\alpha}} = (\Psi^{-1} \circ \mathbf{J}_{\vec{\alpha}} \circ \Phi)'(\vec{0})$. Let $\ddot{+}$ be the action of $W_{\vec{\alpha}}(\rho)$ on $\prod_{j=1}^g B(\alpha_j, \rho)$ from Theorem 6.9. By parts (C) and (E) of Theorem 6.9, for each $\vec{z} \in (\pi_w^\ell \mathcal{O}_w)^g$ we have $\Psi(L_{\vec{\alpha}}(\vec{z})), \mathbf{J}_{\vec{\alpha}}(\Phi(\vec{z})) \in J_w(\rho)$, with $\Psi(L_{\vec{\alpha}}(\vec{z})) \equiv \mathbf{J}_{\vec{\alpha}}(\Phi(\vec{z}))$ (mod $J_w(\eta\rho)$).

First suppose the residue characteristic of F_w is odd. By our choice of N_0, at least two points θ_k^* belong to $B(\alpha_j, r_j)$ for each j. Recall that $r_j = \eta\rho$ for $j = 1, \ldots, g$. Without loss, we can assume that $\theta_1^*, \ldots, \theta_N^*$ are labeled in such a way that θ_j^* and θ_{g+j}^* belong to $B(\alpha_j, \eta\rho)$, for $j = 1, \ldots, g$.

Fix an element $t \in \pi_w^\ell \mathcal{O}_w$ whose image in $\pi_w^\ell \mathcal{O}_w / \pi_w^{\ell+1} \mathcal{O}_w$ is nonzero. Put $\vec{t} = (t, \cdots, t) \in (\pi_w^\ell \mathcal{O}_w)^g$ and set $\Delta = \Psi(L_{\vec{\alpha}}(\vec{t}))$, then define $\theta_1, \ldots, \theta_{2g} \in \mathcal{C}_v(F_w)$ by

$$(\theta_1, \ldots, \theta_g) \;=\; \Delta \ddot{+} (\theta_1^*, \ldots, \theta_g^*) \ , \qquad (\theta_{g+1}, \ldots, \theta_{2g}) \;=\; (-(\delta \dot{+} \Delta)) \ddot{+} (\theta_1^*, \ldots, \theta_g^*) \ .$$

Put $\theta_k = \theta_k^*$ for $k = 2g+1, \ldots, N$, and set

$$D_N \;=\; \sum_{k=0}^{N-1} (\theta_k) - \sum_{i=1}^{m} N s_i(x_i) \ .$$

The divisor D_N is principal, since by (6.56) we have $[(\theta_1) + \cdots + (\theta_g)] = \Delta \dot{+} [(\theta_1^*) + \cdots + (\theta_g^*)]$ and $[(\theta_{g+1}) + \cdots + (\theta_{2g})] = (-(\delta \dot{+} \Delta)) \dot{+} [(\theta_{g+1}^*) + \cdots + (\theta_{2g}^*)]$, which gives

$$\begin{aligned} [D_N] \;&=\; [D_N^*] \dot{+} \Big([\sum_{k=1}^{g} (\theta_k)] \dot{-} [\sum_{k=1}^{g} (\theta_k^*)] \Big) \dot{+} \Big([\sum_{k=g+1}^{2g} (\theta_k)] \dot{-} [\sum_{k=g+1}^{2g} (\theta_k^*)] \Big) \\ &=\; \delta \dot{+} \Delta \dot{+} (-(\delta \dot{+} \Delta)) \;=\; 0 \ . \qquad (6.66) \end{aligned}$$

We will now show that $\theta_1, \ldots, \theta_{2g}$ belong to $E_w \backslash \widetilde{E}_w$, and that they are well-separated from each other and $\theta_{2g+1}, \ldots, \theta_N$.

Put $\vec{\beta} = (\beta_1, \ldots, \beta_g) = (\varphi_1(t), \ldots, \varphi_g(t))$. Then $\beta_j \in \mathcal{C}_v(F_w) \cap B(\alpha_j, \rho)$ for each j, and by Theorem 6.9(B)

$$\mathbf{J}_{\vec{\alpha}}(\vec{\beta}) \;=\; \mathbf{J}_{\vec{\alpha}}(\Phi(\vec{t})) \;\equiv\; \Psi(L_{\vec{\alpha}}(\vec{t})) \;=\; \Delta \quad (\text{mod } W_{\vec{\alpha}}(\eta\rho)) \ .$$

This means there is a $\delta' \in W_{\vec{\alpha}}(\eta\rho)$ such that $\mathbf{J}_{\vec{\alpha}}(\vec{\beta}) = \delta' \dot{+} \Delta$. It follows that $\delta' \ddot{+} (\Delta \ddot{+} \vec{\alpha}) = (\delta' \dot{+} \Delta) \ddot{+} \vec{\alpha} = \mathbf{J}_{\vec{\alpha}}(\vec{\beta}) \ddot{+} \vec{\alpha} = \vec{\beta}$, so $\Delta \ddot{+} \vec{\alpha} = (-\delta') \ddot{+} \vec{\beta}$. Likewise, put $\delta^* = \mathbf{J}_{\vec{\alpha}}((\theta_1^*, \ldots, \theta_g^*)) \in W_{\vec{\alpha}}(\eta\rho)$, so $(\theta_1^*, \ldots, \theta_g^*) = \delta^* \ddot{+} \vec{\alpha}$. By properties of the action $\ddot{+}$ and Theorem 6.9(D)

$$\begin{aligned}(\theta_1, \ldots, \theta_g) \;&=\; \Delta \ddot{+} (\theta_1^*, \ldots, \theta_g^*) \;=\; \Delta \ddot{+} (\delta^* \ddot{+} \vec{\alpha}) \;=\; \delta^* \ddot{+} (\Delta \ddot{+} \vec{\alpha}) \;=\; \delta^* \ddot{+} ((-\delta') \ddot{+} \vec{\beta}) \\ &=\; (\delta^* \dot{-} \delta') \ddot{+} \vec{\beta} \;\in\; \prod_{j=1}^{g} \left(\mathcal{C}_v(F_w) \cap B(\beta_j, \eta\rho)\right) \ .\end{aligned}$$

Similarly, put $\vec{\gamma} = (\gamma_1, \ldots, \gamma_g) = (\varphi_1(-t), \ldots, \varphi_g(-t))$; then $\gamma_j \in \mathcal{C}_v(F_w) \cap B(\alpha_j, \rho)$ for each j. Let $\delta^{**} = \mathbf{J}_{\vec{\alpha}}(\theta_{g+1}^*, \ldots, \theta_{2g}^*) \in W_{\vec{\alpha}}(\eta\rho)$. By computations like those above one sees that there is a $\delta'' \in W_{\vec{\alpha}}(\eta\rho)$ such that $\mathbf{J}_{\vec{\alpha}}(\vec{\gamma}) = \delta'' \dot{+} (-\Delta)$, and that

$$(\theta_{g+1}, \ldots, \theta_{2g}) \;=\; (\delta^{**} \dot{-} \delta'' \dot{-} \delta) \ddot{+} \vec{\gamma} \;\in\; \prod_{j=1}^{g} \left(\mathcal{C}_v(F_w) \cap B(\gamma_j, \eta\rho)\right) \ .$$

For each j the map $\varphi_j : D(0, \rho) \to B(\alpha_j, \rho)$ is an isometric parametrization, so our choice of t means that $\|\beta_j, \alpha_j\|_v = |t|_v = |\pi_w^\ell|_v = \rho$ and $\|\gamma_j, \alpha_j\|_v = |-t|_v = \rho$. Moreover $\|\beta_j, \gamma_j\|_v = |t - (-t)|_v = |2t|_v = \rho$ since the residue characteristic is odd. Thus the balls $B(\alpha_j, \eta\rho)$, $B(\beta_j, \eta\rho)$ and $B(\gamma_j, \eta\rho)$ are pairwise disjoint and contained in $B(\alpha_j, \rho)$. Since $B(\alpha_1, \rho), \ldots, B(\alpha_g, \rho)$ are pairwise disjoint, all the balls

$$B(\alpha_1, \eta\rho), \ldots, B(\alpha_g, \eta\rho),\; B(\beta_1, \eta\rho), \ldots, B(\beta_g, \eta\rho),\; B(\gamma_1, \eta\rho), \ldots, B(\gamma_g, \eta\rho)$$

are pairwise disjoint. Since $\widetilde{E}_w = \left(E_w \backslash \bigcup_{j=1}^g B(\alpha_j, \rho)\right) \cup \left(\bigcup_{j=1}^g (\mathcal{C}_v(F_w) B(\alpha_j, \eta\rho)\right)$, it follows that $\theta_1, \ldots, \theta_{2g} \in E_w \backslash \widetilde{E}_w$. In addition, for each $j = 1, \ldots, 2g$, if $1 \le k \le N$ and $k \neq j$ then

$$\|\theta_j, \theta_k\|_v \;\ge\; \rho \ . \tag{6.67}$$

Next suppose the residue characteristic of F_w is 2. Since $\pi_w^\ell \mathcal{O}_w / \pi_w^{\ell+2} \mathcal{O}_w$ is an abelian 2-group with at least 4 elements, it either has a subgroup isomorphic to $\mathbb{Z}/4\mathbb{Z}$ or a subgroup isomorphic to $(\mathbb{Z}/2\mathbb{Z})^2$. If it has a subgroup isomorphic to $\mathbb{Z}/4\mathbb{Z}$, choose an element $t \in \pi_w^\ell \mathcal{O}_w$ whose image in $\pi_w^\ell \mathcal{O}_w / \pi_w^{\ell+2} \mathcal{O}_w$ has order 4. Then the same construction as in the case of odd residue characteristic applies, but at the very end, in place of (6.67) one gets that for $j = 1, \ldots, 2g$ and $1 \le k \le N$ with $k \neq j$,

$$\|\theta_j, \theta_k\|_v \;\ge\; |\pi_w|_v \cdot \rho \ . \tag{6.68}$$

If the residue characteristic is 2 and $\pi_w^\ell \mathcal{O}_w / \pi_w^{\ell+2} \mathcal{O}_w$ has no elements of order 4, we modify the construction as follows. Let $t_1, t_2 \in \pi_w^\ell \mathcal{O}_w$ be elements whose images in $\pi_w^\ell \mathcal{O}_w / \pi_w^{\ell+2} \mathcal{O}_w$ generate a subgroup isomorphic to $(\mathbb{Z}/2\mathbb{Z})^2$. By our choice of N_0, at least three of the points $\theta_1^*, \ldots, \theta_N^*$ belong to $\mathcal{C}_v(F_w) \cap B(\alpha_j, r_j)$ for each j. Without loss, we can assume that $\theta_1^*, \ldots, \theta_N^*$ are indexed in such a way that

θ_j^*, θ_{g+j}^* and θ_{2g+j}^* belong to $B(\alpha_j, \eta\rho)$ for $j = 1, \ldots, g$. Put $\vec{t_1} = (t_1, \cdots, t_1)$ and $\vec{t_2} = (t_2, \cdots, t_2)$, then set $\Delta_1 = \Psi(L_{\vec{\alpha}}(\vec{t_1}))$, $\Delta_2 = \Psi(L_{\vec{\alpha}}(\vec{t_2}))$. Recall that $\delta = [D_N^*] \in J_w(\eta\rho)$. Define $\theta_1, \ldots, \theta_{3g} \in \mathcal{C}_v(F_w)$ by

$$(\theta_1, \ldots, \theta_g) = \Delta_1 \ddot{+} (\theta_1^*, \ldots, \theta_g^*) , \quad (\theta_{g+1}, \ldots, \theta_{2g}) = \Delta_2 \ddot{+} (\theta_{g+1}^*, \ldots, \theta_{2g}^*) ,$$
$$(\theta_{2g+1}, \ldots, \theta_{3g}) = (-(\delta \dot{+} \Delta_1 \dot{+} \Delta_2)) \ddot{+} (\theta_{2g+1}^*, \ldots, \theta_{3g}^*) ,$$

and put $\theta_k = \theta_k^*$ for $k = 3g+1, \ldots, N$. If we take

$$D_N = \sum_{k=0}^{N-1} (\theta_k) - \sum_{i=1}^{m} N s_i(x_i) ,$$

an argument similar to the one before shows that D_N is principal, that $\theta_1, \ldots, \theta_{3g}$ belong to $E_w \backslash \widetilde{E}_w$, and that for each $j = 1, \ldots, 3g$, if $1 \le k \le N$ and $k \ne j$ then

$$(6.69) \qquad \|\theta_j, \theta_k\|_v \ \ge \ |\pi_w|_v \cdot \rho .$$

Finally, we show that for sufficiently large N divisible by N_0, the divisor D_N satisfies conditions (1), (2), and (3) of Lemma 6.7.

We have already seen that D_N is principal, so condition (1) holds.

For condition (2), let $A = \widetilde{A}$ be the constant from Proposition 6.6(A) for the set $\widetilde{H}_w$. As shown above, for all N divisible by N_0 and all $j \ne k$ with $1 \le j, k \le N$, we have $\|\theta_j^*, \theta_k^*\|_v \ge \widetilde{A}/N$. Suppose in addition that N is large enough that $A/N < |\pi_w|_v \cdot \rho$. For $1 \le j \le 2g$ (resp. $3g$ in the third case), and all $k \ne j$ we have $\|\theta_j, \theta_k\|_v \ge |\pi_w|_v \cdot \rho > A/N$ by (6.67), (6.68) and (6.69). By symmetry, this also holds for $1 \le k \le 2g$ (resp. $3g$) and all $j \ne k$. For $j, k > 2g$ (resp. $3g$ in the third case) with $j \ne k$ we have $\|\theta_j, \theta_k\|_v > A/N$ since $\theta_j = \theta_j^*$, $\theta_k = \theta_k^*$. Thus condition (2) holds.

For condition (3), consider the pseudopolynomials $Q_N(z) = \prod_{j=1}^N [z, \theta_j]_{\mathfrak{X}, \vec{s}}$ and $Q_N^*(z) = \prod_{j=1}^N [z, \theta_j^*]_{\mathfrak{X}, \vec{s}}$. We must show there is a constant B such that for each $z \in B(a, r)$, if $J = J(z)$ is an index for which $\|z, \theta_J\|_v = \min_{1 \le j \le N}(\|z, \theta_j\|_v)$, then

$$(6.70) \qquad -\log_v(|Q_N(z)|_v) \ \le \ N \cdot V_{\mathfrak{X}, \vec{s}}(\widetilde{E}_w) + B + \big(-\log_v(\|z, \theta_J\|_v)\big) .$$

By (6.65), there is a constant $\widetilde{B}$ such that for each $z \in B(a, r)$, if $K = K(z)$ is an index for which $\|z, \theta_K^*\|_v = \min_{1 \le j \le N}(\|z, \theta_j^*\|_v)$, then

$$-\log_v(Q_N^*(z)) \ \le \ N \cdot V_{\mathfrak{X}, \vec{s}}(\widetilde{E}_w) + \widetilde{B} + \big(-\log_v(\|z, \theta_K^*\|_v)\big) .$$

Throughout the discussion below, J and K will have this meaning.

Let G be the number of roots of $Q_N^*(z)$ that were moved, i.e. $G = 2g$ or $G = 3g$. Then

$$Q_N(z) = \prod_{j=1}^{G} \frac{[z, \theta_j]_{\mathfrak{X}, \vec{s}}}{[z, \theta_j^*]_{\mathfrak{X}, \vec{s}}} \cdot Q_N^*(z) .$$

(If $z = \theta_h^*$ for some $h = 1, \ldots, G$, we regard the right side as defined by its limit as $z \to \theta_h^*$.)

For all $z, w \in B(a, r)$ we have $[z, w]_{\mathfrak{X}, \vec{s}} = C_{\mathfrak{X}, \vec{s}} \|z, w\|_v$. Hence if we set

$$(6.71) \quad D(z) = \sum_{j=1}^{G} (-\log_v(\|z, \theta_j\|_v) + \sum_{j=1}^{G} \log_v(\|z, \theta_j^*\|_v) + (-\log_v(\|z, \theta_K^*\|_v)) ,$$

to prove (6.70) it suffices to show there is a constant $\widehat{B}$ such that for all $z \in B(a,r)$,

$$D(z) \;\le\; \widehat{B} + \big(-\log_v(\|z,\theta_J\|_v)\big)\ . \tag{6.72}$$

(If $z = \theta_h^*$ for some $h = 1,\ldots,G$, then $\theta_K^* = \theta_h^*$, and we define $D(z)$ by the sum obtained by omitting the corresponding terms from the right side of (6.71); if $z = \theta_j$ for some j, we regard both sides of (6.71) and (6.72) as being ∞.)

We will prove (6.72) by considering cases. By (6.67), (6.68), (6.69), the balls

$$B(\theta_1,\eta\rho)\ ,\ \ldots\ ,\ B(\theta_G,\eta\rho)\ ,\ B(\alpha_1,r_1)\ ,\ldots\ ,\ B(\alpha_d,r_d)$$

are disjoint. Put $\widehat{r} = \min_{1\le j\le d}(r_j)$ and take $\widehat{B} = G\cdot(-\log_v(\eta\rho)) + (-\log_v(\widehat{r}))$.

First suppose that $z \in B(\theta_h,\eta\rho)$ for some h, $1 \le h \le G$. Then $\theta_J = \theta_h$, and $\|z,\theta_j\|_v > \eta\rho$ for all $j = 1,\ldots,G$ with $j \ne h$. Furthermore, $\|z,\theta_j^*\|_v \le r < 1$ for $j = 1,\ldots,G$ and $\|z,\theta_K^*\|_v > \eta\rho$. Hence

$$\begin{aligned} D(z) &\le (G-1)\cdot(-\log_v(\eta\rho)) + (-\log_v(\|z,\theta_J\|_v) + (-\log_v(\eta\rho)) \\ &\le \widehat{B} + (-\log_v(\|z,\theta_J\|_v). \end{aligned}$$

Next suppose that $z \in B(\alpha_h,r_h)$ for some h, $1 \le h \le d$. In this case $\|z,\theta_j\|_v > \eta\rho$ for $j = 1,\ldots,G$. By our choice of N_0, at least four of the θ_k^* belong to $B(\alpha_h,r_h)$, and at least one remains after $\theta_1^*,\ldots,\theta_G^*$ are moved. This means that $\theta_J = \theta_\ell^*$ for some ℓ with $\theta_\ell^* \in B(\alpha_h,r_h)$, and that $\theta_K^* \in B(\alpha_h,r_h)$. By the definition of K, we have $\|z,\theta_J\|_v = \|z,\theta_\ell^*\|_v \ge \|z,\theta_K^*\|_v$.

If $\|z,\theta_J\|_v > \|z,\theta_K^*\|_v$, then θ_K^* was one of the roots moved out of $B(\alpha_h,r_h)$, and in particular $1 \le K \le G$. Thus there is a term $\log_v(\|z,\theta_j^*\|_v)$ in second sum in (6.71) which cancels the term $-\log_v(\|z,\theta_K^*\|_v)$. (If $z = \theta_K^*$, then $-\log_v(\|z,\theta_K^*\|_v) = \infty$ but we have defined $D(z)$ by omitting these two terms from (6.71).) This gives

$$D(z) \;\le\; G\cdot(-\log_v(\eta\rho)) \;\le\; \widehat{B} + (-\log_v(\|z,\theta_J\|))\ .$$

On the other hand, if $\|z,\theta_J\|_v = \|z,\theta_K^*\|_v$, once more

$$D(z) \;\le\; G\cdot(-\log_v(\eta\rho)) + (-\log_v(\|z,\theta_K^*\|) \;\le\; \widehat{B} + (-\log_v(\|z,\theta_J\|))\ .$$

Finally, take $z \in B(a,r)\backslash\big(\bigcup_{j=1}^G B(\theta_j,\eta\rho)\cup\bigcup_{j=1}^d B(\alpha_j,r_j)\big)$. Clearly $\|z,\theta_j\|_v > \eta\rho$ for $j = 1,\ldots,G$, and $\|z,\theta_K^*\| > \widehat{r}$. Since $\|z,\theta_J\|_v \le r < 1$, again we have

$$D(z) \;\le\; G\cdot(-\log_v(\eta\rho)) + (-\log_v(\widehat{r})) \;\le\; \widehat{B} + (-\log_v(\|z,\theta_J\|))\ .$$

This establishes (6.72), and completes the proof of condition (3) of Lemma 6.7. Applying Lemma 6.7, we obtain Proposition 6.5A. □

We have now completed the proof of Theorem 6.3.

5. Corollaries to the Proof of Theorem 6.3

In this section we note two consequences of the proof of Theorem 6.3 which will be used in §11.4, the local patching construction for K_v-simple sets.

DEFINITION 6.10. Let $E_v \subset \mathcal{C}_v(\mathbb{C}_v)$ be a K_v-simple set with a K_v-simple decomposition $E_v = \bigcup_{\ell=1}^D \big(B(a_\ell,r_\ell)\cap\mathcal{C}_v(F_{w_\ell})\big)$. Let $H_v \subset E_v$ be a K_v-simple set compatible with E_v, with a K_v-simple decomposition $H_v = \bigcup_{h=1}^N \big(B(\theta_h,\rho_h)\cap\mathcal{C}_v(F_{u_h})\big)$ compatible with the decomposition $E_v = \bigcup_{\ell=1}^D \big(B(a_\ell,r_\ell)\cap\mathcal{C}_v(F_{w_\ell})\big)$. We will say the decomposition $H_v = \bigcup_{h=1}^N \big(B(\theta_h,\rho_h)\cap\mathcal{C}_v(F_{u_h})\big)$ is *move-prepared* relative to $B(a_1,r_1),\ldots,B(a_D,r_D)$ if

(*A*) $g(\mathcal{C}) = 0$, or

(*B*) $g = g(\mathcal{C}) > 0$, and for each $\ell = 1, \ldots, D$ there are g indices $h_{\ell 1}, \ldots, h_{\ell g}$ such that

(1) $B(\theta_{h_{\ell 1}}, \rho_{h_{\ell 1}}), \ldots, B(\theta_{h_{\ell g}}, \rho_{h_{\ell g}}) \subset B(a_\ell, r_\ell)$;

(2) there is a number $\overline{r}_\ell$ such that $\rho_{h_{\ell 1}}, \ldots, \rho_{h_{\ell g}} < \overline{r}_\ell < r_\ell$ and $B(\theta_{h_{\ell 1}}, \overline{r}_\ell), \ldots, B(\theta_{h_{\ell g}}, \overline{r}_\ell)$ are disjoint and contained in $B(a_\ell, r_\ell)$;

(3) taking $\vec{\theta}_\ell = (\theta_{h_{\ell 1}}, \ldots, \theta_{h_{\ell g}})$, the Abel map $J_{\vec{\theta}_\ell} : \mathcal{C}_v(\mathbb{C}_v)^g \to \mathrm{Jac}(\mathcal{C}_v)(\mathbb{C}_v)$ is injective on $\prod_{j=1}^g B(\theta_{h_{\ell j}}, \overline{r}_\ell)$, and $W_{\vec{\theta}_\ell}(\overline{r}_\ell) := \mathbf{J}_{\vec{\theta}_\ell}\big(\prod_{j=1}^g B(\theta_{h_{\ell j}}, \overline{r}_\ell)\big)$ is an open subgroup of $\mathrm{Jac}(\mathcal{C}_v)(\mathbb{C}_v)$ with the properties in Theorem 6.9.

We will call $B(\theta_{h_{\ell 1}}, \rho_{h_{\ell 1}}), \ldots, B(\theta_{h_{\ell g}}, \rho_{h_{\ell g}})$ *distinguished balls* corresponding to $B(a_\ell, r_\ell)$ in the decomposition of H_v.

COROLLARY 6.11. *In Theorem 6.3, by choosing N_v sufficiently large, we can arrange that the K_v-simple decomposition $H_v = \bigcup_{h=1}^N B(\theta_h, \rho_h) \cap \mathcal{C}_v(F_{u_h})$ of $H_v := f_v^{-1}(D(0,1)) \cap E_v$ is move-prepared relative to $B(a_1, r_1), \ldots, B(a_D, r_D)$.*

PROOF. When $g(\mathcal{C}_v) = 0$ there is nothing to show. When $g = g(\mathcal{C}_v) > 0$, the corollary follows by tracing through the proof of Theorem 6.3. We note the key points in the argument, below.

The proof begins by using the decomposition $E_v = \bigcup_{\ell=1}\big(B(a_\ell, r_\ell) \cap \mathcal{C}_v(F_{w_\ell})\big)$ to reduce to a single set of the form $E_{w,\ell} = B(a_\ell, r_\ell) \cap \mathcal{C}_v(F_{w_\ell})$; the function f_v is a scaled product of conjugates of functions $f_{w,\ell}$ for representatives of Galois orbits of the balls $B(a_\ell, r_\ell)$ (see (6.30)). Since the Abel map is Galois-equivariant, if the conditions in Definition 6.10 are satisfied for some $B(a_\ell, r_\ell)$, they also hold for its conjugates.

Proposition 6.5A constructs $f_{w,\ell}$ for a single set $E_{w,\ell} = B(a_\ell, r_\ell) \cap \mathcal{C}_v(F_{w_\ell})$. (For notational simplicity, the index ℓ is suppressed in its proof.) The first step in the proof (see (6.61) and (6.62)) is to construct a subset

$$\widetilde{E}_{w,\ell} = \bigcup_{j=1}^{d_\ell} \big(\mathcal{C}_v(F_{w_\ell}) \cap B(\alpha_{\ell j}, r_{\ell j})\big) \subset E_{w,\ell}$$

in which $r_{\ell 1} = \cdots = r_{\ell g} = \eta\rho < r_\ell$, such that $W_{\vec{\alpha}}(\eta\rho) := \mathbf{J}_{\vec{\alpha}}\big(\prod_{j=1}^g B(\alpha_{\ell j}, \eta\rho)\big)$ is an open subgroup of $\mathrm{Jac}(\mathbb{C}_v)$ satisfying the conditions of Theorem 6.9. Put $\overline{r}_\ell = \eta\rho$. In the construction of $f_{w,\ell}$, zeros are initially assigned to the cosets $\mathcal{C}_v(F_{w_\ell}) \cap B(\alpha_{\ell j}, r_{\ell j})$ in proportion to their weights under the $(\mathfrak{X}, \vec{s})$-equilibrium distribution of $\widetilde{E}_{w,\ell}$, giving a nonprincipal divisor D_N^*; then at most $3g$ zeros are moved to obtain a principal divisor D_N, which becomes the divisor of $f_{w,\ell}$. The corollary follows by noting that if N_v (hence N) is sufficiently large, then $f_{w,\ell}$ has zeros in $\mathcal{C}_v(F_{w,\ell}) \cap B(\alpha_{\ell j}, \overline{r}_\ell)$ for each $j = 1, \ldots, g$, and the distance between these zeros is at most $\overline{r}_\ell$ so the corresponding balls $B(\theta_h, \rho_h)$ in the decomposition of H_v have radii less than $\overline{r}_\ell$. If we let $h_{\ell 1}, \ldots, h_{\ell g}$ be indices of zeros with $\theta_{h_{\ell j}} \in B(\alpha_{\ell j}, r_\ell)$, then by Theorem 6.9(D),

$$W_{\vec{\theta}_\ell}(\overline{r}_\ell) = \mathbf{J}_{\vec{\theta}_\ell}\Big(\prod_{j=1}^g B(\theta_{h_{\ell j}}, \overline{r}_\ell)\Big) = \mathbf{J}_{\vec{\alpha}}\Big(\prod_{j=1}^g B(\alpha_{\ell j}, \eta\rho)\Big) . \tag{6.73}$$

When $\mathrm{char}(K_v) = p > 0$, the final step in the construction of f_v replaces the function $f_v(z)$ described above with $\pi_v^{-N_v B/2} S_{p^B, \mathcal{O}_v}(f(z))$ (see (6.31), in order to assure that the leading coefficient $c_{v,i}$ belongs to $K_v(x_i)^{\mathrm{sep}}$ for each i. Since $x = 0$

is a root of the Stirling polynomial $S_{p^B,\mathcal{O}_v}(x)$, the roots of the original $f_v(z)$ remain roots of the new one, and (6.73) still holds.

Thus $H_v = f_v^{-1}(D(0,1)) \cap E_v$ is move-prepared relative to the collection of balls $B(a_1,r_1),\ldots,B(a_D,r_D)$. $\square$

A second consequence of the proof of Theorem 6.3 is

COROLLARY 6.12. *In Theorem 6.3, by choosing N_v sufficiently large and appropriately divisible, we can arrange that $E_v = \bigcup_{\ell=1}^D \big(B(a_\ell,r_\ell)\cap \mathcal{C}_v(F_{w_\ell})\big)$ and $H_v = \bigcup_{h=1}^N \big(B(\theta_h,\rho_h)\cap \mathcal{C}_v(F_{u_h})\big)$ have the property that there is a point $\overline{w}_\ell \in \big(B(a_\ell,r_\ell)\cap \mathcal{C}_v(F_{w_\ell})\big)\backslash H_v$ for each $\ell = 1,\ldots,D$.*

PROOF. This can be seen by tracing through the proof of Theorem 6.3. However, a simple modification at the end of the proof gives the claim directly.

Applying Theorem 6.3 in its stated form, given $\varepsilon > 0$ and K_v-simple decomposition $E_v = \bigcup_{\ell=1}^D \big(B(a_\ell,r_\ell)\cap \mathcal{C}_v(F_{w_\ell})\big)$, there is a K_v-simple set $\widetilde{E}_v \subset E_v$ such that for each $0 < \beta_v \in \mathbb{Q}$ and each K_v-symmetric probability vector $\vec{s} \in \mathcal{P}^m(\mathbb{Q})$, assertions (A) and (B) in the theorem hold.

Fixing $0 < \beta_v < \mathbb{Q}$ and $\vec{s}$, write $\beta_v = \beta_v' + \beta_v''$ with $0 < \beta_v',\beta_v'' \in \mathbb{Q}$. Applying the theorem with β_v replaced by β_v', there is an integer $N_v' \geq 1$ such that for each positive integer N divisible by N_v', there is an $(\mathfrak{X},\vec{s})$-function $f_v' \in K_v(\mathcal{C}_v)$ of degree N such that assertion (B) holds for β_v'; in particular, for each $x_i \in \mathfrak{X}$,

$$\Lambda_{x_i}(f_v',\vec{s}) \;=\; \Lambda_{x_i}(\widetilde{E}_v,\vec{s}) + \beta_v' \ .$$

Furthermore, the zeros $\theta_1',\ldots,\theta_N'$ of f_v' are distinct and belong to E_v, the preimage $(f_v')^{-1}(D(0,1))$ decomposes as $\bigcup_{h=1}^N B(\theta_h',\rho_h')$ where the balls $B(\theta_h',\rho_h')$ are pairwise disjoint and contained in $\bigcup_{\ell=1}^D B(a_\ell,r_\ell)$, and $H_v' := (f_v')^{-1}(D(0,1))\cap E_v$ has a K_v-simple decomposition $H_v' = \bigcup_{h=1}^N \big(B(\theta_h',\rho_h')\cap \mathcal{C}_v(F_{u_h'})\big)$ compatible with the decomposition $E_v = \bigcup_{\ell=1}^D \big(B(a_\ell,r_\ell)\cap \mathcal{C}_v(F_{w_\ell})\big)$. For each h, ρ_h' belongs to $|F_{u_h'}^\times|_v$ and f_v' induces an $F_{u_h'}$-rational scaled isometry from $B(\theta_h',\rho_h')$ onto $D(0,1)$.

Fix a positive integer N_v'' such that $N_v''\cdot\beta_v'' \in \mathbb{Z}$, and put $N_v = N_v' N_v''$. Given an integer N divisible by N_v, let $f_v'(z)$ be as above, and put

$$f_v(z) \;=\; \pi_v^{-N\beta_v''}\cdot f_v'(z)$$

where π_v is a uniformizer for the maximal ideal of $\mathcal{O}_v$. Clearly $\Lambda_{x_i}(f_v',\vec{s}) = \Lambda_{x_i}(\widetilde{E}_v,\vec{s}) + \beta_v$ for each $x_i \in \mathfrak{X}$. The zeros $\theta_1,\ldots,\theta_N$ of $f_v(z)$ are the same as the zeros $\theta_1',\ldots,\theta_N'$ of f_v', so they are distinct and belong to E_v.

For each h, put $\rho_h = |\pi_v^{N\beta_v''}|_v\cdot\rho_h' < \rho_h'$ and put $F_{u_h} = F_{u_h'}$. Since $\pi_v^{-N\beta_v''} \in K_v^\times$ and $\rho_h' \in |F_{u_h'}^\times|_v$, we have $\rho_h \in |F_{u_h}^\times|_v$ and $B(\theta_h,\rho_h) \subsetneq B(\theta_h',\rho_h')$. Since $f_v' : B(\theta_h',\rho_h') \to D(0,1)$ is a scaled isometry, it follows that

$$f_v^{-1}(D(0,1)) \;=\; \bigcup_{h=1}^N B(\theta_h,\rho_h) \;\subset\; \bigcup_{h=1}^N B(\theta_h',\rho_h') \;=\; (f_v')^{-1}(D(0,1))\ ,$$

and $H_v := f_v^{-1}(D(0,1))\cap E_v$ has the K_v-simple decomposition

$$H_v \;=\; \bigcup_{h=1}^N \big(B(\theta_h,\rho_h)\cap \mathcal{C}_v(F_{u_h})\big)$$

compatible with the K_v-simple decomposition $E_v = \bigcup_{\ell=1}^{D} \big(B(a_\ell, r_\ell) \cap \mathcal{C}_v(F_{w_\ell})\big)$.

Now fix ℓ, and take any $1 \le h \le N$ with $B(\theta_h, \rho_h) \subset B(a_\ell, r_\ell)$. Then $F_{u_h} = F_{u'_h} = F_{w_\ell}$, $\theta_h = \theta'_h \in B(a_\ell, r_\ell) \cap \mathcal{C}_v(F_{w_\ell})$, $B(\theta_h, \rho_h) \subsetneq B(\theta_h, \rho'_h) \subset B(a_\ell, r_\ell)$, and $\rho'_h \in |F_{w_\ell}^\times|_v$. Since $f'_v : B(\theta_h, \rho'_h) \to D(0,1)$ is an F_{w_ℓ}-rational scaled isometry, there infinitely many points $\overline{w} \in \mathcal{C}_v(F_{w_\ell})$ with $\|\theta_h, \overline{w}\|_v = \rho'_h$. Such points belong to $B(\theta_h, \rho'_h) \cap \mathcal{C}_v(F_{w_\ell})$ but not $B(\theta_h, \rho_h)$, so they are not in H_v.

Thus for each ℓ there is a point $\overline{w}_\ell \in \big(B(a_\ell, r_\ell) \cap \mathcal{C}_v(F_{w_\ell})\big) \backslash H_v$. □

CHAPTER 7

The Global Patching Construction

In this section we will give the global patching construction for the proof of Theorem 4.2. This argument manages the patching process in such a way that the final patched function is K-rational. Part of the argument specifies the order in which the coefficients are patched, and the way the target coefficients are chosen. In expansions of functions, all coefficients are with respect to the L-rational basis, constructed in §3.3.

The inputs to the global patching argument are the construction of the initial approximating functions, carried out in Chapters 5 and 6 above, and the local patching constructions given in Chapters 8–11 below. The global and local patching constructions are largely independent; we have chosen to present the global argument first in order to provide the reader with an overview of the proof. Nonetheless, in order to understand some aspects of the global patching construction (in particular the need for patching the coefficients in bands and the reason for using different patching coefficient bounds for high, middle, and low-order coefficients), it is necessary to be acquainted with the local constructions. Therefore we encourage the reader to examine the local patching constructions in parallel with the global one. The patching argument for the nonarchimedean RL-domain case when $\mathrm{char}(K) = 0$, given in Chapter 10, is the easiest and will shed light on the issues above.

For the convenience of the reader, we restate Theorem 4.2. The notion of a K_v-simple set is given in Definition 4.1.

Theorem 4.2 (FSZ with Local Rationality, for K_v-simple sets). *Let K be a global field, and let $\mathcal{C}/K$ be a smooth, geometrically integral, projective curve. Let $\mathfrak{X} = \{x_1, \ldots, x_m\} \subset \mathcal{C}(\widetilde{K})$ be a finite set of points stable under* $\mathrm{Aut}(\widetilde{K}/K)$, *and let $\mathbb{E} = \prod_v E_v \subset \prod_v \mathcal{C}_v(\mathbb{C}_v)$ be a K-rational adelic set compatible with $\mathfrak{X}$. Let $S \subset \mathcal{M}_K$ be a finite set of places containing all archimedean v and all nonarchimedean v such that E_v is not $\mathfrak{X}$-trivial.*

Assume that:

(A) for each $v \in S$, E_v is K_v-simple;

(B) $\gamma(\mathbb{E}, \mathfrak{X}) > 1$.

Then there are infinitely many points $\alpha \in \mathcal{C}(\widetilde{K}^{\mathrm{sep}})$ such that for each $v \in \mathcal{M}_K$, the $\mathrm{Aut}(\widetilde{K}/K)$*-conjugates of α all belong to E_v.*

In Chapter 4 we have reduced Theorems 0.4, 1.2, 1.3, 1.4, 1.5, 1.6, and 1.7 to Theorem 4.2. In Chapters 5 and 6 we have constructed the initial approximating functions.

In this chapter we prove Theorem 4.2, assuming the local patching constructions given in Chapters 8–11. In §7.1–§7.3 below we discuss some preliminaries: an adelic version of the Strong Approximation theorem, the existence of a dense set of subunits, and the semi-local theory. In §7.4 we prove Theorem 4.2 when $\mathrm{char}(K) = 0$. First, we specify the patching parameters, then we construct the functions used to initiate the patching process, and finally we carry out the global patching construction. In §7.5 we prove Theorem 4.2 when $\mathrm{char}(K) = p > 0$.

We use the conventions concerning notation and absolute values from §3.1. We assume familiarity with the theory of Green's functions from §3.8, §3.9 (or from [**51**], §3.2, §4.4), and with Green's matrices and the Cantor capacity from §3.10 (or from [**51**],§5.3).

The Green's matrix and the Cantor capacity only depend on values of the Green's functions outside E_v. For a compact set E_v, if $x \notin E_v$, the upper Green's function $\overline{G}(z,x;E_v)$ coincides with the Green's function $G(z,x;E_v)$ from §3.8 for all z, and it coincides with the (lower) Green's function $G(z,x;E_v)$ studied in ([**51**]) for all $z \notin E_v$. Likewise, for an $\mathfrak{X}$-trivial set, or more generally for any algebraically capacitable set, the upper Green's function $\overline{G}(z,x;E_v)$ coincides with the lower Green's function studied in ([**51**]) for all $z \notin E_v$ (see [**51**], Theorem 4.4.4). Hence for sets $\mathbb{E} = \prod_v E_v$ meeting the conditions of Theorem 4.2, the upper Green's matrix $\overline{\Gamma}(\mathbb{E},\mathfrak{X})$ and inner Cantor capacity $\overline{\gamma}(\mathbb{E},\mathfrak{X})$ from §4.10 coincide with the Green's matrix $\Gamma(\mathbb{E},\mathfrak{X})$ and Cantor capacity $\gamma(\mathbb{E},\mathfrak{X})$ from ([**51**], §5.3).

For this reason, for the remainder of this chapter, we will drop the "bar" from $\overline{G}(z,x;E_v)$, $\overline{\Gamma}(\mathbb{E},\mathfrak{X})$, and $\overline{\gamma}(\mathbb{E},\mathfrak{X})$ and simply write $G(z,x;E_v)$, $\Gamma(\mathbb{E},\mathfrak{X})$ and $\gamma(\mathbb{E},\mathfrak{X})$.

Let $S_K = S$ be the set of places of K in Theorem 4.2, and let $\widehat{S}_K = \widehat{S}$ be the set of all places of K where any of the following conditions hold:

(7.1)

(1) $v \in S_K$; in particular if
   - v is archimedean; or
   - $\mathcal{C}$ has bad reduction at v (with respect to the model $\mathfrak{C}$ determined by the given projective embedding of $\mathcal{C}$); or
   - the points in $\mathfrak{X}$ do not specialize to distinct points (mod v); or
   - E_v is not $\mathfrak{X}$-trivial.

(2) $\mathcal{C}$ has good reduction at v, and one or more of the uniformizing parameters $g_{x_i}(z)$ fails to specialize to a well-defined, nonconstant function on the fiber $\mathfrak{C} \pmod v = \mathfrak{C} \times_{\mathcal{O}_K} k_v$ (in the classical terminology, "has bad reduction at v").

(3) $\mathcal{C}$ has good reduction at v, and one or more of the basis functions $\varphi_{ij}(z)$ and φ_λ from Section 3.3 has bad reduction at v.

Note that although there are infinitely many basis functions, only finitely many places are affected by condition (3), because the basis functions belong to a multiplicatively finitely generated set. Thus $\widehat{S}_K$ is finite.

Put $L = K(\mathfrak{X})$ and $\widehat{S}_L$ be the set of places of L above $\widehat{S}_K$. Write $\mathbb{E}_K = \mathbb{E}$ and let $\mathbb{E}_L = \prod_{w \in \mathcal{M}_L} E_w$ be the set obtained from $\mathbb{E}_K$ by base change, identifying $\mathbb{C}_w$ with $\mathbb{C}_v$ and putting $E_w = E_v$ if $w|v$. For each $w \notin \widehat{S}_L$, E_w is $\mathfrak{X}$-trivial and the $g_{x_i}(z)$ have good reduction, so the local Green's matrix $\Gamma(E_w,\mathfrak{X})$ is the zero matrix. Thus

$$\Gamma(\mathbb{E}_K,\mathfrak{X}) = \frac{1}{[L:K]}\Gamma(\mathbb{E}_L,\mathfrak{X}) = \frac{1}{[L:K]} \sum_{w \in \widehat{S}_L} \Gamma(E_w,\mathfrak{X}) \log(q_w) . \tag{7.2}$$

In Theorem 4.2, our hypothesis that $\gamma(\mathbb{E},\mathfrak{X}) > 1$ is equivalent to $\Gamma(\mathbb{E},\mathfrak{X})$ being negative definite. Let $\{\widetilde{E}_v\}_{v \in \widehat{S}_K}$ be another collection of sets for which

$\widetilde{\mathbb{E}} := \prod_{v \in \widehat{S}_K} \widetilde{E}_v \times \prod_{v \notin \widehat{S}_K} E_v$ is K-rational and compatible with $\mathfrak{X}$. By continuity, there are numbers $\varepsilon_v > 0$ for $v \in \widehat{S}_K$ such that $\Gamma(\widetilde{\mathbb{E}}, \mathfrak{X})$ is also negative definite, provided that for each $v \in \widehat{S}_K$,

$$(7.3) \qquad \begin{cases} |G(x_j, x_i; \widetilde{E}_v) - G(x_j, x_i; E_v)| \ < \ \varepsilon_v & \text{for all } i \neq j \ , \\ |V_{x_i}(\widetilde{E}_v) - V_{x_i}(E_v)| \ < \ \varepsilon_v & \text{for all } i \ . \end{cases}$$

1. The Uniform Strong Approximation Theorem

Let F be a global field. Write $\mathcal{M}_F$ for the set of all places of F, and let $|x|_u$ be the absolute value associated to $u \in \mathcal{M}_F$, normalized as in §3.1. Let $\mathbb{A}_F$ and $\mathbb{J}_F$ denote the adele ring and idele group of F, respectively, and write $b = (b_u)_{u \in \mathcal{M}_F}$ for an element of $\mathbb{A}_F$ or $\mathbb{J}_F$. Let

$$\|b\|_F \ = \ \prod_{u \in \mathcal{M}_F} |b_u|_u^{D_u}$$

denote the "size" of b. Recall that $D_u = 2$ if u is archimedean and $K_u \cong \mathbb{C}$, and $D_u = 1$ otherwise.

The following version of the Strong Approximation theorem is well known, but there seems to be no convenient reference for it.

LEMMA 7.1. *Let F be a global field. There is a constant B_F depending only on F such that for each $b \in \mathbb{J}_F$ with $\|b\|_F \geq B_F$, given any $c \in \mathbb{A}_F$, there is an $f \in F$ such that*

$$|f - c_u|_u \ \leq |b_u|_u \quad \textit{for all } u \in \mathcal{M}_F .$$

PROOF. By the Lemma on ([**18**], p.66), there is a constant $A(F) > 0$ such that for any $a \in \mathbb{J}_F$ with $\|a\|_F > A(F)$, there is a $\beta \in F^\times$ satisfying $|\beta|_u \leq |a_u|_u$ for all $u \in \mathcal{M}_F$.

Given any idele $y = (y_u)_{u \in \mathcal{M}_F} \in \mathbb{J}_F$, put

$$V(y) \ = \ \{x = (x_u)_{u \in \mathcal{M}_F} \in \mathbb{A}_F : |x|_u \leq |y_u|_u\} \ .$$

By Corollary 1 of ([**18**], p.65), there is a $d \in \mathbb{J}_F$ such that $V(d)$ contains a fundamental domain for $\mathbb{A}_F/F$: that is, $\mathbb{A}_F = V(d) + F$, where we view F as embedded on the diagonal in $\mathbb{A}_F$. Let $D(F) = \|d\|_F$.

Take $B_F = A(F) \cdot D(F)$. Suppose $b \in \mathbb{J}_F$ is an idele with $\|b\|_F \geq B_F$. Put $a = b \cdot d^{-1}$; then $\|a\|_F \geq A(F)$. Let $\beta \in F^\times$ be such that $|\beta|_u \leq |a_u|_u$ for each $u \in \mathcal{M}_F$; then $|\beta d_u|_u \leq |b_u|_v$ for each u. Since $V(d)$ contains a fundamental domain for $\mathbb{A}_F$, so does $V(\beta d)$; indeed,

$$\begin{aligned} \mathbb{A}_F \ &= \ \beta \cdot \mathbb{A}_F \ = \ \beta \cdot (V(d) + F) \\ &= \ \beta \cdot V(d) + \beta \cdot F \ = \ V(\beta d) + F \ . \end{aligned}$$

Since $V(\beta d) \subseteq V(b)$, it follows that $\mathbb{A}_F = V(b) + F$ as well.

Now, take any adele $c = (c_u)_{u \in \mathcal{M}_F} \in \mathbb{A}_F$. Let $x \in V(b)$ and $f \in F$ be such that $a = x + f$. Then for each $u \in \mathcal{M}_F$,

$$|f - c_u|_u \ = \ |c_u - f|_u \ = \ |x_u|_u \ \leq \ |b_u|_u$$

as desired. □

Restricting to a finite set of places, we have

COROLLARY 7.2. *Let F be a global field, and let $\widehat{S}_F \subset \mathcal{M}_F$ be a nonempty finite set of places. Then there is a constant $B(\widehat{S}_F)$ with the following property. For any set of numbers*

$$\{0 < Q_u \in \mathbb{R} : u \in \widehat{S}_F\}$$

such that $\prod_{u \in \widehat{S}_F} Q_u^{D_u} > B(\widehat{S}_F)$, and any collection of elements $c_u \in F_u$ for $u \in \widehat{S}_F$, there is an $f \in F$ satisfying

$$\begin{cases} |f - c_u|_u \le Q_u & \text{for all } u \in \widehat{S}_F\,, \\ |f|_u \le 1 & \text{for all } u \notin \widehat{S}_F\,. \end{cases}$$

PROOF. Let B_F be the constant from Lemma 7.1, and put

$$B(\widehat{S}_F) \;=\; B_F \cdot \Big(\prod_{\text{nonarchimedean } u \,\in\, \widehat{S}_F} q_u \Big)\,,$$

where q_u is the order of the residue field at u.

Suppose $\prod_{u \in \widehat{S}_F} Q_u^{D_u} \ge B(\widehat{S}_F)$. For each archimedean $u \in \widehat{S}_F$, there is a $b_u \in F_u$ with $|b_u|_u = Q_u$; for each nonarchimedean $u \in \widehat{S}_F$ there is a $b_u \in F_u$ with

$$q_u^{-1} Q_u \;<\; |b_u| \;\le\; Q_u\,.$$

For each $u \notin \widehat{S}_F$, put $b_u = 1$, and let $b = (b_u)_{u \in \mathcal{M}_F} \in \mathbb{J}_F$. Then $\|b\|_F \ge B_F$. Let $c_u \in F_u$ for $u \in \widehat{S}_F$ be the given elements, and put $c_u = 0$ for $u \notin \widehat{S}_F$.

By Lemma 7.1 there is an $f \in F$ with $|f - c_u| \le |b_u|_u$ for all $u \in \mathcal{M}_F$. □

Now consider a global field K and a nonempty finite set of places $\widehat{S}_K$ of K. For any finite extension F/K, let $\widehat{S}_F$ be the set of places of F above $\widehat{S}_K$. The following extension of Corollary 7.2 will be used in the global patching process.

PROPOSITION 7.3 (Uniform Strong Approximation Theorem). *Let K be a global field, and let $\widehat{S}_K$ be a nonempty finite set of places of K. Let H/K be a finite normal extension. Then there is a constant $C_H(\widehat{S}_K) > 0$ with the following property. Let $\{0 < Q_w \in \mathbb{R} : w \in \widehat{S}_H\}$ be a K-symmetric set of numbers with $\prod_{w \in \widehat{S}_H} Q_w^{D_w} > C_H(\widehat{S}_K)$. Let F be any field with $K \subseteq F \subseteq H$, and let $\{c_u \in F_u : u \in \widehat{S}_F\}$ be any set of elements. Then, regarding c_u as an element of H_w for each $w|u$, there is an $f \in F$ satisfying*

$$\begin{cases} |f - c_u|_w \le Q_w & \text{for each } w \in \widehat{S}_H\,, \\ |f|_w \le 1 & \text{for each } w \notin \widehat{S}_H\,. \end{cases} \tag{7.4}$$

Remark. If H/K is finite but not normal, the assumption that $\{Q_w\}_{w \in \widehat{S}_H}$ is K-symmetric can be replaced by the requirement that there is a set of numbers $\{0 < Q_v \in \mathbb{R} : v \subset \widehat{S}_K\}$ such that $Q_w^{D_w} = Q_v^{D_v [H_w : K_v]}$ whenever $w|v$.

PROOF. Define

$$C_H(\widehat{S}_K) \;=\; \max_{K \subseteq F \subseteq H} \big(B(\widehat{S}_F)^{[H:F]}\big)\,. \tag{7.5}$$

Let $\{Q_w\}_{w \in \widehat{S}_L}$, F, and $\{c_u\}_{u \in \widehat{S}_F}$ be as in the proposition. For each $u \in \widehat{S}_F$ and each $w \in \widehat{S}_L$ with $w|u$, define Q_u by $Q_u^{D_u [L_w : K_v]} = Q_w^{D_w}$. Note that Q_u is well-defined since $\{Q_w\}_{w \in \widehat{S}_H}$ is K-symmetric. By our normalization of absolute values, if $x \in F_u$ is regarded as an element of H_w then $|x|_w^{D_w} = (|x|_u^{D_u})^{[H_w:F_u]}$, so $|x|_u \le Q_u$ iff $|x|_w \le Q_w$.

Since $\sum_{w|u}[H_w : F_u] = [H : F]$ we have

$$(\prod_{u\in\widehat{S}_F} Q_u^{D_u})^{[H:F]} \;=\; \prod_{w\in\widehat{S}_L} Q_w^{D_w} \;\ge\; C_L(\widehat{S}_K) \;\ge\; B(\widehat{S}_F)^{[H:F]} ,$$

so $\prod_{u\in\widehat{S}_F} Q_u^{D_u} \ge B(\widehat{S}_F)$. Let $f \in F$ be the element given by Lemma 7.2. For each $u \in \widehat{S}_F$, we have $|f - c_u|_u \le Q_u$, while for each $u \notin \widehat{S}_F$, we have $|f|_u \le 1$. Passing to the extension H/K, for each $w \in \widehat{S}_H$ we have $|f - c_u|_w \le Q_w$ if $w|u$, while for each $w \notin \widehat{S}_H$ we have $|f|_w \le 1$. □

2. S-units and S-subunits

Let F be a global field, and let $\widehat{S}_F$ be a nonempty finite set of places of F containing all the archimedean places. Write $\widehat{S}_F = \widehat{S}_{F,\infty} \bigcup \widehat{S}_{F,0}$ where $\widehat{S}_{F,\infty}$ is the subset of archimedean places, and $\widehat{S}_{F,0}$ is the subset of nonarchimedean places; here, if F is a function field, $\widehat{S}_{F,\infty}$ is empty. The set of $\widehat{S}_F$-units is the group

$$\mathcal{O}_{F,\widehat{S}_F}^\times \;=\; \{f \in F^\times : \mathrm{ord}_u(f) = 0 \text{ for all } u \notin \widehat{S}_F\} .$$

By the S-unit theorem,

$$\mathcal{O}_{F,\widehat{S}_F}^\times \;\cong\; \mathbb{Z}/d\mathbb{Z} \times \mathbb{Z}^{\#(\widehat{S}_F)-1},$$

where $d = d_F$ is the number of roots of unity in F. Furthermore, the homomorphism $\log_{F,\widehat{S}_F} : \mathcal{O}_{F,\widehat{S}_F}^\times \to \mathbb{R}^{\#(S_F)}$

$$\log_{F,\widehat{S}_F}(f) \;=\; (\log_u(|f|_u))_{u\in\widehat{S}_F}$$

maps $\mathcal{O}_{F,\widehat{S}_F}^\times$ onto a $\mathbb{Z}$-lattice which spans the hyperplane

$$\mathcal{H}_{\widehat{S}_F} \;=\; \{\vec{t} \in \mathbb{R}^{\#(\widehat{S}_F)} : \sum_{u\in\widehat{S}_F} t_u \log(q_u) = 0\} \;\subset\; \mathbb{R}^{\#(\widehat{S}_F)} .$$

The kernel of $\log_{F,\widehat{S}}$ is the group of roots of unity μ_d in F.

Note that if $u \in \widehat{S}_{F,0}$ and $f \in F^\times$, then $-\log_u(|f|_u) = \mathrm{ord}_u(f) \in \mathbb{Z}$. The S-unit theorem therefore implies

PROPOSITION 7.4. *Let F be a global field, and let $\widehat{S}_F$ be a finite set of places of F containing $\widehat{S}_{F,\infty}$. Suppose $\vec{t} \in \mathbb{R}^{\#(\widehat{S}_F)}$ satisfies*

$$\sum_{u\in\widehat{S}_F} t_u \log(q_u) \;=\; 0 ,$$

with $t_u \in \mathbb{Q}$ for each $u \in \widehat{S}_{F,0}$. Then for each $\eta > 0$, there are an an integer $m_0 > 0$ and an $\widehat{S}_F$-unit $f_0 \in F^\times$ such that

(1) $\frac{1}{m_0}\log_u(|f_0|_u) = t_u$ *for each* $u \in \widehat{S}_{F,0}$;

(2) $|\frac{1}{m_0}\log_u(|f_0|_u) - t_u| < \eta$ *for each* $u \in \widehat{S}_{F,\infty}$.

When F is a function field, this can be reformulated as follows:

PROPOSITION 7.5. *Let F be a function field, and let $\widehat{S}_F$ be a finite set of places of F. Let $\{c_u \in F_u^\times : u \in \widehat{S}_F\}$ be a collection of elements such that*

$$\sum_{u\in\widehat{S}_F} \log_u(|c_u|_u)\log(q_u) \;=\; 0 \;.$$

Then there are an S_F-unit $f \in \mathcal{O}_{F,\widehat{S}_F}^\times$ and an integer $n_0 > 0$ such that $|c_u^{n_0}|_u = |f|_u$ for each $u \in \widehat{S}_{F,0}$.

PROOF. Take $t_u = \log_u(|c_u|_u)$ for each $u \in \widehat{S}_F$, and let n_0 and f_0 and be the integer m_0 and $\widehat{S}_F$ unit given by Proposition 7.4. □

When F is a number field, there is a stronger version of Proposition 7.5 using the concept of a subunit, introduced by Cantor ([**16**]):

DEFINITION 7.6. An $\widehat{S}_F$-subunit is a vector $\vec{\varepsilon} = (\varepsilon_u)_{u|\infty} \in \bigoplus_{u\in\widehat{S}_{F,\infty}} F_u^\times$ for which there are an integer $n_0 > 0$ and an $\widehat{S}_F$-unit $f \in F^\times$ with $\varepsilon_u^{n_0} = f$ for each $u \in \widehat{S}_{F,\infty}$.

If F is a number field, the group of roots of unity is dense in the unit circle in $\mathbb{C}^\times$, and the units $\{\pm 1\}$ represent both connected components of $\mathbb{R}^\times$. Hence we have

PROPOSITION 7.7. *Let F be a number field, and let $\widehat{S}_F$ be a finite set of places of F containing $\widehat{S}_{F,\infty}$. Let $\{c_u \in F_u^\times : u \in \widehat{S}_F\}$ be a collection of elements such that*

$$\sum_{u\in\widehat{S}_F} \log_u(|c_u|_u)\log(q_u) \;=\; 0 \;.$$

Then for each $\delta > 0$, there are an S_F-unit $f \in \mathcal{O}_{F,\widehat{S}_F}^\times$, an integer $n_0 > 0$, and an S_F-subunit $\vec{\varepsilon} \in \bigoplus_{u\in\widehat{S}_{F,\infty}} F_u^\times$ such that

(1) *$|c_u^{n_0}|_u = |f|_u$ for each $u \in \widehat{S}_{F,0}$;*
(2) *$|\varepsilon_u - c_u|_u < \delta$ and $\varepsilon_u^{n_0} = f$ for each $u \in \widehat{S}_{F,\infty}$.*

PROOF. Apply Proposition 7.4, taking $t_u = \log_u(|c_u|_u)$ for each $u \in \widehat{S}_F$. Let $\eta > 0$ be small enough that for each $u \in \widehat{S}_{F,\infty}$, if $|x - t_u| < \eta$ then $|\exp(x) - |c_u|_u| < \delta/2$. Let m_0 and f_0 and be the positive integer and $\widehat{S}_F$ unit given by Proposition 7.4.

For each $u \in \widehat{S}_{F,\infty}$ we have $||f_0|_u - |c_u|_u| < \delta/2$, and there is a root of unity $\omega_u \in F_u^\times$ such that $|\omega_u f_0 - c_u| < \delta$; put $\varepsilon_u = \omega_u f_0$, and let $\vec{\varepsilon} = (\varepsilon_u)_{u|\infty}$. Let m_1 be the least common multiple of the orders of the ω_u; put $n_0 = m_0 m_1$ and take $f = f_0^{m_1}$. Clearly (1) and (2) hold for this $\vec{\varepsilon}$, n_0 and f. □

3. The Semi-local Theory

Let K be a global field, and let H/K be a finite extension. For each place v of K, there is a canonical isomorphism of topological algebras

$$H \otimes_K K_v \;\cong\; \bigoplus_{w|v} H_w \;. \tag{7.6}$$

(This isomorphism holds even when H/K is not separable; see ([**51**], p.321).) Under this isomorphism $K_v \cong K \otimes_K K_v$ is identified with the set of diagonal elements

$(\kappa_v, \ldots, \kappa_v)$, $\kappa_v \in K_v$. More generally, for any field F with $K \subseteq F \subseteq H$, the algebra $F \otimes_K K_v \cong \bigoplus_{u|v} F_u$ embeds in $H \otimes_K K_v$ in such a way that $\bigoplus_{u|v} h_u \in \bigoplus_{u|v} F_u$ is sent to the quasi-diagonal element $\bigoplus_{w|v} f_w \in \bigoplus_{w|v} H_w$ where $f_w = h_u$ for each $w|u$.

When H/K is normal, the group $\mathrm{Aut}(H/K)$ acts on $H \otimes_K K_v$ through its action on H: for each $\sigma \in \mathrm{Aut}(H/K)$, $f \in H$, and $\kappa_v \in K_v$, we have $\sigma(f \otimes_K \kappa_v) = \sigma(f) \otimes \kappa_v$. When this is interpreted on the right side of (7.6), it says that σ induces a permutation $w \mapsto \sigma(w)$ of the places $w|v$, and a canonical isomorphism $\tau_{\sigma,w} : H_w \to H_{\sigma(w)}$ for each w. That is, the action of σ on $\bigoplus_{w|v} H_w$ is gotten by applying $\tau_{\sigma,w}$ to the w-coordinate, while permuting the coordinates so the w-coordinate goes to the $\sigma(w)$-coordinate. Furthermore, $\mathrm{Aut}(H/K)$ acts transitively on the places w over v. (When H/K is Galois this is well-known. When H/K is merely normal, let H^{sep} be the separable closure of K in H. Then $\mathrm{Aut}(H/K) \cong \mathrm{Gal}(H^{\mathrm{sep}}/K)$ acts transitively on the places w_0 of H^{sep} lying over v, and the assertion follows because there is a unique place w of H over each w_0 of H^{sep}.)

When H/K is Galois, K_v is the sub-algebra fixed by $\mathrm{Gal}(H/K)$; more generally, $F \otimes_K K_v$ is the sub-algebra fixed by $\mathrm{Gal}(H/F)$, for each F with $K \subseteq F \subseteq H$.

We will now apply these facts in the context of Theorem 4.2. Let K be the global field in Theorem 4.2, and put $L = K(\mathfrak{X})$. If K is a number field, take $H = L$; if K is a function field, put $H = L^{\mathrm{sep}}$. Then H/K is Galois. Since $\mathcal{C}/K$ is geometrically integral,

$$H \otimes_K K_v(\mathcal{C}) \cong \bigoplus_{w|v} H_w(\mathcal{C}) \ . \tag{7.7}$$

If $K \subseteq F \subseteq H$, the algebra $F \otimes_K K_v(\mathcal{C})$ embeds quasi-diagonally in $H \otimes_K K_v(\mathcal{C})$.

For each $\sigma \in \mathrm{Gal}(H/K)$ and each w, the isomorphism $\tau_{\sigma,w} : H_w \to H_{\sigma(w)}$ induces an isomorphism $\widehat{\tau}_{\sigma,w} : H_w(\mathcal{C}) \to H_{\sigma(w)}(\mathcal{C})$ fixing $K_v(\mathcal{C})$. As before, the action of $\mathrm{Gal}(H/K)$ on $H \otimes_K K_v(\mathcal{C})$ can be interpreted as applying $\widehat{\tau}_{\sigma,w}$ to the w-component of (7.7), for each w, at the same time permuting the coordinates so the w-component goes to the $\sigma(w)$-component. $K_v(\mathcal{C})$ is the sub-algebra fixed by $\mathrm{Gal}(H/K)$; more generally, $F \otimes_K K_v(\mathcal{C})$ is the sub-algebra fixed by $\mathrm{Gal}(H/F)$, for each F with $K \subseteq F \subseteq H$.

Let J be the number from the construction of the L-rational and L^{sep} rational bases in §3.3, and let $\Lambda_0 = \dim_{\widetilde{K}}(\widetilde{\Gamma}(\sum_{i=1}^m J(x_i))$ be the number of low-order basis elements. Given a probability vector $\vec{s} \in \mathcal{P}^m(\mathbb{Q})$ with positive coordinates, and an integer N such that $N\vec{s} \in \mathbb{Z}^m$, write $N_i = Ns_i$ for $i = 1, \ldots, m$. Assume that N is large enough that $N_i \geq J$ for each i. Suppose we are given an $(\mathfrak{X}, \vec{s})$-function $\phi(z) \in K(\mathcal{C})$ of degree N.

If K is a number field (that is, if $\mathrm{char}(K) = 0$), we can expand $\phi(z)$ using the L-rational basis as

$$\phi(z) = \sum_{i=1}^{m} \sum_{j=0}^{N_i-J+1} a_{ij}\varphi_{i,N_i-j}(z) + \sum_{\lambda=1}^{\Lambda_0} a_\lambda \varphi_\lambda(z) \tag{7.8}$$

with the $a_{ij}, a_\lambda \in L$. Since $\phi(z)$ is K-rational, for each $\sigma \in \mathrm{Gal}(F/K)$ we have $\sigma(\phi)(z) = \phi(z)$. Applying σ to (7.8), and recalling that $\sigma(\varphi_{ij}) = \varphi_{\sigma(i),j}$ and $\sigma(\varphi_\lambda) = \varphi_\lambda$, we find that $\sigma(a_{ij}) = a_{\sigma(i),j}$ for all i, j, and $\sigma(a_\lambda) = a_\lambda$ for all λ. Thus, the a_{ij} are K-symmetric relative to the action of $\mathrm{Gal}(L/K)$ on the x_i, and for each $\sigma \in \mathrm{Gal}(F/K(x_i))$ we have $\sigma(a_{ij}) = a_{\sigma(i),j} = a_{ij}$, so a_{ij} belongs to $K(x_i)$. Likewise, since each φ_λ is K-rational, each a_λ belongs to K.

Similarly, let w be a place of L with $w|v$. If we are given an $(\mathfrak{X}, \vec{s})$ function $\phi_w(z) \in L_w(\mathcal{C})$ of degree N, we can write

$$\phi_w(z) = \sum_{i=1}^{m} \sum_{j=0}^{N_i-J-1} a_{w,ij}\varphi_{i,N_i-j}(z) + \sum_{\lambda=1}^{\Lambda} a_{w,\lambda}\varphi_\lambda(z) \tag{7.9}$$

where each $a_{w,ij} \in L_w$ and each $a_{w,\lambda} \in K_v$. In this context, we have:

PROPOSITION 7.8. *Suppose K is a number field, and let v be a place of K. For each place w of L with $w|v$, let an $(\mathfrak{X}, \vec{s})$-function $\phi_w(z) \in L_w(\mathcal{C})$ be given. Then the following are equivalent:*

(1) *There is a $\phi_v(z) \in K_v(\mathcal{C})$ such that $\phi_w(z) = \phi_v(z)$ for all $w|v$.*

(2) *$\bigoplus_{w|v} \phi_w(z)$ is invariant under the action of $\mathrm{Gal}(L/K)$ on $L \otimes_K K_v(\mathcal{C})$.*

(3) *If each $\phi_w(z)$ is expanded as in (7.9), then*

(a) *for each i, j, each $w|v$, and each $\sigma \in \mathrm{Gal}(L/K)$,*

$$a_{\sigma(w),\sigma(i),j} = \tau_{\sigma,w}(a_{w,ij}) , \quad \textit{and}$$

(b) *for each λ there is an $a_{v,\lambda} \in K_v$ such that $a_{w,\lambda} = a_{v,\lambda}$ for all $w|v$.*

Under these conditions, for each i and j, if we write $F = K(x_i)$ then $\bigoplus_{w|v} a_{w,ij}$ belongs to $\bigoplus_{u|v} F_u$, embedded semi-diagonally in $L \otimes_K K_v$.

PROOF. The equivalence of (1) and (2) follows from the description of the action of $\mathrm{Gal}(L/K)$ on $L \otimes_K K_v(\mathcal{C})$ given above. For the equivalence of (2) and (3), note that if $\bigoplus_{w|v} \phi_w(z)$ is expanded as in (7.9), with $a_{ij} = \bigoplus_{w|v} a_{w,ij}$ and $a_\lambda = \bigoplus_{w|v} a_{w,\lambda}$, then

$$\sigma\left(\sum_{i=1}^{m} \sum_{j=0}^{N_i-(2g+1)} a_{ij}\varphi_{i,N_i-j}(z) + \sum_{\lambda=1}^{\Lambda_0} a_\lambda\varphi_\lambda(z)\right)$$
$$= \sum_{i=1}^{m} \sum_{j=0}^{N_i-(2g+1)} \sigma(a_{ij})\varphi_{\sigma(i),N_i-j}(z) + \sum_{\lambda=1}^{\Lambda_0} \sigma(a_\lambda)\varphi_\lambda(z) .$$

Thus $\bigoplus_{w|v} \phi_w(z)$ is invariant under $\mathrm{Gal}(L/K)$ if and only if $\sigma(a_{ij}) = a_{\sigma(i),j}$ and $\sigma(a_\lambda) = a_\lambda$ for all σ and all i, j, and λ. In view of the description of the action of $\mathrm{Gal}(L/K)$ on $\bigoplus_{w|v} L_w$, this holds if and only if $\tau_{\sigma,w}(a_{w,ij}) = a_{\sigma(w),\sigma(i),j}$ for all σ, w, i, j; and $a_{w,\lambda} = a_{v,\lambda} \in K_v$ for all w, λ. The assertion concerning the F-rationality of $\bigoplus_{w|v} a_{w,ij}$ follows from the discussion at the beginning of the section. □

If K is a function field (so $\mathrm{char}(K) = p > 0$), we can expand $\phi(z)$ using the L^{sep}-rational basis as

$$\phi(z) = \sum_{i=1}^{m} \sum_{j=0}^{N_i-J+1} \widetilde{a}_{ij}\widetilde{\varphi}_{i,N_i-j}(z) + \sum_{\lambda=1}^{\Lambda_0} \widetilde{a}_\lambda\widetilde{\varphi}_\lambda(z) \tag{7.10}$$

with the $\widetilde{a}_{ij}, \widetilde{a}_\lambda \in L^{\mathrm{sep}}$. Again, the $\widetilde{a}_{ij}$ are K-symmetric relative to the action of $\mathrm{Gal}(L^{\mathrm{sep}}/K)$, and each $\widetilde{a}_\lambda$ belongs to K. If w is a place of L^{sep} with $w|v$, and we are given an $(\mathfrak{X}, \vec{s})$ function $\phi_w(z) \in L_w^{\mathrm{sep}}(\mathcal{C})$ of degree N, we can write

$$(7.11) \qquad \phi_w(z) \;=\; \sum_{i=1}^{m} \sum_{j=0}^{N_i-J-1} \widetilde{a}_{w,ij}\widetilde{\varphi}_{i,N_i-j}(z) + \sum_{\lambda=1}^{\Lambda} \widetilde{a}_{w,\lambda}\widetilde{\varphi}_\lambda(z)$$

with each $a_{w,ij} \in L_w^{\mathrm{sep}}$ and each $a_{w,\lambda} \in K_v$. In this case we have

PROPOSITION 7.9. *Suppose K is a function field, and let v be a place of K. For each place w of L^{sep} with $w|v$, let an $(\mathfrak{X}, \vec{s})$-function $\phi_w(z) \in L_w^{\mathrm{sep}}(\mathcal{C})$ be given. Then the following are equivalent:*

(1) *There is a $\phi_v(z) \in K_v(\mathcal{C})$ such that $\phi_w(z) = \phi_v(z)$ for all $w|v$.*

(2) *$\bigoplus_{w|v} \phi_w(z)$ is invariant under the action of $\mathrm{Gal}(L^{\mathrm{sep}}/K)$ on $L^{\mathrm{sep}} \otimes_K K_v(\mathcal{C})$.*

(3) *If each $\phi_w(z)$ is expanded as in (7.11), then*

(a) *for each i, j, each $w|v$, and each $\sigma \in \mathrm{Gal}(L^{\mathrm{sep}}/K)$,*

$$\widetilde{a}_{\sigma(w),\sigma(i),j} \;=\; \tau_{\sigma,w}(\widetilde{a}_{w,ij}) \;, \quad \text{and}$$

(b) *for each λ there is an $\widetilde{a}_{v,\lambda} \in K_v$ such that $\widetilde{a}_{w,\lambda} = \widetilde{a}_{v,\lambda}$ for all $w|v$.*

Under these conditions, for each i and j, if we write $F = K(x_i)^{\mathrm{sep}}$ then $\bigoplus_{w|v} \widetilde{a}_{w,ij}$ belongs to $\bigoplus_{u|v} F_u$, embedded semi-diagonally in $L^{\mathrm{sep}} \otimes_K K_v$.

PROOF. The proof is similar to that of Proposition 7.8. □

4. Proof of Theorem 4.2 when char$(K) = 0$

In this section we will prove Theorem 4.2 when $\mathrm{char}(K) = 0$. Let K, $L = K(\mathfrak{X})$, S_K, and $\mathbb{E}$ be as in Theorem 4.2, and let $\widehat{S}_K \supseteq S_K$ be the finite set of places of K satisfying the conditions (7.1) at the beginning this Chapter. Let $\widehat{S}_L$ be the set of places of L above $\widehat{S}_K$.

The proof has three stages, and will occupy the rest of this section. First, we choose the parameters governing the patching process. Next, we construct a set of "initial approximating functions" $f_v(z)$ for $v \in \widehat{S}_K$, whose roots belong to E_v, and modify them to obtain "coherent approximating functions" $\phi_v(z)$ whose leading coefficients satisfy the conditions of Proposition 7.7. By means of a degree-raising procedure, we use the coherent approximating functions to construct "initial patching functions" $G_v^{(0)}(z)$ whose whose roots also belong to E_v. Finally, we patch the coefficients, creating a sequence of K_v-symmetric functions $G_v^{(1)}(z), G_v^{(2)}(z), \ldots, G_v^{(n)}(z)$ which have more and more coefficients in the global field L, but whose roots still belong to E_v. The final functions $G_v^{(n)}(z)$ have all their coefficients in L, and are K_v-rational but independent of v. Using Proposition 7.8, they can be seen to be K-rational. In this way we construct a function $G^{(n)}(z) \in K(\mathcal{C})$ whose roots belong to E_v for each v.

PROOF OF THEOREM 4.2 WHEN $\mathrm{char}(K) = 0$. We begin by outlining the first two stages of the construction.

First we choose parameters $0 < h_v < r_v < R_v$ for $v \in \widehat{S}_K$, which control the amount of the freedom in the patching process. Using the Green's matrix $\Gamma(\mathbb{E}, \mathfrak{X})$ we construct a K-symmetric probability vector $\vec{s}$ with positive rational coefficients, and a positive integer N, which will be the common degree of the initial approximating functions.

Using the approximation theorems from §5 and §6, we construct the initial approximating functions $\{f_v(z)\}_{v \in \widehat{S}_K}$, which are $(\mathfrak{X}, \vec{s})$-functions of common degree

N with roots belonging to E_v. We then modify the $f_v(z)$ to obtain the coherent approximating functions $\{\phi_v(z)\}_{v \in \widehat{S}_K}$. The key properties of the $\phi_v(z)$ will be

(1) For each $v \in \widehat{S}_K$, $\phi_v(z) \in K_v(\mathcal{C})$ is an $(\mathfrak{X}, \vec{s})$ function of degree N, for which there are a constant $\kappa_v \in K_v^\times$ with $|\kappa_v|_v \geq 1$ and an initial approximating function $f_v(z) \in K_v(\mathcal{C})$ for E_v, such that

$$\phi_v(z) \;=\; \kappa_v f_v(z) \ .$$

Thus the $\phi_v(z)$ inherit the approximation properties of the $f_v(z)$, and their roots are the same as those of the $f_v(z)$.

(2) For each $w \in \widehat{S}_L$, put $\phi_w(z) = \phi_v(z)$ if $w|v$, and view $\phi_w(z)$ as an element of $L_w(\mathcal{C})$. Although the $\phi_w(z)$ for $w|v$ are all the same, for distinct w the points of $\mathfrak{X}$, which are their poles, are identified differently as points of $\mathcal{C}_w(L_w)$. Write $\widetilde{c}_{w,i} = \lim_{z \to x_i} \phi_w(z) \cdot g_{x_i}(z)^{Ns_i}$ for the leading coefficient of $\phi_w(z)$ at x_i.

Then for each i, $\bigoplus_{w|\infty} \widetilde{c}_{w,i}$ is an $\widehat{S}_L$-subunit and

$$\prod_{w \in \widehat{S}_L} |\widetilde{c}_{w,i}|_w^{D_w} \;=\; 1 \ .$$

Our ability to achieve (1) uses that $\gamma(\mathbb{E}, \mathfrak{X}) > 1$. Our ability to achieve (2) depends on the independent variability of the archimedean logarithmic leading coefficients, proved in Theorems 5.1 and 5.2.

For the convenience of the reader, Theorem 7.10 below summarizes Theorems 5.1, 5.2, 6.1, 6.3 and Corollaries 6.11 and 6.12 which construct the initial approximating functions. In broad terms, those theorems say that for each K-symmetric $\vec{s} \in \mathcal{P}^m(\mathbb{Q})$ and each $v \in \widehat{S}_K$, there are a set $\widetilde{E}_v \subset E_v$ and an integer $N_v > 0$ such that for each $N > 0$ divisible by N_v, there is an $(\mathfrak{X}, \vec{s})$-function $f_v(z) \in K_v(\mathcal{C})$ of degree N such that $\frac{1}{N} \log_v(|f_v(z)|_v)$ approximates $\sum_{i=1}^m s_i G(z, x_i; \widetilde{E}_v)$, and $f_v(z)$ has roots in E_v and satisfies certain side conditions.

Recall that if $f_v(z) \in K_v(\mathcal{C})$ is an $(\mathfrak{X}, \vec{s})$-function of degree N, then for each $x_i \in \mathfrak{X}$, the leading coefficient of f_v at x_i is

$$c_{v,i} \;=\; (f_v \cdot g_{x_i}^{Ns_i})|_{x_i} \;=\; \lim_{z \to x_i} f_v(z) \cdot g_{x_i}(z)^{Ns_i} \ ,$$

and that $\Lambda_{x_i}(f_v, \vec{s}) = \frac{1}{N} \log_v(|c_{v,i}|_v)$.

THEOREM 7.10 (Summary of the Initial Approximation Theorems). *Let K, $\mathbb{E}$, and $\mathfrak{X}$ be as in Theorem 4.2. Then for each $v \in \widehat{S}_K$,*

(A) If $K_v \cong \mathbb{C}$ (so $E_v \subset \mathcal{C}_v(\mathbb{C})$ is compact, $\mathbb{C}$-simple, and disjoint from $\mathfrak{X}$), let $U_v = E_v^0$ be the interior of E_v. Then for each $\varepsilon_v > 0$ there is a compact set $\widetilde{E}_v$ contained in E_v^0 such that:

(1) For each $x_i, x_j \in \mathfrak{X}$ with $x_i \neq x_j$,

$$|V_{x_i}(\widetilde{E}_v) - V_{x_i}(E_v)| \;<\; \varepsilon_v \ , \qquad |G(x_i, x_j; \widetilde{E}_v) - G(x_i, x_j; E_v)| \;<\; \varepsilon_v \ .$$

(2) There is a $\delta_v > 0$ with the property that for each $\vec{s} \in \mathcal{P}^m(\mathbb{Q})$, there is an integer $N_v > 0$ such that for each $\vec{\beta}_v = {}^t(\beta_{v,1}, \ldots, \beta_{v,m}) \in [-\delta_v, \delta_v]^m$, and each positive integer N divisible by N_v, there is an $(\mathfrak{X}, \vec{s})$-function $f_v(z) \in K_v(\mathcal{C})$ of degree N satisfying

(a) $\{z \in \mathcal{C}_v(\mathbb{C}) : |f_v(z)|_v \leq 1\} \subset E_v^0$;

(b) *For each* $x_i \in \mathfrak{X}$, $\frac{1}{N}\log_v(|c_{v,i}|_v) = \Lambda_{x_i}(\widetilde{E}_v, \vec{s}) + \beta_{v,i}$.

(B) *If* $K_v \cong \mathbb{R}$ *(so* $E_v \subset \mathcal{C}_v(\mathbb{C})$ *is compact,* $\mathbb{R}$*-simple, and disjoint from* $\mathfrak{X}$*), let* E_v^0 *be the quasi-interior of* E_v*. Then for each* $\varepsilon_v > 0$*, and each open set* $U_v \subset \mathcal{C}_v(\mathbb{C})$ *which is stable under complex conjugation, bounded away from* $\mathfrak{X}$*, and satisfies* $U_v \cap E_v = E_v^0$*, there are a compact set* $\widetilde{E}_v$ *contained in* E_v^0 *such that:*

(1) *For each* $x_i, x_j \in \mathfrak{X}$ *with* $x_i \neq x_j$,

$$|V_{x_i}(\widetilde{E}_v) - V_{x_i}(E_v)| \ < \ \varepsilon_v\,, \quad |G(x_i, x_j; \widetilde{E}_v) - G(x_i, x_j; E_v)| \ < \ \varepsilon_v\,.$$

(2) *For each* $0 < \mathcal{R}_v < 1$*, there is a* $\delta_v > 0$ *with the property that for each* K_v*-symmetric* $\vec{s} \in \mathcal{P}^m(\mathbb{Q})$*, there is an integer* $N_v > 0$ *such that for each* K_v*-symmetric* $\vec{\beta}_v = {}^t(\beta_{v,1}, \ldots, \beta_{v,m}) \in [-\delta_v, \delta_v]^m$ *and each positive integer* N *divisible by* N_v*, there is an* $(\mathfrak{X}, \vec{s})$*-function* $f_v(z) \in K_v(\mathcal{C})$ *of degree* N *satisfying*

(a) $\{z \in \mathcal{C}_v(\mathbb{C}) : |f_v(z)| \leq 1\} \ \subset \ U_v$ *, all the zeros of* $f_v(z)$ *belong to* E_v^0*, and if* $E_{v,i}$ *is a component of* E_v *contained in* $\mathcal{C}_v(\mathbb{R})$ *and* $f_v(z)$ *has* N_i *zeros in* $E_{v,i}$*, then* $f_v(z)$ *oscillates* N_i *times between* $\pm\mathcal{R}_v^N$ *on* $E_{v,i}$.

(b) *For each* $x_i \in \mathfrak{X}$, $\frac{1}{N}\log_v(|c_{v,i}|_v) = \Lambda_{x_i}(\widetilde{E}_v, \vec{s}) + \beta_{v,i}$.

(C) *If* K_v *is nonarchimedean and* $v \in S_K$*, (so* E_v *is compact,* K_v*-simple, and disjoint from* $\mathfrak{X}$*), fix a* K_v*-simple decomposition*

$$E_v \ = \ \bigcup_{\ell=1}^{D_v} B(a_\ell, r_\ell) \cap \mathcal{C}_v(F_{w_\ell}) \tag{7.12}$$

and fix $\varepsilon_v > 0$*. Then there is a* K_v*-simple set* $\widetilde{E}_v \subseteq E_v$ *compatible with* E_v *such that:*

(1) *For each* $x_i, x_j \in \mathfrak{X}$ *with* $x_i \neq x_j$,

$$|V_{x_i}(\widetilde{E}_v) - V_{x_i}(E_v)| \ < \ \varepsilon_v\,, \quad |G(x_i, x_j; \widetilde{E}_v) - G(x_i, x_j; E_v)| \ < \ \varepsilon_v\,.$$

(2) *For each* $0 < \beta_v \in \mathbb{Q}$ *and each* K_v*-symmetric* $\vec{s} \in \mathcal{P}^m(\mathbb{Q})$*, there is an integer* $N_v \geq 1$ *such that for each positive integer* N *divisible by* N_v*, there is an* $(\mathfrak{X}, \vec{s})$*-function* $f_v \in K_v(\mathcal{C}_v)$ *of degree* N *satisfying:*

(a) *The zeros* $\theta_1, \ldots, \theta_N$ *of* f_v *are distinct and belong to* E_v.

(b) $f_v^{-1}(D(0,1)) \subseteq \bigcup_{\ell=1}^{D_v} B(a_\ell, r_\ell)$.

Further, there is a decomposition $f_v^{-1}(D(0,1)) = \bigcup_{h=1}^{N} B(\theta_h, \rho_h)$*, where the balls* $B(\theta_h, \rho_h)$ *are pairwise disjoint and isometrically parametrizable. For each* $h = 1, \ldots, N$*, if* $\ell = \ell(h)$ *is such that* $B(\theta_h, \rho_h) \subseteq B(a_\ell, r_\ell)$*, put* $F_{u_h} = F_{w_\ell}$*; then* $\rho_h \in |F_{u_h}^\times|_v$ *and* f_v *induces an* F_{u_ℓ}*-rational scaled isometry from* $B(\theta_h, \rho_h)$ *to* $D(0,1)$*, with*

$$f_v\big(B(\theta_h, \rho_h) \cap \mathcal{C}_v(F_{u_h})\big) \ = \ \mathcal{O}_{F_{u_h}}\ ,$$

such that $|f_v(z_1) - f_v(z_2)|_v = (1/\rho_h)\|z_1, z_2\|_v$ *for all* $z_1, z_2 \in B(\theta_h, \rho_h)$.

(c) *The set* $H_v := E_v \cap f_v^{-1}(D(0,1))$ *is* K_v*-simple and compatible with* E_v*. Indeed,*

$$H_v \ = \ \bigcup_{h=1}^{N} \big(B(\theta_h, \rho_h) \cap \mathcal{C}_v(F_{u_h})\big)\ , \tag{7.13}$$

and (7.13) *is a* K_v*-simple decomposition of* H_v *compatible with the* K_v*-simple decomposition* (7.12) *of* E_v*, which is move-prepared (see Definition* 6.10*) relative to the balls* $B(a_1, r_1), \ldots, B(a_{D_v}, r_{D_v})$*. For each* ℓ *there is a point* $\overline{w}_\ell \in \big(B(a_\ell, r_\ell) \cap \mathcal{C}_v(F_{w_\ell})\big) \backslash H_v$.

(*d*) *For each* $x_i \in \mathfrak{X}$, $\frac{1}{N}\log_v(|c_{v,i}|_v) \ = \ \Lambda_{x_i}(\widetilde{E}_v, \vec{s}) + \beta_v$.

(*D*) *If* K_v *is nonarchimedean and* $v \notin S_K$, *(so* E_v *is* $\mathfrak{X}$*-trivial and in particular is an* RL*-domain disjoint from* $\mathfrak{X}$*), put* $\widetilde{E}_v = E_v$.

Then for each K_v*-symmetric* $\vec{s} \in \mathcal{P}^m(\mathbb{Q})$*, there is an integer* $N_v \geq 1$ *such that for each positive integer* N *divisible by* N_v*, there is an* $(\mathfrak{X}, \vec{s})$*-function* $f_v \in K_v(\mathcal{C}_v)$ *of degree* N *satisfying*

(*a*) $E_v = \widetilde{E}_v = \{z \in \mathcal{C}_v(\mathbb{C}_v) : |f_v(z)|_v \leq 1\}$;

(*b*) *for each* $x_i \in \mathfrak{X}$, $\frac{1}{N}\log_v(|c_{v,i}|_v) \ = \ \Lambda_{x_i}(\widetilde{E}_v, \vec{s})$.

Note that in (A) and (B) of Theorem 7.10, the number $\delta_v > 0$ depends on ε_v, E_v and U_v, but the numbers $\beta_{v,i}$ for which

$$\frac{1}{N}\log_v(|c_{v,i}|_v) \ = \ \Lambda_{x_i}(\widetilde{E}_v, \vec{s}) + \beta_{v,i}$$

can be specified arbitrarily provided they are K_v-symmetric and satisfy $-\delta_v \leq \beta_{v,i} \leq \delta_v$ for each i. In (C) the number $0 < \beta_v \in \mathbb{Q}$ is the same for all i. In (D), the logarithmic leading coefficients match the $\Lambda_{x_i}(\widetilde{E}_v, \vec{s})$ exactly. For each v, the leading coefficients $c_{v,i}$ of f_v are K_v-symmetric, because $f_v(z)$ is K_v-rational and the $g_{x_i}(z)$ are K_v-symmetric.

We now turn to the details of the proof.

Stage 1. Choices of the sets and parameters.

We begin by making the choices that govern the patching process. Given a number field F containing K, write $\widehat{S}_{F,\infty}$ for the set of archimedean places of F, and $\widehat{S}_{F,0}$ for the set of nonarchimedean places in $\widehat{S}_F$, so $\widehat{S}_F = \widehat{S}_{F,\infty} \cup \widehat{S}_{F,0}$. Similarly, $S_F = S_{F,\infty} \cup S_{F,0}$.

The open sets U_v for $v \in \widehat{S}_{K\infty}$. For each $v \in \widehat{S}_{K,\infty}$ with $K_v \cong \mathbb{C}$, E_v is $\mathbb{C}$-simple, so it has finitely finitely many components, each of which is simply connected, has a piecewise smooth boundary and is the closure of its interior. Let $U_v = E_v^0$ be its interior.

For each $v \in \widehat{S}_{K,\infty}$ with $K_v \cong \mathbb{R}$, E_v is $\mathbb{R}$-simple, so it is stable under complex conjugation and has finitely many components, each of which is an interval of positive length contained in $\mathcal{C}_v(\mathbb{R})$, or is disjoint from $\mathcal{C}_v(\mathbb{R})$ and is simply connected, has a piecewise smooth boundary, and is the closure of its interior. Let $U_v \subset \mathcal{C}_v(\mathbb{C})$ be an open set such that $U_v \cap E_v = E_v^0$, the quasi-interior of E_v. We will choose U_v so that it is stable under complex conjugation, bounded away from $\mathfrak{X}$, and has the same number of connected components as E_v. Thus, U_v is the union of the interiors of the components $E_{v,i}$ in $\mathcal{C}_v(\mathbb{C}) \backslash \mathcal{C}_v(\mathbb{R})$, together with open sets $U_{v,i}$ such that $U_{v,i} \cap E_v$ is the real interior of $E_{v,i}$, for the $E_{v,i} \subset \mathcal{C}_v(\mathbb{R})$.

The K_v-simple decompositions of E_v and the sets U_v, for $v \in S_{K,0}$. For each $v \in S_{K,0}$, the set E_v is compact and K_v-simple (see Definition 4.1).

Choose a K_v-simple decomposition

$$E_v = \bigcup_{\ell=1}^{D_v} B(a_\ell, r_\ell) \cap \mathcal{C}_v(F_{w_\ell}) . \tag{7.14}$$

By refining this decomposition, we can assume that $U_v := \bigcup_{\ell=1}^{D_v} B(a_\ell, r_\ell)$ is disjoint from $\mathfrak{X}$. Such a decomposition will be fixed for the rest of the construction.

The sets $\widetilde{E}_v$ for $v \in \widehat{S}_K$. By hypothesis, $\gamma(\mathbb{E}, \mathfrak{X}) > 1$ in Theorem 4.2, so $\Gamma(\mathbb{E}, \mathfrak{X})$ is negative definite. Suppose $\widetilde{\mathbb{E}} = \prod_{v \in \widehat{S}_K} \widetilde{E}_v \times \prod_{v \notin \widehat{S}_K} E_v$ is another K-rational adelic set compatible with $\mathfrak{X}$. By the discussion leading to (7.3), there are numbers $\varepsilon_v > 0$ for $v \in \widehat{S}_K$ such that such that $\Gamma(\widetilde{\mathbb{E}}, \mathfrak{X})$ is also negative definite, provided that for each $v \in \widehat{S}_K$,

$$\begin{cases} |G(x_j, x_i; \widetilde{E}_v) - G(x_j, x_i; E_v)| < \varepsilon_v & \text{for all } i \neq j , \\ |V_{x_i}(\widetilde{E}_v) - V_{x_i}(E_v)| < \varepsilon_v & \text{for all } i . \end{cases} \tag{7.15}$$

For each $v \in \widehat{S}_K$, we will take $\widetilde{E}_v \subseteq E_v$ to be the set given by Theorem 7.10 for E_v, relative to the number ε_v chosen above (and the set U_v, if $K_v \cong \mathbb{R}$), satisfying (7.15). Put $\widetilde{\mathbb{E}} = \prod_{v \in \widehat{S}_K} \widetilde{E}_v \times \prod_{v \notin \widehat{S}_K} E_v$ with the $\widetilde{E}_v$ chosen above, and let

$$\widetilde{V}_K := V(\widetilde{\mathbb{E}}, \mathfrak{X}) = \mathrm{val}(\Gamma(\widetilde{\mathbb{E}}, \mathfrak{X})) \tag{7.16}$$

be the global Robin constant for $\widetilde{\mathbb{E}}$ and $\mathfrak{X}$. By construction, $\widetilde{V}_K < 0$.

The local parameters η_v, $\mathcal{R}_v$, h_v, r_v, and R_v. Fix a collection of real numbers $\{\eta_v\}_{v \in \widehat{S}_K}$ with $\eta_v > 0$ for each $v \in \widehat{S}_K$ and $\eta_v \in \mathbb{Q}$ for each $v \in \widehat{S}_{K,0}$, such that

$$\sum_{v \in \widehat{S}_K} \eta_v \log(q_v) = |\widetilde{V}_K| = -\widetilde{V}_K . \tag{7.17}$$

The η_v provide the freedom for adjustment needed in the construction of the initial approximating functions, and determine the scaling factors in passing from the initial approximating functions to the coherent approximating functions.

For each $v \in \widehat{S}_{K,\infty}$, fix a number r_v with $1 < r_v < e^{\eta_v}$. Then, choose a set of numbers $\{h_v\}_{v \in \widehat{S}_K}$ with $\prod_{v \in \widehat{S}_K} h_v^{D_v} > 1$, such that

$$\begin{cases} 1 < h_v < r_v & \text{if } v \in \widehat{S}_{K,\infty} , \\ 0 < h_v < 1 & \text{if } v \in \widehat{S}_{K,0} . \end{cases} \tag{7.18}$$

Finally, for each $v \in S_{K,0}$, fix an r_v with $h_v < r_v < 1$, and for each $v \in \widehat{S}_{K,0} \backslash S_{K,0}$ put $r_v = 1$. Note that $D_v = \log(q_v)$ for each archimedean v. Thus $0 < h_v < r_v$ for all v, and

$$1 < \prod_{v \in \widehat{S}_K} h_v^{D_v} < \prod_{v \in \widehat{S}_K} r_v^{D_v} . \tag{7.19}$$

In the patching process, the numbers h_v will control how much the coefficients of the functions being patched can be changed, and the r_v will be "encroachment bounds" which limit how close certain quantities can come to the h_v.

For each $v \in \mathcal{S}_{K,\infty}$ with $K_v \cong \mathbb{C}$, put $\widehat{R}_v = e^{\eta_v}$. For each $v \in \mathcal{S}_{K,\infty}$ with $K_v \cong \mathbb{R}$, fix a number $0 < \mathcal{R}_v < 1$ close enough to 1 that

$$1 < r_v < \mathcal{R}_v \cdot e^{\eta_v} < e^{\eta_v} \tag{7.20}$$

and put $\widehat{R}_v = \mathcal{R}_v \cdot e^{\eta_v}$. The number $\mathcal{R}_v$ specifies the magnitude of the oscillations of the initial approximating functions when $K_v \cong \mathbb{R}$.

In either case, we can choose R_v so that $r_v < R_v < \widehat{R}_v$; thus

$$1 < r_v < R_v < \widehat{R}_v \le e^{\eta_v} . \tag{7.21}$$

For each $v \in \widehat{S}_{K,0}$, put $R_v = q_v^{\eta_v}$. Then $0 < h_v < r_v < R_v$ for each $v \in \widehat{S}_K$, and $R_v \in |\mathbb{C}_v^\times|_v$ for each $v \in \widehat{S}_{K,0}$.

The numbers $\delta_v > 0$ for $v \in \widehat{S}_{K,\infty}$. If $K_v \cong \mathbb{C}$, let δ_v be the number given by Theorem 7.10(A.2) for E_v and $\widetilde{E}_v$. If $K_v \cong \mathbb{R}$, let δ_v be the number given by Theorem 7.10(B.2) for E_v and $\widetilde{E}_v$, relative to the number $\mathcal{R}_v$ chosen in (7.20). For each $v \in \widehat{S}_{K,\infty}$, the number δ_v plays the role of a "radius of independent variability" for the logarithmic leading coefficients at v.

The rational probability vector $\vec{s}$. By construction, the Green's matrix $\Gamma(\widetilde{\mathbb{E}}, \mathfrak{X})$ is negative definite. As above, put $\widetilde{V}_K = V(\widetilde{\mathbb{E}}_K, \mathfrak{X}) < 0$. Let $\widetilde{s} \in \mathcal{P}^m(\mathbb{R})$ be the K-symmetric probability vector given by Proposition 3.33 for which

$$\begin{pmatrix} \widetilde{V}_K \\ \vdots \\ \widetilde{V}_K \end{pmatrix} = \Gamma(\widetilde{\mathbb{E}}_K, \mathfrak{X})\widetilde{s} .$$

The entries of $\widetilde{s}$ are positive, but they need not be rational.

Fix an archimedean place v_0 of K, and let δ_{v_0} be the radius of independent variability for the logarithmic leading coefficients at v_0, constructed above. By continuity, there is a K-symmetric probability vector $\vec{s} \in \mathcal{P}^m(\mathbb{Q})$ close enough to $\widetilde{s}$ that all its entries are positive, and such that for each $i = 1, \ldots, m$, the ith coordinate of $\Gamma(\widetilde{\mathbb{E}}_K, \mathfrak{X})\vec{s}$ satisfies

$$|\widetilde{V}_K - (\Gamma(\widetilde{\mathbb{E}}_K, \mathfrak{X})\vec{s})_i| < \delta_{v_0} \log(q_{v_0}) . \tag{7.22}$$

This $\vec{s}$ will be fixed for the rest of the construction.

Stage 2. Construction of the approximating functions $f_v(z)$ and $\phi_v(z)$.

In this stage, we construct the initial approximating functions $f_v(z)$, then modify them to obtain the coherent approximating functions $\phi_v(z)$. Our goal is to prove the following technical theorem:

THEOREM 7.11. *Let $\mathcal{C}$, K, $\mathbb{E}$, $\mathfrak{X}$, and S_K be as in Theorem 4.2, with* $\mathrm{char}(K) = 0$. *Let $\widehat{S}_K \supseteq S_K$ be the finite set of places satisfying conditions (7.1). For each $v \in \widehat{S}_K$, let $\widetilde{E}_v \subset E_v$ and $0 < h_v < r_v < R_v$ be the set and patching parameters constructed above. For each $v \in \widehat{S}_{K,\infty}$, let $U_v \subset \mathcal{C}_v(\mathbb{C})$ be the chosen set with $U_v \cap E_v = E_v^0$, and let $\delta_v > 0$ be the radius of independent variability for the logarithmic leading coefficients. For each $v \in S_{K,0}$ let $\bigcup_{\ell=1}^{D_v} B(a_\ell, r_\ell) \cap \mathcal{C}_v(F_{w_\ell})$ be the chosen K_v-simple decomposition of E_v. Let $\vec{s} \in \mathcal{P}_m(\mathbb{Q})$ be the chosen rational probability vector with positive coefficients, satisfying (7.22).*

Then there are a positive integer N and $(\mathfrak{X}, \vec{s})$-functions $\phi_v(z) \in K_v(\mathcal{C})$, $v \in \widehat{S}_K$, of common degree N, so that for each v the zeros of $\phi_v(z)$ belong to E_v and

(A) The $\phi_v(z)$ have the following mapping properties:

(1) If $K_v \cong \mathbb{C}$, then $\{z \in \mathcal{C}_v(\mathbb{C}_v) : |\phi_v(z)|_v \le R_v^N\} \subset U_v = E_v^0$.

(2) If $K_v \cong \mathbb{R}$, then $\{z \in \mathcal{C}_v(\mathbb{C}_v) : |\phi_v(z)|_v \le 2R_v^N\} \subset U_v$, and for each component $E_{v,j}$ of E_v contained in $\mathcal{C}_v(\mathbb{R})$, if $\phi_v(z)$ has τ_j zeros in $E_{v,j}$,

then $\phi_v(z)$ oscillates τ_j times between $\pm 2R_v^N$ on $E_{v,j}$.

(3) *If K_v is nonarchimedean and $v \in S_K$, then*

$$r_v^N \ < \ q_v^{-1/(q_v-1)} \ < \ 1 \ , \quad \textit{and} \tag{7.23}$$

(a) *the zeros $\theta_1, \ldots, \theta_N$ of $\phi_v(z)$ are distinct;*

(b) *$\phi_v^{-1}(D(0,1)) = \bigcup_{h=1}^N B(\theta_h, \rho_h)$, where $B(\theta_1, \rho_1), \ldots, B(\theta_N, \rho_N)$ are disjoint, isometrically parametrizable, and contained in $\bigcup_{\ell=1}^{D_v} B(a_\ell, r_\ell)$;*

(c) *$H_v := \phi_v^{-1}(D(0,1)) \cap E_v$ has the K_v-simple decomposition*

$$H_v \ = \ \bigcup_{k=1}^{n} \big(B(\theta_k, \rho_k) \cap \mathcal{C}_v(F_{u_h})\big)$$

compatible with the K_v-simple decomposition $\bigcup_{\ell=1}^{D_v} \big(B(a_\ell, r_\ell) \cap \mathcal{C}_v(F_{w_\ell})\big)$ of E_v, which is move-prepared relative to $B(a_1, r_1), \ldots, B(a_{D_v}, r_{D_v})$. For each ℓ, there is a point $\overline{w}_\ell \in \big(B(a_\ell, r_\ell) \cap \mathcal{C}_v(F_{w_\ell})\big) \backslash H_v$.

(d) *Each F_{u_h}/K_v is separable and finite. For each $\theta_h \in E_v \cap B(a_\ell, r_\ell)$, we have $F_{u_h} = F_{w_\ell}$, $\rho_h \in |F_{w_\ell}^\times|_v$, and $B(\theta_h, \rho_h) \subseteq B(a_\ell, r_\ell)$. Moreover, ϕ_v induces an F_{u_h}-rational scaled isometry from $B(\theta_h, \rho_h)$ onto $D(0,1)$ with $\phi_v(\theta_h) = 0$, which takes $B(\theta_h, \rho_h) \cap \mathcal{C}_v(F_{u_h})$ onto $\mathcal{O}_{u_h}$.*

(4) *If K_v is nonarchimedean and $v \in \widehat{S}_K \backslash S_K$, then*

$$E_v \ = \ \widetilde{E}_v \ = \ \{z \in \mathcal{C}_v(\mathbb{C}_v) : |\phi_v(z)|_v \le R_v^N\} \ .$$

(B) *For each $w \in \widehat{S}_L$, put $\phi_w(z) = \phi_v(z)$ if $w|v$, and regard $\phi_w(z)$ as an element of $L_w(\mathcal{C})$. Each $x_i \in \mathfrak{X}$ is canonically embedded in $\mathcal{C}_w(L_w)$; let $\widetilde{c}_{w,i} = \lim_{z \to x_i} \phi_w(z) \cdot g_{x_i}(z)^{Ns_i}$ be the leading coefficient of $\phi_w(z)$ at x_i. Then for each i,*

$$\sum_{w \in \widehat{S}_L} \log_w(|\widetilde{c}_{w,i}|_w) \log(q_w) \ = \ 0 \ .$$

Moreover, $\bigoplus_{w \in \widehat{S}_L} \widetilde{c}_{w,i}$ is an S_L-subunit: there are an integer n_0 and a K-symmetric set of $\widehat{S}_L$-units $\mu_1, \ldots, \mu_m \in L$, such that for each $i = 1, \ldots, m$,

$$\begin{cases} \widetilde{c}_{w,i}^{\,n_0} = \mu_i \ , & \textit{if } w \in \widehat{S}_{L,\infty} \ , \\ |\widetilde{c}_{w,i}^{\,n_0}|_w = |\mu_i|_w \ , & \textit{if } w \in \widehat{S}_{L,0} \ . \end{cases} \tag{7.24}$$

Necessarily $\mu_i \in K(x_i)$ for each i.

PROOF. The proof has several steps, and consists of carefully choosing a compatible collection of initial approximating functions $f_v(z)$ of common degree N in Theorem 7.10, scaling them, and then modifying their leading coefficients so that they satisfy (7.24).

The choice of N. For each archimedean v with $K_v \cong \mathbb{C}$, let $N_v > 0$ be the integer given by Theorem 7.10(A.2) for E_v, ε_v, $\widetilde{E}_v$, and $\vec{s}$ as chosen above. For each archimedean v with $K_v \cong \mathbb{R}$, let $N_v > 0$ be the integer given by Theorem 7.10(B.2) for E_v, ε_v, $\widetilde{E}_v$, U_v, $\mathcal{R}_v$, and $\vec{s}$ as chosen above. For each nonarchimedean $v \in S_{K,0}$, put $\beta_v = \eta_v$ (where $0 < \eta_v \in \mathbb{Q}$ is as in 7.17) and let $N_v > 0$ be the integer given by Theorem 7.10(C.2) for E_v, $\widetilde{E}_v$, $\vec{s}$, β_v and the K_v-simple decomposition $E_v = \bigcup_{\ell=1}^{D_v} \big(B(a_\ell, r_\ell) \cap \mathcal{C}_v(F_{w_\ell})\big)$ chosen above. For each nonarchimedean $v \in \widehat{S}_{K,0} \backslash S_{K,0}$, let N_v be as given by Theorem 7.10(D) for $E_v = \widetilde{E}_v$ and $\vec{s}$ as chosen above.

Let $N > 0$ be an integer which satisfies the following conditions:

(1) N is divisible by N_v, for each $v \in \widehat{S}_K$;
(2) N is divisible by J, the number from the construction of the L-rational and L^{sep}-rational bases in §3.3;
(3) $N \cdot \eta_v \in \mathbb{N}$ for each $v \in \widehat{S}_{K,0}$, where the η_v are as in (7.17);
(4) N is large enough that

(7.25)
- $Ns_i > J$ for each $i = 1, \ldots, m$;
- $1 < 2R_v^N < \widehat{R}_v^N$ for each archimedean v with $K_v \cong \mathbb{R}$;
- $r_v^N < q_v^{-1/(q_v-1)} < 1$ for each nonarchimedean $v \in S_{K,0}$.

In particular (7.23) holds. This N will be fixed for the rest of the construction.

The choice of the Initial Approximating Functions $f_v(z)$. We will apply Theorem 7.10 with the parameters chosen above. Let v_0 be the archimedean place for which (7.22) holds. We first construct the $f_v(z)$ for $v \in \widehat{S}_K \backslash \{v_0\}$, then choose $f_{v_0}(z)$ to "balance" their leading coefficients, so that (7.29) below will hold.

For each archimedean $v \neq v_0$, take $\vec{\beta}_v = (\beta_{v,1}, \ldots, \vec{\beta}_{v,m}) = \vec{0}$ in Theorem 7.10, and let $f_v(z) \in K_v(\mathcal{C})$ be the $(\mathfrak{X}, \vec{s})$-function of degree N given by Theorem 7.10(A.2) if $K_v \cong \mathbb{C}$, or by Theorem 7.10(B.2) with $\mathcal{R}_v$ as in (7.20), if $K_v \cong \mathbb{R}$. Thus the leading coefficients $c_{v,i}$ of $f_v(z)$ satisfy $\frac{1}{N}\log_v(|c_{v,i}|_v) = \Lambda_{x_i}(\widetilde{E}_v, \vec{s})$ for each i, for such v.

For each $v \in S_{K,0}$, take $\beta_v = \eta_v$ as before (with $0 < \eta_v \in \mathbb{Q}$ as in (7.17)) and let $f_v(z) \in K_v(\mathcal{C})$ be the $(\mathfrak{X}, \vec{s})$-function of degree N given by Theorem 7.10(C.2) with $\frac{1}{N}\log_v(|c_{v,i}|_v) = \Lambda_{x_i}(\widetilde{E}_v, \vec{s}) + \beta_v$ for each i. For each $v \in \widehat{S}_{K,0}\backslash S_{K,0}$, let $f_v(z) \in K_v(\mathcal{C})$ be the $(\mathfrak{X}, \vec{s})$-function of degree N from Theorem 7.10(D), with $\frac{1}{N}\log_v(|c_{v,i}|_v) = \Lambda_{x_i}(\widetilde{E}_v, \vec{s})$ for each i.

To construct $f_{v_0}(z)$, we must first specify $\vec{\beta}_{v_0}$ in Theorem 7.10. For each i, define $\beta_{v_0,i}$ by

$$\beta_{v_0,i}\log(q_{v_0}) \;=\; \widetilde{V}_K - (\Gamma(\widetilde{\mathbb{E}}_K, \mathfrak{X})\vec{s})_i \;; \tag{7.26}$$

then $|\beta_{v_0,i}| < \delta_{v_0}$ by (7.22). The vector $\vec{\beta}_{v_0} := (\beta_{v_0,1}, \ldots, \beta_{v_0,m})$ is K-symmetric since $\Gamma(\widetilde{\mathbb{E}}_K, \mathfrak{X})$ and $\vec{s}$ are; this means that $\beta_{v_0,i} = \beta_{v_0,\sigma(i)}$ for each $\sigma \in \mathrm{Gal}(L/K)$. Let $f_{v_0}(z) \in K_{v_0}(\mathcal{C})$ be the $(\mathfrak{X}, \vec{s})$-function of degree N given by Theorem 7.10(A.2) for $\vec{\beta}_{v_0}$ and $\widetilde{E}_{v_0}$ if $K_{v_0} \cong \mathbb{C}$, or by Theorem 7.10(B.2) with $\mathcal{R}_{v_0}$ as in (7.20), if $K_{v_0} \cong \mathbb{R}$.

Thus, for each $v \in \widehat{S}_K$ the leading coefficients $c_{v,i}$ of the $f_v(z)$ satisfy

$$\frac{1}{N}\log_v(|c_{v,i}|_v) \;=\; \begin{cases} \Lambda_{x_i}(\widetilde{E}_v, \vec{s}) & \text{if } v \in \widehat{S}_K\backslash(S_{K_0} \cup \{v_0\}), \\ \Lambda_{x_i}(\widetilde{E}_{v_0}, \vec{s}) + \beta_{v_0,i} & \text{if } v = v_0, \\ \Lambda_{x_i}(\widetilde{E}_v, \vec{s}) + \eta_v & \text{if } v \in \widehat{S}_{K,0}, \end{cases} \tag{7.27}$$

and each $f_v(z)$ has the mapping properties in Theorem 7.10.

In particular, if $v \in S_{K,0}$, then $H_v := f_v^{-1}(D(0,1)) \cap E_v$ has a K_v-simple decomposition $H_v = \bigcup_{h=1}^N \big(B(\theta_h, \rho_h) \cap \mathcal{C}_v(F_{u_h})\big)$ compatible with the K_v-simple decomposition $E_v = \bigcup_{\ell=1}^{D_v} \big(B(a_\ell, r_\ell) \cap \mathcal{C}_v(F_{w_\ell}))\big)$, which is move-prepared relative to $B(a_1, r_1), \ldots, B(a_{D_v}, r_{D_v})$. Here $\theta_1, \ldots, \theta_N$ are the zeros of $f_v(z)$, $\rho_h \in |F_{u_h}^\times|_v$, and f_v induces an F_{u_h}-rational scaled isometry from $B(\theta_h, \rho_h)$ to $D(0,1)$ which maps $B(\theta_h, \rho_h) \cap \mathcal{C}_v(F_{u_h})$ onto $\mathcal{O}_{u_h}$. For each $\ell = 1, \ldots, D_v$, there is a point $\overline{w}_\ell \in \big(B(a_\ell, r_\ell) \cap \mathcal{C}_v(F_{w_\ell}))\big)\backslash H_v$.

Preliminary choice of the Coherent Approximating Functions $\phi_v(z)$. First we will define functions $\phi_v(z) \in K_v(\mathcal{C})$ for $v \in \widehat{S}_K$, and then we will put $\phi_w(z) = \phi_v(z)$ for all $w|v$ and consider the leading coefficients of the collection $\{\phi_w(z) \in L_w(\mathcal{C})\}_{w \in \widehat{S}_L}$. For archimedean v, the $\phi_v(z)$ will be modified later to make the leading coefficients of the $\phi_w(z)$ subunits.

If v is archimedean, put $\kappa_v = e^{N\eta_v}$, where η_v is as in (7.17). Recall from (7.21) that $K_v \cong \mathbb{C}$ we have $\widehat{R}_v = e^{\eta_v} > R_v$, while if $K_v \cong \mathbb{R}$, we have $\widehat{R}_v = \mathcal{R}_v \cdot e^{\eta_v} > R_v$. Thus $\kappa_v > R_v^N$ for each archimedean v, and $\kappa_v \cdot \mathcal{R}_v^N = \widehat{R}_v^N > R_v^N$ if $K_v \cong \mathbb{R}$.

If v is nonarchimedean and $v \in S_{K,0}$, put $\kappa_v = 1$. If $v \in \widehat{S}_{K,0} \backslash S_{K,0}$, put $\kappa_v = \pi_v^{-N\eta_v}$, where again $0 < \eta_v \in \mathbb{Q}$ as in (7.17). Our choice of N required that $N\eta_v \in \mathbb{N}$, so $\kappa_v \in K_v^\times$ and $|\kappa_v|_v = R_v^N > 1$.

For each $v \in \widehat{S}_K$, define

$$\phi_v(z) \;=\; \kappa_v f_v(z) \;\in\; K_v(\mathcal{C}) \ .$$

For each v and each i, the leading coefficient $\widetilde{c}_{v,i}$ of $\phi_v(z)$ at x_i is given by $\widetilde{c}_{v,i} = \kappa_v c_{v,i}$. By (7.27) and our choice of the κ_v, it follows that

$$(7.28) \qquad \frac{1}{N}\log_v(|\widetilde{c}_{v,i}|_v) \;=\; \begin{cases} \Lambda_{x_i}(\widetilde{E}_v, \vec{s}) + \eta_v & \text{if } v \neq v_0, \\ \Lambda_{x_i}(\widetilde{E}_{v_0}, \vec{s}) + \eta_{v_0} + \beta_{v_0,i} & \text{if } v = v_0 \ . \end{cases}$$

Furthermore, the mapping properties of the $f_v(z)$ from Theorem 7.10, together with our choice of the κ_v, yield the following mapping properties for the $\phi_v(z)$.

(1) If $K_v \cong \mathbb{C}$, then $\{z \in \mathcal{C}_v(\mathbb{C}) : |\phi_v(z)| \le \widehat{R}_v^N\} \subset E_v^0$.

(2) If $K_v \cong \mathbb{R}$, then $\{z \in \mathcal{C}_v(\mathbb{C}) : |\phi_v(z)| \le \widehat{R}_v^N\} \subset U_v$, and for each real component $E_{v,j}$ of E_v, if $\phi_v(z)$ has τ_j zeros in $E_{v,j}$ then $\phi_v(z)$ oscillates τ_j times between $\pm\widehat{R}_v^N$ on $E_{v,j}$.

(3) If $v \in S_{K,0}$ then properties (a)-(d) in Theorem 7.11(A.3) hold for $\phi_v(z)$. Indeed, for $v \in S_{K,0}$ we have $\kappa_v = 1$, so $\phi_v(z) = f_v(z)$ and the mapping properties of $\phi_v(z)$ are inherited from those of $f_v(z)$.

(4) If $v \in \widehat{S}_{K,0} \backslash S_{K,0}$, then $E_v = \widetilde{E}_v = \{z \in \mathcal{C}_v(\mathbb{C}_v) : |\phi_v(z)|_v \le R_v^N\}$.

Coherence of the Leading Coefficients. In order to view $\mathfrak{X}$ as a subset of $\mathcal{C}_v(\mathbb{C}_v)$, for each v we have (noncanonically) fixed an embedding of $\widetilde{K}$ in $\mathbb{C}_v$ (see §3.2), leading to a distinguished choice of a place w_v of L above v. Until now these choices have been a minor concern, since all constructions in the proof have been K_v-symmetric. However, to properly understand the leading coefficients of the $\phi_v(z)$, we must consider them over the fields L_w for $w \in \widehat{S}_L$, since $\mathfrak{X}$ is canonically a subset of $\mathcal{C}(L)$ and of $\mathcal{C}_w(L_w)$ for each w, and the uniformizer $g_{x_i}(z) \in L(\mathcal{C})$ is canonically an element of $L_w(\mathcal{C})$.

For each $w \in \widehat{S}_L$, put $\phi_w(z) = \phi_v(z)$ if $w|v$, and view $\phi_w(z)$ as an element of $L_w(\mathcal{C})$. Although the functions $\phi_w(z)$ for $w|v$ are all the same, the points of $\mathfrak{X}$, which are their poles, are identified differently. For each i and w, let $\widetilde{c}_{w,i} = \lim_{z \to x_i} \phi_w(z) \cdot g_{x_i}(z)^{Ns_i}$ be the leading coefficient of $\phi_w(z)$ at x_i. Let $\sigma_w : L \hookrightarrow \mathbb{C}_v$ be an embedding which induces the place w, and for each $i = 1, \ldots, m$ let $\sigma_w(i)$ be the index j for which $\sigma_w(x_i) = x_j$ (where we identify x_j with its image in $\mathcal{C}_v(\mathbb{C}_v)$ given by the fixed embedding of $\widetilde{K}$ in $\mathbb{C}_v$). Then $\widetilde{c}_{w,i} = \widetilde{c}_{v,\sigma_w(i)}$.

The following proposition is the first step towards making the leading coefficients of the patching functions $\widehat{S}_L$-units:

PROPOSITION 7.12. *For each $i = 1, \ldots, m$,*

$$\sum_{w \in \widehat{S}_L} \log_w(|\widetilde{c}_{w,i}|_w) \log(q_w) \;=\; 0 \;. \tag{7.29}$$

To prove this we will need a lemma. First note that by our normalizations of the absolute values on L and K, if $w|v$ and $x \in \mathbb{C}_v \cong \mathbb{C}_w$, then

$$\log_w(|x|_w) \log(q_w) \;=\; [L_w : K_v] \cdot \log_v(|x|_v) \log(q_v) \;. \tag{7.30}$$

Recall that $\Gamma(\widetilde{\mathbb{E}}, \mathfrak{X}) = \Gamma(\widetilde{\mathbb{E}}_K, \mathfrak{X}) = \frac{1}{[L:K]} \Gamma(\widetilde{\mathbb{E}}_L, \mathfrak{X})$. Using (7.2), this gives

$$[L : K] \cdot \Gamma(\widetilde{\mathbb{E}}_K, \mathfrak{X}) \;=\; \sum_{w \in \widehat{S}_L} \Gamma(\widetilde{E}_w, \mathfrak{X}) \log(q_w) \;, \tag{7.31}$$

LEMMA 7.13. *For each $w \in \widehat{S}_L$ and each $i = 1, \ldots, m$, the ith coordinate of $\Gamma(\widetilde{E}_w, \mathfrak{X})\vec{s}$ satisfies*

$$(\Gamma(\widetilde{E}_w, \mathfrak{X})\vec{s})_i \cdot \log(q_w) \;=\; [L_w : K_v] \cdot \Lambda_{\sigma_w(x_i)}(\widetilde{E}_v, \vec{s}) \log(q_v) \;. \tag{7.32}$$

PROOF. By definition,

$$(\Gamma(\widetilde{E}_w, \mathfrak{X})\vec{s})_i \cdot \log(q_w) \;=\; \sum_{j=1}^{m} \Gamma(\widetilde{E}_w, \mathfrak{X})_{ij} \log(q_w) s_j \;. \tag{7.33}$$

For each $i \neq j$, it follows from (7.30) that

$$\begin{aligned} \Gamma(\widetilde{E}_w, \mathfrak{X})_{ij} \log(q_w) \;&=\; G(x_i, x_j; \widetilde{E}_w) \log(q_w) \\ &=\; [L_w : K_v]\, G(\sigma_w(x_i), \sigma_w(x_j), \widetilde{E}_v) \log(q_v) \;. \end{aligned}$$

Similarly, since the uniformizers have been chosen in such a way that $\sigma_w(g_{x_i})(z) = g_{\sigma_w(x_i)}(z)$, for each i,

$$\Gamma(\widetilde{E}_w, \mathfrak{X})_{ii} \log(q_w) \;=\; V_{x_i}(\widetilde{E}_w) \log(q_w) \;=\; [L_w : K_v] V_{\sigma_w(x_i)}(\widetilde{E}_v) \log(q_v) \;.$$

For compactness of notation, write

$$\widetilde{G}(x_i, x_j; E_v) \;=\; \begin{cases} G(x_i, x_j; \widetilde{E}_v) & \text{if } i \neq j \;, \\ V_{x_i}(\widetilde{E}_v) & \text{if } i = j \;. \end{cases}$$

Since $\vec{s}$ is K-symmetric, we have $s_j = s_{\sigma_w(j)}$ for each j. Hence

$$\begin{aligned} (\Gamma(\widetilde{E}_w, \mathfrak{X})\vec{s})_i \cdot \log(q_w) \;&=\; \sum_{j=1}^{m} \Gamma(\widetilde{E}_w, \mathfrak{X}) s_j \log(q_w) \\ &=\; [L_w : K_v] \cdot \sum_{j=1}^{m} \widetilde{G}(\sigma_w(x_i), \sigma_w(x_j); \widetilde{E}_v) s_{\sigma_w(j)} \log(q_v) \\ &=\; [L_w : K_v] \cdot \Lambda_{\sigma_w(x_i)}(\widetilde{E}_v, \vec{s}) \log(q_v) \end{aligned}$$

as desired. □

PROOF OF PROPOSITION 7.12. For each v, and each $w|v$, fix an embedding $\sigma_w : L \hookrightarrow \mathbb{C}_v$ which induces the place w. Since $\widetilde{c}_{w,i} = \widetilde{c}_{v,\sigma_w(i)}$, it follows from (7.28) that

$$\begin{aligned}\frac{1}{N}\sum_{w\in\widehat{S}_L}\log_w(|\widetilde{c}_{w,i}|_w)\log(q_w) &= \sum_{v\in\widehat{S}_K}\sum_{w|v}[L_w:K_v]\big(\frac{1}{N}\log_v(|\widetilde{c}_{v,\sigma_w(x_i)}|_v)\big)\log(q_v)\\ &= \sum_{v\in\widehat{S}_K}\sum_{w|v}[L_w:K_v]\big(\Lambda_{\sigma_w(x_i)}(\widetilde{E}_v,\vec{s})+\eta_v\big)\log(q_v)\\ &\quad + \sum_{w|v_0}[L_w:K_v]\beta_{v_0,\sigma_w(i)}\log(q_{v_0})\ . \qquad (7.34)\end{aligned}$$

By Lemma 7.13

$$\begin{aligned}\sum_{v\in\widehat{S}_K}\sum_{w|v}[L_w:K_v]\Lambda_{\sigma_w(x_i)}(\widetilde{E}_v,\vec{s})\log(q_v) &= \sum_{w\in\widehat{S}_L}(\Gamma(\widetilde{E}_w,\mathfrak{X})\vec{s})_i\log(q_w)\\ = (\Gamma(\widetilde{\mathbb{E}}_L,\mathfrak{X})\vec{s})_i &= [L:K]\cdot(\Gamma(\widetilde{\mathbb{E}}_K,\mathfrak{X})\vec{s})_i\ . \qquad (7.35)\end{aligned}$$

By our choice of the η_v in (7.17),

$$\sum_{v\in\widehat{S}_K}\sum_{w|v}[L_w:K_v]\eta_v\log(q_v) = [L:K]\sum_{v\in\widehat{S}_K}\eta_v\log(q_v) = -[L:K]\cdot\widetilde{V}_K\ . \qquad (7.36)$$

Finally, since $\vec{\beta}_{v_0}$ is K-symmetric, for each $w|v_0$ we have $\beta_{v_0,\sigma_w(i)} = \beta_{v_0,i}$, so by (7.26)

$$\sum_{w|v_0}[L_w:K_v]\beta_{v_0,\sigma_w(i)}\log(q_{v_0}) = [L:K]\cdot\big(\widetilde{V}_K-(\Gamma(\widetilde{\mathbb{E}}_K,\mathfrak{X})\vec{s})_i\big)\ . \qquad (7.37)$$

Combining (7.34), (7.35), (7.36) and (7.37) gives

$$\sum_{w\in\widehat{S}_L}\log_w(|\widetilde{c}_{w,i}|_w)\log(q_w) = 0$$

as required. □

Adjusting the Leading Coefficients to be $\widehat{S}_L$-subunits. The final step in the proof of Theorem 7.11 involves modifying the archimedean $\phi_v(z)$ so that their leading coefficients become $\widehat{S}_L$-subunits.

By our choices of the R_v, $\widehat{R}_v$, and N we have $R_v^N < \widehat{R}_v^N$ for each archimedean v, and $2R_v^N < \widehat{R}_v^N$ for each archimedean v with $K_v \cong \mathbb{R}$. For each $i = 1, \ldots, m$, put $N_i = Ns_i \in \mathbb{N}$. Noting that our choice of N has required that $N_i > J$, let $\phi_{i,N_i}(z)$ be the corresponding function from the L-rational basis.

If $K_v \cong \mathbb{C}$, the construction of $\phi_v(z)$ has arranged that

$$\{z\in\mathcal{C}_v(\mathbb{C}_v) : |\phi_v(z)| \le \widehat{R}_v^N\} \subset U_v\ ,$$

Since $R_v^N < \widehat{R}_v^N$, by continuity there is a $\delta_v' > 0$ such that if $\Delta_{v,1},\ldots,\Delta_{v,m} \in \mathbb{C}$ satisfy $|\Delta_{v,i}| < \delta_v'$ for each i, and if $\phi_v(z)$ is replaced by $\widehat{\phi}_w(z) = \phi_w(z) + \sum_{i=1}^m \Delta_{w,i}\varphi_{i,N_i}(z)$, then

$$\{z\in\mathcal{C}_v(\mathbb{C}_v) : |\widehat{\phi}_w(z)|_v \le R_v^N\} \subset U_v\ . \qquad (7.38)$$

If $K_v \cong \mathbb{R}$, the construction of $\phi_v(z)$ has arranged that

$$\{z\in\mathcal{C}_v(\mathbb{C}_v) : |\phi_v(z)| \le 2\widehat{R}_v^N\} \subset U_v\ ,$$

and that for each component $E_{v,i}$ of E_v contained in $\mathcal{C}_v(\mathbb{R})$, if $\phi_v(z)$ has τ_i zeros in $E_{v,i}$ then it oscillates τ_i times between $\pm 2\widehat{R}_v^N$ on $E_{v,i}$. Since $2R_v^N < 2\widehat{R}_v^N$, by continuity there is a $\delta_v' > 0$ such that if $\Delta_{v,1}, \ldots, \Delta_{v,m} \in \mathbb{C}$ are a K_v-symmetric set of numbers with $|\Delta_{v,i}| < \delta_v'$ for each i, and if $\phi_v(z)$ is replaced by $\widehat{\phi}_v(z) = \phi_v(z) + \sum_{i=1}^m \Delta_{v,i}\varphi_{i,N_i}(z)$, then

$$\{z \in \mathcal{C}_v(\mathbb{C}_v) : |\widehat{\phi}_v(z)| \le 2R_v^N\} \subseteq U_v \tag{7.39}$$

and for each component $E_{v,i}$ contained in $\mathcal{C}_v(\mathbb{R})$, if $\phi_v(z)$ has τ_i zeros in $E_{v,i}$ then $\widehat{\phi}_v(z)$ oscillates τ_i times between $\pm 2R_v^N$ on $E_{v,i}$.

Let δ' be the minimum of the δ_v', for all $v \in \widehat{S}_{K,\infty}$.

Fix $x_i \in \mathfrak{X}$, and put $F = K(x_i)$. Let $\widehat{S}_F$ be the set of places of F above $\widehat{S}_K$. For each $v \in \widehat{S}_K$, since the $\phi_w(z) \in K_v(\mathcal{C})$ are the same for all $w|v$, Proposition 7.8 tells us that $\bigoplus_{w|v} \widetilde{c}_{w,i} \in \bigoplus_{w|v} L_w$ actually belongs to $\bigoplus_{u|v} F_u$, embedded semi-diagonally in $\bigoplus_{w|v} L_w$. Write $\bigoplus_{u|v} \widetilde{c}_{u,i}$ for the element of $\bigoplus_{u|v} F_u$ that induces it. By (7.29),

$$\sum_{u \in \widehat{S}_F} \log_u(|\widetilde{c}_{u,i}|_u)\log(q_u) \;=\; \frac{1}{[L:F]} \sum_{w \in \widehat{S}_L} \log_w(|\widetilde{c}_{w,i}|_w)\log(q_w) \;=\; 0\ .$$

According to Proposition 7.5 there are an $\widehat{S}_F$-unit $\mu_i \in F$, an integer n_i, and an $\widehat{S}_F$-subunit $\bigoplus_{u \in \widehat{S}_{F,\infty}} \varepsilon_{u,i} \in \bigoplus_{u \in \widehat{S}_{F,\infty}} F_u^\times$ such that $\varepsilon_{u,i}^{n_i} = \mu_i$ for each $u \in \widehat{S}_{F,\infty}$, and

$$\begin{cases} |\widetilde{c}_{u,i} - \varepsilon_{u,i}| < \delta'\,, & \\ |\widetilde{c}_{u,i}^{\,n_i}|_u = |\mu_i|_u\,, & \text{for each } u \in \widehat{S}_{F,0}\ . \end{cases}$$

Since μ_i is an $\widehat{S}_F$-unit (hence also an $\widehat{S}_L$-unit), $\sum_{w \in \widehat{S}_L} \log_u(|\mu_i|_w)\log(q_w) = 0$.

For each archimedean u and each $w|u$, put $\varepsilon_{w,i} = \varepsilon_{u,i}$. Then $\log_w(|\varepsilon_{w,i}|_w) = \frac{1}{n_i}\log_w(|\mu_i|_w)$ for each archimedean $w \in S_L$, and $\log_w(|c_{w,i}|_w) = \frac{1}{n_i}\log_w(|\mu_i|_w)$ for each nonarchimedean $w \in S_L$. It follows that

$$\sum_{w \in S_{L,\infty}} \log_w(\varepsilon_{w,i})\log(q_w) + \sum_{w \in S_{L,0}} \log_w(\widetilde{c}_{w,i})\log(q_w) \;=\; 0\ . \tag{7.40}$$

Now let x_i vary. We next arrange for the μ_i and $\vec{\varepsilon}_{v,i} = \bigoplus_{w|v} \varepsilon_{w,i}$ to be K-symmetric. By Proposition 7.8, the $\bigoplus_{w|v} \widetilde{c}_{w,i}$ are K-symmetric. For each $\mathrm{Gal}(L/K)$-orbit $\mathfrak{X}_\ell \subset \mathfrak{X}$, fix an $x_i \in \mathfrak{X}_\ell$ and put $F = K(x_i)$ as before. By construction, $\vec{\varepsilon}_{v,i} := \bigoplus_{w|v} \varepsilon_{w,i} \in L \otimes_K K_v$ belongs to $F \otimes_K K_v$ for each archimedean v. For each $x_j \in \mathfrak{X}_\ell$, choose $\sigma \in \mathrm{Gal}(L/K)$ with $\sigma(x_i) = x_j$, and replace μ_j with $\sigma(\mu_i)$, $\vec{\varepsilon}_{v,j}$ with $\sigma(\vec{\varepsilon}_{v,i})$. Since $\vec{\varepsilon}_{v,i} \in F \otimes_K K_v$, these objects are independent of the choice of σ with $\sigma(x_i) = x_j$, and are K-symmetric.

After replacing the μ_i with powers of themselves, we can assume there is a number n_0 such that $n_i = n_0$, for all i. The numbers $\mu_1, \ldots, \mu_m$ form a K-symmetric system of $\widehat{S}_L$-units, and the $\varepsilon_{w,i}$ form a K-symmetric system of $\widehat{S}_L$-subunits. For each archimedean w and each i, put

$$\Delta_{w,i} = \varepsilon_{w,i} - c_{w,i}$$

and put $\widehat{\phi}_w(z) = \phi_w(z) + \sum_{i=1}^m \Delta_{w,i}\varphi_{i,N_i}(z)$. Since $\bigoplus_{w|v} \Delta_{w,i}\varphi_{i,N_i} \in L \otimes_K K_v(\mathcal{C})$ and $\bigoplus_{w|v} \phi_w(z) \in L \otimes_K K_v(\mathcal{C})$ are K-symmetric, Proposition 7.8 shows that $\widehat{\phi}_w(z)$ belongs to $K_v(\mathcal{C})$ for each $w|v$, and that $\widehat{\phi}_{w_1}(z) = \widehat{\phi}_{w_2}(z)$ for all $w_1, w_2|v$.

Replace $\phi_w(z)$ with $\widehat{\phi}_w(z)$, for each archimedean v and each $w|v$. The leading coefficients of the new $\phi_w(z)$ are the $\varepsilon_{w,i}$, so we can put $\phi_v(z) = \widehat{\phi}$, for any $w|v$. By (7.40), assertion (B) in the theorem holds. Our construction has established assertions (A1)–(A4), so the proof of Theorem 7.11 is complete. □

Stage 3. The patching construction.

Overview. The patching process has two parts, a global part and a local part. The global part concerns the way the patching coefficients are chosen, managing them so as to achieve global K-rationality for the final patched function. The local part is responsible for assuring K_v-rationality of the partially patched functions, and confining their roots to E_v.

Although this description separates the roles of the global and local parts of patching process, in fact the two interact, and the coefficients are determined recursively, from highest to lowest order. Each local patching construction specifies certain parameters to the global patching process: the number of patching stages it considers high-order and bounds for the size of the patching coefficients it can handle. As patching is carried out, and high-order coefficients chosen by the global process are achieved by the local process, lower-order coefficients are changed as a result. The global process must take these changes into account in determining subsequent coefficients.

The process begins with the coherent approximating functions $\{\phi_v(z)\}_{v \in \widehat{S}_K}$ given by Theorem 7.11. Its goal is to produce a function $G(z) \in K(\mathcal{C})$ independent of v, of much higher degree than the $\phi_v(z)$, whose zeros are points with the properties in Theorem 4.2.

The first step is to compose each $\phi_v(z)$ with a "degree-raising polynomial" $Q_{v,n}(x) \in K_v(x)$. The $Q_{v,n}(x)$ are monic, of common degree n. This allows the leading coefficients to be patched to become $\widehat{S}_L$-units, and makes the degree large enough that certain analytic estimates are satisfied, while keeping the roots in E_v.

For each $v \in \widehat{S}_K$, the local patching process provides $Q_{v,n}(z)$. If $K_v \cong \mathbb{C}$ or if K_v is nonarchimedean and E_v is an RL-domain, then $Q_{v,n}(x) = x^n$. If $K_v \cong \mathbb{R}$, then $Q_{v,n}(x)$ is a composite of two Chebyshev polynomials. If K_v is nonarchimedean and $E_v \subset \mathcal{C}(K_v)$, then $Q_{v,n}(x)$ is the Stirling polynomial of degree n for the ring of integers $\mathcal{O}_v$. For appropriately large and divisible n, this yields the "initial patching functions" $G_v^{(0)}(z) = Q_{v,n}(\phi_v(z))$. Although N is likely quite large, n should be thought of as astronomically larger than N.

Each $G_v^{(0)}(z)$ can be expanded in terms of the L-rational basis functions $\varphi_{i,j}(z)$ and φ_λ, with L_w-rational coefficients for each $w|v$. For notational purposes, it will be useful to deem the basis functions $\varphi_\lambda(z)$ for $\lambda \leq \Lambda_0$ and $\varphi_{i,j}(z)$ with $J < j \leq N_i := Ns_i$, as being "low-order", and list them as φ_λ, $\lambda = 1, \ldots, \Lambda$. Thus

$$G_v^{(0)}(z) = \sum_{i=1}^{m} \sum_{j=0}^{(n-1)N_i-1} A_{v,ij}\varphi_{i,nN_i-j}(z) + \sum_{\lambda=1}^{\Lambda} A_{v,\lambda}\varphi_\lambda .$$

The patching process initially adjusts the leading coefficients of the $G_v^{(0)}(z)$ to be global S_L-units, independent of v. Then, in stages, it inductively constructs functions $G_v^{(1)}(z), \ldots, G_v^{(n)}(z)$, where

$$G_v^{(k)}(z) \;=\; G_v^{(k-1)}(z) + \sum_{i=1}^{m} \sum_{j=(k-1)N_i}^{kN_i-1} \Delta_{v,ij}^{(k)} \vartheta_{v,ij}^{(k)}(z)$$

for $1 \le k \le n-1$ (see Theorem 7.14 for a more precise statement), and

$$G_v^{(n)}(z) \;=\; G_v^{(n-1)}(z) + \sum_{\lambda=1}^{\Lambda} \Delta_{v,\lambda}^{(n)} \varphi_\lambda(z)$$

for $k = n$, in such a way that for each v the coefficients $A_{v,ij}$, $A_{v,\lambda}$ are changed into global S_L-integers A_{ij}, A_λ independent of v. This process is called "patching", because it pieces together a global function out of a collection of local ones.

The "compensating functions" $\vartheta_{v,ij}^{(k)}(z)$, indexed by pairs (i,j) with $(k-1)N_i \le j \le kN_i - 1$ in "bands" for $k = 1, \ldots, n-1$, are determined by the local patching process and have poles are supported on $\mathfrak{X}$. Each $\vartheta_{v,ij}^{(k)}(z)$ has a pole of exact order $nN_i - j$ at x_i, and is chosen so that adding $\Delta_{v,ij}^{(k)}\vartheta_{v,ij}^{(k)}(z)$ to $G_v^{(k)}(z)$ affects only poles of order $nN_i - j$ and below at x_i, and lower order poles outside the band, for $x_{i'} \ne x_i$. It was Fekete and Szegö's insight ([**25**]) that by using compensating functions more complicated than the basis functions $\varphi_{ij}(z)$, one could control movement of the roots of the $G_v^{(k)}(z)$. For each x_i, the $\Delta_{v,ij}^{(k)}$ are chosen by ascending j (decreasing order of the pole), so that coefficients A_{ij} already patched are not changed in subsequent steps.

The global patching process has two concerns.

First, it must choose the $\Delta_{v,ij}^{(k)}$ in such a way that for each (i,j),

$$A_{ij} \;=\; A_{v,ij} + \Delta_{v,ij}^{(k)}$$

is a global $\widehat{S}_L$ integer independent of v. This is accomplished by extending the base to L and choosing the A_{ij} via Proposition 7.3, simultaneously patching the coefficients for all x_i belonging to a given Galois orbit in $\mathfrak{X}$.

Second, it must impose conditions on the sizes of the $\Delta_{v,ij}^{(k)}$ so that the local patching constructions can succeed. The choice of the $\Delta_{v,ij}^{(k)}$ involves tension between the global and local parts of the patching process. The global part is charged with adjusting the $A_{v,ij}$ to make them algebraic numbers in L independent of v. Doing so may require the $\Delta_{v,ij}^{(k)}$ to be fairly large. On the other hand, the local part is charged with assuring that the roots of $G_v^{(k)}(z)$ remain in E_v. For this, it is usually necessary that the $\Delta_{v,ij}^{(k)}$ be fairly small.

If there is a bound $B_v > 0$ such that in the k-th stage of the local patching construction the $\Delta_{v,ij}^{(k)} \in L_v$ can be chosen arbitrarily, provided that $|\Delta_{v,ij}^{(k)}|_v \le B_v$ for all i, j, and the $\Delta_{v,ij}^{(k)}$ are K_v-symmetric, we will say that the coefficients $A_{v,ij}$ for $(k-1)N_i \le j < kN_i$ can be *sequentially patched with freedom* B_v. Equivalently, we will say that *the k-th stage of the local patching process at v can be carried out with freedom* B_v.

As k increases, there is greater and greater freedom in the patching process. However, for small k, balancing the demands of the global and local patching constructions requires care. The leading coefficients are the hardest to patch, and they are controlled through the choice of n. The high-order coefficients are also quite difficult to patch. It turns out that there is a number $\overline{k}$, determined by the sets E_v and initial approximating functions f_v but fortunately independent of n, such that when $1 \le k \le \overline{k}$, the nonarchimedean $\Delta_{v,ij}^{(k)}$ must be very small. To compensate, we must allow the archimedean $\Delta_{v,ij}^{(k)}$ to be quite large.

The archimedean patching procedures accomplish this by exploiting a phenomenon of "magnification" introduced in ([**53**]), by which small changes in $f_v(z)$ create large changes in the leading coefficients of $G_v(z)$. It is shown in Theorems 8.8.1 and 9.9.1 that for any fixed $B_v > 0$, magnification enables us to carry out the first $\overline{k}$ stages of the patching process with freedom B_v.

In combination, the local and global patching processes determine $\overline{k}$, the number of patching stages deemed high-order. For appropriate numbers B_v, we will have

$$|\Delta_{v,ij}^{(k)}|_v \le \begin{cases} B_v & \text{if } k \le \overline{k}, \\ h_v^{kN} & \text{if } k > \overline{k}. \end{cases}$$

If these conditions are met, the local patching constructions will succeed. On the other hand, for global target coefficients $A_{ij} \in L$ to exist, $(\prod_v B_v^{D_v})$ and $(\prod_v h_v^{D_v})^{\overline{k}}$ must be large enough that Proposition 7.3 applies. Achieving this uses condition (7.19) that $\prod_v h_v^{D_v} > 1$, which ultimately depends on the fact that $\gamma(\mathbb{E}, \mathfrak{X}) > 1$.

The final patched functions $G_v^{(n)}(z)$ are K_v-rational but have all their coefficients in L. By Proposition 7.8 there is a global function $G^{(n)}(z) \in K(\mathcal{C})$, independent of v, such that $G^{(n)}(z) = G_v^{(n)}(z)$ for each $v \in \widehat{S}_K$. The local patching constructions assure that its zeros belong to E_v for all $v \in \widehat{S}_K$. For each $v \notin \widehat{S}_K$, the coefficients of $G^{(n)}(z)$ are $\widehat{S}_L$-integers and its leading coefficients are $\widehat{S}_L$-units, so the fact that the basis functions $\varphi_{ij}(z)$ and φ_λ have good reduction outside $\widehat{S}_K$, combined with the fact that E_v is $\mathfrak{X}$-trivial, show that $\{z \in \mathcal{C}_v(\mathbb{C}_v) : |G^{(n)}(z)|_v \le 1\} = E_v$. Thus the zeros of $G^{(n)}(z)$ belong to E_v for all v.

Details. We now give the details of the patching construction. Let K, S_K, $\widehat{S}_K$, $\mathbb{E}$, $\widetilde{\mathbb{E}}$, and the sets U_v for $v \in \widehat{S}_{K,\infty}$ be as in Stage 1 of the proof. Let $\vec{s} \in \mathcal{P}^m(\mathbb{Q})$ be the K-symmetric vector with positive rational coefficients from (7.22), and let h_v, r_v, and R_v be the local patching parameters from (7.18), with $1 < h_v < r_v < R_v$ for archimedean v and $0 < h_v < r_v \le 1 \le R_v$ for nonarchimedean v. Let the natural number N, the coherent $(\mathfrak{X}, \vec{s})$-functions $\{\phi_v(z)\}_{v \in \widehat{S}_K}$ of degree N, and the $\widehat{S}_K$-units μ_i from Theorem 7.11, be as Stage 2 of the proof.

For each v, let w_v be the distinguished place of $L = K(\mathfrak{X})$ over v, induced by the embedding $\widetilde{K} \hookrightarrow \mathbb{C}_v$ chosen in §3.2. This induces an embedding $L_{w_v} \hookrightarrow \mathbb{C}_v$, and allows us to identify $\mathfrak{X}$ with a subset of $\mathcal{C}_v(\mathbb{C}_v)$. We will use these embeddings in comparing coefficients of functions over K and over L.

The ordering $\prec_N$ on $\mathcal{I}$. We will now define an ordering $\prec_N$ on the index set $\mathcal{I} = \{(i,j) \in \mathbb{Z}^2 : 1 \le i \le m, 0 \le j < \infty\}$ which specifies the sequence in which the coefficients are patched.

Let $\mathrm{Gal}(L/K)$ act on $\mathcal{I}$ in a K-symmetric way through its first coordinate, so $\sigma(i,j) = (\sigma(i), j)$ if $\sigma(x_i) = x_{\sigma(i)}$. With $\vec{s}$ as in (7.22) and N as in Theorem 7.11, put $N_i = Ns_i$ for $i = 1, \ldots, m$. For each $(i,j) \in \mathcal{I}$, we can uniquely write $j = (k-1)N_i + r$ with $k, r \in \mathbb{Z}$, $k \ge 1$ and $0 \le r < N_i$; put $k_N(i,j) = k$ and $r_N(i,j) = r$. Let the $\mathrm{Gal}(L/K)$-orbits in $\mathfrak{X} = \{x_1, \ldots, x_m\}$ be $\mathfrak{X}_1, \ldots, \mathfrak{X}_{m_1}$.

Without loss, we can assume the x_i in a given $\mathfrak{X}_\ell$ have consecutive indices. If $x_i \in \mathfrak{X}_\ell$, put $\ell(i,j) = \ell$.

Let $\prec_N$ be the total order on $\mathcal{I}$ defined by $(i_1,j_1) \prec_N (i_2,j_2)$ iff

$$(7.41) \quad \begin{cases} j_1 = j_2 = 0 \text{ and } i_1 < i_2, & \text{or} \\ k_N(i_1,j_1) < k_N(i_2,j_2), & \text{or} \\ k_N(i_1,j_1) = k_N(i_2,j_2),\ \max(j_1,j_2) \geq 1 \text{ and } \ell(i_1,j_1) < \ell(i_2,j_2), & \text{or} \\ k_N(i_1,j_1) = k_N(i_2,j_2),\ \ell(i_1,j_1) = \ell(i_2,j_2), \text{ and } j_1 < j_2, & \text{or} \\ k_N(i_1,j_1) = k_N(i_2,j_2),\ \ell(i_1,j_1) = \ell(i_2,j_2),\ j_1 = j_2 \geq 1, \text{ and } i_1 < i_2. \end{cases}$$

Write $(i_1,j_1) \preceq_N (i_2,j_2)$ iff $(i_1,j_1) \prec_N (i_2,j_2)$ or $(i_1,j_1) = (i_2,j_2)$. Define the "bands" of $\prec_N$, for $k = 1,2,\ldots$ by

$$(7.42) \qquad \mathrm{Band}_N(k) \;=\; \{(i,j) \in \mathcal{I} : k_N(i,j) = k\}\ .$$

Note that the indices $(i,0)$ for the leading coefficients form the initial segment under $\prec_N$, and are contained in $\mathrm{Band}_N(1)$.

Let $\cong_N$ be the equivalence relation on $\mathcal{I}$ defined by $(i_1,j_1) \cong_N (i_2,j_2)$ iff

$$\begin{cases} j_1 = j_2, \text{ and} \\ x_{i_1},\ x_{i_2} \text{ belong to the same Galois orbit } \mathfrak{X}_\ell. \end{cases}$$

Equivalently, $(i_1,j_1) \cong_N (i_2,j_2)$ iff $\sigma(\varphi_{i_1,j_1}) = \varphi_{i_2,j_2}$ for some $\sigma \in \mathrm{Gal}(L/K)$. Define the "Galois blocks" of $\mathcal{I}$ to be the equivalence classes for $\cong_N$, and write

$$\mathrm{Block}(i,j) \;=\; \{(i_1,j_1) \in \mathcal{I} : (i_1,j_1) \cong_N (i,j)\} \;=\; \mathrm{Gal}(L/K)\big((i,j)\big)\ .$$

In patching, coefficients will be adjusted in $\prec_N$ order. This means that the leading coefficients are modified first, then the remaining coefficients are considered band by band. Within each band, they are considered block by block. For each i, they are considered by increasing j. The global patching process simultaneously determines all the coefficients for a given block.

Summary of the Local Patching Theorems. The global patching process interacts with the local patching processes to adjust the coefficients. The following Theorem summarizes the local patching constructions proved in Theorems 8.1, 9.1, 10.1, and 11.1 below.

THEOREM 7.14. *Let K be a number field. Let $\mathcal{C}/K$, $\mathbb{E}$, $\mathfrak{X}$, and S_K be as in Theorem 4.2. Let $\widehat{S}_K \supseteq S_K$ be the finite set of places satisfying conditions (7.1). For each $v \in \widehat{S}_K$, let $\widetilde{E}_v \subset E_v$, and $0 < h_v < r_v < R_v$ be the set and patching parameters constructed in Stage 1 of the proof. For each $v \in \widehat{S}_{K,\infty}$, let $U_v \subset \mathcal{C}_v(\mathbb{C})$ be the chosen open set with $U_v \cap E_v = E_v^0$. For each $v \in S_{K,0}$, let $\bigcup_{\ell=1}^{D_v} B(a_\ell, r_\ell) \cap \mathcal{C}_v(F_{w_\ell})$ be the chosen K_v-simple decomposition of E_v. Let the rational probability vector $\vec{s} \in \mathcal{P}^m(\mathbb{Q})$ be as in (7.22), and let the natural number N and the coherent approximating functions $\{\phi_v(z)\}_{v \in \widehat{S}_K}$ be those constructed in Theorem 7.11 in Stage 2 of the proof.*

Then for each $v \in \widehat{S}_K$, there is a constant $k_v > 0$ determined by the E_v, U_v, and $\phi_v(z)$, representing the minimal number of "high-order" stages in the local patching process for K_v. Let $\overline{k} \geq k_v$ be a fixed integer. If v is nonarchimedean, put $B_v = h_v^{\overline{k}N}$; if v is archimedean, let $B_v > 0$ be arbitrary. Then there is an integer $n_v > 0$, depending on $\overline{k}$ and B_v, such that for each sufficiently large integer n divisible by n_v, one can carry out the local patching process at K_v as follows:

Put $G_v^{(0)}(z) = Q_{v,n}(\phi_v(z))$, *where*

$$\begin{cases} \text{If } K_v \cong \mathbb{C}, \text{ then } Q_{v,n}(x) = x^n. \\ \text{If } K_v \cong \mathbb{R}, \text{ set } \widehat{R}_v = 2^{-1/n_v N} R_v, \text{ write } n = m_v n_v, \text{ and let } T_{m,R}(x) \text{ be the} \\ \quad \text{Chebyshev polynomial of degree } m \text{ for } [-2R, 2R] \text{ (see (9.1)). Then} \\ \qquad Q_{v,n}(x) \ = \ T_{m_v, \widehat{R}_v^{n_v N}}(T_{n_v, R_v^N}(x)). \\ \text{If } K_v \text{ is nonarchimedean and } v \in S_{K,0}, \text{ then } Q_{v,n}(x) = S_{n,v}(x) \\ \quad \text{is the Stirling polynomial of degree } n \text{ for } \mathcal{O}_v \text{ (see (3.56)).} \\ \text{If } K_v \text{ is nonarchimedean and } v \in \widehat{S}_{K,0} \backslash S_{K,0}, \text{ then } Q_{v,n}(x) = x^n. \end{cases}$$

For each k, $1 \le k \le n-1$, *let* $\{\Delta_{v,ij}^{(k)} \in \mathbb{C}_v\}_{(i,j)\in \mathrm{Band}_N(k)}$ *be a* K_v*-symmetric set of numbers, given recursively in* $\prec_N$ *order, subject to the conditions that* $\Delta_{v,i0}^{(1)} = 0$ *for each archimedean* v, *and for all* (i,j),

$$(7.43) \qquad |\Delta_{v,ij}^{(k)}|_v \ \le \ \begin{cases} B_v & \text{if } k \le \overline{k}\ , \\ h_v^{kN} & \text{if } k > \overline{k}\ . \end{cases}$$

For $k = n$, *let* $\{\Delta_{v,\lambda}^{(n)} \in \mathbb{C}_v\}_{1 \le \lambda \le \Lambda}$ *be a* K_v*-symmetric set of numbers satisfying*

$$(7.44) \qquad |\Delta_{v,\lambda}^{(n)}|_v \ \le \ h_v^{nN}\ .$$

Then for each $v \in \widehat{S}_K$, *one can inductively construct* $(\mathfrak{X}, \vec{s})$*-functions* $G_v^{(1)}(z), \ldots, G_v^{(n)}(z)$ *in* $K_v(\mathcal{C})$, *of common degree* Nn, *such that:*

(*A*) *For each* k, $1 \le k \le n$, *there are* K_v*-symmetric functions* $\vartheta_{v,ij}^{(k)}(z) \in L_{w_v}(\mathcal{C})$, *determined recursively in* $\prec_N$ *order, and* $(\mathfrak{X}, \vec{s})$*-functions* $\Theta_v^{(k)}(z) \in K_v(\mathcal{C})$ *of degree at most* $(n-k)N$, *such that*

$$\begin{aligned} G_v^{(k)}(z) &= G_v^{(k-1)}(z) + \sum_{(i,j)\in \mathrm{Band}_N(k)} \Delta_{v,ij}^{(k)} \vartheta_{v,ij}^{(k)}(z) + \Theta_v^{(k)}(z) \quad \text{for } k < n\ , \\ G_v^{(n)}(z) &= G_v^{(n-1)}(z) + \sum_{\lambda=1}^{\Lambda} \Delta_{v,\lambda}^{(n)} \varphi_\lambda(z) \end{aligned}$$

and where for each $k < n$ *and each* (i,j), *if* $\widetilde{c}_{v,i}$ *is the leading coefficient of* $\phi_v(z)$ *at* x_i*:*

(1) $\vartheta_{v,ij}^{(k)}(z)$ *has a pole of order* $nN_i - j > (n-k-1)N_i$ *at* x_i *and leading coefficient* $\widetilde{c}_{v,i}^{\,n-k-1}$, *a pole of order at most* $(n-k-1)N_{i'}$ *at each* $x_{i'} \ne x_i$, *and no other poles;*

(2) $\sum_{i=1}^m \sum_{j=(k-1)N_i}^{kN_i-1} \Delta_{v,i'j}^{(k)} \vartheta_{v,i'j}^{(k)}(z)$ *belongs to* $K_v(\mathcal{C})$;

(3) $\Theta_v^{(k)}(z)$ *is determined by the local patching process at* v *after the coefficients in* $\mathrm{Band}_N(k)$ *have been modified by adding* $\sum_{(i,j)\in \mathrm{Band}_N(k)} \Delta_{v,ij}^{(k)} \vartheta_{v,ij}^{(k)}(z)$ *to* $G_v^{(k)}(z)$; *it has a pole of order at most* $(n-k)N_i$ *at each* x_i *and no other poles, and may be the zero function.*

(*B*) *For each* $k = 0, \ldots, n$,

$$\begin{cases} \text{If } K_v \cong \mathbb{C}, \text{ then } \{z \in \mathcal{C}_v(\mathbb{C}_v) : |G_v^{(k)}(z)|_v \le r_v^{nN}\} \subset U_v = E_v^0. \\ \text{If } K_v \cong \mathbb{R}, \text{ then} \\ \quad (1) \text{ the zeros of } G_v^{(k)}(z) \text{ all belong to } E_v^0, \text{ and for each component } E_{v,i} \text{ of} \\ \quad\quad E_v, \text{ if } \phi_v(z) \text{ has } \tau_i \text{ zeros in } E_{v,i}, \text{ then } G_v^{(k)}(z) \text{ has } T_i = n\tau_i \text{ zeros in } E_{v,i}. \\ \quad (2) \{z \in \mathcal{C}_v(\mathbb{C}_v) : |G_v^{(k)}(z)|_v \le 2r_v^{nN}\} \subset U_v. \\ \quad (3) \text{ On each component } E_{v,i} \text{ contained in } \mathcal{C}_v(\mathbb{R}), \\ \qquad\qquad G_v^{(k)}(z) \text{ oscillates } T_i \text{ times between } \pm 2r_v^{nN} \text{ on } E_{v,i}. \\ \text{If } K_v \text{ is nonarchimedean and } v \in S_{K,0}, \text{ then all the zeros of } G_v^{(k)}(z) \text{ belong} \\ \quad \text{to } E_v, \text{ and for } k = 0 \text{ and } k = n \text{ they are distinct. When } k = n, \\ \qquad\qquad \{z \in \mathcal{C}_v(\mathbb{C}_v) : G_v^{(n)}(z) \in \mathcal{O}_v \cap D(0, r_v^{nN})\} \subset E_v. \\ \text{If } K_v \text{ is nonarchimedean and } v \in \widehat{S}_{K,0} \backslash S_{K,0}, \text{ then} \\ \qquad\qquad \{z \in \mathcal{C}_v(\mathbb{C}_v) : |G_v^{(k)}(z)|_v \le R_v^{nN}\} = E_v. \end{cases}$$

Remark 1. For almost all v and k, we will have $\Theta_v^{(k)}(z) = 0$; the only exception is for one value $k = k_1$ for each $v \in S_{K,0}$, where $\Theta_v^{(k_1)}(z)$ is chosen to "separate the roots" of $G_v^{(k_1)}(z)$. See the discussion after Theorem 11.1, and Phase 3 in the proof of that theorem.

Remark 2. Examining the proofs of the local patching theorems shows that from a local standpoint, the order in which the coefficients are received within a band is immaterial, provided that for each x_i, they are received by increasing j. For any such order, the same changes are produced in lower order coefficients within the band, and the same functions $G_v^{(k)}(z)$ are obtained. For this reason, in the local process at each v, it is permissible to subdivide $\mathrm{Gal}(L/K)$-blocks into $\mathrm{Gal}^c(\mathbb{C}_v/K_v)$-sub-blocks.

In fact, for all bands for nonarchimedean v, and for the bands with $k > \overline{k}$ for archimedean v, the compensating functions $\vartheta_{v,ij}^{(k)}$ are K_v-symmetric and are independent of the $\Delta_{v,ij}^{(k)}$. For archimedean v and bands with $k \le \overline{k}$, patching is carried out by a process called "magnification" (see the proofs of Theorems 8.1 and 9.1), and our description of the $\vartheta_{v,ij}^{(k)}$ in Theorem 7.14 is correct but artificial: rather, for each $\mathrm{Gal}^c(\mathbb{C}_v/K_v)$-stable subset of the indices in a band, the changes in those coefficients produce a canonical K_v-rational change in $G_v^{(k)}(z)$.

The choice of the Parameters $\overline{k}$ and B_v. In Stage 1 of the construction we have chosen a collection of numbers h_v for $v \in \widehat{S}_K$ such that $\prod_{v \in \widehat{S}_K} h_v^{D_v} > 1$. Likewise, for each $v \in \widehat{S}_K$, Theorem 7.14 provides a number k_v, the "minimal number of stages considered high-order" by the local patching process at v.

Let $\overline{k}$ be the smallest integer such that

$$\begin{cases} \overline{k} \ \ge \ k_v \quad \text{ for each } v \in \widehat{S}_K \ , \\ (\prod_{v \in \widehat{S}_K} h_v^{D_v})^{\overline{k}N[L:K]} \ > \ C_L(\widehat{S}_K) \ . \end{cases} \tag{7.45}$$

where $C_L(\widehat{S}_K)$ is the constant from Proposition 7.3.

For each nonarchimedean $v \in \widehat{S}_K$, the choice of $\overline{k}$ determines the constant $B_v = h_v^{\overline{k}N}$ in the local patching process (see Theorem 7.14). For archimedean v, the constants B_v in Theorem 7.14 can be specified arbitrarily. Choose them large enough that

$$(7.46) \qquad \Big(\prod_{v \in \widehat{S}_K} B_v^{D_v}\Big)^{[L:K]} \;>\; C_L(\widehat{S}_K) \ .$$

Given $w \in \widehat{S}_L$, let v be the place of K under w. By our normalization of the absolute values in §3.1, $|x|_w^{D_w} = |x|_v^{D_v[L_w:K_v]}$ for each $x \in \mathbb{C}_w \cong \mathbb{C}_v$. Define h_w by $h_w^{D_w} = h_v^{D_v[L_w:K_v]}$, and define B_w by $B_w^{D_w} = B_v^{D_v[L_w:K_v]}$. Then $|x|_w \le h_w^{kN}$ iff $|x|_v \le h_v^{kN}$, $|x|_w \le B_w$ iff $|x|_v \le B_v$, and

$$\Big(\prod_{w \in \widehat{S}_L^+} h_w^{D_w}\Big)^{\overline{k}N} \;>\; C_L(\widehat{S}_K^+) \ , \qquad \prod_{w \in \widehat{S}_L^+} B_w^{D_w} \;>\; C_L(\widehat{S}_K^+) \ .$$

The choice of the Initial Patching Functions. Theorem 7.11 gives a degree N and a collection of coherent approximating functions $\phi_v(z) \in K_v(\mathcal{C})$ for $v \in \widehat{S}_K$, of common degree N. For each $v \in \widehat{S}_K$, let $Q_{v,n}(x) \in K_v[x]$ be the monic degree-raising polynomial of degree n from Theorem 7.14. For suitable n, we will take the initial patching function at v to be $G_v^{(0)}(z) = Q_{v,n}(\phi_v(z))$. As explained above, our plan is to inductively construct functions $G_v^{(1)}(z), \ldots, G_v^{(n)}(z)$, making more and more coefficients global $\widehat{S}_L$-integers at each stage, until finally the $G_v^{(n)}(z) = G^{(n)}(z)$ are K-rational and independent of v.

At several places in the patching process, it is important to consider the $\phi_v(z)$ and $G_v^{(k)}(z)$ over the fields L_w with $w|v$, rather than over K_v. This has already been seen in Theorem 7.11. However, our ultimate goal is to construct a K-rational function. Hence, the choices made in the local patching constructions must depend only on places v of K, not on the places w of L with $w|v$.

We resolve this by considering the $\phi_v(z)$ and $G_v^{(k)}(z)$ simultaneously over K_v and the L_w. Viewing them over L_w enables us examine their coefficients, which are canonically L_w-rational. Viewing them over K_v assures that any choices in the local patching processes occur in the same way for all $w|v$.

The choice of n. The leading coefficients of the $G_v^{(0)}(z)$ are hardest to patch; we must make them $\widehat{S}_L$-units. The key to this is our choice of n.

Given $v \in \widehat{S}_K$, put $\phi_w(z) = \phi_v(z)$ for each $w \in \widehat{S}_L$ with $w|v$, viewing the $\phi_w(z)$ as functions in $L_w(\mathcal{C})$. By Theorem 7.11 the leading coefficients $\widetilde{c}_{w,i}$ of the $\phi_w(z)$ have the property that there are an integer n_0, and a K-symmetric system of $\widehat{S}_L$-units μ_i, such that for each i,

$$(7.47) \qquad \left\{ \begin{array}{rcll} \widetilde{c}_{w,i}^{\,n_0} & = & \mu_i & \text{for each archimedean } w \in \widehat{S}_L, \\ |\widetilde{c}_{w,i}^{\,n_0}|_w & = & |\mu_i|_w & \text{for each nonarchimedean } w \in \widehat{S}_L. \end{array} \right.$$

For each nonarchimedean $v \in \widehat{S}_K$, all the fields L_w for $w|v$ are isomorphic, and by the structure of the group of units $O_w^\times$ there is an integer $n_v' > 0$ such that for each $x \in \mathcal{O}_w^\times$, and each integer n' divisible by n_v',

$$(7.48) \qquad |x^{n'} - 1|_v \cdot \max_{1 \le i \le m} (|\widetilde{c}_{v,i}|_v^2) \;\le\; B_v \ .$$

Let n_1 be the least common multiple of the n_v'.

For each $v \in \widehat{S}_K$, Theorem 7.14 provides a number n_v such that the local patching process at v will preserve the properties of the roots of $G_v^{(0)}(z)$, provided $n_v|n$ and n is sufficiently large, and the $\Delta_{v,ij}^{(k)}$, $\Delta_{v,\lambda}^{(n)}$ are K_v-symmetric and satisfy the size constraints (7.43), (7.44) relative to h_v, $\overline{k}$ and the B_v chosen above. Let n_2 be the least common multiple of the n_v for $v \in \widehat{S}_K$.

Finally, let n be a positive integer such that

$$n_0 n_1 n_2 | n \ . \tag{7.49}$$

By Theorem 7.14 there is an n_3 such that if $n \geq n_3$, then for each $v \in \widehat{S}_K$ the local patching process can be successfully completed.

Until last step in the proof, $n \geq n_3$ will be a fixed integer satisfying (7.49).

Patching the Leading Coefficients. Given such an n, for each $v \in \widehat{S}_K$ put $G_w^{(0)}(z) = G_v^{(0)}(z)$ for all $w|v$, and expand

$$G_w^{(0)}(z) \ = \ \sum_{i=1}^{m} \sum_{j=0}^{(n-1)N_i-1} A_{w,ij}\varphi_{i,nN_i-j}(z) + \sum_{\lambda=1}^{\Lambda} A_{w,\lambda}\varphi_\lambda \ ,$$

with the $A_{w,ij}, A_{w,\lambda} \in L_w$. Since each $Q_{v,n}(z)$ is monic, for each $w \in \widehat{S}_L$ the leading coefficient of $G_w^{(0)}(z)$ at x_i is

$$A_{w,i0} \ = \ \widetilde{c}_{w,i}^{\,n} \ .$$

For each archimedean v, and all $w|v$, by (7.47) $\widetilde{c}_{w,i}^{\,n} = \mu_i^{n/n_0}$ is a global $\widehat{S}_L$-unit independent of w and v. By construction the μ_i are K-symmetric. In the local patching process at v, take $\Delta_{v,i0}^{(1)} = 0$ for $i = 1, \ldots, m$. Trivially the $\Delta_{v,i0}^{(1)}$ are K_v-symmetric, with $|\Delta_{v,i0}^{(1)}|_v \leq B_v$ for each i. Put $\widehat{G}_v(z) = G_v^{(0)}(z)$.

For nonarchimedean $v \in \widehat{S}_K$, we claim that we can adjust the leading coefficients to be μ_i^{n/n_0}, as well. This depends on the fact that $n_0 n_2 | n$.

Let $v \in \widehat{S}_{K,0}$. For each $w|v$, since $|\widetilde{c}_{w,i}^{\,n_0}|_w = |\mu_i|_w$, it follows that $\mu_i/\widetilde{c}_{w,i}^{\,n_0} \in \mathcal{O}_w^\times$. If we put $\Delta_{w,i0}^{(1)} \ = \ \widetilde{c}_{w,i}^{\,2}\big(\frac{\mu_i^{n/n_0}}{\widetilde{c}_{w,i}^{\,n}} - 1\big)$, then by (7.48),

$$|\Delta_{w,i0}^{(1)}|_v \ = \ |\widetilde{c}_{w,i}^{\,2}|_v \cdot \left| \left(\frac{\mu_i}{\widetilde{c}_{w,i}^{\,n_0}}\right)^{n/n_0} - 1\right|_v \ \leq \ B_v \ . \tag{7.50}$$

Moreover,

$$\mu_i^{n/n_0} \ = \ A_{w,i0} + \Delta_{w,i0}^{(1)}\widetilde{c}_{w,i}^{\,n-2} \ . \tag{7.51}$$

As will be seen below, this is what is needed for the local patching constructions in Theorem 7.14 to change the leading coefficients to μ_i^{n/n_0}.

However, for the local patching process at v we need changes $\Delta_{v,i0}^{(1)}$ independent of w, not the $\Delta_{w,i0}^{(1)}$ which á priori could depend on w. We will now present a "see-saw" argument using Proposition 7.8 which shows that for all $w|v$, the functions $\sum_{i=1}^m \Delta_{w,i0}^{(1)}\varphi_{i,nN_i}(z) \in L_w(\mathcal{C})$ belong to $K_v(\mathcal{C})$, and are independent of w, so we can take $\Delta_{v,i0}^{(1)} = \Delta_{w_v,i0}^{(1)}$ for the distinguished place w_v induced by the embedding $\widetilde{K} \hookrightarrow \mathbb{C}_v$ used to identify $\mathfrak{X}$ with a subset of $\mathcal{C}_v(\mathbb{C}_v)$. A similar argument applies at later steps of the patching process, and in the future we will omit some details.

The $G_w^{(0)}(z)$ with $w|v$ are all the same and belong to $K_v(\mathcal{C})$, so

$$\bigoplus_{w|v} G_w^{(0)}(z) \in \bigoplus_{w|v} L_w(\mathcal{C}) \cong L \otimes_K K_v(\mathcal{C})$$

is $\mathrm{Gal}(L/K)$-invariant in the sense of §7.3. By Proposition 7.8, if we put $F = K(x_i)$, then $\bigoplus_{w|v} A_{w,i0}$ belongs to $\bigoplus_{u|v} F_u$ (embedded semi-diagonally in $\bigoplus_{w|v} L_w \cong L \otimes_K K_v$), and for each $\sigma \in \mathrm{Gal}(L/K)$,

$$\sigma(\bigoplus_{w|v} A_{w,i0}) = \bigoplus_{w|v} A_{w,\sigma(i)0} \ .$$

By a similar argument, $\sigma(\bigoplus_{w|v} \widetilde{c}_{w,i}) = \bigoplus_{w|v} \widetilde{c}_{w,\sigma(i)}$.

On the other hand, by Theorem 7.11, $\mu_i \in K(x_i)$ and $\sigma(\mu_i) = \mu_{\sigma(i)}$. Hence, viewing μ_i as embedded semi-diagonally in $\bigoplus_{w|v} L_w$, we see that

$$\bigoplus_{w|v} \Delta_{w,i0}^{(1)} = \bigoplus_{w|v} (\mu_i^{n/n_0} - A_{w,i0})/\widetilde{c}_{w,i}^{\,n-2}$$

also satisfies $\sigma(\bigoplus_{w|v} \Delta_{w,i0}^{(1)}) = \bigoplus_{w|v} \Delta_{w,\sigma(i)0}^{(1)}$ for each $\sigma \in \mathrm{Gal}(L/K)$.

For the basis functions we have $\sigma(\varphi_{i,j}) = \varphi_{\sigma(i),j}$ by construction. Thus for each Galois orbit $\mathfrak{X}_\ell$

$$\bigoplus_{w|v} \Big(\sum_{x_i \in \mathfrak{X}_\ell} \Delta_{w,i0}^{(1)} \varphi_{i,nN_i}(z)\Big) \in L \otimes K_v(\mathcal{C})$$

is $\mathrm{Gal}(L/K)$-invariant. Applying Proposition 7.8 in reverse, there is a function $H_{v,\ell}(z) = \sum_{x_i \in \mathfrak{X}_\ell} \Delta_{v,i0}^{(1)} \varphi_{i,nN_i} \in K_v(\mathcal{C})$ such that

$$\sum_{x_i \in \mathfrak{X}_\ell} \Delta_{w,i0}^{(1)} \varphi_{i,nN_i}(z) = H_{v,\ell}(z)$$

for each $w|v$. Thus the $\Delta_{v,i0}^{(1)} := \Delta_{w_v,i0}^{(1)}$ are well-defined and K_v-symmetric.

Patch $G_v^{(0)}(z)$ by setting

$$\widehat{G}_v(z) = G_v^{(0)}(z) + \sum_{i=1}^{m} \Delta_{v,i0}^{(1)} \vartheta_{v,i0}^{(1)}(z) \tag{7.52}$$

where the $\vartheta_{v,i0}^{(1)}(z)$ are the compensating functions from Theorem 7.14. The leading coefficient of $\vartheta_{v,i0}^{(1)}(z)$ at x_i is $\widetilde{c}_{v,i}^{\,n-2} = \widetilde{c}_{w_v,i}^{\,n-2}$, so by (7.51) this changes the leading coefficient of $G_w^{(0)}(z)$ at x_i to μ_i^{n/n_0}, for each w and i. (The lower-order coefficients are changed as well, but they will be dealt with in subsequent patching steps.) By Theorem 7.14(A.2), $\widehat{G}_v(z)$ is K_v-rational.

Patching the High-order Coefficients. Next we patch the remaining coefficients for the stage $k = 1$ and inductively carry out the patching process for stages $k = 2, \ldots, \overline{k}$.

Suppose that for some k, we have constructed functions $G_v^{(k-1)}(z) \in K_v(\mathcal{C})$, $v \in \widehat{S}_K$. In the kth stage we patch the coefficients with indices in $\mathrm{Band}_N(k)$ by increasing $\prec_N$ order, patching all the coefficients in a given block at once.

Suppose that after patching a certain number of blocks, we have obtained functions $\widehat{G}_v(z) \in K_v(\mathcal{C})$ for $v \in \widehat{S}_K$. (When $k = 1$, we view patching the high-order coefficients as taking the initial step in passing from $G_v^{(0)}(z)$ to $G_v^{(1)}(z)$.) To lighten notation, we update the coefficients after each step: for each v, write

$$\widehat{G}_v(z) = \sum_{i=1}^{m} \sum_{j=0}^{(n-1)N_i} A_{v,ij}\varphi_{i,nN_i-j}(z) + \sum_{\lambda=1}^{\Lambda} A_{v,\lambda}\,\varphi_\lambda$$

with the $A_{v,ij}, A_{v,\lambda} \in L_{w_v}$. For each $w|v$, put $\widehat{G}_w(z) = \widehat{G}_v(z)$ and regard $\widehat{G}_w(z)$ as belonging to $L_w(\mathcal{C})$. Expand

$$\widehat{G}_w(z) = \sum_{i=1}^{m} \sum_{j=0}^{(n-1)N_i} A_{w,ij}\,\varphi_{i,nN_i-j}(z) + \sum_{\lambda=1}^{\Lambda} A_{w,\lambda}\varphi_\lambda$$

where the $A_{w,ij}$ and $A_{w,\lambda}$ belong to L_w.

Let $(i_0, j_0) \in \mathrm{Band}_N(k)$ be the least index for which the coefficients have not been patched. Thus, for each $(i,j) \prec_N (i_0, j_0)$ there is an $A_{ij} \in L$ such that $A_{w,ij} = A_{ij}$ for all w. To patch the coefficients for the indices $(i,j) \in \mathrm{Block}(i_0, j_0)$, we first determine a target value $A_{i_0j_0} \in K(x_{i_0})$ for the A_{w,i_0j_0}, $w \in \widehat{S}_L$, and then, to preserve Galois equivariance, we define the target values for the other (i,j) in $\mathrm{Block}(i_0, j_0)$ by requiring that if $\sigma \in \mathrm{Gal}(L/K)$ is such that $\sigma(i_0) = i$, then $A_{i,j} = \sigma(A_{i_0j_0})$. This is well-defined, since if $\sigma_1, \sigma_2 \in \mathrm{Gal}(L/K)$ are such that $\sigma_1(i_0) = \sigma_2(i_0)$, then $\sigma_2^{-1}\sigma_1$ fixes x_{i_0}, and so since $A_{i_0j_0} \in K(x_{i_0})$ we have $\sigma_1(A_{i_0j_0}) = \sigma_2(A_{i_0j_0})$.

Consider the vector

$$\vec{A}_{L,i_0j_0} := \bigoplus_{w\in\widehat{S}_L} A_{w,i_0j_0} \in \bigoplus_{w\in\widehat{S}_L} L_w .$$

Put $F = K(x_{i_0})$. For each $v \in \widehat{S}_K$, the $\widehat{G}_w(z) \in K_v(\mathcal{C})$ are the same for all $w|v$, so Proposition 7.8 tells us that $\vec{A}_{L,i_0,j_0}$ belongs to $\bigoplus_{u\in\widehat{S}_F} F_u$, embedded semi-diagonally in $\bigoplus_{w\in\widehat{S}_L} L_w$.

For each $w \in \widehat{S}_L$, put

$$Q_w = B_w \cdot |\widetilde{c}_{w,i_0}^{\,n-k-1}|_w , \tag{7.53}$$

where $\widetilde{c}_{w,i_0}$ is the leading coefficient of $\phi_w(z)$ at x_{i_0}. Theorem 7.11 has arranged that $\prod_{w\in\widehat{S}_L} |\widetilde{c}_{w,i_0}|_w^{D_w} = 1$, so

$$\prod_{w\in\widehat{S}_L} Q_w^{D_w} = \prod_{w\in\widehat{S}_L} B_w^{D_w} > C_L(\widehat{S}_K) .$$

Note that B_w depends only on the place v of K below w, while $|\widetilde{c}_{w,i_0}|_w$ depends only on the place u of F below w, since the $\phi_w(z) \in K_v(\mathcal{C})$ are the same for all $w|v$. Hence Q_w depends only on the place u below w. Similarly the coefficients A_{w,i_0j_0} with $w|u$ belong to F_u and depend only on u.

Thus we can apply Proposition 7.3 to the elements $c_u = A_{w,i_0j_0} \in F_u$, and to the Q_w. By Proposition 7.3, there is an $A_{i_0j_0} \in K(x_{i_0})$ such that

$$\begin{cases} |A_{i_0j_0} - A_{w,i_0j_0}|_w \le Q_w & \text{for each } w \in \widehat{S}_L , \\ |A_{i_0j_0}|_w \le 1 & \text{for each } w \notin \widehat{S}_L . \end{cases}$$

This $A_{i_0j_0}$ will be the target in patching the A_{w,i_0j_0}. For each $(i, j_0) \in \mathrm{Block}(i_0, j_0)$, choose a $\sigma \in \mathrm{Gal}(L/K)$ with $\sigma(x_{i_0}) = x_i$, and put $A_{ij_0} = \sigma(A_{i_0j_0})$. Since $A_{i_0,j_0} \in K(x_{i_0})$, the A_{ij_0} are well-defined and satisfy $\sigma(A_{ij_0}) = A_{\sigma(i)j_0}$ for all $\sigma \in \mathrm{Gal}(L/K)$.

Put $\Delta_{w,ij}^{(k)} = (A_{ij} - A_{w,ij})/\widetilde{c}_{w,i}^{\,n-k-1}$ for each $(i,j) \in \mathrm{Block}(i_0, j_0)$ and each $w \in \widehat{S}_L$. Thus

$$|\Delta_{w,ij}^{(k)}|_w \;\le\; Q_w/|\widetilde{c}_{w,i}^{\,n-k-1}|_w \;=\; B_w \tag{7.54}$$

and

$$A_{ij} \;=\; A_{w,ij} + \Delta_{w,ij}^{(k)}\widetilde{c}_{w,i}^{\,n-k-1}\ . \tag{7.55}$$

By construction the $\bigoplus_{w|v} \Delta_{w,ij}^{(k)}$ are equivariant under $\mathrm{Gal}(L/K)$.

Let $\mathfrak{X}_\ell$ be the Galois orbit of x_{i_0}. By a see-saw argument like the one used in patching the leading coefficients, for each $v \in \widehat{S}_K$ the functions

$$\sum_{x_i \in \mathfrak{X}_\ell} \Delta_{w,ij_0}^{(k)}\varphi_{i,j_0}(z) \;\in\; L_w(\mathcal{C})$$

are independent of $w|v$ and belong to $K_v(\mathcal{C})$. Define the patching coefficients by $\Delta_{v,ij_0}^{(k)} = \Delta_{w_v,ij_0}^{(k)}$, for each $(i, j_0) \in \mathrm{Block}(i_0, j_0)$ and each $v \in \widehat{S}_K$.

For each $v \in \widehat{S}_K$, the local patching construction for v produces functions $\vartheta_{v,ij_0}(z) \in L_{w_v}(\mathcal{C})$ such that $\vartheta_{v,ij}(z)$ has a pole of order $nN_i - j_0$ at x_i, with leading coefficient $\widetilde{c}_{v,i}^{\,n-k-1} = \widetilde{c}_{w_v,i}^{\,n-k-1}$ at x_i, and poles of order $\le (n-k-1)N_{i'}$ for all $i' \ne i$. Theorem 7.14.A2 shows that $\sum_{x_i \in \mathfrak{X}_\ell} \Delta_{v,ij_0}^{(k)}\vartheta_{v,ij_0}(z) \in K_v(\mathcal{C})$. Replace $\widehat{G}_v(z)$ by

$$\check{G}_v(z) \;=\; \widehat{G}_v(z) + \sum_{x_i \in \mathfrak{X}_\ell} \Delta_{v,ij_0}^{(k)}\vartheta_{v,i,j_0}(z)\ .$$

By (7.55), for each $w|v$ and each $(i, j_0) \in \mathrm{Block}(i_0, j_0)$, this changes the coefficient A_{w,ij_0} of $\widehat{G}_w(z)$ to A_{ij_0}, and leaves the coefficients preceding (i_0, j_0) unchanged. Hence, the induction can continue.

When all the coefficients in $\mathrm{Band}_N(k)$ have been patched, the local patching process at v determines an $(\mathfrak{X}, \vec{s})$-function $\Theta_v^{(k)}(z) \in K_v(\mathcal{C})$ with a pole of order at most $(n-k)N_i$ at each x_i. Using the current $\widehat{G}_v(z)$, we set $G_v^{(k)}(z) = \widehat{G}_v(z) + \Theta_v^{(k)}(z)$ and replace k by $k+1$.

Patching the Middle Coefficients. In this stage we carry out the patching process for $k = \overline{k}+1, \ldots, n-1$. The construction is the same as for the high-order coefficients, except that in place of (7.53) we take

$$Q_w \;=\; h_w^{kN} \cdot |\widetilde{c}_{w,i}^{\,n-k-1}|_w\ . \tag{7.56}$$

The construction succeeds because for each $k \ge \overline{k}$,

$$\Big(\prod_{w \in \widehat{S}_L} h_w^{D_w}\Big)^{kN} \;>\; C_L(\widehat{S}_K). \tag{7.57}$$

Patching the Low-order Coefficients. The final stage of the global patching process deals with the coefficients $A_{v,\lambda}$ in the functions

$$G_v^{(n-1)}(z) = \sum_{i=1}^{m} \sum_{j=0}^{(n-1)N_i} A_{v,ij}\varphi_{i,nN_i-j}(z) + \sum_{\lambda=1}^{\Lambda} A_{v,\lambda}\varphi_\lambda .$$

All the $A_{v,\lambda}$ will be patched simultaneously.

For each $w|v$, put $G_w^{(n-1)}(z) = G_v^{(n-1)}(z)$ and expand

$$G_w^{(n-1)}(z) = \sum_{i=1}^{m} \sum_{j=0}^{(n-1)N_i} A_{w,ij}\varphi_{i,nN_i-j}(z) + \sum_{\lambda=1}^{\Lambda} A_{w,\lambda}\varphi_\lambda .$$

By construction, for each $v \in \widehat{S}_K$, the $\bigoplus_{w|v} G_w^{(n-1)}(z) \in L \otimes_K K_v(\mathcal{C})$ are $\mathrm{Gal}(L/K)$ invariant. By Proposition 7.8 this means that for each λ, the coefficient vector $\bigoplus_{w|v} A_{w,\lambda}$ has the same Galois-equivariance properties as $\varphi_\lambda(z)$. In particular, if $K \subset F_\lambda \subset L$ is the smallest field of rationality for $\varphi_\lambda(z)$, then $\bigoplus_{w|v} A_{w,\lambda} \in F_\lambda \otimes_K K_v$.

Since

$$\Big(\prod_{w\in\widehat{S}_L} h_w^{D_w}\Big)^{nN} > C_L(\widehat{S}_K) ,$$

taking $Q_w = h_w^{nN}$ in Proposition 7.3 we can find an $A_\lambda \in F_\lambda$ such that

$$\begin{cases} |A_\lambda - A_{w,\lambda}|_w \le h_w^{nN} & \text{for all } w \in \widehat{S}_L , \\ |A_\lambda|_w \le 1 & \text{for all } w \notin \widehat{S}_L . \end{cases}$$

By working with representatives of Galois orbits as before, we can arrange that for each $\sigma \in \mathrm{Gal}(L/K)$ we have $\sigma(A_\lambda) = A_{\lambda'}$ if $\sigma(\varphi_\lambda) = \varphi_{\lambda'}$.

Put

$$\Delta_{w,\lambda} = A_\lambda - A_{w,\lambda}$$

for each w and λ, and put

$$H_w(z) = \sum_\lambda \Delta_{w,\lambda}\varphi_\lambda(z) .$$

Then $\bigoplus_{w|v} H_w(z) \in L \otimes_K K_v(\mathcal{C})$ is stable under $\mathrm{Gal}(L/K)$, for each $v \in \widehat{S}_K$. It follows that the $H_w(z)$ belong to $K_v(\mathcal{C})$ and are the same for all $w|v$. Let $H_v(z) = H_{w_v}(z)$, and expand $H_v(z) = \sum_{\lambda=1}^{\Lambda} \Delta_{v,\lambda}\varphi_\lambda(z)$. Then

$$|\Delta_{v,\lambda}|_v \le h_v^{nN}$$

for each v and λ.

Patch $G_v^{(n-1)}(z)$ by setting

$$G_v^{(n)}(z) = G_v^{(n-1)}(z) + H_v(z) .$$

This replaces the low-order coefficients of the $G_v^{(n)}(z)$ with the A_λ.

Conclusion of the Patching Argument. The patching process has now arranged that the $G_v^{(n)}(z) \in K_v(\mathcal{C})$ for $v \in \widehat{S}_K$ all coincide with a single function $G^{(n)}(z)$, whose coefficients belong to L. Fix any v, and put $G_w(z) = G(z)$ for all $w|v$; then $\bigoplus_{w|v} G_w(z) \in \bigoplus_{w|v} L(\mathcal{C}) \cong L \otimes_K K(\mathcal{C})$ is invariant under $\mathrm{Gal}(L/K)$, so by Proposition 7.8 it belongs to $K(\mathcal{C})$.

For each $v \in \widehat{S}_K$, our restrictions on the magnitudes of the $\Delta_{v,ij}^{(k)}$ and the $\Delta_{v,\lambda}$ assure that the conclusions of Theorem 7.14 apply. Thus:

$$\begin{cases} \text{If } K_v \cong \mathbb{C}, \text{ then } \{z \in \mathcal{C}_v(\mathbb{C}_v) : |G^{(n)}(z)|_v \le r_v^{Nn}\} \subset U_v = E_v^0. \\ \text{If } K_v \cong \mathbb{R}, \text{ then} \\ \quad (1) \text{ the zeros of } G^{(n)}(z) \text{ all belong to } E_v^0; \text{ for each component } E_{v,i} \text{ of } E_v, \\ \qquad \text{if } \phi_v(z) \text{ has } \tau_i \text{ zeros in } E_{v,i}, \text{ then } G^{(n)}(z) \text{ has } T_i = n\tau_i \text{ zeros in } E_{v,i}. \\ \quad (2)\ \{z \in \mathcal{C}_v(\mathbb{C}_v) : |G^{(n)}(z)|_v \le 2r_v^{nN}\} \subset U_v, \\ \quad (3) \text{ for each component } E_{v,i} \text{ contained in } \mathcal{C}_v(\mathbb{R}), \text{ then} \\ \qquad G^{(n)}(z) \text{ oscillates } T_i \text{ times between } \pm 2r_v^{nN} \text{ on } E_{v,i}. \\ \text{If } K_v \text{ is nonarchimedean and } v \in S_{K,0}, \text{ then the zeros of } G^{(n)}(z) \text{ are distinct} \\ \qquad \text{and belong to } E_v, \text{ and } \{z \in \mathcal{C}_v(\mathbb{C}_v) : G_v^{(n)}(z) \in \mathcal{O}_v \cap D(0, r_v^{nN})\} \subset E_v. \\ \text{If } K_v \text{ is nonarchimedean and } v \in \widehat{S}_{K,0} \backslash S_{K,0}, \text{ then} \\ \qquad \{z \in \mathcal{C}_v(\mathbb{C}_v) : |G^{(n)}(z)|_v \le R_v^{Nn}\} = E_v. \end{cases}$$

On the other hand, for each $v \notin \widehat{S}_K$, our construction has arranged that in

$$G^{(n)}(z) = \sum_{i=1}^{m} \sum_{j=0}^{(n-1)N_i} A_{ij}\varphi_{i,nN_i-j}(z) + \sum_{\lambda=1}^{\Lambda} A_\lambda \varphi_\lambda ,$$

all the coefficients belong to $\widehat{\mathcal{O}}_v$ and the leading coefficients belong to $\widehat{\mathcal{O}}_v^\times$. Our choice of $\widehat{S}_K$ assures that $\mathcal{C}_v$ and the functions $\varphi_{ij}(z)$ and $\varphi_\lambda(z)$ all have good reduction at v, and the x_i specialize to distinct points $\pmod v$. Hence $G^{(n)}(z) \pmod v$ is a nonconstant function with a pole of order $nN_i > 0$ at each x_i.

Thus for each $v \notin \widehat{S}_K$,

$$\{z \in \mathcal{C}_v(\mathbb{C}_v) : |G^{(n)}(z)|_v \le 1\} = \mathcal{C}_v(\mathbb{C}_v) \backslash (\bigcup_{i=1}^{m} B(x_i, 1)^-) = E_v .$$

Construction of the Points in Theorem 4.2. The patching argument holds for each integer $n > n_3$ divisible by $n_0n_1n_2$. For any such n, the zeros of $G^{(n)}(z)$ satisfy the conditions of the theorem. If there are any archimedean $v \in \widehat{S}_K$ with $K_v \cong \mathbb{R}$ such that some component $E_{v,i}$ of E_v is contained in $\mathcal{C}_v(\mathbb{R})$, or if there are any nonarchimedean $v \in \widehat{S}_K$, the construction shows that the zeros of $G^{(n)}(z)$ are distinct. Letting $n \to \infty$, we obtain infinitely many points satisfying the conditions of the theorem.

However, if there are no such v, then since $\prod_{v \in \widehat{S}_K} r_v^{Nn}$ grows arbitrarily large as $n \to \infty$, the number of $\widehat{S}_K$-integers $\kappa \in K$ satisfying $|\kappa|_v \le r_v^{Nn}$ for all $v \in \widehat{S}_K$ also becomes arbitrarily large. For any such κ, the roots of $G^{(n)}(z) = \kappa$ satisfy the conditions of the theorem. Hence there are infinitely many such points.

This completes the proof of Theorem 4.2 when char$(K) = 0$.

□

5. Proof of Theorem 4.2 when Char(K) = $p > 0$

When char$(K) = p > 0$, the proof of 4.2 is similar to that when char$(K) = 0$, but because there are no archimedean places and all the residue fields are lie over the same prime field $\mathbb{F}_p$, many of the details are simpler. On the other hand, there are complications which arise from the fact that L/K may be inseparable. For this reason we carry out the patching process using the L^{sep}-rational basis rather than the L-rational basis; the idea is to preserve separability of the roots at each step.

PROOF OF THEOREM 4.2 WHEN $\text{char}(K) = p > 0$.
Let K be a function field, and let $\mathcal{C}/K$, $\mathfrak{X}$, $\mathbb{E}_K = \mathbb{E} = \prod_v E_v$, and S_K be as in Theorem 4.2. The overall structure of the proof is similar to that when $\text{char}(K) = 0$.

Stage 1. Choices of the sets and parameters.

We begin by making the choices which govern the patching process:

The place v_0. Let $\widehat{S}_K$ be the finite set of places of K containing S_K and satisfying the conditions in (7.1). Fix a place $v_0 \in \mathcal{M}_K \backslash \widehat{S}_K$, which will play the role of a place "at ∞". Put

$$\widehat{S}_K^+ \ = \ \widehat{S}_K \cup \{v_0\} \ .$$

By our choice of $\widehat{S}_K$, the curve $\mathcal{C}$, the uniformizing parameters $g_{x_i}(z)$, and the basis functions $\varphi_{ij}(z), \widetilde{\varphi}_{ij}(z)$, φ_λ, and $\widetilde{\varphi}_\lambda$ all have good reduction at v_0, and the set E_{v_0} is $\mathfrak{X}$-trivial.

Summary of the Initial Approximation Theorems. We will only need the initial approximation theorems concerning K_v-simple sets and RL-domains. Theorem 7.15 below summarizes Theorems 6.1, 6.3 and Corollaries 6.11 and 6.12 in the context of function fields. The main difference from the corresponding results when $\text{char}(K) = 0$ is that we can require that the leading coefficients belong to $K_v(x_i)^{\text{sep}}$, not just $K_v(x_i)$.

THEOREM 7.15. *Let K be a function field with $\text{char}(K) = p > 0$, and let $\mathbb{E}$, and $\mathfrak{X}$ be as in Theorem 4.2. Then for each place v of K,*

(A) If $v \in S_K$ (so E_v is compact, K_v-simple, and disjoint from $\mathfrak{X}$), fix a K_v-simple decomposition

$$E_v \ = \ \bigcup_{\ell=1}^{D_v} B(a_\ell, r_\ell) \cap \mathcal{C}_v(F_{w_\ell}) \ . \tag{7.58}$$

and fix $\varepsilon_v > 0$. Then there is a compact, K_v-simple set $\widetilde{E}_v \subseteq E_v$ compatible with E_v such that

(1) For each $x_i, x_j \in \mathfrak{X}$ with $x_i \neq x_j$,

$$|V_{x_i}(\widetilde{E}_v) - V_{x_i}(E_v)| \ < \ \varepsilon_v \ , \quad |G(x_i, x_j; \widetilde{E}_v) - G(x_i, x_j; E_v)| \ < \ \varepsilon_v \ .$$

(2) For each $0 < \beta_v \in \mathbb{Q}$ and each K_v-symmetric $\vec{s} \in \mathcal{P}^m(\mathbb{Q})$, there is an integer $N_v \geq 1$ such that for each positive integer N divisible by N_v, there is an $(\mathfrak{X}, \vec{s})$-function $f_v \in K_v(\mathcal{C}_v)$ of degree N satisfying:

(a) The zeros $\theta_1, \ldots, \theta_N$ of f_v are distinct and belong to E_v.

(b) $f_v^{-1}(D(0,1)) \subseteq \bigcup_{\ell=1}^{D_v} B(a_\ell, r_\ell)$.

Further, there is a decomposition $f_v^{-1}(D(0,1)) = \bigcup_{h=1}^{N} B(\theta_h, \rho_h)$, where the balls $B(\theta_h, \rho_h)$ are pairwise disjoint and isometrically parametrizable. For each $h = 1, \ldots, N$, if $\ell = \ell(h)$ is such that $B(\theta_h, \rho_h) \subseteq B(a_\ell, r_\ell)$, put $F_{u_h} = F_{w_\ell}$. Then $\rho_h \in |F_{u_h}^\times|_v$ and f_v induces an F_{u_h}-rational scaled isometry from $B(\theta_h, \rho_h)$ to $D(0,1)$, with

$$f_v\big(B(\theta_h, \rho_h) \cap \mathcal{C}_v(F_{u_h})\big) \ = \ \mathcal{O}_{F_{u_h}} \ ,$$

such that $|f_v(z_1) - f_v(z_2)|_v = (1/\rho_h)\|z_1, z_2\|_v$ for all $z_1, z_2 \in B(\theta_h, \rho_h)$.

(*c*) *The set* $H_v := E_v \cap f_v^{-1}(D(0,1))$ *is* K_v*-simple and compatible with* E_v. *Indeed,*

$$H_v \;=\; \bigcup_{h=1}^{N} \left(B(\theta_h,\rho_h) \cap \mathcal{C}_v(F_{u_h})\right) \tag{7.59}$$

is a K_v*-simple decomposition of* H_v *compatible with the* K_v*-simple decomposition* (7.58) *of* E_v*, which is move-prepared relative to* $B(a_1,r_1),\ldots,B(a_{D_v},r_{D_v})$. *(See* Definition 6.10.*) Moreover, for each* $\ell = 1,\ldots,D_v$*, there is a point* $\overline{w}_\ell \in \big(B(a_\ell,r_\ell) \cap \mathcal{C}_v(F_{w_\ell})\big)\big)\backslash H_v$.

(*d*) *For each* $x_i \in \mathfrak{X}$, *the leading coefficient* $c_{v,i} = \lim_{z\to x_i} f_v(z)\cdot g_{x_i}(z)^{Ns_i}$ *belongs to* $K_v(x_i)^{\mathrm{sep}}$*, and* $\frac{1}{N}\log_v(|c_{v,i}|_v) \;=\; \Lambda_{x_i}(\widetilde{E}_v,\vec{s}) + \beta_v$.

(*B*) *If* $v \notin S_K$*, (so* E_v *is* $\mathfrak{X}$*-trivial and in particular is an* RL*-domain disjoint from* $\mathfrak{X}$*), put* $\widetilde{E}_v = E_v$*. Then for each* K_v*-symmetric* $\vec{s} \in \mathcal{P}^m(\mathbb{Q})$*, there is an integer* $N_v \geq 1$ *such that for each positive integer* N *divisible by* N_v*, there is an* $(\mathfrak{X},\vec{s})$*-function* $f_v \in K_v(\mathcal{C}_v)$ *of degree* N *such that*

(*a*) $E_v = \widetilde{E}_v = \{z \in \mathcal{C}_v(\mathbb{C}_v) : |f_v(z)|_v \leq 1\}$;

(*b*) *For each* $x_i \in \mathfrak{X}$, *the leading coefficient* $c_{v,i} = \lim_{z\to x_i} f_v(z)\cdot g_{x_i}(z)^{Ns_i}$ *belongs to* $K_v(x_i)^{\mathrm{sep}}$*, and* $\frac{1}{N}\log_v(|c_{v,i}|_v) \;=\; \Lambda_{x_i}(\widetilde{E}_v,\vec{s}) + \beta_v$.

The K_v-simple Decompositions of E_v and Sets U_v, for $v \in S_K$. For each $v \in S_K$, the set E_v is compact and K_v-simple (see Definition 4.1). Choose a K_v-simple decomposition

$$E_v \;=\; \bigcup_{\ell=1}^{D_v} B(a_\ell,r_\ell) \cap \mathcal{C}_v(F_{w_\ell})\ . \tag{7.60}$$

Refining this decomposition, if necessary, we can assume that $U_v := \bigcup_{\ell=1}^{D_v} B(a_\ell,r_\ell)$ is disjoint from $\mathfrak{X}$. This decomposition will be fixed for the rest of the construction.

The Sets $\widetilde{E}_v$ for $v \in \widehat{S}_K^+$. By hypothesis, $\gamma(\mathbb{E},\mathfrak{X}) > 1$ in Theorem 4.2. This means that the Green's matrix $\Gamma(\mathbb{E},\mathfrak{X})$ is negative definite. Suppose $\widetilde{\mathbb{E}} = \prod_{v\in\widehat{S}_K^+} \widetilde{E}_v \times \prod_{v\notin\widehat{S}_K^+} E_v$ is another K-rational adelic set compatible with $\mathfrak{X}$. By the discussion leading to (7.3), there are numbers $\varepsilon_v > 0$ for $v \in \widehat{S}_K^+$ such that such that $\Gamma(\widetilde{\mathbb{E}},\mathfrak{X})$ is also negative definite, provided that for each $v \in \widehat{S}_K^+$,

$$\begin{cases} |G(x_j,x_i;\widetilde{E}_v) - G(x_j,x_i;E_v)| \;<\; \varepsilon_v & \text{for all } i \neq j\ , \\ |V_{x_i}(\widetilde{E}_v) - V_{x_i}(E_v)| \;<\; \varepsilon_v & \text{for all } i\ . \end{cases} \tag{7.61}$$

For each $v \in S_K$, we will take $\widetilde{E}_v \subseteq E_v$ to be the set given by Theorem 7.15 for E_v, relative to the number ε_v chosen above satisfying (7.61). For each $v \in \widehat{S}_K^+\backslash S_K$ the set E_v is an RL-domain, and we will take $\widetilde{E}_v = E_v$ as in Theorem 7.15. Put $\widetilde{\mathbb{E}}_K = \prod_{v\in\widehat{S}_K^+} \widetilde{E}_v \times \prod_{v\notin\widehat{S}_K^+} E_v$ with the sets $\widetilde{E}_v$ just chosen, and let

$$\widetilde{V}_K \;:=\; V(\widetilde{\mathbb{E}}_K,\mathfrak{X}) \;=\; \mathrm{val}(\Gamma(\widetilde{\mathbb{E}}_K,\mathfrak{X})) \tag{7.62}$$

be the global Robin constant for $\widetilde{\mathbb{E}}_K$ and $\mathfrak{X}$. By construction, $\widetilde{V}_K < 0$.

The Rational Probability Vector $\vec{s}$. By construction, the Green's matrix $\Gamma(\widetilde{\mathbb{E}}_K,\mathfrak{X})$ is K-symmetric and negative definite. However, in a major simplification from the case when char$(K) = 0$, there is a matrix $\Gamma_0 \in M_m(\mathbb{Q})$ such that $\Gamma(\widetilde{\mathbb{E}}_K, \mathfrak{X}) = \Gamma_0 \cdot \log(p)$. This means that the unique probability vector $\vec{s}$ for which the components of $\Gamma(\widetilde{\mathbb{E}}_K, \mathfrak{X})\vec{s}$ are equal has rational coordinates, and that $\widetilde{V}_K \in \mathbb{Q} \cdot \log(p)$.

To see this, note that by (7.2) we have

$$\Gamma(\widetilde{\mathbb{E}}_K, \mathfrak{X}) \;=\; \frac{1}{[L:K]}\Gamma(\widetilde{\mathbb{E}}_L, \mathfrak{X}) \;=\; \frac{1}{[L:K]} \sum_{w \in \widehat{S}_L^+} \Gamma(\widetilde{E}_w, \mathfrak{X}) \log(q_w)\ . \tag{7.63}$$

Since each $\widetilde{E}_w$ is either L_w-simple or is an RL-domain, Proposition 3.28 shows that the entries $G(x_i, x_j; \widetilde{E}_w)$ and $V_{x_i}(\widetilde{E}_w)$ in the local Green's matrices belong to $\mathbb{Q}$. For each w there is a natural number f_w such that $q_w = p^{f_w}$, so $\Gamma(\widetilde{\mathbb{E}}_K, \mathfrak{X}) = \Gamma_0 \cdot \log(p)$ with $\Gamma_0 \in M_m(\mathbb{Q})$ as claimed. Clearly Γ_0 is K-symmetric and negative definite.

By the same argument as in Proposition 3.33, the vector $\vec{s}'$ defined by

$$\vec{s}' \;=\; -\Gamma_0^{-1} \begin{pmatrix} 1 \\ \vdots \\ 1 \end{pmatrix}$$

is K-symmetric with positive entries, and since $\Gamma_0 \in M_m(\mathbb{Q})$ its entries are rational. By suitably scaling $\vec{s}'$ we arrive at a K-symmetric probability vector $\vec{s} \in \mathcal{P}^m(\mathbb{Q})$ for which the entries of $\Gamma_0\vec{s}$ are equal. Evidently we have

$$\Gamma(\widetilde{\mathbb{E}}_K, \mathfrak{X})\,\vec{s} \;=\; \Gamma_0\,\vec{s} \cdot \log(p) \;=\; \begin{pmatrix} \widetilde{V} \\ \vdots \\ \widehat{V} \end{pmatrix} \tag{7.64}$$

for some $\widetilde{V} \in \mathbb{Q} \cdot \log(p)$. By the minimax property defining $\widetilde{V}_K = \mathrm{val}(\Gamma(\widetilde{\mathbb{E}}_K, \mathfrak{X}))$ it must be that $\widetilde{V} = \widetilde{V}_K$, and so $\widetilde{V}_K \in \mathbb{Q} \cdot \log(p)$.

This $\vec{s}$ will be fixed for the rest of the construction.

The Local Parameters η_v, h_v, r_v, and R_v. Since $\widetilde{V}_K \in \mathbb{Q} \cdot \log(p)$ and $\widetilde{V}_K < 0$, and since $\log(q_v) = f_v \log(p)$ for each v, we can choose a collection of numbers $\{\eta_v\}_{v \in \widehat{S}_K^+}$ with $0 < \eta_v \in \mathbb{Q}$ for each v, such that

$$\sum_{v \in \widehat{S}_K^+} \eta_v \log(q_v) \;=\; |\widetilde{V}_K| \;=\; -\widetilde{V}_K\ . \tag{7.65}$$

The η_v provide the freedom for adjustment needed in the construction of the initial approximating functions, and determine the scaling factors in passing from the initial approximating functions to the coherent approximating functions.

For the place v_0, fix a number r_{v_0} such that $1 < r_{v_0} < e^{\eta_{v_0}}$. Then, choose a set of numbers $\{h_v\}_{v \in \widehat{S}_K^+}$ with $\prod_{v \in \widehat{S}_K^+} h_v > 1$, such that

$$\begin{cases} 1 < h_{v_0} < r_{v_0} & \text{if } v = v_0\ , \\ 0 < h_v < 1 & \text{if } v \in \widehat{S}_K\ . \end{cases} \tag{7.66}$$

Finally, for each $v \in S_K$, fix an r_v with $h_v < r_v < 1$, and for each $v \in \widehat{S}_K \backslash S_K$ put $r_v = 1$. For each $v \in \widehat{S}_K^+$, put $R_v = q_v^{\eta_v}$. Then $0 < h_v < r_v < R_v$ for each v, and

$$1 \;<\; \prod_{v \in \widehat{S}_K^+} h_v \;<\; \prod_{v \in \widehat{S}_K^+} r_v\ . \tag{7.67}$$

Furthermore, $R_v \in |\mathbb{C}_v^\times|_v$ for each $v \in \widehat{S}_K^+$.

As in the proof when char$(K) = 0$, the numbers h_v control how much the Laurent coefficients of the patching functions can be changed, and the r_v are "encroachment bounds" which limit how close certain quantities can come to the h_v.

Stage 2. Constructing the coherent approximating functions $\phi_v(z)$.

We will now construct the coherent approximating functions $\phi_v(z)$, modifying the initial approximating functions $f_v(z)$ given by Theorem 7.15 for the sets and parameters chosen above. Let J be the number from the construction of the L-rational and L^{sep}-rational bases in §3.3.

THEOREM 7.16. *Let $\mathcal{C}$, K, $\mathbb{E}$, $\mathfrak{X}$, and S_K be as in Theorem 4.2, where* char$(K) =$ *$p > 0$. Let $\widehat{S}_K^+ \supseteq S_K$ be the finite set of places constructed above. For each $v \in \widehat{S}_K^+$, let $\widetilde{E}_v \subset E_v$ and $0 < h_v < r_v < R_v$ be the set and patching parameters constructed above. For each $v \in S_K$, let $\bigcup_{\ell=1}^{D_v} B(a_\ell, r_\ell) \cap \mathcal{C}_v(F_{w_\ell})$ be the K_v-simple decomposition of E_v chosen above. Let $\vec{s} \in \mathcal{P}_m(\mathbb{Q})$ be the rational probability vector with positive coefficients such that $\widetilde{V}_K = \Gamma(\widetilde{\mathbb{E}}_K, \mathfrak{X})\vec{s}$.*

Then there are a positive integer N and $(\mathfrak{X}, \vec{s})$-functions $\phi_v(z) \in K_v(\mathcal{C})$ for $v \in \widehat{S}_K^+$, of common degree N, such that $N_i := Ns_i$ belongs to $\mathbb{N}$ and is divisible by J, for each $i = 1, \ldots, m$, and:

(A) The $\phi_v(z)$ have the following properties:

(1) If $v \in S_K$, then

$$r_v^N \;<\; q_v^{-1/(q_v-1)} \;<\; 1\,, \quad \textit{and} \tag{7.68}$$

(a) the zeros $\theta_1, \ldots, \theta_N$ of $\phi_v(z)$ are distinct and belong to E_v;

(b) $\phi_v^{-1}(D(0,1)) = \bigcup_{h=1}^N B(\theta_h, \rho_h)$, where $B(\theta_1, \rho_1), \ldots, B(\theta_N, \rho_N)$ are disjoint, isometrically parametrizable, and contained in $\bigcup_{\ell=1}^{D_v} B(a_\ell, r_\ell)$;

(c) $H_v := \phi_v^{-1}(D(0,1)) \cap E_v$ has the K_v-simple decomposition

$$H_v \;=\; \bigcup_{h=1}^{N} \left(B(\theta_h, \rho_h) \cap \mathcal{C}_v(F_{u_h})\right)$$

compatible with the K_v-simple decomposition $\bigcup_{\ell=1}^{D_v}\left(B(a_\ell, r_\ell) \cap \mathcal{C}_v(F_{w_\ell})\right)$ of E_v, which is move-prepared relative to $B(a_1, r_1), \ldots, B(a_{D_v}, r_{D_v})$. For each ℓ, there is a point $\overline{w}_\ell \in \left(B(a_\ell, r_\ell) \cap \mathcal{C}_v(F_{w_\ell})\right)\backslash H_v$.

(d) Each F_{u_h}/K_v is finite and separable. For each $\theta_h \in E_v \cap B(a_\ell, r_\ell)$, we have $F_{u_h} = F_{w_\ell}$, $\rho_h \in |F_{w_\ell}^\times|_v$, and $B(\theta_h, \rho_h) \subseteq B(a_\ell, r_\ell)$. Moreover, ϕ_v induces an F_{u_h}-rational scaled isometry from $B(\theta_h, \rho_h)$ onto $D(0,1)$ with $\phi_v(\theta_h) = 0$, which takes $B(\theta_h, \rho_h) \cap \mathcal{C}_v(F_{u_h})$ onto $\mathcal{O}_{u_h}$.

(2) If $v \in \widehat{S}_K^+ \backslash S_K$, then

$$E_v \;=\; \widetilde{E}_v \;=\; \{z \in \mathcal{C}_v(\mathbb{C}_v) : |\phi_v(z)|_v \le R_v^N\}\,.$$

(B) For each $w \in \widehat{S}_L$, put $\phi_w(z) = \phi_v(z)$ if $w|v$, and regard $\phi_w(z)$ as an element of $L_w(\mathcal{C})$. Each $x_i \in \mathfrak{X}$ is canonically embedded in $\mathcal{C}_w(L_w)$; let $\widetilde{c}_{w,i} = \lim_{z\to x_i} \phi_w(z) \cdot g_{x_i}(z)^{Ns_i}$ be the leading coefficient of $\phi_w(z)$ at x_i. Then for each i

$$\sum_{w \in \widehat{S}_L^+} \log_w(|\widetilde{c}_{w,i}|_w) \log(q_w) \;=\; 0\,. \tag{7.69}$$

Furthermore, there are positive integer n_0, *and a* K*-symmetric system of* $\widehat{S}_L^+$*-units* $\mu_1, \ldots, \mu_m \in L$, *with* $\mu_i \in K(x_i)^{\text{sep}}$ *for each* i, *such that* $|\widetilde{c}_{w,i}^{n_0}|_w = |\mu_i|_w$ *for each* $w \in \widehat{S}_L^+$ *and each* $i = 1, \ldots, m$.

PROOF. Because there are no archimedean places, the proof is simpler than when $\text{char}(K) = 0$. It consists of choosing a collection of initial approximating functions $f_v(z)$ of common degree N using Theorem 7.15, then scaling them so their leading coefficients satisfy (7.69).

The choice of N. For each $v \in S_K$, put $\beta_v = \eta_v$ (where $0 < \eta_v \in \mathbb{Q}$ is the number from (7.65)) and let $N_v > 0$ be the integer given by Theorem 7.15(A.2) for E_v, $\widetilde{E}_v$, $\vec{s}$, β_v and the K_v-simple decomposition $E_v = \bigcup_{\ell=1}^{D_v} \big(B(a_\ell, r_\ell) \cap \mathcal{C}_v(F_{w_\ell})\big)$ chosen above. For each $v \in \widehat{S}_K^+ \backslash S_K$, let N_v be the number given by Theorem 7.15 for $E_v = \widetilde{E}_v$ and $\vec{s}$.

Fix an integer $N > 0$ be an integer which satisfies the following conditions:

(1) N is divisible by N_v, for each $v \in \widehat{S}_K^+$;
(2) $N_i := Ns_i$ belongs to $\mathbb{N}$ and is divisible by J for each $i = 1, \ldots, m$;
(3) $N \cdot \eta_v \in \mathbb{N}$ for each $v \in \widehat{S}_K^+$, where the η_v are as in (7.65);
(4) N is large enough that
- $Ns_i > J$ for each $i = 1, \ldots, m$;
- (7.70) $r_v^N < q_v^{-1/(q_v-1)} < 1$, for each $v \in S_K$.

In particular (7.68) holds.

The choice of the Initial Approximating Functions $f_v(z)$. We will apply Theorem 7.15 with the parameters chosen above. For each $v \in S_K$, take $\beta_v = \eta_v$ (with $0 < \eta_v \in \mathbb{Q}$ as in (7.65)) and let $f_v(z) \in K_v(\mathcal{C})$ be the $(\mathfrak{X}, \vec{s})$-function of degree N given by Theorem 7.15(A.2) with $\frac{1}{N}\log_v(|c_{v,i}|_v) = \Lambda_{x_i}(\widetilde{E}_v, \vec{s}) + \beta_v$ and $c_{v,i} \in K_v(x_i)^{\text{sep}}$ for each i. For each $v \in \widehat{S}_K^+ \backslash S_K$, let $f_v(z) \in K_v(\mathcal{C})$ be the $(\mathfrak{X}, \vec{s})$-function of degree N from Theorem 7.15(B), with $\frac{1}{N}\log_v(|c_{v,i}|_v) = \Lambda_{x_i}(\widetilde{E}_v, \vec{s})$ for each i.

Each $f_v(z)$ has the mapping properties from Theorem 7.15. In particular, for each $v \in S_K$, the set $H_v := f_v^{-1}(D(0,1)) \cap E_v$ has a K_v-simple decomposition $H_v = \bigcup_{h=1}^N \big(B(\theta_h, \rho_h) \cap \mathcal{C}_v(F_{u_h})\big)$ compatible with the K_v-simple decomposition $E_v = \bigcup_{\ell=1}^{D_v} \big(B(a_\ell, r_\ell) \cap \mathcal{C}_v(F_{w_\ell})\big)$, which is move-prepared relative to $B(a_1, r_1), \ldots, B(a_{D_v}, r_{D_v})$. Here $\theta_1, \ldots, \theta_N$ are the zeros of $f_v(z)$, $\rho_h \in |F_{u_h}^\times|_v$, and f_v induces an F_{u_h}-rational scaled isometry from $B(\theta_h, \rho_h)$ to $D(0,1)$ which maps $B(\theta_h, \rho_h) \cap \mathcal{C}_v(F_{u_h})$ onto $\mathcal{O}_{u_h}$. For each $\ell = 1, \ldots, D_v$, there is a point $\overline{w}_\ell \in \big(B(a_\ell, r_\ell) \cap \mathcal{C}_v(F_{w_\ell})\big) \backslash H_v$.

The choice of the Coherent Approximating Functions $\phi_v(z)$. If $v \in S_K$, put $\kappa_v = 1$. If $v \in \widehat{S}_K^+ \backslash S_K$, put $\kappa_v = \pi_v^{-N\eta_v}$, where $0 < \eta_v \in \mathbb{Q}$ is as in (7.65). Our choice of N required that $N\eta_v \in \mathbb{N}$, so $\kappa_v \in K_v^\times$ and $|\kappa_v|_v = R_v^N > 1$. For each $v \in \widehat{S}_K^+$, put

$$\phi_v(z) \ = \ \kappa_v f_v(z) \ \in \ K_v(\mathcal{C}) \ .$$

For each v and each i, the leading coefficient $\widetilde{c}_{v,i}$ of $\phi_v(z)$ at x_i is given by $\widetilde{c}_{v,i} = \kappa_v c_{v,i}$, so $\widetilde{c}_{v,i} \in K_v(x_i)^{\text{sep}}$. By our choices of the β_v and κ_v, for each $v \in \widehat{S}_K^+$ we have

$$\frac{1}{N}\log_v(|\widetilde{c}_{v,i}|_v) \ = \ \Lambda_{x_i}(\widetilde{E}_v, \vec{s}) + \eta_v \ . \tag{7.71}$$

Furthermore, the mapping properties of the $f_v(z)$ from Theorem 7.15, together with our choice of the κ_v, yield the following mapping properties for the $\phi_v(z)$.

(1) If $v \in S_K$ then properties (a)-(d) in Theorem 7.16(A.1) hold for $\phi_v(z)$ and H_v. Indeed, since $\kappa_v = 1$ for $v \in S_K$, then $\phi_v(z) = f_v(z)$ so the mapping properties of $\phi_v(z)$ are inherited from those of $f_v(z)$.

(2) If $v \in \widehat{S}_K^+ \backslash S_K$, then $E_v = \widetilde{E}_v = \{z \in \mathcal{C}_v(\mathbb{C}_v) : |\phi_v(z)|_v \le R_v^N\}$.

Coherence of the Leading Coefficients. To understand the leading coefficients of the $\phi_v(z)$, we must consider them over the fields L_w for $w \in \widehat{S}_L^+$, since $\mathfrak{X}$ is canonically a subset of $\mathcal{C}(L)$ and of $\mathcal{C}_w(L_w)$ for each w, and the uniformizer $g_{x_i}(z) \in L(\mathcal{C})$ is canonically an element of $L_w(\mathcal{C})$.

For each $w \in \widehat{S}_L$, put $\phi_w(z) = \phi_v(z)$ if $w|v$, and view $\phi_w(z)$ as an element of $L_w(\mathcal{C})$. Although the functions $\phi_w(z)$ for $w|v$ are all the same, the points of $\mathfrak{X}$, which are their poles, are identified differently. For each i and w, let $\widetilde{c}_{w,i} = \lim_{z \to x_i} \phi_w(z) \cdot g_{x_i}(z)^{Ns_i}$ be the leading coefficient of $\phi_w(z)$ at x_i. Let $\sigma_w : L \hookrightarrow \mathbb{C}_v$ be an embedding which induces the place w, and for each $i = 1, \ldots, m$ let $\sigma_w(i)$ be the index j for which $\sigma_w(x_i) = x_j$ (where we identify x_j with its image in $\mathcal{C}_v(\mathbb{C}_v)$ given by the fixed embedding of $\widetilde{K}$ in $\mathbb{C}_v$). Then $\widetilde{c}_{w,i} = \widetilde{c}_{v,\sigma_w(i)}$.

Recall that $\Gamma(\widetilde{\mathbb{E}}_K, \mathfrak{X}) = \frac{1}{[L:K]}\Gamma(\widetilde{\mathbb{E}}_L, \mathfrak{X})$. It follows from (7.2) that

$$[L:K] \cdot \Gamma(\widetilde{\mathbb{E}}_K, \mathfrak{X}) \;=\; \sum_{w \in \widehat{S}_L^+} \Gamma(\widetilde{E}_w, \mathfrak{X}) \log(q_w) \; . \tag{7.72}$$

Just as when $\mathrm{char}(K) = 0$, Lemma 7.13 shows that for each $w \in \widehat{S}_L^+$, and each $i = 1, \ldots, m$, the ith coordinate of $\Gamma(\widetilde{E}_w, \mathfrak{X})\vec{s}$ satisfies

$$(\Gamma(\widetilde{E}_w, \mathfrak{X})\vec{s})_i \cdot \log(q_w) \;=\; [L_w : K_v] \cdot \Lambda_{\sigma_w(x_i)}(\widetilde{E}_v, \vec{s}) \log(q_v) \; . \tag{7.73}$$

We can now prove (7.69). Since $\widetilde{c}_{w,i} = \widetilde{c}_{v,\sigma_w(i)}$, it follows from (7.28) that

$$\begin{aligned} \frac{1}{N} \sum_{w \in \widehat{S}_L^+} \log_w(|\widetilde{c}_{w,i}|_w) \log(q_w) \;&=\; \sum_{v \in \widehat{S}_K^+} \sum_{w|v} [L_w : K_v]\big(\frac{1}{N} \log_v(|\widetilde{c}_{v,\sigma_w(x_i)}|_v)\big) \log(q_v) \\ &=\; \sum_{v \in \widehat{S}_K^+} \sum_{w|v} [L_w : K_v]\big(\Lambda_{\sigma_w(x_i)}(\widetilde{E}_v, \vec{s}) + \eta_v\big) \log(q_v) \; . \end{aligned} \tag{7.74}$$

By Lemma 7.13 and our choice of $\vec{s}$ in (7.64),

$$\begin{aligned} \sum_{v \in \widehat{S}_K^+} \sum_{w|v} [L_w : K_v]\Lambda_{\sigma_w(x_i)}(\widetilde{E}_v, \vec{s}) \log(q_v) \;&=\; \sum_{w \in \widehat{S}_L^+} (\Gamma(\widetilde{E}_w, \mathfrak{X})\vec{s})_i \log(q_w) \\ &=\; (\Gamma(\widetilde{\mathbb{E}}_L, \mathfrak{X})\vec{s})_i \;=\; [L:K] \cdot \widetilde{V}_K \; . \end{aligned} \tag{7.75}$$

By our choice of the η_v in (7.65),

$$\sum_{v \in \widehat{S}_K^+} \sum_{w|v} [L_w : K_v]\eta_v \log(q_v) \;=\; [L:K] \sum_{v \in \widehat{S}_K^+} \eta_v \log(q_v) \;=\; -[L:K] \cdot \widetilde{V}_K \; . \tag{7.76}$$

Combining (7.74), (7.75), and (7.76) gives (7.69):

$$\sum_{w \in \widehat{S}_L^+} \log_w(|\widetilde{c}_{w,i}|_w) \log(q_w) \;=\; 0 \; .$$

The final assertion in Theorem 7.16 concerns the existence of a K-symmetric system of S_L^+-units $\mu_1, \ldots, \mu_m$ with $\mu_i \in K(x_i)^{\mathrm{sep}}$ for each i, and a positive integer n_0 such that $|\widetilde{c}_{w,i}^{n_0}|_w = |\mu_i|_w$ for each $w \in \widehat{S}_L^+$ and each $i = 1, \ldots, m$.

To show this, note that by our choice of N (see (7.70)) $N_i := Ns_i$ is an integer divisible by J for each $i = 1, \ldots, m$. By the construction of the L-rational and L^{sep}-rational bases in §3.3, this means that $\varphi_{i,N_i} = \widetilde{\varphi}_{i,N_i}$ for each i. For each $v \in \widehat{S}_K^+$, the leading coefficients $\widetilde{c}_{v,i}$ of $\phi_v(z)$ are K_v-symmetric since ϕ_v is K_v-rational, with each $\widetilde{c}_{v,i} \in K_v(x_i)^{\mathrm{sep}}$ by construction. The L^{sep}-rational basis is K_v-symmetric by construction, so the function

$$\phi_v^0(z) := \sum_{i=1}^{m} \widetilde{c}_{v,i}\varphi_{i,N_i}(z) = \sum_{i=1}^{m} \widetilde{c}_{v,i}\widetilde{\varphi}_{i,N_i}(z)$$

consisting of the leading terms of $\phi_v(z)$, is K_v-rational.

Consider the field $H = L^{\mathrm{sep}}$. Since L/L^{sep} is purely inseparable, for each place of w_0 of H there is a unique place w of L with $w|w_0$, and w is totally ramified over w_0 with ramification index $[L : K]^{\mathrm{insep}}$. Since H/K is Galois, the group $\mathrm{Aut}(L/K) \cong \mathrm{Gal}(H/K)$ acts transitively on the places $w|v$ of L, and the places $w_0|v$ of H. For each $w_0|v$, put $\phi_{w_0}^0(z) = \phi_v^0(z)$, regarding $\phi_{w_0}^0(z) \in K_v(\mathcal{C})$ as an element of $H_{w_0}(\mathcal{C})$. Write $\widetilde{c}_{w_0,i}$, $i = 1, \ldots, m$ for its leading coefficients; thus if w is the place of L over w_0, then $\widetilde{c}_{w_0,i} = \widetilde{c}_{w,i} = \widetilde{c}_{v,\sigma_w(i)}$.

First fix $x_i \in \mathfrak{X}$, and put $F = K(x_i)^{\mathrm{sep}}$. Let $\widehat{S}_F^+$ be the set of places of F above $\widehat{S}_K^+$. For each $v \in \widehat{S}_K^+$, since the $\phi_{w_0}^0(z) \in K_v(\mathcal{C})$ are the same for all $w_0|v$, applying Proposition 7.9 to $\bigoplus_{w_0|v} \phi_{w_0}^0(z)$ tells us that $\bigoplus_{w|v} \widetilde{c}_{w,i} = \bigoplus_{w_0|v} \widetilde{c}_{w_0,i} \in \bigoplus_{w_0|v} H_{w_0}$ actually belongs to $\bigoplus_{u|v} F_u$, embedded semi-diagonally in $\bigoplus_{w_0|v} H_{w_0}$. Write $\bigoplus_{u|v} \widetilde{c}_{u,i}$ for the element of $\bigoplus_{u|v} F_u$ that induces it. Then by (7.69) (and the fact that $[L : K] = \sum_{w|v}[L_w : K_v]$, even when $\mathrm{char}(K) = p > 0$; see ([**51**], p.321)),

$$\begin{aligned}\sum_{u\in\widehat{S}_F^+} \log_u(|\widetilde{c}_{u,i}|_u)\log(q_u) &= \frac{1}{[H:F]} \sum_{w_0\in\widehat{S}_H^+} \log_{w_0}(|\widetilde{c}_{w_0,i}|_{w_0})\log(q_{w_0}) \\ &= \frac{1}{[L:F]} \sum_{w\in\widehat{S}_L^+} \log_w(|\widetilde{c}_{w,i}|_w)\log(q_w) = 0\ .\end{aligned}$$

By Proposition 7.5 there are an $\widehat{S}_F^+$-unit $\mu_i \in F$, and an integer n_i such that $|\widetilde{c}_{u,i}^{n_i}|_u = |\mu_i|_u$ for each $u \in \widehat{S}_F^+$.

Now let x_i vary; we will arrange for the μ_i to be K-symmetric. By Proposition 7.9, the $\bigoplus_{w_0|v} \widetilde{c}_{w_0,i}$ are K-symmetric. For each $\mathrm{Aut}(L/K)$-orbit $\mathfrak{X}_\ell \subset \mathfrak{X}$, fix an $x_{i_\ell} \in \mathfrak{X}_\ell$. For each $x_j \in \mathfrak{X}_\ell$, choose $\sigma \in \mathrm{Aut}(L/K)$ with $\sigma(x_{i_\ell}) = x_j$, and replace μ_j with $\sigma(\mu_{i_\ell})$. These μ_j are independent of the choice of σ with $\sigma(x_{i_\ell}) = x_j$, and are K-symmetric. After further replacing $\mu_1, \ldots, \mu_m$ with appropriate powers of themselves, we can assume there is a number n_0 such that $n_i = n_0$, for all i. Thus $\mu_1, \ldots, \mu_m$ form a K-symmetric system of $\widehat{S}_L^+$-units, with $\mu_i \in K(x_i)^{\mathrm{sep}}$ for each i, and $|\widetilde{c}_{w,i}^{n_0}|_w = |\mu_i|_w$ for each $w \in \widehat{S}_L^+$.

This completes the proof of Theorem 7.16. □

Stage 3. The patching construction.

There are several differences in the patching argument between the function field case and the number field case.

One complication arises from the fact that L/K may be inseparable. In order to preserve the K_v-rationality of the patching functions $G_v^{(k)}(z)$, it is helpful to expand them in terms of the L^{sep}-rational basis $\{\widetilde{\varphi}_{ij}, \widetilde{\varphi}_\lambda\}$ rather than the L-rational basis. However, the several the basis functions $\widetilde{\varphi}_{ij}$ can contribute to poles of the $G_v^{(k)}(z)$ of the same order. To deal with this, instead of patching the coefficients of the $\widetilde{\varphi}_{ij}$ in $\prec_N$ order, we patch all the coefficients in a band simultaneously.

In addition, in the global patching construction we cannot use the same method for patching the high-order coefficients as when $\mathrm{char}(K) = 0$, because there are no archimedean places where a "magnification argument" can apply. Instead, by the choice of n, in the local patching construction we arrange that all the high-order coefficients (apart from the leading coefficient) are 0, so they do not need to be patched. This is possible because $\mathrm{char}(K) = p$.

The following theorem summarizes the local patching constructions proved in Theorems 10.2 and 11.2 below. After stating the theorem, we compare the patching constructions when $\mathrm{char}(K) = 0$ and $\mathrm{char}(K) = p > 0$, and establish some estimates for the coefficients needed to carry out the patching process in bands. We then choose the parameters $\overline{k}$ and n, and give the details of the patching process.

THEOREM 7.17. *Let K be a function field. Let $\mathcal{C}/K$, $\mathbb{E}$, $\mathfrak{X}$, and S_K be as in Theorem 4.2. Let $\widehat{S}_K^+ \supseteq S_K$ be the finite set of places satisfying conditions (7.1), together with v_0. For each $v \in \widehat{S}_K^+$, let $\widetilde{E}_v \subset E_v$, and $0 < h_v < r_v < R_v$ be the set and patching parameters constructed in Stage 1 above. For each $v \in S_K$, let $\bigcup_{\ell=1}^{D_v} B(a_\ell, r_\ell) \cap \mathcal{C}_v(F_{w_\ell})$ be the chosen K_v-simple decomposition of E_v; by construction, $U_v = \bigcup_{\ell=1}^{D_v} B(a_\ell, r_\ell)$ is disjoint from $\mathfrak{X}$. Let the rational probability vector $\vec{s} \in \mathcal{P}^m(\mathbb{Q})$ be as in (7.64), and let the natural number N and the coherent approximating functions $\{\phi_v(z)\}_{v \in \widehat{S}_K^+}$ be the ones constructed in Theorem 7.16. Put $N_i = Ns_i$ for $i = 1, \ldots, m$. By construction $N_i \in \mathbb{N}$ and $J|N_i$ for each i, and for each $v \in \widehat{S}_K^+$ and each i, the leading coefficient $\widetilde{c}_{v,i} = \lim_{z \to x_i} \phi_v(z) \cdot g_{x_i}(z)^{N_i}$ belongs to $K_v(x_i)^{\mathrm{sep}}$.*

For each $v \in \widehat{S}_K^+$, Theorem 10.2 or 11.2 provides a number $k_v > 0$ determined by E_v and $\phi_v(z)$, representing the minimal number of "high-order" stages in the local patching process at v. Let $\overline{k} \ge k_v$ be a fixed integer. Then for each $v \in \widehat{S}_K^+$, there are an integer $n_v > 0$ and a number $0 < B_v < 1$, depending on $\overline{k}$, E_v, and $\phi_v(z)$, such that for each sufficiently large integer n divisible by n_v, the local patching process at v can be carried out as follows:

Put $G_v^{(0)}(z) = Q_{v,n}(\phi_v(z))$, where:

$$\begin{cases} \text{If } v \in S_K, \text{ then } Q_{v,n}(x) = S_{n,v}(x) \\ \quad \text{is the Stirling polynomial of degree } n \text{ for } \mathcal{O}_v \text{ (see (3.56))}. \\ \text{If } v \in \widehat{S}_K^+ \backslash S_K, \text{ then } Q_{v,n}(x) = x^n. \end{cases}$$

Then all the zeros of $G_v^{(0)}(z)$ belong to E_v, and if $v \in S_K$ they are distinct. For each x_i, the leading coefficient of $G_v^{(0)}(z)$ at x_i is $\widetilde{c}_{v,i}^{\,n}$ and when $G_v^{(0)}(z)$ is expanded *in terms of the L-rational basis as*

$$G_v^{(0)}(z) = \sum_{i=1}^{m} \sum_{j=0}^{(n-1)N_i-1} A_{v,ij}\varphi_{i,nN_i-j}(z) + \sum_{\lambda=1}^{\Lambda} A_{v,\lambda}\varphi_\lambda ,$$

then $A_{v,ij} = 0$ *for all* (i,j) *with* $1 \le j < \overline{k}N_i$.

For each k, $1 \le k \le n-1$, *let* $\{\Delta_{v,ij}^{(k)} \in L_{w_v}\}_{(i,j)\in \mathrm{Band}_N(k)}$ *be a* K_v*-symmetric set of numbers satisfying*

$$(7.77) \qquad \begin{cases} |\Delta_{v,i0}^{(1)}|_v \le B_v \text{ and } \Delta_{v,ij}^{(1)} = 0 \text{ for } j = 1,\ldots,N_i-1, & \text{if } k=1 , \\ \Delta_{v,ij}^{(k)} = 0 \text{ for } j = (k-1)N_i,\ldots,kN_i-1, & \text{if } k = 2,\ldots,\overline{k} , \\ |\Delta_{v,ij}^{(k)}|_v \le h_v^{kN}, & \text{if } k = \overline{k}+1,\ldots,n-1 , \end{cases}$$

such that $\Delta_{v,i0}^{(1)} \in K_v(x_i)^{\mathrm{sep}}$ *for each* i *and such that for* $k = \overline{k}+1,\ldots,n-1$

$$(7.78) \qquad \Delta_{v,k}(z) := \sum_{i=1}^{m} \sum_{j=(k-1)N_i}^{kN_i-1} \Delta_{v,ij}^{(k)} \cdot \varphi_{i,(k+1)N_i-j} \in K_v(\mathcal{C}) .$$

For $k = n$, *let* $\{\Delta_{v,\lambda}^{(n)} \in L_{w_v}\}_{1\le\lambda\le\Lambda}$ *be a* K_v*-symmetric set of numbers such that*

$$(7.79) \qquad |\Delta_{v,\lambda}^{(n)}|_v \le h_v^{nN}$$

for each λ, *and such that*

$$(7.80) \qquad \Delta_{v,n}(z) := \sum_{\lambda=1}^{\Lambda} \Delta_{v,\lambda}^{(n)} \cdot \varphi_\lambda \in K_v(\mathcal{C}) .$$

Then one can inductively construct $(\mathfrak{X},\vec{s})$*-functions* $G_v^{(1)}(z),\ldots,G_v^{(n)}(z)$ *in* $K_v(\mathcal{C})$, *of common degree* Nn, *such that:*

(A) For each $k = 1,\ldots,n$, $G_v^{(k)}(z)$ *is obtained from* $G_v^{(k-1)}(z)$ *as follows:*

(1) When $k=1$, *the local patching process at* v *provides a* K_v*-symmetric set of functions* $\widetilde{\theta}_{v,10}^{(1)}(z),\ldots,\widetilde{\theta}_{v,m0}^{(1)}(z) \in L_{w_v}^{\mathrm{sep}}(\mathcal{C})$ *such that*

$$G_v^{(1)}(z) = G_v^{(0)}(z) + \sum_{i=1}^{m} \Delta_{v,i0}^{(1)} \cdot \widetilde{\theta}_{v,i0}^{(1)}(z) ,$$

where for each $i = 1,\ldots,m$, $\widetilde{\theta}_{v,i0}^{(1)}(z) \in K_v(x_i)^{\mathrm{sep}}(\mathcal{C})$ *has the form*

$$\widetilde{\theta}_{v,i0}^{(1)}(z) = \widetilde{c}_{v,i}^{\,n}\varphi_{i,nN_i}(z) + \widetilde{\Theta}_{v,i0}^{(1)}(z)$$

for an $(\mathfrak{X},\vec{s})$*-function* $\widetilde{\Theta}_{v,i0}^{(1)}(z)$ *with a pole of order at most* $(n-\overline{k})N_{i'}$ *at each* $x_{i'}$. *Thus, in passing from* $G_v^{(0)}(z)$ *to* $G_v^{(1)}(z)$, *each of the leading coefficients* $A_{v,i0} = \widetilde{c}_{v,i}^{\,n}$ *is replaced with* $\widetilde{c}_{v,i}^{\,n} + \Delta_{v,i0}^{(1)} \cdot \widetilde{c}_{v,i}^{\,n}$, *and the* $A_{v,ij}$ *for* $1 \le j < \overline{k}N_i$ *remain* 0.

(2) For $k = 2,\ldots,\overline{k}$, *we have* $G_v^{(k)}(z) = G_v^{(k-1)}(z)$.

(3) For $k = \overline{k}+1,\ldots,n-1$, *we have*

$$(7.81) \qquad G_v^{(k)}(z) = G_v^{(k-1)}(z) + \Delta_{v,k}(z) \cdot F_{v,k}(z) + \Theta_v^{(k)}(z) ,$$

where

(a) $\Delta_{v,k}(z) = \sum_{(i,j)\in\mathrm{Band}_N(k)} \Delta_{v,ij}^{(k)}\varphi_{i,(k+1)N_i-j}(z)$ *belongs to* $K_v(\mathcal{C})$ *by (7.78);*

(b) *$F_{v,k}(z) \in K_v(\mathcal{C})$ is an $(\mathfrak{X}, \vec{s})$-function determined by the local patching process using $G_v^{(k-1)}(z)$, whose roots belong to E_v. For each x_i, it has a pole of order $(n-k-1)N_i$ at x_i, and its leading coefficient $d_{v,i} = \lim_{z \to x_i} F_{v,k}(z) \cdot g_{x_i}(z)^{(n-k-1)N_i}$ has $|d_{v,i}|_v = |\widetilde{c}_{v,i}|_v^{n-k-1}$;*

(c) *$\Theta_v^{(k)}(z) \in K_v(\mathcal{C})$ is an $(\mathfrak{X}, \vec{s})$-function determined by the local patching process after the coefficients in* $\mathrm{Band}_N(k)$ *have been modified; it has a pole of order at most $(n-k)N_i$ at each x_i and no other poles, and may be the zero function.*

(4) *For $k = n$,*

$$G_v^{(n)}(z) \;=\; G_v^{(n-1)}(z) + \sum_{\lambda=1}^{\Lambda} \Delta_{v,\lambda}^{(n)} \cdot \varphi_\lambda(z) \ .$$

(B) *For each $v \in \widehat{S}_K^+$ and each $k = 1, \ldots, n$:*

$$\left\{ \begin{array}{l} \text{If } v \in S_K, \text{ then all the zeros of } G_v^{(k)}(z) \text{ belong to } E_v, \\ \quad \text{and for } k=0 \text{ and } k=n \text{ they are distinct. When } k=n, \\ \qquad\qquad \{z \in \mathcal{C}_v(\mathbb{C}_v) : G_v^{(n)}(z) \in \mathcal{O}_v \cap D(0, r_v^{Nn})\} \subset E_v. \\ \text{If } v \in \widehat{S}_K^+ \backslash S_K, \text{ then all the zeros of } G_v^{(k)}(z) \text{ belong to } E_v, \text{ and} \\ \qquad\qquad \{z \in \mathcal{C}_v(\mathbb{C}_v) : |G_v^{(k)}(z)|_v \le R_v^{Nn}\} = E_v. \end{array} \right.$$

Remark. As when $\mathrm{char}(K) = 0$, we will have $\Theta_v^{(k)}(z) = 0$ except for one value $k = k_1$ for each $v \in S_K$, where $\Theta_v^{(k_1)}(z)$ is chosen to "separate the roots" of $G_v^{(k_1)}(z)$. See the discussion after Theorem 11.2, and Phase 3 in the proof of that theorem.

The underlying patching constructions when $\mathrm{char}(K) = 0$ and $\mathrm{char}(K) = p > 0$ are the same: we expand

$$G_v^{(k-1)}(z) \;=\; \sum_{i=1}^{m} \sum_{j=0}^{(n-1)N_i} A_{v,ij}\varphi_{i,nN_i-j}(z) + \sum_{\lambda=1}^{\Lambda} A_{v,\lambda}\varphi_\lambda$$

and we modify the coefficients of $G_v^{(k-1)}(z)$ in $\mathrm{Band}_N(k)$ by setting

$$G_v^{(k)}(z) \;=\; G_v^{(k-1)}(z) + \sum_{i=1}^{m} \sum_{j=(k-1)N_i}^{kN_i-1} \Delta_{v,ij}^{(k)} \vartheta_{v,ij}^{(k)}(z) \ . \tag{7.82}$$

Here $\vartheta_{v,ij}^{(k)}(z)$ has a pole of order $nN_i - j$ at x_i and a pole of order at most $(n-k)N_{i'}$ at $x_{i'}$ for $i' \ne i$. Examining the local patching constructions for nonarchimedean v shows that

$$\vartheta_{v,ij}^{(k)}(z) \;=\; \varphi_{i,r}(z) \cdot F_{v,k}(z) \tag{7.83}$$

where $F_{v,k}(z)$ is K_v-rational, with a pole of order $(n-k-1)N_i$ at each x_i, and $nN_i - j = (n-k-1)N_i + r$, so $N_i + 1 \le r = (k+1)N_i - j \le 2N_i$. (When E_v is an RL-domain, $F_{v,k}(z) = \phi_v(z)^{n-k-1}$; it is more complicated when E_v is K_v-simple.) Since the φ_{ir} are K_v-symmetric, the $\vartheta_{v,ij}^{(k)}(z)$ are K_v-symmetric. Rewriting (7.82) using (7.83) gives

$$
\begin{aligned}
G_v^{(k)}(z) &= G_v^{(k-1)}(z) + \Big(\sum_{i=1}^{m} \sum_{j=(k-1)N_i}^{kN_i-1} \Delta_{v,ij}^{(k)} \varphi_{i,(k+1)N_i-j}(z)\Big) \cdot F_{v,k}(z) \\
&= G_v^{(k-1)}(z) + \Delta_{v,k}(z) \cdot F_{v,k}(z) \ , \qquad (7.84)
\end{aligned}
$$

which is (7.81) before the addition of $\Theta_v^{(k)}(z)$.

However, the patching constructions when $\mathrm{char}(K) = 0$ and $\mathrm{char}(K) = p > 0$ have different aims. When $\mathrm{char}(K) = 0$, the patching construction modifies the coefficients $A_{v,ij}$ one by one in $\prec_N$ order, making them global numbers $A_{ij} \in L$ which depend on the numbers chosen in earlier patching steps. Since each φ_{ij} in the L-rational basis has a pole of different order, patching the coefficients of lower degree basis functions does not change coefficients patched earlier, and since L_{w_v}/K_v is separable, the modification term $\sum_{i=1}^m \sum_{j=(k-1)N_i}^{kN_i-1} \Delta_{v,ij}^{(k)} \vartheta_{v,ij}^{(k)}(z)$ is K_v-rational if the $\Delta_{v,ij}^{(k)}$ are K_v-symmetric.

When $\mathrm{char}(K) = p > 0$, since L_{w_v}/K_v may be inseparable, we cannot simply use Galois equivariance of the $\Delta_{v,ij}^{(k)}$ to deduce the K_v-rationality of the modification term. Instead, we expand $G_v^{(k-1)}(z)$ and $\Delta_{v,k}(z)F_{v,k}(z)$ using the L^{sep}-rational basis, writing

$$
G_v^{(k-1)}(z) = \sum_{i=1}^{m} \sum_{j=0}^{(n-1)N_i} \widetilde{A}_{v,ij} \widetilde{\varphi}_{i,nN_i-j}(z) + \sum_{\lambda=1}^{\Lambda} \widetilde{A}_{v,\lambda} \widetilde{\varphi}_{\lambda} \ , \qquad (7.85)
$$

$$
\Delta_{v,k}(z) F_{v,k}(z) = \sum_{i=1}^{m} \sum_{j=(k-1)N_i}^{kN_i-1} \widetilde{\delta}_{v,ij} \widetilde{\varphi}_{i,nN_i-j}(z) \ + \ \text{lower degree terms} \ , \qquad (7.86)
$$

and patch the coefficients of $G_v^{(k-1)}(z)$ in $\mathrm{Band}_N(k)$ choosing the $\Delta_{v,ij}^{(k)}$ so as to modify the coefficients $\widetilde{A}_{v,ij}$ relative to the L^{sep}-rational basis. By Proposition 7.18 below, $L_{w_v}^{\mathrm{sep}}$-rationality and Galois equivariance for the $\widetilde{\delta}_{v,ij}$ implies the K_v-rationality of $\Delta_{v,k}(z)F_{v,k}(z)$.

Since several terms of the L^{sep}-rational basis can have poles of the same order, this requires us to choose all the patching coefficients in a band simultaneously. By the construction of the L-rational and L^{sep}-rational bases in §3.3, the transition matrix from the L-rational basis to the L^{sep}-rational basis is block-diagonal, with blocks of size J. Since we have required that $J|N_i$ in Theorem 7.17, we can modify the coefficients $\widetilde{A}_{v,ij}$ from $\mathrm{Band}_N(k)$ by patching all the coefficients $A_{v,ij}$ from $\mathrm{Band}_N(k)$ at once, and the patching modifications from later bands do not affect the modifications made in earlier ones.

The patching process must also keep the roots of the $G_v^{(k)}(z)$ in E_v. Doing so requires an analysis of the relationship between patching modifications of the form (7.82) and those of the form (7.86). This is carried out in Proposition 7.18. We study the growth rates of the coefficients, and show that by using modifications of the form (7.84) we can independently vary the L^{sep}-rational coefficients of the $G_v^{(k)}(z)$ in a given band, in a uniform way independent of k. The uniformity ultimately depends on the fact that each $\phi_v(z)$ has its zeros in E_v and its poles in $\mathfrak{X}$, and E_v is bounded away from $\mathfrak{X}$.

To motivate the formulation of Proposition 7.18, note that since $(k-1)N_i \leq j \leq kN_i - 1$ in (7.84), if we replace j by $s = j - (k-1)N_i$ and write $\Delta_{v,is} = \Delta_{v,ij}^{(k)}$, then (7.84) becomes

$$(7.87) \qquad G_v^{(k)}(z) \;=\; G_v^{(k-1)}(z) + \Big(\sum_{i=1}^{m}\sum_{s=0}^{N_i-1} \Delta_{v,is}\cdot \varphi_{i,2N_i-s}(z)\Big) F_{v,k}(z)\ .$$

We will use Proposition 7.18 again in the proof of Theorem 11.2, so we state it in more generality than is needed for Theorem 7.17, letting char(K) be arbitrary and letting $\ell \geq 1$ bands be patched at once. (In the proof of Theorem 7.17 we will take $\ell = 1$.)

PROPOSITION 7.18. *Let* char(K) *be arbitrary, and let v be a nonarchimedean place of K. Then there are numbers $\Lambda_v, \widetilde{\Upsilon}_v > 0$, depending only on E_v, $\mathfrak{X}$, the choice of the L-rational and L^{sep}-rational bases, and the projective embedding of $\mathcal{C}_v$, with the following property.*

Let $r > 0$ be small enough that

(1) $r < \min_{i\neq j}(\|x_i, x_j\|_v)$;
(2) *each ball $B(x_i, r)$ is isometrically parametrizable and disjoint from E_v;*
(3) *for each i, none of the $\varphi_{ij}(z)$ has a zero in $B(x_i, r)$.*

Put $\varpi_v = \min(1, \Lambda_v \cdot r)$, and let ℓ, k be integers with $\ell \geq 1$ and $1 \leq k \leq n-1$.

Let $\vec{s} \in \mathcal{P}^m(\mathbb{Q})$ be a positive rational probability vector; let N be a positive integer such that $N_i = Ns_i \in \mathbb{N}$ and $J|N_i$, for each $i = 1, \ldots, m$. Let $F_v(z) \in \mathbb{C}_v(\mathcal{C})$ be an $(\mathfrak{X}, \vec{s})$-function which has a pole of order $(n-k-1)N_i$ at x_i, for each i, and whose zeros belong to E_v. Let $0 \neq d_{v,i} = \lim_{z\to x_i} F_v(z)\cdot g_{x_i}(z)^{(n-k-1)N_i}$ be its leading coefficient at x_i.

Given $\vec{\Delta} = (\Delta_{v,is})_{1\leq i\leq m, 0\leq s<\ell N_i} \in \mathbb{C}_v^{\ell N}$, let $\Delta_v(z) \in \mathbb{C}_v(\mathcal{C})$ be the $(\mathfrak{X}, \vec{s})$-function

$$(7.88) \qquad \Delta_v(z) \;=\; \sum_{i=1}^{m}\sum_{s=0}^{\ell N_i-1} \Delta_{v,is}\cdot \varphi_{i,(\ell+1)N_i-s}(z)\ .$$

Expand $\Delta_v(z)F_v(z)$ in terms of the L^{sep}-rational basis as

$$\Delta_v(z)F_v(z) \;=\; \sum_{i=1}^{m}\sum_{s=0}^{\ell N_i-1} \widetilde{\delta}_{v,is}\cdot \widetilde{\varphi}_{i,(k+\ell)N_i-s}(z) \;+\; \textit{lower order terms}$$

and write $\widetilde{\delta} = (\widetilde{\delta}_{v,is})_{1\leq i\leq m, 0\leq s<\ell N_i} \in \mathbb{C}_v^{\ell N}$. Let $\Phi_{F_v}^{\mathrm{sep}} : \mathbb{C}_v^{\ell N} \to \mathbb{C}_v^{\ell N}$ be the linear map defined by $\Phi_{F_v}^{\mathrm{sep}}(\vec{\Delta}) = \widetilde{\delta}$.

Then $\Phi_{F_v}^{\mathrm{sep}}$ is an isomorphism and for each $\rho > 0$,

$$(7.89) \qquad \Phi_{F_v}^{\mathrm{sep}}\Big(\bigoplus_{i=1}^{m} D(0,\rho)^{\ell N_i}\Big) \;\supseteq\; \bigoplus_{i=1}^{m} D(0, \widetilde{\Upsilon}_v \varpi_v^{\ell N} |d_{v,i}|_v \rho)^{\ell N_i}$$

and

$$(7.90) \qquad \Phi_{F_v}^{\mathrm{sep}}\Big(\bigoplus_{i=1}^{m}\bigoplus_{s=0}^{\ell N_i-1} D(0, \varpi_v^{-s}\rho)\Big) \;\supseteq\; \bigoplus_{i=1}^{m}\bigoplus_{s=0}^{\ell N_i-1} D(0, \widetilde{\Upsilon}_v \varpi_v^{-s} |d_{v,i}|_v \cdot \rho)\ .$$

Moreover, if $F_v(z)$ is K_v-rational, then for each K_v-symmetric $\widetilde{\delta} \in \left(L_{w_v}^{\mathrm{sep}}\right)^{\ell N}$, the unique solution to $\Phi_{F_v}^{\mathrm{sep}}(\vec{\Delta}) = \widetilde{\delta}$ belongs to $L_{w_v}^{\ell N}$ and is K_v-symmetric, and the corresponding function $\Delta_v(z) = \sum_{i=1}^{m} \sum_{s=0}^{\ell N_i - 1} \Delta_{v,is} \cdot \varphi_{i,(\ell+1)N_i - s}(z)$ is K_v-rational.

The proof of Proposition 7.18 will be given in §7.6 below. We now choose the patching parameters $\overline{k}$ and n in Theorem 7.17, and give the details of the global patching process.

The choice of $\overline{k}$. In Stage 1 we have chosen a collection of numbers h_v for $v \in \widehat{S}_K^+$ such that $\prod_{v \in \widehat{S}_K^+} h_v > 1$. Likewise, for each $v \in \widehat{S}_K^+$, Proposition 7.14 provides a number k_v (the minimal number of stages considered high-order by the patching process at v). Finally, for each $v \in \widehat{S}_K^+$, fix a number $r = r_v > 0$ satisfying the conditions of Proposition 7.18 and small enough that each ball $B(x_i, r_v)$ is disjoint from U_v; then Proposition 7.18 provides numbers $\widetilde{\Upsilon}_v > 0$ and $0 < \varpi_v \leq 1$ (the comparison constants for the transition between the L-rational and L^{sep}-rational bases). Put $\widetilde{\Upsilon} = \prod_{v \in \widehat{S}_K^+} (\widetilde{\Upsilon}_v \varpi_v^N)$ and $h = \prod_{v \in \widehat{S}_K^+} h_v > 1$. Finally, put $H = L^{\mathrm{sep}}$ and let $C_H(\widehat{S}_K^+)$ be the constant from Proposition 7.3.

Let $\overline{k}$ be the smallest integer such that

$$\left\{ \begin{array}{lll} \overline{k} & \geq k_v & \text{for each } v \in \widehat{S}_K^+ \,, \\ (\widetilde{\Upsilon} \cdot h^{\overline{k}N})^{[H:K]} & > & C_H(\widehat{S}_K^+) \,. \end{array} \right. \tag{7.91}$$

The choice of n. As in the patching construction when $\mathrm{char}(K) = 0$, for suitable n we will take the initial patching functions to be $G_v^{(0)}(z) = Q_{v,n}(\phi_v(z))$, where $Q_{v,n}(x) \in \mathcal{O}_v(z)$ is the monic polynomial of degree n given by Theorem 7.17.

One consideration in choosing n is to facilitate patching the leading coefficients of the $G_v^{(0)}(z)$ to be $\widehat{S}_L^+$-units. Given $v \in \widehat{S}_K$, put $\phi_w(z) = \phi_v(z)$ for each $w \in \widehat{S}_L$ with $w|v$, viewing the $\phi_w(z)$ as functions in $L_w(\mathcal{C})$. By Theorem 7.16 the leading coefficients $\widetilde{c}_{w,i}$ of the $\phi_w(z)$ have the property that there are an integer n_0, and a K-symmetric system of $\widehat{S}_L$-units μ_i, such that for each i and each $w \in \widehat{S}_L^+$,

$$|\widetilde{c}_{w,i}^{\,n_0}|_w \;=\; |\mu_i|_w \,. \tag{7.92}$$

For each $v \in \widehat{S}_K^+$, let $0 < B_v < 1$ be the number from Theorem 7.17 controlling the freedom in patching the leading coefficients. All the fields L_w for $w|v$ are isomorphic, and by the structure of the group of units $O_w^\times$ there is an integer $n_v' > 0$ such that for each $x \in \mathcal{O}_w^\times$, and each integer n' divisible by n_v',

$$|x^{n'} - 1|_v \;\leq\; B_v \,. \tag{7.93}$$

Let n_1 be the least common multiple of the n_v'.

Another consideration in choosing n is to assure that various analytic estimates are satisfied. For each $v \in \widehat{S}_K^+$, Theorem 7.17 provides a number n_v such that the local patching process at v will keep the roots of $G_v^{(0)}(z)$ in E_v, provided $n_v|n$ and n is sufficiently large. Let n_2 be the least common multiple of the n_v for $v \in \widehat{S}_K^+$.

Finally, let n be a positive integer such that

$$n_0 n_1 n_2 | n \tag{7.94}$$

and is large enough that

$$(\prod_{v\in\widehat{S}_H^+} h_v)^{nN[H:K]} \;>\; C_H(\widehat{S}_K^+)\ . \tag{7.95}$$

By Theorem 7.17 there is an n_3 such that if $n \ge n_3$, then (7.95) holds and for each $v \in \widehat{S}_K^+$ the local patching process can be successfully completed.

Until last step in the proof, $n \ge n_3$ will be a fixed integer satisfying (7.94).

The order $\prec_N$. The index set $\mathcal{I} = \{(i,j) \in \mathbb{Z}^2 : 1 \le i \le m, 0 \le j < \infty\}$, the order $\prec_N$, the bands $\mathrm{Band}_N(k) = \{(i,j) \in \mathcal{I} : 1 \le i \le m, (k-1)N_i \le j \le kN_i - 1\}$, and the blocks $\mathrm{Block}((i_0,j_0)) = \{(i,j_0) :\ \sigma(x_{i_0}) = x_i \text{ for some } \sigma \in \mathrm{Aut}(L/K)\}$ will be the same as those when $\mathrm{char}(K) = 0$ (see (7.41)).

Patching the Leading Coefficients. For each $v \in \widehat{S}_K^+$, the initial patching function $G_v^{(0)}(z) = Q_{v,n}(\phi_v(z)) \in K_v(\mathcal{C})$ is an $(\mathfrak{X},\vec{s})$-function of degree nN, with a pole of order nN_i and leading coefficient $\widetilde{c}_{v,i}^{\,n}$ at each x_i. Let $\mu_1,\ldots,\mu_m \in L^{\mathrm{sep}}$ and $n_0 \ge 1$ be the K-symmetric set of $\widehat{S}_L^+$-units and integer constructed in Stage 2 above.

Fix $v \in \widehat{S}_K^+$, and view $\mu_1,\ldots,\mu_m$ as embedded in $L_{w_v}^{\mathrm{sep}}$; thus $|\widetilde{c}_{v,i}^{\,n_0}|_v = |\mu_i|_v$. For each $i = 1,\ldots,m$ put

$$\Delta_{v,i0}^{(1)} \;=\; (\mu_i^{n/n_0}/\widetilde{c}_{v,i}^{\,n}) - 1 \;=\; (\mu_i/\widetilde{c}_{v,i}^{\,n_0})^{n/n_0} - 1\ .$$

Since $n_0 n_1 | n$, (7.93) shows that

$$|\Delta_{v,i0}^{(1)}|_v \;\le\; B_v\ . \tag{7.96}$$

Since $\widetilde{c}_{v,i} \in L_{w_v}^{\mathrm{sep}}$, we have $\Delta_{v,i0}^{(1)} \in L_{w_v}^{\mathrm{sep}}$. Since the μ_i and $\widetilde{c}_{v,i}$ are K_v-symmetric, the $\Delta_{v,i0}^{(1)}$ are K_v-symmetric as well. We will take the $\Delta_{v,ij}^{(1)}$ for $j = 1,\ldots,N_i-1$, to be 0.

By Theorem 7.17, when $G_v^{(0)}(z)$ is expanded using the L^{sep}-rational basis as

$$G_v^{(0)}(z) \;=\; \sum_{i=1}^{m}\sum_{j=1}^{(n-1)N_i} \widetilde{A}_{v,ij}\widetilde{\varphi}_{ij}(z) + \sum_{\lambda=1}^{\Lambda} \widetilde{A}_{v,\lambda}\widetilde{\varphi}_\lambda\ ,$$

then for each i we have $\widetilde{A}_{v,i0} = \widetilde{c}_{v,i}^{\,n}$ and $A_{v,ij} = 0$ for $j = 1,\ldots,\overline{k}N_i - 1$. The local patching construction at v provides a K_v-symmetric set of functions $\widetilde{\theta}_{v,i0}^{(1)}(z) \in L_{w_v}^{\mathrm{sep}}(\mathcal{C})$ for $i = 1,\ldots,m$, such that for each i there is an $(\mathfrak{X},\vec{s})$-function $\widetilde{\Theta}_{v,i}(z) \in L_{w_v}^{\mathrm{sep}}(\mathcal{C})$ with a pole of order at most $(n-\overline{k})N_{i'}$ for each i', for which

$$\widetilde{\theta}_{v,i0}^{(1)}(z) \;=\; \widetilde{c}_{v,i}^{\,n}\widetilde{\varphi}_{i,nN_i}(z) + \widetilde{\Theta}_{v,i}(z)\ .$$

(See (10.20) and (11.53); note that $\widetilde{\varphi}_{i,nN_i} = \varphi_{i,nN_i}$ since $J|N_i$.) Thus if we put

$$G_v^{(1)}(z) \;=\; G_v^{(0)}(z) + \sum_{i=1}^{m} \Delta_{v,i0}^{(1)}\cdot\widetilde{\theta}_{v,i0}^{(1)}(z)\ ,$$

then the leading coefficient of $G_v^{(1)}(z)$ at x_i becomes

$$\widetilde{A}_{v,i0} + \Delta_{v,i0}^{(1)}\cdot\widetilde{c}_{v,i}^{\,n} \;=\; \widetilde{c}_{v,i}^{\,n} + \big((\mu_i^{n/n_0}/\widetilde{c}_{v,i}^{\,n}) - 1\big)\cdot\widetilde{c}_{v,i}^{\,n} \;=\; \mu_i^{n/n_0}$$

while the coefficients $\widetilde{A}_{v,ij}$ for $j = 1, \ldots, \overline{k}N_i - 1$ remain 0. Since the $\Delta^{(1)}_{v,i0}$ and $\vartheta^{(1)}_{v,i0}(z)$ are K_v-symmetric and defined over $L^{\mathrm{sep}}_{w_v}$, $G^{(1)}_v(z)$ belongs to $K_v(\mathcal{C})$.

By construction all the zeros of $G^{(0)}_v(z)$ belong to E_v, and since (7.96) holds, the local patching process assures that the zeros of $G^{(1)}_v(z)$ belong to E_v as well.

Patching the High-order Coefficients. No patching is needed for $k = 2, \ldots, \overline{k}$. Since the $\widetilde{A}_{v,ij} = 0$ for $j = 1, \ldots, \overline{k}N_i - 1$, they are already independent of v and belong to L^{sep}. Hence for $k = 2, \ldots, \overline{k}$ we can take $\Delta^{(k)}_{v,ij} = 0$ for all i, j and set $G^{(k)}_v(z) = G^{(k-1)}_v(z) = G^{(1)}_v(z)$.

Patching the Middle Coefficients. For each $k = \overline{k}+1, \ldots, n-1$ we first choose the target coefficients using Proposition 7.9 and a see-saw argument. We then choose the patching coefficients $\Delta^{(k)}_{v,ij}$ using Proposition 7.18, and patch using Theorem 7.17.

As before, let $H = L^{\mathrm{sep}}$. Then L/H is purely inseparable and H/K is Galois, with $\mathrm{Aut}(L/K) \cong \mathrm{Gal}(H/K)$. For each place w_0 of H there is a unique place w of L lying over w_0, and w/w_0 is totally ramified. In the discussion below, we will work primarily with H, and to simplify notation we will write w both for places of L and H. Let $\widehat{S}^+_H$ be the set of places w of H over places $v \in \widehat{S}^+_K$.

Suppose that for some $k > \overline{k}$, we have completed the patching process through stage $k-1$, and have constructed functions $G^{(k-1)}_v(z) \in K_v(\mathcal{C})$ with the properties in Theorem 7.17. For each $v \in \widehat{S}^+_K$, and each $w \in \widehat{S}^+_H$ with $w|v$, put $G^{(k-1)}_w(z) = G^{(k-1)}_v(z)$ and expand $G^{(k-1)}_w(z)$ using the L^{sep}-rational basis as

$$G^{(k-1)}_w(z) = \sum_{i=1}^{m} \sum_{j=0}^{(n-1)N_i-1} \widetilde{A}_{w,ij}\widetilde{\varphi}_{i,nN_i-j}(z) + \sum_{\lambda=1}^{\Lambda} \widetilde{A}_{w,\lambda}\widetilde{\varphi}_{\lambda} .$$

Since $G^{(k-1)}_w(z)$ is rational over K_v and the $\widetilde{\varphi}_{i,nN_i-j}$ and $\widetilde{\varphi}_{\lambda}$ are rational over $H = L^{\mathrm{sep}}$, the $\widetilde{A}_{w,ij}$ and $\widetilde{A}_{w,\lambda}$ belong to H_w and are K_v-symmetric. Since $\widetilde{\varphi}_{i,nN_i-j}$ is rational over $K(x_i)^{\mathrm{sep}}$, by Galois equivariance $\widetilde{A}_{w,ij}$ belongs to $K_v(x_i)^{\mathrm{sep}} \subset H_w$.

We next choose the target coefficients $\widetilde{A}_{ij} \in K(x_i)^{\mathrm{sep}}$. Let $(i_0, j_0) \in \mathrm{Band}_N(k)$ be the least index under $\prec_N$ for which the target coefficient has not been chosen. Because of the way the L^{sep}-rational basis was constructed, $\mathrm{Block}_N((i_0, j_0)) \subset \mathrm{Band}_N(k)$ is the set of indices $(i, j_0) \in \mathcal{I}$ for which there is a $\sigma \in \mathrm{Gal}(L^{\mathrm{sep}}/K)$ such that $\sigma(\widetilde{\varphi}_{i_0,nN_{i_0}-j_0}) = \widetilde{\varphi}_{i,nN_i-j_0}$. We first determine the target coefficient $\widetilde{A}_{i_0,j_0} \in K(x_{i_0})^{\mathrm{sep}}$ and then define the target coefficients for the other (i, j) in $\mathrm{Block}_N((i_0, j_0))$ so as to preserve Galois equivariance.

Consider the vector

$$\vec{A}_{H,i_0j_0} := \bigoplus_{w \in \widehat{S}^+_H} \widetilde{A}_{w,i_0j_0} \in \bigoplus_{w \in \widehat{S}^+_H} H_w$$

and let $F = K(x_{i_0})^{\mathrm{sep}}$. For each $v \in \widehat{S}_K$, the functions $G^{(k-1)}_w(z) \in K_v(\mathcal{C})$ are the same for all places w of H with $w|v$, and the $\widetilde{A}_{w,i_0,j_0}$ belong to H^{sep}_w, so Proposition 7.9 tells us that $\vec{A}_{H,i_0,j_0}$ belongs to $\bigoplus_{u \in \widehat{S}_F} F_u$, embedded semi-diagonally in $\bigoplus_{w \in \widehat{S}_H} H_w$.

Since $k > \overline{k}$, our choice of $\overline{k}$ (see 7.91) assures that

$$\prod_{v \in \widehat{S}_K^+} (\widetilde{\Upsilon}_v \varpi_v^N \cdot h_v^{kN})^{[H:K]} > C_H(\widehat{S}_K^+) .$$

Recalling that for each $w \in \widehat{S}_H^+$ we have $|x|_w = |x|_v^{[H_w:K_v]}$, put $\widetilde{\Upsilon}_w = \widetilde{\Upsilon}_v^{[H_w:K_v]}$, $b_w = \varpi_v^{[H_w:K_v]}$, and $h_w = h_v^{[H_w:K_v]}$. Since $\sum_{w|v}[H_w : K_v] = [H : K]$ for each v, it follows that

$$\prod_{w \in \widehat{S}_H^+} (\widetilde{\Upsilon}_w b_w^N \cdot h_w^{kN}) > C_H(\widehat{S}_K^+) . \tag{7.97}$$

For each each $w \in \widehat{S}_H^+$, put

$$Q_w = \widetilde{\Upsilon}_w b_w^N \cdot h_w^{kN} \cdot |\widetilde{c}_{w,i_0}^{\,n-k-1}|_w , \tag{7.98}$$

where $\widetilde{c}_{w,i_0}$ is the leading coefficient of $\phi_w(z)$ at x_{i_0}. By (7.69), $\prod_{w \in \widehat{S}_H^+} |\widetilde{c}_{w,i_0}|_w = 1$, so (7.97) gives

$$\prod_{w \in \widehat{S}_H^+} Q_w > C_H(S_H^+) . \tag{7.99}$$

Note that $\widetilde{\Upsilon}_w$, b_w and h_w depend only on the place v of K below w, while $|\widetilde{c}_{w,i_0}|_w$ depends only on the place u of F below w, since the $\phi_w(z) \in K_v(\mathcal{C})$ are the same for all $w|v$. Hence Q_w depends only on the place u below w. Similarly the coefficients $\widetilde{A}_{w,i_0 j_0}$ with $w|u$ belong to F_u and depend only on u.

By (7.99) we can apply Proposition 7.3 to the elements $c_u = \widetilde{A}_{w,i_0 j_0} \in F_u$, and to the Q_w. (Note that in the function field case, there are no archimedean places, so the exponents D_w in Proposition 7.3 are all 1.) By Proposition 7.3, there is an $\widetilde{A}_{i_0 j_0} \in K(x_{i_0})^{\text{sep}}$ such that

$$\begin{cases} |\widetilde{A}_{i_0 j_0} - \widetilde{A}_{w,i_0 j_0}|_w \le Q_w & \text{for each } w \in \widehat{S}_H^+ , \\ \quad |\widetilde{A}_{i_0 j_0}|_w \le 1 & \text{for each } w \notin \widehat{S}_H^+ . \end{cases}$$

This $\widetilde{A}_{i_0 j_0}$ is the target in patching the $\widetilde{A}_{w,i_0 j_0}$. For each $(i, j_0) \in \text{Block}((i_0, j_0))$, take $\sigma \in \text{Gal}(H/K)$ with $\sigma(\widetilde{\varphi}_{i_0,nN_{i_0}-j_0}) = \widetilde{\varphi}_{i,nN_i-j_0}$, and put $\widetilde{A}_{ij} = \sigma(\widetilde{A}_{i_0 j_0})$. Since $\widetilde{A}_{i_0,j_0} \in K(x_{i_0})^{\text{sep}}$, the $\widetilde{A}_{ij}$ are well-defined and Galois equivariant.

Repeat this process until target coefficients $\widetilde{A}_{ij}$ have been chosen for all $(i, j) \in \text{Band}_N(k)$. Since the $\widetilde{A}_{ij}$ belong to L^{sep} and are K-symmetric, the function

$$H^{(k)}(z) := \sum_{(i,j) \in \text{Band}_N(k)} \widetilde{A}_{ij} \widetilde{\varphi}_{ij}(z)$$

is K-rational.

We next choose the patching coefficients $\Delta_{v,ij}^{(k)}$ so as to replace the part of the L^{sep}-rational expansions of the $G_v^{(k-1)}(z)$ coming from $\text{Band}_N(k)$ with $H^{(k)}(z)$. Fix $v \in \widehat{S}_K^+$, and let $F_{v,k}(z) \in K_v(\mathcal{C})$ be the $(\mathfrak{X}, \vec{s})$-function of degree $(n - k - 1)N$ provided by Theorem 7.17. View $\mathfrak{X}$ and the $\widetilde{A}_{ij}$ as embedded in L_{w_v}, and let $d_{v,i} \in L_{w_v}$ be the leading coefficient of $F_{v,k}(z)$ at x_i. By hypothesis $|d_{v,i}|_v = |\widetilde{c}_{v,i}^{\,n-k-1}|_v$.

For each $(i,j) \in \mathrm{Band}_N(k)$, put $\widetilde{\delta}_{v,ij}^{(k)} = \widetilde{A}_{ij} - \widetilde{A}_{w_v,ij} \in K_v(x_i)^{\mathrm{sep}} \subseteq H_{w_v}$, and let $Q_v = Q_{w_v}^{1/[H_{w_v}:K_v]} = \widetilde{\Upsilon}_v \varpi_v^N h_v^{kN} \cdot |\widetilde{c}_{v,i}^{n-k-1}|_v$. Since $|\widetilde{\delta}_{v,ij}^{(k)}|_{w_v} = |\widetilde{A}_{ij} - \widetilde{A}_{w_v,ij}|_{w_v} \le Q_{w_v}$, we have

$$|\widetilde{\delta}_{v,ij}^{(k)}|_v \ \le\ Q_v \ =\ \widetilde{\Upsilon}_v \varpi_v^N |d_{v,i}|_v \cdot h_v^{kN} \tag{7.100}$$

and

$$\widetilde{A}_{ij} \ =\ \widetilde{A}_{v,ij} + \widetilde{\delta}_{v,ij}^{(k)} \ . \tag{7.101}$$

The $\widetilde{A}_{v,ij}$ are K_v-symmetric since $G_v^{(k)}(z)$ and $F_{v,k}(z)$ are K_v-rational, and the $\widetilde{A}_{ij}$ are K_v-symmetric since they are K-symmetric. Hence the $\widetilde{\delta}_{v,ij}^{(k)}$ are K_v-symmetric. Put

$$\widetilde{\delta}_v^{(k)} \ =\ (\widetilde{\delta}_{v,ij}^{(k)})_{(i,j)\in \mathrm{Band}_N(k)} \ \in\ (L_{w_v}^{\mathrm{sep}})^N$$

and let $\Phi_{F_{v,k}}^{\mathrm{sep}} : \mathbb{C}_v^N \to \mathbb{C}_v^N$ be the map from Proposition 7.18. By Proposition 7.18 there a unique $\vec{\Delta}_v^{(k)} = (\Delta_{v,ij}^{(k)})_{(i,j)\in\mathrm{Band}_N(k)} \in \mathbb{C}_v^N$ such that $\Phi_{F_{v,k}}^{\mathrm{sep}}(\vec{\Delta}_v^{(k)}) = \widetilde{\delta}_v^{(k)}$. Using (7.100), and applying (7.89) of Proposition 7.18 with $\rho = h_v^{kN}$ and $\ell = 1$, we see that

$$|\Delta_{v,ij}^{(k)}|_v \ \le\ h_v^{kN} \tag{7.102}$$

for all (i,j). Furthermore, Proposition 7.18 tells us that

$$\Delta_{v,k}(z) \ :=\ \sum_{(i,j)\in\mathrm{Band}_N(k)} \Delta_{v,ij}^{(k)} \varphi_{ij}(z)$$

is K_v-rational. By the definition of the map $\Phi_{F_{v,k}}^{\mathrm{sep}}$,

$$\Delta_{v,k}(z) \cdot F_{v,k}(z) \ =\ \sum_{(i,j)\in\mathrm{Band}_N(k)} \widetilde{\delta}_{v,ij}^{(k)} \cdot \widetilde{\varphi}_{ij}(z) \ +\ \text{terms of lower order} \ . \tag{7.103}$$

If we patch $G_v^{(k-1)}(z)$ using the $\Delta_{v,ij}^{(k)}$, then Theorem 7.17 provides a K_v-rational $(\mathfrak{X}, \vec{s})$-function $\widetilde{\Theta}_v^{(k)}(z)$ with a pole of order at most $(n-k)N_i$ at each x_i, such that

$$G_v^{(k)}(z) \ =\ G_v^{(k-1)} + \Delta_{v,k}(z) \cdot F_{v,k}(z) + \widetilde{\Theta}_v^{(k)}(z) \ .$$

By (7.101) and (7.103), for each $(i,j) \in \mathrm{Band}_N(k)$ the coefficient of $\widetilde{\varphi}_{ij}$ in the L^{sep}-rational expansion of $G_v^{(k)}(z)$ becomes $\widetilde{A}_{ij}$. Since (7.102) holds, the roots of $G_v^{(k)}(z)$ belong to E_v. Finally, since $G_v^{(k)}(z)$, $\Delta_{v,k}(z)F_{v,k}(z)$, and $\widetilde{\Theta}_v^{(k)}(z)$ are K_v-rational, so is $G_v^{(k)}(z)$.

Patching the Low-order Coefficients. The final stage of the global patching process deals with the coefficients $\widetilde{A}_{v,\lambda}$ in the expansions

$$G_v^{(n-1)}(z) \ =\ \sum_{i=1}^{m} \sum_{j=0}^{(n-1)N_i} \widetilde{A}_{v,ij} \widetilde{\varphi}_{i,nN_i-j}(z) + \sum_{\lambda=1}^{\Lambda} \widetilde{A}_{v,\lambda} \widetilde{\varphi}_{\lambda} \ .$$

For each $v \in \widehat{S}_K^+$, all the $\widetilde{A}_{v,\lambda}$ will be patched simultaneously. As before, we use a see-saw argument. Let $H = L^{\mathrm{sep}}$. For each w of H with $w|v$, put $G_w^{(n-1)}(z) = G_v^{(n-1)}(z)$ and expand

$$G_w^{(n-1)}(z) \ =\ \sum_{i=1}^{m} \sum_{j=0}^{(n-1)N_i} A_{w,ij} \widetilde{\varphi}_{i,nN_i-j}(z) + \sum_{\lambda=1}^{\Lambda} A_{w,\lambda} \widetilde{\varphi}_{\lambda} \ .$$

By construction, for each $v \in \widehat{S}_K^+$, the vector $\bigoplus_{w|v} G_w^{(n-1)}(z) \in H \otimes_K K_v(\mathcal{C})$ is $\mathrm{Gal}(H/K)$ invariant. By Proposition 7.8 this means that for each λ, the coefficient vector $\bigoplus_{w|v} \widetilde{A}_{w,\lambda}$ has the same Galois-equivariance properties as $\widetilde{\varphi}_\lambda(z)$. In particular, if $K \subset F_\lambda \subset L$ is the smallest field of rationality for $\widetilde{\varphi}_\lambda(z)$, then $\bigoplus_{w|v} A_{w,\lambda} \in F_\lambda \otimes_K K_v$.

By (7.95) in our choice of n, we have

$$\Big(\prod_{w \in \widehat{S}_L} h_w\Big)^{nN} \;>\; C_L(\widehat{S}_K^+) \ ,$$

so taking $Q_w = h_w^{nN}$ in Proposition 7.3 we can find an $\widetilde{A}_\lambda \in F_\lambda$ such that

$$\begin{cases} |\widetilde{A}_\lambda - \widetilde{A}_{w,\lambda}|_w \le h_w^{nN} & \text{for all } w \in \widehat{S}_L \ , \\ |\widetilde{A}_\lambda|_w \le 1 & \text{for all } w \notin \widehat{S}_L \ . \end{cases}$$

By working with representatives of Galois orbits as before, we can arrange that for each $\sigma \in \mathrm{Gal}(H/K)$ we have $\sigma(\widetilde{A}_\lambda) = \widetilde{A}_{\lambda'}$ if $\sigma(\varphi_\lambda) = \varphi_{\lambda'}$.

Put $\widetilde{\Delta}_{w,\lambda} = \widetilde{A}_\lambda - \widetilde{A}_{w,\lambda}$ for each w and λ, and put

$$\Delta_w^{(n)}(z) \;=\; \sum_\lambda \widetilde{\Delta}_{w,\lambda} \widetilde{\varphi}_\lambda(z) \ .$$

Then $\bigoplus_{w|v} \Delta_w(z) \in L \otimes_K K_v(\mathcal{C})$ is stable under $\mathrm{Gal}(L/K)$, for each $v \in \widehat{S}_K$. It follows that the $\Delta_w(z)$ belong to $K_v(\mathcal{C})$ and are the same for all $w|v$. Put $\Delta_{v,n}(z) = H_{w_v}^{(n)}(z)$, and expand $\Delta_{v,n}(z) = \sum_{\lambda=1}^{\Lambda} \Delta_{v,\lambda}\widetilde{\varphi}_\lambda(z)$. Then

$$|\widetilde{\Delta}_{v,\lambda}|_v \;\le\; h_v^{nN}$$

for each v and λ. Patch $G_v^{(n-1)}(z)$ by setting

$$G_v^{(n)}(z) \;=\; G_v^{(n-1)}(z) + \Delta_{v,n}(z)$$

This replaces the low-order coefficients of the $G_v^{(n)}(z)$ with the $\widetilde{A}_\lambda$.

Conclusion of the Patching Argument. The patching process has now arranged that the $G_v^{(n)}(z) \in K_v(\mathcal{C})$ for $v \in \widehat{S}_K$ all coincide with a single function $G^{(n)}(z)$, whose coefficients relative to the L^{sep}-rational basis belong to L^{sep}. Fix any v, and put $G_w(z) = G(z)$ for all places w of L^{sep} with $w|v$; then $\bigoplus_{w|v} G_w(z) \in \bigoplus_{w|v} L^{\mathrm{sep}}(\mathcal{C}) \cong L^{\mathrm{sep}} \otimes_K K(\mathcal{C})$ is invariant under $\mathrm{Gal}(L^{\mathrm{sep}}/K)$, so by Proposition 7.9 it belongs to $K(\mathcal{C})$.

For each $v \in \widehat{S}_K$, our restrictions on the magnitudes of the $\Delta_{v,ij}^{(k)}$ and the $\Delta_{v,\lambda}$ assure that the conclusions of Proposition 7.14 apply. Thus

$$\begin{cases} \text{If } v \in S_K, \text{ so } E_v \text{ is } K_v\text{-simple, the zeros of } G^{(n)}(z) \text{ are distinct and lie in } E_v \ . \\ \text{If } v \in \widehat{S}_K^+ \backslash S_K, \text{ then } \{z \in \mathcal{C}_v(\mathbb{C}_v) : |G^{(n)}(z)|_v \le R_v^{Nn}\} = E_v \ . \end{cases}$$

Also, for each $v \notin \widehat{S}_K^+$ the construction has arranged that in the expansion

$$G^{(n)}(z) \;=\; \sum_{i=1}^{m} \sum_{j=0}^{(n-1)N_i} A_{ij}\varphi_{i,nN_i-j}(z) + \sum_{\lambda=1}^{\Lambda} A_\lambda \varphi_\lambda \ ,$$

all the coefficients belong to $\widehat{\mathcal{O}}_v$ and the leading coefficients belong to $\widehat{\mathcal{O}}_v^\times$. Our choice of $\widehat{S}_K^+$ assures that $\mathfrak{C}_v$ and the $\varphi_{ij}(z)$ and $\varphi_\lambda(z)$ all have good reduction at

v, and the x_i specialize to distinct points $\pmod v$. Hence $G^{(n)}(z) \pmod v$ is a nonconstant function with a pole of order $nN_i > 0$ at each x_i, and for each $v \notin \widehat{S}_K^+$,

$$\{z \in \mathcal{C}_v(\mathbb{C}_v) : |G^{(n)}(z)|_v \le 1\} \;=\; \mathcal{C}_v(\mathbb{C}_v)\backslash(\bigcup_{i=1}^{m} B(x_i,1)^-) \;=\; E_v \;.$$

Construction of the Points in Theorem 4.2. The patching argument holds for each integer $n > n_3$ divisible by $n_0 n_1 n_2$. For any such n, the zeros of $G^{(n)}(z)$ satisfy the conditions of the theorem. If there are any v for which E_v is K_v-simple, the construction shows that the zeros of $G^{(n)}(z)$ are distinct, and letting $n \to \infty$ we obtain the points in the theorem.

However, if there are no such v, then since $\prod_{v \in \widehat{S}_K^+} r_v^{Nn}$ grows arbitrarily large as $n \to \infty$, the number of $\widehat{S}_K^+$-integers $\kappa \in K$ satisfying $|\kappa|_v \le r_v^{Nn}$ for all $v \in \widehat{S}_K^+$ also becomes arbitrarily large. For any such κ, the roots of $G^{(n)}(z) = \kappa$ are points satisfying the conditions of the Theorem. Hence there are infinitely many such points.

This completes the proof of Theorem 4.2 when $\mathrm{char}(K) = p > 0$. □

6. Proof of Proposition 7.18

Fix integers $\ell \ge 1$ and $1 \le k \le n-1$. Let $F_v(z) \in \mathbb{C}_v(\mathcal{C})$ be an $(\mathfrak{X}, \vec{s})$-function with a pole of order $(n-k-1)N_i$ and leading coefficient $d_{v,i} \ne 0$ at each x_i, whose zeros all belong to E_v. In proving Proposition 7.18, it will be useful to introduce a scaled version of the L-rational basis, consisting of the basis functions $\{d_{v,i}\varphi_{ij}, \varphi_\lambda\}$. Write

$$\Delta_v = \sum_{i=1}^{m} \sum_{s=0}^{\ell N_i - 1} \Delta_{v,is} \cdot \varphi_{i,(\ell+1)N_i - s} \;, \tag{7.104}$$

$$\Delta_v F_v = \sum_{i=1}^{m} \sum_{s=0}^{\ell N_i - 1} \delta_{v,is} \cdot d_{v,i}\varphi_{i,(n-k+\ell)N_i - s} \;+\; \text{lower order terms} \;, \tag{7.105}$$

put $\vec{\Delta} = (\Delta_{v,is})_{1 \le i \le m, 0 \le s < N_i}$ $\vec{\delta} = (\delta_{v,is})_{1 \le i \le m, 0 \le s < \ell N_i}$, and let $\Phi_{F_v} : \mathbb{C}_v^{\ell N} \to \mathbb{C}_v^{\ell N}$ be the linear map defined by

$$\Phi_{F_v}(\vec{\Delta}) \;=\; \vec{\delta} \;, \tag{7.106}$$

which takes the coefficients of Δ_v to the high-order coefficients of $\Delta_v F_v$. Note that Φ_{F_v} decomposes as direct sum of maps $\Phi_{F_v,i} : \mathbb{C}_v^{\ell N_i} \to \mathbb{C}_v^{\ell N_i}$, since only the terms in Δ_v involving $\varphi_{i,(\ell+1)N_i - s}$ for $s = 0, \ldots, \ell N_i - 1$ can contribute to poles of $\Delta_v F_v$ with order greater than $(n-k)N_i$ at x_i.

The maps Φ_{F_v}, $\Phi_{F_v,i}$ have an intrinsic interpretation as follows. Put

$$V \;=\; \bigoplus_{i=1}^{m} \bigoplus_{s=0}^{\ell N_i - 1} \mathbb{C}_v \varphi_{i,(\ell+1)N_i - s} \;, \qquad W \;=\; \bigoplus_{i=1}^{m} \bigoplus_{s=0}^{\ell N_i - 1} \mathbb{C}_v d_{v,i}\varphi_{i,(n-k+\ell)N_i - s} \;.$$

Then Φ_{F_v} is the map on coordinates associated to a linear transformation $\Phi_{F_v}^0 : V \to W$, defined as follows. For any divisor D on $\mathcal{C}_v(\mathbb{C}_v)$, let $\Gamma_{\mathbb{C}_v}(D) = \{f \in \mathbb{C}_v(\mathcal{C}) : \mathrm{div}(f) + D \ge 0\}$. Then for $D = \sum_{i=1}^m N_i(x_i)$, the inclusion of W into $\Gamma_{\mathbb{C}_v}((n-k+\ell)D)$ induces an isomorphism

$$\iota : W \;\cong\; \Gamma_{\mathbb{C}_v}\big((n-k+\ell)D\big)/\Gamma_{\mathbb{C}_v}\big((n-k)D\big) \;,$$

and for each function $\Delta_v(z) \in V$,

$$\Phi_{F_v}^0(\Delta_v) \;=\; \iota^{-1}\big(\,\Delta_v F_v \pmod{\Gamma_{\mathbb{C}_v}((n-k)D)}\big)\ .$$

Similarly, for each i, put

$$V_i \;=\; \bigoplus_{s=0}^{\ell N_i-1} \mathbb{C}_v \varphi_{i,(\ell+1)N_i-s}\ , \qquad W_i \;=\; \bigoplus_{s=0}^{\ell N_i-1} \mathbb{C}_v d_{v,i}\varphi_{i,(n-k+\ell)N_i-s}\ ,$$

Then $\Phi_{F_v,i}$ is the map on coordinates associated to a map $\Phi_{F_v,i}^0 : V_i \to W_i$, defined as follows. The inclusion of W_i into $\Gamma_{\mathbb{C}_v}((n-k)D+\ell N_i(x_i))$ induces an isomorphism

$$\iota_i : W_i \;\cong\; \Gamma_{\mathbb{C}_v}\big((n-k)D+\ell N_i(x_i)\big)/\Gamma_{\mathbb{C}_v}\big((n-k)D\big)\ ,$$

and for each function $\Delta_v(z) \in V_i$

$$\Phi_{F_v,i}^0(\Delta_v) \;=\; \iota_i^{-1}\big(\,\Delta_v F_v \pmod{\Gamma_{\mathbb{C}_v}((n-k)D)}\big)\ .$$

As $J|N_i$, the functions in the sets $\{\varphi_{i,(\ell+1)N_i-s}\}_{0\le s<\ell N_i}$ and $\{\widetilde{\varphi}_{i,(\ell+1)N_i-s}\}_{0\le s<\ell N_i}$ are $K(x_i)$-linear combinations of each other, and each set forms a basis for V_i. Similarly, the functions in $\{d_{v,i}\varphi_{i,(n-k+\ell)N_i-s}\}_{0\le s<\ell N_i}$ and $\{\widetilde{\varphi}_{i,(n-k+\ell)N_i-s}\}_{0\le s<\ell N_i}$ are $K(x_i)$-linear combinations of each other, and each set forms a basis for W_i.

The map $\Phi_{F_v}^{\mathrm{sep}}$ in Proposition 7.18 is the coordinate map associated to $\Phi_{F_v}^0$ using the L-rational basis on the source and the L^{sep}-rational basis on the target: if we write

$$\Delta_v F_v \;=\; \sum_{i=1}^{m}\sum_{s=0}^{\ell N_i-1} \widetilde{\delta}_{v,is}\cdot \widetilde{\varphi}_{i,(n-k+\ell)N_i-s} \;+\; \text{lower order terms}\ ,$$

and put $\widetilde{\delta} = (\widetilde{\delta}_{v,is})_{1\le i\le m, 0\le s<\ell N_i}$ then $\Phi_{F_v}^{\mathrm{sep}}(\vec{\Delta}) = \widetilde{\delta}$. Clearly $\Phi_{F_v}^{\mathrm{sep}}$ decomposes as a direct sum of maps $\Phi_{F_v,i}^{\mathrm{sep}} : \mathbb{C}_v^{\ell N_i} \to \mathbb{C}_v^{\ell N_i}$ associated to the $\Phi_{F_v,i}^0 : V_i \to W_i$.

Before proving Proposition 7.18, we will need two lemmas. Recall that the L-rational basis is multiplicatively generated by finitely many functions. This means that collectively, the basis functions $\varphi_{ij}(z)$ have only finitely many distinct zeros.

Lemma 7.19. *Let K_v be nonarchimedean. Then there is a constant $\Lambda_v > 0$, depending only on $\mathfrak{X}$, the choice of the uniformizing parameters $g_{x_i}(z)$, and the projective embedding of $\mathcal{C}_v$, with the following property:*

Let $r > 0$ be small enough that

(1) $r < \min_{i\ne j}(\|x_i, x_j\|_v)$;

(2) *each of the balls $B(x_i, r)$ is isometrically parametrizable;*

(3) *for each i, none of the $\varphi_{ij}(z)$ has a zero in $B(x_i, r)$.*

Put $\varpi_v = \min(1, \Lambda_v \cdot r)$, and let ℓ, k be integers with $\ell \ge 1$ and $1 \le k \le n-1$.

Suppose $F_v(z) \in \mathbb{C}_v(\mathcal{C})$ is an $(\mathfrak{X}, \vec{s})$-function with a pole of order $(n-k-1)N_i$ and leading coefficient $d_{v,i} \ne 0$ at x_i, for each i. Assume $F_v(z)$ has no zeros in $\bigcup_{i=1}^m B(x_i, r)$. Then for each i, and each integer $0 \le s < \ell N_i$, when we expand $\varphi_{i,(\ell+1)N_i-s}(z)F_v(z)$ using the scaled L-rational basis as

$$\varphi_{i,(\ell+1)N_i-s}\cdot F_v \;=\; \sum_{t=0}^{\ell N_i-s-1} C_i(s,t)\cdot d_{v,i}\varphi_{i,(n-k+\ell)N_i-s-t}$$

$$+ \text{ terms with poles of order} \le (n-k)N_{i'} \text{ at each } x_{i'}, \tag{7.107}$$

we have $|C_i(s,t)|_v \le 1/\varpi_v^t$ for each t.

PROOF. For each i, let $g_{x_i}(z)$ be the uniformizing parameter used to normalize the basis functions $\varphi_{ij}(z)$; thus $\lim_{z\to x_i} F_v(z)\cdot g_{x_i}(z)^{(n-k-1)N_i} = d_{v,i}$ and $\lim_{z\to x_i}\varphi_{ij}(z)\cdot g_{x_i}(z)^j = 1$ for each $j > N_i$. Let $\varrho_i : D(0,r)\to B(x_i,r)$ be an isometric parametrization with $\varrho_i(0) = x_i$, and put $b_{v,i} = \lim_{Z\to 0} g_{x_i}(\varrho_i(Z))/Z$.

Take $\Lambda_v = \min_{1\le i\le m}(|b_{v,i}|_v)$, and put

$$\varpi_v \;=\; \min(1,\Lambda_v r) \;=\; \min(1, r|b_{v,1}|_v,\ldots,r|b_{v,m}|_v)\ . \tag{7.108}$$

To establish the bounds in the lemma, first fix i. By abuse of notation, write $F_v(Z)$ for $F_v(\varrho_i(Z))$ and $\varphi_{ij}(Z)$ for $\varphi_{ij}(\varrho_i(Z))$. For compactness of notation, temporarily write $h = n-k-1$. Then

$$\lim_{Z\to 0} F_v(Z)\cdot Z^{hN_i} = \lim_{Z\to 0}\Big(F_v(\varrho_i(Z))\cdot g_{x_i}(\varrho_i(Z))^{hN_i}\Big)\cdot\Big(\frac{Z}{g_{x_i}(\varrho_i(Z))}\Big)^{hN_i} = d_{v,i}b_{v,i}^{-hN_i}\ ,$$

so $F_v(Z)$ has a Laurent expansion of the form

$$F_v(Z) \;=\; d_{v,i}b_{v,i}^{-(n-k-1)N_i}\cdot Z^{-(n-k-1)N_i}\cdot(1+\sum_{j=1}^{\infty} f_j Z^j)\ . \tag{7.109}$$

Since $F_v(Z)$ has no zeros in $D(0,r)$, the theory of Newton Polygons shows that $|f_j|_v < 1/r^j$ for each $j\ge 1$ (see Lemma 3.35 and the discussion before it). Similarly, for each $0\le s<\ell N_i$,

(7.110)

$$\varphi_{i,(\ell+1)N_i-s}(Z)\cdot F_v(Z) \;=\; d_{v,i}b_{v,i}^{-(n-k+\ell)N_i+s}\cdot Z^{-(n-k+\ell)N_i+s}\cdot(1+\sum_{j=1}^{\infty} f_{is,j}Z^j)$$

with $|f_{is,j}|_v < 1/r^j$ for each $j\ge 1$, and for each $0\le t\le \ell N_i - s-1$,

(7.111)

$$\varphi_{i,(n-k+\ell)N_i-s-t}(Z) \;=\; b_{v,i}^{-(n-k+\ell)N_i+s+t}Z^{-(n-k+\ell)N_i+s+t}\cdot(1+\sum_{j=1}^{\infty} c_{i,s+t,j}Z^j)$$

with $|c_{i,s+t,j}|_v < 1/r^j$ for each $j\ge 1$.

To prove the lemma, fix $0\le s<\ell N_i$, insert the expansions (7.111) into (7.107) and compare the coefficients of the resulting series with those in (7.110). Comparing the coefficients of $Z^{-(n-k+\ell)N_i+s}$ we see that when $t=0$,

$$C_i(s,0)\cdot b_{v,i}^{-(n-k+\ell)N_i+s} \;=\; b_{v,i}^{-(n-k+\ell)N_i+s}\ ,$$

so $C_i(s,0) = 1$; trivially, $|C_i(s,0)|_v \le 1/(r|b_{v,i}|_v)^0$. Inductively, take $1\le t\le \ell N_i - s-1$ and assume that $|C_i(s,j)|_v\le 1/(r|b_{v,i}|_v)^j$ for $0\le j\le t-1$. Comparing the coefficients of $Z^{-(n-k+\ell)N_i+s+t}$ we find that

$$\begin{aligned} C_i(s,t)\cdot b_{v,i}^{-(n-k+\ell)N_i+s+t} \;&+\; \sum_{j=0}^{t-1} C_i(s,j)\cdot b_{v,i}^{-(n-k+\ell)N_i+s+j}c_{i,s+j,t-j} \\ &=\; b_{v,i}^{-(n-k+\ell)N_i+s}\cdot f_{is,t}\ , \end{aligned}$$

or equivalently,

$$C_i(s,t) \;=\; b_{v,i}^{-t}\cdot f_{is,t} - \sum_{j=0}^{t-1} C_i(s,j)\cdot b_{v,i}^{j-t}c_{i,s+j,t-j}\ .$$

Since $|f_{is,t}|_v \le 1/r^t$, it follows that $|b_{v,i}^{-t} f_{is,t}|_v \le 1/(r|b_{v,i}|_v))^t$. Similarly, for $0 \le j \le t-1$ we have $|c_{i,s+j,t-j}|_v \le 1/r^{t-j}$, so by induction, for each such j

$$|C_i(s,j) \cdot b_{v,i}^{j-t} c_{i,s+t,t-j}|_v \;\le\; 1/(r|b_{v,i}|_v)^j \cdot |b_{v,i}|_v^{j-t} \cdot 1/r^{t-j} \;=\; 1/(r|b_{v,i}|_v)^t \;.$$

By the ultrametric inequality $|C_i(s,t)|_v \le 1/(r|b_{v,i}|_v)^t$, and the induction can continue.

Now vary i. For each i we have $r|b_{v,i}|_v \ge \Lambda_v r \ge \varpi_v$, and the lemma follows. □

The following lemma gives bounds for the entries of the inverse of a unipotent lower triangular matrix, given a suitable bound for the entries in each subdiagonal. The indices i, j, k, ℓ in the lemma are unrelated to i, j, k and ℓ as used elsewhere.

LEMMA 7.20. *Let $C \in M_k(\mathbb{C}_v)$ be a lower triangular matrix whose diagonal elements are* 1, *and write it as*

$$C \;=\; \begin{pmatrix} 1 & 0 & 0 & \cdots & 0 \\ C_{2,1} & 1 & 0 & \cdots & 0 \\ C_{3,1} & C_{3,2} & 1 & \cdots & 0 \\ \vdots & \vdots & \vdots & \ddots & \vdots \\ C_{k,1} & C_{k,2} & C_{k,3} & \cdots & 1 \end{pmatrix} . \tag{7.112}$$

Let $\varpi_v > 0$ be such that for each $h = 1, \ldots, k$, the elements $C_{i,j}$ belonging to the ℓth subdiagonal (i.e. those with $i - j = \ell$), satisfy $|C_{i,j}|_v \le 1/\varpi_v^\ell$. Then

$$C^{-1} \;=\; \begin{pmatrix} 1 & 0 & 0 & \cdots & 0 \\ c_{2,1} & 1 & 0 & \cdots & 0 \\ c_{3,1} & c_{3,2} & 1 & \cdots & 0 \\ \vdots & \vdots & \vdots & \ddots & \vdots \\ c_{k,1} & c_{k,2} & c_{k,3} & \cdots & 1 \end{pmatrix} \tag{7.113}$$

and for all $c_{i,j}$ in the ℓth subdiagonal, we have $|c_{i,j}|_v \le 1/\varpi_v^\ell$.

PROOF. Clearly C^{-1} exists and has the form (7.113). To show that $|c_{i,j}|_v \le 1/\varpi_v^{i-j}$, we use induction on $\ell = i - j$. When $i - j = 1$, the (i,j) term in $C \cdot C^{-1} = I$ is

$$c_{i,j} + C_{i,j} \;=\; 0 \;,$$

so $|c_{i,j}|_v \;=\; |C_{i,j}|_v \;\le\; 1/\varpi_v$. Now suppose $i - j \;=\; \ell > 1$. The (i,j) term in $C \cdot C^{-1} = I$ is

$$C_{i,j} + C_{i,j+1}c_{j+1,j} + \cdots + C_{i,i-1}c_{i-1,j} + c_{i,j} \;=\; 0 \;.$$

Assuming that $|c_{i',j'}|_v \le 1/\varpi_v^{1'-j'}$ for all (i',j') with $i' - j' < \ell$, and using our hypothesis on C, the ultrametric inequality gives

$$|c_{i,j}|_v \;\le\; \max\big(|C_{i,j+1}|_v |c_{j+1,j}|_v, \cdots, |C_{i,i-1}|_v |c_{i-1,j}|_v, |C_{i,j}|_v\big) \;\le\; 1/\varpi_v^{i-j}$$

as desired. □

We can now prove Proposition 7.18.

PROOF OF PROPOSITION 7.18. Let $F_v(z) \in \mathbb{C}_v(\mathcal{C})$ be an $(\mathfrak{X}, \vec{s})$-function with a pole of order $(n-k-1)N_i$ and leading coefficient $d_{v,i} \neq 0$ at each x_i, whose zeros all belong to E_v. Writing

$$\Delta_v(z) = \sum_{i=1}^{m} \sum_{s=0}^{\ell N_i - 1} \Delta_{v,is} \cdot \varphi_{i,(\ell+1)N_i - s}(z) \ , \tag{7.114}$$

$$\Delta_v(z) F_v(z) = \sum_{i=1}^{m} \sum_{s=0}^{\ell N_i - 1} \delta_{v,is} \cdot d_{v,i}\varphi_{i,(n-k+\ell)N_i - s}(z) \tag{7.115}$$

$$+ \text{ lower order terms} \ ,$$

put

$$\vec{\Delta} \ = \ (\Delta_{v,is})_{\substack{1 \le i \le m \\ 0 \le s < \ell N_i}} \ , \qquad \vec{\delta} \ = \ (\delta_{v,is})_{\substack{1 \le i \le m \\ 0 \le s < \ell N_i}} \ ,$$

and define $\Phi_{F_v} : \mathbb{C}_v^{\ell N} \to \mathbb{C}_v^{\ell N}$ by $\Phi_{F_v}(\vec{\Delta}) = \vec{\delta}$ as in (7.106). If we write $\vec{\Delta}_i = (\Delta_{v,is})_{0 \le s < \ell N_i}$ and $\vec{\delta}_i = (\delta_{v,it})_{0 \le t < \ell N_i}$, then Φ_{F_v} is a direct sum of the maps $\Phi_{F_v,i} : \mathbb{C}_v^{\ell N_i} \to \mathbb{C}_v^{\ell N_i}$ with $\Phi_{F_v,i}(\vec{\Delta}_i) = \vec{\delta}_i$ for each i.

Let $r > 0$ be small enough that

(1) $r < \min_{i \neq j}(\|x_i, x_j\|_v)$;
(2) each ball $B(x_i, r)$ is isometrically parametrizable and disjoint from E_v;
(3) for each i, none of the φ_{ij} has a zero in $B(x_i, r)$.

Let $\Lambda_v > 0$ and $\varpi_v = \min(1, \Lambda_v \cdot r) > 0$ be as in Lemma 7.19.

We begin by showing that Φ_{F_v} is an isomorphism and that for each $\rho > 0$,

$$\Phi_{F_v}\Big(\bigoplus_{i=1}^{m} \bigoplus_{s=0}^{\ell N_i - 1} D(0, \varpi_v^{-s}\rho)\Big) \ \supseteq \ \bigoplus_{i=1}^{m} \bigoplus_{t=0}^{\ell N_i - 1} D(0, \varpi_v^{-t} \cdot \rho) \ . \tag{7.116}$$

For this, it is enough to show that each $\Phi_{F_v,i}$ is an isomorphism, and that

$$\Phi_{F_v,i}\Big(\bigoplus_{s=0}^{\ell N_i - 1} D(0, \varpi_v^{-s}\rho)\Big) \ \supseteq \ \bigoplus_{s=0}^{\ell N_i - 1} D(0, \varpi_v^{-s} \cdot \rho) \ . \tag{7.117}$$

Fix i. As in (7.107), for each $0 \le t < \ell N_i$ the product $\varphi_{i,(\ell+1)N_i - t} \cdot F_v$ can be expanded using the scaled L-rational basis as

$$d_{v,i}\varphi_{i,(n-k+\ell)N_i - s} \quad + \quad \sum_{t=1}^{\ell N_i - s - 1} C_i(s,t) \cdot d_{v,i}\varphi_{i,(n-k+\ell)N_i - s - t}$$

$$+ \text{ terms not contributing to } \Phi_{F_v,i} \ .$$

This means that the matrix of $\Phi_{F_v,i}$ is

$$C_i \ = \ \begin{pmatrix} 1 & 0 & 0 & \cdots & 0 \\ C_i(0,1) & 1 & 0 & \cdots & 0 \\ C_i(0,2) & C_i(1,1) & 1 & \cdots & 0 \\ \vdots & \vdots & \vdots & \ddots & \vdots \\ C_i(0,\ell N_i - 1) & C_i(1,\ell N_i - 2) & \ldots & C_i(\ell N_i - 2, 1) & 1 \end{pmatrix} ; \tag{7.118}$$

in particular, $\Phi_{F_v,i}$ is nonsingular. By Lemma 7.19 we have $C_i(s,t)\varpi_v^{-t}$ for all s,t. Hence by Lemma 7.20,

$$C_i^{-1} = \begin{pmatrix} 1 & 0 & 0 & \cdots & 0 \\ c_i(0,1) & 1 & 0 & \cdots & 0 \\ c_i(0,2) & c_i(1,1) & 1 & \cdots & 0 \\ \vdots & \vdots & \vdots & \ddots & \vdots \\ c_i(0,\ell N_i-1) & c_i(1,\ell N_i-2) & \ldots & c_i(\ell N_i-2,1) & 1 \end{pmatrix},$$

with $c_i(s,t) \le \varpi_v^{-t}$ for all s,t.

Fix $\rho > 0$, and assume that $|\delta_{v,it}|_v \le \varpi_v^{-s}\rho$ for $0 \le t < \ell N_i$. Since $\vec{\Delta}_i = \Phi_{F_v}^{-1}(\vec{\delta}_i) = C_i^{-1}\vec{\delta}_i$, this means that for each s,

$$\Delta_{v,is} = \Big(\sum_{t=0}^{s-1} c_i(t,s-t)\delta_{i,t}\Big) + \delta_{i,s} , \tag{7.119}$$

and the ultrametric inequality shows that $|\Delta_{v,is}|_v \le \varpi_v^{-s}\rho$. This proves (7.117) and (7.116).

We can now prove (7.89) and (7.90) in Proposition 7.18. Write $\Delta_v(z)F_v(z)$ using the L-rational basis as

$$\Delta_v(z)F_v(z) = \sum_{i=1}^{m}\sum_{s=0}^{\ell N_i-1} \widehat{\delta}_{v,is}\varphi_{i,(n-k+\ell)N_i-s}(z) + \text{lower order terms} . \tag{7.120}$$

Comparing (7.120) and (7.115) shows that $|\widehat{\delta}_{v,is}|_v = |d_{v,is}|_v|\delta_{v,is}|_v$ for all i,s. Next, expand $\Delta_v(z)F_v(z)$ in terms of the $L^{\rm sep}$-rational basis as

$$\Delta_v(z)F_v(z) = \sum_{i=1}^{m}\sum_{s=0}^{\ell N_i-1} \widetilde{\delta}_{v,is}\widetilde{\varphi}_{i,(n-k+\ell)N_i-s}(z) + \text{lower order terms} . \tag{7.121}$$

By Proposition 3.3(C), for each $i = 1,\ldots,m$ there is an invertible $J \times J$ matrix $\widetilde{\mathcal{B}}_i$ which expresses each set of J consecutive basis elements $\{\varphi_{ij}\}_{hJ+1\le j\le (h+1)J}$ of the L-rational basis in terms of the corresponding set $\{\widetilde{\varphi}_{ij}\}_{hJ+1\le j\le (h+1)J}$ from the $L^{\rm sep}$-rational basis. Since $J|N_i$ for each i, it follows that there is a constant $\widetilde{\Upsilon}_v > 0$ such that if $|\widetilde{\delta}_{v,is}|_v \le \widetilde{\Upsilon}_v\varpi_v^{-s}|d_{v,i}|_v \cdot \rho$ for all i,s, then $|\widehat{\delta}_{v,is}|_v \le \varpi_v^{-s}|d_{v,i}|_v \cdot \rho$ for all i,s. This in turn means $|\delta_{v,is}|_v \le \varpi_v^{-s}\rho$ for all i,s. By the discussion above, $|\Delta_{v,is}|_v \le \varpi_v^{-s}\rho$ for all i,s.

Since $\Phi_{F_v}^{\rm sep}$ is the coordinate map for $\Phi_{F_v}^0$ using the L-rational basis on the source and the $L^{\rm sep}$-rational basis on the target, we see that

$$\Phi_{F_v}^{\rm sep}\Big(\bigoplus_{i=1}^{m}\bigoplus_{s=0}^{\ell N_i-1} D(0,\varpi_v^{-s}\rho)\Big) \supseteq \bigoplus_{i=1}^{m}\bigoplus_{s=0}^{\ell N_i-1} D(0,\widetilde{\Upsilon}_v\varpi_v^{-s}|d_{v,i}|_v \cdot \rho) , \tag{7.122}$$

which is (7.90). To show (7.89), replace ρ with $\varpi_v^{\ell N}\rho$ in (7.122). Since $0 < \varpi_v \le 1$,

$$\begin{aligned} \Phi_{F_v}^{\rm sep}\Big(\bigoplus_{i=1}^{m}\bigoplus_{s=0}^{\ell N_i-1} D(0,\varpi_v^{\ell N-s}\rho)\Big) &\supseteq \bigoplus_{i=1}^{m}\bigoplus_{s=0}^{\ell N_i-1} D(0,\widetilde{\Upsilon}_v\varpi_v^{\ell N-s}|d_{v,i}|_v \cdot \rho) \\ &\supseteq \bigoplus_{i=1}^{m} D(0,\widetilde{\Upsilon}_v\varpi_v^{\ell N}|d_{v,i}|_v \cdot \rho)^{\ell N_i} . \end{aligned}$$

This yields (7.89) since $D(0,\rho)^{\ell N} \supseteq \bigoplus_{i=1}^{m} \bigoplus_{s=0}^{\ell N_i - 1} D(0, \varpi_v^{\ell N - s}\rho)$.

Finally, we prove the rationality assertions in Proposition 7.18.

Assume that $F_v(z)$ is K_v-rational. Using the L^{sep}-rational basis, we can write

$$\Delta_v(z) = \sum_{i=1}^{m} \sum_{s=0}^{\ell N_i - 1} \widetilde{\Delta}_{v,is} \cdot \widetilde{\varphi}_{i,(\ell+1)N_i - s}(z) \ , \tag{7.123}$$

$$\Delta_v(z) F_v(z) = \sum_{i=1}^{m} \sum_{s=0}^{\ell N_i - 1} \widetilde{\delta}_{v,is} \cdot \widetilde{\varphi}_{i,(n-k+\ell)N_i - s}(z) \ + \text{ lower order terms} \ . \tag{7.124}$$

Put

$$\widetilde{\Delta} \ = \ (\widetilde{\Delta}_{v,is})_{\substack{1 \le i \le m \\ 0 \le s < \ell N_i}} \ , \qquad \widetilde{\delta} \ = \ (\widetilde{\delta}_{v,is})_{\substack{1 \le i \le m \\ 0 \le s < \ell N_i}} \ ,$$

and for each $i = 1, \ldots, m$ put

$$\widetilde{\Delta}_i \ = \ (\widetilde{\Delta}_{v,is})_{0 \le s < \ell N_i} \ , \qquad \widetilde{\delta}_i \ = \ (\widetilde{\delta}_{v,is})_{0 \le s < \ell N_i} \ .$$

Fix i. Then for each $0 \le t < \ell N_i$ the product $\widetilde{\varphi}_{i,(\ell+1)N_i - t} \cdot F_v$ is rational over $K_v(x_i)^{\mathrm{sep}}$ and can be expanded using the L^{sep}-rational basis as

$$\widetilde{\varphi}_{i,(\ell+1)N_i - t} \cdot F_v \ = \ \sum_{s=0}^{\ell N_i - 1} \widetilde{C}_i(s,t) \cdot \varphi_{i,(n-k+\ell)N_i - s} + \text{terms not contributing to } \Phi^0_{F_v,i}.$$

Since each $\widetilde{\varphi}_{i,(\ell+1)N_i - t}$ is rational over $K_v(x_i)^{\mathrm{sep}}$, by Galois equivariance each $\widetilde{C}_i(s,t)$ belongs to $K_v(x_i)^{\mathrm{sep}}$. Thus the matrix for $\Phi^0_{F_v,i}$ using the L^{sep}-rational basis on the source and the target is

$$\widetilde{C}_i \ = \ \begin{pmatrix} \widetilde{C}_i(0,0) & \cdots & \widetilde{C}_i(0, \ell N_i - 1) \\ \vdots & \ddots & \vdots \\ \widetilde{C}_i(\ell N_i - 1, 0) & \cdots & \widetilde{C}_i(\ell N_i - 1, \ell N_i - 1) \end{pmatrix} \ \in \ \mathrm{GL}_{\ell N_i}(K_v(x_i)^{\mathrm{sep}}) \ .$$

Because the L^{sep}-rational basis is K_v-symmetric, the set of matrices $\{\widetilde{C}_1, \ldots, \widetilde{C}_m\}$ is K_v-symmetric.

Suppose $\widetilde{\delta}$ belongs to $(L_{w_v}^{\mathrm{sep}})^{\ell N}$ and is K_v-symmetric. Then $\widetilde{\delta}_i$ belongs to $(K_v(x_i)^{\mathrm{sep}})^{\ell N_i}$ for each i, and the set of vectors $\{\widetilde{\delta}_1, \ldots, \widetilde{\delta}_m\}$ is K_v-symmetric. It follows that $\widetilde{\Delta}_i = \widetilde{C}_i^{-1} \widetilde{\delta}_i$ belongs to $(K_v(x_i)^{\mathrm{sep}})^{\ell N_i}$ for each i, and the set of vectors $\{\widetilde{\Delta}_1, \ldots, \widetilde{\Delta}_m\}$ is K_v-symmetric. Since

$$\Delta_v(z) \ = \ \sum_{i=1}^{m} \sum_{s=0}^{\ell N_i - 1} \widetilde{\Delta}_{v,is} \cdot \widetilde{\varphi}_{i,(\ell+1)N_i - s}(z) \ ,$$

where the $\widetilde{\Delta}_{v,is}$ and $\widetilde{\varphi}_{i,(\ell+1)N_i - s}$ are K_v-symmetric and rational over $L_{w_v}^{\mathrm{sep}}$, it follows that Δ_v is K_v-rational.

If we re-express Δ_v in terms of the L-rational basis as

$$\Delta_v(z) \ = \ \sum_{i=1}^{m} \sum_{s=1}^{\ell N_i} \Delta_{v,is} \cdot \varphi_{i,N_i + s}(z) \ ,$$

then the associated vector $\vec{\Delta} \ = \ (\Delta_{v,is})_{1 \le i \le m, 1 \le s \le \ell N_i}$ is the unique solution to $\Phi^{\mathrm{sep}}_{F_v}(\vec{\Delta}) = \widetilde{\delta}$ in $\mathbb{C}_v^{\ell N}$. Since Δ_v is K_v-rational, necessarily $\vec{\Delta}$ belongs to $L_{w_v}^{\ell N}$ and is K_v-symmetric.

This completes the proof of Proposition 7.18. □

CHAPTER 8

Local Patching when $K_v \cong \mathbb{C}$

In this section we give the confinement argument for Theorem 4.2 when $K_v \cong \mathbb{C}$. Write $\mathbb{C}$ for $\mathbb{C}_v$ and $|x|$ for $|x|_v$. Let w_v be the distinguished place of $L = K(\mathfrak{X})$ determined by the embedding $\widetilde{K} \hookrightarrow \mathbb{C}$ used to identify $\mathfrak{X}$ with a subset of $\mathcal{C}_v(\mathbb{C})$, and identify $L_{w_v} = K_v = \mathbb{C}$.

At several places in this section, we assert that certain objects are K_v-symmetric. Since $\mathrm{Aut}_c(\mathbb{C}_v/K_v) = \mathrm{Aut}(\mathbb{C}/\mathbb{C})$ is trivial, this is vacuous. However, we include it for compatibility with the statements of the results in Chapters 9–11.

Using the construction of the coherent approximating functions in Theorem 7.11, we begin with the following data:

(1) A K_v-symmetric probability vector $\vec{s} \in \mathcal{P}^m(\mathbb{Q})$ with positive rational coefficients.

(2) A $\mathbb{C}$-simple set $E_v \subset \mathcal{C}_v(\mathbb{C})\backslash\mathfrak{X}$: thus, E_v is compact and nonempty with finitely many connected components, each of which is simply connected, has a piecewise smooth boundary, and is the closure of its interior E_v^0.

(3) Parameters h_v, r_v, R_v with $1 < h_v < r_v < R_v$, which govern the freedom in the patching process.

(4) A number N and an $(\mathfrak{X}, \vec{s})$-function $\phi_v(z) \in K_v(\mathcal{C})$ of degree N such that

$$\{z \in \mathcal{C}_v(\mathbb{C}) : |\phi_v(z)| \le R_v^N\} \ \subset \ E_v^0 \ .$$

Put $N_i = Ns_i$ for each i, and write $\widetilde{c}_{v,i} = \lim_{z \to x_i} \phi_v(z) \cdot g_{x_i}(z)^{N_i}$ for the leading coefficient of $\phi_v(z)$ at x_i.

(5) An order $\prec_N$ on the index set $\mathcal{I} = \{(i,j) \in \mathbb{Z}^2 : 1 \le i \le m, 0 \le j\}$ determined by N and $\vec{s}$ as in (7.41), which gives the sequence in which coefficients are patched. The group $\mathrm{Aut}_c(\mathbb{C}/K_v)$ acts on $\mathcal{I}$, and we write $\mathrm{Aut}_c(\mathbb{C}/K_v)(i,j)$ for the orbit of $(i,j) \in \mathcal{I}$.

We will use the L-rational basis $\{\varphi_{ij}, \varphi_\lambda\}$ from §3.3 to expand all functions, and $\Lambda = \dim_K(\Gamma(\sum_{i=1}^m N_i(x_i)))$ will be the number of low-order basis elements, as in the global patching process. The order $\prec_N$ respects the N-bands (7.42), and for each $x_i \in \mathfrak{X}$, specifies the terms to be patched in decreasing pole order.

THEOREM 8.1. *Suppose $K_v \cong \mathbb{C}$. Let $E_v \subset \mathcal{C}_v(\mathbb{C})\backslash\mathfrak{X}$ be $\mathbb{C}$-simple, with interior E_0. Let $\vec{s} \in \mathcal{P}^m(\mathbb{Q})$ be a K_v-symmetric probability vector with positive rational coefficients. Let $1 < h_v < r_v < R_v$ be real numbers.*

Let $\phi_v(z) \in K_v(\mathcal{C})$ be an $(\mathfrak{X}, \vec{s})$-function of degree N satisfying

$$\{z \in \mathcal{C}_v(\mathbb{C}) : |\phi_v(z)| \le R_v^N\} \ \subset \ E_v^0 \ .$$

Let $N_i = Ns_i$ for each x_i, and let $\widetilde{c}_{v,i} = \lim_{z\to x_i} \phi_v(z)\cdot g_{x_i}(z)^{N_i}$ be the leading coefficient of $\phi_v(z)$ at x_i. Put

$$M_v = \max\Big(\max_{\substack{1\le i\le m\\ N_i<j\le 2N_i}} \|\varphi_{ij}\|_{E_v},\ \max_{1\le\lambda\le\Lambda} \|\varphi_\lambda\|_{E_v}\Big) .$$

Let $k_v > 0$ be the least integer such that

$$\frac{2NM_v}{1-(h_v/r_v)^N}\cdot\Big(\frac{h_v}{r_v}\Big)^{k_vN} < \frac{1}{4}, \tag{8.1}$$

and let $\overline{k}\ge k_v$ be a fixed integer. Let $B_v>0$ be an arbitrary constant. Then there is an integer n_v depending on $\phi_v(z)$, $\overline{k}$, B_v, r_v, and R_v, such that for each sufficiently large integer n divisible by n_v, one can carry out the local patching process at K_v as follows:

Put $G_v^{(0)}(z) = \phi_v(z)^n$. For each $k=1,\ldots,n-1$, let $\{\Delta_{v,ij}^{(k)}\in L_{w_v}\}_{(i,j)\in\mathrm{Band}_N(k)}$ be a K_v-symmetric set of numbers given in $\prec_N$ order, subject to the conditions that for each i, we have $\Delta_{v,i0}^{(1)} = 0$ and for each $j>0$

$$|\Delta_{v,ij}^{(k)}| \le \begin{cases} B_v & \text{if } k\le\overline{k},\\ h_v^{kN} & \text{if } k>\overline{k}.\end{cases} \tag{8.2}$$

For $k=n$, let $\{\Delta_{v,\lambda}^{(n)}\in L_{w_v}\}_{1\le\lambda\le\Lambda}$ be a K_v-symmetric set of numbers satisfying

$$|\Delta_{v,\lambda}^{(n)}| \le h_v^{nN} . \tag{8.3}$$

Then one can inductively construct $(\mathfrak{X},\vec{s})$-functions $G_v^{(1)}(z),\ldots,G_v^{(n)}(z)\in K_v(\mathcal{C})$, of common degree nN, having leading coefficient $\widetilde{c}_{v,i}^{\,n}$ at each x_i, and satisfying

(A) For each $k=1,\ldots,n$, there are functions $\vartheta_{v,ij}^{(k)}(z)\in L_{w_v}(\mathcal{C})$, determined recursively in $\prec_N$ order, such that

$$G_v^{(k)}(z) = G_v^{(k-1)}(z) + \sum_{(i,j)\in\mathrm{Band}_N(k)} \Delta_{v,ij}^{(k)}\vartheta_{v,ij}^{(k)}(z) \quad \text{for } k<n ,$$

$$G_v^{(n)}(z) = G_v^{(n-1)}(z) + \sum_{\lambda=1}^{\Lambda} \Delta_{v,\lambda}^{(n)}\varphi_\lambda(z) ,$$

and where for each (i,j):

(1) $\vartheta_{v,ij}^{(k)}(z)$ has a pole of order $nN_i - j > (n-k-1)N_i$ at x_i and leading coefficient $\widetilde{c}_{v,i}^{\,n-k-1}$, a pole of order at most $(n-k-1)N_{i'}$ at each $x_{i'}\ne x_i$, and no other poles;

(2) $\sum_{(i',j)\in\mathrm{Aut}_c(\mathbb{C}_v/K_v)(i,j)} \Delta_{v,i'j}^{(k)}\vartheta_{v,i'j}^{(k)}(z)$ belongs to $K_v(\mathcal{C})$;

(B) For each $k=1,\ldots,n$, $\{z\in\mathcal{C}_v(\mathbb{C}) : |G_v^{(k)}(z)|\le r_v^{Nn}\}\subset E_v^0$.

Remark. A key aspect of Theorem 8.1 is that by choosing n appropriately, the freedom B_v in patching the coefficients for $k\le\overline{k}$ can be made arbitrarily large. The patching procedure accomplishes this by exploiting a phenomenon of "magnification" introduced in ([**53**]). It first raises $\phi_v(z)$ to a power $n_v>\overline{k}$, so that $F_v(z)=\phi_v(z)^{n_v}$ has enough coefficients to adjust independently. It then varies those coefficients 'infinitesimally', preserving the analytic properties of $F_v(z)$. Finally it raises the modified $F_v(z)$ to a further power m_v with $n=m_vn_v$, creating large changes in the coefficients of $G_v^{(0)}(z)$.

PROOF OF THEOREM 8.1. Let $\overline{k} \geq k_v$ and $B_v > 0$ be as in the Theorem. Let $n_v > 0$ be an integer large enough that

$$n_v > \overline{k}, \tag{8.4}$$

$$\widehat{R}_v := 2^{-1/(Nn_v)} R_v > r_v , \tag{8.5}$$

and suppose n is a multiple of n_v, say $n = m_v n_v$ for an appropriate integer m_v.

The construction will be carried out in three phases.

Phase 1. Patching the high-order coefficients.

In this phase we carry out the patching for stages $k = 1, \ldots, \overline{k}$.

Using the basis functions $\varphi_{i,k}$ and φ_λ we can write

$$G_v^{(0)}(z) = \phi_v(z)^n = \sum_{i=1}^{m} \sum_{j=0}^{(n-1)N_i-1} A_{v,ij}\varphi_{i,nN_i-j}(z) + \sum_{\lambda=1}^{\Lambda} A_{v,\lambda}\varphi_\lambda(z) . \tag{8.6}$$

Here $A_{v,i0} = \widetilde{c}_{v,i}^{\,n}$, for each i. The coefficients $A_{v,ij}$ with $j = 1, \ldots, \overline{k}N_i - 1$ will be deemed "high-order".

Put $F_v(z) = \phi_v(z)^{n_v}$. Then $G_v^{(0)}(z) = \phi_v(z)^n = (F_v(z))^{m_v}$. We will patch the high-order coefficients of $G_v^{(0)}(z)$ by sequentially adjusting corresponding coefficients of $F_v(z)$. As will be seen, a small change in the latter produces a large change in the former. Write

$$F_v(z) = \sum_{i=1}^{m} \sum_{j=0}^{(n_v-1)N_i-1} a_{v,ij}\varphi_{i,n_vN_i-j}(z) + \sum_{\lambda=1}^{\Lambda} a_{v,\lambda}\varphi_\lambda(z) .$$

To adjust $G_v^{(0)}(z)$, we will replace $F_v(z)$ with

$$\widehat{F}_v(z) = \phi_v(z)^{n_v} + \sum_{i=1}^{m} \sum_{j=0}^{\overline{k}N_i-1} \eta_{v,ij}\varphi_{i,n_vN_i-j}(z) \tag{8.7}$$

for appropriately chosen $\eta_{v,ij}$.

We will take $\eta_{v,i0} = 0$ for each i, since $\Delta_{v,i0}^{(1)} = 0$ by hypothesis. The remaining $\eta_{v,ij}$ will be determined recursively, in terms of the $\Delta_{v,ij}^{(k)}$, in $\prec_N$ order; in particular, for each x_i the $\eta_{v,ij}$ will be determined in order of increasing j. As $F_v(z)$ is changed stepwise to $\widehat{F}_v(z)$, then $G_v^{(0)}(z) = (F_v(z))^{m_v}$ is changed stepwise to $G_v^{(\overline{k})}(z) = \widehat{F}_v(z)^{m_v}$, passing through $G_v^{(1)}(z)$, $G_v^{(2)}(z)$, ..., $G_v^{(\overline{k}-1)}(z)$ at intermediate steps.

It will be useful to consider what happens as each $\eta_{v,ij}$ is varied in turn. Suppose $\check{F}_v(z)$ is a function obtained at one of the intermediate steps, and at the next step $\check{F}_v(z)$ is replaced by $\check{F}_v'(z) = \check{F}_v(z) + \eta_{v,ij}\varphi_{i,n_vN_i-j}(z)$. Let k be such that $(k-1)N_i \leq j < kN_i$. When $\check{F}_v'(z)^{m_v}$ is expanded using the binomial theorem, the result is

$$\check{F}_v'(z)^{m_v} = \check{F}_v(z)^{m_v} + (m_v\eta_{v,ij}/\widetilde{c}_{v,i}^{\,n_v-k-1}) \cdot \vartheta_{v,ij}^{(k)}(z) \tag{8.8}$$

where

$$\begin{aligned} \vartheta_{v,ij}^{(k)}(z) = {} & \widetilde{c}_{v,i}^{\,n_v-k-1}\Big(\varphi_{i,n_vN_i-j}(z)\check{F}_v(z)^{m_v-1} \\ & + \sum_{t=2}^{m_v} \frac{1}{m_v}\binom{m_v}{t} \eta_{v,ij}^{t-1}\varphi_{i,n_vN_i-j}(z)^t\check{F}_v(z)^{m_v-t}\Big) . \end{aligned} \tag{8.9}$$

The first term on the right has a pole of order $nN_i - j$ at x_i, while all the other terms have poles of lower order at x_i. Likewise, for each $x_{i'} \neq x_i$, the first term has a pole of order $(n - n_v)N_{i'}$ at $x_{i'}$ and all the other terms have poles of lower order.

Since the $\Delta_{v,i0}^{(1)} = 0$, the leading coefficient of each $\check{F}_v(z)$ at x_i is $\widetilde{c}_{v,i}^{n_v}$, the same as that of $\phi_v(z)^{n_v}$. It follows that each $\vartheta_{v,ij}^{(k)}(z)$ has leading coefficient $\widetilde{c}_{v,i}^{n-k-1}$ and meets the conditions of the theorem.

It is essentially trivial by continuity (but will be rigorously proved in Lemma 8.2 below), that there is an $\epsilon_v > 0$ such that if $|\eta_{v,ij}| < \epsilon_v$ for each i, j then

$$
\begin{aligned}
\{z : |\widehat{F}_v(z)| \leq \widehat{R}_v^{n_v N}\} \quad &\subset \quad \{z : |\phi_v(z)^{n_v}| \leq R_v^{n_v N}\} \\
&= \quad \{z : |\phi_v(z)| \leq R_v^N\} \subset E_v^0 . \qquad (8.10)
\end{aligned}
$$

The numbers n_v and $\widetilde{c}_{v,i} \neq 0$ are fixed. Assuming the existence of such an ϵ_v, let B_v be the number in the statement of Theorem 8.1. Suppose n (and hence $m_v = n/n_v$) is large enough that

$$
\frac{B_v}{m_v} \cdot \max(1, \max_{1 \leq i \leq m}(|\widetilde{c}_{v,i}|))^{n_v} \; < \; \epsilon_v . \qquad (8.11)
$$

For all (i, j) with $1 \leq j < \overline{k}N_i$, the $\Delta_{v,ij}^{(k)} \in \mathbb{C}$ satisfy $|\Delta_{v,ij}^{(k)}| \leq B_v$. Hence, taking

$$
\eta_{v,ij} \;=\; \frac{1}{m_v} \Delta_{v,ij}^{(k)} \widetilde{c}_{v,i}^{\,n_v - k - 1}
$$

we have $|\eta_{v,ij}| < \epsilon_v$ and (8.10) holds. On the other hand, (8.8) becomes

$$
\check{F}_v'(z)^{m_v} \;=\; \check{F}_v(z)^{m_v} + \Delta_{v,ij}^{(k)} \cdot \vartheta_{v,ij}^{(k)}(z)
$$

for the chosen $\Delta_{v,ij}^{(k)}$.

In summary, small changes $\eta_{v,ij}$ in the coefficients of $\phi_v(z)^{m_v}$ are "magnified" to large changes $\Delta_{v,ij}^{(k)}$ in the coefficients of $G_v^{(0)}(z)$. If these changes are carried out in $\prec_N$ order, then at appropriate steps in the construction we obtain functions

$$
G_v^{(k)}(z) \;=\; G_v^{(k-1)}(z) + \sum_{i=1}^{m} \sum_{j=(k-1)N_i}^{kN_i - 1} \Delta_{v,ij}^{(k)} \vartheta_{v,ij}^{(k)}(z)
$$

for $k = 1, \ldots, \overline{k}$. Note that the leading coefficients of the $G_v^{(k)}(z)$ are never changed. Since the leading coefficient of $G_v^{(0)}(z)$ at x_i is $\widetilde{c}_{v,i}^{\,n}$, the same is true for each $G_v^{(k)}(z)$. Similarly, the leading coefficient of $\widehat{F}_v(z)$ at x_i is $\widetilde{c}_{v,i}^{\,n_v}$.

To rigorously prove the existence of an ϵ_v for which (8.10) holds, we must use some information about the $\varphi_{ij}(z)$.

Given a function $F(z) \in \mathbb{C}(\mathcal{C})$ and a number $R > 0$, write

$$
\begin{aligned}
W_R &= \{z \in \mathcal{C}_v(\mathbb{C}) : |F(z)| \leq R\} , \\
V_R &= \{z \in \mathcal{C}_v(\mathbb{C}) : |F(z)| \geq R\}
\end{aligned}
$$

(regarding the poles of $F(z)$ as belonging to V_R), and put

$$
\Gamma_R \;=\; \{z \in \mathcal{C}_v(\mathbb{C}) : |F(z)| = R\} .
$$

Then Γ_R is the common boundary of W_R and V_R.

LEMMA 8.2. *Let $F(z) \in \mathbb{C}(\mathcal{C}_v)$ have polar divisor $\mathrm{div}(F)_\infty$. Suppose that $H(z) \in \mathbb{C}(\mathcal{C}_v)$ has polar divisor $\mathrm{div}(H)_\infty \le \mathrm{div}(F)_\infty$, and for some $\delta < 1$ we have $|H(z)| < \delta \cdot R$ on Γ_R. Then*

$$\{z \in \mathcal{C}_v(\mathbb{C}) : |F(z) + H(z)| \le (1-\delta)\cdot R\} \subset W_R .$$

PROOF. Consider $G(z) = H(z)/F(z)$. On Γ_R we have $|G(z)| < \delta$. The hypothesis on the poles implies that $G(z)$ extends to a function holomorphic in V_R. By the Maximum Modulus Principle $|G(z)| < \delta$ on V_R, so $|H(z)| < \delta \cdot |F(z)|$ on V_R. It follows that $|F(z) + H(z)| > (1-\delta)|F(z)| \ge (1-\delta)R$ on V_R, so $\{z \in \mathcal{C}_v(\mathbb{C}) : |F(z) + H(z)| \le (1-\delta)\cdot R\} \subset W_R$. □

To obtain (8.10), put $\widehat{M}_v = \max_{1 \le i \le m}\Big(\max_{1\le j\le \overline{k}N_i} \|\varphi_{i,n_vN_i-j}\|_{E_v}\Big)$ and let $\epsilon_v > 0$ be small enough that

$$\epsilon_v \cdot \overline{k}N\widehat{M}_v \;<\; R_v^{n_vN} - \widehat{R}_v^{n_vN} .$$

Apply Lemma 8.2 with $F(z) = \phi_v(z)^{n_v}$ and $R = R_v^{n_vN}$, taking $\delta = 1-(\widehat{R}_v/R_v)^{n_vN}$. By hypothesis, we have $W_R \subset E_v^0$. Take

$$H(z) \;=\; \sum_{i=1}^{m} \sum_{j=1}^{\overline{k}N_i-1} \eta_{v,ij}\varphi_{i,n_vN_i-j}(z)$$

where $|\eta_{v,ij}| \le \epsilon_v$ for each (i,j). Then $|H(z)| \le \epsilon_v \cdot \overline{k}N\widehat{M}_v < \delta R$ on $\Gamma_R = \{z \in \mathcal{C}_v(z) : |\phi_v(z)^{n_v}| = R_v^{n_vN}\}$, and $\widehat{F}_v(z) = F(z) + H(z)$, while $\widehat{R}_v^{n_vN} = (1-\delta)R$, so (8.10) follows from the lemma.

Let $\widehat{\Gamma}_v$ denote the level curve $\{z : |\widehat{F}_v(z)| = \widehat{R}_v^{n_vN}\}$. By (8.10),

$$\widehat{\Gamma}_v \;\subset\; \{z \in \mathcal{C}_v(\mathbb{C}) : |\phi_v(z)| \le R_v^N\} \;\subset\; E_v^0 . \tag{8.12}$$

The function $\widehat{F}_v(z)$ and the curve $\widehat{\Gamma}_v$ will play a key role in the rest of the construction.

Phase 2. Patching the middle coefficients.

In this phase we carry out the patching process for $k = \overline{k}+1, \ldots, n-1$. For each k we begin with a function $G_v^{(k-1)}(z)$, and we modify the coefficients with $(k-1)N_i \le j < kN_i$, for each i. For each such j we can uniquely write

$$nN_i - j \;=\; r_{ij} + (n-k-1)N_i, \quad \text{with } N_i < r_{ij} \le 2N_i.$$

We can then write

$$n-k-1 \;=\; \ell_1 + \ell_2 n_v, \quad \text{with } 0 \le \ell_1 < n_v,\ 0 \le \ell_2 < m_v,$$

so $nN_i - j = r_{ij} + \ell_1 N_i + \ell_2 n_v N_i$. Put

$$\vartheta_{v,ij}^{(k)}(z) \;=\; \varphi_{i,r_{ij}}(z)\phi_v(z)^{\ell_1}\widehat{F}_v(z)^{\ell_2} .$$

Then $\vartheta_{v,ij}^{(k)}(z)$ has a pole of exact order $nN_i - j$ at x_i, with leading coefficient $\widetilde{c}_{v,i}^{\,n-k-1}$. Its poles at the $x_{i'} \ne x_i$ are of order at most $(n-k-1)N_{i'}$, so it meets the conditions of the theorem.

Modifying the coefficients stepwise in $\prec_N$ order, we put

$$G_v^{(k)}(z) = G_v^{(k-1)}(z) + \sum_{i=1}^{m} \sum_{j=(k-1)N_i+1}^{kN_i} \Delta_{v,ij}^{(k)} \vartheta_{v,ij}^{(k)}(z) \tag{8.13}$$

where $|\Delta_{v,ij}^{(k)}| \le h_v^{kN}$.

We now seek a bound for $|\vartheta_{v,ij}^{(k)}(z)|$ on the level curve $\widehat{\Gamma}_v$. By definition, $|\widehat{F}_v(z)^{\ell_2}| = \widehat{R}_v^{Nn_v\ell_2}$ on $\widehat{\Gamma}_v$. Since $\widehat{R}_v = 2^{-1/(Nn_v)}R_v$ it follows from (8.12) that on $\widehat{\Gamma}_v$, for $0 \le \ell_1 < n_v$,

$$|\phi_v(z)^{\ell_1}| \le 2\widehat{R}_v^{N\ell_1} . \tag{8.14}$$

Finally, $|\varphi_{i,r_{ij}}(z)| \le M_v$ on $\widehat{\Gamma}_v$ for all $N_i + 1 \le r_{ij} \le 2N_i$ (in fact this holds for all $z \in E_v$). Hence

$$|\vartheta_{v,ij}^{(k)}(z)| \le |\varphi_{i,r_{ij}}| \cdot |\phi_v(z)^{\ell_1}| \cdot |\widehat{F}_v(z)^{\ell_2}| \le 2M_v\widehat{R}_v^{N(n-k-1)} .$$

Since there are N terms in the sum (8.13), on $\widehat{\Gamma}_v$,

$$|G_v^{(k)}(z)| \le |G_v^{(k-1)}(z)| + Nh_v^{kN} \cdot 2M_v\widehat{R}_v^{N(n-k-1)} . \tag{8.15}$$

Phase 3. Patching the low-order coefficients.

In the final step we take

$$G_v^{(n)}(z) = G_v^{(n-1)}(z) + \sum_{\lambda=1}^{\Lambda} \Delta_{v,\lambda}^{(n)}\varphi_\lambda$$

where $|\Delta_{v,\lambda}^{(n)}| \le h_v^{nN}$ for each λ.

Since $\Lambda \le N$, and each $|\varphi_\lambda(z)| \le M_v$ on $\widehat{\Gamma}_v$ (indeed on all of E_v), on $\widehat{\Gamma}_v$

$$|G_v^{(n)}(z)| \le |G_v^{(n-1)}(z)| + NM_vh_v^{nN} . \tag{8.16}$$

To complete the proof, we must show that if n is sufficiently large then part (B) of Theorem 8.1 holds. Assume that n is large enough that

$$NM_v \left(\frac{h_v}{r_v}\right)^{nN} < \frac{1}{4} , \tag{8.17}$$

and recall that $\overline{k}$ satisfies

$$\frac{2NM_v}{1-(h_v/r_v)^N} \cdot \left(\frac{h_v}{r_v}\right)^{\overline{k}N} < \frac{1}{4} . \tag{8.18}$$

Consider the total change on $\widehat{\Gamma}_v$ in passing from $G_v^{(\overline{k})}(z) = \widehat{F}_v(z)^{m_v}$ to $G_v^{(n)}(z)$. By (8.15) and (8.16), for each $z \in \widehat{\Gamma}_v$,

$$|G_v^{(k)}(z) - \widehat{F}_v(z)^{m_v}| \le NM_vh_v^{nN} + \frac{2NM_v}{\widehat{R}_v^N} \cdot \widehat{R}_v^{nN} \cdot \sum_{k=\overline{k}}^{n-1} \frac{h_v^{kN}}{\widehat{R}_v^{kN}} . \tag{8.19}$$

Since $\widehat{R}_v > r_v > 1$, by inserting (8.17), and (8.18) in (8.19), we find that on $\widehat{\Gamma}_v$,

$$|G_v^{(n)}(z) - \widehat{F}_v(z)^{m_v}| < \frac{1}{2}\widehat{R}_v^{nN} . \tag{8.20}$$

As $|\widehat{F}_v(z)^{m_v}| = \widehat{R}_v^{nN}$ on $\widehat{\Gamma}_v$, by applying Lemma 8.2 with $F(z) = \widehat{F}_v(z)^{m_v}$ and $H(z) = G_v^{(n)}(z) - \widehat{F}_v(z)^{m_v}$, taking $\delta = \frac{1}{2}$, we see that

$$\{z \in \mathcal{C}_v(\mathbb{C}) : |G_v^{(n)}(z)| \le \frac{1}{2}\widehat{R}_v^{nN}\} \subset \{z \in \mathcal{C}_v(\mathbb{C}) : |\widehat{F}_v(z)^{m_v}| \le \widehat{R}_v^{nN}\}$$

which is contained in E_v^0.

Finally, if n is also large enough that

$$\frac{1}{2}\widehat{R}_v^{nN} > r_v^{nN} , \tag{8.21}$$

then

$$\{z \in \mathcal{C}_v(\mathbb{C}) : |G_v^{(n)}(z)| \le r_v^{nN}\} \subset E_v^0 . \tag{8.22}$$

A similar argument shows that $\{z \in \mathcal{C}_v(\mathbb{C}) : |G_v^{(k)}(z)| \le r_v^{nN}\} \subset E_v^0$ for each $k = 1, \ldots, n$.

In summary if $n > \overline{k}$ is divisible by n_v and large enough that conditions (8.11), (8.17) and (8.21) hold, the construction succeeds. $\square$

CHAPTER 9

Local Patching when $K_v \cong \mathbb{R}$

In this section we give the confinement argument for Theorem 4.2 when $K_v \cong \mathbb{R}$. Write $\mathbb{C}_v$ for $\mathbb{C}$ and $|\ |_v$ for $|\ |$. Let w_v be the distinguished place of $L = K(\mathfrak{X})$ determined by the embedding $\widetilde{K} \hookrightarrow \mathbb{C}_v$ used to identify $\mathfrak{X}$ with a subset of $\mathcal{C}_v(\mathbb{C}_v)$. Identify K_v with $\mathbb{R}$, and L_{w_v} with $\mathbb{R}$ or $\mathbb{C}$ as appropriate.

Using the construction of the coherent approximating functions in Theorem 7.11, we begin with the following data:

(1) A K_v-symmetric probability vector $\vec{s} \in \mathcal{P}^m(\mathbb{Q})$ with positive rational coefficients.

(2) An $\mathbb{R}$-simple set E_v: in particular E_v is nonempty and compact, stable under complex conjugation, and is a union of finitely many pairwise disjoint, nonempty compact sets $E_{v,1}, \ldots, E_{v,\ell}$ such that each $E_{v,i}$ is either
   (a) a closed interval of positive length contained in $\mathcal{C}_v(\mathbb{R})$, or
   (b) is disjoint from $\mathcal{C}_v(\mathbb{R})$, simply connected, has a piecewise smooth boundary, and is the closure of its $\mathcal{C}_v(\mathbb{C})$-interior.

(3) Let E_v^0 be the quasi-interior of E_v, the union of the real interiors of the components $E_{v,i} \subset \mathcal{C}_v(\mathbb{R})$ and the complex interiors of the components $E_{v,i} \subset \mathcal{C}_v(\mathbb{C}) \backslash \mathcal{C}_v(\mathbb{R})$; then we are given a $\mathcal{C}_v(\mathbb{C})$-open set U_v such that
   (a) $U_v \cap E_v = E_v^0$,
   (b) the components of U_v are simply connected, and
   (c) the closure $\overline{U}_v$ is disjoint from $\mathfrak{X}$.

(4) Parameters h_v, r_v, R_v, with $1 < h_v < r_v < R_v$, which govern the freedom in the patching process.

(5) A number N and an $(\mathfrak{X}, \vec{s})$-function $\phi_v(z) \in K_v(\mathcal{C})$ of degree N whose zeros all belong to E_v^0, and which has the following properties:
   (a) $\phi_v^{-1}(D(0, 2R_v^N)) \subset U_v$.
   (b) For each component $E_{v,i} \subset \mathcal{C}_v(\mathbb{R})$, if $\phi_v(z)$ has τ_j zeros in $E_{v,j}$, then $\phi_v(z)$ oscillates τ_j times between $\pm 2R_v^N$ on $U_v \cap E_{v,j}$.
   Put $N_i = Ns_i$ for each i, and write $\widetilde{c}_{v,i} = \lim_{z \to x_i} \phi_v(z) \cdot g_{x_i}(z)^{N_i}$ for the leading coefficient of $\phi_v(z)$ at x_i.

(6) An order $\prec_N$ on the index set $\mathcal{I} = \{(i,j) \in \mathbb{Z}^2 : 1 \le i \le m, 0 \le j\}$ determined by N and $\vec{s}$ as in (7.41), which gives the sequence in which coefficients are patched. The group $\mathrm{Aut}_c(\mathbb{C}/K_v)$ acts on $\mathcal{I}$, and we write $\mathrm{Aut}_c(\mathbb{C}/K_v)(i,j)$ for the orbit of $(i,j) \in \mathcal{I}$.

We will use the L-rational basis $\{\varphi_{ij}, \varphi_\lambda\}$ from §3.3 to expand all functions, and $\Lambda = \dim_K(\Gamma(\sum_{i=1}^m N_i(x_i)))$ will be the number of low-order basis elements, as in §7.4. The order $\prec_N$ respects the N-bands (7.42), and for each $x_i \in \mathfrak{X}$, it specifies the terms to be patched in decreasing pole order.

Let $D(0,R) = \{z \in \mathbb{C} : |z| \le R\}$ be the filled disc, and let

$$E(a,b) \;=\; \{x+iy \in \mathbb{C} : x^2/a^2 + y^2/b^2 \le 1\}$$

be the filled ellipse. Write $C(0,R) = \partial D(0,R)$ and $\partial E(a,b)$ for their boundaries, and note that if $a > b$ then $D(0,b) \subset E(a,b) \subset D(0,a)$.

Let $T_n(z)$ be the Chebyshev polynomial of degree n for the interval $[-2,2]$, defined by $T_n(2\cos(\theta)) = 2\cos(n\theta)$. Equivalently, $T_n(z)$ is the unique polynomial of degree n for which $T_n(z+1/z) = z^n + 1/z^n$. For each $R > 0$, let $T_{n,R}(z) = R^n T_n(z/R)$ be the Chebyshev polynomial for the interval $[-2R, 2R]$. Then $T_{n,R}(z)$ is monic of degree n with coefficients in $\mathbb{R}$, and as noted in ([**48**]),

$$T_{n,R}(z) \;=\; z^n + \sum_{k=1}^{\lfloor n/2 \rfloor} \frac{n}{k}\binom{n-k-1}{k-1} R^{2k} z^{n-2k} \;. \tag{9.1}$$

Furthermore, $T_{n,R}(z)$ has the following mapping properties:

First, $T_{n,R}([-2R,2R]) = [-2R^n, 2R^n]$ and $T_{n,R}^{-1}([-2R^n,2R^n]) = [-2R,2R]$. These facts follow from the identity $T_n(2\cos(\theta)) = 2\cos(n\theta)$, which shows that $T_{n,R}$ oscillates n times between $\pm 2R^n$ on $[-2R,2R]$. Second, for each $t > R$,

$$T_{n,R}\big(E(t+\frac{R^2}{t}, t-\frac{R^2}{t})\big) = E(t^n + \frac{R^{2n}}{t^n}, t^n - \frac{R^{2n}}{t^n}),$$
$$\big(T_{n,R}\big)^{-1}\big(E(t^n + \frac{R^{2n}}{t^n}, t^n - \frac{R^{2n}}{t^n})\big) = E(t+\frac{R^2}{t}, t-\frac{R^2}{t}) \;.$$

Indeed, $T_{n,R}$ maps $E(t+R^2/t, t-R^2/t)$ onto $E(t^n + R^{2n}/t^n, t^n - R^{2n}/t^n)$ and is n-to-1 (counting multiplicities). This follows from the commutativity of the diagram

$$\begin{array}{ccc} C(0,t/R) & \xrightarrow{z^n} & C(0,t^n/R^n) \\ \downarrow\; z+\frac{1}{z} & & \downarrow\; z+\frac{1}{z} \\ \partial E(\frac{t}{R}+\frac{R}{t}, \frac{t}{R}-\frac{R}{t}) & \xrightarrow{T_n(z)} & \partial E(\frac{t^n}{R^n}+\frac{R^n}{t^n}, \frac{t^n}{R^n}-\frac{R^n}{t^n}) \;. \end{array}$$

THEOREM 9.1. *Suppose $K_v \cong \mathbb{R}$. Let $E_v \subset \mathcal{C}_v(\mathbb{C}) \backslash \mathfrak{X}$ be an $\mathbb{R}$-simple set. Let U_v be an open set in $\mathcal{C}_v(\mathbb{C}_v)$ with $U_v \cap E_v = E_v^0$, whose components are simply connected, and whose closure $\overline{U}_v$ is disjoint from $\mathfrak{X}$. Let $\vec{s} \in \mathcal{P}^m(\mathbb{Q})$ be a K_v-symmetric probability vector with positive rational coefficients, and let $1 < h_v < r_v < R_v$ be real numbers.*

Let $\phi_v(z) \in K_v(\mathcal{C})$ be an $(\mathfrak{X}, \vec{s})$-function of degree N whose zeros belong to E_v^0, satisfying

(1) $\varphi^{-1}(D(0, 2R_v^N)) \subset U_v$,

(2) *For each component $E_{v,i}$ contained in $\mathcal{C}_v(\mathbb{R})$, if $\phi_v(z)$ has τ_i zeros in $E_{v,i}$, then $\phi_v(z)$ oscillates τ_i times between $\pm 2R_v^N$ on $E_{v,i}$.*

Let $\widetilde{c}_{v,i}$ be the leading coefficient of $\phi_v(z)$ at x_i, and put

$$M_v \;=\; \max\big(\max_{\substack{1 \le i \le m \\ N_i < j \le 2N_i}} \|\varphi_{ij}\|_{\overline{U}_v}, \max_{1 \le \lambda \le \Lambda} \|\varphi_\lambda\|_{\overline{U}_v}\big) \;.$$

Let $k_v > 0$ be the least integer such that

$$\frac{16 N M_v}{1-(h_v/r_v)^N} \cdot \big(\frac{h_v}{r_v}\big)^{k_v N} \;<\; \frac{1}{4} \;, \tag{9.2}$$

and let $\overline{k} \ge k_v$ be a fixed integer. Let $B_v > 0$ be an arbitrary constant. Then there is an integer n_v, depending on $\phi_v(z)$, $\overline{k}$, B_v, r_v, and R_v, such that for each sufficiently *large integer n divisible by n_v, one can carry out the local patching process at K_v as follows:*

Write $n = m_v n_v$. For suitable $\widehat{R}_1, \widehat{R}_2$ with $r_v < \widehat{R}_2 < \widehat{R}_1 < R_v$, put

$$G_v^{(0)}(z) \;=\; T_{m_v,\widehat{R}_2^{n_v N}}(T_{n_v,\widehat{R}_1^N}(\phi_v(z))) \ .$$

For each k, $1 \le k < n$, let $\{\Delta_{v,ij}^{(k)} \in \mathbb{C}_v\}_{(i,j)\in \mathrm{Band}_N(k)}$ be an arbitrary K_v-symmetric set of numbers given recursively in $\prec_N$ order, subject to the conditions that for each i, we have $\Delta_{v,i0}^{(1)} = 0$ and for each $j > 0$,

$$|\Delta_{v,ij}^{(k)}|_v \;\le\; \begin{cases} B_v & \text{if } k \le \overline{k} \ , \\ h_v^{kN} & \text{if } k > \overline{k} \ . \end{cases} \tag{9.3}$$

For $k = n$, let $\{\Delta_{v,\lambda}^{(n)} \in \mathbb{C}_v\}_{1\le\lambda\le\Lambda}$ be an arbitrary K_v-symmetric set of numbers satisfying

$$|\Delta_{v,\lambda}^{(n)}|_v \;\le\; h_v^{nN} \ . \tag{9.4}$$

Then one can inductively construct $(\mathfrak{X},\vec{s})$-functions $G_v^{(1)}(z),\ldots,G_v^{(n)}(z)$ in $K_v(\mathcal{C})$, of common degree nN, having the following properties:

(A) For each $k = 1,\ldots,n$, there are functions $\vartheta_{v,ij}^{(k)}(z) \in L_{w_v}(\mathcal{C})$, determined recursively in $\prec_N$ order, such that

$$G_v^{(k)}(z) \;=\; G_v^{(k-1)}(z) + \sum_{(i,j)\in \mathrm{Band}_N(k)} \Delta_{v,ij}^{(k)}\vartheta_{v,ij}^{(k)}(z) \quad \text{for } k < n \ ,$$

$$G_v^{(n)}(z) \;=\; G_v^{(n-1)}(z) + \sum_{\lambda=1}^{\Lambda} \Delta_{v,\lambda}^{(n)}\varphi_\lambda(z) \ ,$$

and where for each (i,j),

(1) $\vartheta_{v,ij}^{(k)}(z)$ has a pole of order $nN_i - j > (n-k-1)N_i$ at x_i and leading coefficient $\widetilde{c}_{v,i}^{\,n-k-1}$, a pole of order at most $(n-k-1)N_{i'}$ at each $x_{i'} \ne x_i$, and no other poles;

(2) $\sum_{(i',j)\in \mathrm{Aut}_c(\mathbb{C}_v/K_v)(i,j)} \Delta_{v,i'j}^{(k)}\vartheta_{v,i'j}^{(k)}(z)$ belongs to $K_v(\mathcal{C})$.

(B) For each $k = 1,\ldots,n$,

(1) the zeros of $G_v^{(k)}(z)$ all belong to E_v^0, and for each component $E_{v,i}$ of E_v, if $\phi_v(z)$ has τ_i zeros in $E_{v,i}$, then $G_v^{(k)}(z)$ has $T_i = n\tau_i$ zeros in $E_{v,i}$;

(2) $\{z \in \mathcal{C}_v(\mathbb{C}_v) : |G_v^{(k)}(z)|_v \le 2r_v^{nN}\} \subset U_v$, and

(3) for each component $E_{v,i}$ contained in $\mathcal{C}_v(\mathbb{R})$, $G_v^{(k)}(z)$ oscillates T_i times between $\pm 2r_v^{nN}$ on $E_{v,i}$.

Remark. As in the patching construction when $K_v \cong \mathbb{C}$, a key feature of Theorem 9.1 is that by choosing n appropriately, the freedom B_v in patching the coefficients for $k \le \overline{k}$ can be made arbitrarily large. Again this is accomplished by using "magnification". The degree of $\phi_v(z)$ is raised by a two-stage composition with Chebyshev polynomials.

Recall that E_v is $\mathbb{R}$-simple, so each connected component of E_v is either an interval in $\mathcal{C}_v(\mathbb{R})$, or is contained in $\mathcal{C}_v(\mathbb{C})\backslash\mathcal{C}_v(\mathbb{R})$ and is the closure of its interior.

The argument confining the roots of the $G_v^{(k)}(z)$ to E_v has two parts. One part, which goes back to Fekete and Szegö ([**25**]) and uses the Maximum Modulus Principle, confines the roots to U_v and shows that the number of roots in each component of U_v is preserved. Since $U_v \cap E_v = E_v^0$, if $E_{v,i}$ is a component of E_v which is disjoint from $\mathcal{C}_v(\mathbb{R})$, this means that roots in $E_{v,i}^0$ must remain there. The other part, which goes back to Robinson ([**48**]), is based on oscillation properties of Chebyshev polynomials and the intermediate value theorem. It shows that the number of roots in each component $E_{v,i}$ contained in $\mathcal{C}_v(\mathbb{R})$ is preserved. The mapping properties of Chebyshev polynomials noted before the Theorem enable to us carry out both confinement arguments simultaneously.

PROOF OF THEOREM 9.1. Let $\overline{k} \geq k_v$ and $B_v > 0$ be as in the theorem. Choose $n_v \in \mathbb{N}$ large enough that $n_v > \overline{k}$ and $8^{-1/(n_vN)}R_v > r_v$. Put $\widehat{R}_1 = 2^{-1/(n_vN)}R_v$ and $\widehat{R}_2 = 8^{-1/(n_vN)}R_v$, so $r_v < \widehat{R}_2 < \widehat{R}_1 < R_v$ and $2\widehat{R}_1^{n_vN} = R_v^{n_vN}$, $4\widehat{R}_2^{n_vN} = \widehat{R}_1^{n_vN}$. Set

$$F_v(z) \;=\; T_{n_v,\widehat{R}_1^N}(\phi_v(z)) \;.$$

Let n be a multiple of n_v, write $n = m_v n_v$, and put

$$G_v^{(0)}(z) \;=\; T_{m_v,\widehat{R}_2^{n_vN}}\big(T_{n_v,\widehat{R}_1^N}(\phi_v(z))\big) \;=\; T_{m_v,\widehat{R}_2^{n_vN}}\big(F_v(z)\big) \;.$$

We will begin by investigating the mapping properties of $F_v(z)$ and $G_v^{(0)}(z)$. We first show that all the zeros of $F_v(z)$ belong to E_v^0, and that for each component $E_{v,i}$ of E_v, if ϕ_v has τ_i zeros in $E_{v,i}$, then $F_v(z)$ has $n_v\tau_i$ zeros in $E_{v,i}$ (counted with multiplicities).

Let t_1 be the largest real root of

$$t_1^{n_vN} + \frac{\widehat{R}_1^{2n_vN}}{t_1^{n_vN}} \;=\; 4\widehat{R}_1^{n_vN} \;,$$

so that $t_1^{n_vN} = (2+\sqrt{3})\widehat{R}_1^{n_vN} = ((2+\sqrt{3})/2)R_v^{n_vN}$. Put

$$\begin{aligned} a_1 &= t_1^N + \frac{\widehat{R}_1^{2N}}{t_1^N}\,, \qquad & A_1 &= t_1^{n_vN} + \frac{\widehat{R}_1^{2n_vN}}{t_1^{n_vN}} \;=\; 4\widehat{R}_1^{n_vN} \;=\; 2R_v^{n_vN}\,,\\ b_1 &= t_1^N - \frac{\widehat{R}_1^{2N}}{t_1^N}\,, \qquad & B_1 &= t_1^{n_vN} - \frac{\widehat{R}_1^{2n_vN}}{t_1^{n_vN}} \;=\; 2\sqrt{3}\widehat{R}_1^{n_vN} \;=\; \sqrt{3}R_v^{n_vN}\;; \end{aligned}$$

then

$$T_{n_v,\widehat{R}_1^N}(E(a_1,b_1)) \;=\; E(A_1,B_1) \;.$$

Here $a_1 = g(1/n_v)\cdot R_v^N$ where $g(x) = ((2+\sqrt{3})/2)^x + (2(2+\sqrt{3}))^{-x}$. Using Calculus, one sees that $g(x) < 2$ for $0 < x < 1$, so $a_1 < 2R_v^N$. It follows that

$$E(a_1,b_1) \;\subset\; D(0,2R_v^N)\;, \tag{9.5}$$

$$D(0,\sqrt{3}R_v^{n_vN}) \;\subset\; E(A_1,B_1) \;\subset\; D(0,2R_v^{n_vN})\;. \tag{9.6}$$

As $2\widehat{R}_1^{n_vN} = R_v^{n_vN}$, (9.6) shows that

$$D(0,2\widehat{R}_1^{n_vN}) \;\subset\; E(A_1,B_1)\;. \tag{9.7}$$

Since $\phi_v^{-1}(D(0,2R_v^N)) \subset U_v$ and $T^{-1}_{n_v,\widehat{R}_1^N}(E(A_1,B_1)) = E(a_1,b_1)$, (9.5) gives

$$F_v^{-1}(E(A_1,B_1)) \;\subset\; U_v\;.$$

For each component $E_{v,i}$ of E_v contained in $\mathcal{C}_v(\mathbb{R})$, the function ϕ_v is real-valued and oscillates τ_i times between $\pm 2R_v^N$ on $E_{v,i}$. Since $[-2\widehat{R}_1^N, 2\widehat{R}_1^N] \subset [-2R_v^N, 2R_v^N]$ and $T_{n_v,\widehat{R}_1^N}$ oscillates n_v times between $\pm 2\widehat{R}_1^{n_vN}$ on $[-2\widehat{R}_1^N, 2\widehat{R}_1^N]$, it follows that $F_v(z)$ oscillates $n_v\tau_i$ times between $\pm 2\widehat{R}_1^{n_vN}$ on $E_{v,i}$.

If $E_{v,i}$ is a component of E_v disjoint from $\mathcal{C}_v(\mathbb{R})$, then $U_v \cap E_{v,i} = E_{v,i}^0$. Since $T_{n_v,\widehat{R}_1^N}$ has n_v zeros in $E(a_1, b_1)$, and $E_{v,i}$ is simply connected with a piecewise smooth boundary, the Argument Principle shows that $F_v(z)$ has $n_v\tau_i$ zeros in $E_{v,i}^0$. On the other hand, if $E_{v,i}$ is a component contained in $\mathcal{C}_v(\mathbb{R})$, then by the discussion above $F_v(z)$ has at least $n_v\tau_i$ zeros in $E_{v,i}^0$. Since $\sum_i n_v\tau_i = n_vN$ and $F_v(z)$ has degree n_vN, these zeros account for all the zeros of $F_v(z)$. Thus all the zeros of $F_v(z)$ belong to E_v^0, and $F_v(z)$ has exactly $n_v\tau_i$ zeros in each $E_{v,i}$.

Next, let t_2 be the largest real root of

$$t_2^{n_vN} + \frac{\widehat{R}_2^{2n_vN}}{t_2^{n_vN}} = 4\widehat{R}_2^{n_vN} ,$$

so that $t_2^{n_vN} = (2+\sqrt{3})\widehat{R}_2^{n_vN}$. If we put

$$\begin{aligned} a_2 &= t_2^{n_vN} + \frac{\widehat{R}_2^{2n_vN}}{t_2^{n_vN}} = 4\widehat{R}_2^{n_vN} = \widehat{R}_1^{n_vN} , \\ b_2 &= t_2^{n_vN} - \frac{\widehat{R}_2^{2n_vN}}{t_2^{n_vN}} = 2\sqrt{3}\widehat{R}_2^{n_vN} = \frac{\sqrt{3}}{2}\widehat{R}_1^{n_vN} , \end{aligned}$$

and

$$\begin{aligned} A_2 &= t_2^{m_vn_vN} + \frac{\widehat{R}_2^{2m_vn_vN}}{t_2^{m_vn_vN}} = \left(1 + (2-\sqrt{3})^{2m_v}\right)t_2^{nN} , \\ B_2 &= t_2^{m_vn_vN} - \frac{\widehat{R}_2^{2m_vn_vN}}{t_2^{m_vn_vN}} = \left(1 - (2-\sqrt{3})^{2m_v}\right)t_2^{nN} , \end{aligned}$$

then $T_{m_v,\widehat{R}_2^{n_vN}}(E(a_2, b_2)) = E(A_2, B_2)$.

Since $T_{m_v,\widehat{R}_2^{n_vN}}^{-1}(E(A_2, B_2)) = E(a_2, b_2) \subset D(0, \widehat{R}_1^{n_vN}) \subset E(A_1, B_1)$, it follows that

$$(9.8) \qquad \left(G_v^{(0)}\right)^{-1}(E(A_2, B_2)) \subset F_v^{-1}(E(A_1, B_1)) \subset U_v .$$

Since $[-2\widehat{R}_2^{n_vN}, 2\widehat{R}_2^{n_vN}] \subset E(a_2, b_2)$ and $T_{m_v,\widehat{R}_2^{n_vN}}$ oscillates m_v times between $\pm 2\widehat{R}_2^{nN}$ on $[-2\widehat{R}_2^{n_vN}, 2\widehat{R}_2^{n_vN}]$, an argument similar to the one for $F_v(z)$ shows that $G_v^{(0)}(z)$ oscillates $n\tau_i$ times between $\pm 2\widehat{R}_2^{nN}$ on each $E_{v,i}$ contained in $\mathcal{C}_v(\mathbb{R})$, that all the zeros of $G_v^{(0)}(z)$ belong to E_v^0, and that $G_v^{(0)}(z)$ has $T_i = n\tau_i$ zeros in each $E_{v,i}$ (counting multiplicities).

We now turn to the patching construction, which will be done in three phases.

Phase 1. Patching the high-order coefficients.

In this phase we carry out the patching for stages $k = 1, \ldots, \overline{k}$. The fact that $E(a_2, b_2) \subset D(0, \widehat{R}_1^{n_vN})$, while $D(0, 2\widehat{R}_1^{n_vN}) \subset E(A_1, B_1)$, gives us freedom to adjust $F_v(z)$ while maintaining (9.8), and this is the basis for the magnification argument. Write $\widehat{T}_\ell(z)$ for $T_{\ell,\widehat{R}_2^{n_vN}}(z)$.

Using the basis functions $\varphi_{ij}(z)$, $\varphi_\lambda(z)$ we can write

$$(9.9) \qquad G_v^{(0)}(z) = \sum_{i=1}^{m} \sum_{j=0}^{(n-1)N_i-1} A_{v,ij}\varphi_{i,nN_i-j}(z) + \sum_{\lambda=1}^{\Lambda} A_{v,\lambda}\varphi_\lambda(z) \ .$$

Here $A_{v,i0} = \widetilde{c}_{v,i}^{\,n}$, for each i. The coefficients $A_{v,ij}$ with $j = 0, \ldots, \overline{k}N_i - 1$ will be deemed "high-order". They will be patched a magnification argument similar to the one when $K_v \cong \mathbb{C}$: we will sequentially modify the coefficients of $F_v(z)$, changing it from

$$F_v(z) \ = \ \sum_{i=1}^{m} \sum_{j=0}^{(n_v-1)N_i-1} a_{v,ij}\varphi_{i,n_vN_i-j}(z) + \sum_{\lambda=1}^{\Lambda} a_{v,\lambda}\varphi_\lambda(z)$$

to

$$\widehat{F}_v(z) \ = \ F_v(z) + \sum_{i=1}^{m} \sum_{j=1}^{\overline{k}N_i-1} \eta_{v,ij}\varphi_{i,n_vN_i-j}(z) \ ,$$

thereby stepwise changing $G_v^{(0)}(z) = \widehat{T}_{m_v}(F_v(z))$ to $\widehat{G}_v(z) = \widehat{T}_{m_v}(\widehat{F}_v(z))$. We will require the $\eta_{v,ij}$ to be K_v-symmetric, so $\widehat{F}_v(z)$ is K_v-rational.

We claim there there is an $\epsilon_v > 0$ such that if the $\eta_{v,ij}$ are K_v-symmetric and each $|\eta_{v,ij}|_v < \epsilon_v$, then $\widehat{F}_v$ oscillates $n_v\tau_i$ times between $\pm\widehat{R}_1^{n_vN}$ on each $E_{v,i}$ contained in $\mathcal{C}_v(\mathbb{R})$, and

$$(9.10) \qquad \widehat{F}_v^{-1}(D(0, \widehat{R}_1^{n_vN})) \ \subset U_v \ .$$

To see this, put $\widehat{M}_v \ = \ \max_{1\le i\le m}\Big(\max_{1\le j\le \overline{k}N_i} \|\varphi_{i,n_vN_i-j}\|_{\overline{U}_v}\Big)$ and take ϵ_v small enough that $\epsilon_v \cdot \overline{k}N\widehat{M}_v < \widehat{R}_1^{n_vN}$. Write

$$H_v(z) \ = \ \sum_{i=1}^{m} \sum_{j=1}^{\overline{k}N_i-1} \eta_{v,ij}\varphi_{i,n_vN_i-j}(z) \ ,$$

so that $\widehat{F}_v(z) = F_v(z) + H_v(z)$. Since the $\eta_{v,ij}$ and $\varphi_{i,n_vN_i-j}(z)$ are K_v-symmetric, $H_v(z)$ is K_v-symmetric and, in particular, is real-valued on $\mathcal{C}_v(\mathbb{R})$. At each z where $|F_v(z)|_v = 2\widehat{R}_1^{n_vN}$ we have $|H_v(z)|_v < \varepsilon{\cdot}\overline{k}N\widehat{M}_v < \widehat{R}_1^{n_vN}$, and so $|\widehat{F}_v(z)|_v > \widehat{R}_1^{n_vN}$. If $E_{v,i}$ is a component of E_v contained in $\mathcal{C}_v(\mathbb{R})$, then since $F_v(z)$ oscillates $n_v\tau_i$ times between $\pm2\widehat{R}_1^{n_vN}$ on $E_{v,i}$, it follows that $\widehat{F}_v(z)$ oscillates $n_v\tau_i$ times between $\pm\widehat{R}_1^{n_vN}$ on $E_{v,i}$. It remains to show (9.10). For this, put $\Gamma = \{z \in \mathcal{C}_v(\mathbb{C}) : |F_v(z)|_v = 2\widehat{R}_1^{n_vN}\}$, and apply Lemma 8.2 to $F_v(z)$ and $H_v(z)$, taking $R = 2\widehat{R}_1^{n_vN}$ and $\delta = 1/2$. Since $F_v^{-1}(D(0, 2\widehat{R}_1^{n_vN})) \subset U_v$, we conclude that $\widehat{F}_v^{-1}(D(0, \widehat{R}_1^{n_vN})) \subset U_v$.

Since $[-2\widehat{R}_2^{n_vN}, 2\widehat{R}_2^{n_vN}] \subset E(a_2, b_2) \subset D(0, \widehat{R}_1^{n_vN})$, the same argument as for $G_v^{(0)}(z)$ shows that $\widehat{G}_v(z) = T_{m_v,\widehat{R}_2^{n_vN}}(\widehat{F}_v(z))$ oscillates $n\tau_i$ times between $\pm2\widehat{R}_2^{nN}$ on each $E_{v,i}$ contained in $\mathcal{C}_v(\mathbb{R})$, that all the zeros of $\widehat{G}_v(z)$ belong to E_v^0, that $\widehat{G}_v(z)$ has $T_i = n\tau_i$ zeros in each $E_{v,i}$ (counting multiplicities), and that

$$(9.11) \qquad \widehat{G}_v^{-1}(E(A_2, B_2)) \ \subset U_v \ .$$

Let $B_v > 0$ be the number in the statement of Theorem 9.1. We now show that by choosing the n and the $\eta_{v,ij}$ appropriately, we can achieve freedom B_v in patching the high-order coefficients. That is, the $\Delta_{v,ij}^{(k)}$ with $1 \le j < \overline{k}N_i$ can be can be specified arbitrarily, provided that they are K_v-symmetric and satisfy $|\Delta_{v,ij}^{(k)}|_v \le B_v$.

By (9.1),

$$\begin{aligned} G_v^{(0)}(z) &= \widehat{T}_{m_v}(F_v(z)) \\ &= F_v(z)^{m_v} + \sum_{k=1}^{\lfloor m_v/2 \rfloor} (-1)^k \frac{m_v}{k} \binom{m_v - k - 1}{k - 1} \widehat{R}_2^{2kn_vN} F_v(z)^{m_v - 2k} . \end{aligned}$$

When the right side is expanded in terms of the basis functions, only the pure power $F_v^{m_v}(z)$ can contribute to the coefficients of $\varphi_{i,nN_i-j}(z)$ with $j < 2n_vN_i$; in particular, this holds for $j < \overline{k}N_i$.

Since only $F_v(z)^{m_v}$ contributes to the high-order coefficients, essentially the same argument applies here as in the patching construction over $\mathbb{C}$. By sequentially adjusting the numbers $\eta_{v,ij}$ we will modify the corresponding coefficients in the expansion of $\widehat{T}_{m_v}(F_v(z))$. By (8.8), the change in $A_{v,ij}$ induced by replacing $a_{v,ij}$ with $a_{v,ij} + \eta_{v,ij}$ is

$$\Delta_{v,ij}^{(k)} = m_v \eta_{v,ij} / \widetilde{c}_{v,ij}^{\,n_v - k + 1} .$$

Conversely, if a desired change $\Delta_{v,ij}^{(k)}$ is given, then taking

$$\eta_{v,ij} = \frac{\widetilde{c}_{v,ij}^{\,n_v - k - 1} \Delta_{v,ij}^{(k)}}{m_v} \tag{9.12}$$

will produce that change.

Henceforth we will assume n is large enough that (with $m_v = n/n_v$)

$$\frac{B_v}{m_v} \cdot \max(1, \max_{1 \le i \le m} (|\widetilde{c}_{v,i}|_v))^{n_v} < \epsilon_v . \tag{9.13}$$

If $|\Delta_{v,ij}^{(k)}|_v < B_v$, and $\eta_{v,ij}$ is defined by (9.12), then $|\eta_{v,ij}|_v < \epsilon_v$ so our discussion about the mapping properties of $\widehat{F}_v(z)$ applies.

For $i = 1, \ldots, m$ and $j = 0$, since $\Delta_{v,i0}^{(1)} = 0$ by hypothesis, we have $\eta_{v,i0} = 0$. The remaining $\eta_{v,ij}$ will be determined recursively, in terms of the $\Delta_{v,ij}^{(k)}$, in $\prec_N$ order; in particular, for each x_i the $\eta_{v,ij}$ will be determined in order of increasing j. As $F_v(z)$ is changed stepwise to $\widehat{F}_v(z)$, then $G_v^{(0)}(z) = \widehat{T}_{m_v}(F_v(z))$ is changed stepwise to $G_v^{(\overline{k})}(z) = \widehat{T}_{m_v}(\widehat{F}_v(z))$, passing through $G_v^{(1)}(z)$, $G_v^{(2)}(z)$, …, $G_v^{(\overline{k}-1)}(z)$ at intermediate steps.

Consider what happens as each $\eta_{v,ij}$ is varied in turn. Suppose $\check{F}_v(z)$ is a function obtained at one of the intermediate steps, and at the next step $\check{F}_v(z)$ is replaced by $\check{F}_v'(z) = \check{F}_v(z) + \eta_{v,ij}\varphi_{i,n_vN_i-j}(z)$.

If $x_i \in \mathcal{C}_v(\mathbb{R})$, then $\varphi_{i,n_vN_i-j} \in K_v(\mathcal{C})$ and $\widetilde{c}_{v,i} \in \mathbb{R}$. By hypothesis, $\Delta_{v,ij}^{(k)} \in \mathbb{R}$, so $\eta_{v,ij} \in \mathbb{R}$ and

$$\check{F}_v'(z) := \check{F}_v(z) + \eta_{v,ij}\varphi_{i,n_vN_i-j}(z)$$

is K_v-symmetric. Considering an expansion similar to (8.9) one sees that

$$\widehat{T}_{m_v}(\check{F}_v'(z)) = \widehat{T}_{m_v}(\check{F}_v(z)) + \Delta_{v,ij}^{(k)} \vartheta_{v,ij}(z)$$

for a K_v-rational (hence K_v-symmetric) function $\vartheta_{v,ij}(z)$ meeting the conditions of the theorem.

If $x_i \in \mathcal{C}_v(\mathbb{C})\backslash\mathcal{C}_v(\mathbb{R})$, let $\overline{z}$ denote the complex conjugate of z, and let $\overline{i}$ be the index such that $x_{\overline{i}} = \overline{x_i}$. As observed in the remark after Theorem 7.11, we can assume without loss that (i,j) and $(\overline{i},j)$ are consecutive indices under $\prec_N$. By our hypothesis of K_v-symmetry, $\Delta_{v,\overline{i}j}^{(k)} = \overline{\Delta_{v,ij}^{(k)}}$, and $\widetilde{c}_{v,\overline{i}} = \overline{\widetilde{c}_{v,i}}$, so $\eta_{v,\overline{i}j} = \overline{\eta_{v,ij}}$ and $N_{\overline{i}} = N_i$. Hence, after two steps

$$\check{F}_v''(z) = \check{F}_v(z) + \eta_{v,ij}\varphi_{i,n_vN_i-j}(z) + \eta_{v,\overline{i}j}\varphi_{\overline{i},n_vN_i-j}(z)$$

is K_v-symmetric. After grouping the terms in the multinomial expansions appropriately, one sees that

$$\widehat{T}_{m_v}(\check{F}_v''(z)) = \widehat{T}_{m_v}(\check{F}_v(z)) + \Delta_{v,ij}^{(k)}\vartheta_{v,ij}^{(k)}(z) + \Delta_{v,\overline{i}j}^{(k)}\vartheta_{v,\overline{i}j}^{(k)}(z)$$

where $\vartheta_{v,ij}^{(k)}(z), \vartheta_{v,\overline{i}j}^{(k)}(z) \in L_{w_v}(\mathcal{C})$ meet the conditions of the theorem.

The order $\prec_N$ specifies the $\eta_{v,ij}$ in "N-bands" with $(k-1)N_i \le j < kN_i$, for $k = 1, 2, \ldots, \overline{k}$. When we have completed patching the kth band, we obtain a function $G_v^{(k)}(z)$ which oscillates $n\tau_i$ times between $\pm 2\widehat{R}_2^{nN}$ on each real component $E_{v,i}^0$, and satisfies $\{z : G_v^{(k)}(z) \in E(A_2, B_2)\} \subset U_v$. Since the $\Delta_{v,ij}^{(k)}$ are K_v-symmetric, each $G_v^{(k)}(z)$ will be K_v-rational.

The final function $\widehat{F}_v(z)$ thus obtained, for which

$$G_v^{(\overline{k})}(z) = \widehat{T}_{m_v}(\widehat{F}_v(z)) \ ,$$

will play an important role in the rest of the argument. Note that the leading coefficient of $\widehat{F}_v(z)$ at x_i is $\widetilde{c}_{v,i}^{\,n_v}$.

Phase 2. Patching the middle coefficients.

In this phase we will construct functions $G_v^{(k)}(z)$ for $k = \overline{k}+1, \ldots, n-1$, setting

$$G_v^{(k)}(z) = G_v^{(k-1)}(z) + \sum_{i=1}^{m}\sum_{j=(k-1)N_i}^{kN_i-1} \Delta_{v,ij}^{(k)}\vartheta_{v,ij}(z) \tag{9.14}$$

for the given $\Delta_{v,ij}^{(k)}$ and appropriate functions $\vartheta_{v,ij}(z) \in L_{w_v}(\mathcal{C})$, adjoining the terms in $\prec_N$ order. The conditions of theorem require that the $\Delta_{v,ij}^{(k)} \in \mathbb{C}$ be K_v-symmetric; they also satisfy

$$|\Delta_{v,ij}^{(k)}|_v \le h_v^{kN} \ .$$

The $\vartheta_{v,ij}(z)$ will be $\mathrm{Aut}_c(\mathbb{C}_v/K_v)$-equivariant, so for each (i,j)

$$\sum_{(i',j)\in\mathrm{Aut}_c(\mathbb{C}_v/K_v)(i,j)} \Delta_{v,ij}^{(k)}\vartheta_{v,ij}(z) \in K_v(\mathcal{C}) \ ;$$

consequently, $G_v^{(k)}(z) \in K_v(\mathcal{C})$ as well.

Fix k, and write

$$n - k - 1 = \ell_1 + n_v\ell_2, \quad \text{with } 0 \le \ell_1 < n_v,\ 0 \le \ell_2 < m_v.$$

For each i, and each j with $(k-1)N_i \le j < kN_i$, we can uniquely write

$$nN_i - j = r_{ij} + (n-k-1)N_i \quad \text{where } N_i < r_{ij} \le 2N_i$$

so $nN_i - j = r_{ij} + \ell_1 N_i + \ell_2 n_v N_i$. Put

$$\vartheta_{v,ij}(z) = \varphi_{i,r_{ij}}(z) \cdot T_{\ell_1,R_1^N}(\phi_v(z)) \cdot \widehat{T}_{\ell_2}(\widehat{F}_v(z)) \ .$$

Then $\vartheta_{v,ij}(z)$ has a pole of order $nN_i - j$ at x_i, and leading coefficient $\widetilde{c}_{v,i}^{\,n-k-1}$. Its poles at the $x_{i'} \neq x_i$ are of order at most $(n-k-1)N_{i'}$, so it meets the requirements of the global patching process. Since the $\varphi_{ij}(z)$ are $\mathrm{Aut}_c(\mathbb{C}_v/K_v)$-equivariant, and $\phi_v(z)$ and $\widehat{F}_v(z)$ are K_v-rational, the $\vartheta_{v,ij}(z)$ are $\mathrm{Aut}_c(\mathbb{C}_v/K_v)$-equivariant.

Define

$$\begin{aligned}\widehat{E}_1 &= \{z \in E_v \cap \mathcal{C}_v(\mathbb{R}) : \phi_v(z) \in [-2\widehat{R}_1^{n_v N}, 2\widehat{R}_1^{n_v N}]\} ,\\ \widehat{E}_2 &= \{z \in E_v \cap \mathcal{C}_v(\mathbb{R}) : \widehat{F}_v(z) \in [-2\widehat{R}_2^{n_v N}, 2\widehat{R}_2^{n_v N}]\}\\ &= \{z \in \mathcal{C}_v(\mathbb{R}) : G_v^{(\overline{k})}(z) \in [-2\widehat{R}_2^{nN}, 2\widehat{R}_2^{nN}]\} .\end{aligned}$$

By construction, $\widehat{E}_2 \subset \widehat{E}_1 \subset E_v \cap \mathcal{C}_v(\mathbb{R})$. Likewise, put

$$\begin{aligned}\widehat{W}_1 &= \phi_v^{-1}(E(a_1,b_1)) = F_v^{-1}(E(A_1,B_1)) ,\\ \widehat{W}_2 &= \widehat{F}_v^{-1}(E(a_2,b_2)) = \big(G_v^{(\overline{k})}\big)^{-1}(E(A_2,B_2)) .\end{aligned}$$

Then $\widehat{W}_2 \subset \widehat{W}_1 \subset U_v$.

We will now bound the change in $|G_v^{(k)}(z) - G_v^{(k-1)}(z)|_v$ on $\widehat{E}_2$ and $\widehat{W}_2$. We first bound $\|G_v^{(k)} - G_v^{(k-1)}\|_{\widehat{E}_2}$, and we begin by showing that for each (i,j) occurring in (9.14),

$$\|\vartheta_{v,ij}\|_{\widehat{E}_2} \le 16 M_v \widehat{R}_2^{(n-k-1)N} . \tag{9.15}$$

To see this, note that by the definition of M_v, we have $|\varphi_{i,r_{ij}}(z)|_v \le M_v$ for all $z \in E_v \cap \mathcal{C}_v(\mathbb{R})$. Also, since $\ell_1 < n_v$, for each $z \in \widehat{E}_1$,

$$|T_{\ell_1,R_1^N}(\phi_v(z))|_v \le 2\widehat{R}_1^{\ell_1 N} \le 8\widehat{R}_2^{\ell_1 N} .$$

by the definitions of $\widehat{R}_1$ and $\widehat{R}_2$. Finally, by the properties of Chebyshev polynomials, for each $z \in \widehat{E}_2$,

$$|\widehat{T}_{\ell_2}(\widehat{F}_v(z))|_v \le 2\widehat{R}_2^{\ell_2 n_v N}.$$

Combining these, and using that $\widehat{E}_2 \subset \widehat{E}_1 \subset E_v \cap \mathcal{C}_v(\mathbb{R})$, gives (9.15).

Since each $|\Delta_{v,ij}^{(k)}|_v \le h_v^{kN}$ and $1 < r_v < \widehat{R}_2$, it follows that

$$\begin{aligned}\|G_v^{(k)} - G_v^{(k-1)}\|_{\widehat{E}_2} &\le \sum_{i=1}^{m} \sum_{j=(k-1)N_i}^{kN_i-1} |\Delta_{v,ij}^{(k)}|_v \|\vartheta_{v,ij}(z)\|_{\widehat{E}_2}\\ &\le N \cdot h_v^{kN} \cdot 16 M_v \widehat{R}_2^{(n-k-1)N} = \frac{16NM_v}{\widehat{R}_2^N} \cdot \widehat{R}_2^{nN} \cdot \left(\frac{h_v}{\widehat{R}_2}\right)^{kN} \qquad (9.16)\\ &< 16NM_v \cdot \widehat{R}_2^{nN} \cdot \left(\frac{h_v}{r_v}\right)^{kN} .\end{aligned}$$

We next bound $\|G_v^{(k)} - G_v^{(k-1)}\|_{\widehat{W}_2}$. We claim that for each (i,j) in (9.14),

$$\|\vartheta_{v,ij}\|_{\widehat{W}_2} \le 5 M_v t_2^{(n-k-1)N} . \tag{9.17}$$

To see this, recall that $|\varphi_{i,r_{ij}}(z)|_v \le M_v$ for all $z \in \overline{U}_v$. Also, note that $T_{\ell_1,\widehat{R}_1^N}$ maps $E(a_1,b_1)$ to $E(A^{(\ell_1)}, B^{(\ell_1)})$ where $B^{(\ell_1)} < A^{(\ell_1)}$ and

$$A^{(\ell_1)} = t_1^{\ell_1 N} + \frac{\widehat{R}_1^{2\ell_1 N}}{t_1^{\ell_1 N}} = \big(1 + (2-\sqrt{3})^{2\ell_1}\big) \cdot t_1^{\ell_1 N} < 1.1 \cdot t_1^{\ell_1 N} .$$

Since $t_1^{n_v N} = (2+\sqrt{3})\widehat{R}_1^{n_v N}$, $\widehat{R}_1^{n_v N} = 4\widehat{R}_2^{n_v N}$ and $t_2^{n_v N} = (2+\sqrt{3})\widehat{R}_2^{n_v N}$, we have $t_1^{n_v N} = 4t_2^{n_v N}$ and for $0 \le \ell_1 < n_v$ it follows that $t_1^{\ell_1 N} < 4t_2^{\ell_1 N}$. This means that $E(A^{(\ell_1)}, B^{(\ell_1)}) \subset D(0, 1.1 \cdot 4t_2^{\ell_1 N})$, so for each $z \in \widehat{W}_1 = \phi_v^{-1}(E(a_1, b_1))$,

$$|T_{\ell_1, \widehat{R}_1^N}(\phi_v(z))|_v \;\le\; 1.1 \cdot 4t_2^{\ell_1 N}\;. \tag{9.18}$$

Similarly, since $\widehat{T}_{\ell_2} = T_{\ell_2, \widehat{R}_2^{n_v N}}$ maps $E(a_2, b_2)$ to $E(\widetilde{A}^{(\ell_2)}, \widetilde{B}^{(\ell_2)})$ where $\widetilde{B}^{(\ell_2)} < \widetilde{A}^{(\ell_2)}$ and

$$\widetilde{A}^{(\ell_1)} \;=\; t_2^{\ell_2 n_v N} + \frac{\widehat{R}_2^{2\ell_2 n_v N}}{t_2^{\ell_2 n_v N}} \;=\; \left(1 + (2-\sqrt{3})^{2\ell_2}\right) \cdot t_2^{\ell_2 n_v N} \;<\; 1.1 \cdot t_2^{\ell_1 n_v N}\;,$$

for each $z \in \widehat{W}_2 = \widehat{F}_v^{-1}(E(a_2, b_2))$ we have

$$|\widehat{T}_{\ell_2}(\widehat{F}_v(z))|_v \;\le\; 1.1 \cdot t_2^{\ell_2 n_v N}\;. \tag{9.19}$$

Since $\widehat{W}_2 \subset \widehat{W}_1 \subset \overline{U}_v$ and $4 \cdot (1.1)^2 < 5$, combining (9.18) and (9.19) gives (9.17).

Since each $|\Delta_{v,ij}^{(k)}|_v \le h_v^{kN}$ and $1 < r_v < \widehat{R}_2 < t_2$, it follows that

$$\begin{aligned} \|G_v^{(k)} - G_v^{(k-1)}\|_{\widehat{W}_2} \;&\le\; \sum_{i=1}^{m} \sum_{j=(k-1)N_i}^{kN_i - 1} |\Delta_{v,ij}^{(k)}|_v \|\vartheta_{v,ij}(z)\|_{\widehat{W}_2} \\ &\le\; N \cdot h_v^{kN} \cdot 5M_v t_2^{(n-k-1)N} \;=\; \frac{5NM_v}{t_2^N} \cdot t_2^{nN} \cdot \left(\frac{h_v}{t_2}\right)^{kN} \\ &<\; 5NM_v \cdot t_2^{nN} \cdot \left(\frac{h_v}{r_v}\right)^{kN}\;. \end{aligned} \tag{9.20}$$

Phase 3. Patching the low-order coefficients.

In the final step we take

$$G_v^{(n)}(z) \;=\; G_v^{(n-1)}(z) + \sum_{\lambda=1}^{\Lambda} \Delta_{v,\lambda}^{(n)} \varphi_\lambda$$

with K_v-symmetric $\Delta_{v,\lambda}^{(n)}$ satisfying $|\Delta_{v,\lambda}^{(n)}|_v \le h_v^{nN}$ for each λ. Since $\Lambda \le N$, and each $|\varphi_\lambda|_v \le M_v$ on $\overline{U}_v$, it follows that on $\widehat{E}_2$,

$$\|G_v^{(n)} - G_v^{(n-1)}\|_{\widehat{E}_2} \;\le\; NM_v h_v^{nN} \;<\; 16NM_v \cdot \widehat{R}_2^{nN} \cdot \left(\frac{h_v}{r_v}\right)^{nN}\;, \tag{9.21}$$

while on $\widehat{W}_2$,

$$\|G_v^{(n)} - G_v^{(n-1)}\|_{\widehat{W}_2} \;\le\; NM_v h_v^{nN} \;<\; 5NM_v \cdot t_2^{nN} \cdot \left(\frac{h_v}{r_v}\right)^{nN}\;. \tag{9.22}$$

To conclude the proof, we show that if n is sufficiently large, then $G_v^{(n)}(z)$ has the mapping properties in part (B) of the theorem. A similar argument applies to $G_v^{(k)}(z)$ for each $k = 1, \ldots, n$.

Consider the total change in passing from $G_v^{(\overline{k})}(z) = \widehat{T}_{m_v}(\widehat{F}_v(z))$ to $G_v^{(n)}(z)$. By (9.16) and (9.21), for each $z \in \widehat{E}_2$,

$$|G_v^{(n)}(z) - G_v^{(\overline{k})}(z)|_v \le 16NM_v \cdot \widehat{R}_2^{nN} \cdot \sum_{k=\overline{k}}^{n} \left(\frac{h_v}{r_v}\right)^{kN} < \frac{16NM_v}{1-(h_v/r_v)^N} \cdot \left(\frac{h_v}{r_v}\right)^{\overline{k}N} \cdot \widehat{R}_2^{nN} .$$

Since $\overline{k} \ge k_v$, assumption (9.2) in Theorem 9.1 shows that

$$\|G_v^{(n)} - G_v^{(\overline{k})}\|_{\widehat{E}_2} < \frac{1}{4}\widehat{R}_2^{nN} . \tag{9.23}$$

Similarly, on $\widehat{W}_2$

$$\|G_v^{(n)} - G_v^{(\overline{k})}\|_{\widehat{W}_2} < \frac{1}{4}t_2^{nN} . \tag{9.24}$$

We first show that

$$\{z \in \mathcal{C}_v(\mathbb{C}) : |G_v^{(n)}(z)|_v \le \frac{1}{2}t_2^{nN}\} \subset U_v . \tag{9.25}$$

Let $\widehat{\Gamma}_2 = \partial\widehat{W}_2 = \{z \in \mathcal{C}_v(\mathbb{C}) : G_v^{(\overline{k})}(z) \in \partial E(A_2, B_2)\}$. Since $A_2 > B_2$ and

$$B_2 = (1-(2-\sqrt{3})^{2m_v})t_2^{nN} > 0.9t_2^{nN} ,$$

for each $z \in \widehat{\Gamma}_2$ we have $|G_v^{(\overline{k})}(z)|_v > 0.9t_2^{nN}$. By (9.24), $|G_v^{(n)}(z) - G_v^{(\overline{k})}(z)|_v < 0.25t_2^{nN}$. Applying Lemma 8.2 with $F(z) = G_v^{(\overline{k})}(z)$ and $H(z) = G_v^{(n)}(z) - G_v^{(\overline{k})}(z)$, we see that

$$\{z \in \mathcal{C}_v(\mathbb{C}) : |G_v^{(n)}(z)|_v \le 0.65t_2^{nN}\} \subset \widehat{W}_2 ,$$

which yields (9.25).

If $E_{v,i}$ is a component of E_v which is disjoint from $\mathcal{C}_v(\mathbb{R})$, then $G_v^{(\overline{k})}(z)$ has $n\tau_i$ zeros in $E_{v,i}^0$. Put $\widehat{W}_{2,i} = \widehat{W}_2 \cap E_{v,i}$. Since $\widehat{W}_2 \subset U_v$ and $U_v \cap E_{v,i} = E_{v,i}^0$, it follows that $\widehat{W}_{2,i} \subset E_{v,i}^0$. Applying Rouché's theorem to $G_v^{(\overline{k})}(z)$ and $G_v^{(n)}(z)$ on $\partial\widehat{W}_{2,i}$, we conclude that $G_v^{(n)}(z)$ has $n\tau_i$ zeros in $E_{v,i}^0$ as well.

We next show that if $E_{v,i}$ is a component of E_v contained in $\mathcal{C}_v(\mathbb{R})$, then $G_v^{(n)}(z)$ oscillates $n\tau_i$ times between $\pm(7/4)\widehat{R}_2^{nN}$ on $E_{v,i}$. Recall that $G_v^{(\overline{k})}(z) = \widehat{T}_{m_v}(\widehat{F}_v(z))$ oscillates $n\tau_i$ times between $\pm 2\widehat{R}_2^{nN}$ on $E_{v,i}$. Equation (9.23) shows that at each $z_i \in E_{v,i}$ where $G_v^{(\overline{k})}(z_i) = \pm 2\widehat{R}_2^{nN}$, then $G_v^{(n)}(z_i)$ has the same sign as $G_v^{(\overline{k})}(z_i)$ and $|G_v^{(n)}(z_i)|_v \ge (7/4)\widehat{R}_2^{nN}$. Hence $G_v^{(n)}(z)$ oscillates $n\tau_i$ times between $\pm(7/4)\widehat{R}_2^{nN}$ on $E_{v,i}$ and, in particular, it has $n\tau_i$ zeros in $E_{v,i}^0$.

Since $G_v^{(n)}(z)$ has degree nN and $\sum n\tau_i = nN$, all the zeros of $G_v^{(n)}(z)$ lie in E_v^0. Since $t_2 > \widehat{R}_2 > r_v$, for all sufficiently large n we have $(1/2)t_2^{nN} > 2r_v^{nN}$ and $(7/4)\widehat{R}_2^{nN} > 2r_v^{nN}$. For such n, $G_v^{(n)}(z)$ oscillates $n\tau_i$ times between $\pm 2r_v^{nN}$ on each $E_{v,i}$ contained in $\mathcal{C}_v(\mathbb{R})$, and

$$\{z \in \mathcal{C}_v(\mathbb{C}) : |G_v^{(n)}(z)|_v \le 2r_v^{nN}\} \subset U_v .$$

Thus the construction succeeds for any integer n divisible by n_v which is large enough that condition (9.13) holds and $n > \overline{k}$, $(1/2)t_2^{nN} > 2r_v^{nN}$, and $(7/4)\widehat{R}_2^{nN} > 2r_v^{nN}$. $\square$

CHAPTER 10

Local Patching for Nonarchimedean RL-domains

In this section we give the confinement argument for Theorem 4.2 when K_v is nonarchimedean, and E_v is a K_v-rational RL-domain.

Let q_v be the order of the residue field of K_v. Let w_v be the distinguished place of $L = K(\mathfrak{X})$ determined by the embedding $\widetilde{K} \hookrightarrow \mathbb{C}_v$ used to identify $\mathfrak{X}$ with a subset of $\mathcal{C}_v(\mathbb{C}_v)$, and view L_{w_v} as a subset of $\mathbb{C}_v$. Using the construction of the coherent approximating functions in Theorems 7.11 and 7.16, we begin with the following data:

(1) A K_v-symmetric probability vector $\vec{s} \in \mathcal{P}^m(\mathbb{Q})$ with positive rational coefficients;

(2) An RL-domain E_v, a number N, a number $R_v \in |\mathbb{C}_v^\times|_v$, and an $(\mathfrak{X}, \vec{s})$-function $\phi_v(z) \in K_v(\mathcal{C})$ of degree N such that

$$E_v \;=\; \{z \in \mathcal{C}_v(\mathbb{C}_v) : |\phi_v(z)|_v \le R_v^N\} \ .$$

Put $N_i = N s_i$ for each i, and write $\widetilde{c}_{v,i} = \lim_{z\to x_i} \phi_v(z) \cdot g_{x_i}(z)^{N_i}$ for the leading coefficient of $\phi_v(z)$ at x_i.

If $\mathrm{char}(K_v) = p > 0$, we will assume that the number J from the construction of the L-rational and L^{sep}-rational bases in §3.3 divides N_i, and that the leading coefficient $\widetilde{c}_{v,i}$ of $\phi_v(z)$ at x_i belongs to $K_v(x_i)^{\mathrm{sep}}$, for each $i = 1, \ldots, m$.

(3) Parameters h_v and r_v satisfying $0 < h_v < r_v \le R_v$, which govern the freedom in the patching process;

(4) An order $\prec_N$ on the index set $\mathcal{I} = \{(i,j) \in \mathbb{Z}^2 : 1 \le i \le m, 0 \le j\}$ determined by N and $\vec{s}$ as in (7.41), which gives the sequence in which coefficients are patched. The group $\mathrm{Aut}_c(\mathbb{C}/K_v)$ acts on $\mathcal{I}$, and we write $\mathrm{Aut}_c(\mathbb{C}/K_v)(i,j)$ for the orbit of $(i,j) \in \mathcal{I}$.

We will use the L-rational basis $\{\varphi_{ij}, \varphi_\lambda\}$ from §3.3 to expand functions, and $\Lambda = \dim_K(\Gamma(\sum_{i=1}^m N_i(x_i)))$ will be the number of low-order basis elements, as in §7.4. The order $\prec_N$ respects the N-bands (7.42), and for each $x_i \in \mathfrak{X}$, specifies the terms to be patched in decreasing pole order.

When $\mathrm{char}(K_v) = 0$, we will need the following the following patching theorem.

THEOREM 10.1. *Suppose K_v is nonarchimedean, and that* $\mathrm{char}(K_v) = 0$. *Let $E_v \subset \mathcal{C}_v(\mathbb{C}_v)\backslash\mathfrak{X}$ be an* RL-*domain. Let $\vec{s} \in \mathcal{P}^m(\mathbb{Q})$ be a K_v-symmetric probability vector with positive rational coefficients. Suppose $R_v \in |\mathbb{C}_v^\times|_v$, let $0 < h_v < r_v \le R_v$, and let $\phi_v(z) \in K_v(\mathcal{C})$ be an $(\mathfrak{X}, \vec{s})$-function of degree N satisfying*

$$\{z \in \mathcal{C}_v(\mathbb{C}_v) : |\phi_v(z)|_v \le R_v^N\} \;=\; E_v \ .$$

For each $i = 1, \ldots, m$ *let* $N_i = Ns_i$ *and let* $\widetilde{c}_{v,i} = \lim_{z \to x_i} \phi_v(z) \cdot g_{x_i}(z)^{N_i}$ *be the leading coefficient of* $\phi_v(z)$ *at* x_i. *Put*

$$M_v = \max\Big(\max_{\substack{1 \le i \le m \\ N_i < j \le 2N_i}} \|\varphi_{ij}\|_{E_v}, \max_{1 \le \lambda \le \Lambda} \|\varphi_\lambda\|_{E_v} \Big) .$$

Let $k_v \ge 1$ *be the least integer such that*

$$\left(\frac{h_v}{r_v}\right)^{k_v N} \cdot \frac{M_v}{\min(1, R_v^N)} < 1 , \tag{10.1}$$

and let $\overline{k} \ge k_v$ *be a fixed integer. Put* $n_v = 1$. *Then there is a number* $0 < B_v < 1$ *depending on* $\overline{k}$, E_v, *and* $\phi_v(z)$ *such that for each sufficiently large integer* n, *one can carry out the local patching process at* K_v *as follows :*

Put $G_v^{(0)}(z) = \phi_v(z)^n$. *For each* k, $1 \le k < n$, *let* $\{\Delta_{v,ij}^{(k)} \in L_{w_v}\}_{(i,j) \in \mathrm{Band}_N(k)}$ *be a* K_v*-symmetric set of numbers given recursively in* $\prec_N$ *order, such that for each* (i, j),

$$|\Delta_{v,ij}^{(k)}|_v \le \begin{cases} B_v & \text{if } k \le \overline{k} , \\ h_v^{kN} & \text{if } k > \overline{k} . \end{cases} \tag{10.2}$$

For $k = n$, *let* $\{\Delta_{v,\lambda}^{(n)} \in L_{w_v}\}_{1 \le \lambda \le \Lambda}$ *be a* K_v*-symmetric set of numbers such that for each* λ,

$$|\Delta_{v,\lambda}^{(n)}|_v \le h_v^{nN} . \tag{10.3}$$

Then one can inductively find $(\mathfrak{X}, \vec{s})$*-functions* $G_v^{(1)}(z), \ldots, G_v^{(n)}(z) \in K_v(\mathcal{C})$, *of common degree* nN, *having the following properties:*
(*A*) *For each* $k = 1, \ldots, n-1$,

$$G_v^{(k)}(z) = G_v^{(k-1)}(z) + \sum_{(i,j) \in \mathrm{Band}_N(k)} \Delta_{v,ij}^{(k)} \cdot \vartheta_{v,ij}^{(k)}(z)$$

where $\vartheta_{v,ij}^{(k)}(z) = \varphi_{i,(k+1)N_i - j} \cdot \phi_v(z)^{n-k-1}$, *and for* $k = n$,

$$G_v^{(n)}(z) = G_v^{(n-1)}(z) + \sum_{\lambda=1}^{\Lambda} \Delta_{v,\lambda}^{(n)} \cdot \varphi_\lambda(z) ,$$

and where

(1) *Each* $\vartheta_{v,ij}^{(k)}(z)$ *is in* $K_v(x_i)(\mathcal{C})$, *has a pole of order* $nN_i - j > (n-k-1)N_i$ *at* x_i *with leading coefficient* $\widetilde{c}_{v,i}^{\,n-k-1}$, *has poles of order at most* $(n-k-1)N_{i'}$ *at each* $x_{i'} \ne x_i$, *and has no other poles;*

(2) *The* $\vartheta_{v,ij}^{(k)}(z)$ *are* K_v*-symmetric;*

(3) *In passing from* $G_v^{(0)}(z)$ *to* $G_v^{(1)}(z)$, *each of the leading coefficients* $A_{v,i0} = \widetilde{c}_{v,i}^{\,n}$ *is replaced with* $\widetilde{c}_{v,i}^{\,n} + \Delta_{v,i0}^{(1)} \cdot \widetilde{c}_{v,i}^{\,n-2}$.

(*B*) *For each* $k = 0, \ldots, n$, $\quad \{z \in \mathcal{C}_v(\mathbb{C}_v) : |G_v^{(n)}(z)|_v \le R_v^{Nn}\} = E_v$.

When $\mathrm{char}(K_v) = p > 0$, we will use the following patching theorem instead. The K_v-rationality assumptions in the theorem are addressed by the global patching process.

THEOREM 10.2. *Suppose K_v is nonarchimedean and* $\mathrm{char}(K_v) = p > 0$. *Let $\vec{s} \in \mathcal{P}^m(\mathbb{Q})$ be a K_v-symmetric probability vector with positive rational coefficients. Suppose $R_v \in |\mathbb{C}_v^\times|_v$, let $0 < h_v < r_v \le R_v$, and let $\phi_v(z) \in K_v(\mathcal{C})$ be an $(\mathfrak{X}, \vec{s})$-function of degree N satisfying*

$$\{z \in \mathcal{C}_v(\mathbb{C}_v) : |\phi_v(z)|_v \le R_v^N\} \;=\; E_v \;.$$

Let $N_i = Ns_i$ for each x_i, and let $\widetilde{c}_{v,i} = \lim_{z \to x_i} \phi_v(z) \cdot g_{x_i}(z)^{N_i}$ be the leading coefficient of $\phi_v(z)$ at x_i. Assume that $J|N_i$ and $\widetilde{c}_{v,i} \in K_v(x_i)^{\mathrm{sep}}$, for $i = 1, \ldots, m$. Put

$$M_v \;=\; \max\big(\max_{1 \le i \le m}\ \max_{N_i < j \le 2N_i} \|\varphi_{ij}\|_{E_v},\ \max_{1 \le \lambda \le \Lambda} \|\varphi_\lambda\|_{E_v}\big) \;.$$

Let $k_v > 0$ be the least integer such that

$$\left(\frac{h_v}{r_v}\right)^{k_v N} \cdot \frac{M_v}{\min(1, R_v^N)} \;<\; 1 \;, \tag{10.4}$$

and let $\overline{k} \ge k_v$ be a fixed integer. Then there are an integer $n_v \ge 1$ and a number $0 < B_v < 1$ depending on $\overline{k}$, E_v, and $\phi_v(z)$, such that for each sufficiently large integer n divisible by n_v, one can carry out the local patching process at K_v as follows:

Put $G_v^{(0)}(z) = \phi_v(z)^n$. Then the leading coefficient of $G_v^{(0)}(z)$ at each x_i is $\widetilde{c}_{v,i}^{\,n}$, $\{z \in \mathcal{C}_v(\mathbb{C}_v) : |G_v^{(0)}(z)|_v \le R_v^{nN}\} = E_v$, and when $G_v^{(0)}(z)$ is expanded in terms of the L-rational basis as

$$G_v^{(0)}(z) \;=\; \sum_{i=1}^{m} \sum_{j=0}^{(n-1)N_i - 1} A_{v,ij}\varphi_{i,nN_i-j}(z) + \sum_{\lambda=1}^{\Lambda} A_{v,\lambda}\varphi_\lambda \;,$$

we have $A_{v,ij} = 0$ for all (i,j) with $1 \le j < \overline{k}N_i$.

For each k, $1 \le k \le n-1$, let $\{\Delta_{v,ij}^{(k)} \in L_{w_v}\}_{(i,j) \in \mathrm{Band}_N(k)}$ be a K_v-symmetric set of numbers satisfying

$$\begin{cases} |\Delta_{v,i0}^{(1)}|_v \le B_v \text{ and } \Delta_{v,ij}^{(1)} = 0 \text{ for } j = 1, \ldots, N_i - 1, & \text{if } k = 1 \;, \\ \Delta_{v,ij}^{(k)} = 0 \text{ for } j = (k-1)N_i, \ldots, kN_i - 1, & \text{if } k = 2, \ldots, \overline{k} \;, \\ |\Delta_{v,ij}^{(k)}|_v \le h_v^{kN}, & \text{if } k = \overline{k}+1, \ldots, n-1 \;, \end{cases} \tag{10.5}$$

such that $\Delta_{v,i0}^{(1)} \in K_v(x_i)^{\mathrm{sep}}$ for each i and such that for $k = \overline{k}+1, \ldots, n-1$,

$$\Delta_v^{(k)}(z) \;=\; \sum_{i=1}^{m} \sum_{j=(k-1)N_i}^{kN_i - 1} \Delta_{v,ij}^{(k)} \cdot \varphi_{i,(k+1)N_i - j} \;\in\; K_v(\mathcal{C}) \;. \tag{10.6}$$

For $k = n$, let $\{\Delta_{v,\lambda}^{(n)} \in L_{w_v}\}_{1 \le \lambda \le \Lambda}$ be a K_v-symmetric set of numbers such that

$$|\Delta_{v,\lambda}^{(n)}|_v \;\le\; h_v^{nN} \tag{10.7}$$

for each λ, and

$$\sum_{\lambda=1}^{\Lambda} \Delta_{v,\lambda}^{(n)} \cdot \varphi_\lambda \;\in K_v(\mathcal{C}) \;. \tag{10.8}$$

Then one can inductively find $(\mathfrak{X}, \vec{s})$-functions $G_v^{(1)}(z), \ldots, G_v^{(n)}(z)$ in $K_v(\mathcal{C})$, of common degree Nn, such that:

(A) For each $k = 1, \ldots, n$, $G_v^{(k)}(z)$ *is obtained from* $G_v^{(k-1)}(z)$ *as follows:*

(1) *When* $k = 1$, *there is a* K_v*-symmetric set of functions* $\widetilde{\theta}_{v,10}^{(1)}(z), \ldots, \widetilde{\theta}_{v,m0}^{(1)}(z)$ *in* $L_{w_v}^{\mathrm{sep}}(\mathcal{C})$ *such that*

$$G_v^{(1)}(z) \;=\; G_v^{(0)}(z) + \sum_{i=1}^{m} \Delta_{v,i0}^{(1)} \cdot \widetilde{\theta}_{v,i0}^{(1)}(z) \ ,$$

where for each $i = 1, \ldots, m$, $\widetilde{\theta}_{v,i0}^{(1)}(z) \in K_v(x_i)^{\mathrm{sep}}(\mathcal{C})$ *has the form*

$$\widetilde{\theta}_{v,i0}^{(1)}(z) \;=\; \widetilde{c}_{v,i}^{\,n} \varphi_{i,nN_i}(z) + \widetilde{\Theta}_{v,i0}^{(1)}(z)$$

for an $(\mathfrak{X}, \vec{s})$*-function* $\widetilde{\Theta}_{v,i0}^{(1)}(z)$ *with a pole of order at most* $(n - \overline{k})N_i$ *at each* x_i*. Thus, in passing from* $G_v^{(0)}(z)$ *to* $G_v^{(1)}(z)$*, each leading coefficient* $A_{v,i0} = \widetilde{c}_{v,i}^{\,n}$ *is replaced with* $\widetilde{c}_{v,i}^{\,n} + \Delta_{v,i0}^{(1)} \cdot \widetilde{c}_{v,i}^{\,n}$*, and the coefficients* $A_{v,ij}$ *for* $1 \le j < \overline{k}N_i$ *remain* 0*.*

(2) *For* $k = 2, \ldots, \overline{k}$, *we have*$G_v^{(k)}(z) = G_v^{(k-1)}(z)$*.*

(3) *For* $k = \overline{k} + 1, \ldots, n - 1$, *we have*

$$G_v^{(k)}(z) \;=\; G_v^{(k-1)}(z) + \Delta_v^{(k)}(z) \cdot F_{v,k}(z) + \Theta_v^{(k)}(z) \ , \tag{10.9}$$

where

(a) $\Delta_v^{(k)}(z) = \sum_{(i,j) \in \mathrm{Band}_N(k)} \Delta_{v,ij}^{(k)} \varphi_{i,(k+1)N_i - j}(z)$*.*
(Here $\Delta_v^{(k)}(z)$ *belongs to* $K_v(\mathcal{C})$ *by* (10.6)*).*

(b) $F_{v,k}(z) = \phi_v(z)^{n-k-1}$ *is a* K_v*-rational* $(\mathfrak{X}, \vec{s})$*-function with roots in* E_v*. It has a pole of order* $(n - k - 1)N_i$ *at each* x_i*, and its leading coefficient* $d_{v,i} = \lim_{z \to x_i} F_{v,k}(z) \cdot g_{x_i}(z)^{(n-k-1)N_i}$ *belongs to* $K_v(x_i)^{\mathrm{sep}}$ *and has absolute value* $|d_{v,i}|_v = |\widetilde{c}_{v,i}|_v^{n-k-1}$*.*

(c) $\Theta_v^{(k)}(z) \in K_v(\mathcal{C})$ *is an* $(\mathfrak{X}, \vec{s})$*-function determined by the local patching process at* v *after the coefficients in* $\mathrm{Band}_N(k)$ *have been modified; it has a pole of order at most* $(n - k)N_i$ *at each* x_i *and no other poles, and may be the zero function.*

(4) *For* $k = n$,

$$G_v^{(n)}(z) \;=\; G_v^{(n-1)}(z) + \sum_{\lambda=1}^{\Lambda} \Delta_{v,\lambda}^{(n)} \cdot \varphi_\lambda(z) \ .$$

(B) For each $k = 1, \ldots, n$,

$$\{z \in \mathcal{C}_v(\mathbb{C}_v) : |G_v^{(n)}(z)|_v \le R_v^{nN}\} \;=\; E_v \ .$$

Theorems 10.1 and 10.2 will be proved together. There are some differences in the way the leading and high-order coefficients are treated, but the underlying patching constructions for the middle and low-order coefficients are the same. In Theorem 10.1 the compensating functions are $\vartheta_{v,ij}^{(k)}(z) = \varphi_{i,(k+1)N_i - j}(z) \cdot \phi_v(z)^{n-k-1}$, while in Theorem 10.2 we have $\omega_v^{(k)}(z) \;=\; \sum_{(i,j) \in \mathrm{Band}_N(k)} \Delta_{v,ij}^{(k)} \varphi_{i,(k+1)N_i - j}(z)$ and $F_{v,k}(z) = \phi_v(z)^{n-k-1}$ so, as noted after Theorem 7.17,

$$\sum_{(i,j) \in \mathrm{Band}_N(k)} \Delta_{v,ij}^{(k)} \vartheta_{v,ij}^{(k)}(z) \;=\; \omega_v^{(k)}(z) \cdot F_{v,k}(z) \ .$$

To prove Theorems 10.1 and 10.2, we will need the following nonarchimedean analogue of Lemma 8.2, valid both when $\mathrm{char}(K_v) = 0$ and when $\mathrm{char}(K_v) = p > 0$:

LEMMA 10.3. *Let $F(z) \in \mathbb{C}_v(\mathcal{C}_v)$ be a nonconstant rational function, and let $R > 0$ be an element of the value group of $\mathbb{C}_v^\times$. Put*

$$U \;=\; \{z \in \mathcal{C}_v(\mathbb{C}_v) : |F(z)|_v \le R\} \;.$$

Suppose $H(z) \in \mathbb{C}_v(\mathcal{C}_v)$ is a function such that $|H(z)|_v < R$ for all $z \in U$, and whose polar divisor satisfies $\mathrm{div}_\infty(H) \le \mathrm{div}_\infty(F)$. Then

$$\{z \in \mathcal{C}_v(\mathbb{C}_v) : |F(z) + H(z)|_v \le R\} \;=\; U \;.$$

Lemma 10.3 depends on the following version of the nonarchimedean Maximum Modulus Principle:

PROPOSITION 10.4 (Maximum Principle with Distinguished Boundary). *Let $f(z) \in \mathbb{C}_v(\mathcal{C}_v)$ be a nonconstant rational function, and let $R > 0$ belong to the value group of $\mathbb{C}_v^\times$. Put*

$$\begin{aligned} U \;&=\; \{z \in \mathcal{C}_v(\mathbb{C}_v) : |f(z)|_v \le R\} \;, \\ \partial U(f) \;&=\; \{z \in \mathcal{C}_v(\mathbb{C}_v) : |f(z)|_v = R\} \;. \end{aligned}$$

Let $g(z) \in \mathbb{C}_v(\mathcal{C}_v)$ be a function with no poles in U. Then $|g(z)|_v$ achieves its maximum for $z \in U$ at a point $z_0 \in \partial U(f)$.

PROOF. See ([**51**], Theorem 1.4.2, p.51). □

PROOF OF LEMMA 10.3. For $z \in U$, we have $|F(z) + H(z)|_v \le R$ by the ultrametric inequality, so

$$U \;\subseteq\; \{z \in \mathcal{C}_v(\mathbb{C}_v) : |F(z) + H(z)|_v \le R\} \;.$$

To establish the reverse containment, put

$$V \;=\; \{z \in \mathcal{C}_v(\mathbb{C}_v) : |(1/F)(z)|_v \le 1/R\}$$

regarding $1/F$ as a rational function whose value is 0 on the poles of F. The distinguished boundaries of U and V satisfy

$$\begin{aligned} \partial U(F) \;&=\; \{z \in \mathcal{C}_v(\mathbb{C}_v) : |F(z)|_v = R\} \\ &=\; \{z \in \mathcal{C}_v(\mathbb{C}_v) : |1/F(z)|_v = 1/R\} \;= \partial V(1/F) \;. \end{aligned}$$

Put $G(z) = H(z)/F(z) \in \mathbb{C}_v(\mathcal{C})$. By hypothesis, $G(z)$ has no poles in V, and $|G(z)|_v < 1$ for all $z \in \partial V(1/F) = \partial U(F) \subset U$. By Proposition 10.4 $|G(z)|_v < 1$ for all $z \in V$. Equivalently $|H(z)|_v < |F(z)|_v$ for all $z \in V$ which are not poles of F, so by the ultrametric inequality $|F(z) + H(z)|_v = |F(z)|_v$ for such z. In particular, for $z \notin U$ we have $|F(z) + H(z)|_v = |F(z)|_v > R$ so

$$\{z \in \mathcal{C}_v(\mathbb{C}_v) : |F(z) + H(z)|_v \le R\} \;\subseteq\; U \;.$$ □

PROOF OF THEOREMS 10.1 AND 10.2. The patching construction will be carried out in three phases. The proofs differ only in their treatment of the high-order coefficients.

Phase 1. Patching the high-order coefficients.

In this phase we carry out the patching process for stages $k = 1, \ldots, \overline{k}$.

First assume $\mathrm{char}(K_v) = 0$. Let the $(\mathfrak{X}, \vec{s})$-function $\phi_v(z)$ of degree N, and the numbers k_v, M_v, $0 < h_v < r_v \le R_v$, and $\overline{k} \ge k_v$ be as in Theorem 10.1. Take $n_v = 1$, and put

$$B_v \;=\; \frac{\min(1, R_v)^{(\overline{k}+1)N}}{2M_v} \;.$$

Assume $n > \overline{k}$, and let $G_v^{(0)}(z) = \phi_v(z)^n$.

For each $k = 1, \ldots, \overline{k}$, we begin with a K_v-rational $(\mathfrak{X}, \vec{s})$-function $G_v^{(k-1)}(z)$ satisfying

$$\{z \in \mathcal{C}_v(\mathbb{C}_v) : |G_v^{(k-1)}(z)|_v \le 1\} \;=\; E_v \ .$$

Expand

$$G_v^{(k-1)}(z) \;=\; \sum_{i=1}^{m} \sum_{j=0}^{(n-1)N_i-1} A_{v,ij}\varphi_{i,nN_i-j}(z) + \sum_{\lambda=1}^{\Lambda} A_{v,\lambda}\varphi_\lambda \ .$$

We will patch the coefficients $A_{v,ij}$ with $(k-1)N_i \le j < kN_i$ in $\prec_N$ order. Given (i,j) we can uniquely write

$$nN_i - j \;=\; r_{ij} + (n-k-1)N_i, \quad \text{with } N_i + 1 \le r_{ij} \le 2N_i \ ;$$

thus $r_{ij} = (k+1)N_i - j$. Put

$$\vartheta_{v,ij}^{(k)}(z) \;=\; \varphi_{i,r_{ij}} \cdot \phi_v(z)^{n-k-1} \;=\; \varphi_{i,(k+1)N_i-j} \cdot \phi_v(z)^{n-k-1} \ . \tag{10.10}$$

Then $\vartheta_{v,ij}^{(k)}(z)$ has pole of order $nN_i - j$ at x_i, with leading coefficient $\widetilde{c}_{v,i}^{\,n-k-1}$, and a pole of order at most $(n-k-1)N_{i'}$ at each $x_{i'} \ne x_i$. As a collection, the $\vartheta_{v,ij}^{(k)}(z)$ are K_v-symmetric. By the definition of M_v, for each (i,j),

$$\|\vartheta_{v,ij}^{(k)}(z)\|_{E_v} \;\le\; M_v R_v^{(n-k-1)N} \ . \tag{10.11}$$

Since $1 \le k \le \overline{k}$, the definition of B_v shows that $B_v M_v < R_v^{(k+1)N}$. By hypothesis, $\{\Delta_{v,ij}^{(k)} \in L_{w_v}\}_{(i,j)\in \mathrm{Band}_N(k)}$ is a K_v-symmetric collection of numbers such that $|\Delta_{v,ij}^{(k)}|_v \le B_v$ for each (i,j). Thus

$$\|\Delta_{v,ij}^{(k)}\vartheta_{v,ij}^{(k)}(z)\|_{E_v} \;\le\; B_v M_v R_v^{(n-k-1)N} \;<\; R_v^{nN} \ . \tag{10.12}$$

Put

$$G_v^{(k)}(z) \;=\; G_v^{(k-1)}(z) + \sum_{i=1}^{m} \sum_{j=(k-1)N_i}^{kN_i-1} \Delta_{v,ij}^{(k)} \cdot \vartheta_{v,ij}(z) \ . \tag{10.13}$$

Let $H(z)$ denote the sum on the right in (10.13). $H(z)$ is K_v-rational, since $\mathrm{char}(K_v) = 0$ and the $\widetilde{c}_{v,i}$, $\Delta_{v,ij}^{(k)}$ and $\vartheta_{v,ij}(z)$ are defined over L_{w_v} and are K_v-symmetric. It follows that $G_v^{(k)}(z)$ is a K_v-rational $(\mathfrak{X}, \vec{s})$-function of degree nN. Clearly $\mathrm{div}_\infty(H) \le \mathrm{div}_\infty(G_v^{(k)})$. By (10.12) and the ultrametric inequality, we have $|H(z)|_v < 1$ for each $z \in E_v$. Hence by Lemma 10.3,

$$\{z \in \mathcal{C}_v(\mathbb{C}_v) : |G_v^{(k)}(z)|_w \le R_v^{nN}\} \;=\; E_v \ .$$

This completes the patching process for high-order coefficients when $\mathrm{char}(K_v) = 0$.

Next assume $\mathrm{char}(K_v) = p > 0$. Let the $(\mathfrak{X}, \vec{s})$-function $\phi_v(z)$ of degree N and the numbers k_v, $\overline{k} \ge k_v$, M_v, and $0 < h_v < r_v \le R_v$ be as in Theorem 10.2. Let $J = p^A \ge \max(2g+1, \max_i([K_v(x_i) : K_v]^{\mathrm{insep}}))$ be the number from §3.3 in the construction of the L-rational and L^{sep}-rational bases. By assumption, $J | N_i$ for each i. The leading coefficient of $\phi_v(z)$ at x_i is $\widetilde{c}_{v,i} = \lim_{z\to x_i} \phi_v(z) \cdot g_{x_i}(z)^{N_i}$; by hypothesis, $\widetilde{c}_{v,i} \in K_v(x_i)^{\mathrm{sep}}$ for each i.

We will choose n_v and B_v differently from the way they were chosen when $\mathrm{char}(K_v) = 0$. Recall that k_v is the least integer for which

$$\Big(\frac{h_v}{R_v}\Big)^{k_v} \cdot M_v \; < \; 1 \; ,$$

and that $\overline{k} \geq k_v$ is a a fixed integer (specified by the global patching process). Let $n_v = p^r$ be the least power of p for which

$$p^r \; \geq \; \max(\overline{k}N_1, \ldots, \overline{k}N_m) \; , \tag{10.14}$$

and let

$$B_v \; = \; \frac{R_v^{n_v N}}{2\max_i \big(|\widetilde{c}_{v,i}^{n_v}|_v \|\varphi_{i,n_vN_i}\|_{E_v}\big)} \; . \tag{10.15}$$

We now show that for all sufficiently large n divisible by n_v, we can carry out the patching process imposing the conditions in the theorem. In fact, we will see that we can take $n = n_v \cdot Q = p^r \cdot Q$ for any integer $Q > \max(3, \overline{k})$.

Given such an n, put $G_v^{(0)}(z) = \phi_v(z)^n$. We first show that $G_v^{(0)}(z)$ has the properties in the theorem. It is clear that $E_v = \{z \in \mathcal{C}_v(\mathbb{C}_v) : |G_v^{(0)}(z)|_v \leq R_v^{nN}\}$. For each i, the leading coefficient of $G_v^{(0)}(z)$ at x_i is $\widetilde{c}_{v,i}^{\,n}$, which belongs to $K_v(x_i)^{\mathrm{sep}}$ since $\widetilde{c}_{v,i} \in K_v(x_i)^{\mathrm{sep}}$. If we expand $\phi_v(z)^Q$ using the L-rational basis as

$$\phi_v(z)^Q \; = \; \sum_{i=1}^{m} \sum_{j=0}^{(Q-1)N_i-1} B_{v,ij}\varphi_{i,QN_i-j}(z) \; + \; \sum_{\lambda=1}^{\Lambda} B_\lambda \varphi_\lambda(z) \; ,$$

then since $\mathrm{char}(K_v) = p > 0$ and $n_v = p^r$ it follows that
(10.16)

$$G_v^{(0)}(z) \; = \; (\phi_v(z)^Q)^{n_v} \; = \; \sum_{i=1}^{m} \sum_{j=0}^{(Q-1)N_i-1} B_{v,ij}^{n_v}\varphi_{i,QN_i-j}(z)^{n_v} \; + \; \sum_{\lambda=1}^{\Lambda} B_\lambda^{n_v} \varphi_\lambda^{n_v} \; .$$

Since $J|N_i$ for each i, Proposition 3.3(B) shows that

$$\varphi_{i,QN_i}(z)^{n_v} \; = \; \varphi_{i,nN_i}(z)$$

belongs to the L-rational basis. On the other hand if D is the divisor $\sum_{i=1}^m N_i(x_i)$, then since $n_v \geq \overline{k}N_i$ for each i, and since $Q \geq 2$, all other terms in the expansion (10.16) belong to $\Gamma_{\mathbb{C}_v}((n-\overline{k})D)$. Since $J|N_i$ for each i, this means that when we expand $G_v^{(0)}(z)$ in terms of the L-rational basis as

$$G_v^{(0)}(z) \; = \; \sum_{i=1}^{m} \sum_{j=0}^{(n-1)N_i-1} A_{v,ij}\varphi_{i,nN_i-j}(z) + \sum_{\lambda=1}^{\Lambda} A_{v,\lambda}\varphi_\lambda \; , \tag{10.17}$$

then $A_{v,ij} = 0$ for all (i,j) with $1 \leq j < \overline{k}N_i$.

We next carry out the patching process for the stage $k = 1$. We want to modify the leading coefficients $A_{v,i0}$ and leave the remaining-order high coefficients $A_{v,ij}$ for $1 \leq j < \overline{k}N_i$ (which are 0) unchanged. By assumption, we are given a K_v-symmetric set of numbers $\{\Delta_{v,ij}^{(1)} \in L_{w_v}\}_{1 \leq i \leq m, 0 \leq j < N_i}$, with $|\Delta_{v,i0}^{(1)}|_v \leq B_v$ for each i, and $\Delta_{v,ij}^{(1)} = 0$ for all $j \geq 1$ and all i, such that $\Delta_{v,i0}^{(1)}$ belongs to $K_v(x_i)^{\mathrm{sep}}$ for each i, and we wish to replace $A_{v,i0} = \widetilde{c}_{v,i}^{\,n}$ with $\widetilde{c}_{v,i}^{\,n} + \Delta_{v,i0}\widetilde{c}_{v,i}^{\,n}$ in (10.17).

Recall that $n = n_v Q$. We claim that setting

$$\widetilde{\theta}_{v,i0}^{(1)}(z) \;=\; \widetilde{c}_{v,i}^{n_v} \varphi_{i,n_v N_i} \cdot \phi_v(z)^{n_v(Q-1)} \tag{10.18}$$

in Theorem 10.2 for each $i = 1, \ldots, m$, and then putting

$$G_v^{(1)}(z) \;=\; G_v^{(0)}(z) \;+\; \sum_{i=1}^{m} \Delta_{v,i0}^{(1)} \, \widetilde{\theta}_{v,i0}^{(1)}(z) \;, \tag{10.19}$$

accomplishes what we need. Let $H(z)$ denote the sum on the right side of (10.19).

First, adding $H(z)$ to $G_v^{(0)}(z)$ adds $\Delta_{v,i0}^{(1)} \widetilde{c}_{v,i}^{\,n}$ to $A_{v,i0}$, for each i. This follows from the fact that $\widetilde{c}_{v,i}^{n_v} \varphi_{i,n_v N_i} \cdot \phi_v(z)^{n_v(Q-1)}$ has a pole of order nN_i at x_i with leading coefficient $\widetilde{c}_{v,i}^{\,n}$, and at each $x_{i'} \neq x_i$ its pole has order less than $(n - \overline{k})N_{i'}$.

Second, adding $H(z)$ to $G_v^{(0)}(z)$ leaves $A_{v,ij} = 0$ for $1 \le j < \overline{k} N_i$. This follows by considering an expansion of $\varphi_{i,n_v N_i} \cdot \big(\phi_v(z)^{(Q-1)}\big)^{n_v}$ like the one in (10.16): if we write $\phi_v(z)^{Q-1}$ as

$$\phi_v(z)^{Q-1} \;=\; \sum_{\ell=1}^{m} \sum_{j=0}^{(Q-2)N_\ell - 1} C_{v,\ell j} \varphi_{\ell, QN_\ell - j}(z) \;+\; \sum_{\lambda=1}^{\Lambda} C_\lambda \varphi_\lambda(z) \;,$$

then

$$\begin{aligned} \widetilde{\theta}_{v,i0}^{(1)}(z) \;&=\; \varphi_{i,n_v N_i} \cdot \big(\phi_v(z)^{Q-1}\big)^{n_v} \\ &=\; \sum_{\ell=1}^{m} \sum_{j=0}^{(Q-2)N_\ell - 1} C_{v,\ell j}^{n_v} \cdot \varphi_{i,n_v N_i} \cdot \varphi_{\ell,(Q-1)N_\ell - j}(z)^{n_v} \\ &\qquad +\; \sum_{\lambda=1}^{\Lambda} C_\lambda^{n_v} \cdot \varphi_{i,n_v N_i} \cdot \varphi_\lambda^{n_v} \;, \end{aligned} \tag{10.20}$$

and since $n_v \ge \overline{k} N_i$ and $Q \ge 3$, all the terms in (10.20) besides the one with $(\ell, j) = (i, 0)$ belong to $\Gamma_{\mathbb{C}_v}((n - \overline{k})D)$. On the other hand, since $J | N_i$, Proposition 3.3(B) shows that that term coincides with $\widetilde{c}_{v,i}^{\,n_v(Q-1)} \varphi_{i,nN_i}(z)$. Note that $\varphi_{i,n_v N_i} = \widetilde{\varphi}_{i,n_v N_i}$ is rational over $K_v(x_i)^{\mathrm{sep}}$ by Proposition 3.3(B), since $J | N_i$. Since the $\widetilde{c}_{v,i}$ and $\varphi_{i,n_v N_i}$ are K_v-symmetric, and $\phi_v(z)^{n_v(Q-1)}$ is K_v-rational, the $\widetilde{\theta}_{v,i0}^{(1)}(z)$ are K_v-symmetric. This discussion also shows that each $\widetilde{\theta}_{v,i0}^{(1)}(z)$ belongs to $K_v(x_i)^{\mathrm{sep}}(\mathcal{C})$ and has the form $\widetilde{\theta}_{v,i0}^{(1)}(z) = \widetilde{c}_{v,i}^{\,n} \varphi_{i,nN_i}(z) + \widetilde{\Theta}_{v,i0}(1)(z)$ for an $(\mathfrak{X}, \vec{s})$-function $\widetilde{\Theta}_{v,i0}^{(1)}(z)$ with poles of order at most $(n - \overline{k})N_{i'}$ at each $x_{i'}$, as asserted in Theorem 10.2.

Third, $G_v^{(1)}(z)$ is K_v-rational. Indeed, the $\widetilde{c}_{v,i}^{n_v}$ and $\widetilde{\theta}_{v,i0}^{(1)}(z)$ are $L_{w_v}^{\mathrm{sep}}$-rational and K_v-symmetric, so $H(z)$ is K_v-rational. Since $G_v^{(0)}(z)$ is K_v-rational, so is $G_v^{(1)}(z)$.

Finally, $E_v = \{z \in \mathcal{C}_v(\mathbb{C}_v) : |G_v^{(1)}(z)|_v \le R_v^{nN}\}$. To see this, note that our choice of B_v in (10.15), and the fact that $|\Delta_{v,i0}^{(1)}|_v \le B_v$ for each i, means that $\|H(z)\|_{E_v} \le \frac{1}{2} R_v^{nN}$. Hence the claim follows by applying Lemma 10.3 to $F(z) = G_v^{(0)}(z)$ and $H(z)$, taking $R = R_v^{nN}$.

For $k = 2, \ldots, \overline{k}$, by assumption we have $\Delta_{v,ij}^{(k)} = 0$ for all (i,j), and we take $G_v^{(k)}(z) = G_v^{(k-1)}(z)$.

Phase 2. Patching the middle coefficients.

In this phase we carry out the patching process for $k = \overline{k}+1, \ldots, n-1$. Regardless of $\mathrm{char}(K_v)$, the construction is the same as the construction when $\mathrm{char}(K_v) = 0$ in Phase 1, except that for each k, instead of having $|\Delta_{v,ij}^{(k)}|_v \le B_v$ we have

$$|\Delta_{v,ij}^{(k)}|_v \le h_v^{kN} .$$

Since $k > k_v$, if we take $\vartheta_{v,ij}^{(k)}(z) \;=\; \varphi_{i,(k+1)N_i-j} \cdot \phi_v(z)^{n-k-1}$, then by condition (10.1) (resp. condition (10.4))

$$\|\Delta_{v,ij}^{(k)}\vartheta_{v,ij}(z)\|_{U_v} \;\le\; h_v^{kN} M_v R_v^{(n-k-1)N} \;\le\; \left(\frac{h_v}{r_v}\right)^{kN} \frac{M_v}{R_v^N} \cdot R_v^{nN} \;<\; R_v^{nN} .$$

Hence if

$$G_v^{(k)}(z) \;=\; G_v^{(k-1)}(z) + \sum_{i=1}^{m} \sum_{j=(k-1)N_i}^{kN_i-1} \Delta_{v,ij}^{(k)} \cdot \vartheta_{v,ij}(z) , \tag{10.21}$$

then as before, by Lemma 10.3,

$$E_v \;=\; \{z \in \mathcal{C}_v(\mathbb{C}_v) : |G_v^{(k)}(z)|_v \le R_v^{nN}\} .$$

When $\mathrm{char}(K_v) = 0$, the sum on the right in (10.21) is K_v-rational for the same reasons of K_v-symmetry as in Phase 1. When $\mathrm{char}(K_v) = p > 0$, it can be written

$$\Big(\sum_{i=1}^{m} \sum_{j=(k-1)N_i}^{kN_i-1} \Delta_{v,ij}^{(k)} \cdot \varphi_{v,(k+1)N_i-j}(z)\Big) \cdot \phi_v(z)^{n-k-1}$$

which is K_v-rational by assumption (10.6). Thus $G_v^{(k)}(z)$ is K_v-rational.

Phase 3. Patching the low-order coefficients.

In the final step we take

$$G_v^{(n)}(z) \;=\; G_v^{(n-1)}(z) + \sum_{\lambda=1}^{\Lambda} \Delta_{v,\lambda}^{(n)} \varphi_\lambda$$

with K_v-symmetric $\Delta_{v,\lambda} \in L_{w_v}$ satisfying $|\Delta_{v,\lambda}|_v \le h_v^{nN}$ for each λ. When $\mathrm{char}(K_v) = 0$, the sum $\sum_{\lambda=1}^{\Lambda} \Delta_{v,\lambda}^{(n)} \varphi_\lambda$ is K_v-rational by the K_v-symmetry of the $\Delta_{v,\lambda}^{(n)}$ and the φ_λ. When $\mathrm{char}(K_v) = p > 0$, it is K_v-rational by assumption (10.8). Thus $G_v^{(n)}(z)$ is K_v-rational.

Since $n > k_v$, condition (10.1) (resp. condition (10.4)) shows that for each λ,

$$\|\Delta_{v,ij}^{(n)}\varphi_\lambda(z)\|_{E_v} \;\le\; h_v^{nN} \cdot M_v \;\le\; (h_v/r_v)^{nN} M_v \cdot R_v^{nN} \;<\; R_v^{nN} .$$

Hence by Lemma 10.3, as before,

$$E_v \;=\; \{z \in \mathcal{C}(\mathbb{C}_v) : |G_v^{(n)}(z)|_v \le R_v^{nN}\} .$$

This completes the proof. □

CHAPTER 11

Local Patching for Nonarchimedean K_v-simple Sets

In this section we give the confinement argument for Theorem 4.2 when K_v is nonarchimedean, and $E_v \subset \mathcal{C}_v(\mathbb{C}_v)$ is K_v-simple, hence compact. This construction is the most intricate of the four confinement arguments, and uses results from Appendices C and D. It breaks into two cases, when $\mathrm{char}(K_v) = 0$ and when $\mathrm{char}(K_v) = p > 0$.

Let q_v be the order of the residue field of K_v. Let w_v be the distinguished place of $L = K(\mathfrak{X})$ determined by the embedding $\widetilde{K} \hookrightarrow \mathbb{C}_v$ used to identify $\mathfrak{X}$ with a subset of $\mathcal{C}_v(\mathbb{C}_v)$, and view L_{w_v} as a subset of $\mathbb{C}_v$. Using the construction of the coherent approximating functions in Theorems 7.11 and 7.16, we begin with the following data:

(1) A K_v-symmetric probability vector $\vec{s} \in \mathcal{P}^m(\mathbb{Q})$ with positive rational coefficients.

(2) A compact, K_v-simple set $E_v \subset \mathcal{C}_v(\mathbb{C}_v)$ equipped with a K_v-simple decomposition $E_v = \bigcup_{\ell=1}^{D_v} \big(B(a_\ell, r_\ell) \cap \mathcal{C}_v(F_{w_\ell})\big)$ such that $U_v := \bigcup_{\ell=1}^{D_v} B(a_\ell, r_\ell)$ is disjoint from $\mathfrak{X}$ (to ease notation, we henceforth write D for D_v). In particular,
   - (a) the balls $B(a_1, r_1), \ldots, B(a_D, r_D)$ are pairwise disjoint and isometrically parametrizable; each $B(a_\ell, r_\ell)$ is disjoint from $\mathfrak{X}$;
   - (b) the set of balls $\{B(a_\ell, r_\ell)\}_{1 \le \ell \le D}$ is stable under $\mathrm{Aut}_c(\mathbb{C}_v/K_v)$. For each $\sigma \in \mathrm{Aut}_c(\mathbb{C}_v/K_v)$, if $\sigma(B(a_j, r_j)) = B(a_k, r_k)$, then $\sigma(F_{w_j}) = F_{w_k}$. For each ℓ, F_{w_ℓ}/K_v is a finite and separable, $a_\ell \in \mathcal{C}_v(F_{w_\ell})$, $r_\ell \in |F_{w_\ell}^\times|_v$, and $B(a_\ell, r_\ell)$ has exactly $[F_{w_\ell} : K_v]$ conjugates under $\mathrm{Aut}_c(\mathbb{C}_v/K_v)$.

(3) A number N and an $(\mathfrak{X}, \vec{s})$-function $\phi_v(z) \in K_v(\mathcal{C})$ of degree N such that
   - (a) the zeros $\theta_1, \ldots, \theta_N$ of $\phi_v(z)$ are distinct and belong to E_v;
   - (b) $\phi_v^{-1}(D(0,1)) = \bigcup_{h=1}^N B(\theta_h, \rho_h)$, where $B(\theta_1, \rho_1), \ldots, B(\theta_N, \rho_N)$ are disjoint, isometrically parametrizable, and contained in $\bigcup_{\ell=1}^D B(a_\ell, r_\ell)$.
   - (c) $H_v := \phi_v^{-1}(D(0,1)) \cap E_v$ has the K_v-simple decomposition
   $$H_v = \bigcup_{h=1}^{N} \big(B(\theta_h, \rho_h) \cap \mathcal{C}_v(F_{u_h})\big)$$
   compatible with the K_v-simple decomposition of E_v above. This decomposition is move-prepared relative to $B(a_1, r_1), \ldots, B(a_D, r_D)$. For each $\ell = 1, \ldots, D$, there is a point $\overline{w}_\ell \in \big(B(a_\ell, r_\ell) \cap \mathcal{C}_v(F_{w_\ell})\big) \backslash H_v$ (see Definition 6.10).
   - (d) For each $h = 1, \ldots, N$, ϕ_v induces an F_{u_h}-rational scaled isometry from $B(\theta_h, \rho_h)$ onto $D(0,1)$ which takes $B(\theta_h, \rho_h) \cap \mathcal{C}_v(F_{u_h})$ onto $\mathcal{O}_{u_h}$, with $\phi_v(\theta_h) = 0$,

Put $N_i = Ns_i$ for each i, and write $\widetilde{c}_{v,i} = \lim_{z \to x_i} \phi_v(z) \cdot g_{x_i}(z)^{N_i}$ for the leading coefficient of $\phi_v(z)$ at x_i. If $\mathrm{char}(K_v) = p > 0$, the number J from the construction of the L-rational and L^{sep}-rational bases in §3.3 divides N_i, and the leading coefficient $\widetilde{c}_{v,i}$ belongs to $K_v(x_i)^{\mathrm{sep}}$, for each $i = 1, \ldots, m$.

(4) Parameters $0 < h_v < r_v < 1$ satisfying $h_v^N < r_v^N < q_v^{-1/(q_v-1)} < 1$.

(5) An order $\prec_N$ on the index set $\mathcal{I} = \{(i,j) \in \mathbb{Z}^2 : 1 \le i \le m, 0 \le j\}$ determined by N and $\vec{s}$ as in (7.41), which gives the sequence in which coefficients are patched. The group $\mathrm{Aut}_c(\mathbb{C}/K_v)$ acts on $\mathcal{I}$, and we write $\mathrm{Aut}_c(\mathbb{C}/K_v)(i,j)$ for the orbit of $(i,j) \in \mathcal{I}$.

We will use the L-rational basis $\{\varphi_{ij}, \varphi_\lambda\}$ from §3.3 to expand functions, and $\Lambda = \dim_K(\Gamma(\sum_{i=1}^m N_i(x_i)))$ will be the number of low-order basis elements, as in §7.4. The order $\prec_N$ respects the N-bands (7.42), and for each $x_i \in \mathfrak{X}$, specifies the terms to be patched in decreasing pole order.

As in Definition 3.39, let the Stirling polynomial of degree n for $\mathcal{O}_v$ be

$$S_{n,v}(x) = \prod_{j=0}^{n-1} (x - \psi_v(j)) , \tag{11.1}$$

where $\{\psi_v(j)\}_{0 \le j < \infty}$ is the basic well-distributed sequence in $\mathcal{O}_v$.

When $\mathrm{char}(K_v) = 0$, we will prove the following patching theorem.

THEOREM 11.1. *Suppose K_v is nonarchimedean, and $\mathrm{char}(K_v) = 0$. Let E_v be K_v-simple, with a K_v-simple decomposition $E_v = \bigcup_{\ell=1}^D \big(B(a_\ell, r_\ell) \cap C_v(F_{w_\ell})\big)$ such that $U_v := \bigcup_{\ell=1}^D B(a_\ell, r_\ell)$ is disjoint from $\mathfrak{X}$. Let $\vec{s} \in \mathcal{P}^m(\mathbb{Q})$ be a K_v-symmetric probability vector with positive coefficients, and let $\phi_v(z) \in K_v(\mathcal{C})$ be an $(\mathfrak{X}, \vec{s})$-function of degree N with distinct zeros $\theta_1, \ldots, \theta_N \in E_v$, such that:*

(1) $\phi_v^{-1}(D(0,1)) = \bigcup_{k=1}^N B(\theta_k, \rho_k)$, where the balls $B(\theta_1, \rho_1), \ldots, B(\theta_N, \rho_N)$ are pairwise disjoint, isometrically parametrizable, and contained in $\bigcup_{\ell=1}^D B(a_\ell, r_\ell)$.

(2) The set $H_v := \phi_v^{-1}(D(0,1)) \cap E_v$ has the K_v-simple decomposition

$$H_v = \bigcup_{h=1}^N \big(B(\theta_h, \rho_h) \cap C_v(F_{u_h})\big)$$

compatible with the K_v-simple decomposition $\bigcup_{\ell=1}^D \big(B(a_\ell, r_\ell) \cap C_v(F_{w_\ell})\big)$ of E_v, and is move-prepared relative to $B(a_1, r_1), \ldots, B(a_D, r_D)$. In particular, if $\theta_h \in B(a_\ell, r_\ell) \cap C_v(F_{w_\ell})$, then $F_{u_h} = F_{w_\ell}$, $\rho_h \in |F_{w_\ell}^\times|_v$, and $B(\theta_h, \rho_h) \subseteq B(a_\ell, r_\ell)$. For each $\ell = 1, \ldots, D$, there is a point $\overline{w}_\ell \in \big(B(a_\ell, r_\ell) \cap C_v(F_{w_\ell}))\big) \backslash H_v$.

(3) For each $h = 1, \ldots, N$, ϕ_v induces an F_{u_h}-rational scaled isometry from $B(\theta_h, \rho_h)$ onto $D(0,1)$, which takes $B(\theta_h, \rho_h) \cap C_v(F_{u_h})$ onto $\mathcal{O}_{u_h}$.

Let $0 < h_v < r_v < 1$ be numbers satisfying

$$h_v^N < r_v^N < q_v^{-1/(q_v-1)} < 1 . \tag{11.2}$$

Put

$$M_v = \max\big(\max_{\substack{1 \le i \le m \\ N_i < j \le 2N_i}} \|\varphi_{ij}\|_{U_v}, \max_{1 \le \lambda \le \Lambda} \|\varphi_\lambda\|_{U_v}\big) , \tag{11.3}$$

and let $k_v > 0$ be the least integer such that for all $k \ge k_v$,

$$h_v^{kN} \cdot M_v \;<\; q_v^{-\frac{k+1}{q_v-1}-\log_v(k+1)} \;. \tag{11.4}$$

Let $\overline{k} \ge k_v$ be a fixed integer, and put $B_v = h_v^{\overline{k}N}$. Then there is an integer $n_v \ge 1$ such that for each sufficiently large integer n divisible by n_v, the local patching process at K_v can be carried out as follows:

Put $G_v^{(0)}(z) = S_{n,v}(\phi_v(z))$.

For each k, $1 \le k < n$, let $\{\Delta_{v,ij}^{(k)} \in \mathbb{C}_v\}_{(i,j)\in \mathrm{Band}_N(k)}$ be an arbitrary K_v-symmetric set of numbers given recursively in $\prec_N$ order, subject to the condition that for each (i,j),

$$|\Delta_{v,ij}^{(k)}|_v \;\le\; \begin{cases} B_v & \text{if } k \le \overline{k}\,, \\ h_v^{kN} & \text{if } k > \overline{k}\,. \end{cases} \tag{11.5}$$

For $k = n$, let $\{\Delta_{v,\lambda}^{(n)} \in \mathbb{C}_v\}_{1\le\lambda\le\Lambda}$ be an arbitrary K_v-symmetric set of numbers satisfying

$$|\Delta_{v,\lambda}^{(n)}|_v \;\le\; h_v^{nN} \;. \tag{11.6}$$

Then one can inductively construct $(\mathfrak{X},\vec{s})$-functions $G_v^{(1)}(z),\dots,G_v^{(n)}(z)$ in $K_v(\mathcal{C})$, of common degree Nn, such that

(A) For each $k = 1,\dots,n-1$,

$$G_v^{(k)}(z) \;=\; G_v^{(k-1)}(z) + \sum_{(i,j)\in \mathrm{Band}_N(k)} \Delta_{v,ij}^{(k)} \cdot \vartheta_{v,ij}^{(k)}(z) \;+\; \Theta_v^{(k)}(z)\;,$$

where $\vartheta_{v,ij}^{(k)}(z) = \varphi_{i,(k+1)N_i-j} \cdot F_{v,k}(z)$ with an $(\mathfrak{X},\vec{s})$ function $F_{v,k}(z) \in K_v(\mathcal{C})$ of degree $(n-k-1)$ independent of (i,j) whose roots belong to E_v, and $\Theta_v^{(k)}(z) \in K_v(\mathcal{C})$ has a pole of order at most $(n-k)N_i$ at each x_i and no other poles. For each i, the leading coefficient $d_{v,i}$ of $F_k(z)$ at x_i belongs to $K_v(x_i)$ and has absolute value $|d_{v,i}|_v = |\widetilde{c}_{v,i}|_v^{n-k-1}$. For $k = n$,

$$G_v^{(n)}(z) \;=\; G_v^{(n-1)}(z) + \sum_{\lambda=1}^{\Lambda} \Delta_{v,\lambda}^{(n)} \cdot \varphi_\lambda(z)\;.$$

In particular:

(1) Each $\vartheta_{v,ij}^{(k)}(z)$ is in $K_v(x_i)(\mathcal{C})$, has a pole of order $nN_i - j > (n-k-1)N_i$ at x_i with leading coefficient $\widetilde{c}_{v,i}^{\,n-k-1}$, has poles of order at most $(n-k-1)N_{i'}$ at each $x_{i'} \ne x_i$, and has no other poles.

(2) The $\vartheta_{v,ij}^{(k)}(z)$ are K_v-symmetric.

(3) In passing from $G_v^{(0)}(z)$ to $G_v^{(1)}(z)$, the leading coefficient $\widetilde{c}_{v,i}^{\,n}$ of $G_v^{(0)}(z)$ at x_i is replaced with $\widetilde{c}_{v,i}^{\,n} + \Delta_{v,i0}^{(1)} \cdot \widetilde{c}_{v,i}^{\,n-2}$, for each i.

(B) For each $k = 0,1,\dots,n$, the zeros of $G_v^{(k)}(z)$ belong to E_v, and for $k = 0$ and $k = n$ they are distinct. Furthermore,

$$\{z \in \mathcal{C}_v(\mathbb{C}_v) : G_v^{(n)}(z) \in \mathcal{O}_v \text{ and } |G_v^{(n)}(z)|_v \le r_v^{Nn}\} \;\subseteq\; E_v\;.$$

When $\mathrm{char}(K_v) = p > 0$, we have the following patching theorem. The K_v-rationality assumptions (11.11), (11.13) in the theorem are addressed by the global patching process.

THEOREM 11.2. *Suppose K_v is nonarchimedean, and* $\mathrm{char}(K_v) = p > 0$. *Let E_v be K_v-simple, with a K_v-simple decomposition $E_v = \bigcup_{\ell=1}^{D} \big(B(a_\ell, r_\ell) \cap \mathcal{C}_v(F_{w_\ell})\big)$ such that $U_v := \bigcup_{\ell=1}^{D} B(a_\ell, r_\ell)$ is disjoint from $\mathfrak{X}$. Let $\vec{s} \in \mathcal{P}^m(\mathbb{Q})$ be a K_v-symmetric probability vector with positive coefficients, and let $\phi_v(z) \in K_v(\mathcal{C})$ be an $(\mathfrak{X}, \vec{s})$-function of degree N with distinct zeros $\theta_1, \ldots, \theta_N \in E_v$, such that*

(1) $\phi_v^{-1}(D(0,1)) = \bigcup_{h=1}^{N} B(\theta_h, \rho_h)$, where the balls $B(\theta_1, \rho_1), \ldots, B(\theta_N, \rho_N)$ are pairwise disjoint, isometrically parametrizable, and contained in $\bigcup_{\ell=1}^{D} B(a_\ell, r_\ell)$;

(2) $H_v := \phi_v^{-1}(D(0,1)) \cap E_v$ is K_v-simple, has a K_v-simple decomposition

$$H_v = \bigcup_{h=1}^{N} \big(B(\theta_h, \rho_h) \cap \mathcal{C}_v(F_{u_h})\big)$$

compatible with the K_v-simple decomposition $\bigcup_{\ell=1}^{D} \big(B(a_\ell, r_\ell) \cap \mathcal{C}_v(F_{w_\ell})\big)$ of E_v, and is move-prepared relative to $B(a_1, r_1), \ldots, B(a_D, r_D)$. In particular, if $\theta_h \in B(a_\ell, r_\ell) \cap \mathcal{C}_v(F_{w_\ell})$, then $F_{u_h} = F_{w_\ell}$, $\rho_h \in |F_{w_\ell}^\times|_v$, and $B(\theta_h, \rho_h) \subseteq B(a_\ell, r_\ell)$. For each $\ell = 1, \ldots, D$, there is a point $\overline{w}_\ell \in \big(B(a_\ell, r_\ell) \cap \mathcal{C}_v(F_{w_\ell}))\big) \backslash H_v$.

(3) For each $h = 1, \ldots, N$, ϕ_v induces an F_{u_h}-rational scaled isometry from $B(\theta_h, \rho_h)$ onto $D(0,1)$, which takes $B(\theta_h, \rho_h) \cap \mathcal{C}_v(F_{u_h})$ onto $\mathcal{O}_{u_h}$.

Let $0 < h_v < r_v < 1$ and M_v be numbers satisfying

$$h_v^N < r_v^N < q_v^{-1/(q_v-1)} < 1 , \tag{11.7}$$

$$M_v = \max\Big(\max_{\substack{1 \le i \le m \\ N_i < j \le 2N_i}} \|\varphi_{ij}\|_{U_v}, \max_{1 \le \lambda \le \Lambda} \|\varphi_\lambda\|_{U_v} \Big) . \tag{11.8}$$

Let $k_v > 0$ be the least integer such that for all $k \ge k_v$,

$$h_v^{kN} \cdot M_v < q_v^{-\frac{k+1}{q_v-1} - \log_v(k+1)} . \tag{11.9}$$

Let $\overline{k} \ge k_v$ be a fixed integer. Then there are an integer $n_v \ge 1$ and a number $0 < B_v < 1$ such that for each sufficiently large integer n divisible by n_v, the local patching process at K_v can be carried out as follows:

Put $G_v^{(0)}(z) = S_{n,v}(\phi_v(z))$.

Then the leading coefficient of $G_v^{(0)}(z)$ at x_i is $\widetilde{c}_{v,i}^{\,n}$, the zeros of $G_v^{(0)}(z)$ are distinct and belong to E_v, and when $G_v^{(0)}(z)$ is expanded in terms of the L-rational basis as

$$G_v^{(0)}(z) = \sum_{i=1}^{m} \sum_{j=0}^{(n-1)N_i-1} A_{v,ij} \varphi_{i,nN_i-j}(z) + \sum_{\lambda=1}^{\Lambda} A_{v,\lambda} \varphi_\lambda ,$$

we have $A_{v,ij} = 0$ for all (i,j) with $1 \le j < \overline{k} N_i$.

For each k, $1 \le k \le n-1$, let $\{\Delta_{v,ij}^{(k)} \in L_{w_v}\}_{(i,j) \in \mathrm{Band}_N(k)}$ be a K_v-symmetric set of numbers satisfying

$$\begin{cases} |\Delta_{v,i0}^{(1)}|_v \le B_v \text{ and } \Delta_{v,ij}^{(1)} = 0 \text{ for } j = 1, \ldots, N_i - 1, & \text{if } k = 1 , \\ \Delta_{v,ij}^{(k)} = 0 \text{ for } j = (k-1)N_i, \ldots, kN_i - 1, & \text{if } k = 2, \ldots, \overline{k} , \\ |\Delta_{v,ij}^{(k)}|_v \le h_v^{kN}, & \text{if } k = \overline{k}+1, \ldots, n-1 , \end{cases} \tag{11.10}$$

such that $\Delta_{v,i0}^{(1)} \in K_v(x_i)^{\text{sep}}$ *for each* i *and such that for* $k = \overline{k}+1, \ldots, n-1$,

$$\Delta v, k(z) \ := \ \sum_{i=1}^{m} \sum_{j=(k-1)N_i}^{kN_i-1} \Delta_{v,ij}^{(k)} \cdot \varphi_{i,(k+1)N_i-j} \ \in \ K_v(\mathcal{C}) \ . \tag{11.11}$$

For $k = n$, *let* $\{\Delta_{v,\lambda}^{(n)} \in L_{w_v}\}_{1 \le \lambda \le \Lambda}$ *be a* K_v*-symmetric set of numbers such that*

$$|\Delta_{v,\lambda}^{(n)}|_v \ \le \ h_v^{nN} \tag{11.12}$$

for each λ, *such that*

$$\Delta_{v,n}(z) \ := \ \sum_{\lambda=1}^{\Lambda} \Delta_{v,\lambda}^{(n)} \cdot \varphi_\lambda \ \in K_v(\mathcal{C}) \ . \tag{11.13}$$

Then one can inductively construct $(\mathfrak{X}, \vec{s})$*-functions* $G_v^{(1)}(z), \ldots, G_v^{(n)}(z)$ *in* $K_v(\mathcal{C})$, *of common degree* Nn, *such that*

(*A*) *For each* $k = 1, \ldots, n$, $G_v^{(k)}(z)$ *is obtained from* $G_v^{(k-1)}(z)$ *as follows:*

(1) *When* $k = 1$, *there is a* K_v*-symmetric set of functions* $\widetilde{\theta}_{v,10}^{(1)}(z), \ldots, \widetilde{\theta}_{v,m0}^{(1)}(z)$ *in* $L_{w_v}^{\text{sep}}(\mathcal{C})$ *such that*

$$G_v^{(1)}(z) \ = \ G_v^{(0)}(z) + \sum_{i=1}^{m} \Delta_{v,i0}^{(1)} \cdot \widetilde{\theta}_{v,i0}^{(1)}(z) \ ,$$

where for each $i = 1, \ldots, m$, $\widetilde{\theta}_{v,i0}^{(1)}(z) \in K_v(x_i)^{\text{sep}}(\mathcal{C})$ *has the form*

$$\widetilde{\theta}_{v,i0}^{(1)}(z) \ = \ \widetilde{c}_{v,i}^{\,n} \varphi_{i,nN_i}(z) + \widetilde{\Theta}_{v,i0}^{(1)}(z)$$

for an $(\mathfrak{X}, \vec{s})$*-function* $\widetilde{\Theta}_{v,i0}^{(1)}(z)$ *with a pole of order at most* $(n-\overline{k})N_{i'}$ *at each* $x_{i'}$. *Thus, in passing from* $G_v^{(0)}(z)$ *to* $G_v^{(1)}(z)$, *each leading coefficient* $A_{v,i0} = \widetilde{c}_{v,i}^{\,n}$ *is replaced with* $\widetilde{c}_{v,i}^{\,n} + \Delta_{v,i0}^{(1)} \cdot \widetilde{c}_{v,i}^{\,n}$, *and the coefficients* $A_{v,ij}$ *for* $1 \le j < \overline{k}N_i$ *remain* 0.

(2) *For* $k = 2, \ldots, \overline{k}$, *we have* $G_v^{(k)}(z) = G_v^{(k-1)}(z)$.

(3) *For* $k = \overline{k}+1, \ldots, n-1$, *we have*

$$G_v^{(k)}(z) \ = \ G_v^{(k-1)}(z) + \Delta_{v,k}(z) \cdot F_{v,k}(z) + \Theta_v^{(k)}(z) \ , \tag{11.14}$$

where

(*a*) $\omega_v^{(k)}(z) = \sum_{(i,j) \in \text{Band}_N(k)} \Delta_{v,ij}^{(k)} \varphi_{i,(k+1)N_i-j}(z)$.

(*Here* $\omega_v^{(k)}(z)$ *belongs to* $K_v(\mathcal{C})$ *by* (11.13)).

(*b*) $F_{v,k}(z) = \phi_v(z)^{n-k-1}$ *is a* K_v*-rational* $(\mathfrak{X}, \vec{s})$*-function with roots in* E_v. *It has a pole of order* $(n-k-1)N_i$ *at each* x_i, *and its leading coefficient* $d_{v,i} = \lim_{z \to x_i} F_{v,k}(z) \cdot g_{x_i}(z)^{(n-k-1)N_i}$ *belongs to* $K_v(x_i)^{\text{sep}}$ *and has absolute value* $|d_{v,i}|_v = |\widetilde{c}_{v,i}|_v^{n-k-1}$.

(*c*) $\Theta_v^{(k)}(z) \in K_v(\mathcal{C})$ *is an* $(\mathfrak{X}, \vec{s})$*-function determined by the local patching process at* v *after the coefficients in* $\text{Band}_N(k)$ *have been modified; it has a pole of order at most* $(n-k)N_i$ *at each* x_i *and no other poles, and may be the zero function.*

(4) *For* $k = n$,

$$G_v^{(n)}(z) \ = \ G_v^{(n-1)}(z) + \Delta_{v,n}(z) \ .$$

(B) For each $k = 1, \ldots, n$, the zeros of $G_v^{(k)}(z)$ belong to E_v, and for $k = n$ they are distinct. Furthermore,

$$\{z \in \mathcal{C}_v(\mathbb{C}_v) : G_v^{(n)}(z) \in \mathcal{O}_v \text{ and } |G_v^{(n)}(z)|_v \le r_v^{Nn}\} \subseteq E_v .$$

Remark. In both theorems, we will have $\Theta_v^{(k)}(z) = 0$ for all but one value $k = k_1$. For $k = k_1$, after computing

$$G_v^{(k_1)}(z) = G_v^{(k_1-1)}(z) + \sum_{(i,j)\in \mathrm{Band}_N(k_1)} \Delta_{v,ij}^{(k_1)} \vartheta_{v,ij}^{(k_1)}(z)$$

we "move the roots of $G_v^{(k_1)}(z)$ apart", constructing an $(\mathfrak{X}, \vec{s})$-function $\widehat{G}_v^{(k_1)}(z)$ whose patched coefficients are the same as those of $G_v^{(k_1)}(z)$, but whose roots are well-separated, and replace $G_v^{(k_1)}(z)$ by $\widehat{G}_v^{(k_1)}(z)$. The function $\Theta_v^{(k_1)}(z) = \widehat{G}_v^{(k_1)}(z) - G_v^{(k_1)}(z)$ is chosen to accomplish this change: see Phase 3 in the proof, given in §11.3 below.

Theorems 11.1 and 11.2 will be proved together. There are some differences in the way the leading and high-order coefficients are treated, but the underlying patching constructions for the middle and low-order coefficients are the same.

Given a positive integer n, we begin the patching process by composing $\phi_v(z)$ with the Stirling polynomial $S_{n,v}(x)$. This yields the K_v-rational function

$$G_v^{(0)}(z) := S_{n,v}(\phi_v(z)) = \prod_{j=0}^{n-1} (\phi_v(z) - \psi_v(j)) .$$

An important observation is that $S_{n,v}(\phi_v(z))$ is highly factorized, and by taking products $\prod_{j\in\mathcal{S}}(\phi_v(z) - \psi_v(j))$ corresponding to subsets $\mathcal{S} \subset \{0, \ldots, n-1\}$ we can easily obtain K_v-rational functions dividing $G_v^{(0)}(z)$. On the other hand, by restricting $S_{n,v}(\phi_v(z))$ to one of the isometrically parametrizable balls $B(\theta_h, \rho_h)$ and composing it with a parametrization, we obtain a power series which behaves much like $S_{n,v}(x)$. The key to the construction is the interaction between these global and local ways of viewing $G_v^{(0)}(z)$.

Before we can give the proof, we must develop some machinery.

1. The Patching Lemmas

In this section, we consider aspects of the construction involving power series.

For each zero θ_h of $\phi_v(z)$, let $\sigma_h : D(0, \rho_h) \to B(\theta_h, \rho_h)$ be an F_{u_h}-rational isometric parametrization with $\sigma_h(0) = \theta_h$. Let $d_h \in F_{u_h}^\times$ be such that $|d_h|_v = \rho_h$, and define $\widehat{\sigma}_h : D(0, 1) \to B(\theta_h, \rho_h)$ by

$$\widehat{\sigma}_h(Z) = \sigma_h(d_h Z) .$$

Put $\widehat{\Phi}_h(Z) = \phi_v(\widehat{\sigma}_h(Z))$. Thus $\widehat{\Phi}_h(Z)$ is a power series converging on $D(0, 1)$ which induces an F_{u_h}-rational distance-preserving isomorphism from $D(0, 1)$ to itself and takes $\mathcal{O}_{u_h}$ to $\mathcal{O}_{u_h}$. It satisfies $\widehat{\Phi}_h(0) = 0$, and $|\widehat{\Phi}_h'(0)|_v = 1$. After replacing d_h by $\mu_h d_h$ for an appropriate $\mu_h \in \mathcal{O}_{u_h}^\times$, if necessary, we can assume that $\widehat{\Phi}_h'(0) = 1$. Hence we can expand

$$\widehat{\Phi}_h(Z) = Z + \sum_{i=2}^{\infty} C_{hi} Z^i \in \mathcal{O}_{u_h}[[Z]] ,$$

where $\mathrm{ord}_v(C_{hi}) > 0$ for each $i \geq 2$, and $\mathrm{ord}_v(C_{hi}) \to \infty$ as $i \to \infty$. Let $\widetilde{\Phi}_h(Z)$ be the inverse power series to $\widehat{\Phi}_h(Z)$, which is easily seen to converge on $D(0,1)$. Thus $\widetilde{\Phi}_h(\widehat{\Phi}_h(Z)) = \widehat{\Phi}_h(\widetilde{\Phi}_h(Z)) = Z$.

The restriction of $S_{n,v}(\phi_v(z))$ to $B(\theta_h, \rho_h)$ induces a function $S_{n,v}(\widehat{\Phi}_h(Z))$ on $D(0,1)$. There is a one-to-one correspondence between the zeros θ_{hj} of $S_{n,v}(\phi_v(z))$ in $B(\theta_h, \rho_h)$ and the zeros α_{hj} of $S_{n,v}(\widehat{\Phi}_h(Z))$ in $D(0,1)$, given by $\theta_{hj} = \widehat{\sigma}_h(\alpha_{hj})$; the θ_{hj} belong to $\mathcal{C}_v(F_{u_h})$, and the α_{hj} belong to $\mathcal{O}_{u_h}$. Since the zeros of $S_{n,v}(z)$ are the $\psi_v(j)$ for $j = 0, \ldots, n-1$ we have $\widehat{\Phi}_h(\alpha_{hj}) = \psi_v(j)$ (or equivalently, $\widetilde{\Phi}_h(\psi_v(j)) = \alpha_{hj}$) for each j, $0 \leq j \leq n-1$. Thus the power series

$$G_h^{(0)}(Z) \;:=\; S_{n,v}(\widehat{\Phi}_h(Z)) \;=\; \prod_{j=0}^{n-1}(\widehat{\Phi}_h(Z) - \psi_v(j))$$

can be factored as

$$G_h^{(0)}(Z) \;=\; \prod_{j=0}^{n-1}(Z - \alpha_{hj}) \cdot \mathcal{H}_h(Z)$$

where $\mathcal{H}_h(Z)$ is an invertible power series in $\mathcal{O}_{u_h}[[Z]]$. Write $q = q_v$. By (3.54), since $\widehat{\Phi}_h(Z)$ is distance-preserving, for all $0 \leq j, k < n$ with $j \neq k$,

$$\mathrm{ord}_v(\alpha_{hj} - \alpha_{hk}) \;=\; \mathrm{ord}_v(\psi_v(j) - \psi_v(k)) \;=\; \mathrm{val}_q(|j-k|)\ . \tag{11.15}$$

We will now consider how to modify the functions $G_h^{(0)}(Z)$, while keeping their zeros in $\mathcal{O}_{u_h}$. To do this, we establish a series of lemmas concerning power series, analogous to those proved for polynomials in ([**53**]). For the rest of this section, F_u/K_v will denote a finite extension in $\mathbb{C}_v$, with ring of integers $\mathcal{O}_u$.

DEFINITION 11.3. Let F_u/K_v be finite. Suppose $\mathcal{S} = \{j_1, j_1+1, \ldots, j_1+\ell-1\}$ is a sequence of ℓ consecutive nonnegative integers, and that $\widehat{\Phi} : D(0,1) \to D(0,1)$ is a distance-preserving automorphism defined by an F_u-rational power series, so $\widehat{\Phi}(\mathcal{O}_u) = \mathcal{O}_u$. A *$\psi_v$-regular sequence of length ℓ in $\mathcal{O}_u$ attached to $\mathcal{S}$ and $\widehat{\Phi}(z)$* is a sequence $\{\alpha_j\}_{j\in\mathcal{S}} \subset \mathcal{O}_u$ such that

$$\mathrm{ord}_v(\widehat{\Phi}(\alpha_j) - \psi_v(j)) \;\geq\; \log_v(\ell)$$

(or equivalently, $\mathrm{ord}_v\left(\alpha_j - \widetilde{\Phi}(\psi_v(j))\right) \geq \log_v(\ell)$), for each $j \in \mathcal{S}$.

In particular, for each h, $\{\alpha_{hj}\}_{0\leq j<n}$ is a ψ_v-regular sequence of length n in $\mathcal{O}_{u_h}$ relative to $\widehat{\Phi}_h(Z)$. In the applications, we will have $\widehat{\Phi}(Z) = \widehat{\Phi}_h(Z)$, and $\mathcal{S}$ will be a subsequence of $\{0, 1, \ldots, n-1\}$. However, often the precise power series defining the ψ_v-regular sequence in $\mathcal{O}_u$ is not important (generally all that is used is the fact the power series is an isometry), and we will frequently speak of a ψ_v-regular sequence of length ℓ, or just a ψ_v-regular sequence, if $\mathcal{S}$, $\widehat{\Phi}(Z)$, and $\mathcal{O}_u$ are understood from context. Note that each subsequence of a ψ_v-regular sequence consisting of consecutive elements, is itself a ψ_v-regular sequence.

For the rest of this section, we will assume that $\widehat{\Phi}(Z) \in \mathcal{O}_u[[Z]]$ as in Definition 11.3 has been fixed; let $\widetilde{\Phi}(Z)$ denote its inverse. If $\{\alpha_j\}_{j\in\mathcal{S}}$ is a ψ_v-regular sequence of length ℓ in $\mathcal{O}_u$ for $\widehat{\Phi}(Z)$, then for each $k \neq j \in \mathcal{S}$,

$$\mathrm{ord}_v(\alpha_k - \alpha_j) \;=\; \mathrm{val}_q(|k-j|) \tag{11.16}$$

because

$$\mathrm{ord}_v(\alpha_k - \widetilde{\Phi}(\psi_v(k))) \;\geq\; \log_v(\ell)\ , \quad \mathrm{ord}_v(\alpha_j - \widetilde{\Phi}(\psi_v(j))) \;\geq\; \log_v(\ell)\ ,$$

while by formula (3.54), since $\widetilde{\Phi}(Z)$ is distance-preserving,

$$\begin{aligned} \mathrm{ord}_v(\widetilde{\Phi}(\psi_v(k)) - \widetilde{\Phi}(\psi_v(j))) &= \mathrm{ord}_v(\psi_v(k) - \psi_v(j)) \\ &= \mathrm{val}_q(|k-j|) \ < \ \log_v(\ell) \ . \end{aligned}$$

LEMMA 11.4. *Let $\{\alpha_j\}_{j_1 \le j < j_1+\ell}$ be a ψ_v-regular sequence of length ℓ in $\mathcal{O}_u$. Given $z \in D(0,1)$, let J be an index for which $\mathrm{ord}_v(z - \alpha_J)$ is maximal. Then for each $j \ne J$ with $j_1 \le j < j_1 + \ell$, we have*

$$\mathrm{ord}_v(z - \alpha_j) \ \le \ \mathrm{ord}_v(\alpha_J - \alpha_j) \ = \ \mathrm{val}_q(|J-j|) \ .$$

PROOF. Fix $j \ne J$. By hypothesis $\mathrm{ord}_v(z - \alpha_j) \le \mathrm{ord}_v(z - \alpha_J)$. By the ultrametric inequality,

$$\mathrm{ord}_v(\alpha_J - \alpha_j) \ \ge \ \min(\mathrm{ord}_v(z-\alpha_J), \mathrm{ord}_v(z - \alpha_j)) \ = \ \mathrm{ord}_v(z - \alpha_j) \ .$$

By (11.16), $\mathrm{ord}_v(\alpha_J - \alpha_j) = \mathrm{val}_q(|J-j|)$, so we obtain the result. □

Recall from (3.59) that

$$\sum_{k=1}^{\ell} \mathrm{val}_q(k) \ = \ \frac{\ell}{q-1} - \frac{1}{q-1} \sum_{j \ge 0} d_j(\ell) \ . \tag{11.17}$$

The following lemma generalizes the parts of Proposition 3.40 we will need:

LEMMA 11.5. *Let $\{\alpha_j\}_{j \in \mathcal{S}}$ be a ψ_v-regular sequence of length ℓ in $\mathcal{O}_u$. Put $P_{\mathcal{S}}(Z) = \prod_{j \in \mathcal{S}}(Z - \alpha_j)$. Then*

(A) For each $J \in \mathcal{S}$,

$$\mathrm{ord}_v\Big(\prod_{\substack{j \in \mathcal{S} \\ j \ne J}} (\alpha_J - \alpha_j)\Big) \ < \ \frac{\ell}{q-1} \ .$$

(B) For each $z \in D(0,1)$, if $J \in \mathcal{S}$ is such that $\mathrm{ord}_v(z - \alpha_J)$ is maximal, then

$$\mathrm{ord}_v(P_{\mathcal{S}}(z)) \ < \ \frac{\ell}{q-1} + \mathrm{ord}_v(z - \alpha_J) \ .$$

PROOF. Suppose $\mathcal{S} = \{j_1, \ldots, j_1 + \ell - 1\}$. To prove (A), fix $J \in \mathcal{S}$ and recall that if $j \in \mathcal{S}$ and $j \ne J$ then $\mathrm{ord}_v(\alpha_J - \alpha_j) = \mathrm{val}_q(|J-j|)$. Hence

$$\begin{aligned} \mathrm{ord}_v\Big(\prod_{\substack{j \in \mathcal{S} \\ j \ne J}} (\alpha_J - \alpha_j)\Big) \ &= \ \sum_{j=j_1}^{J-1} \mathrm{val}_q(|J-j|) + \sum_{j=J+1}^{j_1+\ell-1} \mathrm{val}_q(|J-j|) \\ &= \ \sum_{k=1}^{J-j_1} \mathrm{val}_q(k) + \sum_{k=1}^{j_1+\ell-J-1} \mathrm{val}_q(k) \qquad (11.18) \\ &= \ \frac{(J-j_1) - \sum_{j \ge 0} d_j(J - j_1)}{q-1} \\ &\qquad + \frac{(j_1 + \ell - J - 1) - \sum_{j \ge 0} d_j(j_1 + \ell - J - 1)}{q-1} \\ &= \ \frac{\ell}{q-1} - \frac{1 + \sum_{j\ge 0} d_j(J-j_1) + \sum_{j \ge 0} d_j(j_1 + \ell - J - 1)}{q-1} \ < \ \frac{\ell}{q-1} \ . \end{aligned}$$

To prove (B), let $J \in \mathcal{S}$ be such that $\text{ord}_v(z - \alpha_J)$ is maximal, or equivalently $|z - \alpha_J|_v$ is minimal. By Lemma 11.4

$$\text{ord}_v(P_{\mathcal{S}}(z)) \ \le \ \text{ord}_v(z - \alpha_J) + \text{ord}_v(\prod_{j \ne J} (\alpha_J - \alpha_j)) \ .$$

Applying part (A), we get

$$\text{ord}_v(P_{\mathcal{S}}(z)) \ < \ \frac{\ell}{q-1} + \text{ord}_v(z - \alpha_J) \ ,$$

as required. □

We now come to the fundamental lemma governing the patching process. If $\mathcal{Q}(Z) \in \mathbb{C}_v[[Z]]$ converges in $D(0,1)$, write

$$\|\mathcal{Q}\|_{D(0,1)} \ = \ \sup_{Z \in D(0,1)} |\mathcal{Q}(Z)|_v$$

for its sup norm relative to the absolute value $|x|_v$. The lemma will be applied to functions of the form $\mathcal{Q}(Z) = \mathcal{Q}_{\mathcal{S},h}(Z) = \prod_{j \in \mathcal{S}}(\widehat{\Phi}_h(Z) - \psi_v(j))$, for appropriate (usually short) subsequences $\mathcal{S} \subset \{0, 1, \ldots, n-1\}$.

LEMMA 11.6 (Basic Patching Lemma). *Let F_u/K_v be a finite extension in $\mathbb{C}_v$, and let $\mathcal{Q}(Z) \in F_u[[Z]]$ be a power series converging on $D(0,1)$, with $\|\mathcal{Q}\|_{D(0,1)} = 1$. Suppose that $\mathcal{Q}(Z)$ has exactly ℓ roots in $D(0,1)$, and that these roots form a ψ_v-regular sequence $\{\alpha_j\}_{j \in \mathcal{S}}$ of length ℓ in $\mathcal{O}_u$ with respect to $\widehat{\Phi}(Z)$. Let $M \ge \log_v(\ell)$ be given. Then for any power series $\Delta(Z) \in F_u[[Z]]$ converging on $D(0,1)$, with* sup *norm*

$$\|\Delta\|_{D(0,1)} \ \le \ q_v^{-\frac{\ell}{q-1} - M} \ ,$$

the roots α_j^* of $\mathcal{Q}^*(Z) = \mathcal{Q}(Z) + \Delta(Z)$ again form a ψ_v-regular sequence of length ℓ in $\mathcal{O}_u$ with respect to $\widehat{\Phi}(Z)$. They can be uniquely labeled in such a way that

$$\text{ord}_v(\alpha_j^* - \alpha_j) \ > \ M$$

for each $j \in \mathcal{S}$.

PROOF. If $\mathcal{Q}(Z) = \sum_{i=0}^\infty B_i Z^i$, then since $\|\mathcal{Q}\|_{D(0,1)} = 1$ we have $\text{ord}_v(B_i) \ge 0$ for all i, and $\text{ord}_v(B_i) = 0$ for some i. Since $\mathcal{Q}(Z)$ converges in the disc $D(0,1)$,

$$\lim_{i \to \infty} \text{ord}_v(B_i) \ = \ \infty \ .$$

Hence there is a largest index K for which $\text{ord}_v(B_K) = 0$, and the theory of Newton Polygons shows that $K = \ell$ is the number of roots of $\mathcal{Q}(Z)$ in $D(0,1)$ (see Lemma 3.35).

Similarly, if $\Delta(Z) = \sum_{i=0}^\infty \Delta_i Z^i$, the fact that $\Delta(Z)$ converges in $D(0,1)$, with $\|\Delta\|_{D(0,1)} \le q_v^{-\frac{\ell}{q-1} - M}$, tells us that $\text{ord}_v(\Delta_i) \ge \frac{\ell}{q-1} + M$ for all i.

Now consider $\mathcal{Q}^*(Z) = \mathcal{Q}(Z) + \Delta(Z)$. Writing $\mathcal{Q}^*(Z) = \sum_{i=0}^\infty C_i Z^i$, we have $C_i = B_i + \Delta_i$, so $\text{ord}_v(C_i) \ge 0$ for all i, $\text{ord}_v(C_\ell) = 0$, and $\text{ord}_v(C_i) > 0$ for all $i > \ell$. By the theory of Newton Polygons, $\mathcal{Q}^*(Z)$ also has exactly ℓ roots in $D(0,1)$.

By Lemma 3.35,

$$\mathcal{Q}(Z) \ = \ B \cdot \mathcal{P}(Z) \cdot \mathcal{H}(Z),$$

where $B \in F_u$ is a constant, $\mathcal{P}(Z) \in F_u(Z)$ is the polynomial

$$\mathcal{P}(Z) \;=\; \mathcal{P}_{\mathcal{S}}(Z) \;=\; \prod_{j \in \mathcal{S}} (Z - \alpha_j),$$

and $\mathcal{H}(Z) \in \mathcal{O}_u[[Z]]$ is an invertible power series with constant term 1.

Since $\mathcal{Q}(Z)$ has only finitely many roots α_j, there is a point $z_0 \in D(0,1)$ with $|z_0 - \alpha_j|_v = 1$ for all j. At each such point $|\mathcal{Q}(z_0)|_v = \|\mathcal{Q}\|_{D(0,1)} = 1$. Using $|\mathcal{P}(z_0)|_v = |\mathcal{H}(z_0)|_v = 1$, we find that $|B|_v = 1$.

Now fix a root α_J, and expand $\mathcal{Q}(Z)$ and $\Delta(Z)$ as power series in $Z - \alpha_J$:

$$\begin{aligned} \mathcal{Q}(Z) &= B_0^{(J)} + B_1^{(J)}(Z - \alpha_J) + B_2^{(J)}(Z - \alpha_J)^2 + \cdots \;, \\ \Delta(Z) &= \Delta_0^{(J)} + \Delta_{v,1}^{(J)}(Z - \alpha_J) + \Delta_2^{(J)}(Z - \alpha_J)^2 + \cdots \;. \end{aligned}$$

We will use the theory of Newton Polygons to show that $\mathcal{Q}^*(Z)$ has a unique root $\alpha_J^* \in \mathcal{O}_u$ satisfying $\mathrm{ord}_v(\alpha_J^* - \alpha_J) > M$.

By Lemma 3.35, the initial part of the Newton Polygon of $\mathcal{Q}(Z)$ (expanded about α_J) coincides with that of the polynomial $\mathcal{P}(Z)$, while its remaining sides have slope ≥ 0. Here

$$\mathcal{P}(Z) \;=\; \prod_{j \in \mathcal{S}} ((Z - \alpha_J) - (\alpha_j - \alpha_J)) \;=\; \sum_{i=1}^{\ell} A_i^{(J)}(Z - \alpha_J)^i \;.$$

Up to sign, the $A_i^{(J)}$ are elementary symmetric polynomials in the $\alpha_j - \alpha_J$. In particular, $A_0^{(J)} = 0$ and

$$A_1^{(J)} \;=\; \pm \prod_{j \neq J} (\alpha_j - \alpha_J) \;.$$

Since $\{\alpha_j\}_{j \in \mathcal{S}}$ is a ψ_v-regular sequence of length ℓ, Lemma 11.5 tells us that

$$\mathrm{ord}_v(A_1^{(J)}) \;<\; \frac{\ell}{q-1} \;.$$

For each $i \geq 2$, considering the expansions of $A_1^{(J)}$ and $A_i^{(J)}$ as elementary symmetric functions in the $\alpha_j - \alpha_J$ gives

$$A_i^{(J)} \;=\; \pm A_1^{(J)} \cdot \left(\sum_{\substack{0 \leq j_1 < \cdots < j_{i-1} < \ell \\ \text{each } j_\ell \neq J}} \frac{1}{(\alpha_{j_1} - \alpha_J) \cdots (\alpha_{j_{i-1}} - \alpha_J)} \right) \;.$$

For each $j \neq J$, we have $\mathrm{ord}_v(\alpha_j - \alpha_J) = \mathrm{val}_q(|j - J|) < \log_v(\ell)$. Hence

$$\mathrm{ord}_v(A_i^{(J)}) \;>\; \mathrm{ord}_v(A_1^{(J)}) - (i-1)\log_v(\ell) \;.$$

Returning to the Newton polygon of $\mathcal{Q}(Z)$, we see that $B_0^{(J)} = 0$,

$$\mathrm{ord}_v(B_1^{(J)}) \;=\; \mathrm{ord}_v(A_1^{(J)}) \;<\; \frac{\ell}{q-1} \;,$$

and for each $i \geq 2$,

$$\mathrm{ord}_v(B_i^{(J)}) \;>\; \mathrm{ord}_v(B_1) - (i-1)\log_v(\ell) \;.$$

(This holds trivially for $i \geq \ell$.)

For the power series $\Delta(Z)$, elementary estimates show that for each i

$$\mathrm{ord}_v(\Delta_{i,i}^{(J)}) \geq \frac{\ell}{q-1} + M \ .$$

Thus $\mathrm{ord}_v(\Delta_{v,1}^{(J)}) \geq \frac{\ell}{q-1} > \mathrm{ord}_v(B_1)$, and $\mathrm{ord}_v(\Delta_i^{(J)}) > \mathrm{ord}_v(B_1) - (i-1)\log_v(\ell)$ for each $i \geq 2$. Now consider the Newton polygon of $\mathcal{Q}^*(Z) = \mathcal{Q}(Z) + \Delta(Z)$, expanded about α_J:

$$\mathcal{Q}^*(Z) \ = \ C_0^{(J)} + C_1^{(J)}(Z-\alpha_J) + C_2^{(J)}(Z-\alpha_J)^2 + \cdots \ .$$

By the discussion above,

$$\begin{aligned} \mathrm{ord}_v(C_0^{(J)}) &= \mathrm{ord}_v(\Delta_0^{(J)}) \ \geq \ \frac{\ell}{q-1} + M \ , \\ \mathrm{ord}_v(C_1^{(J)}) &= \mathrm{ord}_v(B_1^{(J)}) \ < \ \frac{\ell}{q-1} \ , \end{aligned}$$

and for each $i \geq 2$,

$$\mathrm{ord}_v(C_i^{(J)}) \ > \ \mathrm{ord}_v(B_1^{(J)}) - (i-1)\log_v(\ell) \ .$$

Since $M \geq \log_v(\ell)$, the Newton polygon of $\mathcal{Q}^*(Z)$ has a break at $i = 1$, and its initial segment has slope $< -M$. Hence $\mathcal{Q}^*(Z)$ has a unique root α_J^* satisfying

$$\mathrm{ord}_v(\alpha_J^* - \alpha_J) \ > \ M \ .$$

Since $\mathcal{Q}^*(Z) \in F_u[[Z]]$, the theory of Newton Polygons shows that $Z - \alpha_J^*$ is a linear factor in the Weierstrass factorization of $\mathcal{Q}^*(Z)$ over F_u (see Proposition 3.36(B)). Thus $\alpha_J^* \in \mathcal{O}_u$.

This applies for each J. Since $M \geq \log_v(\ell)$, the ℓ roots $\{\alpha_j^*\}_{j\in\mathcal{S}}$ are distinct and form a ψ_v-regular sequence of length ℓ in $\mathcal{O}_u$ attached to $\mathcal{S}$. □

A rough sketch of the way Lemma 11.6 will be applied is as follows:

We begin with the function $G_v^{(0)}(z) = S_{n,v}(\phi_v(z)) = \prod_{j=0}^{n-1}(\phi_v(z) - \psi_v(j))$. The early stages of the patching process seek to preserve this factorization as much as possible. Suppose that at the beginning of the kth stage, $k \geq 1$, there is a sequence of $k+1$ consecutive integers

$$\mathcal{S}_k \ = \ \{j_1, j_1+1, \ldots, j_1+k\} \ \subset \ \{0, 1, \ldots, n-1\} \ ,$$

such that $G_v^{(k-1)}(z) \ = \ \prod_{j\in\mathcal{S}_k}(\phi_v(z) - \psi_v(j)) \cdot F_{v,k}(z)$ for some $(\mathfrak{X}, \vec{s})$-function $F_{v,k}(z) \in K_v(\mathcal{C})$. Expand

$$G_v^{(k-1)}(z) \ = \ \sum_{i=1}^{m} \sum_{j=0}^{(n-1)N_i} A_{v,ij}\varphi_{i,nN_i-j}(z) + \sum_{\lambda=1}^{\Lambda} A_{v,\lambda}\varphi_\lambda(z) \ ,$$

where the $A_{v,ij}, A_{v,\lambda} \in L_{w_v}$. We must patch the coefficients $A_{v,ij}$ in the range $(k-1)N_i \leq j < kN_i$.

Given (i,j) with $(k-1)N_i \leq j < kN_i$, write $nN_i - j = (n-k-1)N_i + r_{ij}$ where $N_i < r_{ij} \leq 2N_i$ (so $r_{ij} = (k+1)N_i - j$), and put

$$\vartheta_{v,ij}^{(k)}(z) \ = \ \varphi_{i,r_{ij}}(z) \cdot F_{v,k}(z) \ .$$

The function $\vartheta_{v,ij}^{(k)}(z)$ has a pole of order $nN_i - j$ at x_i and a pole of order at most $(n-k-1)N_{i'}$ at each $x_{i'} \neq x_i$. By construction, the $\vartheta_{v,ij}^{(k)}(z)$ are K_v-symmetric.

Let $\{\Delta_{v,ij}^{(k)}\}$ for $1 \le i \le m$, $(k-1)N_i + 1 \le j < kN_i$ be a K_v-symmetric set of numbers belonging to L_{w_v}, as given by the global patching process. Patch $G_v^{(k-1)}(z)$ by setting

$$G_v^{(k)}(z) \;=\; G_v^{(k-1)}(z) + \sum_{i=1}^{m} \sum_{j=(k-1)N_i}^{kN_i-1} \Delta_{v,ij}^{(k)} \vartheta_{v,ij}^{(k)}(z) \ .$$

Then the coefficients $A_{v,ij}$ with $j < (k-1)N_i$ are unchanged, the coefficients $A_{v,ij}$ with $(k-1)N_i < j \le kN_i$ are adjusted by the $\Delta_{v,ij}$, and the coefficients $A_{v,ij}$ with $j \ge kN_i$ are changed in complicated ways that are unimportant to us.

If $\mathrm{char}(K_v) = 0$, then since the $\Delta_{v,ij}^{(k)}$ are K_v-symmetric, the sum

$$\sum_{i=1}^{m} \sum_{j=(k-1)N_i}^{kN_i-1} \Delta_{v,ij}^{(k)} \vartheta_{v,ij}^{(k)}(z)$$

is K_v-rational. If $\mathrm{char}(K_v) = p > 0$, then

$$\sum_{i=1}^{m} \sum_{j=(k-1)N_i}^{kN_i-1} \Delta_{v,ij}^{(k)} \vartheta_{v,ij}^{(k)}(z) \;=\; \Big(\sum_{i=1}^{m} \sum_{j=(k-1)N_i}^{kN_i-1} \Delta_{v,ij}^{(k)} \varphi_{i,(k+1)N_i-j}(z)\Big) \cdot F_{v,k}(z)$$

is K_v-rational by assumption (11.11) and the fact that $F_{v,k}(z)$ is K_v-rational. It follows that $G_v^{(k)}(z)$ is K_v-rational.

Put $G_k(z) = \prod_{j \in \mathcal{S}_k} (\phi_v(z) - \psi_v(j))$, so $G_v^{(k-1)}(z) = G_k(z) \cdot F_{v,k}(z)$. By our choice of the $\vartheta_{v,ij}^{(k)}(z)$,

$$G_v^{(k)}(z) \;=\; \left(G_k(z) + \sum_{i=1}^{m} \sum_{j=(k-1)N_i+1}^{kN_i} \Delta_{v,ij} \cdot \varphi_{i,r_{ij}}(z) \right) \cdot F_{v,k}(z) \ .$$

Thus the changes in the roots of $G_v^{(k)}(z)$ have been localized to $G_k(z)$. Put $\Delta_{v,k}(z) = \sum_{i=1}^{m} \sum_{j=(k-1)N_i}^{kN_i-1} \Delta_{v,ij} \cdot \varphi_{i,r_{ij}}(z)$, and let $G_k^*(z) = G_k(z) + \Delta_{v,k}(z)$.

Observe that $G_k(z) = \prod_{j \in \mathcal{S}_k} (\phi_v(z) - \psi_v(j))$ has sup norm 1 on each ball $B(\theta_h, \rho_h)$. Fixing h, when we compose $G_k(z)$ with $\widehat{\sigma}_h : D(0,1) \to B(\theta_h, \rho_h)$, we obtain a function $\mathcal{Q}_{k,h}(Z)$ whose roots in $D(0,1)$ form a ψ_v-regular sequence of length $k+1$ in $\mathcal{O}_{u_h}$ attached to $\mathcal{S}_k$. Taking $\Delta_{k,h}(Z) = \Delta_{v,k}(\widehat{\sigma}_h(Z))$, we can apply Lemma 11.6 to $G_{k,h}^*(Z) = \mathcal{Q}_{k,h}(Z) + \Delta_{k,h}(Z)$. If the $|\Delta_{v,ij}^{(k)}|_v$ are small enough, the roots of $\mathcal{Q}_{k,h}^*(Z)$ will form a ψ_v-regular sequence of length $k+1$ in $\mathcal{O}_{u_h}$. Since $\widehat{\sigma}_h(Z)$ is F_{u_h}-rational, the roots of $G_v^{(k)}(z)$ in $B(a_h, \rho_h)$ belong to $\mathcal{C}(F_{u_h})$.

This was a rough sketch; the actual patching argument will be more complicated, because we eventually we will run out of "new" subsequences $\mathcal{S}_k$ to use in patching, and we must deal with roots which have been previously patched.

In the later steps of the construction, we will use the following lemma:

LEMMA 11.7 (Refined Patching Lemma). *Let $\mathcal{Q}(Z) \in F_u[[Z]]$ be a power series converging on $D(0,1)$, with* sup *norm $\|\mathcal{Q}\|_{D(0,1)} = 1$. Suppose the roots $\{\alpha_j\}$ of $\mathcal{Q}(Z)$ in $D(0,1)$ belong to $\mathcal{O}_u$ and can be partitioned into r disjoint ψ_v-regular sequences in $\mathcal{O}_u$ attached to index sets $\mathcal{S}_1, \dots, \mathcal{S}_r$ of respective lengths $\ell_1, \dots, \ell_r$.*

Put $\ell = \sum_{k=1}^{r} \ell_k$. *Suppose also that there is a bound* $T \geq \max_i(\log_v(\ell_i))$ *such that*

$$\mathrm{ord}_v(\alpha_j - \alpha_k) \ \leq \ T$$

for all $j \neq k$.

Then for any $M \geq T$, *and any power series* $\Delta(Z) \in F_u[[Z]]$ *converging on* $D(0,1)$ *which satisfies*

$$\|\Delta\|_{D(0,1)} \ \leq \ q_v^{-\frac{\ell}{q-1}-(r-1)T-M} \ ,$$

the roots $\{\alpha_j^*\}$ *of* $\mathcal{Q}^*(Z) = \mathcal{Q}(Z)+\Delta(Z)$ *in* $D(0,1)$ *again form a union of* ψ_v*-regular sequences in* $\mathcal{O}_u$ *attached to* $\mathcal{S}_1, \ldots, \mathcal{S}_r$. *They can uniquely be labeled so that*

$$\mathrm{ord}_v(\alpha_j^* - \alpha_j) \ > \ M$$

for each $j \in \bigcup_{i=1}^{r} S_i$.

PROOF. By the Weierstrass Preparation Theorem (Lemma 3.35), we can write $\mathcal{Q}(Z) = B{\cdot}\mathcal{P}(Z){\cdot}\mathcal{H}(Z)$ where $B \in F_u$ is a constant, $\mathcal{P}(Z) \in \mathcal{O}_u[Z]$ is the polynomial

$$\mathcal{P}(Z) \ = \ \prod_{k=1}^{r} \prod_{j \in \mathcal{S}_k} (Z - \alpha_j) \ ,$$

and $\mathcal{H}(Z) \in \mathcal{O}_u[[Z]]$ is an invertible power series with constant term 1. As before, $|B|_v = 1$. We are concerned with the roots of

$$\mathcal{Q}^*(Z) \ = \ \mathcal{Q}(Z) + \Delta(Z) \ .$$

As in the proof of the Basic Patching Lemma, the theory of Newton Polygons shows that $\mathcal{Q}(Z)$ and $\mathcal{Q}^*(Z)$ both have exactly ℓ roots in $D(0,1)$.

Fix a root α_J, and expand $\mathcal{Q}(Z)$ and $\Delta(Z)$ as power series in $Z - \alpha_J$:

$$\begin{aligned} \mathcal{Q}(Z) &= B_0^{(J)} + B_1^{(J)}(Z-\alpha_J) + B_2^{(J)}(Z-\alpha_J)^2 + \cdots \ , \\ \Delta(Z) &= \Delta_0^{(J)} + \Delta_{v,1}^{(J)}(Z-\alpha_J) + \Delta_2^{(J)}(Z-\alpha_J)^2 + \cdots \ . \end{aligned}$$

The initial part of the Newton Polygon of $\mathcal{Q}(Z)$ (expanded about α_J) coincides with that of $\mathcal{P}(Z)$, while its remaining sides have slope > 0. Here

$$\mathcal{P}(Z) \ = \ \prod_{j} ((Z-\alpha_J) - (\alpha_j - \alpha_J)) \ = \ \sum_{i=0}^{\ell} A_i^{(J)} (Z-\alpha_J)^i \ .$$

The coefficients $A_i^{(J)}$ are symmetric polynomials in the $\alpha_j - \alpha_J$. In particular, $A_0^{(J)} = 0$ and

$$A_1^{(J)} \ = \ \pm \prod_{j \neq J} (\alpha_j - \alpha_J) \ .$$

Suppose $J \in \mathcal{S}_i$. By part (A) of Lemma 11.5,

$$\mathrm{ord}_v\Big(\prod_{\substack{j \in \mathcal{S}_i \\ j \neq J}} (\alpha_J - \alpha_j)\Big) \ < \ \frac{\ell_i}{q-1} \ .$$

For each $k \neq i$, by part (B) of Lemma 11.5,

$$\mathrm{ord}_v\Big(\prod_{j \in \mathcal{S}_k} (\alpha_J - \alpha_j)\Big) \ < \ \frac{\ell_k}{q-1} + T$$

since $\max_{j\in\mathcal{S}_k}(\mathrm{ord}_v(\alpha_J-\alpha_j)) \le T$. Summing these, we see that

$$\mathrm{ord}_v(A_1^{(J)}) \ < \ \frac{\ell}{q-1} + (r-1)T \ .$$

For each $i \ge 2$,

$$A_i^{(J)} \ = \ \pm A_1^{(J)} \cdot \left(\sum_{\substack{j_1<\cdots<j_{i-1} \\ \text{each } j_k \ne J}} \frac{1}{(\alpha_{j_1}-\alpha_J)\cdots(\alpha_{j_{i-1}}-\alpha_J)} \right)$$

just as in the Basic Patching Lemma. We have $\mathrm{ord}_v(\alpha_j-\alpha_J) \le T$ for each $j \ne J$. Hence

$$\mathrm{ord}_v(A_i^{(J)}) \ \ge \ \mathrm{ord}_v(A_1^{(J)}) - (i-1)T \ .$$

Returning to the Newton polygon of $\mathcal{Q}(Z)$, we see that $B_0^{(J)} = 0$,

$$\mathrm{ord}_v(B_1^{(J)}) \ = \ \mathrm{ord}_v(A_1^{(J)}) \ < \ \frac{\ell}{q-1} + (r-1)T \ ,$$

and for each $i \ge 2$,

$$\mathrm{ord}_v(B_i^{(J)}) \ \ge \ \mathrm{ord}_v(B_1^{(J)}) - (i-1)T \ .$$

(Note that $T > 0$ unless $\ell = 0$, in which case there is nothing to prove.)

For the power series $\Delta(Z)$, elementary estimates give

$$\mathrm{ord}_v(\Delta_i^{(J)}) \ \ge \ \frac{\ell}{q-1} + (r-1)T + M$$

for each i. In particular, $\mathrm{ord}_v(\Delta_{v,1}^{(J)}) > \frac{\ell}{q-1} + (r-1)T$, and $\mathrm{ord}_v(\Delta_i^{(J)}) > \mathrm{ord}_v(B_1) - (i-1)T$ for each $i \ge 2$. Now consider the Newton polygon of $\mathcal{Q}^*(Z) = \mathcal{Q}(Z) + \Delta(Z)$, expanded about α_J:

$$\mathcal{Q}^*(Z) \ = \ C_0^{(J)} + C_1^{(J)}(Z-\alpha_J) + C_2^{(J)}(Z-\alpha_J)^2 + \cdots \ .$$

By the discussion above,

$$\begin{aligned} \mathrm{ord}_v(C_0^{(J)}) \ &= \ \mathrm{ord}_v(\Delta_0^{(J)}) \ \ge \ \frac{k}{q-1} + (r-1)T + M \ , \\ \mathrm{ord}_v(C_1^{(J)}) \ &= \ \mathrm{ord}_v(B_1^{(J)}) \ < \ \frac{k}{q-1} + (r-1)T \ , \end{aligned}$$

and for each $i \ge 2$,

$$\mathrm{ord}_v(C_i^{(J)}) \ \ge \ \mathrm{ord}_v(B_1^{(J)}) - (i-1)T \ .$$

Since $M \ge T$, the Newton polygon of $\mathcal{Q}^*(Z)$ has a break at $i = 1$, and its initial segment has slope $< -M$. Hence $\mathcal{Q}^*(Z)$ has a unique root α_J^* satisfying

$$\mathrm{ord}_v(\alpha_J^* - \alpha_J) \ > \ M \ .$$

Since $\mathcal{Q}^*(Z) \in F_u[[Z]]$, the theory of Newton Polygons shows that $Z - \alpha_J^*$ is a linear factor in the Weierstrass factorization of $\mathcal{Q}^*(Z)$ over F_u (see Proposition 3.36(B)). Thus $\alpha_J^* \in \mathcal{O}_u$.

This applies for each J. Since $M \ge T$, the roots α_j^* are distinct and form a union of ψ_v-regular sequences in $\mathcal{O}_u$ attached to $\mathcal{S}_1, \ldots, \mathcal{S}_r$. □

2. Stirling Polynomials when $\mathrm{Char}(K_v) = p > 0$

In this section, assume that $\mathrm{char}(K_v) = p > 0$.

In Proposition 11.9 below, we will show that by requiring that n be divisible by a sufficiently high power of p, we can make arbitrarily many high-order coefficients of the Stirling polynomial $S_{n,v}(z) = \prod_{j=0}^{n-1}(z - \psi_v(j))$ be 0. This fact plays a key role in the degree-raising argument in the proof of Theorem 11.2 in §11.3.

We begin with a lemma concerning homogeneous products of linear forms over a finite field.

LEMMA 11.8. *Let $\mathbb{F}_q$ be the finite field with q elements, let $r \geq 1$ be an integer, and put*

$$Q_r(z; T_1, \ldots, T_r) \;=\; \prod_{a_1,\ldots,a_r \in \mathbb{F}_q} \big(z - \sum_{i=1}^{r} a_i T_i\big) \; . \tag{11.19}$$

Then Q_r has the form

$$Q_r(z; T_1, \ldots, T_r) \;=\; z^{q^r} + \sum_{\ell=1}^{r} P_{r,\ell}(T_1, \ldots, T_r) \cdot z^{q^{r-\ell}} \tag{11.20}$$

where $P_{r,\ell}(T_1, \ldots, T_r)$ is homogeneous of degree $q^r - q^{r-\ell}$ in $\mathbb{F}_q[T_1, \ldots, T_r]$, and is symmetric in $T_1, \ldots, T_r$, for each $\ell = 1, \ldots, r$.

PROOF. Since Q_r is symmetric in $T_1, \ldots, T_r$ and is homogeneous of degree q^r, if Q_r has the form (11.20), necessarily the $P_{r,\ell}(T_1, \ldots, T_r)$ are symmetric in $T_1, \ldots, T_r$ and homogeneous of degree $q^r - q^{r-\ell}$. We now prove (11.20) by induction on r. When $r = 1$,

$$\begin{aligned} Q_1(z; T) \;&=\; \prod_{a \in \mathbb{F}_q} (z - aT) \;=\; T^q \cdot \prod_{a \in \mathbb{F}_q} (\frac{z}{T} - a) \\ &=\; T^q \cdot \big((\frac{z}{T})^q - \frac{z}{T}\big) \;=\; z^q - T^{q-1} \cdot z \; . \end{aligned} \tag{11.21}$$

Suppose that (11.20) holds for some r. Then for $r+1$,

$$\begin{aligned} Q_{r+1}(z; T_1, \ldots, T_{r+1}) \;&=\; \prod_{a_1 \in \mathbb{F}_q} \Big(\prod_{a_2 \in \mathbb{F}_q} \cdots \prod_{a_{r+1} \in \mathbb{F}_q} \Big((z - a_1 T_1) - \sum_{i=2}^{r+1} a_i T_i\Big)\Big) \\ &=\; \prod_{a_1 \in \mathbb{F}_q} Q_r(z - a_1 T_1; T_2, \ldots, T_{r+1}) \; . \end{aligned}$$

Using (11.20) for $Q_r(z - a_1 T_1; T_2, \ldots, T_{r+1})$, noting that $(z - a_1 T_1)^q = z^q - a_1 T_1^q$, and applying (11.21), we see that $Q_{r+1}(z; T_1, \ldots, T_{r+1})$ has the form

$$\begin{aligned} &Q_{r+1}(z; T_1, \ldots, T_{r+1}) \\ &= \prod_{a_1 \in \mathbb{F}_q} \Big((z^{q^r} - a_1 T_1^{q^r}) + \sum_{\ell=1}^{r} (z^{q^{r-\ell}} - a_1 T_1^{q^{r-\ell}}) P_{r,\ell}(T_2, \ldots, T_{r+1}) \Big) \\ &= \prod_{a_1 \in \mathbb{F}_q} \Big(Q_r(z; T_2, \ldots, T_{r+1}) - a_1 Q_r(T_1; T_2, \ldots, T_r) \Big) \\ &= Q_r(z; T_2, \ldots, T_{r+1})^q - Q_r(z; T_2, \ldots, T_{r+1}) \cdot Q_r(T_1; T_2, \ldots, T_r)^{q-1} \; . \end{aligned}$$

Using (11.20) for $Q_r(z; T_2, \ldots, T_{r+1})$, then expanding the q-th power and collecting terms, we obtain

$$Q_{r+1}(z; T_1, \ldots, T_{r+1}) \;=\; z^{q^{r+1}} + \sum_{\ell=1}^{r+1} P_{r+1,\ell}(T_1, \ldots, T_{r+1}) \cdot z^{q^{r+1-\ell}} . \qquad \square$$

Now let π_v be a uniformizer for the maximal ideal of $\mathcal{O}_v$. Let $q = q_v = p^{f_v}$ be the order of the residue field $\mathcal{O}_v/\pi_v\mathcal{O}_v$. By the structure theory of local fields in positive characteristic, $\mathcal{O}_v \cong \mathbb{F}_q[[\pi_v]]$ and $K_v \cong \mathbb{F}_q((\pi_v))$. The following proposition depends on the fact that in the basic well-distributed sequence $\{\psi_v(k)\}_{0 \le k < \infty}$ for $\mathcal{O}_v$, the representatives $\psi_v(0), \ldots, \psi_v(q-1)$ for the cosets of $\mathcal{O}_v/\pi_v\mathcal{O}_v$ are the Teichmüller representatives, the elements of $\mathbb{F}_q$.

PROPOSITION 11.9. *Let $K_v \cong \mathbb{F}_q((\pi_v))$ be a local field of characteristic $p > 0$. If $q^r | n$ for some $r > 0$, then $S_{n,v}(z)$ can be expanded as*

$$S_{n,v}(z) \;=\; z^n + \sum_{j=q^r-q^{r-1}}^{n} C_j \, z^{n-j} \tag{11.22}$$

with each $C_j \in \mathcal{O}_v$.

PROOF. First suppose $n = q^r$ for some $r > 0$. The numbers $\psi_v(k)$, for $k = 0, \ldots, q^r - 1$, run over all possible sums $a_0 + a_1\pi_v + \cdots + a_{r-1}\pi_v^{r-1}$ with $a_0, \ldots, a_{r-1} \in \mathbb{F}_q$. By Lemma 11.8,

$$S_{q^r,v}(z) \;=\; Q_r(z; 1, \pi_v, \ldots, \pi_v^{r-1}) \;=\; z^{q^r} + A_r z^{q^{r-1}} + \text{terms of lower degree} ,$$

where $A_r = P_{r,1}(1, \pi_v, \ldots, \pi_v^{r-1})$. Now suppose $n = q^r\ell$. If $0 \le k < \ell$ and we write $k = jq^r + s$ with $0 \le j < \ell$, $0 \le s < q^r$, then $\psi_v(k) = \psi_v(j)\pi_v^r + \psi_v(s)$. It follows that

$$\begin{aligned} S_{n,v}(z) &= \prod_{j=0}^{\ell-1}\prod_{s=0}^{q^r-1} (z - \psi_v(j)\pi_v^r - \psi_v(s)) \;=\; \prod_{j=0}^{\ell-1} S_{q^r,v}\big(z - \psi_v(j)\pi_v^r\big) \\ &= \prod_{j=0}^{\ell-1} \Big(\big(z - \psi_v(j)\pi_v^r\big)^{q^r} + A_r \cdot \big(z - \psi_v(j)\pi_v^r\big)^{q^{r-1}} + \cdots \Big) . \end{aligned} \tag{11.23}$$

Since $(z - \psi_v(j)\pi_v^r)^q = z^q - \psi_v(j)^q\pi_v^{qr}$, upon multiplying out (11.23) we see that

$$S_{n,v}(z) \;=\; (z^{q^r})^\ell + \ell \cdot (z^{q^r})^{\ell-1}(A_r z^{q^{r-1}}) \;+\; \text{terms of lower degree} ,$$

which yields (11.22). $\square$

3. Proof of Theorems 11.1 and 11.2

In this section we prove Theorems 11.1 and 11.2. The construction is a generalization of those in ([**52**]) and ([**53**]).

We begin the construction with

$$\begin{aligned} G_v^{(0)}(z) &= S_{n,v}(\phi_v(z)) \;=\; \prod_{j=0}^{n-1} (\phi_v(z) - \psi_v(j)) \\ &= \sum_{i=1}^{m} \sum_{j=0}^{(n-1)N_i-1} A_{v,ij}\varphi_{i,nN_i-j} + \sum_{\lambda=1}^{\Lambda} A_{v,\lambda}\varphi_\lambda , \end{aligned}$$

whose leading coefficient at x_i is $A_{v,i0} = \widetilde{c}_{v,i}^n$.

The roots $\{\theta_{hj}\}_{1 \le h \le N, 0 \le j < n}$ of $G_v^{(0)}(z)$ belong to $H_v := E_v \cap \big(\bigcup_{h=1}^N B(\theta_h, \rho_h)\big)$, and are distinct. Indeed, as noted at the beginning of §11.1, for each h there is a $1-1$ correspondence between the roots θ_{hj} of $S_{n,v}(\phi_v(z))$ in $B(\theta_h, \rho_h)$ and the zeros α_{hj} of $S_{n,v}(\widehat{\Phi}_h(Z))$ in $D(0,1)$, given by $\theta_{hj} = \widehat{\sigma}_h(\alpha_{hj})$ for $j = 0, \ldots, n-1$. Moreover, the α_{hj} belong to $\mathcal{O}_{u_h}$, and form a ψ_v-regular sequence of length n in $\mathcal{O}_{u_h}$. Put

$$U_v^0 \ := \ \bigcup_{h=1}^{N} B(\theta_h, \rho_h) \ \subset \ U_v \ = \ \bigcup_{\ell=1}^{D} B(a_\ell, r_\ell) \ . \tag{11.24}$$

For suitable n, the patching process will inductively construct K_v-rational $(\mathfrak{X}, \vec{s})$-functions $G_v^{(k)}(z)$, for $k = 1, \ldots, n$, whose roots belong to H_v for all k. For each k, there will be a natural one-to-one correspondence between the roots of $G_v^{(k-1)}(z)$ and $G_v^{(k)}(z)$, and the roots of $G_v^{(n)}(z)$ will be distinct. The conditions on n needed for the construction to succeed will be noted as they arise; they will all require that n be sufficiently large, or that it be divisible by a certain integer. Only a finite number of conditions will be imposed, so there is an $n_v \ge 1$ such that the construction will succeed if n is sufficiently large and divisible by n_v.

Recall that

$$M_v \ = \ \max\big(\max_i \max_{N_i < j \le 2N_i} (\|\varphi_{i,j}\|_{U_v}), \max_\lambda(\|\varphi_\lambda\|_{U_v})\big) \ . \tag{11.25}$$

We have defined k_v to be the smallest integer such that for all $k \ge k_v$,

$$h_v^{Nk} \cdot M_v \ < \ q_v^{-(\frac{k+1}{q_v - 1} + \log_v(k+1))} \ . \tag{11.26}$$

The global patching process specifies a number $\overline{k} \ge k_v$, the number of bands of coefficients considered "high-order".

Phase 1. Patching the leading and high-order coefficients, for $k = 1, \ldots, \overline{k}$.

The patching constructions for the leading and high-order coefficients are different when $\mathrm{char}(K_v) = 0$ and when $\mathrm{char}(K_v) = p > 0$.

Case A. Suppose $\mathrm{char}(K_v) = 0$.

In this case, the leading coefficient is patched along with the other high-order coefficients. The bound for the high-order patching coefficients in Theorem 11.1 is $B_v = h_v^{\overline{k}N}$. Since $0 < h_v < 1$ and $\overline{k} \ge k_v$, it follows from (11.26) that

$$B_v \cdot M_v \ = \ h_v^{\overline{k}N} \cdot M_v \ \le \ q_v^{-\frac{\overline{k}+1}{q_v-1} - \log_v(\overline{k}+1)} \ \le \ q_v^{-\frac{k+1}{q_v-1} - \log_v(k+1)} \tag{11.27}$$

for each $k = 1, \ldots, \overline{k}$.

When $k = 1$, we are given a K_v-symmetric set $\{\Delta_{v,ij}^{(1)} \in L_{w_v}\}_{(i,j) \in \mathrm{Band}_N(1)}$, determined recursively in $\prec_N$ order, such that for each (i,j),

$$|\Delta_{v,ij}^{(1)}|_v \ \le \ B_v \ . \tag{11.28}$$

Put $\mathcal{S}_1 = \{0, 1\}$, and let

$$P_1(x) \ = \ \prod_{j \in \mathcal{S}_1} (x - \psi_v(j)), \qquad \widehat{P}_1(x) \ = \ \prod_{j=2}^{n-1} (x - \psi_v(j))$$

so $S_{n,v}(x) = P_1(x) \cdot \widehat{P}_1(x)$. Set $Q_1(z) = P_1(\phi_v(z))$; then

$$G_v^{(0)}(z) = Q_1(z) \cdot \widehat{P}_1(\phi(z)) .$$

For each $(i,j) \in \mathrm{Band}_N(1)$, put

$$\vartheta_{v,ij}^{(1)}(z) = \varphi_{i,2N_i-j}(z) \cdot \widehat{P}_1(\phi_v(z)) .$$

Thus $\vartheta_{v,ij}^{(1)}(z)$ has a pole of order $nN_i - j$ at x_i with leading coefficient $\widetilde{c}_{v,i}^{\,n-2}$, and a pole of order $(n-2)N_{i'}$ at each $x_{i'} \neq x_i$. By construction the $\vartheta_{v,ij}^{(1)}(z)$ are K_v-symmetric. Since the $\Delta_{v,ij}^{(1)}$ are K_v-symmetric as well, for each (i,j),

$$\sum_{x_{i'} \in \mathrm{Aut}_c(\mathbb{C}_v/K_v)(x_i)} \Delta_{v,i'j}^{(1)} \vartheta_{v,i'j}^{(1)} \in K_v(\mathcal{C}) .$$

This assures that when the global patching process determines the patching coefficients $\Delta_{v,ij}^{(k)}(z)$ recursively in $\prec_N$ order, the partially patched function $G_v^{(0)}(z)$ is K_v-rational after each $\mathrm{Aut}(\widetilde{K}/K)$-orbit of coefficients is chosen.

Patch $G_v^{(0)}(z)$ by setting

$$G_v^{(1)}(z) = G_v^{(0)}(z) + \sum_{i=1}^{m} \sum_{j=0}^{N_i-1} \Delta_{v,ij}^{(1)} \cdot \vartheta_{v,ij}^{(1)}(z) .$$

Then $G_v^{(1)}(z)$ is a K_v-rational $(\mathfrak{X}, \vec{s})$-function of degree Nn.

Now consider how the roots $\{\theta_{hj}\}$ change in passing from to $G_v^{(0)}(z)$ to $G_v^{(1)}(z)$. If we write $\Delta_{v,1}(z) = \sum_{i=1}^m \sum_{j=0}^{N_i-1} \Delta_{v,ij}^{(1)} \cdot \varphi_{i,2N_i-j}(z)$, and put

$$G_1^*(z) = Q_1(z) + \Delta_{v,1}(z) ,$$

then by our choice of the $\vartheta_{v,ij}^{(1)}(z)$,

$$G_v^{(1)}(z) = G_1^*(z) \cdot \widehat{P}_1(\phi_v(z)) .$$

Hence the roots θ_{hj} with $j \geq 2$ are all preserved.

For each ball $B(\theta_h, \rho_h)$, let $\widehat{\sigma}_h : D(0,1) \to B(\theta_h, \rho_h)$ be the F_{u_h}-rational parametrization from §11.1; recall that $\theta_{hj} = \widehat{\sigma}_h(\alpha_{hj})$. By abuse of language, we will refer to both the θ_{hj} and the α_{hj} as "roots". By (11.25), (11.27) and (11.28),

$$\|\Delta_{v,1}(z)\|_{U_v^0} \leq B_v \cdot M_v \leq q^{-(2/(q-1)+\log_v(2))} .$$

Put $\mathcal{Q}_{1,h}(Z) = Q_1(\widehat{\sigma}_h(Z)) = P_1(\widehat{\Phi}_h(Z)))$ and $\Delta_{1,h}(Z) = \Delta_{v,1}(\widehat{\sigma}_h(Z))$. The roots $\{\alpha_{h0}, \alpha_{h1}\}$ of $\mathcal{Q}_{1,h}(Z)$ form a ψ_v-regular sequence of length 2 in $\mathcal{O}_{u_h}$, and $\|\Delta_{1,h}\|_{D(0,1)} \leq q^{-(2/(q-1)+\log_v(2))}$. By Lemma 11.6 the roots of $\mathcal{Q}_{1,h}^*(Z) = \mathcal{Q}_{1,h}(Z) + \Delta_{1,h}(Z)$ form a ψ_v-regular sequence $\{\alpha_{h0}^*, \alpha_{h1}^*\}$ of length 2 in $\mathcal{O}_w$, with

$$\mathrm{ord}_v(\alpha_{h0}^* - \alpha_{h0}) > \log_v(2), \quad \mathrm{ord}_v(\alpha_{h1}^* - \alpha_{h1}) > \log_v(2) .$$

Thus the roots of $G_v^{(1)}(\widehat{\sigma}_h(Z))$ in $D(0,1)$ are

$$\{\alpha_{h0}^*, \alpha_{h1}^*, \alpha_{h2}, \ldots, \alpha_{h,n-1}\} ,$$

a union of a ψ_v-regular sequence of length 2 and the remaining $n-2$ elements of the original ψ_v-regular sequence of length n. Transferring this back to $G_v^{(1)}(z)$, put

$\theta_{h0}^* = \widehat{\sigma}_h(\alpha_{h0}^*)$, $\theta_{h1}^* = \widehat{\sigma}_h(\alpha_{h1}^*)$. Then $\theta_{h0}^*, \theta_{h1}^* \in B(\theta_h, \rho_h) \cap \mathcal{C}(F_{u_h}) \subset \widetilde{E}_v$. The roots of $G_v^{(1)}(z)$ in $B(\theta_h, \rho_h)$ are

$$\{\theta_{h0}^*, \theta_{h1}^*, \theta_{h2}, \ldots, \theta_{h,n-1}\} \ .$$

Let $\varepsilon_{v,i} = 1 + \Delta_{v,i0}^{(1)}$. Since the $\Delta_{v,i0}^{(1)}$ are K_v-symmetric, with $|\Delta_{v,i0}^{(1)}|_v < 1$ for each i, the $\varepsilon_{v,i}$ form a K_v-symmetric system of units in $\mathcal{O}_{w_v}^\times$. The leading coefficient $A_{v,i0} = \widetilde{c}_{v,i}^{\,n}$ of $G_v^{(0)}(z)$ at x_i is changed to $\varepsilon_{v,i}\widetilde{c}_{v,i}^{\,n}$ in $G_v^{(1)}(z)$. (From the global patching process, we know that $\varepsilon_{v,i}\widetilde{c}_{v,i}^{\,N} = \mu_i^{n/n_0}$ where the μ_i are the $\widehat{S}$-units from Theorem 7.11; however, from a local standpoint, this is irrelevant.) The leading coefficient of $Q_1^*(z)$ at x_i is $\varepsilon_{v,i}$.

Next we construct the $G_v^{(k)}(z)$ for $k = 2, \ldots, \overline{k}$. Each $G_v^{(k)}(z)$ will be a K_v-rational $(\mathfrak{X}, \vec{s})$-function of degree nN, having a pole of order nN_i and leading coefficient $\varepsilon_{v,i}\widetilde{c}_{v,i}^{\,n}$ at x_i, and with a factorization into K_v-rational $(\mathfrak{X}, \vec{s})$-functions of the form

$$G_v^{(k)}(z) \ = \ Q_1^*(z) Q_2^*(z) \cdots Q_k^*(z) \cdot \widehat{P}_k(\phi_v(z)) \ .$$

The properties of the $Q_k^*(z)$ will be discussed below.

Recall that $\mathcal{S}_1 = \{0, 1\}$. Put $\mathcal{S}_2 = \{2, 3, 4\}$, $\mathcal{S}_3 = \{5, 6, 7, 8\}$, and so on, through $\mathcal{S}_{\overline{k}}$, where $\mathcal{S}_k$ consists of the next $k+1$ integers in $\{0, 1, \ldots, n-1\}$ after $\mathcal{S}_{k-1}$. Explicitly,

$$\mathcal{S}_k \ = \ \{j_k, \ldots, j_k + k\} \tag{11.29}$$

where $j_k = (k^2 + k - 2)/2$. For this to be possible we need

$$n \ \geq \ (\overline{k}^2 + 3\overline{k} - 2)/2 \tag{11.30}$$

which we henceforth assume.

In the kth step, the roots corresponding to $\mathcal{S}_k$ will be moved. Put

$$P_k(x) = \prod_{j \in \mathcal{S}_k} (x - \psi_v(j)), \quad \widehat{P}_k(x) = \prod_{0 \leq j \leq n-1, \ j \notin \mathcal{S}_1 \cup \ldots \mathcal{S}_k} (x - \psi_v(j)) \ , \tag{11.31}$$

so that $S_{n,v}(x) = P_1(x) \cdots P_k(x) \cdot \widehat{P}_k(x)$ and $\widehat{P}_{k-1}(x) = P_k(x) \cdot \widehat{P}_k(x)$. Set

$$Q_k(z) \ = \ P_k(\phi_v(z)) \ . \tag{11.32}$$

For each k, $2 \leq k \leq \overline{k}$, the function $Q_k^*(z)$ will have the following properties:

(1) $Q_k^*(z)$ is obtained by perturbing $Q_k(z)$;
(2) $Q_k^*(z)$ is a K_v-rational $(\mathfrak{X}, \vec{s})$-function of degree $(k+1)N$, with a pole of order $(k+1)N_i$ and leading coefficient $\widetilde{c}_{v,i}^{\,k+1}$ at x_i;
(3) for each ball $B(\theta_h, \rho_h)$, the roots $\{\alpha_{h,j_k}^*, \ldots, \alpha_{h,j_k+k}^*\}$ of $Q_k^*(\widehat{\sigma}_h(Z))$ in $D(0,1)$ form a ψ_v-regular sequence $\{\alpha_{hj}\}_{j \in \mathcal{S}_k}$ of length $k+1$ in $\mathcal{O}_{u_h}$ relative to $\widehat{\Phi}_h(Z)$, and for each $j \in \mathcal{S}_k$,

$$\mathrm{ord}_v(\alpha_{hj}^* - \alpha_{hj}) \ > \ \frac{k+1}{q-1} + \log_v(k+1);$$

(4) for each ball $B(\theta_h, \rho_h)$, $\|Q_k^*\|_{B(\theta_h, \rho_h)} = 1$.

Inductively, suppose $G_v^{(k-1)}(z)$ has been constructed, with

$$\begin{aligned} G_v^{(k-1)}(z) &= Q_1^*(z)\cdots Q_{k-1}^*(z)\cdot \widehat{P}_{k-1}(\phi_v(z)) \\ &= Q_1^*(z)\cdots Q_{k-1}^*(z)\cdot Q_k(z)\cdot \widehat{P}_k(\phi_v(z)) \ . \end{aligned} \tag{11.33}$$

Expand

$$G_v^{(k-1)}(z) = \sum_{i=1}^{m}\sum_{j=0}^{(n-1)N_i-1} A_{v,ij}\varphi_{i,nN_i-j}(z) + \sum_{\lambda=1}^{\Lambda} A_\lambda\varphi_\lambda(z) \ .$$

The $A_{v,ij}$ and $A_{v,\lambda}$ belong to L_{w_v} and are K_v-symmetric.

We are given a K_v-symmetric set of numbers $\{\Delta_{v,ij}^{(k)} \in L_{w_v}\}_{(i,j)\in \mathrm{Band}_N(k)}$, determined recursively in $\prec_N$ order, such that for each (i,j),

$$|\Delta_{v,ij}^{(k)}|_v \ \le \ B_v \ . \tag{11.34}$$

For each (i,j), we have $(k-1)N_i \le j < kN_i$. Let $r_{ij} = (k+1)N_i - j$, so $N_i < r_{ij} \le 2N_i$ and $nN_i - j = (n-k-1)N_i + r_{ij}$, and put

$$\vartheta_{v,ij}^{(k)}(z) = \varepsilon_{v,i}^{-1}\cdot \varphi_{i,r_{ij}}(z)\cdot \prod_{\ell=1}^{k-1} Q_\ell^*(z)\cdot \widehat{P}_k(\phi_v(z)) \ . \tag{11.35}$$

Then $\vartheta_{v,ij}^{(k)}(z)$ has a pole of order $nN_i - j$ at x_i. Its leading coefficient at x_i is $\widetilde{c}_{v,i}^{\,n-k-1}$, because the leading coefficient of $Q_1^*(z)$ is $\varepsilon_{v,i}$, while for $2 \le \ell \le k-1$ the leading coefficient of $Q_\ell^*(z)$ is $\widetilde{c}_{v,i}^{\,\ell+1}$. It has a pole of order at most $(n-k-1)N_{i'}$ at each $x_{i'} \ne x_i$.

Since the $\widetilde{c}_{v,i}$, $\varepsilon_{v,i}$, and $\varphi_{ij}(z)$ are K_v-symmetric, and $\widehat{P}_k(\phi_v(z))$ is K_v-rational, the $\vartheta_{v,ij}^{(k)}(z)$ are K_v-symmetric. Thus for each (i,j),

$$\sum_{x_{i'}\in \mathrm{Aut}_c(\mathbb{C}_v/K_v)(x_i)} \Delta_{v,i'j}^{(k)}\vartheta_{v,i'j}^{(k)} \ \in \ K_v(\mathcal{C}) \ .$$

We patch $G_v^{(k-1)}(z)$ by setting

$$G_v^{(k)}(z) = G_v^{(k-1)}(z) + \sum_{i=1}^{m}\sum_{j=(k-1)N_i}^{kN_i-1} \Delta_{v,ij}^{(k)}\vartheta_{v,ij}^{(k)}(z) \ . \tag{11.36}$$

The coefficients $A_{v,ij}$ with $j < (k-1)N_i$ remain unchanged. In particular, $G_v^{(k)}(z)$ has the same leading coefficients $\varepsilon_{v,i}\widetilde{c}_{v,i}^{\,n}$ as $G_v^{(1)}(z)$. Each $G_v^{(k)}(z)$ is a K_v-rational $(\mathfrak{X},\vec{s})$-function of degree nN, so its roots are K_v-symmetric.

Put

$$\Delta_{v,k}(z) = \sum_{i=1}^{m}\sum_{j=(k-1)N_i+1}^{kN_i} \Delta_{v,ij}^{(k)}\cdot \varepsilon_{v,i}^{-1}\varphi_{i,r_{ij}}(z) \ , \tag{11.37}$$

$$Q_k^*(z) = Q_k(z) + \Delta_{v,k}(z) \ . \tag{11.38}$$

Then $\Delta_{v,k}(z)$ and $Q_k^*(z)$ are K_v rational. Since $Q_k(z)$ has a pole of order $(k+1)N_i$ at each x_i, with leading coefficient $\widetilde{c}_{v,i}^{\,k+1}$, while $\Delta_{v,k}(z)$ has a pole of order at most $2N_i$ at x_i, $Q_k^*(z)$ has the same leading coefficients as $Q_k(z)$.

By (11.33) and the definition of the $\vartheta_{v,ij}^{(k)}(z)$, changes in the roots of $G_v^{(k-1)}(z)$ are localized to the factor $Q_k(z) = P_k(\phi_v(z))$:

$$(11.39) \qquad G_v^{(k)}(z) \;=\; Q_1^*(z)Q_2^*(z)\cdots Q_k^*(z)\cdot \widehat{P}_k(\phi_v(z)) \;.$$

Furthermore, by (11.25), (11.27), and (11.34)

$$\|\Delta_{v,k}(z)\|_{U_v^0} \;\le\; B_v\cdot M_v \;\le\; q^{-\frac{k+1}{q-1}-\log_v(k+1)} \;.$$

For each ball $B(\theta_h,\rho_h)$, we have $\|\Delta_{v,k}(z)\|_{B(\theta_h,\rho_h)} < 1$ while $\|Q_k(z)\|_{B(\theta_h,\rho_h)} = 1$, so $\|Q_k^*(z)\|_{B(\theta_h,\rho_h)} = 1$.

Let $\widehat{\sigma}_h : D(0,1) \to B(\theta_h,\rho_h)$ be the chosen F_{u_h}-rational parametrization; recall that $\theta_{hj} = \widehat{\sigma}_h(\alpha_{hj})$. Apply Lemma 11.6 to $\mathcal{Q}_{k,h}(Z) = Q_k(\widehat{\sigma}_h(Z)) = P_k(\widehat{\Phi}_h(Z))$ and $\Delta_{k,h}(Z) = \Delta_{v,k}(\widehat{\sigma}_h(Z))$. The roots of $\mathcal{Q}_{k,h}(Z)$ in $D(0,1)$ are $\{\alpha_{hj}\}_{j\in\mathcal{S}_k}$, which is a ψ_v-regular sequence of length $k+1$ in $\mathcal{O}_{u_h}$ in attached to $\mathcal{S}_k$, and $\|\Delta_{k,h}\|_{D(0,1)} \le q^{-(k+1)/(q-1)-\log_v(k+1)}$. By Lemma 11.6 the roots $\{\alpha_{hj}^*\}_{j\in\mathcal{S}_k}$ of $\mathcal{Q}_{k,h}^*(Z) = \mathcal{Q}_{k,h}(Z) + \Delta_{k,h}(Z)$ form a ψ_v-regular sequence of length $k+1$ in $\mathcal{O}_{u_h}$, with

$$(11.40) \qquad \operatorname{ord}_v(\alpha_{hj}^* - \alpha_{hj}) \;>\; \log_v(k+1)$$

for each $j \in \mathcal{S}_k$.

Thus the roots of $G_v^{(k)}(\widehat{\sigma}_h(Z))$ in $D(0,1)$ are

$$\{\alpha_{h0}^*, \alpha_{h1}^*,\ \alpha_{h2}^*, \alpha_{h3}^*, \alpha_{h4}^*,\ \ldots, \alpha_{h,j_k+k}^*,\ \ \alpha_{h,j_{k+1}}, \ldots, \alpha_{h,n-1}\} \;,$$

which is a union of k ψ_v-regular sequences of lengths $2,3,4,\ldots,k+1$, together with the remainder of the original ψ_v-regular sequence of length n. Transferring this back to $G_v^{(k)}(z)$, put $\theta_{hj}^* = \widehat{\sigma}_h(\alpha_{hj}^*)$ for each $j \in \mathcal{S}_k$. The roots of $G_v^{(k)}(z)$ in $B(\theta_h,\rho_h)$ are

$$\{\theta_{h0}^*, \theta_{h1}^*,\ \theta_{h2}^*, \theta_{h3}^*, \theta_{h4}^*,\ \ldots, \theta_{h,j_k+k}^*,\ \ \theta_{h,j_{k+1}}, \ldots, \theta_{h,n-1}\} \;,$$

and they all belong to $\mathcal{C}_v(F_{u_h}) \cap B(\theta_h,\rho_h)$.

Case B. Suppose $\operatorname{char}(K_v) = p > 0$.

In this case the goals of the high-order patching process are different. First, we need to choose n_v so that if $n_v|n$, then when $G_v^{(0)}(z) = S_{n,v}(\phi_v(z))$ is expanded using the L-rational basis as

$$(11.41) \qquad G_v^{(0)}(z) \;=\; \sum_{i=1}^{m}\sum_{j=0}^{(n-1)N_i-1} A_{v,ij}\varphi_{i,nN_i-j}(z) + \sum_{\lambda=1}^{\Lambda} A_{v,\lambda}\varphi_\lambda(z) \;,$$

we have $A_{v,ij} = 0$ for all (i,j) with $1 \le i \le m$, $1 \le j < \overline{k}N_i$. Second, when we patch the leading coefficients $A_{v,i0} = \widetilde{c}_{v,i}^{\,n}$, we must carry out the patching process in such a way that the $A_{v,ij}$ with $1 \le i \le m$, $1 \le j < \overline{k}N_i$ remain 0.

To accomplish these goals, we will choose n_v and B_v differently from how they were chosen when $\operatorname{char}(K_v) = 0$. Recall that $q = q_v = p^{f_v}$ is the order of the residue field of $\mathcal{O}_v$, and that k_v is the least integer for such that for all $k \ge k_v$,

$$(11.42) \qquad h_v^{Nk}\cdot M_v \;<\; q^{-(\frac{k+1}{q-1}+\log_v(k+1))} \;.$$

Recall also that $J = p^A$ is the least power of p such that

$$p^A \;\ge\; \max\big(2g+1, \max_{1\le i\le m}([K(x_i):K]^{\text{insep}})\big) \;,$$

(see §3.3) and that we have chosen N in such a way that $J|N_i$ for each i. By the construction of the L-rational and L^{sep}-rational bases, this means that $\varphi_{i,nN_i}(z) = \widetilde{\varphi}_{i,nN_i}(z)$ is L^{sep}-rational for each $i = 1, \ldots, m$.

Let $\overline{k} \geq k_v$ is a fixed integer specified by the global patching process. We will take $n_v = q^r$ to be the least power of q such that

$$\left\{ \begin{array}{lcl} q^r & \geq & \max(\overline{k}N_1, \ldots, \overline{k}N_m) , \\ q^r - q^{r-1} & \geq & \overline{k} , \end{array} \right. \tag{11.43}$$

and take

$$B_v = \min\left(\frac{1}{2}, \frac{q^{-\left(\frac{n_v}{q-1}+\log_v(n_v)\right)}}{\max_i\left(|\widetilde{c}_{v,i}^{n_v}|_v \|\varphi_{i,n_vN_i}\|_{U_v}\right)} \right) . \tag{11.44}$$

Note that $J|n_v$, since $n_v \geq \overline{k}N_1 \geq J$ and both n_v and J are powers of p.

We will that for see all sufficiently large n divisible by n_v, we can carry out the patching process described in Theorem 11.2.

Given n divisible by n_v, write $n = n_vQ = q^r \cdot Q$, and put $G_v^{(0)}(z) = S_{n,v}(\phi_v(z))$. We first show when $G_v^{(0)}(z)$ is expanded as in (11.41), then the high-order coefficients $A_{v,ij}$ are 0 for $1 \leq i \leq m$, $1 \leq j < \overline{k}N_i$, as required. For this, we apply Proposition 11.9, which says that $S_{n,v}(x)$ has the form

$$S_{n,v}(x) = x^n + \sum_{j=q^r-q^{r-1}}^{n} C_j x^{n-j} \tag{11.45}$$

with each $C_j \in \mathcal{O}_v$.

For each i, the leading coefficient of $G_v^{(0)}(z)$ at x_i is $\widetilde{c}_{v,i}^{\,n}$, which belongs to $K_v(x_i)^{\text{sep}}$ since $\widetilde{c}_{v,i} \in K_v(x_i)^{\text{sep}}$ by the construction of $\phi_v(z)$. If we expand $\phi_v(z)^Q$ using the L-rational basis as

$$\phi_v(z)^Q = \sum_{i=1}^{m} \sum_{j=0}^{(Q-1)N_i-1} B_{v,ij}\varphi_{i,QN_i-j}(z) + \sum_{\lambda=1}^{\Lambda} B_\lambda \varphi_\lambda(z) ,$$

then since $\text{char}(K_v) = p > 0$ and n_v is a power of p, it follows that

$$\phi_v(z)^n = (\phi_v(z)^Q)^{n_v} = \sum_{i=1}^{m} \sum_{j=0}^{(Q-1)N_i-1} B_{v,ij}^{n_v}\varphi_{i,QN_i-j}(z)^{n_v} + \sum_{\lambda=1}^{\Lambda} B_\lambda^{n_v} \varphi_\lambda(z)^{n_v} .$$

Here for each i, since $J|N_i$, Proposition 3.3(B) shows that $\varphi_{i,QN_i}^{n_v} = \varphi_{i,nN_i}$, and since the leading coefficient of $\phi_v(z)$ is $\widetilde{c}_{v,i}$, we have $B_{v,ij}^{n_v} = \widetilde{c}_{v,i}^{\,n}$. Each term $B_{v,ij}^{n_v}\varphi_{i,QN_i-j}(z)^{n_v}$ for $j \geq 1$ has a pole of order at most $nN_i - n_v$ at x_i and no other poles, and the $B_\lambda^{n_v}\varphi_\lambda(z)^{n_v}$ have poles of order at most $n_vN_{i'}$ at each $x_{i'}$. Thus if $n \geq n_v + 1$, which we henceforth assume, then

$$\phi_v(z)^n = \sum_{i=1}^{m} \widetilde{c}_{v,i}^{\,n}\varphi_{i,nN_i}(z) + \text{terms with poles of order} \leq (n-n_v)N_i \text{ at each } x_i . \tag{11.46}$$

Now consider

$$G_v^{(0)}(z) = S_{n,v}(\phi_v(z)) = \phi_v(z)^n + \sum_{j=q^r-q^{r-1}}^{n} C_j\phi_v(z)^{n-j} . \tag{11.47}$$

By (11.43), (11.45) and (11.46), when $G_v^{(0)}(z)$ is expanded as in (11.41) the coefficients $A_{v,ij}$ for $1 \le i \le m$, $1 \le j < \overline{k}N_i$ are all 0.

We next carry out the patching process for the stage $k = 1$. We want to modify the leading coefficients $A_{v,i0}$ and leave the remaining-order high coefficients $A_{v,ij}$ for $1 \le j < \overline{k}N_i$ (which are 0) unchanged. By assumption, we are given a K_v-symmetric set of numbers $\{\Delta_{v,i0}^{(1)} \in L_{w_v}\}_{1 \le i \le m, 0 \le j < N_i}$, with $|\Delta_{v,i0}^{(1)}|_v \le B_v$ for each i, and $\Delta_{v,ij}^{(1)} = 0$ for all $j \ge 1$ and all i, such that $\Delta_{v,i0}^{(1)}$ belongs to $K_v(x_i)^{\mathrm{sep}}$ for each i, and we wish to replace $A_{v,i0} = \widetilde{c}_{v,i}^{\,n}$ with $\widetilde{c}_{v,i}^{\,n} + \Delta_{v,i0}\widetilde{c}_{v,i}^{\,n}$ in (11.41).

We claim that taking

$$\widetilde{\theta}_{v,i0}^{(1)}(z) \;=\; \widetilde{c}_{v,i}^{\,n_v}\varphi_{i,n_vN_i}(z) \cdot S_{n-n_v,v}(\phi_v(z)) \tag{11.48}$$

in Theorem 11.2 for each $i = 1, \ldots, m$, and then putting

$$G_v^{(1)}(z) \;=\; G_v^{(0)}(z) \;+\; \sum_{i=1}^{m} \Delta_{v,i0}^{(1)}\,\widetilde{\theta}_{v,i0}^{(1)}(z) \;, \tag{11.49}$$

accomplishes what we need. Let $H(z)$ denote the sum on the right side of (11.49).

First, adding $H(z)$ to $G_v^{(0)}(z)$ adds $\Delta_{v,i0}^{(1)}\widetilde{c}_{v,i}^{\,n}$ to $A_{v,i0} = \widetilde{c}_{v,i}^{\,n}$, for each i. This follows from the fact that $\widetilde{c}_{v,i}^{\,n_v}\varphi_{i,n_vN_i}(z) \cdot S_{n-n_v,v}(\phi_v(z))$ has a pole of order nN_i at x_i with leading coefficient $\widetilde{c}_{v,i}^{\,n}$, and at each $x_{i'} \ne x_i$ its pole has order less than $(n - \overline{k})N_{i'}$.

Put $\varepsilon_{v,i} = 1 + \Delta_{v,i0}^{(1)}$; then the leading coefficient of $G_v^{(1)}(z)$ at x_i is $\varepsilon_{v,i}\widetilde{c}_{v,i}^{\,n}$. Since the $\Delta_{v,i0}^{(1)}$ are K_v-symmetric, with $\Delta_{v,i0}^{(1)} \in K_v(x_i)^{\mathrm{sep}}$ and $|\Delta_{v,i0}^{(1)}|_v \le B_v < 1$ for each i, the $\varepsilon_{v,i}$ form a K_v-symmetric system of $\mathcal{O}_{w_v}^{\mathrm{sep}}$-units with $\varepsilon_{v,i} \in K_v(x_i)^{\mathrm{sep}}$ for each i. Since $\widetilde{c}_{v,i}$ belongs to $K_v(x_i)^{\mathrm{sep}}$, so does $\varepsilon_{v,i}\widetilde{c}_{v,i}^{\,n}$.

Second, adding $H(z)$ to $G_v^{(0)}(z)$ leaves $A_{v,ij} = 0$ for $1 \le i \le m$, $1 \le j < \overline{k}N_i$. To see this, note that if $n = n_vQ$ then $n - n_v = n_v(Q-1)$. Since $n_v|(n - n_v)$, Proposition 11.9 and an argument like the one which gave (11.47) show that

$$S_{n-n_v,v}(\phi_v(z)) \;=\; \phi_v(z)^{n-n_v} + \sum_{j=q^r-q^{r-1}}^{n-n_v} C_j'\phi_v(z)^{n-n_v-j} \tag{11.50}$$

for certain $C_j' \in \mathcal{O}_v$. Likewise, by an argument similar to the one which gave (11.46), if $n \ge n_v + 2$ (which we henceforth assume), then

$$\begin{aligned} \phi_v(z)^{n-n_v} &= \sum_{i=1}^{m} \widetilde{c}_{v,i}^{\,n-n_v}\varphi_{i,(n-n_v)N_i}(z) \\ &\quad + \text{terms with poles of order} \le (n-2n_v)N_i \text{ at each } x_i \;. \end{aligned} \tag{11.51}$$

Finally, since $J|N_i$ for each i, Proposition 3.3(B) shows that for each basis function $\varphi_{ij}(z)$ we have

$$\varphi_{i,n_vN_i}(z) \cdot \varphi_{ij}(z) \;=\; \varphi_{i,n_vN_i+j}(z) \;. \tag{11.52}$$

Combining (11.50), (11.51) and (11.52), and using that $q^r - q^{r-1} \ge \overline{k}$, we see that

(11.53)

$$\begin{aligned} \widetilde{\theta}_{v,i0}^{(1)}(z) &= \widetilde{c}_{v,i}^{\,n_v}\varphi_{i,n_vN_i}(z) \cdot S_{n-n_v,v}(\phi_v(z)) \\ &= \widetilde{c}_{v,i}^{\,n}\varphi_{i,nN_i}(z) + \text{terms with poles of order} \le (n-\overline{k})N_{i'} \text{ at each } x_i' \;. \end{aligned}$$

Thus $\widetilde{\theta}_{v,i0}^{(1)}(z) = \widetilde{c}_{v,i}^{n}\varphi_{i,nN_i}(z) + \widetilde{\Theta}_{v,i0}^{(1)}(z)$ for an $(\mathfrak{X}, \vec{s})$-function $\widetilde{\Theta}_{v,i0}^{(1)}(z)$ with poles of order at most $(n-\overline{k})N_{i'}$ at each $x_{i'}$, as asserted in Theorem 11.2.

Third, $\widetilde{\theta}_{v,i0}^{(1)}(z)$ is rational over $K_v(x_i)^{\text{sep}}$ for each i, and the $\widetilde{\theta}_{v,i0}^{(1)}(z)$ are K_v-symmetric. To see this, note that $\varphi_{i,n_vN_i}(z) = \widetilde{\varphi}_{i,n_vN_i}(z)$ is rational over $K_v(x_i)^{\text{sep}}$ by Proposition 3.3(B), since $J|N_i$. Furthermore, $\widetilde{c}_{v,i} \in K_v(x_i)^{\text{sep}}$ by hypothesis, and $S_{n-n_v,v}(\phi_v(z))$ is K_v-rational, so $\widetilde{\theta}_{v,i0}^{(1)}(z)$ is $K_v(x_i)^{\text{sep}}$-rational. The $\widetilde{\theta}_{v,i0}^{(1)}(z)$ are K_v-symmetric since the $\widetilde{c}_{v,i}$ and $\varphi_{i,n_vN_i}(z)$ are K_v-symmetric and $S_{n-n_v,v}(\phi_v(z))$ is K_v-rational.

Fourth, $G_v^{(1)}(z)$ is K_v-rational. To see this, note that $\widetilde{c}_{v,i}^{n_v}$ belongs to $K_v(x_i)^{\text{sep}}$, and the $\widetilde{\theta}_{v,i0}^{(1)}(z)$ are L_{w_v}-rational and K_v-symmetric as remarked above. It follows that $H(z)$ is K_v-rational. Since $G_v^{(0)}(z)$ is K_v-rational, $G_v^{(1)}(z)$ is K_v-rational as well.

We will now show that the roots of $G_v^{(1)}(z)$ belong to E_v. Let

$$\mathcal{S}_1 \ = \ \{n-n_v, n-n_v+1, \ldots, n-1\} \ ,$$

and put $P_1(x) = \prod_{j\in\mathcal{S}_1}(x-\psi_v(j))$, $\widehat{P}_1(x) = S_{n-n_v,v}(x) = \prod_{j=0}^{n-n_v-1}(x-\psi_v(j))$, so

$$S_{n,v}(x) \ = \ P_1(x)\cdot S_{n-n_v,v}(x) \ = \ P_1(x)\cdot\widehat{P}_1(x) \ .$$

Since $G_v^{(0)}(z) = S_{n,v}(\phi_v(z))$, using (11.48) we can write (11.49) as

$$\begin{aligned} G_v^{(1)}(z) \ &= \ G_v^{(0)}(z) \ + \ \Big(\sum_{i=1}^{m}\Delta_{v,i0}^{(1)}\cdot\widetilde{c}_{v,i}^{n_v}\varphi_{i,n_vN_i}(z)\Big)\cdot S_{n-n_v,v}(\phi_v(z)) \\ &= \ \Big(P_1(\phi_v(z)) + \Delta_{v,1}(z)\Big)\cdot\widehat{P}_1(\phi_v(z)) \\ &= \ Q_1^*(z)\cdot\widehat{P}_1(\phi_v(z)) \ , \end{aligned}$$

where $\Delta_{v,1}(z) = \sum_{i=1}^m \Delta_{v,i0}^{(1)}\cdot\widetilde{c}_{v,i}^{n_v}\varphi_{i,n_vN_i}(z)$ and $Q_1^*(z) = P_1(\phi_v(z)) + \Delta_{v,1}(z)$. Since $|\Delta_{v,i0}^{(1)}|_v \le B_v$ for each i, by (11.44) and the ultrametric inequality we have

$$\|\Delta_{v,1}(z)\|_{U_v} \ \le \ q^{-\frac{n_v}{q-1}-\log_v(n_v)} \ .$$

On each ball $B(\theta_h,\rho_h)$, we have $\|\Delta_{v,1}(z)\|_{B(\theta_h,\rho_h)} \le q^{-n_v/(q-1)-\log_v(n_v)} < 1$. Since $\|P_1(\phi_v(z))\|_{B(\theta_h,\rho_h)} = 1$, it follows that $\|Q_1^*(z)\|_{B(\theta_h,\rho_h)} = 1$.

For each $h = 1,\ldots,N$, let $\widehat{\sigma}_h : D(0,1) \to B(\theta_h,\rho_h)$ be the F_{u_h}-rational parametrization chosen at the beginning of §11.1; recall that the roots of $S_{n,v}(\phi_v(z))$ in $B(\theta_h,\rho_h)$ are θ_{hj} for $j = 0,\ldots,n-1$ and that $\theta_{hj} = \widehat{\sigma}_h(\alpha_{hj})$ where $\alpha_{hj} = \widetilde{\Phi}_h(\psi_v(j)) \in \mathcal{O}_{u_h}$.

Just as when $\text{char}(K_v) = 0$, we can apply Lemma 11.6 to $\mathcal{Q}_{1,h}(Z) = Q_1(\widehat{\sigma}_h(Z))$ and $\Delta_{1,h}(Z) = \Delta_{v,1}(\widehat{\sigma}_h(Z))$. The roots of $\mathcal{Q}_{1,h}(Z)$ in $D(0,1)$ are $\{\alpha_{hj}\}_{j\in\mathcal{S}_1}$, which is a ψ_v-regular sequence of length n_v in $\mathcal{O}_{u_h}$ in attached to $\mathcal{S}_1$. By Lemma 11.6 the roots $\{\alpha_{hj}^*\}_{j\in\mathcal{S}_1}$ of $\mathcal{Q}_{1,h}^*(Z) = \mathcal{Q}_{1,h}(Z) + \Delta_{1,h}(Z)$ form a ψ_v-regular sequence of length n_v in $\mathcal{O}_{u_h}$, with

$$\text{ord}_v(\alpha_{hj}^* - \alpha_{hj}) \ > \ \log_v(n_v) \tag{11.54}$$

for each $j \in \mathcal{S}_1$.

Thus the roots of $G_v^{(1)}(\widehat{\sigma}_h(Z))$ in $D(0,1)$ are

$$\{\alpha_{h0}, \alpha_{h1}, \ldots, \alpha_{h,n-n_v-1}, \ \alpha_{h,n-n_v}^*, \ldots, \alpha_{h,n-1}^*\} \ ,$$

which is the union of the initial part of the original ψ_v-regular sequence of length n and a ψ_v-regular sequence of length n_v. Transferring this back to $G_v^{(k)}(z)$, put $\theta_{hj}^* = \widehat{\sigma}_h(\alpha_{hj}^*)$ for each $j \in \mathcal{S}_1$. Then the roots of $G_v^{(1)}(z)$ in $B(\theta_h, \rho_h)$ are

$$\{\theta_{h0}, \theta_{h1}, \ldots, \theta_{h,n-n_v-1}, \ \theta_{h,n-n_v}^*, \ldots, \theta_{h,n-1}^*\} ,$$

and they all belong to $\mathcal{C}_v(F_{u_h}) \cap B(\theta_h, \rho_h) \subseteq E_v$.

For $k = 2, \ldots, \overline{k}$, we have $\Delta_{v,ij}^{(k)} = 0$ for all (i,j), and we take $G_v^{(k)}(z) = G_v^{(k-1)}(z)$. For notational compatibility with Case A, for each $k = 2, \ldots, \overline{k}$ let $\mathcal{S}_k$ be the empty set and put $Q_k^*(z) = 1$, $\widehat{P}_k(z) = \widehat{P}_1(z)$. Thus for each k,

$$G_v^{(k)}(z) \ = \ Q_1^*(z)Q_2^*(z)\cdots Q_k^*(z) \cdot \widehat{P}_k(\phi_v(z)) \ .$$

Note that the leading coefficient of $Q_1^*(z)$ at x_i is $\varepsilon_{v,i}\widetilde{c}_{v,i}^{n_v}$.

Phase 2. Patching for $k = \overline{k}+1, \ldots, k_1$.

Since $\overline{k} \geq k_v$, for each $k \geq \overline{k}+1$ we have

$$h_v^{kN} \cdot M_v \ < \ q^{-\frac{k+1}{q-1} - \log_v(k+1)} \ .$$

Let k_1 be the least integer such that for all $k \geq k_1$,

$$h_v^{kN} \cdot M_v \ < \ q^{-\frac{k+1}{q-1} - \log_v(n)} \ . \tag{11.55}$$

If n is large enough that

$$q^{-\frac{\overline{k}+1}{q-1} - \log_v(n)} \ < \ h_v^{\overline{k}N} \cdot M_v \ , \tag{11.56}$$

which we henceforth assume, then $k_1 > \overline{k}$. Since $h_v^N < q^{-1/(q-1)}$, there is a constant $A_1 > 0$ such that

$$k_1 \ \leq \ A_1 \log_v(n) \ ,$$

so if n is sufficiently large, then $n > k_1$ which we also assume.

The purpose of Phase 2 is simply to "carry on" until k is large enough that (11.55) holds, at which point Lemma 11.6 preserves the position of the roots α_{hj} within balls of size $q_v^{-\lceil \log_v(n) \rceil}$. Equivalently, in patching steps for $k > k_1$, each pair of patched and unpatched roots will satisfy

$$\mathrm{ord}_v(\alpha_{hj}^* - \alpha_{hj}) \ \geq \ \log_v(n) \ .$$

The patching process in Phase 2 is the same as the one for steps $k \geq 2$ in Case A of Phase 1, except that in place of (11.34) we require that each $\Delta_{v,ij}^{(k)}$ satisfy

$$|\Delta_{v,ij}^{(k)}|_v \ \leq \ h_v^{kN}. \tag{11.57}$$

When $\mathrm{char}(K_v) = 0$, for each $k = \overline{k}+1, \ldots, k_1$ we put $\mathcal{S}_k = \{j_k, \ldots, j_k + k\}$ as in (11.29). When $\mathrm{char}(K_v) = p > 0$, we put $\mathcal{S}_{\overline{k}+1} = \{0, 1, \ldots, \overline{k}\}$ and for $k = \overline{k}+2, \ldots, k_1$ we let $\mathcal{S}_k$ be the set consisting of the next $k+1$ integers after $\mathcal{S}_{k-1}$. These $\mathcal{S}_k$ will be subsequences of $\{0, 1, \ldots, n-1\}$ if n is large enough that

$$n \ > \ n_v + (A_1 \log_v(n) + 1)(\ A_1 \log_v(n) + 2)/2 \ , \tag{11.58}$$

which we henceforth assume.

Let the $P_k(z)$, $\widehat{P}_k(z)$, $Q_k(z)$, and $Q_k^*(z)$ be as in (11.31), (11.32), and (11.38). Inductively, suppose $G_v^{(k-1)}(z)$ has been constructed with

$$\begin{aligned} G_v^{(k-1)}(z) &= Q_1^*(z)\cdots Q_{k-1}^*(z)\cdot \widehat{P}_{k-1}(\phi_v(z)) \\ &= Q_1^*(z)\cdots Q_{k-1}^*(z)\cdot P_k(\phi_v(z))\cdot \widehat{P}_k(\phi_v(z)). \end{aligned} \tag{11.59}$$

When $\mathrm{char}(K_v)=0$ we patch $G_v^{(k-1)}(z)$ by setting

$$\begin{aligned} G_v^{(k)}(z) &= G_v^{(k-1)}(z) + \sum_{i=1}^{m}\sum_{j=(k-1)N_i}^{kN_i-1} \Delta_{v,ij}^{(k)}\vartheta_{v,ij}^{(k)}(z) \\ &= Q_1^*(z)\cdots Q_{k-1}^*(z)\cdot\big(P_k(\phi_v(z))+\Delta_{v,k}(z)\big)\cdot \widehat{P}_k(\phi_v(z))\ , \end{aligned}$$

where the compensating functions are

$$\vartheta_{v,ij}^{(k)}(z) = \varepsilon_{v,i}^{-1}\varphi_{i,(k+1)N_i-j}(z)\cdot Q_1^*(z)\cdots Q_{k-1}^*(z)\cdot \widehat{P}_k(\phi_v(z))\ ,$$

and where as in (11.37),

$$\Delta_{v,k}(z) = \sum_{i=1}^{m}\sum_{j=(k-1)N_i}^{kN_i-1} \Delta_{v,ij}^{(k)}\cdot\varepsilon_{v,i}^{-1}\varphi_{i,(k+1)N_i-j}(z)\ .$$

It is easy to see that the $\vartheta_{v,ij}^{(k)}(z)$ have the properties required by Theorem 11.1. Including the factor of $\varepsilon_{v,i}^{-1}$ makes the leading coefficient of $\vartheta_{v,ij}^{(k)}(z)$ at x_i be $\widetilde{c}_{v,i}^{\,n-k-1}$.

When $\mathrm{char}(K_v)=p>0$, we put

$$\Delta_{v,k}(z) = \sum_{i=1}^{m}\sum_{j=(k-1)N_i}^{kN_i-1} \Delta_{v,ij}^{(k)}\cdot\varphi_{i,(k+1)N_i-j}(z)\ ,$$

which is K_v-rational by the hypotheses of Theorem 11.2, and let

$$F_{v,k}(z) = Q_1^*(z)\cdots Q_{k-1}^*(z)\cdot \widehat{P}_k(\phi_v(z))\ .$$

We patch $G_v^{(k-1)}(z)$ by setting

$$\begin{aligned} G_v^{(k)}(z) &= G_v^{(k-1)}(z) + \Delta_{v,k}(z)\cdot F_{v,k}(z) \\ &= Q_1^*(z)\cdots Q_{k-1}^*(z)\cdot\big(P_k(\phi_v(z))+\Delta_{v,k}(z)\big)\cdot \widehat{P}_k(\phi_v(z)). \end{aligned}$$

By induction $F_{v,k}(z)$ is K_v-rational, its roots belong in E_v, and it has a pole of order $(n-k-1)N_i$ with leading coefficient $d_{v,i}=\varepsilon_{v,i}\widetilde{c}_{v,i}^{\,n-k-1}$ at each x_i. In particular, $|d_{v,i}|_v = |\widetilde{c}_{v,i}^{\,n-k-1}|_v$. Thus $F_{v,k}(z)$ satisfies the conditions of Theorem 11.2.

The two patching constructions are the same except for a minor difference in the choice of $\Delta_{v,k}(z)$, and in both cases

$$\|\Delta_{v,k}\|_{U_v^0} \le \max_{i,j}(\Delta_{|v,ij}|_v)\cdot\max_{i,j}(\|\varphi_{(k+1)N_i-j}\|_{U_v^0}) \le h_v^{kN}\cdot M_v < q^{-\frac{k}{q-1}-\log_v(k+1)}\ .$$

The argument from Phase 1, using Lemma 11.6, shows $Q_k^*(z)=P_k(\phi_v(z))+\Delta_{v,k}(z)$ is K_v-rational and has its roots in E_v. For each $h=1,\ldots,N$, the roots of $Q_k^*(z)$ in $B(\theta_h,\rho_h)$ are $\theta_{hj}^*=\widehat{\sigma}_h(\alpha_{hj}^*)$, for $j\in\mathcal{S}_k$. Since for each h,

$$\|\Delta_{v,k}\|_{B(\theta,\rho_h)} < q^{-\frac{k+1}{q-1}-\log_v(k+1)}\ ,$$

the α_{hj}^* form a ψ_v-regular sequence of length $k+1$ in $\mathcal{O}_{u_h}$, and for each $j\in\mathcal{S}_k$,

$$\mathrm{ord}_v(\alpha_{hj}^*-\alpha_{hj}) > \log_v(k+1)\ . \tag{11.60}$$

After patching, we have a factorization

$$G_v^{(k)}(z) \;=\; Q_1^*(z)\cdots Q_k^*(z)\cdot \widehat{P}_k(\phi_v(z)) \ ,$$

so the induction in (11.59) can continue.

Phase 3. Moving the roots apart.

At this point we have obtained a K_v-rational $(\mathfrak{X},\vec{s})$-function $G_v^{(k_1)}(z)$ of degree nN whose coefficients $A_{v,ij}$ with $1 \le i \le m$, $0 \le j < k_1 N_i$ have been patched, and which has the factorization

$$G_v^{(k_1)}(z) \;=\; Q_1^*(z)\cdots Q_{k_1}^*(z)\cdot \widehat{P}_{k_1}(\phi_v(z)) \ .$$

Its zeros belong to E_v, and it has n roots in each ball $B(\theta_h,\rho_h)$: when $\mathrm{char}(K_v)=0$, these are

$$\{\theta_{h0}^*,\theta_{h1}^*,\ \theta_{h2}^*,\theta_{h3}^*,\theta_{h4}^*,\ \ldots,\ \theta_{h,j_{k_1}+k_1}^*,\ \theta_{h,j_{k_1}+k_1+1},\ldots,\theta_{h,n-1}\} \ .$$

When the zeros in $B(\theta_h,\rho_h)$ are pulled back to $D(0,1)$ using $\widehat{\sigma}_h$, they become

$$\{\alpha_{h0}^*,\alpha_{h1}^*,\ \alpha_{h2}^*,\alpha_{h3}^*,\alpha_{h4}^*,\ \ldots,\ \alpha_{h,j_{k_1}+k_1}^*,\ \alpha_{h,j_{k_1}+k_1+1},\ldots,\alpha_{h,n-1}\} \ ,$$

a union of the ψ_v-regular sequences in $\mathcal{O}_{u_h}$ corresponding to $\mathcal{S}_1,\ldots,\mathcal{S}_{k_1}$, together with the unpatched roots. For each $k=1,\ldots,k_1$ the "patched" roots $\{\alpha_{hj}^*\}_{j\in\mathcal{S}_k}$ satisfy

$$\mathrm{ord}_v(\alpha_{hj}^*-\alpha_{hj}) \;>\; \log_v(\#(\mathcal{S}_k))$$

relative to the original "unpatched" roots α_{hj} for $j\in\mathcal{S}_k$.

Although the patched roots corresponding to $\mathcal{S}_k$ are well-separated from each other, they may have come close to (or even coincide with) roots corresponding to some other $\mathcal{S}_\ell$ or to unpatched roots. The purpose of Phase 3 is to move the patched roots α_{hj}^* to new points α_{hj}^{**} which are well-separated from each other and the unpatched roots, while preserving the coefficients of $G_v^{(k_1)}(z)$ that have already been patched.

We accomplish this in two steps. First, we move the patched roots away from any roots they have come too near to. This changes the high-order coefficients. Second, we restore the high-order coefficients to their original values, by repatching using a new sequence of roots. We allow these new roots only to move in such a way that they stay well-separated from the other roots, which in turn limits how far the original α_{hj}^* can be moved. A computation shows that for a suitable constant A_4, the roots α_{hj}^{**} of the repatched function $\widehat{G}_v^{(k_1)}(z)$ can be required to satisfy $|\alpha_{hj}^{**}-\alpha_{h\ell}^{**}|_v \ge n^{-A_4}$ (or equivalently $\mathrm{ord}_v(\alpha_{hj}^{**}-\alpha_{h\ell}^{**}) \le A_4\log_v(n)$) for all h and all $j\ne\ell$. We then replace $G_v^{(k_1)}(z)$ with $\widehat{G}_v^{(k_1)}(z)$.

There are several obstacles to carrying this out. Moving the roots α_{hj}^* generally results in a nonprincipal divisor. To compensate, we choose a collection of roots in "good position" which are well-separated from the patched roots, and move them to regain a principal divisor. (It is here that we use the assumption that H_v is "move-prepared".) In Lemma 11.10 below, we construct a function $Y(z)\in K_v(\mathcal{C})$ whose zeros are the moved roots, whose poles are the original roots, and which is very close to 1 outside U_v. Multiplying $G_v^{(k_1)}(z)$ by $Y(z)$ yields a function $\overline{G}_v^{(k_1)}(z)$ with the desired new roots. By standard estimates for Laurent coefficients, the amount the high-order coefficients are changed in passing from $G_v^{(k_1)}(z)$ to $\overline{G}_v^{(k_1)}(z)$

depends on how close $Y(z)$ is to 1 at the points x_i, which in turn depends on how far the roots were moved.

To restore the high-order coefficients to their previous values, we apply the basic patching lemma but use a carefully chosen, previously unpatched sequence of roots to absorb the resulting movement in the roots. We prove an estimate giving the "cost" of independently adjusting each coefficient. This tells us how far the α_{hj}^* can be moved.

We now give the details of the construction, postponing the proofs of the three Moving Lemmas 11.10, 11.11, and 11.12 to §11.4. Put

$$\delta_n = q_v^{-\lceil \log_v(n) \rceil} .$$

For each $\alpha \in D(0,1)$, we will call the disc $D(\alpha, \delta_n)$ the "δ_n-coset" of α, and we will refer to elements of $\{0, 1, \ldots, n-1\}$ as "indices".

Let $\mathcal{S}^\triangle = \mathcal{S}_1 \cup \mathcal{S}_2 \cup \cdots \cup \mathcal{S}_{k_1}$ be the collection of indices which have already been used in the patching process; we will call it the set of "patched" indices. We will call its complement $\mathcal{S}^\diamondsuit = \{0, 1, \ldots, n-1\} \backslash \mathcal{S}^\triangle$ the set of "unpatched" indices. Let $\mathcal{S}^\dagger \subset \mathcal{S}^\diamondsuit$ be the collection of unpatched indices which Steps 1 and 2 of the patching process have "endangered": the set of indices $\ell \in \mathcal{S}^\diamondsuit$ for which some patched root α_{hj}^* has moved too close to an unpatched root $\alpha_{h\ell}$:

$$\begin{aligned} \mathcal{S}^\dagger \quad = \quad & \{\ell \in \mathcal{S}^\diamondsuit : \operatorname{ord}_v(\alpha_{hj}^* - \alpha_{h\ell}) \geq \log_v(n) \text{ for some } j \in \mathcal{S}_1 \cup \mathcal{S}_2 \cup \cdots \cup \mathcal{S}_{k_1} \\ & \text{and some } h,\ 1 \leq h \leq N\} . \end{aligned}$$

This set of indices must be "protected" until later in the patching process; we will call the corresponding roots "endangered". Finally, put

$$\mathcal{S}^\heartsuit \ = \ \mathcal{S}^\diamondsuit \backslash \mathcal{S}^\dagger .$$

This is the set of indices which are "safe" to use in repatching: for each $\ell \in \mathcal{S}^\heartsuit$, there is no h for which any α_{hj}^* belongs to the δ_n-coset of $\alpha_{h\ell}$.

We have $\#(\mathcal{S}^\triangle) \leq A_2 \cdot (\log_v(n))^2$ for an appropriate constant A_2. Since each patched root α_{hj}^* can belong to the δ_n-coset of at most one unpatched root $\alpha_{h\ell}$, it follows that

$$\#(\mathcal{S}^\triangle \cup \mathcal{S}^\dagger) \ \leq \ (N+1) A_2 (\log_v(n))^2 . \tag{11.61}$$

We view $\mathcal{S}^\triangle \cup \mathcal{S}^\dagger$ as a collection of marked indices which partitions its complement in $\{0, 1, \ldots, n-1\}$ (the set of "safe" indices $\mathcal{S}^\heartsuit$) into a collection of sequences of consecutive integers. Let $\mathcal{S}^0$ be the longest such sequence (to be specific, the first one, if there are two or more of the same length). The Pigeon-hole Principle shows that by taking n sufficiently large, we can make S^0 arbitrarily long. If

$$n \ \geq \ ((N+1) A_2 (\log_v(n))^2 + 1) \cdot (A_1 \log_v(n) + 2) , \tag{11.62}$$

which we will henceforth assume, then

$$\#(\mathcal{S}^0) \ \geq \ A_1 \log_v(n) + 2 \ \geq \ k_1 + 2 . \tag{11.63}$$

We will call S^0 the "long safe sequence". Write $\mathcal{S}^0 = \{j_0, j_1, \ldots, j_L\}$. The index j_0 will be used to provide the roots in "good position" needed to recover principal divisors when the α_{hj}^* are moved in the first step of the process. Let

$$\mathcal{S}^0[k_1] \ = \ \{j_1, \ldots, j_{k_1+1}\}$$

be the subsequence of $\mathcal{S}^0$ consisting of the next k_1+1 integers. This is the sequence of indices that will be used for "repatching" in the second step.

Our first lemma describes the properties of the divisor $\mathcal{D}$ and the function $Y(z)$. It relates ε, the distance we can move the roots, to $C_2\varepsilon$, a bound for the size of $|Y(z)-1|_v$.

LEMMA 11.10 (First Moving Lemma). *Let $\delta_n = q_v^{-\lceil \log_v(n)\rceil}$, $\mathcal{S}^\triangle$, j_0, and $\{\theta_{hj}^* \in \mathcal{C}_v(F_{u_h})\}_{1\le h\le N, j\in S^\triangle}$ be as above. Then there are an $\varepsilon_1>0$ and constants $C_1, C_2 \ge 1$ (depending on $\phi_v(z)$, E_v, H_v, and their K_v-simple decompositions, but not on n), with the following property:*

For each $0<\varepsilon<\varepsilon_1$ such that

$$C_1\varepsilon \;\le\; \delta_n \cdot \min_{1\le h\le N}(\rho_h)\ , \tag{11.64}$$

given any K_v-symmetric set $\{\theta_{hj}^{**} \in \mathcal{C}_v(F_{u_h}) \cap B(\theta_h,\rho_h)\}_{1\le h\le N, j\in S^\triangle}$ with the property that $\|\theta_{hj}^{**},\theta_{hj}^*\|_v<\varepsilon$ for all (h,j), there is a K_v-symmetric collection of points $\{\theta_{h,j_0}^{**} \in \mathcal{C}_v(F_{u_h}) \cap B(\theta_h,\rho_h)\}_{1\le h\le N}$ satisfying

$$\|\theta_{h,j_0}^{**},\theta_{h,j_0}\|_v \;\le\; C_1\varepsilon \;\le\; \delta_n\rho_h$$

for each h, such that:

(A) The divisor

$$\mathcal{D} \;=\; \sum_{j\in S^\triangle}\sum_{h=1}^{N}((\theta_{hj}^{**})-(\theta_{hj}^*)) + \sum_{h=1}^{N}((\theta_{h,j_0}^{**})-(\theta_{h,j_0}))$$

is K_v-rational and principal.

(B) Writing $U_v^0=\bigcup_{h=1}^N B(\theta_h,\rho_h) \subset U_v$ as before, we have

(1) $|Y(z)|_v = 1$ for all $z \in \mathcal{C}_v(\mathbb{C}_v)\backslash U_v^0$;

(2) $|Y(z)-1|_v \le C_2\varepsilon$ for all $z\in\mathcal{C}_v(\mathbb{C}_v)\backslash U_v^0$.

For now, let $\varepsilon>0$ and $\{\theta_{hj}^{**}\}_{1\le h\le N, j\in S^\triangle\cup\{j_0\}}$ be any number and collection of points satisfying the conditions of Lemma 11.10, and let $\mathcal{D}$ and $Y(z)$ be the corresponding divisor and function. We will explain the rest of the construction, then make the final choice of ε and the θ_{hj}^{**} at the end.

Put $\overline{G}_v^{(k_1)}(z) = Y(z)G_v^{(k_1)}(z)$. We first consider how the coefficients of $G_v^{(k_1)}(z)$ change in passing from $G_v^{(k_1)}(z)$ to $\overline{G}_v^{(k_1)}(z)$. Write

$$G_v^{(k_1)}(z) \;=\; \sum_{i=1}^{m}\sum_{j=0}^{(n-1)N_i-1} A_{v,ij}\varphi_{i,nN_i-j}(z) + \sum_{\lambda=1}^{\Lambda} A_\lambda\varphi_\lambda(z)\ , \tag{11.65}$$

$$\overline{G}_v^{(k_1)}(z) \;=\; \sum_{i=1}^{m}\sum_{j=0}^{(n-1)N_i-1} \overline{A}_{v,ij}\varphi_{i,nN_i-j}(z) + \sum_{\lambda=1}^{\Lambda} \overline{A}_\lambda\varphi_\lambda(z)\ . \tag{11.66}$$

Because $G_v^{(k_1)}(z)$ and $\overline{G}_v^{(k_1)}(z)$ are K_v-rational, the $A_{v,ij}$ and $\overline{A}_{v,ij}$ belong to L_{w_v} and are K_v-symmetric.

LEMMA 11.11 (Second Moving Lemma). *There are constants $\varepsilon_2>0$ and $C_3, C_4\ge 1$ (depending on E_v, $\mathfrak{X}$, the choices of the L-rational and L^{sep}-rational bases, the uniformizers $g_{x_i}(z)$, and the projective embedding of $\mathcal{C}_v$), such that if ε*

and $Y(z)$ are as in Lemma 11.10, and in addition $\varepsilon < \varepsilon_2$ and n is sufficiently large, then for all $i = 1, \ldots, m$ and all $0 \le j < k_1 N_i$, we have

$$|\overline{A}_{v,ij} - A_{v,ij}|_v \ \le \ C_3 C_4^j (|\widetilde{c}_{v,i}|_v)^n \cdot \varepsilon \ .$$

We next ask about the largest change in the coefficients $A_{v,ij}$ we can correct for, by repatching. Write

$$\overline{G}_v^{(k_1)}(z) \ = \ Y(z) \cdot G_v^{(k_1)}(z) \ = \ Y(z) Q_1^*(z) \cdots Q_{k_1}^*(z) \cdot \widehat{P}_{k_1}(\phi_v(z)) \ .$$

Multiplying $G_v^{(k_1)}(z)$ by $Y(z)$ moves the θ_{hj}^* for $j \in \mathcal{S}^\triangle$ to the θ_{hj}^{**}, moves the θ_{hj_0} to the $\theta_{hj_0}^{**}$, and leaves all other zeros unchanged. Recalling that $x - \psi_v(j_0)$ is a factor of $\widehat{P}_{k_1}(x)$, write

$$\widehat{P}_{k_1}(x) \ = \ (x - \psi_v(j_0)) \cdot \widehat{P}_{k_1,j_0}(x)$$

and put

$$\overline{Q}_{k_1}(z) \ = \ Y(z) Q_1^*(z) \cdots Q_{k_1}^*(z)(\phi_v(z) - \psi_v(j_0))$$

so that

$$\overline{G}_v^{(k_1)}(z) = \overline{Q}_{k_1}(z) \cdot \widehat{P}_{k_1,j_0,}(\phi_v(z)) \ . \tag{11.67}$$

The function $\overline{Q}_{k_1}(z)$ accounts for the change in passing from $G_v^{(0)}(z)$ to $\overline{G}_v^{(k_1)}(z)$. It has the following properties.

First, $\overline{Q}_{k_1}(z)$ is a K_v-rational $(\mathfrak{X}, \vec{s})$-function, and it extends to a function defined and finite on $B(\theta_h, \rho_h)$, for each $1 \le h \le N$. Indeed, the zeros of $\overline{Q}_{k_1}(z)$ are those of $Y(z)$, and the poles of $\overline{Q}_{k_1}(z)$ are those of $Q_1^*(z) \cdots Q_{k_1}^*(z)(\phi_v(z) - \psi_v(j_0))$: the poles of $Y(z)$ cancel with the zeros of $Q_1^*(z) \cdots Q_{k_1}^*(z)(\phi_v(z) - \psi_v(j_0))$.

Second, $\|\overline{Q}_{k_1}\|_{B(\theta_h,\rho_h)} = 1$ for each $1 \le h \le N$. To see this, fix h and restrict $\overline{Q}_{k_1}(z)$ to $B(\theta_h, \rho_h)$. Let $\{\eta_{h\ell}\}_{1 \le \ell \le T_h}$ be a list of the zeros and poles of $Y(z)$ and the $Q_k^*(z)$ in $B(\theta_h, \rho_h)$, and write $\eta_{h\ell} = \widehat{\sigma}_h(\tau_{h\ell})$ for each h, ℓ. For all $z \in B(\theta_h, \rho_h) \backslash \big(\bigcup_{\ell=1}^{T_h} B(\eta_{h\ell}, \rho_h)^-\big)$ we have $|\overline{Q}_{k_1}(z)|_v = 1$, since this is true for each of the factors in its definition. Pulling this back to $D(0,1)$, we see that $\overline{Q}_{k_1}(\widehat{\sigma}_h(Z))$ is a power series converging in $D(0,1)$, with absolute value 1 except on the finitely many subdiscs $D(\tau_{h\ell}, 1)^-$. By the Maximum Modulus Principle for power series, $\|\overline{Q}_{k_1}\|_{B(\theta_h,\rho_h)} = 1$.

Third, when the zeros of $\overline{Q}_{k_1}(z)$ in $B(\theta_h, \rho_h)$ are pulled back to $D(0,1)$ using $\widehat{\sigma}_h(Z)$, they form a union of ψ_v-regular sequences of lengths $1, \#(\mathcal{S}_1), \ldots, \#(\mathcal{S}_{k_1})$ attached to the sets $\{j_0\}, \mathcal{S}_1, \ldots, \mathcal{S}_{k_1}$. This holds by construction, since the zeros of $\phi_v(z) - \psi_v(j_0)$ and $Q_1^*(z), \ldots, Q_{k_1}^*(z)$ have this property, and multiplying by $Y(z)$ moves each root only by an amount which preserves its position in its ψ_v-regular sequence.

Fourth, $\overline{Q}_{k_1}(z)$ has degree Nt, where $t = \#(\mathcal{S}_1 \cup \cdots \cup \mathcal{S}_{k+1})$. For each $x_i \in \mathfrak{X}$, the leading coefficient of $\overline{Q}_{k_1}(z)$ at x_i has the form $\mu_{v,i} \cdot \vec{c}_{v,i}^{\,t}$ where $\mu_{v,i} \in \mathcal{O}_{w_v}^\times$. Indeed, the leading coefficient of $(\phi_v(z) - \psi_v(j_0)) Q_1^*(z) \ldots Q_{k_1}^*(z)$ at x_i is $\varepsilon_{v,i} \cdot \vec{c}_{v,i}^{\,t}$, while $Y(x_i) \in \mathcal{O}_{w_v}^\times$ by Lemma 11.10(B2).

Our plan is to replace $\overline{G}_v^{(k_1)}(z)$ with a new function $\widehat{G}_v^{(k_1)}(z)$ with the same high-order coefficients as the original $G_v^{(k_1)}(z)$, and whose zeros are well-separated from each other. To do this, we will use the basic patching lemma via the sequence of "safe" indices $\mathcal{S}^0[k_1]$ of length $k_1 + 1$.

Put

$$P^0_{k_1}(x) = \prod_{j \in \mathcal{S}^0[k_1]} (x - \psi_v(j)) \ , \quad \widehat{P}^0_{k_1}(x) = \prod_{j \in \mathcal{S}^\diamond \backslash (\{j_0\} \cup \mathcal{S}^0[k_1])} (x - \psi_v(j)) \ ,$$

noting that $\widehat{P}_{k_1,j_0}(x) = P^0_{k_1}(x) \cdot \widehat{P}^0_{k_1}(x)$. Then write

$$\overline{F}_{v,k_1}(z) \ = \ \overline{Q}_{k_1}(z) \cdot \widehat{P}^0_{k_1}(\phi_v(z)) \ , \tag{11.68}$$

so that

$$\overline{G}_v^{(k_1)}(z) \ = \ P^0_{k_1}(\phi_v(z)) \cdot \overline{F}_{v,k_1}(z) \ . \tag{11.69}$$

Thus $\overline{F}_{v,k_1}(z)$ is a K_v-rational $(\mathfrak{X}, \vec{s})$-function of degree $N \cdot (n - k_1 - 1)$, whose roots form a union of ψ_v-regular sequences accounting for the indices in $\{0, \ldots, n-1\} \backslash \mathcal{S}^0[k_1]$. It has a pole of order $(n - k_1 - 1)N_i$ at each x_i, with leading coefficient $\mu_{v,i}\widetilde{c}_{v,i}^{\,n-k_1-1}$. Furthermore, $\|\overline{F}_{v,k_1}\|_{U_v^0} \le 1$, since $\|\overline{Q}_{k_1}\|_{U_v^0} = 1$ and $\|\widehat{P}^0_{k_1}\|_{U_v^0} = 1$.

LEMMA 11.12 (Third Moving Lemma). *There are constants $\varepsilon_3 > 0$ and $C_6, C_7 \ge 1$ (depending only on $\phi_v(z)$, E_v, H_v, their K_v-simple decompositions $\bigcup_{\ell=1}^D \big(B(a_\ell, r_\ell) \cap \mathcal{C}_v(F_{w_\ell})\big)$ and $\bigcup_{h=1}^N \big(B(\theta_h, \rho_h) \cap \mathcal{C}_v(F_{u_h})\big)$, the choices of the L-rational and L^{sep}-rational bases, the uniformizers $g_{x_i}(z)$, and the projective embedding of $\mathcal{C}_v$), such that if $0 < \varepsilon < \varepsilon_3$, then there is a K_v-rational $(\mathfrak{X}, \vec{s})$-function $\overline{\Delta}_{v,k_1}(z)$ of the form*

$$\overline{\Delta}_{v,k_1}(z) \ = \ \sum_{i=1}^{m} \sum_{j=0}^{k_1 N_i - 1} \overline{\Delta}_{v,ij} \varphi_{i,(k_1+1)N_i - j}(z) \ ,$$

and for which

$$\|\overline{\Delta}_{v,ij}(z)\|_{U_v^0} \ \le \ C_6 C_7^{k_1} \varepsilon \ ,$$

such that when $\overline{G}_v^{(k_1)}(z)$ from Lemma 11.11 is replaced with

$$\widehat{G}_v^{(k_1)}(z) \ = \ \overline{G}_v^{(k_1)}(z) + \overline{\Delta}_{v,k_1}(z) \overline{F}_{v,k_1}(z) \ , \tag{11.70}$$

then for each (i, j) with $1 \le i \le m$, $0 \le j < k_1 N_i$, the coefficient $\overline{A}_{v,ij}$ of $\overline{G}_v^{(k_1)}(z)$ is restored to the coefficient $A_{v,ij}$ of $G_v^{(k_1)}(z)$ in $\widehat{G}_v^{(k_1)}(z)$.

Finally, we consider the amount of movement in the roots caused by (11.70). Since

$$\widehat{G}_{k_1}(z) \ = \ \big(P^0_{k_1}(\phi_v(z)) + \overline{\Delta}_{v,k_1}(z)\big) \cdot \overline{F}_{v,k_1}(z) \ ,$$

the movement is isolated to the roots of $P^0_{k_1}(\phi_v(z))$.

Write $Q^0_{k_1}(z) = P^0_{k_1}(\phi_v(z))$, put

$$Q^{0*}_{k_1}(z) \ := \ Q^0_{k_1}(z) + \overline{\Delta}_{v,k_1}(z) \ , \tag{11.71}$$

and consider $Q^0_{k_1}(z)$ and $Q^{0*}_{k_1}(z)$ on each ball $B(\theta_h, \rho_h)$. Pull them back to $D(0,1)$ using $\widehat{\sigma}_h(Z)$, and apply the Basic Patching Lemma (Lemma 11.6). The roots of $Q^0_{k_1}(\widehat{\sigma}_h(Z)))$ in $D(0,1)$ form a ψ_v-regular sequence of length $k_1 + 1$ in $\mathcal{O}_{u_h}$ attached to $\mathcal{S}^0[k_1]$, namely $\{\alpha_{hj}\}_{j \in \mathcal{S}^0[k_1]}$. We have chosen $\mathcal{S}^0[k_1]$ to be a "safe" sequence, which means these α_{hj} are the only roots within their δ_n-cosets.

If we can arrange that the roots only move within their δ_n-cosets, they will remain separated from all the other roots. To assure this it is enough to require

$$C_6 C_7^{k_1} \varepsilon \ \le \ q^{-\frac{k_1+1}{q-1} - \log_v(n)} \ . \tag{11.72}$$

Under this condition, Lemma 11.6 shows that for each h the roots $\{\alpha_{hj}^*\}_{j\in\mathcal{S}^0[k_1]}$ of $\overline{Q}_{k_1}^{0*}(\widehat{\sigma}_h(Z)))$ form a ψ_v-regular sequence of length k_1+1 in $\mathcal{O}_{u_h}$, satisfying

$$\mathrm{ord}_v(\alpha_{hj}^* - \alpha_{hj}) \ \geq \ \log_v(n)$$

for each $j \in \mathcal{S}^0[k_1]$.

We will now specify ε and the θ_{hj}^{**}.

We want ε to be as large as possible. For the construction to succeed, we must have $\varepsilon < \min(\varepsilon_1, \varepsilon_2, \varepsilon_3)$, and (11.64) and (11.72) must hold:

$$C_1\varepsilon \le q^{-\lceil \log_v(n)\rceil} \cdot \min_{1\le h\le N}(\rho_h), \qquad C_6 C_7^{k_1}\varepsilon \ \le \ q^{-\frac{k_1+1}{q-1}-\log_v(n)} \ .$$

Since $k_1 \le A_1 \log_v(n)$, then for an appropriate constant A_3 we can choose ε so that

$$-\log_v(\varepsilon) \ = \ A_3 \log_v(n) \ , \tag{11.73}$$

provided n is sufficiently large.

We next choose the θ_{hj}^{**} for $1 \le h \le N$, $j \in \mathcal{S}^{\triangle}$.

Fixing h, let $\widehat{\sigma}_h : D(0,1) \to B(\theta_h, \rho_h)$ be the parametrization used before; specifying the θ_{hj}^{**} is equivalent to specifying numbers $\alpha_{hj}^{**} \in \mathcal{O}_{u_h}$ such that $\widehat{\sigma}_h(\alpha_{hj}^{**}) = \theta_{hj}^{**}$. Recall that $\|\widehat{\sigma}_h(x), \widehat{\sigma}_h(y)\|_v = \rho_h|z-x|_v$ for all $x, y \in D(0,1)$. Put $\overline{\rho} = \max_{1\le h\le N}(\rho_h)$ and let $\varepsilon_0 = \varepsilon/\overline{\rho}$, noting that $\varepsilon_0 \le \delta_n$ by (11.64). In Lemma 11.10 we can move the α_{hj}^* for $j \in \mathcal{S}^{\triangle}$ to arbitrary points $\alpha_{hj}^{**} \in \mathcal{O}_{u_h}$ such that $|\alpha_{hj}^{**} - \alpha_{hj}^*|_v \le \varepsilon_0$ (provided the collection $\{\alpha_{hj}^{**}\}$ is K_v-symmetric), while only moving the α_{hj_0} within their δ_n-cosets. However, we need to choose the α_{hj}^{**} in such a way that they become well-separated from each other and from the unpatched roots.

For a given $\alpha \in \mathcal{O}_{u_h}$, consider the ε_0-coset of α in $\mathcal{O}_{u_h}$,

$$D_{u_h}(\alpha, \varepsilon_0) \ := \ \mathcal{O}_{u_h} \cap D(\alpha, \varepsilon_0) \ = \ \{z \in \mathcal{O}_{u_h} : |z-\alpha|_v \le \varepsilon_0\} \ .$$

The roots α_{hj}^* (which correspond to the indices $j \in \mathcal{S}^{\triangle} = \mathcal{S}_1 \cup \cdots \cup \mathcal{S}_{k_1}$), form a union of ψ_v-regular subsequences of respective lengths $\#(\mathcal{S}_1), \ldots, \#(\mathcal{S}_{k_1})$. At most one α_{hj}^* from each subsequence can belong to $D_{u_h}(\alpha, \varepsilon_0)$. Since the original sequence $\{\alpha_{hj}\}_{0\le j\le n-1}$ was a ψ_v-regular sequence of length n in $\mathcal{O}_{u_h}$ there is at most one unpatched root $\alpha_{hj'}$ in $D_{u_h}(\alpha, \varepsilon_0)$. Thus $D_{u_h}(\alpha, \varepsilon_0)$ contains at most k_1+1 roots. Furthermore, if $\alpha_{h,j_0}^{**} \in D_{u_h}(\alpha, \varepsilon_0)$, then since $\varepsilon_0 \le \delta_n$, our choice of j_0 means that $D_{u_h}(\alpha, \varepsilon_0)$ does not contain any of the α_{hj}^*.

Put $\delta_0 = q^{-\lceil \log_v(k_1+1)\rceil}$. Since $\log_v(k_1) \le \log_v(A_1\log_v(n))$, it follows from (11.73) that

$$-\log_v(\delta_0\varepsilon_0) \ = \ -\log_v(\delta_0\varepsilon/\overline{\rho}) \ \le \ A_4 \log_v(n) \tag{11.74}$$

for an appropriate constant A_4 . There are at least $q^{\lceil \log_v(k_1+1)\rceil} \ge k_1+1$ distinct $\delta_0\varepsilon_0$-cosets $D_{u_h}(\beta, \delta_0\varepsilon_0) \subset D_{u_h}(\alpha, \delta_0)$ with $\beta \in \mathcal{O}_{u_h}$. By simultaneously adjusting all the α_{hj}^* belonging to $D_{u_h}(\alpha, \varepsilon_0)$ we can choose new roots $\alpha_{hj}^{**} \in D_{u_h}(\alpha, \varepsilon_0) = \mathcal{O}_{u_h} \cap D(\alpha, \varepsilon_0)$ which are separated from each other and from the unpatched root $\alpha_{hj'}$ (if it exists), by a distance at least $\delta_0\varepsilon_0$. Do this for each h and each $D_{u_h}(\alpha, \varepsilon_0)$, making the choices for different h in a Galois-equivariant way.

It follows that for each h, and each $j \in \mathcal{S}_1 \cup \cdots \cup \mathcal{S}_{k_1}$, we can choose the $\alpha_{hj}^{**} \in \mathcal{O}_{u_h}$ so that $|\alpha_{hj}^{**} - \alpha_{hj}^*|_v \le \varepsilon_0$ and

(1) for each $\ell \in \mathcal{S}_1 \cup \cdots \cup \mathcal{S}_{k_1}$ with $\ell \neq j$,

$$\operatorname{ord}_v(\alpha_{hj}^{**} - \alpha_{h\ell}^{**}) \ \leq \ A_4 \log_v(n) \ ,$$

(2) for each $\ell \in \{j_0\} \cup \mathcal{S}^0[k_1]$

$$\operatorname{ord}_v(\alpha_{hj}^{**} - \alpha_{h\ell}^{*}) \ \leq \ A_4 \log_v(n) \ ,$$

(3) for each $\ell \in \{0, 1, \ldots, n-1\} \backslash (\{j_0\} \cup \mathcal{S}_1 \cup \cdots \cup \mathcal{S}_{k_1} \cup \mathcal{S}^0[k_1])$

$$\operatorname{ord}_v(\alpha_{hj}^{**} - \alpha_{h\ell}) \ \leq \ A_4 \log_v(n) \ .$$

Property (2) holds because $\{j_0\} \bigcup \mathcal{S}_{k_1}^0 \subset \mathcal{S}^{\heartsuit}$. Fix such a choice of the α_{hj}^{**}, and define $\widehat{G}_v^{(k_1)}(z)$ by means of Lemma 11.12.

Given $\alpha, \alpha' \in \mathbb{C}_v$, we will say that α and α' are "logarithmically separated by at least T" if

$$\operatorname{ord}_v(\alpha' - \alpha) \ \leq \ T$$

or, equivalently, if $|\alpha' - \alpha|_v \geq q_v^{-T}$.

To ease the notation in subsequent steps, relabel the roots α_{hj}^{**} as α_{hj}^{*}, and replace $G_v^{(k_1)}(z)$ with $\widehat{G}_v^{(k_1)}(z)$. In the statements of Theorems 11.1 and 11.2, this is accounted for by adding $\Theta_v^{(k_1)}(z) := \widehat{G}_v^{(k_1)}(z) - G_v^{(k_1)}(z)$ to $G_v^{(k_1)}(z)$. Thus, the new function $G_v^{(k_1)}(z)$ has the same high-order coefficients as the old one, and its roots are logarithmically separated from each other by at least $A_4 \log_v(n)$.

Phase 4. Patching with the long safe sequence for $k = k_1 + 1, \ldots, k_2$.

We have now arrived at a function $G_v^{(k_1)}(z)$ with "patched" roots α_{hj}^{*} for $j \in \mathcal{S}_1 \cup \cdots \cup \mathcal{S}_{k_1} \cup \{j_0\} \cup \mathcal{S}^0[k_1]$, and "unpatched" roots α_{hj} for all other j. By the construction in Phase 3, for all h, and all $j \neq \ell$, the roots satisfy

$$\operatorname{ord}_v(\alpha_{hj}^{*} - \alpha_{h\ell}^{*}),\ \operatorname{ord}_v(\alpha_{hj}^{*} - \alpha_{h\ell}),\ \operatorname{ord}_v(\alpha_{hj} - \alpha_{h\ell}) \ \leq A_4 \log_v(n)$$

as appropriate for each j, ℓ.

The number k_1 has the property that for all $k \geq k_1$,

$$h_v^{kN} \ \leq \ q^{-\frac{k}{q-1} - \log_v(n)} \ . \tag{11.75}$$

The purpose of Phase 4 is to carry on the patching process until h_v^{kN} is so much smaller than $q^{-k/(q-1)}$ that the "endangered" roots α_{hj} for $j \in \mathcal{S}^{\dagger}$ can be included in the patching process; this will allow us to apply the Refined Patching Lemma (Lemma 11.7) in Phase 5.

Let k_2 be the least integer such that for all $k \geq k_2$,

$$h_v^{kN} \cdot M_v \ \leq \ q^{-\frac{k+1}{q-1} - 3A_4 \log_v(n)} \ . \tag{11.76}$$

Thus, for an appropriate constant A_5,

$$k_2 \ \leq \ A_5 \log_v(n) \ . \tag{11.77}$$

As in Phase 3, we will use the "long safe sequence" of roots $\mathcal{S}^0$ in patching. We will now impose a condition on n which means that $\mathcal{S}^0$ is actually much longer than was required by (11.63). By (11.75), patched roots only move within their δ_n-cosets, so they maintain their position within a ψ_v-regular sequence of length n. This means that instead of choosing a new ψ_v-regular subsequence of length k to use in patching at each step, we can simply extend the previous one.

Recall (11.62). If n is large enough that

$$n \ \geq \ ((N+1)A_2(\log_v(n))^2+1)\cdot(A_5\log_v(n)+2) \ . \tag{11.78}$$

which we will henceforth assume, then by the Pigeon-hole Principle, the long safe sequence satisfies

$$\#(\mathcal{S}^0) \ \geq \ A_5\log_v(n)+2 \ \geq \ k_2+2 \ .$$

Recall that we write $\mathcal{S}^0=\{j_0,j_1,\ldots,j_L\}$. For each $k=k_1+1,\ldots,k_2$, put

$$\mathcal{S}^0[k] \ = \ \{j_1,j_2,\ldots,j_{k+1}\} \ .$$

Also recall that $\mathcal{S}^\diamondsuit=\{0,1,\ldots,n-1\}\backslash(\mathcal{S}_1\cup\cdots\cup\mathcal{S}_{k_1})$, and that

$$G_v^{(k_1)}(z) \ = \ \overline{Q}_{k_1}(z)\cdot Q_{k_1}^{0*}(z)\cdot \widehat{P}_{k_1}^0(\phi_v(z)) \ .$$

For $k=k_1+1,\ldots,k_2$ we will patch $G_v^{(k-1)}(z)$ to $G_v^{(k)}(z)$ as follows. Noting that $\mathcal{S}^0[k]=\mathcal{S}^0[k-1]\cup\{j_{k+1}\}$, define

$$\widehat{P}_k^0(x) \ = \prod_{j\in\mathcal{S}^\diamondsuit\backslash(\{j_0\}\cup\mathcal{S}^0[k])}(x-\psi_v(j)) \ = \ \widehat{P}_{k-1}^0(x)/(x-\psi_v(j_{k+1})) \ .$$

Then $\widehat{P}_{k-1}^0(x)=(x-\psi_v(j_{k+1}))\cdot\widehat{P}_k^0(x)$, and if we set

$$Q_k^0(z) \ = \ Q_{k-1}^{0*}(z)\cdot(\phi_v(z)-\psi_v(j_{k+1})) \ ,$$

then

$$G_v^{(k-1)}(z) \ = \ \overline{Q}_{k_1}(z)\cdot Q_k^0(z)\cdot\widehat{P}_k^0(\phi_v(z)) \ . \tag{11.79}$$

By construction, when the roots of $Q_k^0(z)$ in $B(\theta_h,\rho_h)$ are pulled back to $D(0,1)$ using $\widehat{\sigma}_h(Z)$, they form a ψ_v-regular sequence of length $k+1$ in $\mathcal{O}_{u_h}$. For notational simplicity, we will relabel these roots (the α_{hj}^* for $j\in\mathcal{S}^0[k-1]$, together with $\alpha_{h,j_{k+1}}$) as $\{\dot{\alpha}_{hj}\}_{j\in\mathcal{S}^0[k]}$.

When $\mathrm{char}(K_v)=0$, we are given a K_v-symmetric collection of numbers $\{\Delta_{v,ij}\in L_{w_v}\}_{(i,j)\in\mathrm{Band}_N(k)}$ determined recursively in $\prec_N$ order, and we patch $G_v^{(k-1)}(z)$ by setting

$$\begin{aligned} G_v^{(k)}(z) \ &= \ G_v^{(k-1)}(z)+\sum_{i=1}^{m}\sum_{j=(k-1)N_i}^{kN_i-1}\Delta_{v,ij}^{(k)}\vartheta_{v,ij}^{(k)}(z) \\ &= \ Q_1^*(z)\cdots Q_{k-1}^*(z)\cdot\big(P_k(\phi_v(z))+\Delta_{v,k}(z)\big)\cdot\widehat{P}_k(\phi_v(z)) \ , \end{aligned}$$

where the compensating functions are

$$\vartheta_{v,ij}^{(k)}(z) \ = \ \varepsilon_{v,i}^{-1}\varphi_{i,(k+1)N_i-j}(z)\cdot Q_1^*(z)\cdots Q_{k-1}^*(z)\cdot\widehat{P}_k(\phi_v(z)) \ ,$$

and where as in (11.37),

$$\Delta_{v,k}(z) \ = \ \sum_{i=1}^{m}\sum_{j=(k-1)N_i}^{kN_i-1}\Delta_{v,ij}^{(k)}\cdot\varepsilon_{v,i}^{-1}\varphi_{i,(k+1)N_i-j}(z) \ .$$

We claim that the leading coefficient of $\vartheta_{v,ij}^{(k)}(z)$ at x_i is $\widetilde{c}_{v,i}^{\,n-k-1}$. To see this, note that the leading coefficient of $G_v^{(k-1)}(z)$ at x_i is $A_{v,i0}=\varepsilon_{v,i}\widetilde{c}_{v,i}^{\,n}$, while the leading coefficient of $\widehat{P}_k^0(\phi_v(z))$ at x_i is $\widetilde{c}_{v,i}^{\,k+1}$. By (11.79), the leading coefficient of $\overline{Q}_{k_1}(z)\cdot\widehat{P}_k(\phi_v(z))$ is $\varepsilon_{v,i}\widetilde{c}_{v,i}^{\,n-k-1}$, which gives what we want. Note also that

$$\|\vartheta_{v,ij}^{(k)}\|_{U_v^0} \ \leq \ M_v \ .$$

Indeed, $\|\overline{Q}_{k_1}(z)\|_{U_v^0} \le 1$ and $\|\widehat{P}_k^0(\phi_v(z))\|_{U_v^0} \le 1$, while $\|\varphi_{i,r_{ij}}\|_{U_v^0} \le M_v$.

Clearly the $\vartheta_{v,ij}^{(k)}(z)$ are K_v-symmetric. It follows that $\Delta_{v,k}(z)$ and $G_v^{(k)}(z)$ are K_v-rational, and for each (i,j),

$$\sum_{x_{i'} \in \mathrm{Aut}_c(\mathbb{C}_v/K_v)(x_i)} \Delta_{v,i'j}^{(k)} \vartheta_{v,i'j}^{(k)},$$

is K_v-rational.

When $\mathrm{char}(K_v) = p > 0$, let

$$F_{v,k}(z) \;=\; Q_1^*(z) \cdots Q_{k-1}^*(z) \cdot \widehat{P}_k(\phi_v(z)) \;.$$

By arguments similar to those above, $F_{v,k}(z)$ is K_v-rational, its roots belong in E_v, and it has a pole of order $(n-k-1)N_i$ with leading coefficient $d_{v,i} = \varepsilon_{v,i}\widetilde{c}_{v,i}^{\,n-k-1}$ at each x_i. In particular, $|d_{v,i}|_v = |\widetilde{c}_{v,i}^{\,n-k-1}|_v$. Thus $F_{v,k}(z)$ satisfies the conditions of Theorem 11.2.

By the hypotheses of Theorem 11.2 we are given a K_v-symmetric collection of numbers $\{\Delta_{v,ij} \in L_{w_v}\}_{(i,j)\in \mathrm{Band}_N(k)}$ such that the function

$$\Delta_{v,k}(z) \;=\; \sum_{i=1}^{m} \sum_{j=(k-1)N_i}^{kN_i-1} \Delta_{v,ij}^{(k)} \cdot \varphi_{i,(k+1)N_i-j}(z) \;,$$

is K_v-rational. We patch $G_v^{(k-1)}(z)$ by setting

$$\begin{aligned} G_v^{(k)}(z) \;&=\; G_v^{(k-1)}(z) + \Delta_{v,k}(z) \cdot F_{v,k}(z) \\ &=\; Q_1^*(z) \cdots Q_{k-1}^*(z) \cdot \big(P_k(\phi_v(z)) + \Delta_{v,k}(z)\big) \cdot \widehat{P}_k(\phi_v(z)). \end{aligned}$$

The two patching constructions differ only in the choice of $\Delta_{v,k}(z)$, and in both cases $\|\Delta_{v,k}\|_{U_v^0} \le h_v^{kN} \cdot M_v < q^{-(k/(q+1)-\log_v(n))}$. Write

$$Q_k^{0*}(z) \;=\; Q_k^0(z) + \Delta_{v,k}(z) \;.$$

The change in passing from $G_v^{(k-1)}(z)$ to $G_v^{(k)}(z)$ is localized to $G_k^{0*}(z)$, and

$$G_v^{(k)}(z) \;= \overline{Q}_{k_1}(z) \cdot Q_k^{0*}(z) \cdot \widehat{P}_k(\phi_v(z)) \;.$$

By Lemma 11.6, when the roots of $Q_k^{0*}(z)$ in $B(\theta_h, \rho_h)$ are pulled back to $D(0,1)$ using $\widehat{\sigma}_h(Z)$, they form a ψ_v-regular sequence $\{\alpha_{hj}^*\}$ of length $k+1$ in $\mathcal{O}_{u_h}$ attached to $\mathcal{S}^0[k]$. Since $k > k_1$ we have $h_v^{kN} \cdot M_v \le q^{-\frac{k+1}{q-1}-\log_v(n)}$, which means that

$$\mathrm{ord}_v(\alpha_{hj}^* - \dot{\alpha}_{hj}) \;\ge\; \log_v(n)$$

for each j. The fact that $\mathcal{S}^0$ consists only of "safe" indices means that the α_{hj}^* for $j \in \mathcal{S}^0[k]$ remain the only roots within their δ_n-cosets, and have not moved nearer to any of the other roots. Thus

$$\{\alpha_{hj}^*\}_{j\in\mathcal{S}^0[k]} \cup \{\alpha_{hj}\}_{\mathcal{S}^0\backslash\mathcal{S}^0[k]}$$

is again a ψ_v-regular sequence attached to $\mathcal{S}^0$, and the induction can continue.

Phase 5. Patching using the remaining unpatched indices.

At this point we have constructed a K_v-rational $(\mathfrak{X}, \vec{s})$-function

$$G_v^{(k_2)}(z) \;=\; \overline{Q}_{k_1}(z) \cdot Q_{k_2}^{0*}(z) \cdot \widehat{P}_{k_2}^0(\phi_v(z))$$

whose coefficients $A_{v,ij}$ have been patched for all (i,j) with $1 \le i \le m$, $0 \le j < k_2 N_i$. When the roots of $G_v^{(k_2)}(z)$ in $B(\theta_h, \rho_h)$ are pulled back to $D(0,1)$ using $\widehat{\sigma}_h(Z)$, they form a union of ψ_v-regular sequences consisting of patched roots α_{hj}^* corresponding to the sets $\{j_0\}$, $\mathcal{S}_1, \ldots, \mathcal{S}_{k_1}$, and $\mathcal{S}^0[k_2]$, together with unpatched roots α_{hj} for j in the set

$$\mathcal{S}^\Diamond \backslash (\{j_0\} \cup \mathcal{S}^0[k]) \;=\; \{0,1,\ldots,n-1\} \backslash (\{j_0\} \cup \mathcal{S}_1 \cup \cdots \cup \mathcal{S}_{k_1} \cup \mathcal{S}^0[k_2]) \;.$$

As the roots of $G_v^{(k_1)}(z)$ were logarithmically separated by at least $A_4 \log_v(n)$ and since Phase 4 only patched using "safe" roots, the roots of $G_v^{(k_2)}(z)$ remain logarithmically separated by at least $A_4 \log_v(n)$.

Note that $\mathcal{S}^\Diamond = \{0,1,\ldots,n-1\} \backslash (\mathcal{S}_1 \cup \cdots \cup \mathcal{S}_{k_1})$ is a sequence of consecutive indices, both when $\mathrm{char}(K_v) = 0$ and when $\mathrm{char}(K_v) = p > 0$. Since $\{j_0\} \cup \mathcal{S}^0[k_2] \subset \mathcal{S}^0 \subset \mathcal{S}^\Diamond$ is also a sequence of consecutive indices, its complement in $\mathcal{S}^\Diamond$ consists of at most two sequences of consecutive indices. Recalling that $\{j_0\} \cup \mathcal{S}^0[k_2] = \{j_0, j_1, \ldots, j_{k_2+1}\}$, put $k_3 = \#(\mathcal{S}^\Diamond) - 3$ and list the elements of $\mathcal{S}^\Diamond \backslash (\{j_0\} \cup \mathcal{S}^0[k_2])$ in increasing order as $\{j_{k_2+2}, \ldots, j_{k_3+1}\}$. For each k with $k_2 < k \le k_3$, put

$$\mathcal{S}^\Diamond[k] \;=\; \mathcal{S}^0[k_2] \cup \{j_{k_2+2}, \ldots, j_{k+1}\} \;.$$

By the discussion above, $\mathcal{S}^\Diamond[k]$ is a union of most 3 subsequences of consecutive indices.

Recall from §11.1 the following:

Lemma 11.7 (Refined Patching Lemma). *Let $\mathcal{Q}(Z) \in K_v[[Z]]$ be a power series converging on $D(0,1)$, with* sup *norm $\|\mathcal{Q}\|_{D(0,1)} = 1$. Suppose the roots $\{\alpha_j\}$ of $\mathcal{Q}(Z)$ in $D(0,1)$ can be partitioned into r disjoint ψ_v-regular sequences in $\mathcal{O}_w$ attached to index sets $\mathcal{S}_1, \ldots, \mathcal{S}_r$ of respective lengths $\ell_1, \ldots, \ell_r$. Put $\ell = \sum_{k=1}^r \ell_k$. Suppose further that there is a bound $T \ge \max_i(\log_v(\ell_i))$ such that*

$$\mathrm{ord}_v(\alpha_j - \alpha_k) \;\le\; T$$

for all $j \ne k$.

Then for any $M \ge T$, and any power series $\Delta(Z) \in K_v[[z]]$ converging on $D(0,1)$ with

$$\|\Delta\|_{D(0,1)} \;\le\; q^{-\frac{\ell}{q-1}-(r-1)T-M}$$

the roots $\{\alpha_j^*\}$ of $\mathcal{Q}^*(Z) = \mathcal{Q}(Z) + \Delta(Z)$ in $D(0,1)$ again form a union of ψ_v-regular sequences in $\mathcal{O}_w$ attached to $\mathcal{S}_1, \ldots, \mathcal{S}_r$. They can uniquely be labeled in such a way that

$$\mathrm{ord}_v(\alpha_j^* - \alpha_j) \;>\; M$$

for each j.

The number k_2 was chosen so that if $r = 3$ and $T = A_4 \log_v(n)$, then for all $k \ge k_2$,

$$h_v^{kN} \cdot M_v \;\le\; q^{-\frac{k+1}{q-1}-rT} \;. \tag{11.80}$$

This means that when we apply the Refined Patching Lemma using at most 3 sequences of roots, all roots remain logarithmically separated by at least $A_4 \log_v(n)$.

When $k = k_2$ write

$$Q_{k_2}^{\Diamond*}(z) \;=\; Q_{k_2}^{0*}(z) \;, \qquad \widehat{P}_{k_2}^\Diamond(z) \;=\; \widehat{P}_{k_2}^0(z) \;,$$

so that with this notation

$$G_v^{(k_2)}(z) \;=\; \overline{Q}_{k_1}(z)\cdot Q_{k_2}^{\Diamond *}(z)\cdot \widehat{P}_{k_2}^{\Diamond}(\phi_v(z))\ .$$

For $k = k_2+1,\ldots,k_3$, inductively suppose that

$$G_v^{(k-1)}(z) \;=\; \overline{Q}_{k_1}(z)\cdot Q_{k-1}^{\Diamond *}(z)\cdot \widehat{P}_{k-1}^{\Diamond}(\phi_v(z))$$

where the roots of $Q_{k-1}^{\Diamond *}(z)$ correspond to $\mathcal{S}^{\Diamond}[k-1]$. Put

$$\widehat{P}_k^{\Diamond}(z) \;=\; \prod_{j\in\mathcal{S}^{\Diamond}\backslash(\{j_0\}\cup\mathcal{S}^{\Diamond}[k])} (z-\psi_v(j)) \;=\; \widehat{P}_{k-1}^{\Diamond}(z)/(z-\psi_v(j_{k+1}))$$

and put

$$Q_k^{\Diamond}(z) \;=\; Q_{k-1}^{\Diamond *}(z)\cdot(\phi_v(z)-\psi_v(j_{k+1}))$$

so that

$$G_v^{(k-1)}(z) \;=\; \overline{Q}_{k_1}(z)\cdot Q_k^{\Diamond}(z)\cdot \widehat{P}_k^{\Diamond}(\phi_v(z))\ .$$

The patching argument in Phase 5 is very similar to that in Phase 4.

When $\mathrm{char}(K_v) = 0$, we are given a K_v-symmetric collection of numbers $\{\Delta_{v,ij} \in L_{w_v}\}_{(i,j)\in \mathrm{Band}_N(k)}$ determined recursively in $\prec_N$ order, and we patch $G_v^{(k-1)}(z)$ by setting

$$(11.81)\qquad G_v^{(k)}(z) \;=\; G_v^{(k-1)}(z) + \sum_{i=1}^{m}\sum_{j=(k-1)N_i}^{kN_i-1} \Delta_{v,ij}^{(k)}\vartheta_{v,ij}^{(k)}(z)$$
$$\qquad\qquad = \;\overline{Q}_{k_1}(z)\cdot\big(P_k(\phi_v(z))+\Delta_{v,k}(z)\big)\cdot \widehat{P}_k^{\Diamond}(\phi_v(z))\ ,$$

where the compensating functions are

$$(11.82)\qquad \vartheta_{v,ij}^{(k)}(z) \;=\; \varepsilon_{v,i}^{-1}\varphi_{i,(k+1)N_i-j}(z)\cdot\overline{Q}_{k_1}(z)\cdot \widehat{P}_k^{\Diamond}(\phi_v(z))$$

and where as in (11.37),

$$\Delta_{v,k}(z) \;=\; \sum_{i=1}^{m}\sum_{j=(k-1)N_i}^{kN_i-1} \Delta_{v,ij}^{(k)}\cdot\varepsilon_{v,i}^{-1}\varphi_{i,(k+1)N_i-j}(z)\ .$$

As in Phase 4, the $\vartheta_{v,ij}^{(k)}(z)$ are K_v-symmetric. Each $\vartheta_{v,ij}^{(k)}(z)$ has a pole of order nN_i-j at x_i with leading coefficient $\widetilde{c}_{v,i}^{\,n-k-1}$, and poles of order at most $(n-k-1)N_{i'}$ at each $x_{i'}\neq x_i$; furthermore, $\|\vartheta_{v,ij}^{(k)}\|_{U_v^0} \le M_v$. $G_v^{(k)}(z)$ is K_v-rational by the K_v-symmetry of the $\Delta_{v,ij}^{(k)}$ and $\vartheta_{v,ij}^{(k)}(z)$, and for each (i,j),

$$\sum_{x_{i'}\in\mathrm{Aut}_c(\mathbb{C}_v/K_v)(x_i)} \Delta_{v,i'j}^{(k)}\vartheta_{v,i'j}^{(k)}$$

is K_v-rational.

When $\mathrm{char}(K_v)=p>0$, let

$$F_{v,k}(z) \;=\; \overline{Q}_{k_1}(z)\cdot \widehat{P}_k^{\Diamond}(\phi_v(z))\ .$$

By arguments similar to those above, $F_{v,k}(z)$ is K_v-rational, its roots belong in E_v, and it has a pole of order $(n-k-1)N_i$ with leading coefficient $d_{v,i}=\varepsilon_{v,i}\widetilde{c}_{v,i}^{\,n-k-1}$ at each x_i. In particular, $|d_{v,i}|_v = |\widetilde{c}_{v,i}^{\,n-k-1}|_v$. Thus $F_{v,k}(z)$ satisfies the conditions of Theorem 11.2.

By the hypotheses of Theorem 11.2 we are given a K_v-symmetric collection of numbers $\{\Delta_{v,ij} \in L_{w_v}\}_{(i,j)\in \mathrm{Band}_N(k)}$ such that the function

$$\Delta_{v,k}(z) \;=\; \sum_{i=1}^{m} \sum_{j=(k-1)N_i}^{kN_i-1} \Delta_{v,ij}^{(k)} \cdot \varphi_{i,(k+1)N_i-j}(z) \;,$$

is K_v-rational. We patch $G_v^{(k-1)}(z)$ by setting

$$\begin{aligned} G_v^{(k)}(z) &= G_v^{(k-1)}(z) + \Delta_{v,k}(z) \cdot F_{v,k}(z) \\ &= Q_1^*(z) \cdots Q_{k-1}^*(z) \cdot \big(P_k(\phi_v(z)) + \Delta_{v,k}(z)\big) \cdot \widehat{P}_k(\phi_v(z)). \end{aligned}$$

The two patching constructions differ only in the choice of $\Delta_{v,k}(z)$, and in both cases $\|\Delta_{v,k}\|_{U_v^0} \le h_v^{kN} \cdot M_v \le q^{-\frac{k+1}{q-1} - 3A_4 \log_v(n)}$. Let

$$Q_k^{\Diamond *}(z) \;=\; Q_k^{\Diamond}(z) + \Delta_{v,k}(z) \;.$$

In passing from $G_v^{(k-1)}(z)$ to $G_v^{(k)}(z)$, the change is isolated in the factor $Q_k^{\Diamond}(z)$, and we have

$$G_v^{(k)}(z) \;= \overline{Q}_{k_1}(z) \cdot Q_k^{\Diamond *}(z) \cdot \widehat{P}_k^{\Diamond}(\phi_v(z)) \;.$$

When the roots of $Q_k^{\Diamond *}(z)$ in $B(\theta_h, \rho_h)$ are pulled back to $D(0,1)$ using $\widehat{\sigma}_h(Z)$, then Lemma 11.7, applied with $r \le 3$ and

$$M \;=\; -\log(h_v^{kN} \cdot M_v) - \frac{k+1}{q-1} - (r-1)) \;\ge\; T,$$

shows they form a union of at most three ψ_v-regular sequences in $\mathcal{O}_{u_h}$ attached to $\mathcal{S}^{\Diamond}[k]$ and

$$\mathrm{ord}_v(\alpha_{hj}^* - \alpha_{hj}) \;\ge\; T \;=\; A_4 \log_v(n)$$

for each j. Hence the roots of $Q_k^{\Diamond *}(z)$ have not moved closer to any of the other roots of $G_v^{(k-1)}(z)$, and the induction can continue.

Phase 6. Completing the patching process

We have now obtained a function $G_v^{(k_3)}(z)$ whose roots have all been patched. When the roots in each ball $B(\theta_h, \rho_h)$ are pulled back to $D(0,1)$ using $\ddot{\sigma}_h(Z)$, they form a union of at most $r := k_1 + 4$ ψ_v-regular sequences in $\mathcal{O}_{u_h}$, with total length n. These roots are logarithmically separated from each other by at least $T = A_4 \log_v(n)$.

We must now include all the roots in the patching process. To be able to apply Lemma 11.7, we need that for all $k > k_3$,

$$h_v^{kN} \cdot M_v \;\le\; q_v^{-\frac{n}{q-1} - (k_1+4)T} \;. \tag{11.83}$$

However, k_3 is quite large: $k_3 = \#(\mathcal{S}^{\Diamond}) - 3 \;\ge\; n - A_2(\log_v(n))^2 - 3$. Thus (11.83) will hold if

$$n \;\ge\; A_6 \cdot (\log_v(n))^2 \tag{11.84}$$

for a suitable constant A_6, which we henceforth assume.

When $\mathrm{char}(K_v) = 0$, for $k = k_3 + 1, \ldots, n-1$ we patch as follows. For each (i,j) with $1 \le i \le m$, $(k-1)N_i \le j < kN_i$, put

$$\vartheta_{v,ij}^{(k)}(z) \;=\; \varphi_{i,(k+1)N_i-j}(z) \cdot \prod_{j=1}^{n-k-1} (\phi_v(z) - \psi_v(j)) \;. \tag{11.85}$$

The $\vartheta_{v,ij}^{(k)}$ are K_v-symmetric. It is easy to see that each $\vartheta_{v,ij}^{(k)}$ has a pole of order $nN_i - j$ at x_i with leading coefficient $\widetilde{c}_{v,i}^{\,n-k-1}$, and has a pole of order at most $(n-k-1)N_{i'}$ at each $x_{i'} \neq x_i$, with $\|\vartheta_{v,ij}^{(k)}(z)\|_{U_v^0} \leq M_v$.

By Theorem 11.1 we are given a K_v-symmetric set $\{\Delta_{v,ij}^{(k)} \in L_{w_v}\}_{(i,j)\in \mathrm{Band}_N(k)}$, determined recursively in $\prec_N$ order, such that $|\Delta_{v,ij}^{(k)}|_v \leq h_v^{kN}$ for each i, j. Put

$$G_v^{(k)}(z) \;=\; G_v^{(k-1)}(z) + \sum_{i=1}^{m} \sum_{j=(k-1)N_i+1}^{kN_i} \Delta_{v,ij}^{(k)} \vartheta_{v,ij}^{(k)}(z) \; .$$

Here $G_v^{(k)}(z)$ is K_v-rational by the K_v-symmetry of the $\Delta_{v,ij}^{(k)}$ and $\vartheta_{v,ij}^{(k)}(z)$, and for each (i,j),

$$\sum_{x_{i'} \in \mathrm{Aut}_c(\mathbb{C}_v/K_v)(x_i)} \Delta_{v,i'j}^{(k)} \vartheta_{v,i'j}^{(k)} \;\in\; K_v(\mathcal{C}) \; .$$

Furthermore, $h_v^{kN} M_v \leq q^{-\frac{n}{q-1}-rT}$, so Lemma 11.7 shows that the roots of $G_v^{(k)}(z)$ belong to E_v and have the same separation properties as those of $G_v^{(k-1)}(z)$.

When $k = n$, we are given a K_v-symmetric set of numbers $\{\Delta_{v,\lambda}\}_{1\leq \lambda \leq \Lambda}$ with $|\Delta_{v,\lambda}|_v \leq h_v^{nN}$ for each λ. Put

$$G_v^{(n)}(z) \;=\; G_v^{(n-1)}(z) + \sum_{\lambda=1}^{\Lambda} \Delta_{v,\lambda}\varphi_\lambda \; .$$

Clearly $G_v^{(n)}(z)$ is K_v-rational. Since $\|\varphi_\lambda\|_{U_v^0} \leq M_v$ for each λ, Lemma 11.7 shows that the roots of $G_v^{(n)}(z)$ belong to E_v and have the same separation properties as those of $G_v^{(n-1)}(z)$. In particular, they are distinct.

When $\mathrm{char}(K_v) = p > 0$, for $k = k_3+1, \ldots, n-1$ we patch as follows. For each k, put

$$F_{v,k}(z) \;=\; \prod_{j=1}^{n-k-1} (\phi_v(z) - \psi_v(j)) \; . \tag{11.86}$$

Then $F_{v,k}(z)$ is K_v-rational, its roots belong in E_v, and it has a pole of order $(n-k-1)N_i$ with leading coefficient $d_{v,i} = \widetilde{c}_{v,i}^{\,n-k-1}$ at each x_i. In particular, $|d_{v,i}|_v = |\widetilde{c}_{v,i}^{\,n-k-1}|_v$. Thus $F_{v,k}(z)$ satisfies the conditions of Theorem 11.2.

By the hypotheses of Theorem 11.2 we are given a K_v-symmetric collection of numbers $\{\Delta_{v,ij} \in L_{w_v}\}_{(i,j)\in \mathrm{Band}_N(k)}$ such that the function

$$\Delta_{v,k}(z) \;=\; \sum_{i=1}^{m} \sum_{j=(k-1)N_i}^{kN_i-1} \Delta_{v,ij}^{(k)} \cdot \varphi_{i,(k+1)N_i-j}(z) \; ,$$

is K_v-rational. We patch $G_v^{(k-1)}(z)$ by setting

$$G_v^{(k)}(z) \;=\; G_v^{(k-1)}(z) + \Delta_{v,k}(z) \cdot F_{v,k}(z).$$

Here $G_v^{(k)}(z)$ is K_v-rational since $\Delta_{v,k}(z)$ and $F_{v,k}(z)$ are K_v-rational. Using the bounds $\|\Delta_{v,k}\|_{U_v^0} \leq h_v^{kN} M_v \leq q^{-\frac{n}{q-1}-rT}$ and $\|F_{v,k}\|_{U_v^0} \leq 1$, it follows from Lemma 11.7 that the roots of $G_v^{(k)}(z)$ belong to E_v and have the same separation properties as those of $G_v^{(k-1)}(z)$.

When $k = n$, by the hypotheses of Theorem 11.2 we are given a K_v-symmetric set of numbers $\{\Delta_{v,\lambda}\}_{1 \le \lambda \le \Lambda}$ with $|\Delta_{v,\lambda}|_v \le h_v^{nN}$ for each λ, such that

$$\Delta_{v,n}(z) \;=\; \sum_{\lambda=1}^{\Lambda} \Delta_{v,\lambda}\varphi_\lambda(z)$$

is K_v-rational. Put

$$G_v^{(n)}(z) \;=\; G_v^{(n-1)}(z) + \Delta_{v,n}(z)\ .$$

Clearly $G_v^{(n)}(z)$ is K_v-rational. Since $\|\varphi_\lambda\|_{U_v^0} \le M_v$ for each λ, Lemma 11.7 shows that the roots of $G_v^{(n)}(z)$ belong to E_v and have the same separation properties as those of $G_v^{(n-1)}(z)$. In particular, they are distinct.

The final assertion in Theorems 11.1 and 11.2 is that

$$\{z \in \mathcal{C}_v(\mathbb{C}_v) : G_v^{(n)}(z) \in \mathcal{O}_v \cap D(0, r_v^{nN}\} \;\subseteq\; E_v\ .$$

To assure this, we must assume that

$$r_v^{nN} \;<\; q_v^{-\frac{n}{q-1}-(k_1+4)T}\ , \tag{11.87}$$

which holds for all sufficiently large n since $r_v^N < q_v^{-1/(q-1)}$ by (11.2) and (11.7).

Given (11.87), for each $h = 1, \ldots, N$ when we restrict $G_v^{(n)}(z)$ to $B(\theta_h, \rho_h)$ and pull it back to $D(0,1)$ using $\widehat{\sigma}_h(Z)$, the Refined Patching Lemma (Lemma 11.7) shows that for each $\kappa_v \in \mathcal{O}_v$ with $|\kappa_v|_v \le r_v^{nN}$, the function $G_v^{(n)}(\widehat{\sigma}_h(Z)) - \kappa_v$ has n distinct roots in $\mathcal{O}_{u_h}$. Correspondingly $G_v^{(n)}(z) = \kappa_v$ has n distinct solutions in $\mathcal{C}_v(F_{u_h}) \cap B(\theta_h, \rho_h)$. Since there are N balls $B(\theta_h, \rho_h)$ and $G_v^{(n)}(z)$ has degree nN, this accounts for all the solutions to $G_v^{(n)}(z) = \kappa_v$ in $\mathcal{C}_v(\mathbb{C}_v)$. It follows that

$$\{z \in \mathcal{C}_v(\mathbb{C}_v) : G_v^{(n)}(z) \in \mathcal{O}_v \cap D(0, r_v^{nN})\} \;\subset\; \bigcup_{h=1}^{N} \mathcal{C}_v(F_{u_h}) \cap B(\theta_h, \rho_h) \;\subseteq\; E_v\ .$$

This completes the proof of Theorems 11.1 and 11.2, subject to the proofs of the three Moving Lemmas below.

4. Proofs of the Moving Lemmas

In this section we give the proofs of the three Moving Lemmas. Our notation and assumptions are the same as in §11.3. For the convenience of the reader, before giving the proofs we restate the lemmas making the hypotheses more explicit.

Lemma 11.10 (First Moving Lemma). *Let E_v and $\phi_v(z)$ be as in Theorems* 11.1 *and* 11.2 : *E_v has the K_v-simple decomposition $E_v = \bigcup_{\ell=1}^{D} \big(B(a_\ell, r_\ell) \cap \mathcal{C}_v(F_{w_\ell})\big)$ such that $U_v := \bigcup_{\ell=1}^{D} B(a_\ell, r_\ell)$ is disjoint from $\mathfrak{X}$, and $H_v := \phi_v^{-1}(D(0,1))$ has the K_v-simple decomposition $H_v = \bigcup_{h=1}^{N} \big(B(\theta_h, \rho_h) \cap \mathcal{C}_v(F_{u_h})\big)$ which is compatible with $\bigcup_{\ell=1}^{D} \big(B(a_\ell, r_\ell) \cap \mathcal{C}_v(F_{w_\ell})\big)$ and move-prepared with respect to $B(a_1, r_1), \ldots,$ $B(a_D, r_D)$. For each ℓ, there is a point $\overline{w}_\ell \in \big(\mathcal{C}_v(F_{w_\ell}) \cap B(a_\ell, r_\ell)\big) \backslash H_v$.*

Let $S^\triangle = \mathcal{S}_1 \cup \ldots \cup \mathcal{S}_{k_1}$, so the set of patched roots from Phases 1 *and* 2 *is $\{\theta_{hj}^*\}_{1 \le h \le N, j \in S^\triangle}$. Let j_0 be as in Phase* 3, *so $\{\theta_{hj_0} \in \mathcal{C}_v(F_{u_h}) \cap B(\theta_h, \rho_h)\}_{1 \le h \le N}$ is a set of "safe" roots.*

Then there are constants $\varepsilon_1 > 0$ and $C_1, C_2 \ge 1$ (depending on $\phi_v(z)$, E_v, H_v, and their K_v-simple decompositions), with the following property:

Put $\delta_n = q_v^{-\lceil \log_v(n) \rceil}$. *Then for each* $0 < \varepsilon < \varepsilon_1$ *small enough that*

$$C_1 \varepsilon \ \le \ \delta_n \cdot \min_{1 \le h \le N} (\rho_h) \ ,$$

given any K_v*-symmetric set* $\{\theta_{hj}^{**} \in \mathcal{C}_v(F_{u_h}) \cap B(\theta_h, \rho_h)\}_{1 \le h \le N, j \in S^{\triangle}}$ *with the property that* $\|\theta_{hj}^{**}, \theta_{hj}^{*}\|_v < \varepsilon$ *for all* (h, j)*, there is a* K_v*-symmetric collection of points* $\{\theta_{h,j_0}^{**} \in \mathcal{C}_v(F_{u_h}) \cap B(\theta_h, \rho_h)\}_{1 \le h \le N}$ *satisfying*

$$\|\theta_{h,j_0}^{**}, \theta_{h,j_0}\|_v \ \le \ C_1 \varepsilon \ \le \ \delta_n \rho_h$$

for each h*, such that:*

(*A*) *The divisor*

$$\mathcal{D} \ = \ \sum_{j \in S^{\triangle}} \sum_{h=1}^{N} ((\theta_{hj}^{**}) - (\theta_{hj}^{*})) + \sum_{h=1}^{N} ((\theta_{h,j_0}^{**}) - (\theta_{h,j_0})) \tag{11.88}$$

is K_v*-rational and principal.*

(*B*) *Writing* $U_v^0 = \bigcup_{h=1}^{N} B(\theta_h, \rho_h) \subset U_v$ *as before, we have*

(1) $|Y(z)|_v = 1$ *for all* $z \in \mathcal{C}_v(\mathbb{C}_v) \backslash U_v^0$*;*

(2) $|Y(z) - 1|_v \le C_2 \varepsilon$ *for all* $z \in \mathcal{C}_v(\mathbb{C}_v) \backslash U_v^0$.

Remark. Although the statement of this lemma is rather technical, it is a deep result which depends on the theory of the Universal Function developed in Appendix C and the local action of the Jacobian studied in Appendix D. It is the key to the local patching construction for nonarchimedean K_v-simple sets.

PROOF. Consider the decompositions $E_v = \bigcup_{\ell=1}^{D} \big(B(a_\ell, r_\ell) \cap \mathcal{C}_v(F_{w_\ell})\big)$ and $H_v = \bigcup_{h=1}^{N} \big(B(\theta_h, \rho_h) \cap \mathcal{C}_v(F_{u_h})\big)$. Since these decompositions are K_v-simple and compatible, we have $F_{u_h} = F_{w_\ell}$ for each h and ℓ such that $B(\theta_h, \rho_h) \subset B(a_\ell, r_\ell)$.

We first reduce the lemma to a similar assertion for a single ball $B(a_\ell, r_\ell)$. For each $\ell = 1, \ldots, D$, let $\mathcal{I}_\ell$ be the set of indices $1 \le h \le N$ such that $B(\theta_h, \rho_h) \subseteq B(a_\ell, r_\ell)$. Suppose that for each ℓ there are constants $\varepsilon_1^{(\ell)} > 0$ and $C_1^{(\ell)}, C_2^{(\ell)} \ge 1$ such that if $0 < \varepsilon < \varepsilon_1^{(\ell)}$ and

$$C_1^{(\ell)} \cdot \varepsilon \ \le \ \delta_n \cdot \min_{h \in \mathcal{I}_\ell} (\rho_h) \ ,$$

then given any set of points $\{\theta_{hj}^{**} \in \mathcal{C}_v(F_{w_\ell}) \cap B(\theta_h, \rho_h)\}_{h \in \mathcal{I}_\ell, j \in S^{\triangle}}$ with the property that $\|\theta_{hj}^{**}, \theta_{hj}\|_v \le \varepsilon$ for all h, j, there are points $\{\theta_{h,j_0}^{**} \in \mathcal{C}_v(F_{w_\ell}) \cap B(\theta_h, \rho_h)\}_{h \in \mathcal{I}_\ell}$ with $\|\theta_{h,j_0}^{**}, \theta_{h,j_0}\|_v \le C_1^{(\ell)} \varepsilon \le \delta_n \rho_h$ such that the F_{w_ℓ}-rational divisor

$$\mathcal{D}_\ell \ = \ \sum_{h \in \mathcal{I}_\ell, j \in S^{\triangle}} ((\theta_{hj}^{**}) - (\theta_{hj}^{*})) \ + \ \sum_{h \in \mathcal{I}_\ell} \big((\theta_{h,j_0}^{**}) - (\theta_{h,j_0})\big) \tag{11.89}$$

is principal, and if we put $U_{v,\ell}^0 = \bigcup_{h \in \mathcal{I}_\ell} B(\theta_h, \rho_h)$, then there is an F_{w_ℓ}-rational function $Y_\ell(z)$ with divisor $\mathcal{D}_\ell$ such that

(1_ℓ) $|Y_\ell(z)|_v = 1$ for all $z \in \mathcal{C}_v(\mathbb{C}_v) \backslash U_{v,\ell}^0$;

(2_ℓ) $|Y_\ell(z) - 1|_v \le C_2^{(\ell)} \varepsilon$ for all $z \in \mathcal{C}_v(\mathbb{C}_v) \backslash U_{v,\ell}^0$.

Put $\varepsilon_1 = \min_\ell(\varepsilon_1^{(\ell)})$, $C_1 = \max_\ell(C_1^{(\ell)})$ and $C_2 = \max_\ell(C_2^{(\ell)})$. Let $\overline{\rho} = \min_{1 \le h \le N}(\rho_h)$. Take $0 < \varepsilon \le \varepsilon_1$ small enough that $C_1 \varepsilon \le \delta_n \overline{\rho}$, and let $\{\theta_{hj}^{**} \in \mathcal{C}_v(F_{u_h}) \cap B(\theta_h, \rho_h)\}_{1 \le h \le N, j \in S^{\triangle}}$ be a K_v-symmetric set of points with the property that $\|\theta_{hj}^{**}, \theta_{hj}^{*}\|_v < \varepsilon$ for all h, j.

We now construct the K_v-symmetric set $\{\theta_{h,j_0}^{**} \in \mathcal{C}_v(F_{u_h}) \cap B(\theta_h, \rho_h)\}_{1 \le h \le N}$, the divisor $\mathcal{D}$, and the function $Y(z)$ in the lemma, by using Galois equivariance: we keep the divisors $\mathcal{D}_\ell$ and functions $Y_\ell(z)$ for a set of representatives of the Galois orbits for the balls $B(a_\ell, r_\ell)$, then throw away the others and replace them with the Galois conjugates for the representatives.

To be precise, write $\widetilde{K}_v^{\mathrm{sep}}$ for the maximal separable extension of K_v. Since $E_v = \bigcup_{\ell=1}^D \big(\mathcal{C}_v(F_{w_\ell}) \cap B(a_\ell, r_\ell)\big)$ is a K_v-simple decomposition, F_{w_ℓ}/K_v is separable for each ℓ, and the orbit of $B(a_\ell, r_\ell)$ under $\mathrm{Gal}(\widetilde{K}_v^{\mathrm{sep}}/K_v)$ has exactly $d_\ell := [F_{w_\ell} : K_v]$ elements; this means there is an action of $\mathrm{Gal}(\widetilde{K}_v^{\mathrm{sep}}/K_v)$ on the index set $\{\ell \in \mathbb{N} : 1 \le \ell \le D\}$ such that $B(a_{\sigma(\ell)}, r_{\sigma(\ell)}) = \sigma(B(a_\ell, r_\ell))$ and $F_{w_{\sigma(\ell)}} = \sigma(F_{w_\ell})$ for each ℓ and each $\sigma \in \mathrm{Gal}(\widetilde{K}_v^{\mathrm{sep}}/K_v)$. Similarly, since $H_v = \bigcup_{h=1}^N \big(\mathcal{C}_v(F_{u_h}) \cap B(\theta_h, \rho_h)\big)$ is a K_v-simple decomposition, there is an action of $\mathrm{Gal}(\widetilde{K}_v^{\mathrm{sep}}/K_v)$ on $\{h \in \mathbb{N} : 1 \le h \le N\}$ such that $B(\theta_{\sigma(h)}, \rho_{\sigma(h)}) = \sigma(B(\theta_h, \rho_h))$ and $F_{u_{\sigma(h)}} = \sigma(F_{u_h})$ for each h and σ. The fact that the θ_{hj}^{**} are K_v-symmetric implies that $\theta_{\sigma(h),j}^{**} = \sigma(\theta_{hj}^{**})$ for all h, σ. The compatibility of the decompositions of H_v and E_v means that $\mathcal{I}_{\sigma(\ell)} = \{\sigma(h) : h \in \mathcal{I}_\ell\}$ for each σ, and that for each h such that $B(\theta_h, \rho_h) \subseteq B(a_\ell, r_\ell)$, we have $\sigma(h) = h$ if and only if $\sigma(\ell) = \ell$.

Let $\mathcal{L} = \{\ell_1, \ldots, \ell_r\}$ be a set of representatives for the distinct Galois orbits of the balls $B(a_1, r_1), \ldots, B(a_D, r_D)$. For each $\ell_k \in \mathcal{L}$, we have $F_{u_h} = F_{w_{\ell_k}}$ for all $h \in \mathcal{I}_{\ell_k}$. Let

$$\{\theta_{hj}^{**} \in \mathcal{C}_v(F_{w_{\ell_k}}) \cap B(\theta_h, \rho_h)\}_{h \in \mathcal{I}_{\ell_k}, j \in \mathcal{S}^\triangle}$$

be the corresponding subset of $\{\theta_{hj}^{**} \in \mathcal{C}_v(F_{u_h}) \cap B(\theta_h, \rho_h)\}_{1 \le h \le N, j \in \mathcal{S}^\triangle}$. By hypothesis, there is a collection of points $\{\theta_{h,j_0}^{**} \in \mathcal{C}_v(F_{w_{\ell_k}}) \cap B(\theta_h, \rho_h)\}_{h \in \mathcal{I}_{\ell_k}}$ with $\|\theta_{h,j_0}^{**}, \theta_{h,j_0}\|_v \le C_1^{(\ell_k)} \varepsilon \le \delta_n \rho_h$ such that the $F_{w_{\ell_k}}$-rational divisor

$$\mathcal{D}_{\ell_k} = \sum_{h \in \mathcal{I}_{\ell_k}, j \in S^\triangle} \big((\theta_{hj}^{**}) - (\theta_{hj}^{*})\big) + \sum_{h \in \mathcal{I}_{\ell_k}} \big((\theta_{h,j_0}^{**}) - (\theta_{h,j_0})\big)$$

is principal; let $Y_{\ell_k}(z) \in F_{w_{\ell_k}}(\mathcal{C})$ be the corresponding function. For an arbitrary $1 \le \ell \le D$ there are an $\ell_k \in \mathcal{L}$ and a $\sigma \in \mathrm{Gal}(\widetilde{K}_v^{\mathrm{sep}}/K_v)$ such that $\ell = \sigma(\ell_k)$; redefine

$$D_\ell = \sigma(D_{\ell_k}) , \qquad Y_\ell(z) = \sigma(Y_{\ell_k})(z)$$

and redefine the θ_{h,j_0}^{**} for $h \in \mathcal{I}_\ell$ by putting $\theta_{\sigma(h),j_0}^{**} = \sigma(\theta_{h,j_0}^{**}) \in \mathcal{C}_v(F_{w_\ell}) \cap B(a_\ell, r_\ell)$ for each $h \in \mathcal{I}_{\ell_k}$. By the discussion above, all of these are well-defined, the set $\{\theta_{h,j_0}^{**}\}_{1 \le h \le N}$ is K_v-symmetric, and for all ℓ and σ we have $D_{\sigma(\ell)} = \sigma(D_\ell)$, $Y_{\sigma(\ell)}(z) = \sigma(Y_\ell)(z)$.

Finally, define

$$\mathcal{D} = \sum_{\ell=1}^D \mathcal{D}_\ell , \qquad Y(z) = \prod_{\ell=1}^D Y_\ell(z) .$$

Since $\mathcal{D}$ is $\widetilde{K}_v^{\mathrm{sep}}$-rational and is fixed by $\mathrm{Gal}(\widetilde{K}_v^{\mathrm{sep}}/K_v)$, it is K_v-rational; similarly $Y(z)$ is K_v-rational. Clearly $\mathrm{div}(Y(z)) = \mathcal{D}$. By construction, $\{\theta_{h,j_0}^{**}\}_{1 \le h \le N}$ is K_v-symmetric, and by Galois equivariance, $\theta_{h,j_0}^{**} \in \mathcal{C}_v(F_{u_h}) \cap B(\theta_h, \rho_h)$ and $\|\theta_{h,j_0}^{**}, \theta_{h,j_0}\|_v \le C_1^{(\ell)} \varepsilon \le \delta_n \rho_h$ for each h.

Since the original sets $\{\theta_{hj}^{**}\}_{1 \le h \le N, j \in S^\triangle}$, $\{\theta_{hj}\}_{1 \le h \le N, j \in S^\triangle}$, and $\{\theta_{h,j_0}^{**}\}_{1 \le h \le N}$ were K_v-symmetric, $\mathcal{D}$ has the form (11.88). Clearly $U_{v,\ell}^0 \subset B(a_\ell, r_\ell) \subseteq U_v$ for each

ℓ. For each $z \notin U_{v,\ell}^0$, we have $|Y_\ell(z)|_v = 1$, so for each $z \notin U_v^0 = \bigcup_{\ell=1}^D U_{v,\ell}^0$ we have $|Y(z)|_v = 1$, and (B1) in the lemma holds; similarly, for each $z \notin U_{v,\ell}^0$, we have $|Y_\ell(z) - 1|_v \le C_2\varepsilon$, so since

$$Y(z) - 1 \;=\; \sum_{\ell=1}^{D}(Y_\ell(z) - 1)\cdot\left(\prod_{k=\ell+1}^{D} Y_k(z)\right) ,$$

the ultrametric inequality shows that for each $z \notin U_v^0$ we have $|Y(z) - 1|_v \le C_2\varepsilon$, and (B2) holds.

Now fix ℓ; we will construct $\mathcal{D}_\ell$ and $Y_\ell(z)$ for $B(a_\ell, r_\ell)$, and show they satisfy properties (1_ℓ) and (2_ℓ). The proof has two steps: first we use the local action of the Jacobian, from Appendix D, to construct the principal divisor $\mathcal{D}_\ell$; then we use the theory of the Universal Function, from Appendix C, to construct $Y_\ell(z)$. Put $E_{v,\ell} = E_v \cap B(a_\ell, r_\ell) = \mathcal{C}_v(F_{w_\ell}) \cap B(a_\ell, r_\ell)$ and put $H_{v,\ell} = H_v \cap B(a_\ell, r_\ell) \subset E_{v,\ell}$. As noted above, we have $F_{u_h} = F_{w_\ell}$ for each $h \in \mathcal{I}_\ell$, so $H_{v,\ell} = \bigcup_{h\in\mathcal{I}_\ell}\left(\mathcal{C}_v(F_{w_\ell}) \cap B(\theta_h, \rho_h)\right) = \mathcal{C}_v(F_{w,\ell}) \cap U_{v,\ell}^0$. By hypothesis, there is a point $\overline{w}_\ell \in \left(\mathcal{C}_v(F_{w_\ell}) \cap B(a_\ell, r_\ell)\right)\backslash H_{v,\ell}$; clearly $\overline{w}_\ell \notin U_{v,\ell}^0$.

We begin by constructing $\mathcal{D}_\ell$.

First assume that $g = g(C_v) > 0$. By hypothesis, the K_v-simple decomposition $H_v = \bigcup_{h=1}^N \left(\mathcal{C}_v(F_{u_h}) \cap B(\theta_h, \rho_h)\right)$ is move-prepared for $B(a_1, r_1), \ldots, B(a_D, r_D)$. For simplicity, relabel the roots $\theta_1, \ldots, \theta_N$ of $\phi_v(z)$ so that $B(\theta_1, \rho_1), \ldots, B(\theta_M, \rho_M)$ are contained in $B(a_\ell, r_\ell)$; thus $\mathcal{I}_\ell = \{1, \ldots, M\}$. Without loss we can assume that $B(\theta_1, \rho_1), \ldots, B(\theta_g, \rho_g)$ are the distinguished balls corresponding to $B(a_\ell, r_\ell)$ in the definition of move-preparedness (Definition 6.10). This means there is a number $\overline{r}_\ell$ with $\rho_1, \ldots, \rho_g < \overline{r}_\ell < r_\ell$ such that $B(\theta_1, \overline{r}_\ell), \ldots, B(\theta_g, \overline{r}_\ell)$ are pairwise disjoint and contained in $B(a_\ell, r_\ell)$, and if we put $\vec{\theta}_\ell = (\theta_1, \ldots, \theta_g)$ then

$$W_{\vec{\theta}_\ell}(\overline{r}_\ell) \;:=\; \mathbf{J}_{\vec{\theta}_\ell}\Big(\prod_{h=1}^{g} B(\theta_h, \overline{r}_\ell)\Big)$$

is an open subgroup of $\mathrm{Jac}(\mathcal{C}_v)(\mathbb{C}_v)$ satisfying the properties in Theorem 6.9.

Since $E_{v,\ell}$ is compact, by Proposition D.3 there are constants $\varepsilon_0^{(\ell)} > 0$, $C_0^{(\ell)} > 0$ such that if $0 < \varepsilon \le \varepsilon_0^{(\ell)}$, then for all $x, y \in E_{v,\ell}$ with $\|x, y\|_v \le \varepsilon$, the divisor class $\mathbf{j}_x(y) = [(y) - (x)]$ belongs to $W_{\vec{\theta}_\ell}(C_0^{(\ell)}\varepsilon \cdot \overline{r}_\ell)$.

In the Lemma, we will take

$$\varepsilon_1^{(\ell)} = \varepsilon_0^{(\ell)} , \qquad C_1^{(\ell)} = \max(1, C_0^{(\ell)}\overline{r}_\ell) .$$

Put $\overline{\rho}_\ell = \min_{h\in\mathcal{I}_\ell}(\rho_h)$, and let $0 < \varepsilon \le \varepsilon_1^{(\ell)}$ be small enough that $C_1^{(\ell)}\varepsilon \le \delta_n\overline{\rho}_\ell$. Thus $B(\theta_{hj}, C_1^{(\ell)}\varepsilon) \subseteq B(\theta_{hj}, \delta_n\rho_h)$ for all $1 \le h \le M$, $1 \le j \le n$. The balls $B(\theta_{hj}, \delta_n\rho_h)$ are pairwise disjoint and isometrically parametrizable, hence the same is true for the balls $B(\theta_{hj}, C_1^{(\ell)}\varepsilon)$. Using the "safe" index j_0, put $\vec{\theta}_{\ell,j_0} = (\theta_{1,j_0}, \ldots, \theta_{g,j_0})$. By Theorem 6.9(D),

$$W_\ell(C_1^{(\ell)}\varepsilon) \;:=\; W_{\vec{\theta}_{\ell,j_0}}(C_1^{(\ell)}\varepsilon) \;=\; \mathbf{J}_{\vec{\theta}_{\ell,j_0}}\Big(\prod_{h=1}^{g} B(\theta_{h,j_0}, C_1^{(\ell)}\varepsilon)\Big)$$

is an open subgroup of $W_{\vec{\theta}_\ell}(\overline{r}_\ell)$. By our choice of $C_1^{(\ell)}$ it contains $W_{\vec{\theta}_\ell}(C_0^{(\ell)}\varepsilon \cdot \overline{r}_\ell)$.

Suppose we are given a set of points $\{\theta_{hj}^{**} \in \mathcal{C}_v(F_{w_\ell}) \cap B(\theta_h, \rho_h)\}_{1\le h\le M, j\in\mathcal{S}^\triangle}$ with $\|\theta_{hj}^{**}, \theta_{hj}^*\|_v \le \varepsilon$ for each (h,j). Since θ_{hj}^* belongs to $B(\theta_h, \rho_h)$ and $\varepsilon \le \rho_h$, we have $\theta_{hj}^{**} \in B(\theta_h, \rho_h)$ as well. Thus, $\theta_{hj}^{**} \in \mathcal{C}_v(F_{w_\ell}) \cap B(\theta_h, \rho_h) \subset H_{v,\ell}$ for each h, j.

Using the action $\ddot{+}$ of the group $W_{\vec{\theta}_{\ell,j_0}}(C_1^{(\ell)}\varepsilon)$ on $\prod_{h=1}^g B(\theta_{h,j_0}, C_1^{(\ell)}\varepsilon)$ from Theorem 6.9, we will construct points $\theta_{h,j_0}^{**} \in \mathcal{C}_v(F_{w_\ell}) \cap B(\theta_{h,j_0}, C_1^{(\ell)}\varepsilon)$, for $h = 1, \ldots, g$, such that

$$\mathcal{D}_\ell \ := \ \sum_{j\in S^\triangle}\sum_{h\in\mathcal{I}_\ell}((\theta_{hj}^{**}) - (\theta_{hj}^*)) + \sum_{h=1}^{g}((\theta_{h,j_0}^{**}) - (\theta_{h,j_0}))$$

is F_{w_ℓ}-rational and principal. Consider the divisor class

$$x \ = \ \sum_{j\in S^\triangle}\sum_{h\in\mathcal{I}_\ell}[(\theta_{hj}^{**}) - (\theta_{hj}^*)] \ \in \ \mathrm{Jac}(\mathcal{C}_v)(F_{w_\ell}) \ .$$

As noted above, the θ_{hj}^{**} and θ_{hj}^* belong to $H_{v,\ell} \subset U_{v,\ell}^0$. By our choice of ε, we have $[(\theta_{hj}^{**}) - (\theta_{hj}^*)] \in W_{\vec{\theta}_\ell}(C_0^\ell\varepsilon \cdot \overline{r}_\ell) \subseteq W_{\vec{\theta}_{\ell,j_0}}(C_1^{(\ell)}\varepsilon)$ for all h, j. Since $W_{\vec{\theta}_{\ell,j_0}}(C_1^{(\ell)}\varepsilon)$ is a group, it follows that $x \in W_{\vec{\theta}_{\ell,j_0}}(C_1^{(\ell)}\varepsilon) \cap \mathrm{Jac}(\mathcal{C}_v)(F_{w_\ell})$. Define

$$(\theta_{1,j_0}^{**}, \ldots, \theta_{g,j_0}^{**}) \ = \ (-x) \ \ddot{+} \ (\theta_{1,j_0}, \ldots, \theta_{g,j_0}) \ \in \ \prod_{h=1}^{g} B(\theta_{h,j_0}, C_1^{(\ell)}\varepsilon) \ .$$

By Theorem 6.9(C),

$$\sum_{h=1}^{g}[(\theta_{h,j_0}^{**}) - (\theta_{h,j_0})] \ = \ \mathbf{J}_{\vec{\theta}_{\ell,j_0}}((\theta_{1,j_0}^{**}, \ldots, \theta_{g,j_0}^{**})) \ = \ -x \ ,$$

so $\mathcal{D}_\ell$ is principal. By Theorem 6.9(E), the action $\ddot{+}$ preserves F_{w_ℓ} rationality, so each $\theta_{h,j_0}^{**} \in \mathcal{C}_v(F_{w_\ell})$, and $\mathcal{D}_\ell$ is F_{w_ℓ}-rational. Finally, our choice of ε required that $C_1^{(\ell)}\varepsilon \le \delta_n\rho_h$, so $\theta_{h,j_0}^{**} \in \mathcal{C}_v(F_{w_\ell}) \cap B(\theta_{h,j_0}, \delta_n\rho_h)$ for each $h = 1, \ldots, g$.

For $h = g+1, \ldots, M$, put $\theta_{h,j_0}^{**} = \theta_{h,j_0}$.

We next construct $Y_\ell(z)$. For this, it will be useful to relabel the θ_{hj}^* and θ_{hj}^{**} by gathering them in groups of size g. For simplicity, first assume that g divides $M \cdot \#(\mathcal{S}^\triangle)$. Put $T = M \cdot \#(\mathcal{S}^\triangle)/g$ and write

$$\{\theta_{hj}^*\}_{1\le h\le M, j\in\mathcal{S}^\triangle} \ = \ \{c_k^{(t)}\}_{1\le k\le g, 1\le t\le T} \ .$$

Using the same correspondence between indices, write

$$\{\theta_{hj}^{**}\}_{1\le h\le M, j\in\mathcal{S}^\triangle} \ = \ \{\widehat{c}_k^{(t)}\}_{1\le k\le g, 1\le t\le T} \ .$$

Put $\vec{c}^{(t)} = (c_1^{(t)}, \ldots, c_g^{(t)})$, $\widehat{c}^{(t)} = (\widehat{c}_1^{(t)}, \ldots, \widehat{c}_g^{(t)})$. Clearly

$$\sum_{j\in S^\triangle}\sum_{h=1}^{M}[(\theta_{hj}^{**}) - (\theta_{hj}^*)] \ = \ \sum_{t=1}^{T}\sum_{k=1}^{g}[(\widehat{c}_k^{(t)}) - (c_k^{(t)})] \ . \tag{11.90}$$

If g does not divide $M \cdot \#(\mathcal{S}^\triangle)$, put $T = \lceil M \cdot \#(\mathcal{S}^\triangle)/g \rceil$ and set $r = T \cdot g - M \cdot \#(\mathcal{S}^\triangle)$. Fix an element $j_1 \in \mathcal{S}^\triangle$ and augment the lists $\{\theta_{hj}^*\}$ and $\{\theta_{hj}^{**}\}$ by adjoining r copies of θ_{1,j_1}^* at the end of each, then break down the lists into groups of size g as before. In this way the final vectors $\vec{c}^{(T)}$, $\widehat{c}^{(T)}$ have their last r components equal to θ_{1,j_1}^*, and (11.90) still holds.

Put $\vec{d}^{(0)} = \vec{\theta}_{\ell,j_0} = (\theta_{1,j_0}, \ldots, \theta_{g,j_0})$ and write $\vec{d}^{(0)} = (d_1^{(0)}, \ldots, d_g^{(0)})$. We will inductively construct vectors $\vec{d}^{(t)} = (d_1^{(t)}, \ldots, d_g^{(t)}) \in \prod_{h=1}^{g} \big(\mathcal{C}_v(F_{w_\ell}) \cap B(\theta_{h,j_0}, C_1^{(\ell)}\varepsilon)\big)$ such that for each $t = 1, \ldots, T$, the divisor

$$\mathcal{D}_\ell^{(t)} \;=\; \sum_{h=1}^{g}((\widehat{c}_h^{(t)}) - (c_h^{(t)})) + \sum_{h=1}^{g}((d_h^{(t)}) - (d_h^{(t-1)}))$$

is F_{w_ℓ}-rational and principal, and such that

$$\mathcal{D}_\ell \;=\; \sum_{t=1}^{T} \mathcal{D}_\ell^{(t)} \ .$$

Suppose $\vec{d}^{(t-1)}$ has been constructed. Put

$$x^{(t)} \;=\; \sum_{j=1}^{g}[(\widehat{c}_j^{(t)}) - (c_j^{(t)})] \;\in\; \mathrm{Jac}(\mathcal{C}_v)(F_{w_\ell}) \cap W_{\vec{\theta}_{\ell,j_0}}(C_1^{(\ell)}\varepsilon) \ .$$

Using the operation $\ddot{+}$ of the group $W_{\vec{\theta}_{\ell,j_0}}(C_1^{(\ell)}\varepsilon)$ on $\prod_{h=1}^{g} B(\theta_{h,j_0}, C_1^{(\ell)}\varepsilon)$ in Theorem 6.9, define

$$\vec{d}^{(t)} \;=\; (-x^{(t)}) \ddot{+} \vec{d}^{(t-1)} \;\in\; \prod_{h=1}^{g} B(\theta_{h,j_0}, C_1^{(\ell)}\varepsilon) \ .$$

By Theorem 6.9(C),

$$\sum_{j=1}^{g}[(d_j^{(t)}) - (d_j^{(t-1)})] \;=\; -x^{(t)} \;=\; -\sum_{j=1}^{g}[(\widehat{c}_j^{(t)}) - (c_j^{(t)})]$$

so $\mathcal{D}^{(t)}$ is principal. Since $C_1^{(\ell)}\varepsilon \le \delta_n \overline{\rho}_\ell$, Theorem 6.9(E) shows that $\vec{d}^{(t)}$ belongs to $\prod_{h=1}^{g} \big(\mathcal{C}_v(F_{w_\ell}) \cap B(\theta_{h,j_0}, \delta_n\rho_h)\big)$, and $\mathcal{D}^{(t)}$ is F_{w_ℓ}-rational.

Since $x^{(1)} + \cdots + x^{(T)} = x$, the fact that $\ddot{+}$ is an action assures that when $t = T$, we have $\vec{d}^{(T)} = (\theta_{1,j_0}^{**}, \ldots, \theta_{g,j_0}^{**})$ with the points θ_{h,j_0}^{**} constructed earlier. Thus the divisor class

$$\sum_{t=1}^{T}\Big(\sum_{j=1}^{g}[(d_j^{(t)}) - (d_j^{(t-1)})]\Big)$$

telescopes to $\sum_{h=1}^{g}[(\theta_{h,j_0}^{**}) - (\theta_{h,j_0})]$, and $\mathcal{D}_\ell = \sum_{t=1}^{T} \mathcal{D}_\ell^{(t)}$ as claimed.

If $g(\mathcal{C}) = 0$, we can again assume the roots of $\phi_v(z)$ are labeled so that $\mathcal{I}_\ell = \{1, \ldots, M\}$. The divisor $\mathcal{D}_\ell := \sum_{j \in \mathcal{S}^\triangle} \sum_{h=1}^{M} (\theta_{hj}^{**}) - (\theta_{hj}^{*})$ is already principal, so we can take $\theta_{h,j_0}^{**} = \theta_{h,j_0}$ for each $h = 1, \ldots, M$. For compatibility with the notation above, put $T = M \cdot \#(\mathcal{S}^\triangle)$, and relabel the sets $\{\theta_{hj}\}_{1 \le h \le M, j \in \mathcal{S}^\triangle}$, $\{\theta_{hj}^{**}\}_{1 \le h \le M, j \in \mathcal{S}^\triangle}$, as $\{c^{(t)}\}_{1 \le t \le T}$, $\{\widehat{c}^{(t)}\}_{1 \le t \le T}$, respectively. For each $t = 1, \ldots, T$, put $\mathcal{D}_\ell^{(t)} = (\widehat{c}^{(t)}) - (c^{(t)})$. Then each $\mathcal{D}_\ell^{(t)}$ is F_{w_ℓ}-rational, and $\mathcal{D}_\ell = \sum_{t=1}^{T} \mathcal{D}^{(t)}$.

We can now construct $Y_\ell(z)$. Recall that

$$C_1^{(\ell)}\varepsilon \;\le\; \delta_n \cdot \overline{\rho}_\ell \;=\; \delta_n \cdot \min_{h \in \mathcal{I}_\ell} \rho_h \ .$$

For each (h, j) with $1 \le h \le M$ and $j \in \mathcal{S}^\triangle$ we have $\|\theta_{hj}^{**}, \theta_{hj}^{*}\|_v \le \varepsilon$ and in particular $\theta_{hj}^{**} \in B(\theta_{hj}, \delta_n\rho_h)$ since $C_1^{(\ell)} \ge 1$; while for $j = j_0$ and $h = 1, \ldots, M$, we have $\|\theta_{h,j_0}^{**}, \theta_{h,j_0}\|_v \le C_1^{(\ell)}\varepsilon$, hence $\theta_{h,j_0}^{**} \in B(\theta_{h,j_0}, \delta_n\rho_h)$.

We now apply Theorem C.2 of Appendix C to the set $H_{v,\ell}$, taking $d=\max(1,2g)$, $r=\overline{\rho}_\ell$, and replacing ε in the theorem with $C_1^{(\ell)}\varepsilon \le \delta_n\overline{\rho}_\ell$. Let $D(H_{v,\ell},d)$ be the constant from the Theorem, and take

$$C_2^{(\ell)} = C_1^{(\ell)} \cdot \frac{D(H_{v,\ell},d)}{(\overline{\rho}_\ell)^d}$$

in the lemma. Let $\overline{w}_\ell \in \big(\mathcal{C}_v(F_{w_\ell}) \cap B(a_\ell,r_\ell)\big)\backslash H_v = \big(\mathcal{C}_v(F_{w_\ell}) \cap B(a_\ell,r_\ell)\big)\backslash U_{v,\ell}^0$ be the point from the statement of the lemma. For each $p \in U_{v,\ell}^0$ we have $\|p,\overline{w}_\ell\|_v > r = \overline{\rho}_\ell$.

For each $t=1,\ldots,T$, by specializing the Universal Function $f(z,w;\vec{p},\vec{q})$ of degree d in Theorem C.1 of Appendix C, taking $w=\overline{w}_\ell$, and letting $\vec{p}$ (resp. $\vec{q}$) be vectors consisting of the zeros (resp. poles) of $\mathcal{D}_\ell^{(t)}$, we obtain a function $Y_\ell^{(t)}(z)$ for which $\operatorname{div}(Y_\ell^{(t)}) = \mathcal{D}_\ell^{(t)}$ and $Y_\ell^{(t)}(\overline{w}_\ell)=1$. Each $Y_\ell^{(t)}(z)$ is F_{w_ℓ}-rational, since $\mathcal{D}_\ell^{(t)}$ is F_{w_ℓ}-rational and $\overline{w}_\ell \in \mathcal{C}_v(F_{w_\ell})$. The sets $(\bigcup_{j=1}^d B(p_j,r_j)^-)\cup(\bigcup_{j=1}^d B(q_j,r_j)^-))$ and $(\bigcup_{j=1}^d B(p_j,r)^-)\cup(\bigcup_{j=1}^d B(q_j,r)^-)$ from Theorem C.2(A,B) are both contained in $U_{v,\ell}^0 = \bigcup_{h\in\mathcal{I}_\ell} B(\theta_h,\rho_h)$. Hence for each $t=1,\ldots,T$,

$(1_{\ell,t})$ $|Y_\ell^{(t)}(z)|_v = 1$ for all $z \in \mathcal{C}_v(\mathbb{C}_v)\backslash U_{v,\ell}^0$;

$(2_{\ell,t})$ $|Y_\ell^{(t)}(z)-1|_v \le C_2^{(\ell)}\varepsilon$ for all $z \in \mathcal{C}_v(\mathbb{C}_v)\backslash U_{v,\ell}^0$.

Put $Y_\ell(z) = \prod_{t=1}^T Y^{(t)}(z) \in F_{w_\ell}(\mathcal{C}_v)$. Then $\operatorname{div}(Y_\ell)=\mathcal{D}_\ell$, and $|Y_\ell(z)|_v = 1$ for all $z \notin U_{v,\ell}^0$. Thus assertion $(1_{\ell,t})$ holds. Since

$$Y_\ell(z)-1 = \sum_{t=1}^T \big(Y_\ell^{(t)}(z)-1\big)\cdot \prod_{s=t+1}^T Y_\ell^{(s)}(z)$$

and $(1_{\ell,t})$ and $(2_{\ell,t})$ above hold for all t, the ultrametric inequality shows that $|Y_\ell(z)-1|_v \le C_2^{(\ell)}\varepsilon$ for all $z \notin U_{v,\ell}^0$. Thus assertion $(2_{\ell,t})$ holds. □

Lemma 11.11 (Second Moving Lemma). *There are constants $\varepsilon_2 > 0$ and $C_3, C_4 > 1$ (depending on E_v, $\mathfrak{X}$, the choices of the L-rational and L^{sep}-rational bases, the uniformizers $g_{x_i}(z)$, and the projective embedding of $\mathcal{C}_v$), such that if $0<\varepsilon<\varepsilon_1$ and $Y(z)$ are as in Lemma 11.10, and in addition $\varepsilon<\varepsilon_2$ and n is sufficiently large, then when we expand*

$$G_v^{(k_1)}(z) = \sum_{i=1}^m \sum_{j=0}^{(n-1)N_i-1} A_{v,ij}\varphi_{i,nN_i-j}(z) + \sum_{\lambda=1}^{\Lambda} A_\lambda\varphi_\lambda(z)\ ,$$

$$\overline{G}_v^{(k_1)}(z) := Y(z)\cdot G_v^{(k_1)}(z) = \sum_{i=1}^m \sum_{j=0}^{(n-1)N_i-1} \overline{A}_{v,ij}\varphi_{i,nN_i-j}(z) + \sum_{\lambda=1}^{\Lambda} \overline{A}_\lambda\varphi_\lambda(z)\ ,$$

for all $i=1,\ldots,m$ and all $0\le j<k_1N_i$, we have

$$|\overline{A}_{v,ij} - A_{v,ij}|_v \le C_3C_4^j(|\widetilde{c}_{v,i}|_v)^n\varepsilon\ .$$

For the proof, we will need the following lemma concerning power series, which is closely related to Lemma 7.19:

LEMMA 11.13. *Let r belong to the value group of $\mathbb{C}_v$, and let $b \in \mathbb{C}_v^\times$. Suppose that for each $j \geq J_0$, $\Phi_j(Z) = b^{-j}Z^{-j}\big(1 + \sum_{\ell=1}^{\infty} C_\ell^{(j)} Z^\ell\big) \in \mathbb{C}_v((Z))$ is a Laurent series with leading coefficient b^{-j}, which converges in $D(0,r)\backslash\{0\}$ and has no zeros there. Let*

$$G(Z) \ = \ Z^{-M}\big(g_0 + \sum_{j=1}^{\infty} g_j Z^j\big) \in \mathbb{C}_v((Z))$$

be another Laurent series converging in $D(0,r)\backslash\{0\}$ and having no zeros there, with $g_0 \neq 0$.

We can uniquely expand $G(Z)$ as a linear combination of the $\Phi_j(Z)$ and a residual series in Z, writing

$$G(Z) \ = \ \sum_{j=0}^{M-J_0} B_j \Phi_{M-j}(Z) + Z^{-J_0+1}\big(\sum_{j=0}^{\infty} B_j' Z^j\big) \ . \tag{11.91}$$

Suppose $Y(Z) = \sum_{\ell=0}^{\infty} h_\ell Z^\ell$ is a power series converging in $D(0,r)$, and there is an ε with $0 < \varepsilon < 1$ such that $|Y(Z) - 1|_v \leq \varepsilon$ for all $Z \in D(0,r)$.

Consider the product $Y(Z)G(Z) = \sum_{j=0}^{\infty} \overline{g}_j Z^{-M+j}$. If we expand

$$Y(Z)G(Z) \ = \ \sum_{j=0}^{M-J_0} \overline{B}_j \Phi_{M-j}(Z) + Z^{-J_0+1}\big(\sum_{j=0}^{\infty} \overline{B}_j' Z^j\big) \tag{11.92}$$

then $|\overline{B}_j - B_j|_v \leq \varepsilon \cdot |g_0|_v \cdot (r|b|_v)^{-j}$ for each j in the range $0 \leq j \leq M - J_0$.

PROOF. After replacing Z by bZ, and $D(0,r)$ by $D(0,r|b|_v)$, we can assume without loss that $b = 1$. In particular, we can assume that each $\Phi_j(Z)$ has leading coefficient 1.

Under this hypothesis, we will first show that for each $j \geq 0$,

$$|\overline{g}_j - g_j|_v \ \leq \ \varepsilon \cdot \frac{|g_0|_v}{r^j} \ .$$

Multiplying $G(Z)$ by Z^M, we obtain a power series converging in $D(0,r)$, having no zeros in $D(0,r)$, whose Taylor coefficients are the g_j. The theory of Newton Polygons shows that $|g_j|_v \leq |g_0|/r^j$ for all j (see Lemma 3.35 and the discussion preceding it; in fact, strict inequality holds when $j \geq 1$). On the other hand, by the Maximum Modulus Principle for power series, since $|Y(Z) - 1|_v \leq \varepsilon$ for all $Z \in D(0,r)$, we have $|h_0 - 1|_v \leq \varepsilon$ and $|h_\ell|_v \leq \varepsilon/r^\ell$ for all $\ell \geq 1$.

In the product $Y(Z)G(Z)$ we have $\overline{g}_j = \sum_{k=0}^{j} h_k g_{j-k}$ for each j. Hence

$$\begin{aligned}
|\overline{g}_j - g_j|_v &= |(h_0 - 1)g_j + h_1 g_{j-1} + \ldots + h_j g_0|_v \\
&\leq \max(|h_0 - 1|_v |g_j|_v, |h_1|_v |g_{j-1}|_v, \ldots, |h_j|_v |g_0|_v) \\
&\leq \max(\varepsilon \cdot |g_0|_v / r^j, \varepsilon/r \cdot |g_0|_v / r^{j-1}, \ldots, \varepsilon/r^j \cdot |g_0|_v) \\
&= \varepsilon \cdot |g_0|_v / r^j \ .
\end{aligned}$$

Now consider the expansions (11.91) and (11.92). Clearly $B_0 = g_0$ and $\overline{B}_0 = \overline{g}_0$, so $|\overline{B}_0 - B_0|_v \leq \varepsilon |B_0|_v$. Since the $\Phi_j(Z)$ have no zeros in $D(0,r)$ and have leading coefficient 1, the theory of Newton Polygons shows that $|C_\ell^{(j)}|_v \leq 1/r^\ell$ for each j and ℓ, as before.

Suppose inductively that for some $J \le M - J_0$, we have shown that $|\overline{B}_j - B_j|_v \le \varepsilon |B_0|_v / r^j$ for all $0 \le j < J$. Using (11.91) and (11.92) we have

$$\begin{aligned} G(Z) - \sum_{j=0}^{J-1} B_j \Phi_{M-j}(Z) &= \sum_{k=0}^{\infty} \delta_k Z^{-M+J+k} , \\ Y(Z)G(Z) - \sum_{j=0}^{J-1} \overline{B}_j \Phi_{M-j}(Z) &= \sum_{k=0}^{\infty} \overline{\delta}_k Z^{-M+J+k} , \end{aligned}$$

for certain numbers $\delta_k, \overline{\delta}_k \in \mathbb{C}_v$. Inserting the Laurent expansions for $G(Z)$, $Y(Z)G(Z)$ and the $\Phi_{M-j}(Z)$, we see that for each k,

$$\begin{aligned} \delta_k &= g_{J+k} - B_0 C_{k+J}^{(M)} - B_1 C_{k+J-1}^{(M-1)} - \cdots - B_{J-1} C_{k+1}^{(M-J+1)} , \\ \overline{\delta}_k &= \overline{g}_{J+k} - \overline{B}_0 d_{J+k}^{(M)} - \cdots - \overline{B}_{J-1} C_{k+1}^{(M-J+1)} . \end{aligned}$$

By the ultrametric inequality and the estimates above,

$$\begin{aligned} |\overline{\delta}_k - \delta_k|_v &\le \max(\varepsilon |B_0|_v / r^{J+k}, \varepsilon |B_0|_v \cdot 1/r^{J+k}, \ldots, \varepsilon |B_0|_v / r^{J-1} \cdot 1/r^{k+1}) \\ &= \varepsilon |B_0|_v / r^{J+k} . \end{aligned}$$

When $k = 0$, the fact that $\Phi_{M-J}(Z)$ has leading term Z^{-M+J} shows that $B_J = \delta_0$ and $\overline{B}_J = \overline{\delta}_0$. Hence $|\overline{B}_j - B_j|_v \le \varepsilon |B_0| / r^J$ and the induction can continue.

When $J = M - J_0$, the induction stops because there is no function $\Phi_{J_0 - 1}(Z)$. □

Proof of Lemma 11.11. Since $\mathfrak{X}$ is disjoint from E_v, and since the basis functions $\varphi_{ij}(z)$ and φ_λ belong to a multiplicatively finitely generated set, there is a radius $r > 0$ in the value group of $\mathbb{C}_v^\times$ such that

(1) $r < \min_{i \ne j}(\|x_i, x_j\|_v)$;
(2) each ball $B(x_i, r)$ is isometrically parametrizable and disjoint from E_v;
(3) for each i, nonc of the $\varphi_{ij}(z)$ has a zero in $B(x_i, r)$.

Fixing $x_i \in \mathfrak{X}$, let $\varrho_i : D(0, r) \to B(x_i, r)$ be an L_{w_0}-rational isometric parametrization of $B(x_i, r)$ with $\varrho_i(0) = x_i$. If Z is the coordinate on $D(0, r)$, then we can expand $G_v^{(k_1)}(z)$, $Y(z)$, and the $\varphi_{ij}(z)$ as Laurent series in Z converging in $D(0, r) \backslash \{0\}$, putting $G(Z) = G_v^{(k_1)}(\varrho_i(Z))$, $H(Z) = Y(\varrho_i(Z))$, and $\Phi_j(z) = \varphi_{ij}(\varrho_i(Z))$.

With respect to the uniformizer $g_{x_i}(z)$ the $\varphi_{ij}(z)$ are monic. That is,

$$\lim_{z \to x_i} \varphi_{ij}(z) \cdot g_{x_i}(z)^j = 1 .$$

When $g_{x_i}(z)$ is expanded in terms of Z, its leading coefficient will be some $b_{v,i} \in K_v(x_i)^\times$:

$$b_{v,i} = \lim_{Z \to 0} \frac{g_{x_i}(\varrho_i(Z))}{Z} .$$

It follows that $\varphi_{ij}(\varrho_i(Z))$ has the leading term $(b_{v,i})^{-j} Z^{-j}$.

We now apply Lemma 11.13 with $\Phi_j(Z)$, $G(Z)$ and $H(Z)$ as above, taking $b = b_{v,i}$ and $J_0 = k_1 N_i$. Because only the $\varphi_{ij}(z)$ with $j \ge (n - k_1) N_i$ have poles of order $(n-k_1)N_i$ or more at x_i, we can write

$$\begin{aligned} G(Z) &= \sum_{j=0}^{k_1N_i-1} A_{v,ij}\varphi_{i,nN_i-j}(\varrho_i(Z)) + \sum_{j=(n-k_1)N_i}^{\infty} A'_{ij}Z^{-nN_i+j}, \\ Y(Z)G(Z) &= \sum_{j=0}^{k_1N_i-1} \overline{A}_{v,ij}\varphi_{i,nN_i-j}(\varrho_i(Z)) + \sum_{j=(n-k_1)N_i}^{\infty} \overline{A}'_{ij}Z^{-nN_i+j} \end{aligned}$$

where the $A_{v,ij}$ and $\overline{A}_{v,ij}$ are the same as in (11.65), (11.66). Thus $g_0 = A_{v,i0}$ in Lemma 11.13. Recall that $|A_{v,i0}|_v = |\widetilde{c}_{v,i}|_v^n$ and that $|Y(Z)-1|_v = |Y(z)-1|_v \le C_2\varepsilon$ for all $Z \in D(0,r)$, where C_2 is the constant from Lemma 11.10. Put $C_3 = C_2$ in Lemma 11.11. By Lemma 11.13,

$$|A_{v,ij} - \overline{A}_{v,ij}|_v \le C_3\varepsilon \cdot \frac{(|\widetilde{c}_{v,i}|_v)^n}{(r|b_{v,i}|_v)^j} \tag{11.93}$$

for each $j = 0, \ldots, k_1N_i$.

Letting x_i vary, Lemma 11.11 holds with

$$C_4 = \frac{1}{\min(1, r|b_{v,1}|_v, \ldots, r|b_{v,m}|_v)} . \tag{11.94}$$

□

Lemma 11.12 (Third Moving Lemma). *There are constants $\varepsilon_3 > 0$ and $C_6, C_7 \ge 1$ (depending only on $\phi_v(z)$, E_v, H_v, their K_v-simple decompositions $\bigcup_{\ell=1}^D \big(B(a_\ell, r_\ell) \cap \mathcal{C}_v(F_{w_\ell})\big)$ and $\bigcup_{h=1}^N \big(B(\theta_h, \rho_h) \cap \mathcal{C}_v(F_{u_h})\big)$, the choices of the L-rational and L^{sep}-rational bases, the uniformizers $g_{x_i}(z)$, and the projective embedding of $\mathcal{C}_v$), such that if $0 < \varepsilon < \varepsilon_3$, and if $\overline{F}_{v,k_1}(z)$ is as in (11.68) and $U_v^0 = \bigcup_{h=1}^N B(\theta_h, \rho_h)$, then there is a K_v-rational $(\mathfrak{X}, \vec{s})$-function $\overline{\Delta}_{v,k_1}(z)$ of the form*

$$\overline{\Delta}_{v,k_1}(z) = \sum_{i=1}^{m} \sum_{j=0}^{k_1N_i-1} \overline{\Delta}_{v,ij}\varphi_{i,(k_1+1)N_i-j}(z) ,$$

satisfying

$$\|\overline{\Delta}_{v,ij}\|_{U_v^0} \le C_6C_7^{k_1}\varepsilon ,$$

such that when $\overline{G}_v^{(k_1)}(z)$ from Lemma 11.11 is replaced with

$$\widehat{G}_v^{(k_1)}(z) = \overline{G}_v^{(k_1)}(z) + \overline{\Delta}_{v,k_1}(z)\overline{F}_{v,k_1}(z) ,$$

then for each (i,j) with $1 \le i \le m$, $0 \le j < k_1N_i$, the coefficient $\overline{A}_{v,ij}$ of $\overline{G}_v^{(k_1)}(z)$ is restored to the coefficient $A_{v,ij}$ of $G_v^{(k_1)}(z)$ in $\widehat{G}_v^{(k_1)}(z)$.

PROOF. This is a consequence of Proposition 7.18. Using the L-rational basis, expand

$$\begin{aligned} G_v^{(k_1)}(z) &= \sum_{i=1}^{m} \sum_{j=0}^{(n-1)N_i-1} A_{v,ij}\varphi_{i,nN_i-j}(z) + \sum_{\lambda=1}^{\Lambda} A_\lambda\varphi_\lambda(z) , \\ \overline{G}_v^{(k_1)}(z) &= \sum_{i=1}^{m} \sum_{j=0}^{(n-1)N_i-1} \overline{A}_{v,ij}\varphi_{i,nN_i-j}(z) + \sum_{\lambda=1}^{\Lambda} \overline{A}_\lambda\varphi_\lambda(z) , \end{aligned}$$

and using the L^{sep}-rational basis, write

$$G_v^{(k_1)}(z) = \sum_{i=1}^{m} \sum_{j=0}^{(n-1)N_i-1} \widetilde{a}_{v,ij}\widetilde{\varphi}_{i,nN_i-j}(z) + \sum_{\lambda=1}^{\Lambda} \widetilde{a}_\lambda \widetilde{\varphi}_\lambda(z) ,$$

$$\overline{G}_v^{(k_1)}(z) = \sum_{i=1}^{m} \sum_{j=0}^{(n-1)N_i-1} \overline{a}_{v,ij}\widetilde{\varphi}_{i,nN_i-j}(z) + \sum_{\lambda=1}^{\Lambda} \overline{a}_\lambda \varphi_\lambda(z) .$$

By Proposition 3.3(C), the transition matrix from the L-rational basis to the L^{sep}-rational basis is block diagonal with blocks of size J, and for a given i the same $J \times J$ matrix $\mathcal{B}_{i,jk}$ occurs for each block. By Lemma 11.11 we have $|A_{v,ij} - \overline{A}_{v,ij}|_v \le C_3 C_4^j (|\widetilde{c}_{v,i}|_v)^n \varepsilon$ for all $1 \le i \le m$, $0 \le j < k_1 N_i$. Since $J | N_i$ for each i, there is a constant C_5 such that

$$|\widetilde{a}_{v,ij} - \overline{a}_{v,ij}|_v \le C_5 C_4^j (|\widetilde{c}_{v,i}|_v)^n \varepsilon \tag{11.95}$$

for all $1 \le i \le m$, $0 \le j < k_1 N_i$. Indeed, putting $B_v = \max_{1 \le i \le m, 1 \le j,k \le J} |\mathcal{B}_{i,jk}|_v$, we can take $C_5 = B_v C_3 C_4^{J-1}$.

For each i, j put $\widetilde{\delta}_{v,ij} = \widetilde{a}_{v,ij} - \overline{a}_{v,ij}$. Since $G_v^{(k_1)}(z)$ and $\overline{G}_v^{(k_1)}(z)$ are K_v-rational, the $\widetilde{\delta}_{v,ij}$ belong to $L_{w_v}^{\text{sep}}$ and are K_v-symmetric (Corollary 3.5).

We now apply Proposition 7.18 taking $\ell = k_1$, $F_v(z) = \overline{F}_{v,k_1}(z)$, and

$$\widetilde{\delta} = (\widetilde{\delta}_{v,ij})_{1 \le i \le m, 0 \le j < k_1 N_i} .$$

We will take r in Proposition 7.18 to be the same number as in the proof of Lemma 11.11. Comparing (7.108) and (11.94) shows that if ϖ_v is the constant from Proposition 7.18 and C_4 is the constant from Lemma 11.11, then $C_4 = \varpi_v^{-1}$. Letting $\widetilde{\Upsilon}_v$ be the constant from Proposition 7.18, and recalling from the discussion after (11.69) that the leading coefficient $d_{v,i}$ of $\overline{F}_{v,k_1}(z)$ at x_i has absolute value $|d_{v,i}|_v = |\widetilde{c}_{v,i}|_v^{n-k_1-1}$, we will take ρ in Proposition 7.18 to be

$$\rho = \frac{C_5 |\widetilde{c}_{v,i}|_v^{k_1+1}}{\widetilde{\Upsilon}_v} \cdot \varepsilon . \tag{11.96}$$

Then for all (i, j) with $1 \le i \le m, 0 \le j < k_1 N_i$ we have

$$|\widetilde{\delta}_{v,ij}|_v = |\widetilde{a}_{v,ij} - \overline{a}_{v,ij}|_v \le C_5 C_4^j (|\widetilde{c}_{v,i}|_v)^n \varepsilon = \widetilde{\Upsilon}_v \varpi_v^{-j} |d_{v,i}|_v \rho . \tag{11.97}$$

By Proposition 7.18 there is a unique $\vec{\Delta} = (\overline{\Delta}_{v,is})_{1 \le i \le m, 0 \le s < k_1 N_i} \in (L_{w_v}^{\text{sep}})^{k_1 N}$ for which

$$\Phi_{\overline{F}_{v,k_1}}^{\text{sep}}(\vec{\Delta}) = \widetilde{\delta} ;$$

moreover, the $\overline{\Delta}_{v,ij}$ are K_v-symmetric and

$$\overline{\Delta}_{v,k_1}(z) := \sum_{i=1}^{m} \sum_{j=0}^{k_1 N_i - 1} \overline{\Delta}_{v,ij} \varphi_{(k_1+1)N_i - j}(z)$$

is K_v-rational. The fact that $\Phi_{\overline{F}_{v,k_1}}^{\text{sep}}(\vec{\Delta}) = \widetilde{\delta}$ and each $\widetilde{\delta}_{v,ij} = \widetilde{a}_{v,ij} - \overline{a}_{v,ij}$ means that

$$\overline{\Delta}_{v,k_1}(z) \overline{F}_{v,k_1}(z) = \sum_{i=1}^{m} \sum_{j=0}^{\ell N_i - 1} (\widetilde{a}_{v,ij} - \overline{a}_{v,ij}) \cdot \widetilde{\varphi}_{i,(k+\ell)N_i - j}(z) + \text{lower order terms} .$$

Consequently, when we expand $\overline{\Delta}_{v,k_1}(z)\overline{F}_{v,k_1}(z)$ using the L-rational basis,

$$\overline{\Delta}_{v,k_1}(z)\overline{F}_{v,k_1}(z) = \sum_{i=1}^{m} \sum_{j=0}^{\ell N_i - 1} (A_{v,ij} - \overline{A}_{v,ij}) \cdot \varphi_{i,(k+\ell)N_i - j}(z) + \text{lower order terms} \ .$$

This means that when $\overline{G}_v^{(k_1)}(z)$ is replaced with

$$\widehat{G}_v^{(k_1)}(z) \ = \ \overline{G}_v^{(k_1)}(z) + \overline{\Delta}_{v,k_1}(z)\overline{F}_{v,k_1}(z) \ ,$$

then for each (i,j) with $1 \le i \le m$, $0 \le j < k_1 N_i$, the coefficient $\overline{A}_{v,ij}$ of $\overline{G}_v^{(k_1)}(z)$ is changed to $A_{v,ij}$ in $\widehat{G}_v^{(k_1)}(z)$.

Finally, by (11.96) and (11.97), and by (7.90) of Proposition 7.18, for each (i,j) with $1 \le i \le m, 0 \le j < k_1 N_i$ we have

$$|\overline{\Delta}_{v,ij}|_v \ \le \ \varpi_v^{-j}\rho \ = \ \frac{C_5}{\widetilde{\Upsilon}_v} C_4^j \cdot (|\widetilde{c}_{v,i}|_v)^{k_1+1} \cdot \varepsilon \ .$$

On the other hand, by Proposition 3.7 there is a constant C_v^0 such that $\|\varphi_{ij}(z)\|_{U_v^0} \le (C_v^0)^j$ for all i and j. Without loss, we can assume that $C_v^0 \ge 1$. It follows that

$$\begin{aligned} \|\overline{\Delta}_{v,k_1}(z)\|_{U_v^0} \ &\le \ \max_{1 \le i \le m, 0 \le j < k_1 N_i} \left(|\overline{\Delta}_{v,ij}|_v \cdot \|\varphi_{i,(k_1+1)N_i-j}\|_{U_v^0}\right) \\ &\le \ \max_{1 \le i \le m, 0 \le j < k_1 N_i} \left(\frac{C_5}{\widetilde{\Upsilon}_v} C_4^j \cdot (|\widetilde{c}_{v,i}|_v)^{k_1+1} \cdot (C_v^0)^{(k_1+1)N_i - j} \cdot \varepsilon\right) \\ &\le \ C_6 C_7^{k_1} \cdot \varepsilon \ , \end{aligned}$$

where

$$C_6 \ = \ \max\left(1, \ \frac{C_5}{\widetilde{\Upsilon}_v} \cdot \max_i(|\widetilde{c}_{v,i}|_v) \cdot (C_v^0)^{\max_i(N_i)}\right)$$

and

$$C_7 \ = \ \max\left(1, \ \max_i(|\widetilde{c}_{v,i}|_v) \cdot \max\left(C_v^0, C_4\right)^{\max_i(N_i)}\right) \ .$$

This completes the proof. □

APPENDIX A

$(\mathfrak{X}, \vec{s})$-Potential Theory

In this appendix we study potential theory for the $(\mathfrak{X}, \vec{s})$-canonical distance. In section A.1 we discuss the basic facts of $(\mathfrak{X}, \vec{s})$-potential theory for compact sets concerning potential functions, equilibrium distributions, the transfinite diameter, and the Chebyshev constant. In section A.2, which concerns the archimedean case, we derive bounds for the mass the $(\mathfrak{X}, \vec{s})$-equilibrium distribution of H can give to "small" subsets of H. In section A.3, which concerns the nonarchimedean case, we determine the $(\mathfrak{X}, \vec{s})$-equilibrium distributions for a class of well-behaved sets.

Fix a projective embedding of $\mathcal{C}/K$. Given a place v of K, let $\|z, w\|_v$ be the corresponding spherical metric on $\mathcal{C}_v(\mathbb{C}_v)$. As in §3.2, if v is nonarchimedean let q_v be the order of the residue field of K_v, and let $\log_v(x)$ be the logarithm to the base q_v. If v is archimedean, put $\log_v(x) = \ln(x)$.

Let $\mathfrak{X} = \{x_1, \ldots, x_m\} \subset \mathcal{C}(\tilde{K})$ be the finite, Galois-stable set of points from §3.2. For each $x_i \in \mathfrak{X}$, let $g_{x_i}(z) \in K(\mathcal{C})$ be the uniformizer at x_i chosen in §3.2, and let the canonical distance $[z, w]_{x_i}$ be normalized so that for each $w \neq x_i$,

$$\lim_{z \to x_i} ([z, w]_{x_i} \cdot |g_{x_i}(z)|_v) \;=\; 1 \; .$$

Let $\vec{s} = (s_1, \ldots, s_m) \in \mathcal{P}^m$ be a probability vector. As in §3.5, we define the $(\mathfrak{X}, \vec{s})$-canonical distance by

$$[z, w]_{\mathfrak{X},\vec{s}} \;=\; \prod_{i=1}^{n} ([z, w]_{x_i})^{s_i} \; .$$

1. $(\mathfrak{X}, \vec{s})$-Potential Theory for Compact Sets

Let $H \subset \mathcal{C}_v(\mathbb{C}_v) \backslash \mathfrak{X}$ be a compact set. In this section we will define analogues of the classical logarithmic capacity, transfinite diameter, Chebyshev constant, potential functions, and Green's functions, relative to the kernel $[z, w]_{\mathfrak{X},\vec{s}}$.

We will study these objects and their relation with the corresponding objects when $\mathfrak{X}$ consists of a single point. The proofs of all the results below are classical and (with minor modifications) are the same as those in ([**51**], §3 and §4), so we only sketch them.

We first define the $(\mathfrak{X}, \vec{s})$-capacity. For any probability measure ν supported on H, the $(\mathfrak{X}, \vec{s})$-energy is

$$\text{(A.1)} \qquad I_{\mathfrak{X},\vec{s}}(\nu) \;=\; \iint_{H \times H} -\log_v([z, w]_{\mathfrak{X},\vec{s}})\, d\nu(z) d\nu(w)$$

and the $(\mathfrak{X}, \vec{s})$-potential function is

$$\text{(A.2)} \qquad u_{\mathfrak{X},\vec{s}}(z, \nu) \;=\; \int_H -\log_v([z, w]_{\mathfrak{X},\vec{s}})\, d\nu(w) \; .$$

The $(\mathfrak{X}, \vec{s})$-Robin constant is defined by

$$(A.3) \qquad V_{\mathfrak{X},\vec{s}}(H) = \inf_{\substack{\text{probability measures} \\ \nu \text{ supported on } H}} I_{\mathfrak{X},\vec{s}}(\nu) ,$$

and the $(\mathfrak{X}, \vec{s})$-capacity is given by

$$(A.4) \qquad \gamma_{\mathfrak{X},\vec{s}}(H) = \begin{cases} e^{-V_{\mathfrak{X},\vec{s}}(H)} & \text{if } v \text{ is archimedean,} \\ q_v^{-V_{\mathfrak{X},\vec{s}}(H)} & \text{if } v \text{ is nonarchimedean.} \end{cases}$$

We next define the $(\mathfrak{X}, \vec{s})$-transfinite diameter. For $N = 2, 3, ...$ let

$$(A.5) \qquad d_N(H) = \sup_{q_1,\dots,q_N \in H} \Big(\prod_{\substack{i,j=1 \\ i \neq j}}^{N} [q_i, q_j]_{\mathfrak{X},\vec{s}} \Big)^{1/N^2} ;$$

then the $(\mathfrak{X}, \vec{s})$-transfinite diameter is

$$(A.6) \qquad d_{\mathfrak{X},\vec{s}}(H) = \lim_{N \to \infty} d_N(H) .$$

The existence of the limit follows by a classical argument, given in ([**51**], p.150 and pp.203-204) for the kernel $[z, w]_\zeta$. There the exponent $1/N^2$ in (A.5) is replaced by $1/N(N-1)$, and the $d_N(H)$ are shown to be monotonically decreasing. Our modification to the exponent does not affect the convergence in (A.6), or the value of the limit.

Finally, we define the restricted $(\mathfrak{X}, \vec{s})$-Chebyshev constant $\mathrm{CH}^*_{\mathfrak{X},\vec{s}}(H)$. Given points $a_1, \dots, a_N \in \mathcal{C}_v(\mathbb{C}) \backslash \mathfrak{X}$, consider the $(\mathfrak{X}, \vec{s})$-pseudopolynomial (see §3.6)

$$P(z; a_1, \dots, a_N) = \prod_{i=1}^{N} [z, a_i]_{\mathfrak{X},\vec{s}} .$$

Writing $\|P\|_H = \sup_{z \in H} P(z)$, put

$$(A.7) \qquad \mathrm{CH}^*_N(H) = \inf_{a_1,\dots,a_N \in H} (\|P(z; a_1, \dots, a_N)\|_H)^{1/N} ,$$

and then define

$$(A.8) \qquad \mathrm{CH}^*_{\mathfrak{X},\vec{s}}(H) = \lim_{N \to \infty} \mathrm{CH}^*_N(H) .$$

The existence of the limit in (A.8) follows from arguments similar to those in ([**51**], p.151 and pp.203-304). We call $\mathrm{CH}^*_{\mathfrak{X},\vec{s}}(H)$ the restricted Chebyshev constant since the points $a_1, \dots, a_N$ are required to be in H; lifting that restriction, it is also possible to define an unrestricted Chebyshev constant $\mathrm{CH}_{\mathfrak{X},\vec{s}}(H)$, whose value turns out to be the same as the restricted one.

The following theorems summarize the main facts concerning these objects:

THEOREM A.1. *Let $H \subset \mathcal{C}_v(\mathbb{C}_v) \backslash \mathfrak{X}$ be compact. Then for each probability vector $\vec{s} \in \mathcal{P}^m$,*

$$(A.9) \qquad \gamma_{\mathfrak{X},\vec{s}}(H) = d_{\mathfrak{X},\vec{s}}(H) = \mathrm{CH}^*_{\mathfrak{X},\vec{s}}(H) ,$$

and these quantities are 0 *if and only if H has capacity* 0 *in the sense of Definition* 3.14.

PROOF. The proofs are similar to those of [**51**], Theorems 3.1.18 and 4.1.19. □

THEOREM A.2. *Let $H \subset \mathcal{C}_v(\mathbb{C}_v)\backslash\mathfrak{X}$ be compact, with positive capacity. Then for each probability vector $\vec{s} \in \mathcal{P}^m$:*

(A) If H has positive capacity, then there is a unique probability measure $\mu = \mu_{\mathfrak{X},\vec{s}}$ on H, called the $(\mathfrak{X}, \vec{s})$-equilibrium distribution of H, for which

$$V_{\mathfrak{X},\vec{s}}(H) = I_{\mathfrak{X},\vec{s}}(\mu) .$$

(B) For this measure $\mu_{\mathfrak{X},\vec{s}}$, the potential function

$$u_{\mathfrak{X},\vec{s}}(z, H) := u_{\mathfrak{X},\vec{s}}(z, \mu_{\mathfrak{X},\vec{s}}) = \int_H -\log_v([z, w]_{\mathfrak{X},\vec{s}})\, d\mu_{\mathfrak{X},\vec{s}}(w)$$

satisfies $u_{\mathfrak{X},\vec{s}}(z, H) \le V_{\mathfrak{X},\vec{s}}(H)$ for all $z \in \mathcal{C}_v(\mathbb{C}_v)$, with $u_{\mathfrak{X},\vec{s}}(z, H) = V_{\mathfrak{X},\vec{s}}(H)$ for all $z \in H$ except possibly an F_σ-set $e_{\mathfrak{X},\vec{s}} \subset H$ of inner capacity 0. Moreover, $u_{\mathfrak{X},\vec{s}}(z, H)$ is continuous on $\mathcal{C}_v(\mathbb{C}_v)\backslash e_{\mathfrak{X},\vec{s}}$.

In the archimedean case, $u_{\mathfrak{X},\vec{s}}(z, H) < V_{\mathfrak{X},\vec{s}}(H)$ on each component of $\mathcal{C}_v(\mathbb{C})\backslash H$ which contains a point $x_i \in \mathfrak{X}$ with $s_i > 0$, and $u_{\mathfrak{X},\vec{s}}(z, H) = V_{\mathfrak{X},\vec{s}}(H)$ on all other components of $\mathcal{C}_v(\mathbb{C})\backslash H$. The exceptional set $e_{\mathfrak{X},\vec{s}}$ is contained in $\partial H_{\mathfrak{X},\vec{s}}$, the part of the boundary of H shared by the components of $\mathcal{C}_v(\mathbb{C})\backslash H$ on which $u_{\mathfrak{X},\vec{s}}(z, H) < 0$, and H and $\partial H_{\mathfrak{X},\vec{s}}$ have the same capacity, potential function, and equilibrium distribution with respect to $[z, w]_{\mathfrak{X},\vec{s}}$. Furthermore, $u_{\mathfrak{X},\vec{s}}(z, H)$ is superharmonic on $\mathcal{C}_v(\mathbb{C})\backslash\mathfrak{X}$, subharmonic on $\mathcal{C}_v(\mathbb{C})\backslash H$, and harmonic on $\mathcal{C}_v(\mathbb{C})\backslash(H \bigcup \mathfrak{X})$. At each $x_i \in \mathfrak{X}$, $u_{\mathfrak{X},\vec{s}}(z, H) + s_i \log(|z - x_i|)$ extends to a function harmonic in a neighborhood of x_i.

In the nonarchimedean case, $u_{\mathfrak{X},\vec{s}}(z, H) < V_{\mathfrak{X},\vec{s}}(H)$ for all $z \in \mathcal{C}_v(\mathbb{C}_v)\backslash H$, and $u_{\mathfrak{X},\vec{s}}(z, H) + s_i \log_v(|z - x_i|_v)$ has a finite limit at each $x_i \in \mathfrak{X}$.

Remark. When H is clear from the context, we will often write $u_{\mathfrak{X},\vec{s}}(z)$ for $u_{\mathfrak{X},\vec{s}}(z, H)$.

PROOF. In the classical theory, the assertions in Theorems A.1 and A.2 are the main consequences of Maria's Theorem and Frostman's theorem. They are established for the kernel $[z, w]_\zeta$ in Theorems 3.1.6, 3.1.7, 3.1.12, and 3.1.18 of ([**51**], §3.1) in the archimedean case, and in Theorems 4.1.11, 4.1.19, and 4.1.22 of ([**51**], §4.1) in the nonarchimedean case.

In the archimedean case, the continuity/harmonicity properties of $\log([z, w]_{\mathfrak{X},\vec{s}})$ shown in Proposition 3.11, together with the Maximum Principle for harmonic functions, allow the proofs in ([**51**], §3.1) to be carried over for $[z, w]_{\mathfrak{X},\vec{s}}$. The property of the canonical distance needed for those proofs is that for each $q \in H$, and each disc $D(q, r)$ with $\zeta \notin \overline{D(q, r)}$, if we fix a coordinate chart on $D(q, r)$, then there is a constant C (depending on ζ and the choice of coordinates) such that for all $z \ne w \in D(q, r)$

$$-\log([z, w]_\zeta) - C \le -\log(|z - w|) \le -\log([z, w]_\zeta) - C$$

(see [**51**], p.139). By Proposition 3.11 this holds for $[z, w]_{\mathfrak{X},\vec{s}}$.

In the nonarchimedean case, the proofs given in ([**51**], §4.1) use two properties of the canonical distance. First, for each $q \in H$, each $\zeta \in \mathfrak{X}$, and each isometrically parametrizable ball $B(q, r)$ disjoint from $\mathfrak{X}$ (see Definition 3.8), there is a constant $C = C_{q,\zeta}$ such that $[z, w]_\zeta = C\|z, w\|_v$ for all $z, w \in B(q, r)$. Since $[z, w]_{\mathfrak{X},\vec{s}}$ is a weighted product of the $[z, w]_{x_i}$, with the weights summing to 1, Proposition 3.11 shows that this property holds for $[z, w]_{\mathfrak{X},\vec{s}}$ as well. Second, for each pair of points $w, \zeta \in \mathcal{C}_v(\mathbb{C})$, and each isometrically parametrizable ball $B(w, r)$ not containing ζ, there is a function $f(z) \in \mathbb{C}_v(\mathcal{C})$ of degree N say, having all its zeros in $B(w,r)$ and having poles only at ζ, such that

$$[z,w]_\zeta \;=\; (|f(z)|_v)^{1/N}$$

for all $z \notin B(w,r)$ (see [**51**], Proposition 2.1.6). For $[z,w]_{\mathfrak{X},\vec{s}}$ the analogue of this is that if $B(w,r)$ is an isometrically parametrizable ball disjoint from $\mathfrak{X}$, then there are functions $f_i(z) \in \mathbb{C}_v(\mathcal{C})$ of degree N_i say, having all their zeros in $B(w,r)$ and such that $f_i(z)$ has poles only at x_i, for which

$$-\log_v([z,w]_{\mathfrak{X},\vec{s}}) \;=\; -\sum_{i=1}^m s_i \cdot \frac{1}{N_i}\log_v(|f_i(z)|_v)$$

for all $z \notin B(w,r)$. □

The following proposition can often be used to show that the exceptional set $e_{\mathfrak{X},\vec{s}}$ in Theorem A.2 is empty. By an arc, we mean a homeomorphic image of the segment $[0,1]$.

PROPOSITION A.3. *Let all assumptions be as in Theorem* A.2.

(A) If K_v is archimedean, then $u_{\mathfrak{X},\vec{s}}(z,H) = V_{\mathfrak{X},\vec{s}}(H)$ at each point $z_0 \in H$ for which there is an arc $A \subset H$ with $z_0 \in A$.

(B) If K_v is nonarchimedean with $\mathrm{char}(K_v) = 0$, *and if p is the rational prime lying under v, then $u_{\mathfrak{X},\vec{s}}(z,H) = V_{\mathfrak{X},\vec{s}}(H)$ at each point $z_0 \in H$ for which, for some $r > 0$, there is an isometric parametrization $f_{z_0} : D(0,r) \to B(z_0,r) \subset \mathcal{C}_v(\mathbb{C}_v)$ with $f_{z_0}(0) = z_0$, such that $f(\mathbb{Z}_p \cap D(0,r)) \subset H$. If K_v is nonarchimedean with* $\mathrm{char}(K_v) = p > 0$, *the analogous assertion holds with $\mathbb{Z}_p$ replaced by $\mathbb{F}_p[[T]]$.*

PROOF. In the archimedean case, this is a classical consequence of the existence of a "barrier". The proof is given ([**51**], Theorem 3.1.9) when $\mathfrak{X}$ is a single point, and the argument, which is purely local, carries over unchanged in the general case.

In the nonarchimedean case, the proof uses the monotonicity of upper Green's functions of compact sets ([**51**], Proposition 4.4.1(A)). We give the argument when $\mathrm{char}(K_v) = 0$; the proof when $\mathrm{char}(K_v) > 0$ is similar. After shrinking r if necessary, we can assume that $B(z_0,r) \cap \mathfrak{X} = \phi$. By Proposition 3.11(B1), there is a constant $C > 0$ such that $[z,w]_{\mathfrak{X},\vec{s}} = C\|z,w\|_v$ for all $z,w \in B(z_0,r)$. Since f_{z_0} is an isometric parametrization, we have $[f_{z_0}(x), f_{z_0}(y)]_{\mathfrak{X},\vec{s}} = C|x-y|_v$ for all $x,y \in D(0,r)$.

Since a set $e \subset D(0,r)$ has positive inner capacity if and only if it supports a probability measure ν for which $\iint -\log_v(|x-y|_v)\,d\nu(x)d\nu(y) < \infty$, by pushing forward or pulling back appropriate measures one sees that f_{z_0} takes sets of positive inner capacity to sets of positive inner capacity, and sets of inner capacity 0 to sets of inner capacity 0.

Put $H_0 = H \cap B(z_0,r)$, and let $\nu_0 = \mu_{\mathfrak{X},\vec{s}}|_{H_0}$. Since H_0 contains $f_{z_0}(\mathbb{Z}_p \cap D(0,r))$, it has positive capacity. This means that $\nu_0(H_0) > 0$. Let $u(z,\nu_0) = \int -\log_v(\|z,w\|_v)\,d\nu_0(w)$. Since $[z,w]_{\mathfrak{X},\vec{x}}$ is constant on pairwise disjoint isometrically parametrizable balls in $\mathcal{C}_v(\mathbb{C}_v)\backslash\mathfrak{X}$, there is a constant D such that for all $z \in B(z_0,r)$,

$$u_{\mathfrak{X},\vec{s}}(z,H) \;=\; D + u(z,\nu_0)\,. \tag{A.10}$$

Pull back $[z,w]_{\mathfrak{X},\vec{s}}$, $u_{\mathfrak{X},\vec{x}}(z,H)$, H_0 and ν_0 to $D(0,r)$ using f_{z_0}. Let $E = f_{z_0}^{-1}(H_0)$ and put $\nu = (1/\nu_0(H_0))f_{z_0}^*(\nu_0)$; then ν is a probability measure supported on E. Consider the potential function

$$u_\infty(x,\nu) \ := \ \int -\log_v(|x-y|_v)\,d\nu(y)$$

on $\mathbb{P}^1(\mathbb{C}_v)$. Since $\mu_{\mathfrak{X},\vec{s}}$ is the equilibrium measure of H, we have $u_{\mathfrak{X},\vec{s}}(z,H) \le V_{\mathfrak{X},\vec{s}}(H)$ for all z, with equality on H except on a set of inner capacity 0. By (A.10) there is a constant V such that $u_\infty(z,\nu_0) \le V$ on E, with equality except on a set of inner capacity 0. It follows from ([**51**], Proposition 4.1.23) that ν is the equilibrium measure of E. This means that $G(x,\infty;H) := V - u_\infty(x,\nu)$ is the upper Green's function of H with respect to ∞. Since $\mathbb{Z}_p \cap D(0,r) \subseteq H$, by ([**51**], Proposition 4.4.1(A)) we have

$$G(x,\infty;\mathbb{Z}_p \cap D(0,r)) \ \ge \ G(x,\infty;H)$$

for all $x \in \mathbb{C}_v$. The explicit computation in Proposition 2.1 gives $G(0,\infty;\mathbb{Z}_p \cap D(0,r)) = 0$, so $G(0,\infty;H) = 0$ as well. This means that $u_\infty(0,\nu) = V$, and hence by (A.10) that $u_{\mathfrak{X},\vec{s}}(z_0,H) = V_{\mathfrak{X},\vec{s}}(H)$. □

The following result sometimes lets us determine the equilibrium distribution.

PROPOSITION A.4. *Let μ_0 be a probability measure on H for which there is a constant $V < \infty$ such that the potential function $u_{\mathfrak{X},\vec{s}}(z,\mu_0)$ equals V on H, except possibly on a set of inner capacity* 0. *Then $V_{\mathfrak{X},\vec{s}}(H) = V$, and $\mu_{\mathfrak{X},\vec{s}} = \mu_0$.*

PROOF. By the same argument as in ([**51**], Lemmas 3.1.4 and 4.17), a positive measure ν for which $I_{\mathfrak{X},\vec{s}}(\nu) < \infty$ cannot charge sets of inner capacity 0. In particular, this applies to μ_0 and $\mu_{\mathfrak{X},\vec{s}}$. Hence by Theorem A.2(B) and Fubini-Tonelli,

$$\begin{aligned} V \ &= \ \int_H u_{\mathfrak{X},\vec{s}}(z,\mu_0)\,d\mu_{\mathfrak{X},\vec{s}}(z) \ = \ \iint_{H\times H} -\log_v([z,w]_{\mathfrak{X},\vec{s}})\,d\mu_0(w)d\mu_{\mathfrak{X},\vec{s}}(z) \\ &= \ \int_H u_{\mathfrak{X},\vec{s}}(w,\mu_{\mathfrak{X},\vec{s}})\,d\mu_0(w) \ = \ V_{\mathfrak{X},\vec{s}}(H)\ . \end{aligned}$$

Consequently,

$$\begin{aligned} I_{\mathfrak{X},\vec{s}}(\mu_0) \ &= \ \iint_{H\times H} -\log_v([z,w]_{\mathfrak{X},\vec{s}})\,d\mu_0(w)d\mu_0(z) \\ &= \ \int_H u_{\mathfrak{X},\vec{s}}(z,\mu_0)\,d\mu_0(z) \ = \ V_{\mathfrak{X},\vec{s}}(H)\ . \end{aligned}$$

Since $\mu_{\mathfrak{X},\vec{s}}$ is the unique probability measure minimizing the $(\mathfrak{X},\vec{s})$-energy integral (Theorem A.2(A)), it follows that $\mu_{\mathfrak{X},\vec{s}} = \mu_0$. □

We define the Green's function $G_{\mathfrak{X},\vec{s}}(z;H)$ to be

$$G_{\mathfrak{X},\vec{s}}(z;H) \ = \ V_{\mathfrak{X},\vec{s}}(H) - u_{\mathfrak{X},\vec{s}}(z)\ .$$

For each $x_i \in \mathfrak{X}$, and each positive measure ν supported on $\mathcal{C}_v(\mathbb{C}_v)\backslash\{x_i\}$, put

$$u_{x_i}(z,\nu) \ = \ \int_{\mathcal{C}_v(\mathbb{C}_v)} -\log_v([z,w]_{x_i})\,d\nu(w)\ .$$

Let μ_i be the equilibrium distribution of H with respect to $[z,w]_{x_i}$, and write

$$u_{x_i}(z) \ = \ u_{x_i}(z,\mu_i) \ = \ \int_H -\log_v([z,w]_{x_i})\,d\mu_i(w)\ .$$

As in §3.9, for each $x_i \in \mathfrak{X}$ the Green's function $G(z, x_i; H)$ is defined by

$$G(z, x_i; H) \;=\; V_{x_i}(H) - u_{x_i}(z) \; .$$

We can express $\mu_{\mathfrak{X},\vec{s}}$ and $G_{\mathfrak{X},\vec{s}}(z; H)$ in terms of the μ_i and $G(z, x_i; H)$:

PROPOSITION A.5. *Suppose $H \subset \mathcal{C}_v(\mathbb{C}_v)\backslash\mathfrak{X}$ is compact with positive capacity. Then*

$$\mu_{\mathfrak{X},\vec{s}} \;=\; \sum_{i=1}^{m} s_i \mu_i \; , \tag{A.11}$$

$$G_{\mathfrak{X},\vec{s}}(z; H) \;=\; \sum_{i=1}^{m} s_i G(z, x_i; H) \; , \tag{A.12}$$

and

$$V_{\mathfrak{X},\vec{s}}(H) \;=\; \sum_{i,j,k=1}^{m} s_i s_j s_k \iint_{H\times H} -\log_v([z, w]_{x_i}) \, d\mu_j(z) d\mu_k(w) \; . \tag{A.13}$$

In particular, $V_{\mathfrak{X},\vec{s}}(H)$ is a continuous function of $\vec{s}$.

PROOF (Archimedean Case). Formula (A.12) follows by applying the strong form of the Maximum Principle for harmonic functions, to

$$G_{\mathfrak{X},\vec{s}}(z, H) - \sum s_i G(z, x_i; H)$$

on $\mathcal{C}_v(\mathcal{C})\backslash(H \bigcup \mathfrak{X})$ (see [**51**], Proposition 3.1.1). Formula (A.11) is a consequence of (A.12), since by the Riesz decomposition theorem (see [**51**], Theorem 3.1.11) a measure can be recovered from its potential function. In modern terminology, applying the dd^c operator on $\mathcal{C}_v(\mathbb{C})\backslash\mathfrak{X}$, one has $\mu_{\mathfrak{X},\vec{s}} = dd^c(-u_{\mathfrak{X},\vec{s}}(z))$ and $\mu_i = dd^c(-u_{x_i}(z)) = dd^c(G(z, x_i; H))$ for each i.

For the assertion about $V_{\mathfrak{X},\vec{s}}(H)$, note that

$$\begin{aligned} V_{\mathfrak{X},\vec{s}}(H) &= \iint_{H\times H} -\log([z, w]_{\mathfrak{X},\vec{s}}) \, d\mu_{\mathfrak{X},\vec{s}}(z) d\mu_{\mathfrak{X},\vec{s}}(w) \\ &= \sum_{i,j,k=1}^{m} s_i s_j s_k \left(\iint_{H\times H} -\log([z, w]_{x_i} \, d\mu_j(z) d\mu_k(w) \right) . \end{aligned}$$

(Nonarchimedean Case). The proof is more complicated in this case, because of the absence of a Laplacian operator on $\mathcal{C}_v(\mathbb{C}_v)$. (Actually, a suitable Laplacian has been defined in the context of Berkovich Spaces, for $\mathbb{P}^1$ by Baker and Rumely ([**7**]), and for curves of arbitrary genus by Thuillier ([**64**]). However, introducing that theory would take us too far afield.) Instead, we use the approximability of Green's functions by algebraic functions, the $[z, w]_{\mathfrak{X},\vec{s}}$-factorization of pseudopolynomials (3.31), and the nonarchimedean Maximum Modulus Principle.

First suppose $\vec{s} \in \mathcal{P}^m \bigcap \mathbb{Q}^m$. Choose decreasing sequences of numbers $r_n > 0$ and $\varepsilon_n > 0$ with $\lim_{n\to\infty} r_n = \lim_{n\to\infty} \varepsilon_n = 0$. For each n, put $W_n = \{z \in \mathcal{C}_v(\mathbb{C}_v) : \|z, w\|_v \le r_n$ for some $w \in H\}$. Thus, the W_n form a decreasing sequence of neighborhoods of H with $\bigcap_{n=1}^{\infty} W_n = H$.

For each i and n, Proposition 4.1.5 of ([**51**]) provides a function $f_i^{(n)}(z) \in \mathbb{C}_v(\mathcal{C})$, with poles only at x_i and normalized so that $|f_i^{(n)}(z)|_v \cdot |g_{x_i}(z)^N|_v = 1$ if

$N = \deg(f_i^{(n)})$, such that for all $z \in \mathcal{C}_v(\mathbb{C}_v) \backslash (W_n \cup \{x_i\})$,

$$\left| u_{x_i}(z) - \left(\frac{-1}{N} \log_v |f_i^{(n)}(z)|_v \right) \right|_v < \varepsilon_n . \tag{A.14}$$

By raising the $f_i^{(n)}(z)$ to appropriate powers, we can assume without loss that for a given n, they have common degree N_n. Let the zeros of $f_i^{(n)}(z)$ (with multiplicity) be $a_{ij}^{(n)}$, $j = 1, \ldots, N_n$. Let $\nu_i^{(n)}$ be the probability measure which gives weight $1/N_n$ to each $a_{ij}^{(n)}$. By the construction in ([**51**], Proposition 4.1.5), for each i the $\nu_i^{(n)}$ converge weakly to μ_i.

The normalization of the $f_i^{(n)}(z)$ means that for each n and i,

$$|f_i^{(n)}(z)|_v = \prod_{j=1}^{N_n} [z, a_{ij}^{(n)}]_{x_i} ,$$

and so

$$u_{x_i}(z, \nu_i^{(n)}) = \frac{1}{N_n} \cdot \sum_{j=1}^{N_n} - \log_v([z, a_{ij}^{(n)}]_{x_i}) = -\frac{1}{N_n} \log_v(|f_i^{(n)}(z)|_v) . \tag{A.15}$$

In particular, for each $z \notin H$, by (A.14),

$$\lim_{n \to \infty} u_{x_i}(z, \nu_i^{(n)}) = u_{x_i}(z, \mu_i) = u_{x_i}(z) . \tag{A.16}$$

Let M be a common denominator for the s_i and for each n put

$$F^{(n)}(z) = \prod_{i=1}^{m} f_i^{(n)}(z)^{M s_i} . \tag{A.17}$$

Then $F^{(n)}(z)$ is an $(\mathfrak{X}, \vec{s})$-function in the sense of Definition 3.12. Let the zeros $F^{(n)}(z)$, listed with multiplicities, be $a_1^{(n)}, \ldots, a_{MN_n}^{(n)}$; these are of course just the $a_{ij}^{(n)}$, repeated certain numbers of times. By formula (3.31) there is a constant C_n such that for all $z \in \mathcal{C}_v(\mathbb{C}_v)$,

$$|F^{(n)}(z)|_v = C_n \cdot \prod_{j=1}^{MN_n} [z, a_j^{(n)}]_{\mathfrak{X}, \vec{s}} . \tag{A.18}$$

Let $\omega^{(n)}$ be the probability measure which gives mass $1/(MN_n)$ to each of the points $a_j^{(n)}$; clearly

$$\omega^{(n)} = \sum_{i=1}^{m} s_i \nu_i^{(n)} .$$

Combining (A.15), (A.17) and (A.18), we see that

$$\begin{aligned} u_{\mathfrak{X},\vec{s}}(z,\omega^{(n)}) &= \int -\log_v([z,w]_{\mathfrak{X},\vec{s}})\, d\omega^{(n)}(w) \\ &= \frac{1}{MN_n}\log_v(C_n) - \frac{1}{MN_n}\log_v(|F^{(n)}(z)|_v) \\ &= \frac{1}{MN_n}\log_v(C_n) + \sum_{i=1}^m s_i\left(-\frac{1}{N_n}\log_v(|f_i^{(n)}(z)|_v)\right) \\ &= \frac{1}{MN_n}\log_v(C_n) + \sum_{i=1} s_i u_{x_i}(z,\nu_i^{(n)}) \ . \end{aligned} \tag{A.19}$$

Since the $\nu_i^{(n)}$ converge weakly to the μ_i, the $\omega^{(n)}$ converge weakly to $\omega = \sum s_i\mu_i$. Hence, for all $z \notin H$,

$$\lim_{n\to\infty} u_{\mathfrak{X},\vec{s}}(z,\omega^{(n)}) \ = \ u_{\mathfrak{X},\vec{s}}(z,\omega) \ . \tag{A.20}$$

Comparing (A.16), (A.20) and (A.19) it follows that $C := \lim_{n\to\infty}\frac{1}{NM_n}\log_v(C_n)$ exists, and that for all $z \notin H$,

$$u_{\mathfrak{X},\vec{s}}(z,\omega) \ = \ C + \sum_{i=1} s_i u_{x_i}(z) \ . \tag{A.21}$$

Using ([**51**], Lemma 4.1.3) and its $(\mathfrak{X}, \vec{s})$-analogue, we conclude that $\omega = \sum s_i\mu_i$. Hence (A.21) holds for $z \in H$ as well.

By ([**51**], Theorem 4.1.11), for each i there is a set $e_i \subset H$ of capacity 0 such that $u_{x_i}(z) = V_{x_i}(H)$ for all $z \in H\backslash e_i$. Moreover, since $u_{x_i}(z) \le V_{x_i}(H)$ for all z, ([**51**], Lemma 4.1.9) shows that $u_{x_i}(z)$ is continuous at each $z \in H\backslash e_i$. Put $e = \bigcup_{i=1}^m e_i$. By ([**51**], Lemma 4.1.9), for all $z \in H\backslash e$,

$$\begin{aligned} u_{\mathfrak{X},\vec{s}}(z,\omega) &= \lim_{\substack{w\to z \\ w\notin H}} \Big(C + \sum_{i=1} s_i u_{x_i}(w)\Big) \\ &= C + \sum_{i=1} s_i V_{x_i}(H) \ . \end{aligned}$$

Moreover, by ([**51**], Corollary 4.1.15) e has inner capacity 0. Thus $u_{\mathfrak{X},\vec{s}}(z,\omega)$ is constant on H, except on a set of inner capacity 0.

By Proposition A.4, ω is the equilibrium distribution $\mu_{\mathfrak{X},\vec{s}}$ of H. That is,

$$\mu_{\mathfrak{X},\vec{s}} \ = \ \sum_{i=1}^m s_i\mu_i \ . \tag{A.22}$$

From this (A.12) and (A.13) follow at once.

Now consider the general case where possibly $\vec{s} \notin \mathbb{Q}^m$. Let $f(\vec{s})$ be the function given by the right side of (A.13). Fix $\vec{s}$, and choose a sequence of probability vectors $\vec{s}^{(n)} \in \mathbb{Q}^m$ approaching $\vec{s}$. Let $\mu_0 := \mu_{\mathfrak{X},\vec{s}}$ be the $(\mathfrak{X}, \vec{s})$-equilibrium distribution of H. Then

$$\begin{aligned} V_{\mathfrak{X},\vec{s}}(H) &= \iint_{H\times H} -\log_v([z,w]_{\mathfrak{X},\vec{s}})\, d\mu_0(z)d\mu_0(w) \\ &= \sum_{i=1}^m s_i \iint_{H\times H} -\log_v([z,w]_{x_i})\, d\mu_0(z)d\mu_0(w) \ . \end{aligned} \tag{A.23}$$

Suppose $V_{\mathfrak{X},\vec{s}}(H) < f(\vec{s})$. By the continuity f and of the right side of (A.23) in $\vec{s}$, for sufficiently large n,

$$\iint_{H \times H} -\log_v([z,w]_{\mathfrak{X},\vec{s}^{(n)}})\, d\mu_0(z) d\mu_0(w) \ < \ f(\vec{s}^{(n)}) \ .$$

However, this contradicts that $f(\vec{s}^{(n)}) = V_{\mathfrak{X},\vec{s}^{(n)}}(H)$ is the minimal value of the energy integral for $[z,w]_{\mathfrak{X},\vec{s}^{(n)}}$.

Thus $V_{\mathfrak{X},\vec{s}}(H) \geq f(\vec{s})$. But $\sum s_i\mu_i$ is a probability measure for which the $(\mathfrak{X},\vec{s})$-energy integral equals $f(\vec{s})$. Hence this is the minimal value of the energy integral, and by the uniqueness of the $(\mathfrak{X},\vec{s})$-equilibrium distribution $\mu_{\mathfrak{X},\vec{s}} = \sum_{i=1}^m s_i\mu_i$. □

2. Mass Bounds in the Archimedean Case

Throughout this section we assume K_v is archimedean, and we identify $\mathbb{C}_v$ with $\mathbb{C}$. The results proved here will be used in Theorem 5.2, the construction of the initial local approximating functions when $K_v \cong \mathbb{R}$.

Suppose $H \subset \mathcal{C}_v(\mathbb{C})\backslash\mathfrak{X}$ can be decomposed as $H = H_1 \cup e$, where H_1 and e are closed and disjoint; we think of e as being "small". The case of interest is when $K_v \cong \mathbb{R}$, $H \subset \mathcal{C}_v(\mathbb{R})$, and e is a short interval.

If $\mu_{\mathfrak{X},\vec{s}}$ is the $(\mathfrak{X},\vec{s})$-equilibrium distribution of H, we seek upper and lower bounds for $\mu_{\mathfrak{X},\vec{s}}(e)$ in terms of the Robin constants and Green's functions of H_1 and e.

LEMMA A.6. *Let $H \subset \mathcal{C}_v(\mathbb{C})\backslash\mathfrak{X}$ be compact with positive capacity. Let M be a constant such that for each $x_i \in \mathfrak{X}$,*

$$\max_{z,w \in H} [z,w]_{x_i} \ < \ M \ ;$$

put $C = \log(M)$. Suppose $H = H_1 \bigcup e$ where H_1 and e are closed and disjoint. Given $\vec{s} \in \mathcal{P}^m$, let $\mu_{\mathfrak{X},\vec{s}}$ be the $(\mathfrak{X},\vec{s})$-equilibrium distribution of H. Then

$$\text{(A.24)} \qquad \mu_{\mathfrak{X},\vec{s}}(e) \ \leq \ \frac{V_{\mathfrak{X},\vec{s}}(H) + C}{V_{\mathfrak{X},\vec{s}}(e) + C} \ .$$

PROOF. By our hypothesis, $[z,w]_{\mathfrak{X},\vec{s}} < M$ for all $z, w \in H$. If $M = 1$, by the same argument as in the proof of ([**51**], formula (15), p.148) one obtains

$$\mu_{\mathfrak{X},\vec{s}}(e) \ \leq \ \frac{V_{\mathfrak{X},\vec{s}}(H)}{V_{\mathfrak{X},\vec{s}}(e)} \ .$$

In the general case, if we renormalize $[z,w]_{\mathfrak{X},\vec{s}}$ by replacing it with $\frac{1}{M}[z,w]_{\mathfrak{X},\vec{s}}$, then for each compact set $X \subset H$, $V_{\mathfrak{X},\vec{s}}(X)$ is replaced by $V_{\mathfrak{X},\vec{s}}(X) + \log(M)$. The result follows. □

To obtain a lower bound, we need information about the potential-theoretic separation between H_1 and e.

LEMMA A.7. *Let $H = H_1 \bigcup e \subset \mathcal{C}_v(\mathbb{C})\backslash\mathfrak{X}$, and the constants M and C, be as in Lemma* A.6. *Let $m > 0$ be such that for each $x_i \in \mathfrak{X}$, and all $z \in e$,*

$$\text{(A.25)} \qquad G(z, x_i; H_1) \ \geq \ m \ .$$

Fixing $\vec{s}$, let $\mu_{\mathfrak{X},\vec{s}}$ be the $(\mathfrak{X}, \vec{s})$-equilibrium distribution of H. Suppose $V_{\mathfrak{X},\vec{s}}(e) \geq V_{\mathfrak{X},\vec{s}}(H_1)$. Then

$$\mu_{\mathfrak{X},\vec{s}}(e) \ \geq \ \frac{m^2}{2(V_{\mathfrak{X},\vec{s}}(H_1)+C)(V_{\mathfrak{X},\vec{s}}(e)+C+2m)} \ . \tag{A.26}$$

PROOF. If e has capacity 0, the result is trivial since $\mu_{\mathfrak{X},\vec{s}}(e) = 0$ and $V_{\mathfrak{X},\vec{s}}(e) = \infty$. Hence without loss we can assume that e has positive capacity. First suppose $M = 1$.

Write μ (resp. μ_1, resp. μ_2) for the $(\mathfrak{X}, \vec{s})$-equilibrium distribution of H (resp. H_1, resp. e). Put $V_1 = V_{\mathfrak{X},\vec{s}}(H_1) = I_{\mathfrak{X},\vec{s}}(\mu_1)$, $V_2 = V_{\mathfrak{X},\vec{s}}(e) = I_{\mathfrak{X},\vec{s}}(\mu_2)$, and let

$$I_{\mathfrak{X},\vec{s}}(\mu_1,\mu_2) \ = \ \iint -\log([z,w]_{\mathfrak{X},\vec{s}})\, d\mu_1(z) d\mu_2(w) \ .$$

Then for each $0 \leq t \leq 1$,

$$\begin{aligned} V_{\mathfrak{X},\vec{s}}(H) \ &= \ I_{\mathfrak{X},\vec{s}}(\mu) \ \leq \ I_{\mathfrak{X},\vec{s}}((1-t)\mu_1 + t\mu_2) \\ &= \ (1-t)^2 I_{\mathfrak{X},\vec{s}}(\mu_1) + 2t(1-t) I_{\mathfrak{X},\vec{s}}(\mu_1,\mu_2) + t^2 I_{\mathfrak{X},\vec{s}}(\mu_2) \ . \end{aligned} \tag{A.27}$$

By our hypothesis, $G_{\mathfrak{X},\vec{s}}(z; H_1) = \sum_{i=1}^m s_i G(z, x_i; H_1) \geq m$ on e; hence the potential function

$$u_{\mathfrak{X},\vec{s}}(z; H_1) \ = \ V_1 - G_{\mathfrak{X},\vec{s}}(z; H_1)$$

satisfies $u_{\mathfrak{X},\vec{s}}(z; H_1) \ \leq \ V_1 - m$ on e. Thus

$$I_{\mathfrak{X},\vec{s}}(\mu_1,\mu_2) \ = \ \int_e u_{\mathfrak{X},\vec{s}}(z; H_1) d\mu_2(z) \ \leq \ V_1 - m \ .$$

Since $I_{\mathfrak{X},\vec{s}}(\mu_1) = V_1$ and $I_{\mathfrak{X},\vec{s}}(\mu_2) = V_2$, (A.27) gives

$$V_{\mathfrak{X},\vec{s}}(H) \ \leq \ (V_2 + 2m - V_1)t^2 - 2mt + V_1 \ . \tag{A.28}$$

The minimum of the right side occurs at $t = m/(V_2 + 2m - V_1)$, which lies in the interval $[0,1]$ because we have assumed $V_2 \geq V_1$. Inserting this in (A.28), we get

$$V_{\mathfrak{X},\vec{s}}(H) \ \leq \ V_1 - \frac{m^2}{V_2 + 2m - V_1} \ . \tag{A.29}$$

Put $\beta = \mu_{\mathfrak{X},\vec{s}}(e)$. Because $-\log([z,w]_{\mathfrak{X},\vec{s}}) \geq 0$ on H, we have

$$\begin{aligned} V_{\mathfrak{X},\vec{s}}(H) \ &= \ \iint_{H \times H} -\log([z,w]_{\mathfrak{X},\vec{s}})\, d\mu(z) d\mu(w) \\ &\geq \ \iint_{H_1 \times H_1} -\log([z,w]_{\mathfrak{X},\vec{s}})\, d\mu(z) d\mu(w) \ = \ I_{\mathfrak{X},\vec{s}}(\mu|_{H_1}) \ . \end{aligned}$$

Since $\frac{1}{1-\beta}\mu|_{H_1}$ is a probability measure on H_1, upon dividing by $(1-\beta)^2$, we get

$$\frac{V_{\mathfrak{X},\vec{s}}(H)}{(1-\beta)^2} \ \geq \ I_{\mathfrak{X},\vec{s}}(\frac{1}{1-\beta}\mu|_{H_1}) \ \geq \ I_{\mathfrak{X},\vec{s}}(\mu_1) \ = \ V_1 \ . \tag{A.30}$$

Combining (A.29) and (A.30) gives

$$(1-\beta)^2 \ \leq \ 1 - \frac{m^2}{V_1(V_2 + 2m - V_1)} \ \leq \ 1 - \frac{m^2}{V_1(V_2 + 2m)} \ .$$

Taking square roots, we see that

$$1 - \beta \ \leq \ 1 - \frac{m^2}{2V_1(V_2 + 2m)} \ ,$$

which is equivalent to (A.26).

The general case follows upon scaling $[z,w]_{\mathfrak{X},\vec{s}}$ by $1/M$. $\square$

Fix a local coordinate patch $U \subset \mathcal{C}_v(\mathbb{C})\backslash\mathfrak{X}$, with coordinate function z say. We can describe subsets of U, such as intervals or discs, in terms of the coordinate function z. Our last result concerns the behavior of the $(\mathfrak{X},\vec{s})$-Robin constant of an interval, as its length goes to 0.

LEMMA A.8. *Let $U \subset \mathcal{C}_v(\mathbb{C})\backslash\mathfrak{X}$ be a local coordinate patch, and let I be a compact subset of U. Then there is a constant A depending only I (and the choice of the local coordinate function z) such that for any interval $e_a(h) = [a-h, a+h] \subset I$, and any probability vector $\vec{s} \in \mathcal{P}^m$, we have*

$$-\log(h) - A \ \leq\ V_{\mathfrak{X},\vec{s}}(e_a(h)) \ \leq\ -\log(h) + A\ . \tag{A.31}$$

PROOF. There is a constant A_0 such that for each $x_i \in \mathfrak{X}$, and all $z, w \in I$ with $z \neq w$, $-\log(|z-w|) - A_0 \leq -\log([z,w]_{x_i}) \leq -\log(|z-w|) + A_0$. Hence for all $z, w \in I$ with $z \neq w$, and all $\vec{s} \in \mathcal{P}^m$.

$$-\log(|z-w|) - A_0 \ \leq\ -\log([z,w]_{\mathfrak{X},\vec{s}}) \ \leq\ -\log(|z-w|) + A_0\ .$$

Fix $\vec{s}$, and let $\mu = \mu_{\mathfrak{X},\vec{s}}$ be the $(\mathfrak{X},\vec{s})$-equilibrium distribution of $e_a(h)$. Also, let μ_0 be the equilibrium distribution of $e_a(h)$ considered as a subset of $\mathbb{C}$, via the local coordinate function z. Then by the energy minimizing property of μ_0 and the fact that the classical Robin constant of a segment of length L is $-\log(L/4)$,

$$\begin{aligned}
V_{\mathfrak{X},\vec{s}}(e_a(h)) &= \iint_{e_a(h)\times e_a(h)} -\log([z,w]_{\mathfrak{X},\vec{s}})\, d\mu(z)d\mu(w) \\
&\geq \iint_{e_a(h)\times e_a(h)} (-\log(|z-w|) - A_0)\, d\mu(z)d\mu(w) \\
&\geq \iint_{e_a(h)\times e_a(h)} -\log(|z-w|)\, d\mu_0(z)d\mu_0(w) - A_0 \\
&= -\log(h/2) - A_0\ .
\end{aligned}$$

Similarly, using the energy minimizing property of $\mu_{\mathfrak{X},\vec{s}}$,

$$\begin{aligned}
-\log(h/2) + A_0 &= \iint_{e_a(h)\times e_a(h)} (-\log(|z-w|) + A_0)\, d\mu_0(z)d\mu_0(w) \\
&\geq \iint_{e_a(h)\times e_a(h)} -\log([z,w]_{\mathfrak{X},\vec{s}})\, d\mu(z)d\mu(w) \\
&= V_{\mathfrak{X},\vec{s}}(e_a(h))\ .
\end{aligned}$$

Putting $A = A_0 + \log(2)$, we obtain the result. $\square$

3. Description of $\mu_{\mathfrak{X},\vec{s}}$ in the Nonarchimedean Case

Throughout this section we assume that K_v is nonarchimedean. Our goal is to determine $\mu_{\mathfrak{X},\vec{s}}$ for a class of well-behaved compact sets. The results proved here will be used in Theorem 11.1, the construction of the initial local approximating functions in the nonarchimedean compact case.

For the remainder of this section, $F_w \subset \mathbb{C}_v$ will be a fixed finite extension of K_v, with ramification index $e = e_{w/v}$ and residue degree $f = f_{w/v}$.

We begin by considering the special case when $\mathcal{C}_v = \mathbb{P}^1/K_v$ and $\zeta = \infty$. Identify $\mathbb{P}^1_v(\mathbb{C}_v)\backslash\{\infty\}$ with $\mathbb{C}_v$ and normalize the canonical distance so that $[z,x]_\infty = |z-x|_v$. The equilibrium distribution and potential function can be determined explicitly when H is a coset of $\mathcal{O}_w$. The following result is a mild generalization of ([**51**], Example 4.1.24):

LEMMA A.9. *Let F_w/K_v be a finite extension with ramification index e and residue degree f. Suppose $H = a + b\mathcal{O}_w$, where $a \in \mathbb{C}_v$ and $b \in \mathbb{C}_v^\times$. Then the equilibrium distribution μ of H relative to $[x,y]_\infty = |x-y|_v$ is the pushforward of additive Haar measure on $\mathcal{O}_w$ by the affine map $x = a + bz$, and if $|b|_v = r$, the potential function $u_\infty(x) = u_\infty(x, H)$ is given by*

$$u_\infty(x) = \begin{cases} -\log_v(r) + \frac{1}{e(q_v^f-1)} & \text{for } x \in H, \\ -\log_v(|x-a|_v) & \text{for } x \notin D(a,r)\ . \end{cases} \tag{A.32}$$

PROOF. After a change of coordinates, we can assume without loss that $H = \mathcal{O}_w$. Since H and $|x-y|_v$ are invariant under translation by elements of $\mathcal{O}_w$, the uniqueness of the equilibrium distribution shows that it must be translation-invariant as well. It follows that μ is the additive Haar measure μ_w on $\mathcal{O}_w$.

Let π_w be a generator for the maximal ideal $\mathfrak{m}_w$ of $\mathcal{O}_w$. We can compute $u_\infty(0)$ directly:

$$u_\infty(0) = \sum_{\ell=0}^\infty \int_{\mathfrak{m}_w^\ell\backslash\mathfrak{m}_w^{\ell+1}} -\log(|0-y|_v)\,d\mu(y) \tag{A.33}$$

$$= \sum_{\ell=0}^\infty \ell\cdot\frac{1}{e}\left(\frac{1}{q_v^{f\ell}} - \frac{1}{q_v^{f(\ell+1)}}\right) = \frac{1}{e(q_v^f-1)}\ . \tag{A.34}$$

By translation invariance, $u_\infty(x) = u_\infty(0)$ for all $x \in \mathcal{O}_v$.

For $x \notin D(0,1)$, the ultrametric inequality gives

$$\begin{aligned} u_\infty(x) &= \int_{\mathcal{O}_v} -\log(|x-y|_v)\,d\mu(y) \\ &= \int_{\mathcal{O}_v} -\log(|x|_v)\,d\mu(y) = -\log_v(|x|_v)\,. \end{aligned}$$

It is not hard to give a formula for $u_\infty(x)$ when $x \in D(0,1)\backslash\mathcal{O}_w$ (see [**51**], Example 4.1.26), but we will not need this. □

Now let $\mathcal{C}_v/K_v$ be arbitrary. Suppose $a \in \mathcal{C}_v(F_w)$, and let $B(a,r)$ be an isometrically parametrizable ball disjoint from $\mathfrak{X}$, whose radius r belongs to $|F_w^\times|_v$. Take $H = \mathcal{C}_v(F_w)\bigcap B(a,r)$. By Theorem 3.9, there is an F_w-rational isometric parametrization $\Lambda : D(0,r) \to B(a,r)$ with $\Lambda(0) = a$, and if $b \in F_w^\times$ is such that $|b|_v = r$, then $\Lambda(b\mathcal{O}_w) = H$.

Fix $\vec{s} \in \mathcal{P}^m$. By Proposition 3.11 there is a constant $C_a(\vec{s})$, which belongs to the value group of $\mathbb{C}_v^\times$ if $\vec{s} \in \mathcal{P}^m \cap \mathbb{Q}^m$, such that $[z,w]_{\mathfrak{X},\vec{s}} = C_a(\vec{s})\|z,w\|_v$ for all $z, w \in B(a,r)$. Using Lemma A.9, we obtain:

COROLLARY A.10. *Suppose $a \in \mathcal{C}_v(F_w)$, $r \in |F_w^\times|_v$, and that $B(a,r) \subset \mathcal{C}_v(\mathbb{C}_v)$ is an isometrically parametrizable ball disjoint from $\mathfrak{X}$. Let $\vec{s} \in \mathcal{P}^m$ be arbitrary, and take $H = \mathcal{C}_v(F_w)\bigcap B(a,r)$.*

Then the $(\mathfrak{X},\vec{s})$-equilibrium distribution of H is the pushforward $\Lambda_*(\mu_w)$ of additive Haar measure on $b\mathcal{O}_w$, normalized to have total mass 1, and the $(\mathfrak{X},\vec{s})$-potential function of H satisfies

$$(A.35) \qquad u_{\mathfrak{X},\vec{s}}(z,H) \ = \ \begin{cases} -\log_v(C_a(\vec{s})\cdot r) + \frac{1}{e(q_v^f-1)} & \text{for all } z \in H, \\ -\log_v([z,a]_{\mathfrak{X},\vec{s}}) & \text{for all } z \notin B(a,r). \end{cases}$$

PROOF. Write $z = \Lambda(x)$, $w = \Lambda(y)$ for $z, w \in B(a,r)$ and $x, y \in D(0,r)$. Then $\|z,w\|_v = |x-y|_v$, and $[z,w]_{\mathfrak{X},\vec{s}} = C_a(\vec{s})|x-y|_v$.

Write μ_w for the additive Haar measure on $b\mathcal{O}_w = F_w \cap D(0,r)$, normalized to have total mass 1, and put $\mu_0 = \Lambda_*(\mu_w)$. By Lemma A.9, for each $z \in H$,

$$\begin{aligned} u_{\mathfrak{X},\vec{s}}(z,\mu_0) \ &= \ \int_H -\log_v([z,w]_{\mathfrak{X},\vec{s}})\, d\mu_0(w) \\ &= \ \int_{b\mathcal{O}_w} -\log_v(C_a(\vec{s})|x-y|_v)\, d\mu_w(y) \\ &= \ -\log_v(C_a(\vec{s})) - \log_v(r) + \frac{1}{e(q_v^f-1)} \ . \end{aligned}$$

In particular, $u_{\mathfrak{X},\vec{s}}(z,\mu_0)$ is constant on H. By Proposition A.4, $\mu_{\mathfrak{X},\vec{s}} = \mu_0$.

If $z \notin B(a,r)$ then Proposition 3.11 gives $[z,w]_{\mathfrak{X},\vec{s}} = [z,a]_{\mathfrak{X},\vec{s}}$ for all $w \in B(a,r)$, so

$$u_{\mathfrak{X},\vec{s}}(z,\mu_0) \ = \ \int_H -\log_v([z,w]_{\mathfrak{X},\vec{s}})\, d\mu_0(w) \ = \ -\log_v([z,a]_{\mathfrak{X},\vec{s}}) \ . \qquad \square$$

Note that in Corollary A.10, the equilibrium measure $\mu_{\mathfrak{X},\vec{s}}$ of H is independent of $\vec{s}$. This always holds, for compact subsets of isometrically parametrizable balls:

LEMMA A.11. *Let $B(a,r) \subset \mathcal{C}_v(\mathbb{C}_v)$ be an isometrically parametrizable ball disjoint from $\mathfrak{X}$, and let $H \subset B(a,r)$ be compact with positive capacity. Then the equilibrium distribution $\mu_{\mathfrak{X},\vec{s}}$ of H is a probability measure μ^* independent of $\vec{s}$.*

PROOF. For a given $\vec{s}$, the equilibrium distribution $\mu_{\mathfrak{X},\vec{s}}$ is the unique probability measure μ supported on H which minimizes the energy integral

$$I_{\mathfrak{X},\vec{s}}(\mu) \ = \ \iint_{H\times H} -\log_v([z,w]_{\mathfrak{X},\vec{s}})\, d\mu(z) d\mu(w) \ .$$

Fix an isometric parametrization of $B(a,r)$. By Proposition 3.11, for each $\vec{s}$ there is a constant $C_a(\vec{s})$ such that for all $z, w \in B(a,r)$,

$$-\log_v([z,w]_{\mathfrak{X},\vec{s}}) \ = \ -\log_v(\|z,w\|_v) - \log_v(C_a(\vec{s})) \ .$$

Hence the same measure μ^* minimizes the energy integral, for all $\vec{s}$. $\square$

Now let $B(a_\ell, r_\ell)$ for $\ell = 1, \ldots, D$ be isometrically parametrizable balls in $\mathcal{C}_v(\mathbb{C}_v)$, disjoint from each other and from $\mathfrak{X}$. Suppose $H = \bigcup_{\ell=1}^D H_\ell$, where $H_\ell \subset B(a_\ell, r_\ell)$ is compact and has positive capacity for each ℓ. Let μ_ℓ^* be the $(\mathfrak{X},\vec{s})$-equilibrium distribution of H_ℓ, which is independent of $\vec{s}$ by Lemma A.11. Let

$$u_{\mathfrak{X},\vec{s}}(z,H_\ell) \ = \ \int_{H_\ell} -\log([z,w]_{\mathfrak{X},\vec{s}})\, d\mu_\ell^*(w)$$

be the $(\mathfrak{X},\vec{s})$-potential function of H_ℓ.

PROPOSITION A.12. *Let $H = \bigcup_{\ell=1}^D H_\ell$ be as above. For each $\vec{s}$, there are weights $w_\ell(\vec{s}) > 0$ with $\sum_{\ell=1}^D w_\ell(\vec{s}) = 1$ such that the equilibrium distribution $\mu_{\mathfrak{X},\vec{s}}$ of H satisfies*

$$\mu_{\mathfrak{X},\vec{s}} = \sum_{\ell=1}^{D} w_\ell(\vec{s})\, \mu_\ell^* \ , \tag{A.36}$$

and the potential function is given by

$$u_{\mathfrak{X},\vec{s}}(z, H) = \sum_{\ell=1}^{D} w_\ell(\vec{s})\, u_{\mathfrak{X},\vec{s}}(z, H_\ell) \ . \tag{A.37}$$

PROOF. Let $w_\ell(\vec{s}) = \mu_{\mathfrak{X},\vec{s}}(H_\ell)$. Since H_ℓ has positive capacity, necessarily $w_\ell(\vec{s}) > 0$ (see [**51**], Lemma 4.1.7). Since the balls $B(a_\ell, r_\ell)$ are pairwise disjoint and do not meet $\mathfrak{X}$, Proposition 3.11 shows that $[z, w]_{\mathfrak{X},\vec{s}}$ is constant for $z \in B(a_\ell, r_\ell)$, $w \in B(a_k, r_k)$, if $\ell \neq k$. Hence the same arguments as in ([**51**], Proposition 4.1.27) yield (A.36) and (A.37). □

Remark. Using Proposition A.5, one sees that there are constants $W_{i\ell} > 0$ such that $w_\ell(\vec{s}) = \sum_{i=1}^m s_i W_{i\ell}$ for all $\vec{s}$ and all ℓ.

THEOREM A.13. *Suppose $B(a_1, r_1), \ldots, B(a_D, r_D)$ are pairwise disjoint isometrically parametrizable balls in $\mathcal{C}_v(\mathbb{C}_v)$, whose union is disjoint from $\mathfrak{X}$. For each ℓ, let $H_\ell \subset B(a_\ell, r_\ell)$ be a compact set of positive capacity, and let $H = \bigcup_{\ell=1}^D H_\ell$.*

Given $\vec{s} \in \mathcal{P}^m$, let $V_{\mathfrak{X},\vec{s}}(H_\ell)$ be the $(\mathfrak{X}, \vec{s})$-Robin constant of H_ℓ. Let μ_ℓ^* be the $(\mathfrak{X}, \vec{s})$-equilibrium distribution of H_ℓ (which is independent of $\vec{s}$, by Lemma A.11*). Then the Robin constant $V = V_{\mathfrak{X},\vec{s}}(H)$ and the weights $w_\ell = w_\ell(\vec{s}) = \mu_{\mathfrak{X},\vec{s}}(H_\ell) > 0$ such that $\mu_{\mathfrak{X},\vec{s}} = \sum_{\ell=1}^D w_\ell(\vec{s})\, \mu_\ell^*$ (given by Proposition* A.12*) are uniquely determined by the $D + 1$ linear equations*

$$\begin{cases} 1 = 0 \cdot V + \sum_{\ell=1}^D w_\ell \ , \\ 0 = V + w_j \cdot (-V_{\mathfrak{X},\vec{s}}(H_j)) + \displaystyle\sum_{\substack{\ell=1 \\ \ell \neq j}}^{D} w_\ell \cdot \log_v([a_\ell, a_j]_{\mathfrak{X},\vec{s}}) \\ \qquad \textit{for } j = 1, \ldots, D. \end{cases} \tag{A.38}$$

In particular, if $\vec{s} \in \mathcal{P}^m \cap \mathbb{Q}^m$ and $V_{\mathfrak{X},\vec{s}}(H_\ell) \in \mathbb{Q}$ for each ℓ, then $V_{\mathfrak{X},\vec{s}}(H)$ and the $w_\ell(\vec{s})$ belong to $\mathbb{Q}$.

PROOF. By Theorem A.2 and Proposition A.4, $u_{\mathfrak{X},\vec{s}}(z, H)$ takes the constant value V on H, except possibly on an exceptional set $e \subset H$ of inner capacity 0. Similarly, each $u_{\mathfrak{X},\vec{s}}(z, H_j)$ takes the constant value $V_{\mathfrak{X},\vec{s}}(H_j)$ on H_j, except possibly on an exceptional set $e_j \subset H_j$ of inner capacity 0. Since H_j has positive capacity, $H_j \backslash (e \cup e_j)$ is nonempty. For each j, let $a_j^* \in H_j$ be a point where $u_{\mathfrak{X},\vec{s}}(a_j^*, H) = V$ and $u_{\mathfrak{X},\vec{s}}(a_j^*, H_j) = V_{\mathfrak{X},\vec{s}}(H_j)$. The first equation in (A.38) follows from Proposition A.12. Using Corollary A.10 and evaluating $u_{\mathfrak{X},\vec{s}}(H)$ at each a_j^*, we obtain the last D equations in (A.38) with the a_ℓ, a_j replaced by the a_ℓ^*, a_j^*. However, if $\ell \neq j$, then $\log_v([x, y]_{\mathfrak{X},\vec{s}})$ is constant for $(x, y) \in B(a_\ell, r_\ell) \times B(a_j, r_j)$ by Proposition 3.11. Hence $\log_v([a_\ell^*, a_j^*]_{\mathfrak{X},\vec{s}}) = \log_v([a_\ell, a_j]_{\mathfrak{X},\vec{s}})$.

Conversely, we claim that the system of linear equations (A.38) in the variables V and w_ℓ is nonsingular. To see this, first note that the values $V = V_{\mathfrak{X},\vec{s}}(H)$

and $w_\ell = \mu_{\mathfrak{X},\vec{s}}(H_\ell)$ provide one solution to this system, with positive w_ℓ. On the other hand, any solution to the system, with positive w_ℓ, determines a probability measure on H having the properties of the equilibrium distribution. If the system were singular, there would be other solutions arbitrarily close to the one given above, contradicting the uniqueness of the equilibrium distribution. Thus the equations (A.38) uniquely determine V and the w_ℓ.

If $\vec{s} \in \mathcal{P}^m \cap \mathbb{Q}^m$ and the $V_{\mathfrak{X},\vec{s}}(H_j) \in \mathbb{Q}$, then the coefficients in (A.38) are rational, since for all $\ell \neq j$ we have $\log_v([a_\ell, a_j]_{\mathfrak{X},\vec{s}}) = \sum_{i=1}^m s_i \log_v([a_\ell, a_j]_{x_i}) \in \mathbb{Q}$. Hence V and the w_ℓ must be rational as well. $\square$

We now apply the preceding results to sets $H = \bigcup_{\ell=1}^D H_\ell$ of a special form:

COROLLARY A.14. *Let $B(a_1, r_1), \ldots, B(a_D, r_D) \subset \mathcal{C}_v(\mathbb{C}_v)$ be pairwise disjoint isometrically parametrizable balls whose union is disjoint from $\mathfrak{X}$. For each ℓ, let F_{w_ℓ}/K_v be a finite extension in $\mathbb{C}_v$, with residue degree $f_\ell = f_{w_\ell/v}$ and ramification index $e_\ell = e_{w_\ell/v}$. Assume that each $a_\ell \in \mathcal{C}_v(F_{w_\ell})$ and each $r_\ell \in |F_{w_\ell}^\times|_v$, and put $H_\ell = \mathcal{C}_v(F_{w_\ell}) \bigcap B(a_\ell, r_\ell)$. Let $H = \bigcup_{\ell=1}^D H_\ell$.*

Given $\vec{s} \in \mathcal{P}^m$, let $C_{a_\ell}(\vec{s})$ be the constant such that $[z, w]_{\mathfrak{X},\vec{s}} = C_{a_\ell}(\vec{s}) \cdot \|z, w\|_v$ for $z, w \in B(a_\ell, r_\ell)$. Let μ_ℓ^* be the $(\mathfrak{X}, \vec{s})$-equilibrium distribution of H_ℓ (which is independent of $\vec{s}$, by Lemma A.11, and is given by a pushforward of additive Haar measure on $F_w \cap D(0, r_\ell)$, normalized to have mass 1, by Corollary A.10).

Then the Robin constant $V = V_{\mathfrak{X},\vec{s}}(H)$ and the weights $w_\ell = w_\ell(\vec{s}) = \mu_{\mathfrak{X},\vec{s}}(H_\ell) > 0$ such that $\mu_{\mathfrak{X},\vec{s}} = \sum_{\ell=1}^D w_\ell \, \mu_\ell^*$ (given by Proposition A.12) are uniquely determined by the $D+1$ linear equations

(A.39)
$$\begin{cases} 1 = 0 \cdot V + \sum_{\ell=1}^D w_\ell \,, \\ 0 = V + w_j \cdot \Big(\log_v(C_{a_j}(\vec{s}) \cdot r_j) - \frac{1}{e_j(q_v^{f_j}-1)}\Big) + \sum_{\substack{\ell=1 \\ \ell \neq j}}^D w_\ell \cdot \log_v\big([a_\ell, a_j]_{\mathfrak{X},\vec{s}}\big) \\ \qquad \text{for } j = 1, \ldots, D. \end{cases}$$

If $\vec{s} \in \mathcal{P}^m \cap \mathbb{Q}^m$, then $V_{\mathfrak{X},\vec{s}}(H)$ and the $w_\ell(\vec{s})$ belong to $\mathbb{Q}$.

PROOF. This follows by combining Theorem A.13, Proposition A.12, and Corollary A.10. Note that if $\vec{s} \in \mathcal{P}^m \cap \mathbb{Q}^m$, then for each j we have $\log_v(C_{a_j}(\vec{s})) \in \mathbb{Q}$ by Proposition 3.11(B2), so $V_{\mathfrak{X},\vec{s}}(H_j) = -1/(e_j(q_v^{f_j} - 1)) - \log_v(C_{a_j}(\vec{s}) \cdot r_j) \in \mathbb{Q}$. $\square$

As a consequence, we show that for sets H of form in Corollary A.14, then for each $\zeta \notin H$ the Green's function $G(x, \zeta; H)$ and the Robin constant $V_\zeta(H)$ take on rational values.

COROLLARY A.15. *Let $B(a_1, r_1), \ldots, B(a_D, r_D) \subset \mathcal{C}_v(\mathbb{C}_v)$ be pairwise disjoint isometrically parametrizable balls whose union is disjoint from $\mathfrak{X}$. For each ℓ, let F_{w_ℓ}/K_v be a finite extension in $\mathbb{C}_v$. Assume that each $a_\ell \in \mathcal{C}_v(F_{w_\ell})$ and each $r_\ell \in |F_{w_\ell}^\times|_v$, and put $H_\ell = \mathcal{C}_v(F_{w_\ell}) \bigcap B(a_\ell, r_\ell)$.*

Let $H = \bigcup_{\ell=1}^D H_\ell$. Then for each $\zeta \in \mathcal{C}_v(\mathbb{C}_v) \backslash H$, we have $V_\zeta(H) \in \mathbb{Q}$, and for each $x \in \mathcal{C}_v(\mathbb{C}_v) \backslash \{\zeta\}$ we have $G(x, \zeta; H) \in \mathbb{Q}$.

PROOF. We apply the preceding results, taking $\mathfrak{X} = \{\zeta\}$ and $\vec{s} = (1)$. Fix a uniformizing parameter $g_\zeta(z) \in \mathbb{C}_v(\mathcal{C})$ and normalize the canonical distance by

$\lim_{x \to \zeta}[x, y]_\zeta \cdot |g_\zeta(z)|_v = 1$ as usual. By Proposition A.12 for each z we have

$$u_\zeta(z, H) = \sum_{\ell=1}^{D} w_\ell \cdot u_\zeta(z, H_\ell) . \tag{A.40}$$

By Corollary A.14, under our hypotheses on H, the Robin constant $V_\zeta(H) = V_{\mathfrak{X},\vec{s}}(H)$ and the weights $w_1, \ldots, w_D$ belong to $\mathbb{Q}$. By the definition of the Green's function, for each z,

$$G(z, \zeta; H) = V_\zeta(H) - u_\zeta(z, H) .$$

First, suppose $x \in H$. The set H satisfies the hypotheses of Proposition A.3, so we have $u_\zeta(x, H) = V_\zeta(H)$ and $G(x, \zeta; H) = 0 \in \mathbb{Q}$. Next, take $x \in \mathcal{C}_v(\mathbb{C}_v) \backslash \Big(\big(\bigcup_{\ell=1}^D B(a_\ell, r_\ell) \big) \cup \{\zeta\} \Big)$. By Corollary A.10, for each $z \notin B(a_\ell, r_\ell)$ we have $u_\zeta(z, H_\ell) = -\log_v([z, a_\ell]_\zeta)$. Inserting this in (A.40) gives

$$u_\zeta(x, H) = -\sum_{\ell=1}^{D} w_\ell \cdot \log_v([x, a_\ell]_\zeta) .$$

By Proposition 3.11(B1) we have $\log_v([x, a_\ell]_\zeta) \in \mathbb{Q}$ for each ℓ, so $u_\zeta(x, H) \in \mathbb{Q}$. Hence $G(x, \zeta; H) = V_\zeta(H) - u_\zeta(x, H) \in \mathbb{Q}$. Finally, let $x \in \mathcal{C}_v(\mathbb{C}_v) \backslash \big(H \cup \{\zeta\} \big)$ be arbitrary. By replacing the cover $B(a_1, r_1), \ldots, B(a_D, r_D)$ of H with a finer cover, we can arrange that $x \notin \bigcup_{\ell=1}^D B(a_\ell, r_\ell)$. Applying the argument above to this new cover, we see that $G(x, \zeta; H) \in \mathbb{Q}$. □

We close with a proposition which says that deleting small balls from a set H of positive capacity does not significantly change its $(\mathfrak{X}, \vec{s})$-Robin constant or potential function.

PROPOSITION A.16. *Let $H \subset \mathcal{C}_v(\mathbb{C}_v) \backslash \mathfrak{X}$ be a compact set of positive capacity, and fix $q_1, \ldots, q_d \in H$. For each $r > 0$, write*

$$\check{H}(r) = H \backslash \left(\bigcup_{\ell=1}^{d} B(q_\ell, r) \right) .$$

Let W be a neighborhood of H. Then for each $\varepsilon > 0$, there is an $R > 0$ such that, uniformly for compact sets H' with $\check{H}(R) \subseteq H' \subseteq H$ and for $\vec{s} \in \mathcal{P}^m$,

$$V_{\mathfrak{X},\vec{s}}(H) \le V_{\mathfrak{X},\vec{s}}(H') < V_{\mathfrak{X},\vec{s}}(H) + \varepsilon ,$$

and for all $z \in \mathcal{C}_v(\mathbb{C}_v) \backslash (\mathfrak{X} \cup W)$,

$$|u_{\mathfrak{X},\vec{s}}(z, H) - u_{\mathfrak{X},\vec{s}}(z, H')| < \varepsilon .$$

PROOF. After shrinking W if necessary, we can assume that W has the form $W = \bigcup_{\ell=1}^D B(a_\ell, r_\ell)$, where the balls $B(a_\ell, r_\ell)$ are isometrically parametrizable, pairwise disjoint, and do not meet $\mathfrak{X}$.

We will first prove analogous assertions for the $V_{x_i}(H')$ and $u_{x_i}(z, H')$, and then consider the situation for general $\vec{s}$. We begin by showing that for each $x_i \in \mathfrak{X}$,

$$\lim_{r \to 0} V_{x_i}(\check{H}(r)) = V_{x_i}(H) . \tag{A.41}$$

A finite set has capacity 0, so for each x_i, by ([**51**], Corollary 4.1.15),

$$\gamma_{x_i}(H) \ = \ \overline{\gamma}_{x_j}(H\backslash\{q_1,\ldots,q_d\}) \ := \ \sup_{\substack{\text{compact} \\ A\subset H\backslash\{q_1,\ldots,q_d\}}} \gamma_{x_i}(A) \ .$$

As $r \to 0$, the $\check{H}(r)$ form an increasing sequence of compact sets whose union is $H\backslash\{q_1,\ldots,q_d\}$. Hence

$$\lim_{r\to 0} \gamma_{x_i}(\check{H}(r)) \ = \ \overline{\gamma}_{x_i}(H\backslash\{q_1,\ldots,q_d\}) \ = \ \gamma_{x_i}(H)$$

which is equivalent to (A.41). As the $\check{H}(r)$ increase, the $V_{x_i}(\check{H}(r))$ decrease.

We next show weak convergence of the equilibrium distributions. Again fix x_i, and let μ_i be the equilibrium distribution of H with respect to x_i. Take a sequence $r_1 > r_2 > \cdots > 0$ with $r_j \to 0$, and put $H_j = \check{H}(r_j)$. We can assume r_1 is small enough that each H_j has positive capacity. Let $\mu_i^{(j)}$ be the equilibrium distribution of H_j with respect to x_i. As shown above, the Robin constants $V_{x_i}(H_j)$ decrease monotonically to $V_{x_i}(H)$. Since the equilibrium measure of H with respect to x_i is unique ([**51**], Theorem 4.1.22), the argument on ([**51**], p.190) shows that the $\mu_i^{(j)}$ converge weakly to μ_i.

Since $[z,w]_{x_i}$ is constant for z, w belonging to disjoint isometrically parametrizable balls in $\mathcal{C}_v(\mathbb{C}_v)\backslash\{x_i\}$, for each $z \in \mathcal{C}_v(\mathbb{C}_v)\backslash(\{x_i\} \cup W)$ we have

$$\begin{aligned} u_{x_i}(z,H) \ &= \ \sum_{\ell=1}^{D} -\log_v([z,a_\ell]_{x_i}) \cdot \mu_i(B(a_\ell,\rho_\ell)) \ , \\ u_{x_i}(z,\check{H}(r_j)) \ &= \ \sum_{\ell=1}^{D} -\log_v([z,a_\ell]_{x_i}) \cdot \mu_i^{(j)}(B(a_\ell,\rho_\ell)) \ . \end{aligned}$$

Let $g_{x_i}(z)$ be the uniformizing parameter which determines the normalization of $[z,w]_{x_i}$. By the same argument as in the proof of ([**51**], Proposition 4.1.5), there is an isometrically parametrizable ball $B(x_i,\delta)$ such that $u_{x_i}(z,H) = u_{x_i}(z,\check{H}(r_j)) = \log_v(|g_{x_i}(z)|_v)$ for all $z \in B(x_i,\delta)\backslash\{x_i\}$ and all j. Furthermore, by ([**51**], Proposition 4.1.1) there is a finite bound B such that $|\log_v([z,w]_{x_i})| \le B$ for each $w \in W$ and each $z \in \mathcal{C}_v(\mathbb{C}_v)\backslash(W \cup B(x_i,\delta))$. Since the $\mu_i^{(j)}$ converge weakly to μ_i as $j \to \infty$, it follows that the $u_{x_i}(z,\check{H}(r_j))$ converge uniformly to $u_{x_i}(z,H)$ on $\mathcal{C}_v(\mathbb{C}_v)\backslash(\{x_i\} \cup W)$.

The argument above applies for each x_i. Given $\varepsilon > 0$, take j large enough that for all $i = 1,\ldots,m$, we have $|V_{x_i}(H) - V_{x_i}(\check{H}(r_j))| < \varepsilon/3$ and $|u_{x_i}(z,H) - u_{x_i}(\check{H}(r_j))| < \varepsilon/3$ for all $z \in \mathcal{C}_v(\mathbb{C}_v)\backslash(\{x_i\} \cup W)$. Put $R_0 = r_j$.

Let H' be a compact set with $\check{H}(R_0) \subseteq H' \subseteq H$. By the monotonicity of the Robin constant, for each x_i,

$$\text{(A.42)} \qquad V_{x_i}(H) \ \le \ V_{x_i}(H') \ \le \ V_{x_i}(\check{H}(R_0)) \ < \ V_{x_i}(H) + \varepsilon/3 \ .$$

We claim as well that $|u_{x_i}(z,H) - u_{x_i}(z,H')| < \varepsilon$ outside W. This follows from a 3-ε's argument, using monotonicity of the upper Green's functions. Recall that $G(z,x_i,H) = V_{x_i}(H) - u_{x_i}(z,H)$, and similarly for $G(z,x_i,H')$ and $G(z,x_i\check{H}(R_0))$. By ([**51**], Proposition 4.4.1), for all $z \in \mathcal{C}_v(\mathbb{C}_v)\backslash\{x_i\}$,

$$G(z,x_i;H) \ \le \ G(z,x_i;H') \ \le \ G(z,x_i;\check{H}(R_0)) \ .$$

It follows that for all $z \in \mathcal{C}_v(\mathbb{C}_v) \backslash (\{x_i\} \cup W)$,

$$\begin{aligned} & |u_{x_i}(z,H) - u_{x_i}(z,H')| \\ \text{(A.43)} \quad = \; & |(V_{x_i}(H) - G(z,x_i;H)) - (V_{x_i}(H') - G(z,x_i;H')| \\ \leq \; & |V_{x_i}(H) - V_{x_i}(H')| + |G(z,x_i;H) - G(z,x_i;H')| \\ \leq \; & |V_{x_i}(H) - V_{x_i}(\check{H}(R_0))| + |G(z,x_i;H) - G(z,x_i;\check{H}(R_0))| \\ \leq \; & 2|V_{x_i}(H) - V_{x_i}(\check{H}(R_0))| + |u_{x_i}(z,H) - u_{x_i}(z,\check{H}(R_0))| \; < \; \varepsilon \; . \end{aligned}$$

Now fix $\vec{s} \in \mathcal{P}^m$. Recall that $W = \bigcup_{\ell=1}^D B(a_\ell, r_\ell)$. For each $\ell = 1, \ldots, D$, put $H_\ell = H \cap B(a_\ell, r_\ell)$. We can assume without loss that each H_ℓ has positive capacity, since removing a set of capacity 0 from H does not change its Robin constant or potential function with respect to any x_i ([**51**], Corollary 4.1.15). Applying the first part of the proof to each H_ℓ with respect to its cover $W_\ell := B(a_\ell, r_\ell)$, there is an $R > 0$ such that for all ℓ, if H'_ℓ is a compact set with $\check{H}_\ell(R) \subset H'_\ell \subset H_\ell$, then for each x_i and for all $z \in \mathcal{C}_v(\mathbb{C}_v) \backslash (\{x_i\} \cup B(a_\ell, r_\ell))$,

$$\text{(A.44)} \qquad V_{x_i}(H_\ell) \;\leq\; V_{x_i}(H'_\ell) \;<\; V_{x_i}(H_\ell) + \varepsilon \;, \quad \text{and}$$

$$\text{(A.45)} \qquad |u_{x_i}(z,H_\ell) - u_{x_i}(z,H'_\ell)| \;<\; \varepsilon \; .$$

We can assume $R \leq R_0$, so (A.42) and (A.43) continue to hold.

Fix a compact set H' with $\check{H}(R) \subseteq H' \subseteq H$, and for each ℓ put $H'_\ell = H' \cap B(a_\ell, r_\ell)$. As above, we can assume that each H'_ℓ has positive capacity. Fix $\vec{s} \in \mathcal{P}^m$. We will now show that

$$\text{(A.46)} \qquad V_{\mathfrak{X},\vec{s}}(H) \;\leq\; V_{\mathfrak{X},\vec{s}}(H') \;<\; V_{\mathfrak{X},\vec{s}}(H) + \varepsilon \; .$$

By Lemma A.11, for each ℓ the equilibrium distribution μ_ℓ^* of H_ℓ is independent of x_i, so the expression (A.13) for $V_{\mathfrak{X},\vec{s}}(H_\ell)$ in Proposition A.5 simplifies to

$$\text{(A.47)} \qquad V_{\mathfrak{X},\vec{s}}(H_\ell) \;=\; \sum_{i=1}^m s_i V_{x_i}(H_\ell) \; .$$

Similarly the equilibrium distribution $\mu_\ell'^*$ of H'_ℓ is independent of x_i, so

$$\text{(A.48)} \qquad V_{\mathfrak{X},\vec{s}}(H'_\ell) \;=\; \sum_{i=1}^m s_i V_{x_i}(H'_\ell) \; .$$

It follows from (A.44), (A.47), and (A.48) that

$$\text{(A.49)} \qquad V_{\mathfrak{X},\vec{s}}(H_\ell) \;\leq\; V_{\mathfrak{X},\vec{s}}(H'_\ell) \;<\; V_{\mathfrak{X},\vec{s}}(H_\ell) + \varepsilon \; .$$

By Proposition A.12, there are weights $w_k(\vec{s}) > 0$ with $\sum_{k=1}^D w_k(\vec{s}) = 1$ such that the $(\mathfrak{X}, \vec{s})$-equilibrium distribution of H is given by $\mu_{\mathfrak{X},\vec{s}} = \sum_k w_k(\vec{s}) \mu_k^*$.

Consider the probability measure $\nu := \sum_k w_\ell(\vec{s}) \mu_k'^*$ on H'. Although ν may not be the equilibrium measure $\mu'_{\mathfrak{X},\vec{s}}$ of H', we will show that

$$\text{(A.50)} \qquad u_{\mathfrak{X},\vec{s}}(z,\nu) \;<\; V_{\mathfrak{X},\vec{s}}(H) + \varepsilon$$

for all $z \in \bigcup_{\ell=1}^D B(a_\ell, r_\ell)$. By the definition of $V_{\mathfrak{X},\vec{s}}(H')$, it follows that

$$V_{\mathfrak{X},\vec{s}}(H') \;\leq\; I_{\mathfrak{X},\vec{s}}(\nu) \;=\; \int_{H'} u_{\mathfrak{X},\vec{s}}(z,\nu)\, d\nu(z) \;<\; V_{\mathfrak{X},\vec{s}}(H) + \varepsilon \; .$$

This will yield (A.46) since the lower bound there is trivial.

By Lemma A.11, for each $\ell = 1, \ldots, D$,

$$u_{\mathfrak{X},\vec{s}}(z, H_\ell) = \sum_{i=1}^{m} s_i u_{x_i}(z, H_\ell) , \qquad u_{\mathfrak{X},\vec{s}}(z, H'_\ell) = \sum_{i=1}^{m} s_i u_{x_i}(z, H'_\ell) .$$

Hence by (A.45), for each $z \in \mathcal{C}_v(\mathbb{C}_v)\backslash(\mathfrak{X} \cup B(a_\ell, r_\ell))$

$$|u_{\mathfrak{X},\vec{s}}(z, H_\ell) - u_{\mathfrak{X},\vec{s}}(z, H'_\ell)| < \varepsilon . \tag{A.51}$$

Fix ℓ and note that $u_{\mathfrak{X},\vec{s}}(z, H'_\ell) \le V_{\mathfrak{X},\vec{s}}(H'_\ell)$ for all z, hence in particular, for $z \in B(a_\ell, r_\ell)$. Thus for each $k \ne \ell$, $u_{\mathfrak{X},\vec{s}}(z, H_k)$ and $u_{\mathfrak{X},\vec{s}}(z, H'_k)$ are constant on $B(a_\ell, r_\ell)$, since $[z, w]_{\mathfrak{X},\vec{s}}$ is constant on pairwise disjoint isometrically parametrizable balls not meeting $\mathfrak{X}$. This means that for any $z \in B(a_\ell, r_\ell)$,

$$V_{\mathfrak{X},\vec{s}}(z, H) = w_\ell(\vec{s}) V_{\mathfrak{X},\vec{s}}(H_\ell) + \sum_{k \ne \ell} w_k(\vec{s}) u_{\mathfrak{X},\vec{s}}(z, H_k) . \tag{A.52}$$

On the other hand, by the definition of ν and (A.49), (A.51), and (A.52), on $B(a_\ell, r_\ell)$ we have

$$\begin{aligned} u_{\mathfrak{X},\vec{s}}(z, H') &= \sum_{k=1}^{D} w_k(\vec{s}) u_{\mathfrak{X},\vec{s}}(z, H'_k) \\ &\le w_\ell(\vec{s})(V_{\mathfrak{X},\vec{s}}(H) + \varepsilon) + \sum_{k \ne \ell} w_k(\vec{s})(u_{\mathfrak{X},\vec{s}}(z, H_k) + \varepsilon) \\ &= V_{\mathfrak{X},\vec{s}}(H) + \varepsilon , \end{aligned}$$

which gives (A.50) as ℓ varies.

Finally, we show that $|u_{\mathfrak{X},\vec{s}}(z, H) - u_{\mathfrak{X},\vec{s}}(z, H')| < 3\varepsilon$ for all $z \notin \mathfrak{X} \cup W$. Indeed, for such z, by (A.42), (A.43), (A.46), and Proposition A.5,

$$\begin{aligned} &|u_{\mathfrak{X},\vec{s}}(z, H) - u_{\mathfrak{X},\vec{s}}(z, H')| \\ &\quad = |(V_{\mathfrak{X},\vec{s}}(H) - G_{\mathfrak{X},\vec{s}}(z, H)) - (V_{\mathfrak{X},\vec{s}}(H') - G_{\mathfrak{X},\vec{s}}(z, H'))| \\ &\quad \le |V_{\mathfrak{X},\vec{s}}(H) - V_{\mathfrak{X},\vec{s}}(H')| + \sum_{i=1}^{m} s_i |G(z, x_i; H) - G(z, x_i; H')| \\ &\quad \le |V_{\mathfrak{X},\vec{s}}(H) - V_{\mathfrak{X},\vec{s}}(H')| + \sum_{i=1}^{m} s_i |V_{x_i}(H) - V_{x_i}(H')| \\ &\qquad\qquad + \sum_{i=1}^{m} s_i |u_{x_i}(z, H) - u_{x_i}(z, H')| < 3\varepsilon . \end{aligned}$$

Since $\varepsilon > 0$ is arbitrary, this proves the Proposition. □

APPENDIX B

The Construction of Oscillating Pseudopolynomials

The purpose of this appendix is to construct $(\mathfrak{X}, \vec{s})$-pseudopolynomials with all their roots in H, and having large oscillations on $H \cap \mathcal{C}_v(\mathbb{R})$, in the case when $K_v \cong \mathbb{R}$. This is accomplished in Theorem B.18, providing the potential-theoretic input to Theorem 5.2.

Throughout this appendix we assume that K_v is archimedean. We specialize to the case $K_v \cong \mathbb{R}$ only in section B.7. Let $\mathcal{C}_v/K_v$ be a smooth, connected projective curve, so $\mathcal{C}_v(\mathbb{C})$ is a Riemann surface.

We keep the notation and assumptions of §3.2. Let $\mathfrak{X} = \{x_1, \ldots, x_m\} \subset \mathcal{C}_v(\widetilde{K}) \subset \mathcal{C}_v(\mathbb{C})$ be a finite, $\mathrm{Aut}(\widetilde{K}/K)$-stable set of points. As usual, $H \subset \mathcal{C}_v(\mathbb{C})\backslash\mathfrak{X}$ will be compact. Let the canonical distances $[z, w]_{x_i}$ be normalized so that $\lim_{z \to x_i} [z, w]_{x_i} \cdot g_{x_i}(z) = 1$, with the $g_{x_i}(z)$ as in §3.2. As in Chapter 5, if $\vec{s} = (s_1, \ldots, s_m) \in \mathcal{P}^m$ is a K_v-symmetric probability vector, the $(\mathfrak{X}, \vec{s})$-canonical distance is defined by

$$[z, w]_{\mathfrak{X},\vec{s}} = \prod_{i=1}^{m} ([z, w]_{x_i})^{s_i} .$$

Recall that an $(\mathfrak{X}, \vec{s})$-pseudopolynomial is a function of the form

$$P(z) = \prod_{i=1}^{N} [z, \alpha_i]_{\mathfrak{X},\vec{s}}$$

where each $\alpha_i \in \mathcal{C}_v(\mathbb{C})\backslash\mathfrak{X}$. For simplicity, we will often just speak of pseudopolynomials, rather than $(\mathfrak{X}, \vec{s})$-pseudopolynomials.

The motivation for the construction is the classical fact that the Chebyshev polynomial $\widetilde{P}_N(z)$ of degree N for an interval $[a, b]$ oscillates N times between $\pm M^N$, where M^N is the sup norm $\|\widetilde{P}_N\|_{[a,b]}$. Thus $|\widetilde{P}_N(z)|$ varies N times from M^N to 0 to M^N on $[a, b]$. One could hope to prove the same thing for an $(\mathfrak{X}, \vec{s})$-Chebyshev pseudopolynomial (that is, a pseudopolynomial having minimal sup norm on H), but this seems difficult because while $[z, w]_{\mathfrak{X},\vec{s}}$ is locally well-understood, it is globally quite mysterious. Furthermore, for disconnected sets H, Chebyshev pseudopolynomials need not have all their roots in H.

Instead, we consider *restricted* Chebyshev pseudopolynomials for H, which by definition have all their roots in H. The result which makes things work is Proposition B.16, which asserts that restricted Chebyshev pseudopolynomials for sufficiently short intervals ("short" is made precise in Definition B.15) have an oscillation property like that of classical Chebyshev polynomials.

This suggests that when $H = \bigcup_{\ell=1}^{D} H_\ell$ is a disjoint union of short intervals, we could obtain the function we want by taking the product of the Chebyshev pseudopolynomials for the intervals H_ℓ. However, this would only be partially right, because the pseudopolynomials from intervals H_j with $j \neq \ell$ would affect the magnitude of the oscillations on H_ℓ. Instead, we need to take the product of appropriate "weighted" Cheybshev pseudopolynomials for the sets H_ℓ, incorporating the background function coming from the other H_j from the very start. This leads to the topic of "weighted potential theory", or "potential theory in the presence of an external field", a subject which goes back at least as far as Gauss.

We refer the reader to the book by Saff and Totik ([**57**]) for an exposition of weighted potential theory for subsets of $\mathbb{C}$. In the classical unweighted case, it is known that under appropriate hypotheses on H, the discrete probability measures associated to Cheybshev polynomials converge weakly to the equilibrium distribution of H. A variant of this for weighted Chebyshev polynomials is given in ([**57**], Theorem III.4.2). Thus, one expects that the correct weight function to use in defining the weighted $(\mathfrak{X}, \vec{s})$-Chebyshev pseudopolynomials for the sets H_ℓ, should be the exponential of the part of the potential function for H coming from $H \backslash H_\ell$.

This appendix works out that idea. To reach the goal, it is necessary to develop a considerable amount of machinery. Fortunately, most of the arguments are direct adaptations of classical proofs to the $(\mathfrak{X}, \vec{s})$-context on curves. Sections B.1–B.5 below "turn the crank" of a standard potential-theoretic machine, developing the background needed to prove three key facts: Theorems B.9, B.12 and B.13. Theorem B.9 says that in the case of interest to us, the weighted equilibrium distribution for each H_ℓ coincides with the restriction to H_ℓ of the unweighted $(\mathfrak{X}, \vec{s})$-equilibrium distribution for H. Theorem B.12 shows that the weighted Chebyshev constant for H_ℓ coincides with the unweighted $(\mathfrak{X}, \vec{s})$-capacity of H. And, Theorem B.13 proves the convergence of the weighted discrete Chebyshev measures to the equilibrium distribution. If one accepts these results, the construction of oscillating pseudopolynomials can be carried out quickly. Proposition B.16 in section B.6 establishes the oscillation property of weighted Chebyshev pseudopolynomials for short intervals, and the main Theorem B.18 in section B.7 assembles the pieces. We now briefly outline the contents of sections B.1–B.5. Sections B.1, B.2 and B.3 introduce the weighted $(\mathfrak{X}, \vec{s})$-capacity, weighted $(\mathfrak{X}, \vec{s})$-Chebyshev constant, and weighted $(\mathfrak{X}, \vec{s})$-transfinite diameter and prove their existence. Our definition of these quantities is somewhat different from that in [**57**]. The detailed convergence proof for the $(\mathfrak{X}, \vec{s})$-Chebyshev constant is needed because a strong notion of convergence for the finite-level Chebyshev constants to the asymptotic one is required in Theorem B.18. A peculiar aspect of classical potential theory is the "rock-paper-scissors" nature of the capacity, Chebyshev constant, and transfinite diameter; although all three are equal, this is seen only after one proves inequalities between them in a cyclic manner. Section B.5 notes that when the weight function is trivial, the weighted objects constructed in sections B.1–B.3 coincide with the unweighted objects studied in Appendix A. This observation has the consequence that in the case of interest, the weighted equilibrium distribution is unique (Theorem B.9).

In our path through this material, we establish only what is needed to prove the main Theorem B.18. Many standard facts, which could be established if we were developing the theory for its own sake, are not touched upon. For a more complete treatment of the theory in classical case, see [**57**].

1. Weighted $(\mathfrak{X}, \vec{s})$-Capacity Theory

In this section we introduce the weighted $(\mathfrak{X}, \vec{s})$-capacity.

Motivation. Consider the energy minimization problem in the definition of the logarithmic capacity for compact sets $H \subset \mathbb{C}$: for each probability measure ν supported on H, put $I(\nu) = \iint -\log|z-w|\, d\mu(z)d\mu(w)$; then the Robin constant is defined by

$$V_\infty(H) \ = \ \inf_\nu I(\nu) \tag{B.1}$$

and the capacity is $\gamma(H) = e^{-V_\infty(H)}$.

If $\gamma(H) > 0$, there is a unique probability measure μ on H, called the equilibrium distribution, for which

$$\iint_{H\times H} -\log(|z-w|)\, d\mu(z)d\mu(w) \ = \ V_\infty(H) \ . \tag{B.2}$$

The equilibrium potential of H is then defined by

$$u_\infty(z, H) \ = \ \int_H -\log(|z-w|)\, d\mu(w) \ .$$

Now suppose $H = H_1 \bigcup H_2$, where the H_i are closed and disjoint, and put $\mu_1 = \mu|_{H_1}$, $\mu_2 = \mu|_{H_2}$. Let

$$\widehat{u}(z) \ = \ \int_{H_2} -\log(|z-w|)\, d\mu_2(w)$$

be the part of the equilibrium potential coming from H_2. Then

$$\begin{aligned} V_\infty(H) \ &= \ \iint_{H\times H} -\log(|z-w|)\, d\mu(z)d\mu(w) \\ &= \ \iint_{H_1\times H_1} -\log(|z-w|)\, d\mu_1(z)d\mu_1(w) + 2\int_{H_1} \widehat{u}(z)d\mu_1(z) \\ &\qquad + \iint_{H_2\times H_2} -\log(|z-w|)\, d\mu_2(z)d\mu_2(w) \ . \end{aligned} \tag{B.3}$$

If one thinks of μ_2 on H_2 as given, then the minimization problem (B.1) can be viewed as asking for a positive measure on H_1 of mass $\sigma = \mu(H_1)$, which minimizes the sum of the first two terms in (B.3). We now generalize this situation, replacing H_1 by the full set H and allowing an arbitrary weight function $W(z)$.

To simplify notation, throughout the discussion below we fix $\mathfrak{X}$ and $\vec{s}$ and suppress the $(\mathfrak{X}, \vec{s})$-dependence in all quantities, though that dependence is present.

Definition of the Weighted $(\mathfrak{X}, \vec{s})$-Capacity. Let H be a compact subset of $\mathcal{C}_v(\mathbb{C})\backslash\mathfrak{X}$, and let $W(z) : \mathcal{C}_v(\mathbb{C}) \to [0, \infty]$ be a function which is positive, bounded and continuous on a neighborhood of H.

Fix a number $\sigma > 0$, and put

$$\widehat{u}(z) \ = \ -\log(W(z)) \ .$$

Given a positive Borel measure ν on H with total mass σ, define the energy

$$I_\sigma(\nu, W) \ = \ \iint_{H\times H} -\log([z, w]_{\mathfrak{X},\vec{s}})\, d\nu(z)d\nu(w) + 2\int_H \widehat{u}(z)\, d\nu(z) \ .$$

Let the weighted Robin constant be

$$V_\sigma(H,W) \;=\; \inf_{\substack{\nu\geq 0\\ \nu(H)=\sigma}} I_\sigma(\nu,W)\ , \tag{B.4}$$

where the inf is taken over positive Borel measures on H with total mass σ. Then, define the weighted capacity

$$\gamma_\sigma(H,W) \;=\; e^{-V_\sigma(H,W)}\ . \tag{B.5}$$

Because $W(z)$ is finite and bounded away from 0, and $[z,w]_{\mathfrak{X},\vec{s}}/\|z,w\|_v$ is bounded for $z \neq w \in H$, it is easy to see that $\gamma_\sigma(H,W) > 0$ if and only if H has positive capacity in the sense of Definition 3.14.

If ν is a positive Borel measure on H, define the potential function

$$\begin{aligned} u_\nu(z) \;&=\; \int_H -\log([z,w]_{\mathfrak{X},\vec{s}})\, d\nu(w) \\ &=\; \lim_{t\to\infty}\int_H -\log^{(t)}([z,w]_{\mathfrak{X},\vec{s}})\, d\nu(w) \end{aligned}$$

where $-\log^{(t)}([z,w]_{\mathfrak{X},\vec{s}}) := \max(t, -\log([z,w]_{\mathfrak{X},\vec{s}})$. Then $u_\nu(z)$ is harmonic in $\mathcal{C}_v(\mathbb{C})\backslash(H\cup\mathfrak{X})$ and superharmonic in $\mathcal{C}_v(\mathbb{C})\backslash\mathfrak{X}$. If $w \notin \mathfrak{X}$, $u_\nu(z) + r\log([z,w]_{\mathfrak{X},\vec{s}})$ extends to a function harmonic in $\mathcal{C}_v(\mathbb{C})\backslash(H\cup\{w\})$.

Suppose H has positive capacity. In this section we show the existence of a measure $\mu = \mu_{\sigma,H,W}$ achieving $V_\sigma(H,W)$, which we call an equilibrium distribution. We will prove uniqueness later, and only in the cases of interest to us (Theorems B.7 and B.9). However, it seems likely that the equilibrium distribution is always unique (this holds in the classical case: see [**57**], p.27).

The proof of the existence of μ is standard. Take a sequence of measures $\nu_k \geq 0$ on H, with $\nu_k(H) = \sigma$ for all k, such that

$$\lim_{k\to\infty} I_\sigma(\nu_k,W) \;=\; V_\sigma(H,W)\ .$$

After passing to a subsequence, we can assume the ν_k converge weakly to a measure μ on H, which is necessarily positive and satisfies $\mu(H) = \sigma$. We claim that $I_\sigma(\mu,W) = V_\sigma(H,W)$.

Tautologically $I_\sigma(\mu,W) \geq V_\sigma(H,W)$. For the reverse inequality, note that

$$I_\sigma(\mu,W) \;=\; \lim_{t\to\infty}\Big(\iint_{H\times H} -\log^{(t)}([z,w]_{\mathfrak{X},\vec{s}})\, d\mu(z)d\mu(w) + 2\int_H \widehat{u}(z)d\mu(z)\Big)$$

$$= \lim_{t\to\infty}\lim_{k\to\infty}\Big(\iint_{H\times H} -\log^{(t)}([z,w]_{\mathfrak{X},\vec{s}})\, d\nu_k(z)d\nu_k(w) + 2\int_H \widehat{u}(z)d\nu_k(z)\Big) \tag{B.6}$$

$$\leq \liminf_{k\to\infty}\lim_{t\to\infty}\Big(\iint_{H\times H} -\log^{(t)}([z,w]_{\mathfrak{X},\vec{s}})\, d\nu_k(z)d\nu_k(w) + 2\int_H \widehat{u}(z)d\nu_k(z)\Big) \tag{B.7}$$

$$= \liminf_{k\to\infty} I_\sigma(\nu_k,W) \;=\; V_\sigma(H,W)\ .$$

The second equality follows from the weak convergence of the ν_k and the continuity of $-\log^{(t)}([z,w]_{\mathfrak{X},\vec{s}})$ on H. The interchange of limits between (B.6) and (B.7) is valid because the kernels $-\log^{(t)}([z,w]_{\mathfrak{X},\vec{s}})$ are increasing with t.

Let $H_\mu^* \subset H$ be the carrier of μ:

$$H_\mu^* \;=\; \Big\{z \in H : \mu(B(z,\delta)\bigcap H) > 0 \text{ for each } \delta > 0\Big\},$$

where the balls $B(a,\delta) = \{z \in \mathcal{C}_v(\mathbb{C}) : \|z,a\|_v \le \delta\}$ are computed relative to a fixed spherical metric on $\mathcal{C}_v(\mathbb{C})$. Note that H_μ^* is closed.

We now show that $u_\mu(z)$ satisfies an analogue of Frostman's theorem (compare [**57**], Theorem 1.3, p.27):

THEOREM B.1. *Suppose H has positive capacity, and let μ be any equilibrium distribution for H relative to $W(z)$. Then there exist a constant $\mathcal{V}_\mu$ and an F_σ set e_μ of inner capacity 0 contained in H such that for all $z \in H_\mu^* \backslash e_\mu$,*

$$u_\mu(z) + \widehat{u}(z) \;=\; \mathcal{V}_\mu \;,$$

and for all $z \in H\backslash e_\mu$,

$$u_\mu(z) + \widehat{u}(z) \;\ge\; \mathcal{V}_\mu \;.$$

Moreover,

$$\sigma \cdot \mathcal{V}_\mu + \int_H \widehat{u}(z)\, d\mu(z) \;=\; V_\sigma(H,W) \;. \tag{B.8}$$

PROOF. Write $f(z) = u_\mu(z) + \widehat{u}(z)$ and set

$$\mathcal{V}_\mu \;:=\; \sup_{z \in H_\mu^*} f(z) \;. \tag{B.9}$$

We will now show that $f(z) \ge \mathcal{V}_\mu$ for 'almost all' $z \in H$. Put

$$e_n \;=\; \{z \in H : f(z) \le \mathcal{V}_\mu - \frac{1}{n}\} \quad \text{for } n = 1,2,3,\ldots,$$

and

$$e_\mu \;=\; \{z \in H : f(z) < \mathcal{V}_\mu\} \;,$$

so $e_\mu = \bigcup_{n=1}^\infty e_n$. Here, each e_n is closed, since $u_\mu(z)$ is superharmonic and hence lower semi-continuous. We will show that each e_n has capacity 0, which implies that e_μ is an F_σ set of inner capacity 0.

Suppose to the contrary that some e_n had positive capacity. Then there would be a probability measure η supported on e_n such that

$$I(\eta) \;=\; \int_{H\times H} -\log([z,w]_{\mathfrak{X},\vec{s}})\, d\eta(z) d\eta(w) \;<\; \infty.$$

By the definition of $\mathcal{V}_\mu$, there is a $q \in H^*$ where $f(q) > \mathcal{V}_\mu - \frac{1}{2n}$. By the lower semi-continuity of $u_\mu(z)$, there is also a $\delta > 0$ such that for all $z \in B(q, 2\delta)$

$$f(z) \;>\; \mathcal{V}_\mu - \frac{1}{2n} \;.$$

Put $A = H \bigcap B(q,\delta)$, noting that A and e_n are closed and disjoint, hence bounded away from each other. Since $q \in H^*$,

$$\mu(A) \;>\; 0 \;.$$

Write $a = \mu(A)$, and put

$$\Delta = \begin{cases} a \cdot \eta & \text{on } e_n , \\ -\mu & \text{on } A . \end{cases}$$

Then $\Delta(H) = 0$, and it is easy to see that $I(\Delta) = \iint -\log([z,w]_{\mathfrak{X},\vec{s}})\, d\Delta(z)d\Delta(w)$ is finite: when it is expanded as a sum of three integrals, the diagonal terms are finite by hypothesis and the cross term is finite because A and e_n are bounded apart.

For $0 \le t \le 1$ put

$$\mu_t = \mu + t\Delta .$$

Then μ_t is a positive measure of total mass σ supported on H. Comparing the integrals $I_\sigma(\mu_t, W)$ and $V_\sigma(\mu, W)$, one finds

$$I_\sigma(\mu_t, W) - I_\sigma(\mu, W) = 2t \int_H u_\mu(z) + \widehat{u}(z)\, d\Delta(z) + t^2 I(\Delta) .$$

By construction

$$\begin{aligned} \int_H u_\mu(z) + \widehat{u}(z)\, d\Delta(z) &= a \cdot \int_{e_n} f(z)\, d\eta(z) - \int_A f(z)\, d\mu(z) \\ &\le a \cdot (\mathcal{V}_\mu - \frac{1}{n}) - a \cdot (\mathcal{V}_\mu - \frac{1}{2n}) = -\frac{a}{2n} . \end{aligned}$$

Thus $I_\sigma(\mu_t, W) - I_\sigma(\mu, W) < 0$ for sufficiently small $t > 0$, contradicting the minimality of $I_\sigma(\mu, W)$. It follows that $f(z) = u_\mu(z) + \widehat{u}(z) \ge \mathcal{V}_\mu$ on $H \backslash e_\mu$.

By (B.9), $f(z) \le \mathcal{V}_\mu$ on H_μ^*. Thus $f(z) = \mathcal{V}_\mu$ on $H_\mu^* \backslash e_\mu$. Necessarily $\mu(e_\mu) = 0$, since $I(\mu) < \infty$ while e_μ has inner capacity 0. Hence

$$\begin{aligned} V_\sigma(H, W) &= \iint_{H \times H} -\log([z,w]_{\mathfrak{X},\vec{s}})\, d\mu(w)d\mu(z) + 2 \int_H \widehat{u}(z)\, d\mu(z) \\ &= \int_H u_\mu(z) + \widehat{u}(z)\, d\mu(z) + \int_H \widehat{u}(z)\, d\mu(z) \\ &= \sigma \cdot \mathcal{V}_\mu + \int_H \widehat{u}(z)\, d\mu(z) . \end{aligned}$$

□

2. The Weighted Cheybshev Constant

Motivation. Consider the classical restricted Chebyshev constant, which is defined for compact sets $H \subset \mathbb{C}$ by first putting

$$\text{(B.10)} \qquad \mathrm{CH}_N^*(H) = \inf_{\substack{\text{monic } P(z) \in \mathbb{C}[z], \\ \text{degree } N, \\ \text{roots in } H}} (\|P(z)\|_E)^{1/N}$$

for $N = 1, 2, 3, \ldots$, and then setting $\mathrm{CH}(H) = \lim_{N\to\infty} \mathrm{CH}_N^*(H)$.

Since H is compact, for each N there is a polynomial $P_N(z) = \prod_{i=1}^N (z - \alpha_i)$ which achieves the inf in (B.10); it is called a (restricted) Chebyshev polynomial. It is known that as $N \to \infty$, the discrete measures $\omega_N = \frac{1}{N} \sum \delta_{\alpha_i}(z)$ associated to the $P_N(z)$ converge weakly to the equilibrium measure μ of H.

Now suppose $H = H_1 \bigcup H_2$, where the H_i are closed, nonempty, and disjoint. Put

$$\widehat{u}(z) = \int_{H_2} -\log(|z - w|)\, d\mu(z)$$

and set $W(z) = \exp(-\widehat{u}(z))$. Label the roots of $P_N(z)$ so $\alpha_1, \ldots, \alpha_n \in H_1$ and $\alpha_{n+1}, \ldots, \alpha_N \in H_2$. For large N, one has $\prod_{i=n+1}^N |z - a_i| \cong \exp(-N\widehat{u}(z))$ outside H_2. Thus, on H_1,

$$|P_N(z)| \cong \prod_{i=1}^n |z - \alpha_i| \cdot W(z)^N . \tag{B.11}$$

If one thinks of $W(z)$ as given, then the minimization problem (B.10) can be thought of as varying $\alpha_1, \ldots, \alpha_n$ over H_1 to achieve the minimum in (B.10).

Definition of the Weighted $(\mathfrak{X}, \vec{s})$-Chebyshev Constant. Let H be a compact subset of $\mathcal{C}_v(\mathbb{C}) \backslash \mathfrak{X}$, and let $W(z) : \mathcal{C}_v(\mathbb{C}) \to [0, \infty]$ be a function which is positive, bounded and continuous on a neighborhood of H. Let $[z, w]_{\mathfrak{X},\vec{s}}$ be as before. By a weighted $(\mathfrak{X}, \vec{s})$-pseudopolynomial of bidegree (n, N) (relative to $W(z)$) we mean a function of the form

$$P(z) = P_{(n,N)}(z, W) = \prod_{i=1}^n [z, \alpha_i]_{\mathfrak{X},\vec{s}} \cdot W(z)^N .$$

The α_i will be called the roots of $P(z)$.

To simplify notation, for the rest of this section we will generally omit the $(\mathfrak{X}, \vec{s})$-dependence of the quantities discussed.

Since $[z, w]_{\mathfrak{X},\vec{s}}$ and $W(z)$ are continuous on H, and H is compact, among the weighted pseudopolynomials $P_{(n,N)}(z, W)$ of bidegree (n, N) whose roots belong to H, there is at least one with minimal minimal sup norm on H (it need not be unique). Fix one, and write $\widetilde{P}_{(n,N)}(z, W)$ for it. We will call it a weighted, restricted Chebyshev pseudopolynomial.

Put

$$\begin{aligned} \mathrm{CH}^*_{(n,N)} &= \mathrm{CH}^*_{(n,N)}(H, W) := \min_{\substack{P_{(n,N)} \\ \text{roots in } H}} (\|P_{(n,N)}(z, W)\|_H)^{1/N} \\ &= \left(\|\widetilde{P}_{(n,N)}\|_H\right)^{1/N} . \end{aligned} \tag{B.12}$$

For each $\sigma > 0$, define the weighted Chebyshev constant $\mathrm{CH}^*_\sigma(H, W)$ by

$$\mathrm{CH}^*_\sigma(H, W) = \lim_{\substack{n/N \to \sigma \\ N \to \infty}} \mathrm{CH}^*_{(n,N)} \tag{B.13}$$

provided the limit exists, that is, if for each $\varepsilon > 0$, there are a $\delta > 0$ and an N_0 such that if $|\frac{n}{N} - \sigma| < \delta$ and $N \geq N_0$, then $|\mathrm{CH}^*_{(n,N)} - \mathrm{CH}^*_\sigma(H, W)| < \varepsilon$.

THEOREM B.2. *For each* $\sigma > 0$, $\mathrm{CH}^*_\sigma(H, W)$ *exists. The function* $g(r) = \mathrm{CH}^*_r(H, W)$ *is continuous for* $r > 0$. *If* $g(r_0) = 0$ *for one* r_0, *then* $g(r) \equiv 0$. *Moreover,* $g(r) \equiv 0$ *if and only if* H *has capacity* 0.

Recall that H has capacity 0 iff for some $\zeta \in \mathcal{C}_v(\mathbb{C}) \backslash H$, we have $\gamma_\zeta(H) = 0$ in the sense of ([**51**], §3.1); this holds for one $\zeta \notin H$ iff it holds for all $\zeta \notin H$.

PROOF. First fix $0 < r \in \mathbb{Q}$, and take (n, N) with $n/N = r$. Given a rational number $\kappa > 0$ such that $\kappa n, \kappa N \in \mathbb{Z}$, we can write

$$\begin{aligned} \kappa n &= kn + a \qquad \text{where } k = \lfloor \kappa \rfloor \in \mathbb{Z} \text{ and } 0 \leq a < n , \\ \kappa N &= kN + A \qquad \text{where } 0 \leq A = a/r < N . \end{aligned}$$

By the compactness of H and our assumptions on $W(z)$, there is a constant $C > 1$ such that $1/C \le W(z) \le C$ for all $z \in H$. Put $D = \max(1, \max_{z,w\in H}[z,w]_{\mathfrak{X},\vec{s}})$. If $\widetilde{P}_{(n,N)}(z,W) = \prod_{i=1}^{n}[z,\alpha_i]_{\mathfrak{X},\vec{s}} \cdot W(z)^N$, then

$$\begin{aligned}
(\mathrm{CH}^*_{(\kappa n,\kappa N)})^{\kappa N} &\le \max_{z\in H}\left(\prod_{i=1}^{n}[z,\alpha_i]^k_{\mathfrak{X},\vec{s}} \cdot \prod_{i=1}^{a}[z,\alpha_i]_{\mathfrak{X},\vec{s}} \cdot W(z)^{\kappa N}\right) \\
&= \max_{z\in H}\left(\prod_{i=1}^{n}[z,\alpha_i]_{\mathfrak{X},\vec{s}} \cdot W(z)^N\right)^k \cdot \prod_{i=1}^{a}[z,\alpha_i]_{\mathfrak{X},\vec{s}} \cdot W(z)^A \\
&\le (\mathrm{CH}^*_{(n,N)})^{kN} \cdot D^a \cdot C^A .
\end{aligned}$$

Thus

$$\mathrm{CH}^*_{(\kappa n,\kappa N)} \le (\mathrm{CH}^*_{(n,N)})^{\frac{k}{\kappa}} \cdot D^{\frac{a}{\kappa N}} \cdot C^{\frac{A}{\kappa N}} .$$

Letting $\kappa \to \infty$ gives

$$\mathrm{CH}^*_{(n,N)} \ge \limsup_{\kappa\to\infty} \mathrm{CH}^*_{(\kappa n,\kappa N)} ,$$

then replacing (n,N) by $(\kappa n, \kappa N)$ and again letting $\kappa \to \infty$ gives

$$\liminf_{\kappa\to\infty} \mathrm{CH}^*_{(\kappa n,\kappa N)} \ge \limsup_{\kappa\to\infty} \mathrm{CH}^*_{(\kappa n,\kappa N)} .$$

It follows that as κ passes through all rational numbers with $\kappa n, \kappa N \in \mathbb{Z}$, the limit

$$g(r) := \lim_{\kappa\to\infty} \mathrm{CH}^*_{(\kappa n,\kappa N)} \tag{B.14}$$

exists, and the same limit is obtained when κ passes through *any* sequence of rationals with $\kappa \to \infty$ and $\kappa n, \kappa N \in \mathbb{Z}$.

We now compare $g(r)$ and $g(s)$, when $0 < r, s$ are rationals with $r \ne s$. Fix positive integers n, N with $r = n/N$ and m, M with $s = m/M$. Let λ be a positive integer such that $\lambda n \ge m$. Write

$$\lambda n = \ell m + b \qquad \text{where } \ell = \left\lfloor \frac{\lambda n}{m} \right\rfloor \text{ and } 0 \le b < m .$$

If $\widetilde{P}_{(m,M)}(z,W) = \prod_{i=1}^{m}[z,\alpha_i] \cdot W(z)^M$, we have

$$\begin{aligned}
(\mathrm{CH}^*_{(\lambda n,\lambda N)})^{\lambda N} &\le \max_{z\in H}\left(\prod_{i=1}^{m}[z,\alpha_i]^\ell_{\mathfrak{X},\vec{s}} \cdot \prod_{i=1}^{b}[z,\alpha_i]_{\mathfrak{X},\vec{s}} \cdot W(z)^{\lambda N}\right) \\
&= \max_{z\in H}\left(\prod_{i=1}^{m}[z,\alpha_i]_{\mathfrak{X},\vec{s}} \cdot W(z)^M\right)^\ell \cdot \prod_{i=1}^{b}[z,\alpha_i]_{\mathfrak{X},\vec{s}} \cdot W(z)^{\lambda N-\ell M} \\
&\le (\mathrm{CH}^*_{(m,M)})^{\ell M} \cdot D^b \cdot C^{|\lambda N-\ell M|} .
\end{aligned}$$

The absolute value appears in the last term because we use $W(z) \le C$ to obtain the final inequality if $\lambda N - \ell M > 0$, and $W(z) > 1/C$ if $\lambda N - \ell M < 0$. Consequently,

$$(\mathrm{CH}^*_{(\lambda n,\lambda N)})^{\frac{\lambda N}{\lambda n}} \le CH_{(m,M)}^{\frac{\ell M}{\ell m+b}} \cdot D^{\frac{b}{\ell m+b}} \cdot C^{|\frac{\lambda N}{\lambda n} - \frac{\ell M}{\ell m+b}|} . \tag{B.15}$$

Letting $\lambda \to \infty$ (so $\ell \to \infty$ as well), we get

$$g(r)^{1/r} \le (\mathrm{CH}^*_{(m,M)})^{\frac{M}{m}} \cdot C^{|\frac{1}{r} - \frac{M}{m}|} ; \tag{B.16}$$

then replacing (m,M) by $(\kappa m, \kappa M)$ and letting $\kappa \to \infty$,

$$g(r)^{1/r} \le g(s)^{1/s} \cdot C^{|\frac{1}{r} - \frac{1}{s}|} . \tag{B.17}$$

Reversing the roles of r and s, we similarly obtain

$$g(s)^{1/s} \leq g(r)^{1/r} \cdot C^{|\frac{1}{s}-\frac{1}{r}|} . \tag{B.18}$$

From (B.17) and (B.18) it follows that if $g(r_0) = 0$ for one r_0, then $g(r) \equiv 0$.

Suppose $g(r) \neq 0$. Taking logarithms in (B.17) and (B.18) we see that

$$\left|\frac{1}{r}\log(g(r)) - \frac{1}{s}\log(g(s))\right| \leq \left|\frac{1}{r} - \frac{1}{s}\right| \log(C) . \tag{B.19}$$

Thus if $\{r_i\}_{1 \leq i < \infty}$ is a Cauchy sequence of rationals converging to some real $r > 0$, then $\{g(r_i)\}$ is also a Cauchy sequence; given two Cauchy sequences converging to the same r, the sequences of values $g(r_i)$ converge to the same value. Thus, we can extend $g(r)$ to a function defined on all positive reals. It is easy to see that the extended function satisfies (B.19) for all real $r, s > 0$ and hence is continuous.

Trivially $g(r)$ can be extended continuously to all $r > 0$ if $g(r) \equiv 0$ on $\mathbb{Q}$.

We will now show that the limit (B.13) exists, and that

$$\lim_{\substack{n/N \to r \\ N \to \infty}} \mathrm{CH}^*_{(n,N)} = g(r) . \tag{B.20}$$

First suppose $g(r) \neq 0$. Then (B.20) is equivalent to

$$\lim_{\substack{n/N \to r \\ N \to \infty}} \frac{N}{n} \log(\mathrm{CH}^*_{(n,N)}) = \frac{1}{r}\log(g(r)) . \tag{B.21}$$

To prove (B.21), fix $\varepsilon > 0$ and take $r_1, r_2 \in \mathbb{Q}$ with $0 < r_1 < r < r_2$ such that

$$\left|\frac{1}{r_1}\log(g(r_1)) - \frac{1}{r}\log(g(r))\right| < \frac{\varepsilon}{5} , \tag{B.22}$$

$$\left|\frac{1}{r_1} - \frac{1}{r_2}\right| \log(C) < \frac{\varepsilon}{5} . \tag{B.23}$$

By (B.14), there is a pair (m_0, M_0) with $m_0/M_0 = r_1$, such that

$$\left|\frac{M_0}{m_0}\log(\mathrm{CH}^*_{(m_0,M_0)}) - \frac{1}{r_1}\log(g(r_1))\right| < \frac{\varepsilon}{5} . \tag{B.24}$$

Fix an integer $N_0 \geq M_0$ large enough that for each integer $\ell \geq \lfloor N_0/M_0 \rfloor$,

$$\frac{1}{\ell}\log(D) < \frac{\varepsilon}{5} , \tag{B.25}$$

$$\left|\frac{\ell M_0}{\ell m_0 + m_0} - \frac{M_0}{m_0}\right| \cdot |\log(\mathrm{CH}^*_{(m_0,M_0)})| < \frac{\varepsilon}{5} , \tag{B.26}$$

$$\frac{\ell M_0}{\ell m_0 + m_0} > \frac{1}{r_2} . \tag{B.27}$$

Consider any pair (n, N) with $N \geq N_0$ and $r_1 < \frac{n}{N} < r_2$.

Applying (B.16), with r replaced by r_1 and (m, M) replaced by (n, N), then taking logarithms, we get

$$\frac{N}{n}\log(\mathrm{CH}^*_{(n,N)}) \geq \frac{1}{r_1}\log(g(r_1)) - \left|\frac{1}{r_1} - \frac{N}{n}\right| \log(C) . \tag{B.28}$$

Combining (B.28), (B.22) and (B.23) gives

$$\frac{N}{n}\log(\mathrm{CH}^*_{(n,N)}) > \frac{1}{r}\log(g(r)) - \frac{2\varepsilon}{5} . \tag{B.29}$$

For the opposite equality, note that since $m_0/M_0 = r_1 < n/N < r_2$ and $N \geq N_0 > M_0$, we have $n \geq m_0$. Apply (B.15), taking $(m, M) = (m_0, M_0)$ and $\lambda = 1$, so that $n = \ell m_0 + b$, with $\ell = \lfloor n/m_0 \rfloor \geq \lfloor N_0/M_0 \rfloor$ and $0 \leq b < m_0$; this gives

$$\frac{N}{n}\log(\mathrm{CH}^*_{(n,N)})$$

$$\text{(B.30)} \qquad \leq \frac{\ell M_0}{\ell m_0 + b}\log(\mathrm{CH}^*_{(m_0,M_0)}) + \frac{b}{\ell m_0 + b}\log(D) + \left|\frac{N}{n} - \frac{\ell M_0}{\ell m_0 + b}\right|\log(C) .$$

Using (B.25) we see that

$$\text{(B.31)} \qquad \frac{b}{\ell m_0 + b}\log(D) \leq \frac{1}{\ell}\log(D) < \frac{\varepsilon}{5} .$$

Likewise, $1/r_1 > N/n > 1/r_2$, and $1/r_1 > \ell M_0/(\ell m_0 + b) > 1/r_2$ by (B.27), so (B.23) yields

$$\text{(B.32)} \qquad \left|\frac{N}{n} - \frac{\ell M_0}{\ell m_0 + b}\right|\log(C) < \frac{\varepsilon}{5}.$$

Combining (B.30), (B.22), (B.23), (B.26), (B.31), and (B.32) gives

$$\frac{N}{n}\log(\mathrm{CH}^*_{(n,N)}) < \frac{1}{r}\log(g(r)) + \varepsilon .$$

In the case where $g(r) \equiv 0$, first note that if H is finite, then $\mathrm{CH}^*_{(n,N)} = 0$ whenever $n \geq \#(H)$, so (B.20) is trivial. If H is infinite, then $\mathrm{CH}^*_{(n,N)} > 0$ for all (n, N). Take $0 < \varepsilon < 1$. Fix $r_1, r_2 \in \mathbb{Q}$ with $0 < r_1 < r < r_2$ such that

$$\text{(B.33)} \qquad C^{|\frac{1}{r_1} - \frac{1}{r_2}|} < 2$$

and fix (m_0, M_0) such that $m_0/M_0 = r_1$ and $(\mathrm{CH}^*_{(m_0,M_0)})^{M_0/m_0} < \varepsilon/5$. Let N_0 be large enough that for each integer $\ell \geq \lfloor N_0/M_0 \rfloor$ we have

$$\text{(B.34)} \qquad (\mathrm{CH}^*_{(m_0,M_0)})^{\frac{\ell M_0}{\ell m_0 + m_0}} < \frac{\varepsilon}{4} ,$$

$$\text{(B.35)} \qquad D^{\frac{1}{\ell}} < 2 , \qquad \frac{\ell M_0}{\ell m_0 + m_0} > \frac{1}{r_2} .$$

Consider any pair (n, N) with $r_1 < n/N < r_2$ and $N \geq N_0$. Again apply (B.15), taking $(m, M) = (m_0, M_0)$ and $\lambda = 1$, so that $n = \ell m_0 + b$, with $\ell = \lfloor n/m_0 \rfloor \geq \lfloor N_0/M_0 \rfloor$ and $0 \leq b < m_0$; this gives

$$\text{(B.36)} \qquad (\mathrm{CH}^*_{(n,N)})^{\frac{N}{n}} \leq (\mathrm{CH}^*_{(m_0,M_0)})^{\frac{\ell M_0}{\ell m_0 + b}} \cdot D^{\frac{b}{\ell m_0 + b}} \cdot C^{|\frac{N}{n} - \frac{\ell M_0}{\ell m_0 + b}|} .$$

By (B.33) and (B.35),

$$\text{(B.37)} \qquad D^{\frac{b}{\ell m_0 + b}} \cdot C^{|\frac{N}{n} - \frac{\ell M_0}{\ell m_0 + b}|} < 4 .$$

Combining (B.36), (B.34) and (B.37) gives

$$(\mathrm{CH}^*_{(n,N)})^{\frac{N}{n}} < \varepsilon .$$

Finally, we show that $g(r) \equiv 0$ if and only if H has capacity 0. If instead of $[z, w]_{\mathfrak{X},\vec{s}}$ and $W(z)$ we had used another distance function

$$[z, w]^*_{\mathfrak{X},\vec{s}} = \prod_{i=1}^{m^*}([z, w]_{x_i^*})^{s_i^*}$$

and another continuous, positive background function $W^*(z)$, then there would be a constant $A > 0$ such that

$$1/A \ < \ [z,w]^*_{\mathfrak{X},\vec{s}}/[z,w]_{\mathfrak{X},\vec{s}} \ < \ A \ ,$$
$$1/A \ < \ W^*(z)/W(z) \ < \ A \ ,$$

for all $z \neq w \in H$. Let $\mathrm{CH}^*_{(n,N)}(N, W^*)$ and $\mathrm{CH}^*_r(H, W^*)$ denote the weighted Chebyshev constants computed relative to $[z,w]^*_{\mathfrak{X},\vec{s}}$ and $W^*(z)$. Taking $\alpha_1, \ldots, \alpha_n \in H$ such that $\widetilde{P}_{(n,N)}(z, W) = \prod_{i=1}^n [z, \alpha_i]_{\mathfrak{X},\vec{s}} \cdot W(z)^N$, we get

$$\mathrm{CH}^*_{(n,N)}(H, W^*) \ \geq \ \mathrm{CH}^*_{(n,N)} \cdot A^{-\frac{n}{N}-1}.$$

A similar inequality holds with $\mathrm{CH}^*_{(n,N)}(H, W^*)$ and $\mathrm{CH}^*_{(n,N)}$ reversed, and it follows that $\mathrm{CH}^*_r(H, W^*) \neq 0$ iff $\mathrm{CH}^*_r(H, W) \neq 0$.

Taking $r = 1$, $W^*(z) = 1$ and $[z,w]^*_{\mathfrak{X},\vec{s}} = [z,w]_\zeta$ for a fixed $\zeta \in \mathcal{C}_v(\mathbb{C}) \backslash H$, we have $\mathrm{CH}^*_1(H, W^*) = \gamma_\zeta(H)$ by ([**51**], Theorem 3.1.18). Thus $g(r) \not\equiv 0$ if and only if for some (hence any) $\zeta \notin H$, we have $\gamma_\zeta(H) > 0$. □

3. The Weighted Transfinite Diameter

Motivation. The classical transfinite diameter for compact sets $H \subset \mathbb{C}$ is defined by first putting

$$\text{(B.38)} \qquad d_N(H) \ = \ \left(\sup_{z_1,\ldots,z_N \in H} \prod_{\substack{i,j=1 \\ i \neq j}}^N |z_i - z_j| \right)^{1/N^2} ,$$

and then setting $d_\infty(H) = \lim_{N\to\infty} d_N(H)$. (Usually the exponent $1/N^2$ in (B.38) is replaced by $1/N(N-1)$, but this does not affect the value of the limit.)

Since H is compact, for each N there exist points $\alpha_1, \ldots, \alpha_N \in H$ realizing the supremum in (B.38): $d_N(H) = (\prod_{i,j=1, i\neq j}^N |\alpha_i - \alpha_j|)^{1/N^2}$. The collection $\{\alpha_1, \ldots, \alpha_n\}$ is called a set of Fekete points.

Now suppose $H = H_1 \bigcup H_2$, where the H_i are closed, nonempty, and disjoint. We can write

$$\text{(B.39)} \quad d_N(H) \ = \ \prod_{\substack{\alpha_i,\alpha_j \in H_1 \\ i \neq j}} |\alpha_i - \alpha_j| \cdot \prod_{\alpha_i \in H_1} \prod_{\alpha_j \in H_2} |\alpha_i - \alpha_j|^2 \cdot \prod_{\substack{\alpha_i,\alpha_j \in H_2 \\ i \neq j}} |\alpha_i - \alpha_j| \ .$$

Let $\nu_N = \sum_{i=1}^N \frac{1}{N} \delta_{\alpha_i}(z)$ be the probability measure equally supported on the α_i. It is known that the ν_N converge weakly to the equilibrium distribution μ of H. As before, put

$$\widehat{u}(z) \ = \ \int_{H_2} -\log([z,w]_{\mathfrak{X},\vec{s}}) \, d\mu(w)$$

and set $W(z) = \exp(-\widehat{u}(z))$. Then

$$\prod_{\alpha_i \in H_1} \prod_{\alpha_j \in H_2} (|\alpha_i - \alpha_j|)^2 \ \cong \ \prod_{\alpha_i \in H_1} W(\alpha_i)^{2N} \ .$$

Label the Fekete points so that $\alpha_1, \ldots, \alpha_n \in H_1$ and $\alpha_{n+1}, \ldots, \alpha_N \in H_2$. If one is interested in H_1 and thinks of $\alpha_{n+1}, \ldots, \alpha_N$ as given, then the maximization problem (B.38) can be thought of as varying $z_1, \ldots, z_n$ over H_1 so as to maximize

$$\prod_{\substack{i,j=1 \\ i \neq j}}^{n} |z_i - z_j| \cdot \prod_{i=1}^{n_1} W(z_i)^{2N} .$$

Note that as $N \to \infty$, then $n/N \to \sigma := \mu(H_1)$.

Definition of the Weighted Transfinite Diameter $d_\sigma(H, W)$.
Let $H \subset \mathcal{C}_v(\mathbb{C}) \backslash \mathfrak{X}$ be compact, and let $W(z) : \mathcal{C}_v(\mathbb{C}) \to [0, \infty]$ be a function which is positive, bounded and continuous on a neighborhood of H. Given positive integers n and N, define

(B.40)
$$d_{(n,N)} = d_{(n,N)}(H, W) = \max_{z_1, \ldots, z_n \in H} \left(\prod_{\substack{i,j=1 \\ i \neq j}}^{n} [z_i, z_j]_{\mathfrak{X}, \vec{s}} \cdot \prod_{i=1}^{n} W(z_i)^{2N} \right)^{1/N^2} .$$

Then, for each $0 < \sigma \in \mathbb{R}$ let the weighted $(\mathfrak{X}, \vec{s})$-transfinite diameter be

(B.41)
$$d_\sigma(H, W) = \lim_{\substack{n/N \to \sigma \\ N \to \infty}} d_{(n,N)}$$

provided the limit exists. This is understood to mean that for each $\varepsilon > 0$, there exist a $\delta > 0$ and an N_0 such that if $|\frac{n}{N} - r| < \delta$ and $N \geq N_0$, then $|d_{(n,N)} - d_\sigma(H, W)| < \varepsilon$. For notational simplicity we suppress the $(\mathfrak{X}, \vec{s})$-dependence in $d_{(n,N)}$ and $d_\sigma(H, W)$.

A set of points $\alpha_1, \ldots, \alpha_n \in H$ which achieves the maximum in (B.40) will be called a set of (n, N)-Fekete points for H relative to the weight $W(z)$.

THEOREM B.3. *For each $\sigma > 0$, $d_\sigma(H, W)$ exists. The function $f(r) = d_r(H, W)$ is continuous for $r > 0$; and if $f(r_0) = 0$ for one r_0, then $f(r) \equiv 0$. Moreover, $f(r) \equiv 0$ if and only if H has capacity 0.*

PROOF. First consider what happens to $d_{(n,N)}$ when (n, N) is replaced by $(\lambda n, \lambda N)$ for some rational $\lambda \geq 1$ for which λn and λN are integers. If $\alpha_1, \ldots, \alpha_{\lambda n} \in H$ are chosen to maximize (B.40) for $d_{(\lambda n, \lambda N)}$, then

(B.42)
$$d_{(\lambda n, \lambda N)}^{(\lambda N)^2} = \prod_{\substack{i,j=1 \\ i \neq j}}^{\lambda n} [\alpha_i, \alpha_j]_{\mathfrak{X}, \vec{s}} \cdot \prod_{i=1}^{\lambda n} W(\alpha_i)^{2\lambda N} .$$

For any n-element subset $\alpha_{k_1}, \ldots, \alpha_{k_n}$ we have

$$d_{(n,N)}^{N^2} \geq \prod_{\substack{i,j=1 \\ i \neq j}}^{n} [\alpha_{k_i}, \alpha_{k_j}]_{\mathfrak{X}, \vec{s}} \cdot \prod_{i=1}^{n} W(\alpha_{k_i})^{2N} .$$

There are $\binom{\lambda n}{n}$ such subsets; each pair $\{\alpha_i, \alpha_j\}$ belongs to $\binom{\lambda n-2}{n-2}$ subsets, and each α_i belongs to $\binom{\lambda n-1}{n-1}$ subsets. If $C > 1$ is such that $1/C \le W(z) \le C$ on H, then

$$\begin{aligned}
(d_{(n,N)}^{N^2})^{\binom{\lambda n}{n}} &\ge \left(\prod_{\substack{i,j=1\\ i\neq j}}^{\lambda n} [\alpha_i, \alpha_j]_{\mathfrak{X},\vec{s}}\right)^{\binom{\lambda n-2}{n-2}} \cdot \left(\prod_{i=1}^{\lambda n} W(\alpha_i)^{2N}\right)^{\binom{\lambda n-1}{n-1}} \\
&= \left(d_{(\lambda n,\lambda N)}\right)^{(\lambda N)^2\binom{\lambda n-2}{n-2}} \cdot \left(\prod_{i=1}^{\lambda n} W(\alpha_i)\right)^{2N\binom{\lambda n-1}{n-1} - 2\lambda N\binom{\lambda n-2}{n-2}} \\
&\ge \left(d_{(\lambda n,\lambda N)}\right)^{(\lambda N)^2\binom{\lambda n-2}{n-2}} \cdot C^{-\lambda n\left(2N\binom{\lambda n-1}{n-1} - 2\lambda N\binom{\lambda n-2}{n-2}\right)} . \qquad \text{(B.43)}
\end{aligned}$$

Simplifying exponents, we find

$$d_{(n,N)}^{\frac{\lambda n-1}{\lambda(n-1)}} \cdot C^{\frac{2n(\lambda-1)}{\lambda N(n-1)}} \ge d_{(\lambda n,\lambda N)} . \qquad \text{(B.44)}$$

This holds for any (n, N) and λ satisfying the conditions above. Given $0 < \sigma \in \mathbb{Q}$, take n, N with $\sigma = n/N$. Letting $\lambda \to \infty$ in (B.44) (where λ runs over all rationals such that $\lambda n, \lambda N \in \mathbb{Z}$), we find

$$d_{(n,N)}^{\frac{n}{n-1}} \cdot C^{\frac{2n}{N(n-1)}} \ge \limsup_{\lambda\to\infty} d_{(\lambda n,\lambda N)} . \qquad \text{(B.45)}$$

Replacing (n, N) by $(\kappa n, \kappa N)$ on the left side of (B.45) and again letting $\kappa \to \infty$ gives

$$\liminf_{\kappa\to\infty} d_{(\kappa n,\kappa N)} \ge \limsup_{\lambda\to\infty} d_{(\lambda n,\lambda N)} .$$

Hence

$$f(r) := \lim_{\lambda\to\infty} d_{(\lambda n,\lambda N)} \qquad \text{(B.46)}$$

is well-defined; moreover, the same limit is obtained when λ passes through any sequence of values such that $\lambda \to \infty$ and $\lambda n, \lambda N$ are integers.

We now seek to compare $f(r)$ and $f(s)$, when $0 < r, s$ are rationals with $r \neq s$. Fix positive integers n, N with $r = n/N$ and m, M with $s = m/M$. Let λ be a positive integer such that $\lambda n \ge m$. As before, let $\alpha_1, \ldots, \alpha_{\lambda n} \in H$ realize the maximum in (B.40) for $d_{(\lambda n,\lambda N)}$. Then as in (B.43),

$$\begin{aligned}
d_{(m,M)}^{M^2\binom{\lambda n}{m}} &\ge \left(d_{(\lambda n,\lambda N)}\right)^{(\lambda N)^2\binom{\lambda n-2}{m-2}} \cdot \left(\prod_{i=1}^{\lambda n} W(\alpha_i)\right)^{\left(2M\binom{\lambda n-1}{m-1} - 2\lambda N\binom{\lambda n-2}{m-2}\right)} \\
&\ge \left(d_{(\lambda n,\lambda N)}\right)^{(\lambda N)^2\binom{\lambda n-2}{m-2}} \cdot C^{-\left|\lambda n\left(2M\binom{\lambda n-1}{m-1} - 2\lambda N\binom{\lambda n-2}{m-2}\right)\right|} .
\end{aligned}$$

The absolute value in the last term occurs since we use $W(z) > 1/C$ to obtain the final inequality if the exponent in the previous line is positive, and $W(z) < C$ if it is negative. Simplifying exponents gives

$$d_{(m,M)}^{\frac{M}{m}\cdot\frac{M}{m-1}} \ge d_{(\lambda n,\lambda N)}^{\frac{\lambda N}{\lambda n}\cdot\frac{\lambda N}{\lambda n-1}} \cdot C^{-2\left|\frac{M}{m-1} - \frac{\lambda N}{\lambda n-1}\right|} . \qquad \text{(B.47)}$$

Letting $\lambda \to \infty$ in (B.47) gives

$$d_{(m,M)}^{\frac{M}{m}\cdot\frac{M}{m-1}} \ge f(r)^{1/r^2} \cdot C^{-2\left|\frac{M}{m-1} - \frac{1}{r}\right|} ; \qquad \text{(B.48)}$$

then, replacing (m, M) by $(\kappa m, \kappa M)$ and letting $\kappa \to \infty$ in (B.48), we find

$$f(s)^{1/s^2} \geq f(r)^{1/r^2} \cdot C^{-2|\frac{1}{s}-\frac{1}{r}|} . \tag{B.49}$$

Interchanging the role of r and s in (B.49), also

$$f(r)^{1/r^2} \geq f(s)^{1/s^2} \cdot C^{-2|\frac{1}{r}-\frac{1}{s}|} . \tag{B.50}$$

It follows from (B.49) and (B.50) that $f(r) \equiv 0$ if and only if $f(r_0) = 0$ for one r_0. Suppose $f(r) \not\equiv 0$. Taking logarithms in (B.49) and (B.50), we find

$$\left|\frac{1}{s^2}\log(f(s)) - \frac{1}{r^2}\log(f(r))\right| \leq 2\left|\frac{1}{s} - \frac{1}{r}\right|\log(C) . \tag{B.51}$$

From (B.51) we see that if $\{r_i\}_{1\leq i<\infty}$ is a Cauchy sequence of rationals converging to some $r > 0$, then $\{f(r_i)\}$ is also a Cauchy sequence; and given two Cauchy sequences converging to the same r, the sequences of values $f(r_i)$ converge to the same value. Thus, we can extend $f(r)$ to a function defined on all positive reals, which is easily seen to satisfy (B.51) for all real $r, s > 0$ and hence is continuous. Note that trivially $f(r)$ can be extended by continuity if $f(r) \equiv 0$ on $\mathbb{Q}$.

We will now show that the limit (B.41) exists. Fix $0 < r \in \mathbb{R}$. We claim that

$$\lim_{\substack{n/N\to r\\ N\to\infty}} d_{(n,N)} = f(r) . \tag{B.52}$$

To show this, first suppose $f(r) \neq 0$. Then (B.52) is equivalent to

$$\lim_{\substack{n/N\to r\\ N\to\infty}} \frac{N}{n}\frac{N}{n-1}\log\left(d_{(n,N)}\right) = \frac{1}{r^2}\log(f(r)) . \tag{B.53}$$

Given $\varepsilon > 0$, fix $r_1, r_2 \in \mathbb{Q}$ with $0 < r_1 < r < r_2$ such that

$$\left|\frac{1}{r_1^2}\log(f(r_1)) - \frac{1}{r^2}\log(f(r))\right| < \frac{\varepsilon}{6} , \tag{B.54}$$

$$2\left|\frac{1}{r_1} - \frac{1}{r_2}\right|\log(C) < \frac{\varepsilon}{6} . \tag{B.55}$$

Then, using (B.46) and (B.55), take (n_0, N_0) with $n_0/N_0 = r_1$ and N_0 large enough that

$$\left|\frac{N_0}{n_0}\frac{N_0}{n_0-1}\log(d_{(n_0,N_0)}) - \frac{1}{r_1^2}\log(f(r_1))\right| < \frac{\varepsilon}{6} , \tag{B.56}$$

$$2\left|\frac{N_0}{n_0-1} - \frac{1}{r_1}\right|\log(C) < \frac{\varepsilon}{6} . \tag{B.57}$$

Consider a pair (n, N) with $r_1 < n/N < r_2$ and $N \geq N_0$. First, replacing (m, M) by (n, N) and r by r_1 in (B.48), we have

$$\frac{N}{n}\frac{N}{n-1}\log(d_{(n,N)}) \geq \frac{1}{r_1^2}\log(f(r_1)) - 2\left|\frac{N}{n-1} - \frac{1}{r_1}\right|\log(C) . \tag{B.58}$$

Note that since $r_1 < n/N < r_2$ and $N \geq N_0$, also

$$\frac{N_0}{n_0-1} \geq \frac{N}{n-1} > \frac{1}{r_2} .$$

Hence (B.55) and (B.57) give

$$\begin{aligned} 2\left|\frac{N}{n-1}-\frac{1}{r_1}\right|\log(C) &< \left(2\left|\frac{N}{n-1}-\frac{1}{r_2}\right|+2\left|\frac{1}{r_1}-\frac{1}{r_2}\right|\right)\log(C) \\ &< \left(2\left|\frac{N_0}{n_0-1}-\frac{1}{r_2}\right|+2\left|\frac{1}{r_1}-\frac{1}{r_2}\right|\right)\log(C) \\ &< \left(2\left|\frac{N_0}{n_0-1}-\frac{1}{r_1}\right|+2\left|\frac{1}{r_1}-\frac{1}{r_2}\right|+2\left|\frac{1}{r_1}-\frac{1}{r_2}\right|\right)\log(C) \\ &< \frac{3\varepsilon}{6}\ . \end{aligned} \tag{B.59}$$

Combining (B.58), (B.54) and (B.59) gives

$$\frac{N}{n}\frac{N}{n-1}\log(d_{(n,N)}) \ \geq\ \frac{1}{r^2}\log(f(r))-\frac{4\varepsilon}{6}\ . \tag{B.60}$$

For the opposite inequality, replace (m, M) by (n_0, N_0) in (B.47), and take $\lambda = 1$ (which is permissible since $n \geq n_0$ under our hypotheses). This yields

(B.61)

$$\frac{N}{n}\frac{N}{n-1}\log(d_{(n,N)}) \ \leq\ \frac{N_0}{n_0}\frac{N_0}{n_0-1}\log(d_{(n_0,N_0)})+2\left|\frac{N}{n-1}-\frac{N_0}{n_0-1}\right|\log(C)\ .$$

Combining (B.61), (B.54), (B.56), (B.57), and (B.59) gives

$$\frac{N}{n}\frac{N}{n-1}\log(d_{(n,N)}) \ <\ \frac{1}{r^2}\log(f(r))+\varepsilon\ . \tag{B.62}$$

In the case where $f(r) \equiv 0$, we need only to show that

$$\lim_{\substack{n/N\to r\\ N\to\infty}} d_{(n,N)}^{\frac{N}{n}\cdot\frac{N}{n-1}} \ =\ 0\ . \tag{B.63}$$

If H is finite, then $d_{(n,N)} = 0$ whenever $n > \#(H)$, and hence (B.63) holds trivially. If H is infinite, then each $d_{(n,N)} > 0$. Fix $\varepsilon > 0$ and take $r_1, r_2 \in \mathbb{Q}$ with $0 < r_1 < r < r_2$, such that

$$C^{2\left|\frac{1}{r_1}-\frac{1}{r_2}\right|} \ \leq\ 2\ . \tag{B.64}$$

Take (n_0, N_0) with $n_0/N_0 = r_1$. Since $\lim_{\lambda\to\infty} d_{(\lambda n_0,\lambda N_0)} = 0$ we can assume N_0 is large enough that

$$d_{(n_0,N_0)}^{\frac{N_0}{n_0}\cdot\frac{N_0}{n_0-1}} \ <\ \varepsilon/3\ .$$

In view of (B.64) we can also assume that N_0 is large enough that for all (n, N) with $r_1 < \frac{n}{N} < r_2$ and $N \geq N_0$, then

$$C^{2\left(\frac{N_0}{n_0-1}-\frac{N}{n-1}\right)} \ \leq\ 3\ .$$

Taking $\lambda = 1$ and $(m, M) = (n_0, N_0)$ in (B.47), for all such (n, N) we have

$$d_{(n,N)}^{\frac{N}{n}\cdot\frac{N}{n-1}} \ <\ \varepsilon\ .$$

Finally, we show that $f(r) \equiv 0$ if and only if H has capacity 0. If instead of $[z, w]_{\mathfrak{X},\vec{s}}$ and $W(z)$ we had used another distance function

$$[z, w]_{\mathfrak{X},\vec{s}}^* \ =\ \prod_{i=1}^{m^*}([z, w]_{x_i^*})^{s_i^*}$$

and another continuous, positive background function $W^*(z)$, then there is a constant $A > 0$ such that

$$\begin{aligned} 1/A &< [z,w]^*_{\mathfrak{X},\vec{s}}/[z,w]_{\mathfrak{X},\vec{s}} < A \ , \\ 1/A &< W^*(z)/W(z) < A \ , \end{aligned}$$

for all $z \neq w \in H$. Let $d^*_r(H, W^*)$ denote the weighted transfinite diameter computed relative to $[z,w]^*_{\mathfrak{X},\vec{s}}$ and $W^*(z)$, and write $d^*_{(n,N)}(H, W^*)$ for the finite terms in its definition. Taking points $\alpha_1, \ldots, \alpha_n \in H$ which realize the maximum in (B.40), we see that

$$d^*_{(n,N)}(H, W^*) \ \geq \ d_{(n,N)}(H, W) \cdot A^{-\left(\frac{n}{N}\frac{n-1}{N} + \frac{2n}{N}\right)}.$$

Passing to a limit as $N \to \infty$ and $n/N \to r$ gives

$$d^*_r(H, W^*) \ \geq \ d_r(H, W) \cdot A^{-(r^2+2r)} \ .$$

A similar inequality holds with $d^*_r(H, W^*)$ and $d_r(H, W)$ reversed, so $d_r(H, W) \neq 0$ if and only if $d^*_r(H, W^*) \neq 0$.

Taking $r = 1$, $W^*(z) = 1$, and $[z,w]^*_{\mathfrak{X},\vec{s}} = [z,w]_\zeta$ for a fixed $\zeta \in \mathcal{C}_v(\mathbb{C})\backslash H$, by ([**51**], Theorem 3.1.18) we have $d^*_1(H, W^*) = \gamma_\zeta(H)$. Thus $f(r) \not\equiv 0$ if and only if for some (hence any) $\zeta \notin H$, $\gamma_\zeta(H) > 0$. □

4. Comparisons

In this section we will compare the weighted Chebyshev constant, the weighted transfinite diameter, and the weighted capacity (for fixed $(\mathfrak{X}, \vec{s})$). Fix a compact set $H \subset \mathcal{C}_v(\mathbb{C})\backslash\mathfrak{X}$, and let $W(z) : \mathcal{C}_v(\mathbb{C}) \to [0, \infty]$ be a function which is positive and continuous on a neigbhorhood of H. Put $\widehat{u}(z) = -\log(W(z))$.

The Weighted Capacity and the Weighted Transfinite Diameter. We first prove an inequality between the weighted Robin constant and the weighted transfinite diameter.

PROPOSITION B.4. *For each* $\sigma > 0$,

$$V_\sigma(H, W) \ \leq \ -\log(d_\sigma(H, W)) \ . \tag{B.65}$$

Before giving the proof, we will need a lemma. Let U be a neighborhood of H, bounded away from $\mathfrak{X}$, on which $\widehat{u}(z)$ is continuous and bounded.

For any set F with $F \subset U$, define the inner weighted Robin constant and inner weighted capacity by

$$\begin{aligned} \overline{V}_\sigma(F, W) &= \inf_{\substack{K \subset F \\ K \text{ compact}}} V_\sigma(K, W) \ , \\ \overline{\gamma}_\sigma(F, W) &= \sup_{\substack{K \subset F \\ K \text{ compact}}} \gamma_\sigma(K, W) \ = \ \exp(-\overline{V}_\sigma(F, W)) \ . \end{aligned}$$

Note that for any compact $K \subset U$, we have $\overline{V}_\sigma(K, W) = V_\sigma(K, W)$, and that if $F_1 \subset F_2 \subset U$, then trivially $\overline{V}_\sigma(F_1, W) \geq \overline{V}_\sigma(F_2, W)$.

LEMMA B.5. *Let* $H \subset U$ *and* $W(z)$ *be as above. For any* $\sigma > 0$, *and any* $\varepsilon > 0$, *there is a neighborhood* $\widetilde{U}$ *of* H *contained in* U *such that*

$$V_\sigma(H, W) \ \geq \ \overline{V}_\sigma(\widetilde{U}, W) \ \geq \ V_\sigma(H, W) - \varepsilon \ .$$

PROOF. Take a sequence of open sets $U_1 \supset U_2 \supset \ldots \supset H$ whose closures are compact and satisfy $\overline{U}_i \subset U$, with $\bigcap_{k=1}^{\infty} \overline{U}_k = H$. We claim that

$$\lim_{k \to \infty} V_\sigma(\overline{U}_k, W) \;=\; V_\sigma(H, W) \;. \tag{B.66}$$

To see this, for each k let μ_k be an equilibrium distribution for $\overline{U}_k$ and W with mass σ, so $I_\sigma(\mu_k, W)) = V_\sigma(\overline{U}_k, W)$. After passing to a subsequence if necessary, we can assume that the μ_k converge weakly to a measure μ_∞. Clearly μ_∞ is positive, supported on H, and has total mass σ. Then

$$\begin{aligned}
V_\sigma(H, W) \;&\leq\; I_\sigma(\mu_\infty, W) \\
&=\; \lim_{t \to \infty} \Big(\iint_{H \times H} -\log^{(t)}([z, w]_{\mathfrak{X},\vec{s}}) \, d\mu_\infty(z) d\mu_\infty(w) \\
&\qquad\qquad + 2 \int_H \widehat{u}(z) d\mu_\infty(z) \Big) \\
&=\; \lim_{t \to \infty} \lim_{k \to \infty} \Big(\iint_{\overline{U}_k \times \overline{U}_k} -\log^{(t)}([z, w]_{\mathfrak{X},\vec{s}}) \, d\mu_k(z) d\mu_k(w) \\
&\qquad\qquad + 2 \int_{\overline{U}_k} \widehat{u}(z) d\mu_k(z) \Big) \qquad \text{(B.67)} \\
&\leq\; \liminf_{k \to \infty} \lim_{t \to \infty} \Big(\iint_{\overline{U}_k \times \overline{U}_k} -\log^{(t)}([z, w]_{\mathfrak{X},\vec{s}}) \, d\mu_k(z) d\mu_k(w) \\
&\qquad\qquad + 2 \int_{\overline{U}_k} \widehat{u}(z) d\mu_k(z) \Big) \qquad \text{(B.68)} \\
&=\; \liminf_{k \to \infty} I_\sigma(\mu_k, W) \;=\; \liminf_{k \to \infty} V_{\sigma,H}(\overline{U}_k, W) \;. \qquad \text{(B.69)}
\end{aligned}$$

The interchange of limits between (B.67) and (B.68) is valid because the kernels $-\log^{(t)}([z, w]_{\mathfrak{X},\vec{s}})$ are increasing with t.

If H has capacity 0, then $V_\sigma(H, W) = \infty$, so $\lim_{k \to \infty} V_\sigma(\overline{U}_k, W) = \infty$. If H has positive capacity, then $V_\sigma(H, W)$ is finite. Since $V_\sigma(H, W) \geq V_\sigma(\overline{U}_k, W)$ for all k, (B.69) gives $\lim_{k \to \infty} V_\sigma(\overline{U}_k, W) = V_\sigma(H, W)$. In either case (B.66) holds, and we obtain the lemma by taking $\widetilde{U} = U_k$ for sufficiently large k. □

Proof of Proposition B.4. By Theorem B.3 and the remarks after (B.5),

$$V_\sigma(H, W) = \infty \quad \text{iff} \quad -\log(d_\sigma(H, W)) = \infty \quad \text{iff} \quad H \text{ has capacity } 0,$$

so we can assume that $V_\sigma(H, W)$ is finite, $d_\sigma(H, W) > 0$, and H has positive capacity.

Fix σ, and fix $\varepsilon > 0$. Let U be a neighborhood of H on which $\widehat{u}(z)$ is continuous and bounded. By Lemma B.5, there is a neighborhood $\widetilde{U}$ of H, whose closure is contained in U, such that $\overline{V}_\sigma(\widetilde{U}, W) > V_\sigma(H, W) - \varepsilon$. We can assume that H is covered by a finite number of local coordinate patches, each of which contained in $\widetilde{U}$. Then there is an $R_0 > 0$ such that for each $x \in H$, there is a coordinate patch which contains the closed disc $D(x, R_0) = \{z : |z - w| \leq R_0\}$ relative to that coordinate. For each $x \in H$ fix such a coordinate patch, and given $0 < R \leq R_0$, write $D(x, R)$ for the corresponding closed disc of radius R.

To prove (B.65) we use a familiar construction involving "smearing out" Fekete measures. Take a sequence of pairs (n_k, N_k) with $n_k/N_k \to \sigma$ and $N_k \to \infty$.

Given k, let $\alpha_1, \ldots, \alpha_{n_k} \in H$ be points where the maximum in the definition of $d_{(n_k,N_k)}(H,W)$ is achieved; let

$$\nu_k = \sum_{i=1}^{n_k} \frac{1}{N_k} \delta_{\alpha_i}(z) \tag{B.70}$$

be the associated measure of mass n_k/N_k on H; it will be called a Fekete measure. After passing to a subsequence, if necessary, we can assume that the ν_k converge weakly to a measure ν on H.

Fix k, and write $(n,N) = (n_k,N_k)$, $\nu_k = \sum_{i=1}^{n} \frac{1}{N}\delta_{\alpha_i}(z)$. Without loss, we can assume that k (hence N) is large enough that $\frac{1}{\sqrt{\pi N}} \le R_0$. Let dm_i be the measure which coincides with Lebesgue measure on the disc $D_i = D(\alpha_i, \frac{1}{\sqrt{\pi N}})$, and is 0 outside that disc. Thus, dm_i has total mass $\frac{1}{N}$. Put $F_k = \bigcup_{i=1}^{n} D_i$, and let

$$\widetilde{\nu}_k = \frac{r}{n/N} \sum_{i=1}^{n} dm_i \ .$$

Then $F_k \subset \widetilde{U}$, and $\widetilde{\nu}_k$ is a positive measure of mass σ on F_k. It follows that

$$\begin{aligned} V_\sigma(H,W) - \varepsilon \ &\le\ \overline{V}_\sigma(\widetilde{U},W) \ \le\ V_\sigma(F_k,W) \ \le\ I_\sigma(\widetilde{\nu}_k,W) \\ &= \frac{\sigma}{n/N} \sum_{i,j=1}^{n} \iint_{D_i \times D_j} -\log([z,w]_{\mathfrak{X},\vec{s}})\, dm_i(z) dm_j(w) \\ &\qquad + \frac{\sigma}{n/N} \cdot 2 \int_{F_k} \widehat{u}(z)\, d\nu_k(z) \ . \end{aligned} \tag{B.71}$$

For each fixed z, the function $-\log([z,w]_{\mathfrak{X},\vec{s}})$ is superharmonic in w. Hence, using polar coordinates in D_j, we have

$$\begin{aligned} \int_{D_j} -\log([z,w]_{\mathfrak{X},\vec{s}})\, dm_j(w) &= \int_0^{\frac{1}{\sqrt{\pi N}}} \int_0^{2\pi} -\log([z, \alpha_j + te^{i\theta}]_{\mathfrak{X},\vec{s}})\, d\theta\, t\, dt \\ &\le -\frac{1}{N} \log([z,\alpha_j]_{\mathfrak{X},\vec{s}}) \ . \end{aligned}$$

If $i \ne j$, then since $-\frac{1}{N}\log([z,\alpha_j]_{\mathfrak{X},\vec{s}})$ is superharmonic in w, this gives

$$\begin{aligned} \iint_{D_i \times D_j} -\log([z,w]_{\mathfrak{X},\vec{s}})\, dm_i(z) dm_j(w) &\le \int_{D_i} -\frac{1}{N} \log([z,\alpha_j]_{\mathfrak{X},\vec{s}})\, dm_i(z) \\ &\le -\frac{1}{N^2} \log([\alpha_i,\alpha_j]_{\mathfrak{X},\vec{s}}) \ . \end{aligned} \tag{B.72}$$

If $i = j$, then since $-\log([z,w]_{\mathfrak{X},\vec{s}}) = -\log(|z-w|) + \eta(z,w)$ for a $\mathcal{C}^\infty$ function $\eta(z,w)$, we see that

$$\begin{aligned} \iint_{D_i \times D_i} -\log([z,w]_{\mathfrak{X},\vec{s}}) dm_i(z) dm_i(w) &\le \int_{D_i} -\frac{1}{N} \log([z,\alpha_i]_{\mathfrak{X},\vec{s}})\, dm_i(z) \\ &= \frac{1}{N} \left(\int_0^{\frac{1}{\sqrt{\pi N}}} \int_0^{2\pi} -\log(t)\, d\theta\, t\, dt + \int_{D_i} \eta(z,\alpha_i)\, dm_i(z) \right) \\ &= O(\frac{1}{N^2} \log(N)) \ . \end{aligned} \tag{B.73}$$

Thus

$$I_{\mathfrak{X},\vec{s}}(\widetilde{\nu}_k) \;=\; \sum_{i,j=1}^{n} \iint_{D_i\times D_j} -\log([z,w]_{\mathfrak{X},\vec{s}})\, dm_i(z)dm_j(w)$$

$$\text{(B.74)}\qquad \leq\; \frac{1}{N^2}\sum_{\substack{i,j=1\\ i\neq j}}^{n} -\log([\alpha_i,\alpha_j]_{\mathfrak{X},\vec{s}}) + O(\frac{\log(N)}{N})\ .$$

On the other hand, $W(z)$, and hence $\widehat{u}(z)$, is continuous on the closure of $\widetilde{U}$. Hence as $k\to\infty$,

$$\text{(B.75)}\qquad \left|\int \widehat{u}(z)d\widetilde{\nu}_k(z) - \int \widehat{u}(z)d\nu_k\right| \;=\; o(1)\ .$$

Inserting (B.74) and (B.75) in (B.71), we obtain

$$\text{(B.76)}\qquad V_\sigma(H,W)-\varepsilon \;\leq\; -\log\left(d_{(n_k,N_k)}(H,W)\right) + o(1)\ .$$

Passing to the limit as $k\to\infty$, and using that $\varepsilon>0$ is arbitrary gives

$$V_\sigma(H,W) \;\leq\; -\log(d_\sigma(H,W))\ .$$

□

The Weighted Transfinite Diameter and Chebyshev Constant. Fix $\sigma>0$, and consider a sequence of pairs of positive integers (n_k,N_k) with $N_k\to\infty$ and $n_k/N_k\to\sigma$. For each k, let ν_k be the associated Fekete measure, as in (B.70).

PROPOSITION B.6. *With the notation above, if ν is any weak limit of the Fekete measures ν_k, then*

$$\text{(B.77)}\qquad -\log(d_\sigma(H,W)) \;\leq\; \sigma\cdot(-\log(\mathrm{CH}^*_\sigma(H,W))) + \int_H \widehat{u}(z)d\nu(z)\ .$$

PROOF. If H has capacity 0, both $d_\sigma(H,W)$ and $\mathrm{CH}^*_\sigma(H,W)$ are 0, and the inequality is trivial. Hence we can assume H has positive capacity, so $d_\sigma(H,W)$ and $\mathrm{CH}^*_\sigma(H,W)$ are positive.

Fix k and write $(n,N)=(n_k,N_k)$. As before, let $\alpha_1,\ldots,\alpha_n\in H$ be points where $d_{(n,N)}(H,W)$ is achieved. Fixing $\alpha_2,\ldots\alpha_n$, consider

$$\text{(B.78)}\qquad F(z) \;:=\; \left(\prod_{i=2}^{n}[z,\alpha_i]_{\mathfrak{X},\vec{s}}\cdot W(z)^N\right)^2 \cdot \prod_{\substack{2\leq i,j\leq n\\ i\neq j}} [\alpha_i,\alpha_j]_{\mathfrak{X},\vec{s}}\cdot\prod_{i=2}^{n} W(\alpha_i)^{2N}\ .$$

By hypothesis $F(z)$ takes its maximum value on H when $z=\alpha_1$. Thus the pseudopolynomial

$$P(z) \;=\; \prod_{i=2}^{n}[z,\alpha_i]_{\mathfrak{X},\vec{s}}\cdot W(z)^N$$

achieves its maximum on H at $z=\alpha_1$. By definition, if

$$\widetilde{P}_{(n-1,N)}(z) \;=\; \prod_{i=2}^{n}[z,\beta_i]_{\mathfrak{X},\vec{s}}W(z)^N$$

is the pseudopolynomial with minimal sup norm on H, then

$$(\mathrm{CH}^*_{(n-1,N)})^N \;:=\; \|\widetilde{P}_{(n-1,N)}\|_H \;\leq\; P(\alpha_1)\ .$$

Thus $(\mathrm{CH}^*_{n-1,N})^N \le \prod_{i=2}^n [\alpha_1, \alpha_i]_{\mathfrak{X},\vec{s}} \cdot W(\alpha_1)^N$. The same argument applies for each α_k, so

$$\begin{aligned} d_{(n,N)}(H,W)^{N^2} &= \prod_{i=1}^{n} \left(\prod_{\substack{j=1 \\ j \ne i}}^{n} [\alpha_i, \alpha_j]_{\mathfrak{X},\vec{s}} \cdot W(\alpha_i)^N \right) \cdot \left(\prod_{i=1}^{n} W(\alpha_i)^N \right) \\ &\ge (\mathrm{CH}^*_{(n-1,N)})^{Nn} \cdot \prod_{i=1}^{n} W(\alpha_i)^N . \end{aligned}$$

Taking logarithms, dividing by N^2, and recalling that all quantities involved depend on k, we get

$$\text{(B.79)} \quad -\log(d_{(n_k,N_k)}(H,W)) \le \frac{n_k}{N_k} \left(-\log(\mathrm{CH}^*_{(n_k-1,N_k)}) \right) + \sum_{i=1}^{n_k} \widehat{u}(\alpha_i) \frac{1}{N_k} .$$

Now let $k \to \infty$. The fractions $(n_k - 1)/N_k$ converge to σ, so by Theorem B.2 the weighted Chebyshev constant satisfies

$$\mathrm{CH}^*_\sigma(H,W) = \lim_{k\to\infty} \mathrm{CH}^*_{(n_k-1,N_k)} .$$

After passing to a subsequence, if necessary, we can assume that the measures $\nu_k = \sum_{i=1}^{n_k} \frac{1}{N_k} \delta_{\alpha_i}(z)$ converge weakly to ν. Hence by (B.79) and Theorem B.3,

$$\begin{aligned} \text{(B.80)} \quad -\log(d_\sigma(H,W)) &= \lim_{k\to\infty} -\log(d_{(n_k,N_k)}(H,W)) \\ &\le \sigma \cdot (-\log(\mathrm{CH}^*_\sigma(H,W)) + \int_H \widehat{u}(z) d\nu(z) . \end{aligned}$$

□

5. Particular Cases of Interest

In this section we first consider the "classical" case where the weight function $W(z)$ is trivial, and we then specialize to the situation of interest for our applications. As in the previous sections, we assume $(\mathfrak{X}, \vec{s})$ has been fixed.

THEOREM B.7. *Let $H \subset \mathcal{C}_v(\mathbb{C}) \backslash \mathfrak{X}$ be compact. Suppose $\sigma = 1$ and $W(z) \equiv 1$. Then $\gamma_1(H,W) = \gamma_{\mathfrak{X},\vec{s}}(H)$, $\mathrm{CH}^*_1(H,W) = \mathrm{CH}^*_{\mathfrak{X},\vec{s}}(H)$, and $d_1(H,W) = d_{\mathfrak{X},\vec{s}}(H)$, so*

$$\gamma_1(H,W) = d_1(H,W) = \mathrm{CH}^*_1(H,W) = \gamma_{\mathfrak{X},\vec{s}}(H) .$$

Furthermore, if H has positive capacity, then the equilibrium distribution μ for H relative to $W(z)$ is unique and coincides with $\mu_{\mathfrak{X},\vec{s}}$. Finally, the constant $\mathcal{V}_\mu$ in Theorem B.1 *coincides with $V_{\mathfrak{X},\vec{s}}(H)$.*

PROOF. For the first assertion, it suffices to note that the optimization problems (B.4), (A.3) defining $V_1(H,W)$ and $V_{\mathfrak{X},\vec{s}}(H)$ are the same; for each N, the optimization problems (B.12), (A.7) defining $\mathrm{CH}^*_{(N,N)}(H,W)$ and $\mathrm{CH}^*_N(H)$ are the same; and for each N, the optimization problems (B.40), (A.5) defining $d_{(N,N)}(H,W)$ and $d_N(H)$ are the same.

The second assertion follows from Theorem A.1. The last assertions follow from the fact that $V_1(H,W) = V_{\mathfrak{X},\vec{s}}(H)$, and the uniqueness of $\mu_{\mathfrak{X},\vec{s}}$ in Theorem A.2. □

Consider a sequence of pairs (n_k, N_k) with $n_k/N_k \to 1$ and $N_k \to \infty$. For a given k, let $\alpha_1, \ldots, \alpha_{n_k}$ achieve the maximum in the definition of $d_{(n_k,N_k)}(H, W)$ for $W(z) \equiv 1$. Let

$$\nu_k = \sum_{i=1}^{n_k} \frac{1}{N_k} \delta_{\alpha_i}(z)$$

be the associated Fekete measure of mass n_k/N_k on H.

We will now see that, just as in the classical case, the Fekete measures converge weakly to the equilibrium distribution $\mu_{\mathfrak{X},\vec{s}}$ of H.

Corollary B.8. *Let $H \subset \mathcal{C}_v(\mathbb{C})\backslash\mathfrak{X}$ be compact with positive capacity, and suppose $W(z) \equiv 1$. Then for any sequence of pairs (n_k, N_k) with $n_k/N_k \to 1$ and $N_k \to \infty$, the associated sequence of Fekete measures $\{\nu_k\}$ converges weakly to $\mu_{\mathfrak{X},\vec{s}}$.*

Proof. Let ν_∞ be a weak limit of a subsequence of the ν_k. Upon passing to that subsequence, we see that

$$V_{\mathfrak{X},\vec{s}}(H) = \lim_{k\to\infty} -\log(d_{(n_k,N_k)}(H, W)) .$$

Let $\widetilde{\nu}_k$ be a sequence of measures "smearing out" the ν_k as in the proof of Proposition B.4; then the $\widetilde{\nu}_k$ also converge weakly to ν_∞. By the argument leading to (B.76), as $k \to \infty$,

$$I_{\mathfrak{X},\vec{s}}(\widetilde{\nu}_k) = I_{n_k/N_k}(\widetilde{\nu}_k, W) \le -\log(d_{(n_k,N_k)}(H, W)) + o(1) .$$

On the other hand, by (B.71), for any $\varepsilon > 0$ and sufficently large k,

$$V_{\mathfrak{X},\vec{s}}(H) - \varepsilon \le I_{\mathfrak{X},\vec{s}}(\widetilde{\nu}_k) .$$

Thus, by Theorem B.7,

$$\begin{aligned} V_{\mathfrak{X},\vec{s}}(H) &\le I_{\mathfrak{X},\vec{s}}(\nu_\infty) \le \liminf_{k\to\infty} I_{n_k/N_k}(\widetilde{\nu}_k, H) \\ &\le \liminf_{k\to\infty} -\log(d_{(n_k,N_k)}(H, W)) = V_{\mathfrak{X},\vec{s}}(H) . \end{aligned}$$

Since $\mu_{\mathfrak{X},\vec{s}}$ is the unique probability measure on H which minimizes the energy integral (Theorem A.2), it follows that $\nu_\infty = \mu_{\mathfrak{X},\vec{s}}$. □

We now come to the case of interest for our application.

Suppose $H \subset \mathcal{C}_v(\mathbb{C})\backslash\mathfrak{X}$ is compact, and H_ℓ is a connected component of H. We will assume that H_ℓ has positive capacity. Let $\mu = \mu_{\mathfrak{X},\vec{s}}$ be the equilibrium distribution of H. Put

$$\begin{aligned} \sigma_\ell &= \mu_{\mathfrak{X},\vec{s}}(H_\ell) , \\ \widehat{u}_\ell(z) &= \int_{H\backslash H_\ell} -\log([z, w]_{\mathfrak{X},\vec{s}})\, d\mu_{\mathfrak{X},\vec{s}}(w) , \\ W_\ell(z) &= \exp(-\widehat{u}_\ell(z)) . \end{aligned} \tag{B.81}$$

(We take $\widehat{u}_\ell(z) \equiv 0$ and $W_\ell(z) \equiv 1$ if $H\backslash H_\ell$ is empty.) We will use the results of previous sections to study weighted potential theory for H_ℓ, σ_ℓ, and $W_\ell(z)$. The problem is a somewhat subtle one: to show that in the weighted case for H_ℓ, with the weight $W_\ell(z)$ coming from the unweighted case for $H\backslash H_\ell$, the extremal objects for H_ℓ are the restrictions of the global unweighted objects for H.

The key results are Theorems B.9, B.12 and B.13.

THEOREM B.9. *Let H, H_ℓ, $\sigma_\ell = \mu_{\mathfrak{X},\vec{s}}(H_\ell)$ and $W_\ell(z)$ be as in* (B.81). *Then the equilibrium distribution $\mu_{\sigma_\ell,H_\ell,W_\ell}$ of H_ℓ relative to $W_\ell(z)$ with mass σ_ℓ is unique, and is given by*

$$\mu_{\sigma_\ell,H_\ell,W_\ell} \;=\; \mu_{\mathfrak{X},\vec{s}}|_{H_\ell} \;. \tag{B.82}$$

PROOF. Put $\mu_\ell = \mu_{\mathfrak{X},\vec{s}}|_{H_\ell}$, $\dot{\mu}_\ell = \mu_{\mathfrak{X},\vec{s}}|_{H\backslash H_\ell}$, and let $\widetilde{\mu}_\ell$ be any equilibrium distribution for H_ℓ relative to $W_\ell(z)$ with mass σ_ℓ. By definition, $\widetilde{\mu}_\ell$ minimizes the weighted energy

$$I_{\sigma_\ell}(\nu, W_\ell) \;=\; I_{\mathfrak{X},\vec{s}}(\nu) + 2\int_H \widehat{u}_\ell(z)\,d\nu(z)$$

among all positive measures of mass σ_ℓ on H_ℓ, so $I_{\sigma_\ell}(\widetilde{\mu}_\ell, W_\ell) = V_{\sigma_\ell}(H_\ell, W_\ell)$. Thus,

$$I_{\sigma_\ell}(\mu_\ell, W_\ell) \;\ge\; I_{\sigma_\ell}(\widetilde{\mu}_\ell, W_\ell) \;. \tag{B.83}$$

On the other hand, by Theorem A.2, $\mu := \mu_{\mathfrak{X},\vec{s}}$ is the unique probability measure minimizing

$$I_{\mathfrak{X},\vec{s}}(\mu) \;=\; \iint_{H\times H} -\log([z,w]_{\mathfrak{X},\vec{s}})\,d\mu(z)d\mu(w) \;.$$

Thus, if we put $\widetilde{\mu} = \widetilde{\mu}_\ell + \dot{\mu}_\ell$, then $I_{\mathfrak{X},\vec{s}}(\widetilde{\mu}) \ge I_{\mathfrak{X},\vec{s}}(\mu)$. Expanding this, we see that

$$\begin{aligned} I_{\mathfrak{X},\vec{s}}(\widetilde{\mu}_\ell) + 2\int_{H_\ell} \widehat{u}_\ell(z)d\widetilde{\mu}_\ell(z) + I_{\mathfrak{X},\vec{s}}(\dot{\mu}_\ell) & \\ \ge\; I_{\mathfrak{X},\vec{s}}(\mu_\ell) + 2\int_{H_\ell} \widehat{u}_\ell(z)d\mu_\ell(z) + I_{\mathfrak{X},\vec{s}}(\dot{\mu}_\ell) \;, & \end{aligned}$$

which implies that $I_{\sigma_\ell}(\widetilde{\mu}_\ell, W_\ell) \ge I_{\sigma_\ell}(\mu_\ell, W_\ell)$. Combining this with (B.83) gives $I_{\sigma_\ell}(\widetilde{\mu}_\ell, W_\ell) = I_{\sigma_\ell}(\mu_\ell, W_\ell)$. Consequently,

$$I_{\mathfrak{X},\vec{s}}(\widetilde{\mu}) \;=\; I_{\sigma_\ell}(\widetilde{\mu}_\ell, W_\ell) + I_{\mathfrak{X},\vec{s}}(\dot{\mu}_\ell) \;=\; I_{\sigma_\ell}(\mu_\ell, W_\ell) + I_{\mathfrak{X},\vec{s}}(\dot{\mu}_\ell) \;=\; I_{\mathfrak{X},\vec{s}}(\mu) \;.$$

From the uniqueness of μ, we conclude that $\widetilde{\mu} = \mu_{\mathfrak{X},\vec{s}}$, hence $\widetilde{\mu}_\ell = \mu_\ell$. □

Given measures ν_1, ν_2, write

$$I_{\mathfrak{X},\vec{s}}(\nu_1, \nu_2) \;=\; \iint -\log([z,w]_{\mathfrak{X},\vec{s}})\,d\nu_1(z)d\nu_2(z) \;,$$

provided the integral is defined. Also, given a measure ν, put

$$\widehat{I}_{\mathfrak{X},\vec{s}}(\nu) = \iint_{H\times H\backslash\{\text{diagonal}\}} -\log([z,w]_{\mathfrak{X},\vec{s}})\,d\nu(z)d\nu(w) \;,$$

when the integral is defined. Then $I_{\mathfrak{X},\vec{s}}(\nu_1, \nu_2)$ is symmetric, and is bilinear when all relevant terms are defined and finite. If ν is a positive measure, then $I_{\mathfrak{X},\vec{s}}(\nu) = I_{\mathfrak{X},\vec{s}}(\nu, \nu)$, and if in addition ν does not charge the diagonal, then $I_{\mathfrak{X},\vec{s}}(\nu) = \widehat{I}_{\mathfrak{X},\vec{s}}(\nu)$.

We will now show that in the situation of (B.81), the weighted transfinite diameter and the weighted capacity coincide. As in the proof of Proposition B.9, put

$$\mu_\ell = \mu_{\mathfrak{X},\vec{s}}|_{H_\ell} \;, \qquad \dot{\mu}_\ell = \mu_{\mathfrak{X},\vec{s}}|_{H\backslash H_\ell} \;. \tag{B.84}$$

Note that $I_{\mathfrak{X},\vec{s}}(\mu_\ell, \dot{\mu}_\ell) = \int_{H_\ell} \widehat{u}_\ell(z)\,d\mu_\ell(z)$.

THEOREM B.10. *Let H, H_ℓ, $\sigma_\ell = \mu_{\mathfrak{X},\vec{s}}(H_\ell)$, and $W_\ell(z)$ be as in Theorem* B.9. *Then*

$$-\log(d_{\sigma_\ell}(H_\ell, W_\ell)) \ = \ V_{\sigma_\ell}(H_\ell, W_\ell) \ = \ \sigma_\ell \cdot V_{\mathfrak{X},\vec{s}}(H) + \int_{H_\ell} \widehat{u}_\ell(z)\, d\mu_\ell(z) \ . \tag{B.85}$$

PROOF. The second equality in (B.85) is easy: by Proposition B.9

$$\begin{aligned} V_{\sigma_\ell}(H_\ell, W_\ell) &= I_{\sigma_\ell}(\mu_\ell, W_\ell) \\ &= \int_{H_\ell \times H_\ell} -\log([z,w]_{\mathfrak{X},\vec{s}})\, d\mu_\ell(z) d\mu_\ell(w) + 2\int_{H_\ell} \widehat{u}_\ell(z)\, d\mu_\ell(z) \\ &= I_{\mathfrak{X},\vec{s}}(\mu_\ell, \mu_\ell) + 2I_{\mathfrak{X},\vec{s}}(\mu_\ell, \dot{\mu}_\ell) \ = \ I_{\mathfrak{X},\vec{s}}(\mu_{\mathfrak{X},\vec{s}}, \mu_\ell) + I_{\mathfrak{X},\vec{s}}(\mu_\ell, \dot{\mu}_\ell) \\ &= \int_{H_\ell} u_{\mathfrak{X},\vec{s}}(z)\, d\mu_\ell(z) + I_{\mathfrak{X},\vec{s}}(\mu_\ell, \dot{\mu}_\ell) \\ &= \sigma_\ell \cdot V_{\mathfrak{X},\vec{s}}(H) + \int_{H_\ell} \widehat{u}_\ell(z)\, d\mu_\ell(z) , \end{aligned}$$

where the last inequality holds since $u_{\mathfrak{X},\vec{s}}(z)$ takes the constant value $V_{\mathfrak{X},\vec{s}}(H)$ on H, except possibly on a set of inner capacity 0; and the exceptional set necessarily has μ_ℓ-measure 0.

We now turn to the first equality. We have $V_{\sigma_\ell}(H_\ell, W_\ell) \le -\log(d_{\sigma_\ell}(H_\ell, W_\ell))$ by Proposition B.4, so we need only show the reverse inequality. For this, we use the fact that the unweighted Fekete measures for H converge weakly to the unweighted equilibrium distribution of H, as shown in Corollary B.8. Given an integer $N > 0$, let $\alpha_\ell^*, \ldots, \alpha_N^*$ be points maximizing the transfinite diameter $d_{(N,N)}(H, W)$ for $W(z) \equiv 1$. Label them so that $\alpha_\ell^*, \ldots, \alpha_{n_N}^* \in H_\ell$ and $\alpha_{n_{N+1}}^*, \ldots, \alpha_N^* \in H \backslash H_\ell$, and put

$$D_{n_N,N}^{(1)} \ = \ \left(\prod_{\substack{i,j=1 \\ i \ne j}}^{n_N} [\alpha_i^*, \alpha_j^*]_{\mathfrak{X},\vec{s}} \cdot \prod_{i=1}^{n_N} W_\ell(\alpha_i^*)^{2N} \right)^{1/N^2} .$$

By the definition of the weighted transfinite diameter, $D_{n_N,N}^{(1)} \le d_{(n_N,N)}(H_\ell, W_\ell)$. Also put

$$\nu^{(N)} \ = \ \sum_{i=1}^{N} \frac{1}{N} \delta_{\alpha_i^*}(z)$$

and put $\nu_\ell^{(N)} = \nu^{(N)}|_{H_\ell}$, $\dot{\nu}_\ell^{(N)} = \nu^{(N)}|_{H \backslash H_\ell}$; then

$$-\log(D_{n,N}^{(1)}) \ = \ \widehat{I}_{\mathfrak{X},\vec{s}}(\nu_\ell^{(N)}) + 2I_{\mathfrak{X},\vec{s}}(\nu_\ell^{(N)}, \dot{\nu}_\ell^{(N)}) \ .$$

It follows easily by weak convergence that

$$\lim_{N \to \infty} I_{\mathfrak{X},\vec{s}}(\nu_\ell^{(N)}, \dot{\nu}_\ell^{(N)}) \ = \ I_{\mathfrak{X},\vec{s}}(\mu_\ell, \dot{\mu}_\ell) \ . \tag{B.86}$$

We will show below that

$$\lim_{N \to \infty} \widehat{I}_{\mathfrak{X},\vec{s}}(\nu_\ell^{(N)}) \ = \ I_{\mathfrak{X},\vec{s}}(\mu_\ell) \ . \tag{B.87}$$

Granting (B.87), by Proposition B.4 we then have

$$\begin{aligned} I_{\mathfrak{X},\vec{s}}(\mu_\ell) &+ 2I_{\mathfrak{X},\vec{s}}(\mu_\ell,\dot{\mu}_\ell) = V_{\sigma_\ell}(H_\ell,W_\ell) \le -\log(d_{\sigma_\ell}(H_\ell,W_\ell)) \\ &= \lim_{N\to\infty} -\log(d_{(n_N,N)}(H_\ell,W_\ell)) \le \limsup_{N\to\infty} -\log(D^{(1)}_{n_N,N}) \\ &= \limsup_{N\to\infty} \widehat{I}_{\mathfrak{X},\vec{s}}(\nu_\ell^{(N)}) + 2I_{\mathfrak{X},\vec{s}}(\nu_\ell^{(n_N)},\dot{\nu}_\ell^{(N)}) = I_{\mathfrak{X},\vec{s}}(\mu_\ell) + 2I_{\mathfrak{X},\vec{s}}(\mu_\ell,\dot{\mu}_\ell) , \end{aligned}$$

so equalities hold throughout, yielding the theorem.

To prove (B.87), note that when $W(z) \equiv 1$, by the definition of $d_{(N,N)}(H,W)$ we have that for each N,

$$-\log(d_{(N,N)}(H,W)) = \widehat{I}_{\mathfrak{X},\vec{s}}(\nu_\ell^{(N)}) + 2I_{\mathfrak{X},\vec{s}}(\nu_\ell^{(N)},\dot{\nu}_\ell^{(N)}) + \widehat{I}_{\mathfrak{X},\vec{s}}(\dot{\nu}_\ell^{(N)}) .$$

Passing to the limit as $N \to \infty$ and using Theorem B.7 and Corollary B.8, we have

$$\begin{aligned} I_{\mathfrak{X},\vec{s}}(\mu_\ell) &+ 2I_{\mathfrak{X},\vec{s}}(\mu_\ell,\dot{\mu}_\ell) + I_{\mathfrak{X},\vec{s}}(\dot{\mu}_\ell) = I_{\mathfrak{X},\vec{s}}(\mu) = V_{\mathfrak{X},\vec{s}}(H) \\ &= \lim_{N\to\infty} -\log(d_{(N,N)}(H,W)) \\ &= \lim_{N\to\infty} \left(\widehat{I}_{\mathfrak{X},\vec{s}}(\nu_\ell^{(N)}) + 2I_{\mathfrak{X},\vec{s}}(\nu_\ell^{(N)},\dot{\nu}_\ell^{(N)}) + \widehat{I}_{\mathfrak{X},\vec{s}}(\nu_\ell^{(N)})\right) \\ &= 2I_{\mathfrak{X},\vec{s}}(\mu_\ell,\dot{\mu}_\ell) + \lim_{N\to\infty} \left(\widehat{I}_{\mathfrak{X},\vec{s}}(\nu_\ell^{(N)}) + \widehat{I}_{\mathfrak{X},\vec{s}}(\nu_\ell^{(N)})\right) \end{aligned}$$

and hence

$$\text{(B.88)} \qquad I_{\mathfrak{X},\vec{s}}(\mu_\ell) + I_{\mathfrak{X},\vec{s}}(\dot{\mu}_\ell) = \lim_{N\to\infty} \left(\widehat{I}_{\mathfrak{X},\vec{s}}(\nu_\ell^{(N)}) + \widehat{I}_{\mathfrak{X},\vec{s}}(\dot{\nu}_\ell^{(N)})\right) .$$

On the other hand, consider the "smearing out" $\widetilde{\nu}_\ell^{(N)}$ of $\nu_\ell^{(N)}$ as in Proposition B.4. The same argument which gave (B.74) gives

$$I_{\mathfrak{X},\vec{s}}(\widetilde{\nu}_\ell^{(N)}) \le \widehat{I}_{\mathfrak{X},\vec{s}}(\nu_\ell^{(N)}) + O\left(\frac{\log(N)}{N}\right) .$$

Since the $\nu_\ell^{(N)}$ convege weakly to μ_ℓ, so do the $\widetilde{\nu}_\ell^{(N)}$, and therefore

$$\text{(B.89)} \qquad I_{\mathfrak{X},\vec{s}}(\mu_\ell) = \lim_{N\to\infty} I_{\mathfrak{X},\vec{s}}(\widetilde{\nu}_\ell^{(N)}) \le \liminf_{N\to\infty} \widehat{I}_{\mathfrak{X},\vec{s}}(\nu_\ell^{(N)}) .$$

Symmetrically,

$$\text{(B.90)} \qquad I_{\mathfrak{X},\vec{s}}(\dot{\mu}_\ell) \le \liminf_{N\to\infty} \widehat{I}_{\mathfrak{X},\vec{s}}(\dot{\nu}_\ell^{(N)}) .$$

Combining (B.88), (B.89) and (B.90) gives (B.87). □

Just as in the unweighted case, it follows that the Fekete measures for the weighted transfinite diameter associated to H_ℓ and $W_\ell(z)$ converge weakly to the equilibrium distribution. More precisely, consider a sequence of pairs (n_k, N_k) with $n_k/N_k \to \sigma_\ell$ and $N_k \to \infty$. For a given k, let $\alpha_1, \ldots, \alpha_{n_k}$ achieve the maximum in the definition of $d_{(n_k,N_k)}(H_\ell, W_\ell)$, and let

$$\nu_k = \sum_{i=1}^{n_k} \frac{1}{N_k} \delta_{\alpha_i}(z)$$

be the associated Fekete measure of mass n_k/N_k on H_ℓ.

COROLLARY B.11. *Let H, H_ℓ, $\sigma_\ell = \mu_{\mathfrak{X},\vec{s}}(H_\ell)$, and $W_\ell(z)$ be as in Theorem* B.9. *Then for any sequence of pairs (n_k, N_k) with $n_k/N_k \to \sigma_\ell$ and $N_k \to \infty$, the corresponding sequence of Fekete measures $\{\nu_k\}$ for H_ℓ relative to $W_\ell(z)$ converges weakly to $\mu_{\sigma_\ell,H_\ell,W_\ell} = \mu_{\mathfrak{X},\vec{s}}|_{H_\ell}$.*

PROOF. The proof is similar to that of Corollary B.8, using Theorems B.9 and B.10. □

We now determine the weighted Chebyshev constant, in the situation of Theorem B.9.

THEOREM B.12. *Let H, H_ℓ, $\sigma_\ell = \mu_{\mathfrak{X},\vec{s}}(H_\ell)$, and $W_\ell(z)$ be as in Theorem* B.9. *Then*

$$\mathrm{CH}^*_{\sigma_\ell}(H_\ell, W_\ell) \;=\; \gamma_{\mathfrak{X},\vec{s}}(H)\ . \tag{B.91}$$

PROOF. Since the Fekete measures converge weakly to the weighted equilibrium distribution μ_ℓ, Proposition B.6 gives

$$-\log(d_{\sigma_\ell}(H_\ell, W_\ell)) \;\le\; \sigma_\ell \cdot (-\log(\mathrm{CH}^*_{\sigma_\ell}(H_\ell, W_\ell))) + \int_{H_\ell} \widehat{u}_\ell(z)\, d\mu_\ell(z)\ .$$

Comparing this with (B.85) gives

$$V_{\mathfrak{X},\vec{s}}(H) \;\le\; -\log(\mathrm{CH}^*_{\sigma_\ell}(H_\ell, W_\ell))\ . \tag{B.92}$$

We will now show that equality holds in (B.92). If not, then there would be an $\varepsilon > 0$ such that

$$V_{\mathfrak{X},\vec{s}}(H) + \varepsilon \;<\; -\log(\mathrm{CH}^*_{\sigma_\ell}(H_\ell, W_\ell))\ . \tag{B.93}$$

Take a sequence of pairs (n_k, N_k) with $n_k/N_k \to \sigma_\ell$ and $N_k \to \infty$. Given k, let $\widetilde{P}_k(z) := \widetilde{P}_{(n_k,N_k)}(z, W_\ell)$ be the corresponding Chebyshev pseudopolynomial for H_ℓ relative to $W_\ell(z)$; put $\mathrm{CH}^*_k = \mathrm{CH}^*_{(n_k,N_k)} = (\|\widetilde{P}_k\|_{H_\ell})^{1/N_k}$. Also, let ω_k be the usual discrete measure of mass n_k/N_k supported on the roots of $\widetilde{P}_k(z)$.

Our assumption (B.93) implies that on H_ℓ, for each sufficiently large k,

$$\begin{aligned} V_{\mathfrak{X},\vec{s}}(H) + 3\varepsilon/4 \;&<\; -\log(\mathrm{CH}^*_k) \\ &\le\; -\frac{1}{N_k}\log(\widetilde{P}_k(z)) \;=\; u_{\omega_k}(z) + \widehat{u}_\ell(z)\ . \end{aligned}$$

Let

$$\widetilde{\omega}_k \;=\; \frac{\sigma_\ell}{n_k/N_k}\omega_k\ ,$$

renormalizing ω_k to have mass σ_ℓ. Then for all sufficiently large k, we have

$$V_{\mathfrak{X},\vec{s}}(H) + \varepsilon/2 \;\le\; u_{\widetilde{\omega}_k}(z) + \widehat{u}_\ell(z) \tag{B.94}$$

on H_ℓ. But $\widehat{u}_\ell(z) = u_{\dot{\mu}_\ell}(z)$ and by Theorem A.2,

$$u_{\mu_\ell}(z) + u_{\dot{\mu}_\ell}(z) \;=\; u_{\mathfrak{X},\vec{s}}(z) \;\le\; V_{\mathfrak{X},\vec{s}}(H) \tag{B.95}$$

on H. By (B.94) and (B.95),

$$u_{\mu_\ell}(z) + \varepsilon/2 \;\le\; u_{\widetilde{\omega}_k}(z) \tag{B.96}$$

on H_ℓ. However, $u_{\widetilde{\omega}_k}(z) - u_{\mu_\ell}(z)$ extends to a function harmonic on $\mathcal{C}_v(\mathbb{C})\backslash H_\ell$. By the Maximum Principle, it follows that (B.96) holds throughout $\mathcal{C}_v(\mathbb{C})$.

We can obtain a contradiction from this as follows. Put $\nu_k = \widetilde{\omega}_k + \dot{\mu}_\ell$; then ν_k and $\mu = \mu_\ell + \dot{\mu}_\ell$ are probability measures supported on H. By (B.95), $u_{\nu_k}(z) =$

$u_{\widetilde{\omega}_k}(z) + u_{\dot{\mu}_\ell}(z) \ge u_\mu(z) + \varepsilon/2$ on H. Since $\int_H u_\mu(z)\,d\mu(z) = I_{\mathfrak{X},\vec{s}}(\mu) = V_{\mathfrak{X},\vec{s}}(H)$, it follows from the Fubini-Tonelli theorem and the fact that $u_\mu(z) \le V_{\mathfrak{X},\vec{s}}(H)$ for all $z \in H$ (Theorem A.2), that

$$\begin{aligned} V_{\mathfrak{X},\vec{s}}(H) + \varepsilon/2 &\le \int_H u_{\nu_k}\,d\mu(z) \\ &= \int_H u_\mu(z)\,d\nu_k(z) \ \le\ V_{\mathfrak{X},\vec{s}}(H)\ . \end{aligned}$$

This contradiction shows that $V_{\mathfrak{X},\vec{s}}(H) = -\log(\mathrm{CH}^*_{\sigma_\ell}(H_\ell, W_\ell))$, which is equivalent to the assertion in the theorem. □

Finally, we show that under appropriate hypotheses, the discrete measures attached to Chebyshev pseudopolynomials converge weakly to the equilibrium distribution. As before, take a sequence of pairs (n_k, N_k) with $n_k/N_k \to \sigma_\ell$ and $N_k \to \infty$. Let the Chebyshev measure ω_k be the discrete measure of mass n_k/N_k supported equally on the roots of $\widetilde{P}_{(n_k,N_k)}(z)$ for H_ℓ relative to $W_\ell(z)$.

THEOREM B.13. *Let H, H_ℓ, $\sigma_\ell = \mu_{\mathfrak{X},\vec{s}}(H_\ell)$, and $W_\ell(z)$ be as in Theorem* B.9. *Assume also that $\mathcal{C}_v(\mathbb{C})\backslash H_\ell$ is connected, and that H_ℓ has empty interior. Then for any sequence of pairs (n_k, N_k) with $n_k/N_k \to \sigma_\ell$ and $N_k \to \infty$, the corresponding sequence of Chebyshev measures $\{\omega_k\}$ for H_ℓ relative to $W_\ell(z)$ converges weakly to the equilibrium distribution $\mu_{\sigma_\ell,H_\ell,W_\ell} = \mu_{\mathfrak{X},\vec{s}}|_{H_\ell}$.*

PROOF. Recall that $\mu_\ell = \mu_{\mathfrak{X},\vec{s}}|_{H_\ell}$, $\dot{\mu}_\ell = \mu_{\mathfrak{X},\vec{s}}|_{H_2}$. Let the numbers CH^*_k, the Chebyshev pseudopolynomials $\widetilde{P}_k(z)$, and the measures $\widetilde{\omega}_k$, $\nu_k = \widetilde{\omega}_k + \dot{\mu}_\ell$ be as in the proof of Theorem B.12. Let ω be a weak limit of the ω_k; after passing to a subsequence, if necessary, we can assume that the full sequence converges weakly to it. Clearly ω is also the weak limit of the $\widetilde{\omega}_k$.

Fix $\varepsilon > 0$. Since $\mathrm{CH}^*_{\sigma_\ell}(H_\ell, W_\ell) = \mathrm{CH}^*_{\mathfrak{X},\vec{s}}(H)$ by Theorem B.12, an argument similar to the one in the proof of Theorem B.12 shows that for sufficiently large k,

$$V_{\mathfrak{X},\vec{s}}(H) - \varepsilon \ <\ u_{\nu_k}(z) \ =\ u_{\widetilde{\omega}_k}(z) + u_{\dot{\mu}_\ell}(z)$$

on H_ℓ. On the other hand,

$$u_{\mu_\ell}(z) + u_{\dot{\mu}_\ell}(z) \ =\ u_{\mathfrak{X},\vec{s}}(z) \ \le\ V_{\mathfrak{X},\vec{s}}(H)\ .$$

Subtracting, we see that

$$u_{\widetilde{\omega}_k}(z) \ \ge\ u_{\mu_\ell}(z) - \varepsilon \tag{B.97}$$

on H_ℓ. Since $u_{\widetilde{\omega}_k}(z) - u_{\mu_\ell}(z)$ extends to a function harmonic in all of $\mathcal{C}_v(\mathbb{C})\backslash H_\ell$, the Maximum Principle for harmonic functions shows (B.97) holds in all of $\mathcal{C}_v(\mathbb{C})$.

As the $u_{\widetilde{\omega}_k}(z)$ converge uniformly to $u_\omega(z)$ on compact subsets of $\mathcal{C}_v(\mathbb{C})\backslash H_\ell$ and $\varepsilon > 0$ is arbitrary, it follows from (B.97) that

$$u_\omega(z) \ \ge\ u_{\mu_\ell}(z) \tag{B.98}$$

for all $z \in \mathcal{C}_v(\mathbb{C})\backslash H_\ell$.

Suppose equality in (B.98) failed to hold for some $z_0 \in \mathcal{C}_v(\mathbb{C})\backslash H_\ell$. Since $\mathcal{C}_v(\mathbb{C})\backslash H_\ell$ is connected, the Maximum Principle implies that $u_\omega(z) > u_{\mu_\ell}(z)$ for all $z \in \mathcal{C}_v(\mathbb{C})\backslash H_\ell$. In particular, there would be a $\delta > 0$ such that

$$u_\omega(z) \ \ge\ u_{\mu_\ell}(z) + \delta$$

on H_2. Now put $\nu = \omega + \dot{\mu}_\ell$; then

$$\begin{aligned} u_\nu(z) &= u_\omega(z) + u_{\dot{\mu}_\ell}(z) \\ &\geq u_{\mu_\ell}(z) + u_{\dot{\mu}_\ell}(z) \;=\; u_{\mathfrak{X},\vec{s}}(z) \end{aligned}$$

on $\mathcal{C}_v(\mathbb{C})\backslash H_\ell$, with $u_\nu \geq u_{\mathfrak{X},\vec{s}}(z) + \delta$ on H_2.

Let $e = e_{\mathfrak{X},\vec{s}}$ be the exceptional subset of H inner capacity 0 where $u_{\mathfrak{X},\vec{s}}(z) < V_{\mathfrak{X},\vec{s}}(H)$, given by Theorem A.2. We claim that $u_\nu(z) \geq V_{\mathfrak{X},\vec{s}}(H)$ on $H\backslash e$. To see this, for each $\eta > 0$ let

$$U_\eta \;=\; \{z \in \mathcal{C}_v(\mathbb{C}) : u_{\mathfrak{X},\vec{s}}(z) > V_{\mathfrak{X},\vec{s}}(H) - \eta\} \ .$$

This is an open set, since $u_{\mathfrak{X},\vec{s}}(z)$ is lower semi-continuous.

Furthermore, at each point of $H\backslash e$, $u_{\mathfrak{X},\vec{s}}(z)$ is continuous and equal to $V_{\mathfrak{X},\vec{s}}(H)$ (Theorem A.2) so apart from a subset of inner capacity 0, ∂U_η is contained in $\mathcal{C}_v(\mathbb{C})\backslash H$. In particular on $\partial U_\eta\backslash e$ we have

$$u_\nu(z) \;\geq\; u_{\mathfrak{X},\vec{s}}(z) \;=\; V_{\mathfrak{X},\vec{s}}(H) - \eta \ .$$

Since $u_\nu(z)$ is superharmonic and bounded from below on U_η, the strong form of the Maximum Principle (see [**51**], Proposition 3.1.1) shows that $u_\nu(z) \geq V_{\mathfrak{X},\vec{s}}(H) - \eta$ on U_η. Since $\eta > 0$ is arbitrary, and $H\backslash e \subset U_\eta$, it follows that $u_\nu(z) \geq V_{\mathfrak{X},\vec{s}}(H)$ on $H\backslash e$.

Because $u_{\mathfrak{X},\vec{s}}(z) = V_{\mathfrak{X},\vec{s}}(H)$ on $(H\backslash H_\ell)\backslash e$, we have $u_\nu(z) \geq V_{\mathfrak{X},\vec{s}}(H) + \delta$ on $(H\backslash H_\ell)\backslash e$. However, a set of inner capacity 0 necessarily has μ-measure 0 ([**51**], Lemma 3.1.4). By Fubini-Tonelli,

$$\begin{aligned} V_{\mathfrak{X},\vec{s}}(H) + \mu_{\mathfrak{X},\vec{s}}(H\backslash H_\ell)\cdot\delta \;&\leq\; \int_H u_\nu(z)\,d\mu_{\mathfrak{X},\vec{s}}(z) \\ &=\; \int_H u_{\mathfrak{X},\vec{s}}(z)\,d\nu(z) \;\leq\; V_{\mathfrak{X},\vec{s}}(H) \ . \end{aligned} \tag{B.99}$$

We claim that $\mu_{\mathfrak{X},\vec{s}}(H\backslash H_\ell) > 0$. Otherwise $u_{\mathfrak{X},\vec{s}}$ would be supported on H_ℓ, and the fact that $\mathcal{C}_v(\mathbb{C})\backslash H_\ell$ is connected and contains $\mathfrak{X}$ would mean that $u_{\mathfrak{X},\vec{s}}(z) < V_{\mathfrak{X},\vec{s}}(H)$ on $\mathcal{C}_v(\mathbb{C})\backslash H_\ell$. However, $u_{\mathfrak{X},\vec{s}}(z) = V_{\mathfrak{X},\vec{s}}(H)$ for all $z \in H$ except possibly a set of inner capacity 0, which contradicts that $H\backslash H_\ell \subset \mathcal{C}_v(\mathbb{C})\backslash H_\ell$ has positive capacity. Thus (B.99) is impossible.

We conclude that in (B.98), we have $u_\nu(z) = u_{\mathfrak{X},\vec{s}}(z)$ for all $z \notin H_\ell$. Moreover, we have shown that $u_\nu(z) \geq V_{\mathfrak{X},\vec{s}}(H)$ for all $z \in H\backslash e$. We now claim that $u_\nu(z) \leq V_{\mathfrak{X},\vec{s}}(H)$ for all z. Suppose to the contrary that $u_\nu(z_0) > V_{\mathfrak{X},\vec{s}}(H)$ for some z_0. Then since $u_\nu(z)$ is lower semi-continuous,

$$U \;:=\; \{z \in \mathcal{C}_v(\mathbb{C}) : u_\nu(z) > V_{\mathfrak{X},\vec{s}}(H)\}$$

would be a nonempty open set. Since H_ℓ has no interior, U contains points of $\mathcal{C}_v(\mathbb{C})\backslash H_\ell$. However, at these points $u_\nu(z) = u_{\mathfrak{X},\vec{s}}(z) \leq V_{\mathfrak{X},\vec{s}}(H)$, contradicting the definition of U.

It follows that $u_\nu(z)$ coincides with $u_{\mathfrak{X},\vec{s}}(z)$ except possibly on the exceptional set e of inner capacity 0. However, a superharmonic function which is bounded below is determined by its values on the complement any set of inner capacity 0 ([**65**], Theorem III.28, p.78). Thus $u_\nu(z) = u_{\mathfrak{X},\vec{s}}(z)$ for all z. By the Riesz Decomposition theorem ([**51**], Theorem 3.1.11) we can recover the measure from the potential function, so $\nu = \mu_{\mathfrak{X},\vec{s}}$, and hence $\omega = \mu_\ell$. □

Remark. Some hypotheses on H_ℓ are necessary in Theorem B.13; it is not always true that the Chebyshev measures converge to the equilibrium distribution. For example, in the classical case consider the unit disc $H_\ell = D(0,1) \subset \mathbb{C}$, with weight $W(z) \equiv 1$, relative to $\mathfrak{X} = \{\infty\}$. The Chebyshev polynomial of degree n for H_ℓ is z^n, with roots only at the origin. However, the equilibrium distribution of H_ℓ is the uniform measure supported on $\partial H_\ell = C(0,1)$.

6. Chebyshev Pseudopolynomials for Short Intervals

In this section we will show that when H is a "sufficiently short" interval, the weighted $(\mathfrak{X}, \vec{s})$-Chebyshev pseudopolynomials for H have oscillation properties like those of classical Chebyshev polynomials. The notion of "shortness" depends on the location of H relative to $\mathfrak{X}$ and the existence of a suitable system of coordinates, but is independent of the choice of the weight function.

The motivating case is case when $K_v \cong \mathbb{R}$ and $H \subset \mathcal{C}_v(\mathbb{R}) \backslash \mathfrak{X}$ is a closed interval. However, since any analytic arc becomes an interval in suitable coordinates, the results apply more generally.

Fix a local coordinate patch $U \subset \mathcal{C}_v(\mathbb{C})$, with coordinate function z say. Thus, z gives a holomorphic isomorphism between U and a simply connected open set $z(U) \subset \mathbb{C}$. We can decompose the coordinate function into its real and imaginary parts, $z = u + iv$, and speak of the real and imaginary coordinates of points in U. For us, the case of interest is when $z(U) \cap \mathbb{R}$ is nonempty; to simplify notation, we will assume that is the situation, and that U and z have been chosen so that $v = 0$ on $z^{-1}(z(U) \cap \mathbb{R})$.

By a real interval $H \subset U$, we mean a set of the form $H = z^{-1}([a,b])$ where $[a,b] \subset z(U) \cap \mathbb{R}$. By abuse of notation, we will simply write $H = [a,b]$. Similarly, we can speak of a disc $D(t,r) \subset U$. Using the coordinate function to identify U with $z(U) \subset \mathbb{C}$, we can speak of translating a point $p \in U$ by a number $c \in \mathbb{C}$: $p \mapsto p + c$, provided that both points involved belong to U. (Formally, if $p \in U$, then $p + c$ means the point $z^{-1}(z(p) + c)$, if $z(p) + c \in z(U)$.)

Recall that $\mathfrak{X} = \{x_1, \ldots, x_m\}$. By Proposition 3.11, relative to the given local coordinate, for each $x_j \in \mathfrak{X}$ there is a $\mathcal{C}^\infty$ function $\eta_j(z,w)$ on $(U \backslash \{x_j\}) \times (U \backslash \{x_j\})$ (which is harmonic in each variable separately) such that for all $z, w \in U \backslash \{x_j\}$

$$-\log([z,w]_{x_j}) \;=\; -\log(|z-w|) + \eta_j(z,w) \ .$$

Writing $z = u_1 + iv_1$, $w = u_2 + iv_2$, we can speak of the partial derivatives of $\eta_j(z,w)$ relative to u_1, v_1, u_2 and v_2.

LEMMA B.14. *Let $U \subset \mathcal{C}_v(\mathbb{C})$ be a local coordinate patch, and let $H = [a,b] \subset U$ be a real interval disjoint from $\mathfrak{X}$.*

Let C be a bound such that, uniformly for all $x_j \in \mathfrak{X}$ and all $z, w \in H$,

$$\left| \frac{\partial \eta_j}{\partial u_2}(z,w) \right| \;\le\; C \ , \qquad \left| \frac{\partial^2 \eta_j}{\partial u_2^2}(z,w) \right| \le C \ . \tag{B.100}$$

Then for each $x_j \in \mathfrak{X}$, all $z, c, d \in H$, and each $0 < \varepsilon \in \mathbb{R}$ such that $c \pm \varepsilon, d \pm \varepsilon$ belong to H,

(1) $|\eta_j(z, c-\varepsilon) - \eta_j(z,c) + \eta_j(z, d+\varepsilon) - \eta_j(z,d)| \;<\; C|d-c|\varepsilon + 2C\varepsilon^2$ *;*
(2) $|\eta_j(z, c-\varepsilon) + \eta_j(z, c+\varepsilon) - 2\eta(z,c)| \;<\; 2C\varepsilon^2$ *;*
(3) $|\eta_j(z, c+\varepsilon) - \eta_j(z,c)| \;<\; C\varepsilon$ *;*
(4) $|\eta_j(z, d-\varepsilon) - \eta_j(z,d)| \;<\; C\varepsilon$ *.*

PROOF. All of these follow from the Mean Value Theorem. For example, we prove (1): for appropriate $c^* \in (c-\varepsilon, c)$, $d^* \in (d, d+\varepsilon)$ and $e^* \in (c^*, d^*)$,

$$\begin{aligned}
&|\eta(z, c-\varepsilon) - \eta(z,c) + \eta(z, d+\varepsilon) - \eta(z,d)| \\
&= \left|-\varepsilon(\frac{\partial \eta}{\partial u_2}(z, c^*)) + \varepsilon(\frac{\partial \eta}{\partial u_2}(z, d^*))\right| = \left|\varepsilon \cdot (d^* - c^*)(\frac{\partial^2 \eta}{\partial u_2^2}(z, e^*))\right| \\
&< \varepsilon(|d-c| + 2\varepsilon) \cdot C \ .
\end{aligned}$$

Clearly (2) is a special case of (1), and (3) and (4) are easy. □

Given a probability vector $\vec{s} = (s_1, \ldots, s_m)$, on $U \backslash \mathfrak{X}$ we have

$$-\log([z,w]_{\mathfrak{X},\vec{s}}) = -\log(|z-w|) + \eta(z,w) \ .$$

where $\eta(z,w) = \eta_{\mathfrak{X},\vec{s}}(z,w) = \sum_{j=1}^m s_j \eta_j(z,w)$. By the triangle inequality and the fact that $\sum s_j = 1$, the bounds in Lemma B.14 hold with $\eta_j(z)$ replaced by $\eta(z)$.

DEFINITION B.15. Let $U \subset \mathcal{C}_v(\mathbb{C})$ be a coordinate patch, with coordinate function z. A real interval $H = [a,b] \subset U$ is *short* (relative to $\mathfrak{X}$ and the coordinate function z on U) if it is disjoint from $\mathfrak{X}$ and $|b-a| < \min(1/C(H,\mathfrak{X}), 1/\sqrt{2C(H,\mathfrak{X})})$, where

$$C(H,\mathfrak{X}) = \max_{x_j \in \mathfrak{X}} \max_{z,w \in H} \left(\left|\frac{\partial \eta_j}{\partial u_2}(z,w)\right|, \left|\frac{\partial^2 \eta_j}{\partial u_2^2}(z,w)\right| \right) \ . \tag{B.101}$$

While this notion of "shortness" is ugly, it is easy to apply. In practice, we will be given a coordinate patch U whose closure $\overline{U}$ is disjoint from $\mathfrak{X}$. Defining $C(U)$ as in (B.101) with H replaced by $\overline{U}$, one sees that any real interval $[a,b] \subset U$ with $|b-a| < \max(1/C(U), 1/\sqrt{2C(U)})$ is "short".

We will now show that restricted, weighted $(\mathfrak{X},\vec{s})$-Chebyshev pseudopolynomials for short real intervals behave like classical Chebyshev polynomials.

PROPOSITION B.16. *Let $U \subset \mathcal{C}_v(\mathbb{C})$ be a coordinate patch with coordinate function z. Suppose $H = [a,b] \subset U \backslash \mathfrak{X}$ is a short real interval.*

Fix $\vec{s} \in \mathcal{P}^m$, and let $W(z)$ be weight function which is continuous, positive and bounded on a neighborhood of H. Then for any pair (n,N) with $n \geq 1$, each Chebyshev pseudopolynomial $\widetilde{P}_{(n,N)}(z,W) = \prod_{i=1}^n [z,\alpha_i]_{\mathfrak{X},\vec{s}} \cdot W(z)^N$ for H relative to $W(z)$ has the following properties. Assume the roots α_i are labeled in increasing order.

(1) *$\widetilde{P}_{(n,N)}(z,W)$ has distinct roots, and they lie in the interior of $[a,b]$.*

(2) *If α_1 is the leftmost root of $\widetilde{P}_{(n,N)}(z,W)$, there is a point $z_0 \in [a,\alpha_1]$ where*

$$\widetilde{P}_{(n,N)}(z_0,W) = \|\widetilde{P}_{(n,N)}\|_H \ .$$

(3) *For each pair of consecutive roots α_i, α_{i+1} there is a point $z_i \in (\alpha_i, \alpha_{i+1})$ where*

$$\widetilde{P}_{(n,N)}(z_i,W) = \|\widetilde{P}_{(n,N)}\|_H \ .$$

(4) *If α_n is the rightmost root, there is a point $z_n \in (\alpha_n, b]$ where*

$$\widetilde{P}_{(n,N)}(z_n,W) = \|\widetilde{P}_{(n,N)}\|_H \ .$$

PROOF. Write $C = C(H, \mathfrak{X})$, where $C(H, \mathfrak{X})$ is as in Definition B.15. The hypothesis that H is "short" implies that

$$\frac{1}{b-a} > C \ , \qquad \frac{1}{(b-a)^2} > 2C \ . \tag{B.102}$$

We will now show that in the presence of the bounds (B.102), the classical arguments concerning oscillation properties of Chebyshev polynomials carry over. The weight function plays a negligible role; it cancels out in all the ratios below.

Fix (n, N). We already know that $\widetilde{P}_{(n,N)}(z, W)$ exists; the problem is to show it has the properties above. Write $\widetilde{P}(z) = \widetilde{P}_{(n,N)}(z, W)$, $M = \|\widetilde{P}(z)\|_H$.

First, suppose $\widetilde{P}(z)$ had a root at $z = a$. Then $\widetilde{P}(a) = 0$ and so there would be an interval $[a, a+\delta]$ on which $\widetilde{P}(z) < M$. We claim that by replacing the factor $[z, a]_{\mathfrak{X},\vec{s}}$ in $\widetilde{P}(z)$ by $[z, a+\varepsilon]_{\mathfrak{X},\vec{s}}$ for an appropriately small ε, we could reduce $\|\widetilde{P}\|_H$. Let $P_\varepsilon(z)$ be the pseudopolynomial thus obtained. By the continuity of $[z, w]_{\mathfrak{X},\vec{s}}$, for sufficiently small $\varepsilon > 0$ we would still have $P_\varepsilon(z) < M$ for $z \in [a, a+\delta]$. For $z \in (a+\delta, b]$,

$$\frac{P_\varepsilon(z)}{\widetilde{P}(z)} = \frac{[z, a+\varepsilon]_{\mathfrak{X},\vec{s}}}{[z, a]_{\mathfrak{X},\vec{s}}}$$

and

$$-\log\left(\frac{[z, a+\varepsilon]_{\mathfrak{X},\vec{s}}}{[z, a]_{\mathfrak{X},\vec{s}}}\right) = -\log\left(\frac{|z-a-\varepsilon|}{|z-a|}\right) + \eta(z, a+\varepsilon) - \eta(z, a) \ .$$

By Lemma B.14, for all $z \in H$, $|\eta(z, a+\varepsilon) - \eta(z, a)| < C\varepsilon$. On the other hand, for $z \in (a+\delta, b]$ and $0 < \varepsilon < \delta$,

$$-\log\left(\frac{|z-a-\varepsilon|}{|z-a|}\right) = -\log\left(|1 - \frac{\varepsilon}{z-a}|\right) > \frac{\varepsilon}{|z-a|} \ .$$

Since $[a, b]$ is short enough that $1/(b-a) > C$, also $1/(z-a) > C$ and so

$$-\log\left(\frac{[z, a+\varepsilon]_{\mathfrak{X},\vec{s}}}{[z, a]_{\mathfrak{X},\vec{s}}}\right) > 0 \, .$$

Thus $[z, a+\varepsilon]_{\mathfrak{X},\vec{s}} < [z, a]_{\mathfrak{X},\vec{s}}$ for all $z \in [a+\delta, b]$. It follows that $\|P_\varepsilon(z)\|_H < \|\widetilde{P}\|_H$, contradicting the minimality of $\|\widetilde{P}\|_H$. A similar argument shows $\widetilde{P}(z)$ cannot have a root at $z = b$.

Let α_1 be the leftmost root of $\widetilde{P}(z)$. If $\widetilde{P}(z)$ did not achieve its maximum in $[a, \alpha_1]$, an argument like the one above shows we could reduce $\|\widetilde{P}\|_H$ by moving α_1 slightly to the right. For similar reasons, if α_n is the rightmost root and $\widetilde{P}(z)$ did not achieve its maximum in $[\alpha_n, b]$, we could reduce $\|\widetilde{P}\|_H$ by moving α_n to the left.

Next, suppose $\widetilde{P}(z)$ had a double root at α_i, say. As shown above, $\alpha_i \in (a, b)$. Since $\widetilde{P}(\alpha_i) = 0$, there is a $\delta > 0$ such that $\widetilde{P}(z) < M$ for $z \in [\alpha_i - \delta, \alpha_i + \delta]$. If we define $P_\varepsilon(z)$ by replacing the factor $[z, \alpha_i]_{\mathfrak{X},\vec{s}}^2$ in $\widetilde{P}(z)$ by $[z, \alpha_i - \varepsilon]_{\mathfrak{X},\vec{s}}[z, \alpha_i + \varepsilon]_{\mathfrak{X},\vec{s}}$, then by the continuity of $[z, w]_{\mathfrak{X},\vec{s}}$, for sufficiently small ε we will have $P_\varepsilon(z) < M$ for $z \in [\alpha_i - \delta, \alpha_i + \delta]$.

For $z \in H \backslash [\alpha_i - \delta, \alpha_i + \delta]$,

$$\frac{P_\varepsilon(z)}{\widetilde{P}(z)} = \frac{[z, \alpha_i - \varepsilon]_{\mathfrak{X},\vec{s}}[z, \alpha_i + \varepsilon]_{\mathfrak{X},\vec{s}}}{[z, \alpha_i]_{\mathfrak{X},\vec{s}}^2}$$

and

$$
\begin{aligned}
&-\log\left(\frac{[z,\alpha_i-\varepsilon]_{\mathfrak{X},\vec{s}}[z,\alpha_i+\varepsilon]_{\mathfrak{X},\vec{s}}}{[z,\alpha_i]^2_{\mathfrak{X},\vec{s}}}\right)\\
&= \ -\log\left(\frac{(z-\alpha_i+\varepsilon)(z-\alpha_i-\varepsilon)}{(z-\alpha_i)^2}\right)+\eta(z,\alpha_i-\varepsilon)+\eta(z,\alpha_i+\varepsilon)-2\eta(z,\alpha_i)\ .
\end{aligned}
$$

We can assume $0<\varepsilon<\delta$, so since $|z-\alpha_i|>\delta$,

$$
-\log\left(\frac{(z-\alpha_i+\varepsilon)(z-\alpha_i-\varepsilon)}{(z-\alpha_i)^2}\right) \ = \ -\log\left(1-\frac{\varepsilon^2}{(z-\alpha_i)^2}\right) \ > \ \frac{\varepsilon^2}{(z-\alpha_i)^2}\ .
$$

On the other hand, by Lemma B.14, $|\eta(z,\alpha_i-\varepsilon)+\eta(z,\alpha_i+\varepsilon)-2\eta(z,\alpha_i)|<2C\varepsilon^2$. As $[a,b]$ is short enough that $1/(b-a)^2>2C$, then $\varepsilon^2/(z-\alpha_i)^2>2C\varepsilon^2$ for $z\in H\backslash[\alpha_i-\varepsilon,\alpha_i+\varepsilon]$, so

$$
-\log\left(\frac{[z,\alpha_i-\varepsilon]_{\mathfrak{X},\vec{s}}[z,\alpha_i+\varepsilon]_{\mathfrak{X},\vec{s}}}{[z,\alpha_i]^2_{\mathfrak{X},\vec{s}}}\right) \ > \ 0\ .
$$

Hence $[z,\alpha_i-\varepsilon]_{\mathfrak{X},\vec{s}}[z,\alpha_i+\varepsilon]_{\mathfrak{X},\vec{s}}<[z,\alpha_i]^2_{\mathfrak{X},\vec{s}}$ for all $z\notin[\alpha_i-\delta,\alpha_i+\delta]$, which implies that $\|P_\varepsilon(z)\|_H<\|\widetilde{P}\|_H$. This contradicts the minimality of $\|\widetilde{P}\|_H$.

Finally, suppose $\widetilde{P}(z)$ did not take on the value M between two consecutive roots α_i,α_{i+1}. There is a $\delta>0$ such that $\widetilde{P}(z)<M$ for all $z\in[\alpha_i-\delta,\alpha_{i+1}+\delta]$. Define $P_\varepsilon(z)$ by replacing the product $[z,\alpha_i]_{\mathfrak{X},\vec{s}}[z,\alpha_{i+1}]_{\mathfrak{X},\vec{s}}$ in $\widetilde{P}(z)$ with $[z,\alpha_i-\varepsilon]_{\mathfrak{X},\vec{s}}[z,\alpha_{i+1}+\varepsilon]_{\mathfrak{X},\vec{s}}$. By the continuity of $[z,w]_{\mathfrak{X},\vec{s}}$, for sufficiently small $\varepsilon>0$ we have $P_\varepsilon(z)<M$ for $z\in[\alpha_i-\delta,\alpha_{i+1}+\delta]$. For $z\neq\alpha_i,\alpha_{i+1}$ we have

$$
\frac{P_\varepsilon(z)}{\widetilde{P}(z)} \ = \ \frac{[z,\alpha_i-\varepsilon]_{\mathfrak{X},\vec{s}}[z,\alpha_{i+1}+\varepsilon]_{\mathfrak{X},\vec{s}}}{[z,\alpha_i]_{\mathfrak{X},\vec{s}}[z,\alpha_{i+1}]_{\mathfrak{X},\vec{s}}}\ .
$$

Furthermore, for $z\in[a,b]$ but $z\notin[\alpha_i-\delta,\alpha_{i+1}+\delta]$,

$$
\begin{aligned}
-\log\left(\frac{[z,\alpha_i-\varepsilon]_{\mathfrak{X},\vec{s}}[z,\alpha_{i+1}+\varepsilon]_{\mathfrak{X},\vec{s}}}{[z,\alpha_i]_{\mathfrak{X},\vec{s}}[z,\alpha_{i+1}]_{\mathfrak{X},\vec{s}}}\right) \ &= \ -\log\left(\frac{(z-\alpha_i+\varepsilon)(z-\alpha_{i+1}-\varepsilon}{(z-\alpha_i)(z-\alpha_{i+1})}\right)\\
&\quad +\ \eta(z,\alpha_i-\varepsilon)+\eta(z,\alpha_{i+1}+\varepsilon)-\eta(z,\alpha_i)-\eta(z,\alpha_{i+1})\ .
\end{aligned}
$$

Here $|\eta(z,\alpha_i-\varepsilon)+\eta(z,\alpha_{i+1}+\varepsilon)-\eta(z,\alpha_i)-\eta(z,\alpha_{i+1})|<C(\alpha_{i+1}-\alpha_i)\varepsilon+2C\varepsilon^2$ by Lemma B.14. On the other hand, for small enough ε and for $z\notin[\alpha_i-\delta,\alpha_{i+1}+\delta]$,

$$
\begin{aligned}
-\log\left(\frac{(z-\alpha_i+\varepsilon)(z-\alpha_{i+1}-\varepsilon)}{(z-\alpha_i)(z-\alpha_{i+1})}\right) \ &= \ -\log\left(1-\frac{\varepsilon(\alpha_{i+1}-\alpha_i)+\varepsilon^2}{(z-\alpha_i)(z-\alpha_{i+1})}\right)\\
&> \ \frac{\varepsilon(\alpha_{i+1}-\alpha_i)}{(z-\alpha_i)(z-\alpha_{i+1})}\ .
\end{aligned}
$$

As $1/(b-a)^2>2C$, then $\frac{\varepsilon(\alpha_{i+1}-\alpha_i)}{(z-\alpha_i)(z-\alpha_{i+1})}>2C(\alpha_{i+1}-\alpha_i)\varepsilon$. Thus for sufficiently small $\varepsilon>0$,

$$
-\log\left(\frac{[z,\alpha_i-\varepsilon]_{\mathfrak{X},\vec{s}}[z,\alpha_{i+1}+\varepsilon]_{\mathfrak{X},\vec{s}}}{[z,\alpha_i]_{\mathfrak{X},\vec{s}}[z,\alpha_{i+1}]_{\mathfrak{X},\vec{s}}}\right) \ > \ 0\ ,
$$

whence $[z,\alpha_i-\varepsilon]_{\mathfrak{X},\vec{s}}[z,\alpha_{i+1}+\varepsilon]_{\mathfrak{X},\vec{s}}<[z,\alpha_i]_{\mathfrak{X},\vec{s}}[z,\alpha_{i+1}]_{\mathfrak{X},\vec{s}}$ for all z as above. This shows that $\|P_\varepsilon(z)\|_H<\|\widetilde{P}\|_H$ for small enough ε, once more contradicting the minimality of $\|\widetilde{P}\|_H$. □

Remark. Because of the presence of the weight function, $\widetilde{P}_{(n,N)}(z,W)$ need not take on its maximum value at a and b (as holds classically). However, $\widetilde{P}_{(n,N)}(z,W)$ does "vary n times from M to 0 to M" on H, which is enough for the application in the next section.

7. Oscillating Pseudopolynomials

In this section we specialize to the case $K_v \cong \mathbb{R}$. Thus, $\mathcal{C}_v(\mathbb{R})$ has meaning, and there is an action of complex conjugation $z \mapsto \overline{z}$ on $\mathcal{C}_v(\mathbb{C})$.

We will assume that $\mathfrak{X}$ and $H \subset \mathcal{C}_v(\mathbb{C})\backslash\mathfrak{X}$ are stable under complex conjugation, that H is compact, and that H has finitely many connected components $H_1, \ldots, H_D$, where no H_ℓ is reduced to a point, and each H_ℓ is simply connected. Under these hypotheses $\mathcal{C}_v(\mathbb{C})\backslash H_\ell$ is connected for each ℓ, and $\mathcal{C}_v(\mathbb{C})\backslash H$ is connected. Since H is stable under complex conjugation, for each ℓ there is an index $\overline{\ell}$ for which $\overline{H_\ell} = H_{\overline{\ell}}$; possibly $H_\ell = H_{\overline{\ell}}$. We will say that a component H_ℓ is a "short interval" if it satisfies the following condition:

(B.103) $H_\ell = [a_\ell, b_\ell] \subset \mathcal{C}_v(\mathbb{R})\backslash\mathfrak{X}$ is short relative to $\mathfrak{X}$ and a suitable local coordinate function z_ℓ, in the sense of Definition B.15.

Recall that a probability vector $\vec{s} \in \mathcal{P}^m$ is K_v-symmetric if $s_j = s_k$ whenever $\overline{x_j} = x_k$. We will say that a vector $\vec{n} = (n, \ldots, n_D) \in \mathbb{N}^D$ is K_v-symmetric if $n_j = n_k$ whenever $\overline{H_j} = H_k$ (i.e., when $\overline{j} = k$).

Let $\vec{s} \in \mathcal{P}^m$ be a K_v-symmetric probability vector. Let $\vec{n} \in \mathbb{N}^D$ be K_v-symmetric, and put $N = \sum_{\ell=1}^{D} n_\ell$. For each such $\vec{n}$ we will construct an $(\mathfrak{X}, \vec{s})$-pseudopolynomial $P_{\vec{n}}(z) = P_{(\mathfrak{X},\vec{s}),\vec{n}}(z)$ whose roots belong to H, which satisfies $P_{\vec{n}}(z) = P_{\vec{n}}(\overline{z})$ for all z, which has large oscillations on the sets H_ℓ which are short intervals, and whose normalized logarithm $(-1/N)\log(P_{\vec{n}}(z))$ approximates $u_{\mathfrak{X},\vec{s}}(z,H)$ outside a neighborhood of H.

Most of the roots of $P_{\vec{n}}(z)$ will be roots of the weighted Chebyshev polynomials, or weighted Fekete points, for the sets H_ℓ. Some care is needed to assure that $P_{\vec{n}}(z) = P_{\vec{n}}(\overline{z})$.

Let $\mu_{\mathfrak{X},\vec{s}}$ be the equilibrium distribution of H relative to $\mathfrak{X}$ and $\vec{s}$. For each ℓ put

$$\widehat{u}_\ell(z) = \int_{H\backslash H_\ell} -\log([z,w]_{\mathfrak{X},\vec{s}})\, d\mu_{\mathfrak{X},\vec{s}}(w)$$

and let $W_\ell(z) = \exp(-\widehat{u}_\ell(z))$. Since $\mathfrak{X}$ and H are stable under complex conjugation, and $\vec{s}$ is K_v-symmetric, we have $\widehat{u}_{\overline{\ell}}(z) = \widehat{u}_\ell(\overline{z})$ and $W_{\overline{\ell}}(z) = W_\ell(\overline{z})$.

If H_ℓ is a short interval, let

$$\widetilde{P}_{\ell,(n_\ell,N)}(z) = \prod_{i=1}^{n_\ell} [z, \alpha_{\ell,i}]_{\mathfrak{X},\vec{s}} \cdot W_\ell(z)^N \tag{B.104}$$

be the weighted Chebyshev pseudopolynomial for H_ℓ with weight $W_\ell(z)$. We are interested in its roots $\alpha_{\ell,1}, \ldots, \alpha_{\ell,n_\ell}$, which belong to H_ℓ.

If H_ℓ is not a short interval and $H_\ell \neq H_{\overline{\ell}}$, let $\alpha_{\ell,1}, \ldots, \alpha_{\ell,n_\ell} \in H_\ell$ be a set of (n_ℓ, N)-Fekete points for H_ℓ relative to the weight $W_\ell(z)$, that is, a set of points achieving the maximum value

$$d_{(n_\ell,N)}(H_\ell, W_\ell) \;=\; \left(\prod_{\substack{i,j=1 \\ i \neq j}}^{n_\ell} [\alpha_{\ell,i}, \alpha_{\ell,j}]_{\mathfrak{X},\vec{s}} \cdot \prod_{i=1}^{n_\ell} W(z_i)^{2N} \right)^{1/N^2}$$

in (B.40). We take the Fekete points $\alpha_{\overline{\ell},i}$ for $H_{\overline{\ell}}$ to be the complex conjugates of the $\alpha_{\ell,i}$ for H_ℓ.

Finally, suppose H_ℓ is not a short interval, but $H_\ell = H_{\overline{\ell}}$. We first show[1] that H_ℓ contains a point β_ℓ fixed by complex conjugation, that is, a point in $\mathcal{C}_v(\mathbb{R})$. Let S be $\mathcal{C}_v(\mathbb{C})$, viewed as a topological surface, and let S_0 be the quotient of S under the action of complex conjugation. Write $\pi : S \to S_0$ for the quotient map. Let $\dot{S} = \mathcal{C}_v(\mathbb{C})\backslash(\mathcal{C}_v(\mathbb{R}) \cup \mathfrak{X})$ and put $\dot{S}_0 = \pi(\dot{S}) \subset S_0$. Then S_0 is a compact, connected (possibly nonorientable) surface with boundary, $\dot{S}_0$ consists of the interior of S_0 with a finite number of points removed, and $\dot{S}$ is a 2-to-1 unramified cover of $\dot{S}_0$. Suppose $H_\ell \cap \mathcal{C}_v(\mathbb{R}) = \phi$. Then $H_\ell \subset \dot{S}$. Choose a point $P \in H_\ell$, and let $\overline{P} \in H_\ell$ be its image under complex conjugation. By hypothesis $P \neq \overline{P}$. Since H_ℓ is simply connected, and in particular, path connected, there is a path α from P to $\overline{P}$ in H_ℓ. Let $\overline{\alpha}$ be the conjugate path; then the concatenation $\overline{\alpha} * \alpha$ is a loop in H_ℓ (by a loop, we mean a continuous image of the unit circle). Since H_ℓ is simply connected, $\overline{\alpha} * \alpha$ is homotopic in H_ℓ to a point. Put $\alpha_0 = \pi(\alpha) = \pi(\overline{\alpha}) \subset \dot{S}_0$. Then $\alpha_0 * \alpha_0$ is homotopic in $\dot{S}_0$ to a point. However, $\pi(P) = \pi(\overline{P})$, so α_0 itself is a loop, and the fundamental group of a surface with at least one puncture is a free group and in particular is torsion-free. This means α_0 is homotopic in $\dot{S}_0$ to a point; let σ be such a homotopy. Since $\dot{S}$ is an unramified cover of $\dot{S}_0$, we can lift σ to a homotopy of loops in $\dot{S}$ whose initial element is the pre-image α of α_0. Thus α is a loop, and so $P = \overline{P}$, a contradiction.

Still assuming H_ℓ not a short interval but $H_\ell = H_{\overline{\ell}}$, put $m_\ell = \lfloor n_\ell/2 \rfloor$, $M = \lfloor N/2 \rfloor$, and let $\alpha_{\ell,1}, \ldots, \alpha_{\ell,m_\ell} \in H_\ell$ be a set of (m_ℓ, M)-Fekete points for H_ℓ relative to $W_\ell(z)$. If $n = 2m_\ell$ is even, put $\alpha_{\ell,i} = \overline{\alpha_{\ell,i-m_\ell}}$ for $i = m_\ell + 1, \ldots, 2m_\ell$. If $n_\ell = 2m_\ell + 1$ is odd, define $\alpha_{\ell,m_\ell+1}, \ldots, \alpha_{\ell,2m_\ell}$ as above and put $\alpha_{\ell,2m_\ell+1} = \beta_\ell$. Now define

$$P_{\vec{n}}(z) \;:=\; \prod_{\ell=1}^{D} \prod_{i=1}^{n_\ell} [z, \alpha_{\ell,i}]_{\mathfrak{X},\vec{s}} \; . \tag{B.105}$$

Then $P_{\vec{n}}(z)$ is an $(\mathfrak{X}, \vec{s})$-pseudopolynomial of degree N for H, satisfying $P_{\vec{n}}(z) = P_{\vec{n}}(\overline{z})$ for all z. Let

$$\mathcal{M}_{\vec{n}} \;=\; \Big(\|P_{\vec{n}}(z)\|_H \Big)^{1/N} \; .$$

If there are components H_ℓ which are short intervals, let $\rho_{\vec{n}} > 0$ be the largest number such that $P_{\vec{n}}(z)$ varies n_ℓ times from $(\rho_{\vec{n}})^N$ to 0 to $(\rho_{\vec{n}})^N$ each of those components; otherwise put $\rho_{\vec{n}} = \mathcal{M}_{\vec{n}}$.

Let

$$\omega_{\vec{n}} \;=\; \frac{1}{N} \sum_{\ell=1}^{D} \sum_{i=1}^{n_\ell} \delta_{\alpha_{\ell,i}}(z)$$

---

[1]The author thanks Will Kazez for this argument.

be the discrete measure of mass 1 associated to $P_{\vec{n}}(z)$. By construction, it is stable under complex conjugation.

For each $\ell = 1, \ldots, D$, put $\sigma_\ell = \mu_{\mathfrak{X},\vec{s}}(H_\ell) > 0$, and let $\vec{\sigma} = (\sigma_1, \ldots, \sigma_D)$. Then $\vec{\sigma}$ is K_v-symmetric: $\sigma_{\overline{\ell}} = \sigma_\ell$ for each ℓ. Given a sequence of K_v-symmetric vectors $\vec{n}_k = (n_{k,1}, \ldots, n_{k,D}) \in \mathbb{N}^D$ for $k = 1, 2, 3, \ldots$, put $N_k = \sum_{\ell=1}^{D} n_{k,\ell}$ for each k.

PROPOSITION B.17. *Suppose $K_v \cong \mathbb{R}$. Assume that $\mathfrak{X}$ and $H \subset \mathcal{C}_v(\mathbb{C})\backslash\mathfrak{X}$ are stable under complex conjugation, and that H is compact and has finitely many connected components $H_1, \ldots, H_D$, where no H_ℓ is reduced to a point, and each H_ℓ is simply connected. Let $\vec{s} \in \mathcal{P}^m$ be a K_v-symmetric probability vector. Then for each K_v-symmetric vector $\vec{n} \in \mathbb{N}^D$, the $(\mathfrak{X}, \vec{s})$-pseudopolynomial $P_{\vec{n}}(z)$ has all its roots in H, with has exactly n_ℓ roots in each H_ℓ. It satisfies $P_{\vec{n}}(z) = P_{\vec{n}}(\overline{z})$ for all $z \in \mathcal{C}_v(\mathbb{C})$. In addition, for each $H_\ell = [a_\ell, b_\ell]$ which is a short interval, the roots of $P_{\vec{n}}(z)$ in H_ℓ are distinct, and $P_{\vec{n}}(z)$ varies n_ℓ times from $\rho_{\vec{n}}^N$ to 0 to $\rho_{\vec{n}}^N$ on H_ℓ, where $N = \deg(P_{\vec{n}_k}) = \sum_{\ell=1}^{d} n_\ell$.*

For any sequence of K_v-symmetric vectors $\vec{n}_k \in \mathbb{N}^D$ such that $N_k \to \infty$ and $\vec{n}_k/N_k \to \vec{\sigma}$, the following assertions hold: as $k \to \infty$:

(1) *The discrete measures $\omega_{\vec{n}_k}$ associated to the $P_{\vec{n}_k}$ converge weakly to the equilibrium distribution $\mu_{\mathfrak{X},\vec{s}}$ of H.*

(2) *For any open neighborhood U of H, the functions $-\frac{1}{N_k}\log(|P_{\vec{n}_k}(z)|)$ converge uniformly to the equilibrium potential $u_{\mathfrak{X},\vec{s}}(z, H)$ on $\mathcal{C}_v(\mathbb{C})\backslash(U \cup \mathfrak{X})$.*

(3) $\lim_{k\to\infty} \mathcal{M}_{\vec{n}_k} = \lim_{k\to\infty} \rho_{\vec{n}_k} = \gamma_{\mathfrak{X},\vec{s}}(H)$.

PROOF. By construction, $P_{\vec{n}}(z) = P_{\vec{n}}(\overline{z})$ for all $z \in \mathcal{C}_v(\mathbb{C})$, the roots of $P_{\vec{n}}(z)$ belong to H, and $P_{\vec{n}}$ has exactly n_ℓ roots in H_ℓ. For each H_ℓ which is a short interval, the roots of $P_{\vec{n}}(z)$ in H_ℓ are distinct by Proposition B.16, and $P_{\vec{n}}(z)$ varies n_ℓ times from $\rho_{\vec{n}}$ to 0 to $\rho_{\vec{n}}$ on H_ℓ by the definition of $\rho_{\vec{n}}$.

Now consider a sequence $\{\vec{n}_k\}_{k\in\mathbb{N}}$ with $N_k \to \infty$ and $n_{k,\ell}/N_k \to \sigma_\ell$ for each ℓ. Put

$$\omega_{\vec{n}_k,\ell} = \sum_{i=1}^{n_\ell} \frac{1}{N_k}\delta_{\alpha_{\ell,i}}(z) .$$

By Corollary B.11 and Theorem B.13, for each ℓ, as $k \to \infty$, the measures $\omega_{\vec{n}_k,\ell}$ converge weakly to $\mu_{\mathfrak{X},\vec{s}}|_{H_\ell}$. (The presence of the points β_ℓ do not affect this.) Hence the $\omega_{\vec{n}_k} = \sum_{\ell=1}^{d} \omega_{\vec{n}_k,\ell}$ converge weakly to $\mu_{\mathfrak{X},\vec{s}}$.

This implies that outside any neighborhood U of H, the potential functions $u_{\mathfrak{X},\vec{s}}(z, \omega_{\vec{n}_k})$ converge uniformly to $u_{\mathfrak{X},\vec{s}}(z, H)$. Indeed, since ∂U and H are compact and disjoint, $-\log([z, w]_{\mathfrak{X},\vec{s}})$ is uniformly continuous on $\partial U \times H$. Since the $\omega_{\vec{n}_k}$ converge weakly to $\mu_{\mathfrak{X},\vec{s}}$, as $k \to \infty$,

$$u_{\mathfrak{X},\vec{s}}(z, \omega_{\vec{n}_k}) = \int_H -\log([z, w]_{\mathfrak{X},\vec{s}})\, d\omega_{\vec{n}_k}(w) = -\frac{1}{N_k}\log(P_{\vec{n}_k}(z))$$

converges uniformly to $u_{\mathfrak{X},\vec{s}}(z, H) = \int_H -\log([z, w]_{\mathfrak{X},\vec{s}})\, d\mu_{\mathfrak{X},\vec{s}}(w)$ on ∂U. However, for each k, by Theorem A.2 the function $u_{\mathfrak{X},\vec{s}}(z, \omega_{\vec{n}_k}) - u_{\mathfrak{X},\vec{s}}(z, H)$ on $\mathcal{C}_v(\mathbb{C})\backslash(U \cup \mathfrak{X})$ extends to a function harmonic on a neighborhood of each $x_i \in \mathfrak{X}$. By the Maximum Principle, $u_{\mathfrak{X},\vec{s}}(z, \omega_{\vec{n}_k}) = -\frac{1}{N_k}\log(P_{\vec{n}_k}(z))$ converges uniformly to $u_{\mathfrak{X},\vec{s}}(z, H)$ on $\mathcal{C}_v(\mathbb{C})\backslash(U \cup \mathfrak{X})$.

We will next show that $\lim_{k\to\infty} \mathcal{M}_{n_k} = \gamma_{\mathfrak{X},\vec{s}}(H)$, or equivalently, that

$$\lim_{k\to\infty} -\log(\mathcal{M}_{\vec{n}_k}) = V_{\mathfrak{X},\vec{s}}(H) . \tag{B.106}$$

Recall that a Hausdorff space X is simply connected if and only if any two points are joined by an arc and every loop in X is homotopic in X to a point. Since each component of H is simply connected and no component is a point, it follows from Theorem A.2 and Proposition A.3 that the potential function $u_{\mathfrak{X},\vec{s}}(z,H)$ is continuous, with $u_{\mathfrak{X},\vec{s}}(z,H) = V_{\mathfrak{X},\vec{s}}(H)$ for all $z \in H$. Since $\mathcal{C}_v(\mathbb{C})\backslash H$ is connected, in addition $u_{\mathfrak{X},\vec{s}}(z,H) < V_{\mathfrak{X},\vec{s}}(H)$ for all $z \notin H$.

We first claim that $-\log(\mathcal{M}_{n_k}) \le V_{\mathfrak{X},\vec{s}}(H)$ for each k. For each $z \in H$, $u_{\mathfrak{X},\vec{s}}(z,\omega_{\vec{n}_k})$ satisfies

$$u_{\mathfrak{X},\vec{s}}(z,\omega_{\vec{n}_k}) \;=\; -\frac{1}{N_k}\log(P_{\vec{n}_k}(z)) \;\ge\; -\log(\mathcal{M}_{\vec{n}_k})\ .$$

By Fubini-Tonelli, it follows that

$$V_{\mathfrak{X},\vec{s}}(H) \;=\; \int_H u_{\mathfrak{X},\vec{s}}(z,H)\,d\omega_{\vec{n}_k}(z) \;=\; \int_H u_{\mathfrak{X},\vec{s}}(z,\omega_{\vec{n}_k})\,d\mu_{\mathfrak{X},\vec{s}}(z) \;\ge\; -\log(\mathcal{M}_{\vec{n}_k})$$

as asserted.

Now fix $\varepsilon > 0$, and put

$$U_\varepsilon \;=\; \{z \in \mathcal{C}_v(\mathbb{C}) : u_{\mathfrak{X},\vec{s}}(z,H) > V_{\mathfrak{X},\vec{s}}(H) - \varepsilon\}\ .$$

By the properties of $u_{\mathfrak{X},\vec{s}}(z,H)$ noted above, U_ε is open and $H \subset U_\varepsilon$. Since the $\omega_{\vec{n}_k}$ converge weakly to $\mu_{\mathfrak{X},\vec{s}}$, the functions $u_{\mathfrak{X},\vec{s}}(z,\omega_{\vec{n}_k})$ converge uniformly to $u_{\mathfrak{X},\vec{s}}(z,H)$ on $\mathcal{C}_v(\mathbb{C})\backslash(U_\varepsilon \cup \mathfrak{X})$. Thus, there is a k_0 such that for all $k \ge k_0$ and all $z \in \mathcal{C}_v(\mathbb{C})\backslash(U_\varepsilon \cup \mathfrak{X})$ we have

$$|u_{\mathfrak{X},\vec{s}}(z,H) - u_{\mathfrak{X},\vec{s}}(z,\omega_{\vec{n}_k})| \;<\; \varepsilon\ .$$

This means that for $k \ge k_0$ and $z \in \partial U_\varepsilon$,

$$u_{\mathfrak{X},\vec{s}}(z,\omega_{\vec{n}_k}) \;>\; V_{\mathfrak{X},\vec{s}}(z) - 2\varepsilon\ . \tag{B.107}$$

However, $u_{\mathfrak{X},\vec{s}}(z,\omega_{\vec{n}_k})$ is superharmonic on U_ε so by the Maximum Principle for superharmonic functions, (B.107) holds throughout U_ε. In particular, $-\log(\mathcal{M}_{\vec{n}_k}) > V_{\mathfrak{X},\vec{s}}(z) - 2\varepsilon$. Since $\varepsilon > 0$ is arbitrary, (B.106) holds.

Finally, we show that $\lim_{k\to\infty}\rho_{\vec{n}_k} = \gamma_{\mathfrak{X},\vec{s}}(H)$. Recall that $W_\ell(z) = \exp(-\widehat{u}_\ell(z))$ where

$$\widehat{u}_\ell(z) \;=\; \int_{H\backslash H_\ell} -\log([z,w]_{\mathfrak{X},\vec{s}})\,d\mu_{\mathfrak{X},\vec{s}}(w)\ .$$

Write $\widetilde{P}_{k,\ell}(z)$ for the weighted Chebyshev pseudopolynomial $\widetilde{P}_{\ell,(n_{k,\ell},N_k)}(z,W_\ell)$ as in (B.104), and write $\mathrm{CH}^*_{k,\ell}$ for the weighted Chebyshev constant $\mathrm{CH}^*_{(n_{k,\ell},N_k)}(H_\ell,W_\ell)$ $= \|\widetilde{P}_{k,\ell}\|_{H_\ell}^{1/N_k}$. Also, for each k and ℓ, put

$$\widehat{u}_{k,\ell}(z) \;:=\; \int_{H\backslash H_\ell} -\log([z,w]_{\mathfrak{X},\vec{s}})\,d\omega_{\vec{n}_k}(w)\ .$$

By Theorem B.2 and Theorem B.12, as $k \to \infty$ and $\vec{n}_k/N_k \to \vec{\sigma}$, the $\mathrm{CH}^*_{k,\ell}$ converge to $\mathrm{CH}^*_{\sigma_\ell}(H_\ell,W_\ell) = \gamma_{\mathfrak{X},\vec{s}}(H)$.

Fix $\varepsilon > 0$. Since the measures $\omega_{\vec{n}_k}$ converge weakly to $\mu_{\mathfrak{X},\vec{s}}$, if k is sufficiently large, then by Theorem B.12, for each ℓ,

$$|\widehat{u}_{k,\ell}(z) - \widehat{u}_\ell(z)| < \varepsilon \quad \text{on} \quad H_\ell \tag{B.108}$$

and

$$e^{-\varepsilon}\cdot\gamma_{\mathfrak{X},\vec{s}}(H) \;<\; \mathrm{CH}^*_{k,\ell}\ . \tag{B.109}$$

Put $\rho = e^{-2\varepsilon}\gamma_{\mathfrak{X},\vec{s}}(H)$.

Suppose H_ℓ is a short interval. Since $\|\widetilde{P}_{k,\ell}\|_{H_\ell} = (\mathrm{CH}^*_{k,\ell})^{N_k}$, Proposition B.16 shows that $\widetilde{P}_{k,\ell}(z)$ oscillates $n_{k,\ell}$ times from $(\mathrm{CH}^*_{k,\ell})^{N_k}$ to 0 to $(\mathrm{CH}^*_{k,\ell})^{N_k}$ on H_ℓ. At each point in H_ℓ where $\widetilde{P}_{k,\ell}(z) = (\mathrm{CH}^*_{\ell,\vec{n}})^{N_k}$, we have

$$P_{\vec{n}_k}(z) \ > \ (e^{-\varepsilon}\,\mathrm{CH}^*_{k,\ell})^{N_k} \ > \ \rho^{N_k}$$

and at each point where $\widetilde{P}_{k,\ell}(z) = 0$, also $P_{\vec{n}_k}(z) = 0$.

Thus $P_{\vec{n}_k}(z)$ oscillates $n_{k,\ell}$ times from ρ^{N_k} to 0 to ρ^{N_k} on each H_ℓ which is a short interval, so $\rho_{\vec{n}_k} \ge \rho = e^{-2\varepsilon}\gamma_{\mathfrak{X},\vec{s}}(H)$. Trivially $\rho_{\vec{n}_k} \le \mathcal{M}_{\vec{n}_k}$, so since $\lim_{k\to\infty} \mathcal{M}_{\vec{n}_k} = \gamma_{\mathfrak{X},\vec{s}}(H)$, and since $\varepsilon > 0$ is arbitrary, assertion (3) in the proposition follows. □

Remark. In constructing the functions $P_{\vec{n}}(z)$, the use of the Chebyshev points for the components which are short intervals is important. However, the use of the Fekete points for the other components is a matter of convenience; all that is needed is that the associated discrete measures be stable under complex conjugation, and converge weakly to $\mu_{\mathfrak{X},\vec{s}}$.

On some curves $\mathcal{C}_v/\mathbb{R}$ there exist connected sets $H \subset \mathcal{C}_v(\mathbb{C})\backslash\mathcal{C}_v(\mathbb{R})$ which are stable under complex conjugation, but are not simply connected. For example, consider an elliptic curve $\mathcal{E}/\mathbb{R}$ such as the one defined by $y^2 = x^3 - 1$, for which the real locus $\mathcal{E}(\mathbb{R})$ is homeomorphic to a loop. Such a curve which has two real 2-torsion points $P_0 = O$ and P_1, and two conjugate nonreal 2-torsion points P_2, P_3. The set $H = P_2 + \mathcal{E}(\mathbb{R})$ (where addition is under the group law on $\mathcal{E}$) is stable under complex conjugation, and complex conjugation acts on H via translation by $P_1 = P_3 - P_2$, so it has no fixed points in H.

In Proposition B.17, one can replace the hypothesis that the components H_ℓ are simply connected with two additional hypotheses: first, that each connected component of $\mathcal{C}_v(\mathbb{C})\backslash H$ must contain at least one x_i for which $s_i > 0$; and second, that for each component H_ℓ which is not a short interval but satisfies $H_\ell = H_{\overline{\ell}}$, the number n_ℓ is even.

We can reformulate Proposition B.17 in a useful way, as follows. For each $\vec{n}$, put

$$Q_{\vec{n}}(z) \ = \ \frac{1}{\mathcal{M}_{\vec{n}}^N} P_{\vec{n}}(z)\ , \tag{B.110}$$

and put $R_{\vec{n}} = \rho_{\vec{n}}/\mathcal{M}_{\vec{n}} \le 1$. Then $\|Q_{\vec{n}}\|_H = 1$, $Q_{\vec{n}}(z) = Q_{\vec{n}}(\overline{z})$ for all z, and $Q_{\vec{n}}(z)$ varies N times from $R_{\vec{n}}^N$ to 0 to $R_{\vec{n}}^N$ on each H_ℓ which is a short interval.

Recall that the Green's function of H is $G_{\mathfrak{X},\vec{s}}(z,H) = V_{\mathfrak{X},\vec{s}}(H) - u_{\mathfrak{X},\vec{s}}(z,H)$. As in §5.1, define the logarithmic leading coefficient of H at x_i by

$$\begin{aligned} \Lambda_{x_i}(H,\vec{s}) \ &= \ \lim_{z\to x_i} G_{\mathfrak{X},\vec{s}}(z;H) + s_i \log(|g_{x_i}(z)|) \\ &= \ V_{x_i}(H) + \sum_{j\ne i} G(x_i,x_j;H)\ . \end{aligned} \tag{B.111}$$

Likewise, as in §5.2, for each $x_i \in \mathfrak{X}$, define the logarithmic leading coefficient of $Q_{\vec{n}}(z)$ at x_i by

$$\Lambda_{x_i}(Q_{\vec{n}},\vec{s}) \ = \ \lim_{z\to x_i} \frac{1}{N}\log(Q_{\vec{n}}(z)) + s_i \log(|g_{x_i}(z)|)\ . \tag{B.112}$$

Recalling that for any sequence $\{\vec{n}_k\}_{k\in\mathbb{N}}$ with $N_k \to \infty$ and $\vec{n}_k/N_k \to \vec{\sigma}$ we have

$$\lim_{k\to\infty} \mathcal{M}_{\vec{n}_k} \;=\; \gamma_{\mathfrak{X},\vec{s}}(H) \;=\; e^{-V_{\mathfrak{X},\vec{s}}(H)} \ ,$$

by Proposition B.17 we have:

THEOREM B.18. *Suppose $K_v \cong \mathbb{R}$. Assume that $\mathfrak{X}$ and $H \subset \mathcal{C}_v(\mathbb{C})\backslash\mathfrak{X}$ are stable under complex conjugation, that H is compact, and that H has finitely many connected components $H_1, \ldots, H_D$, where no H_ℓ is reduced to a point and each H_ℓ is simply connected.*

Fix a K_v-symmetric probability vector $\vec{s} \in \mathcal{P}^m$. For each $\ell = 1, \ldots, D$ put $\sigma_\ell = \mu_{\mathfrak{X},\vec{s}}(H_\ell)$ and let $\vec{\sigma} = (\sigma_1, \ldots, \sigma_D)$. For each K_v-symmetric vector $\vec{n} \in \mathbb{N}^D$ write $N = N_{\vec{n}_k} = \sum_\ell n_\ell$. Then the collection of $(\mathfrak{X}, \vec{s})$-pseudopolynomials

$$Q_{\vec{n}}(z) \;=\; \frac{1}{\mathcal{M}_{\vec{n}}^N} \prod_{\ell=1}^{D} \prod_{i=1}^{n_\ell} [z, \alpha_{\ell,i}]_{\mathfrak{X},\vec{s}} \ ,$$

constructed above for $\vec{n} \in \mathbb{N}^D$, has the following properties:

(1) *For each $\vec{n}$, $Q_{\vec{n}}$ satisfies $\|Q_{\vec{n}}\|_H = 1$, with $Q_{\vec{n}}(z) = Q_{\vec{n}}(\overline{z})$ for all $z \in \mathcal{C}_v(\mathbb{C})$. The roots of $Q_{\vec{n}}$ all belong to H, with n_ℓ roots in each H_ℓ. Put $R_{\vec{n}} = \rho_{\vec{n}}/\mathcal{M}_{\vec{n}}$, so $0 < R_{\vec{n}} \le 1$. For each H_ℓ which is a short interval in the sense of* (B.103), *the roots of $Q_{\vec{n}}$ in H_ℓ are distinct, and $Q_{\vec{n}}$ varies n_ℓ times from $R_{\vec{n}}^N$ to* 0 *to $R_{\vec{n}}^N$ on H_ℓ.*

(2) *Let $\{\vec{n}_k\}_{k\in\mathbb{N}}$ be any sequence with $N_{\vec{n}_k} \to \infty$ and $\vec{n}_k/N_{\vec{n}_k} \to \vec{\sigma}$. Then $\lim_{k\to\infty} R_{\vec{n}_k} = 1$, and the discrete measures $\omega_{\vec{n}_k}$ associated to the $Q_{\vec{n}_k}$ converge weakly to the equilibrium distribution $\mu_{\mathfrak{X},\vec{s}}$ of H. For each neighborhood U of H, the functions $\frac{1}{N_{\vec{n}_k}} \log(Q_{\vec{n}_k}(z))$ converge uniformly to $G_{\mathfrak{X},\vec{s}}(z, H)$ on $\mathcal{C}_v(\mathbb{C}_v)\backslash(U \cup \mathfrak{X})$, and for each $x_i \in \mathfrak{X}$,*

$$\lim_{k\to\infty} \Lambda_{x_i}(Q_{\vec{n}_k}, \vec{s}) \;=\; \Lambda_{x_i}(H, \vec{s}) \ .$$

APPENDIX C

The Universal Function

In this appendix, we construct a parametrization of rational functions of degree d on a curve by their zeros and poles. We then establish a v-adic bound for how much a rational function changes when its zeros and poles are moved slightly. This is used in §11.3, in Step 4 of the patching process in the nonarchimedean compact case, with $d = \max(1, 2g)$, in moving the roots of the partially patched function away from each other.

Let F be a field, and let $\overline{F}$ be a fixed algebraic closure of F. When $\mathcal{C} = \mathbb{P}^1/F$, let z, w, p, q be independent variables and consider the crossratio

$$\chi(z, w; p, q) \quad = \quad \frac{(z-p)}{(z-q)} \frac{(w-q)}{(w-p)} \ ,$$

which extends to a rational function on $(\mathbb{P}^1)^4$. Now specialize p, q to $\mathbb{P}^1(\overline{F})$, and take $w \in \mathbb{P}^1(\overline{F})$ distinct from p, q. Then $f_{w,p,q}(z) = \chi(z, w; p, q)$ is a rational function on $\mathbb{P}^1$ with divisor $(p) - (q)$, normalized by the condition that $f_{w,p,q}(w) = 1$.

More generally, for arbitrary $p_1, \ldots, p_d, q_1, \ldots, q_d \in \mathbb{P}^1(\overline{F})$, if $w \in \mathbb{P}^1(\overline{F})$ is distinct from the p_i and q_i, then

$$\text{(C.1)} \qquad f_{w,\vec{p},\vec{q}}(z) \ := \ f(z, w; \vec{p}, \vec{q}) \ := \ \prod_{i=1}^{d} \chi(z, w; p_i, q_i)$$

is the unique rational function on $\mathbb{P}^1$ with divisor $\mathrm{div}_{\mathcal{C}}(f_{w,\vec{p},\vec{q}}) = \sum(p_i) - \sum(q_i)$, for which $f_{w,\vec{p},\vec{q}}(w) = 1$. Conversely, for any nonconstant rational function $h(z) \in \overline{F}(z)$, there is a point $w \in \mathbb{P}^1(\overline{F})$ where $h(w) = 1$. In this way, we obtain a parametrization of all rational functions of degree d on $\mathbb{P}^1$ by means of their zeros and poles and a normalizing point. In this Appendix, we show the existence of similar parametrizations for curves $\mathcal{C}/F$ of arbitrary genus.

Let $\mathcal{C}/F$ be a smooth, projective, geometrically integral curve of genus $g > 0$. Given $D \in \mathrm{Div}^d_{\mathcal{C}/F}(\overline{F})$, let $[D] \in \mathrm{Pic}^d_{\mathcal{C}/F}(\overline{F})$ be the linear equivalence class of D. Likewise, given $\vec{p} = (p_1, \ldots, p_d) \in \mathcal{C}(\overline{F})^d$, write $[\vec{p}] \in \mathrm{Pic}^d_{\mathcal{C}/F}(\overline{F})$ for the linear equivalence class of $\sum(p_i)$. Let $\mathrm{Jac}(\mathcal{C})/F$ be the Jacobian of $\mathcal{C}$, and let $\Phi : \mathcal{C}^d \times \mathcal{C}^d \to \mathrm{Jac}(\mathcal{C})$ be the F-rational map defined by $\Phi(\vec{p}, \vec{q}) = [\vec{p}] - [\vec{q}]$ for $\vec{p}, \vec{q} \in \mathcal{C}^d(\overline{F})$. Put $Y = \Phi^{-1}(0)$. Then Y is the F-rational subvariety of $\mathcal{C}^d \times \mathcal{C}^d$ for which

$$Y(\overline{F}) \ = \ \{(\vec{p}, \vec{q}) \in \mathcal{C}(\overline{F})^d \times \mathcal{C}(\overline{F})^d : \sum(p_i) - \sum(q_i) \text{ is principal}\} \ .$$

We will construct a "universal rational function" $f(z, w; \vec{p}, \vec{q})$ on $\mathcal{C} \times \mathcal{C} \times Y$ which parametrizes normalized rational functions of degree d on $\mathcal{C}$ in the sense above.

We will then specialize to the case where F is a nonarchimedean local field K_v, and study the continuity properties of $f(z, w; \vec{p}, \vec{q})$.

THEOREM C.1. *Let F be a field, and let $\mathcal{C}/F$ be a smooth, projective, geometrically integral curve of genus $g \geq 0$. Fix $d \geq \max(1, 2g-1)$, and let Y be the F-rational subvariety of $\mathcal{C}^d \times \mathcal{C}^d$ for which*

$$Y(\overline{F}) = \{(\vec{p}, \vec{q}) \in \mathcal{C}^d(\overline{F}) \times \mathcal{C}^d(\overline{F}) : \sum(p_i) \sim \sum(q_i)\} .$$

Then Y is irreducible, and there is an F-rational function $f(z, w; \vec{p}, \vec{q})$ on $\mathcal{C} \times \mathcal{C} \times Y$ uniquely defined by the property that for each $(\vec{p}, \vec{q}) \in Y(\overline{F})$ and each $w \in \mathcal{C}(\overline{F})$ distinct from the p_i, q_i, the function $f_{w,\vec{p},\vec{q}}(z) := f(z, w; \vec{p}, \vec{q}) \in \overline{F}(\mathcal{C})$ satisfies $\text{div}_{\mathcal{C}}(f_{w,\vec{p},\vec{q}}) = \sum(p_i) - \sum(q_i)$ and $f_{w,\vec{p},\vec{q}}(w) = 1$.

Moreover, if $F \subseteq M \subseteq \overline{F}$ is such that $w \in \mathcal{C}(M)$ and $\sum(p_i) - \sum(q_i)$ is rational over M, then $f_{w,\vec{p},\vec{q}}(z) \in M(\mathcal{C})$.

PROOF. The following proof is due to Robert Varley.[1] It applies even if d is not in the stable range $d \geq \max(1, 2g-1)$, though in that case Y may be reducible.

Let F be a field, and let $\mathcal{C}/F$ be a smooth, projective, geometrically integral curve of genus $g \geq 0$. Fix $d \geq 1$ and let $\widetilde{Y} \subset Y$ be a reduced, irreducible, locally closed F-rational subvariety. (If d is in the stable range, then Y is irreducible.) We claim there is a function $\widetilde{f}(z, w; \vec{p}, \vec{q}) \in F(\mathcal{C} \times \mathcal{C} \times \widetilde{Y})$ with the properties in Theorem C.1.

Let $U \subset \mathcal{C} \times \widetilde{Y}$ be the open F-rational subvariety for which

$$U(\overline{F}) = \{(w, \vec{p}, \vec{q}) \in \mathcal{C}(\overline{F}) \times \widetilde{Y}(\overline{F}) : w \text{ is distinct from } p_1, \ldots, p_d, q_1, \ldots, q_d\} .$$

Define sections $\varphi, \sigma_i, \tau_i : U \to \mathcal{C} \times U \hookrightarrow \mathcal{C} \times \mathcal{C} \times \widetilde{Y}$ by

$$\varphi(w, \vec{p}, \vec{q}) = (w, w, \vec{p}, \vec{q}), \quad \sigma_i(w, \vec{p}, \vec{q}) = (p_i, w, \vec{p}, \vec{q}), \quad \tau_i(w, \vec{p}, \vec{q}) = (q_i, w, \vec{p}, \vec{q})$$

for $i = 1, \ldots, d$. Let W be the subvariety $\varphi(U) \subset \mathcal{C} \times U$, and let $\mathcal{D}$ and $\mathcal{E}$ be the Cartier divisors on $\mathcal{C} \times U$ corresponding to the F-rational Weil divisors $\sum_{i=1}^d \sigma_i(U)$, $\sum_{i=1}^d \tau_i(U)$ respectively. Consider the line bundles $\mathcal{O}_{\mathcal{C}\times U}(\mathcal{E} - \mathcal{D})$ and $\mathcal{O}_W(\mathcal{E} - \mathcal{D})$, the restriction map $r : \mathcal{O}_{\mathcal{C}\times U}(\mathcal{E} - \mathcal{D}) \to \mathcal{O}_W(\mathcal{E} - \mathcal{D})$, the projection $p_2 : \mathcal{C} \times U \to U$, and the direct images $(p_2)_*(\mathcal{O}_{\mathcal{C}\times U}(\mathcal{E} - \mathcal{D}))$ and $(p_2)_*(\mathcal{O}_W(\mathcal{E} - \mathcal{D}))$. Then

(1) $H^0(U, (p_2)_*(\mathcal{O}_{\mathcal{C}\times U}(\mathcal{E} - \mathcal{D})))$ embeds in a natural way in the function field $F(\mathcal{C} \times \mathcal{C} \times \widetilde{Y})$;
(2) $(p_2)_*(\mathcal{O}_{\mathcal{C}\times U}(\mathcal{E} - \mathcal{D})) \cong \mathcal{O}_U$;
(3) the function $\widetilde{f}(z, w; \vec{p}, \vec{q}) \in F(\mathcal{C} \times \mathcal{C} \times \widetilde{Y})$ corresponding to the canonical section $1 \in H^0(U, \mathcal{O}_U)$ has the desired properties.

Indeed, (1) follows from a standard interpretation of sections of $\mathcal{O}_{\mathcal{C}\times U}(\mathcal{E} - \mathcal{D})$ as elements of the function field $F(\mathcal{C} \times U) = F(\mathcal{C} \times \mathcal{C} \times \widetilde{Y})$. Assertion (2) follows from two subclaims: first, $(p_2)_*(\mathcal{O}_{\mathcal{C}\times U}(\mathcal{E} - \mathcal{D})) \cong (p_2)_*(\mathcal{O}_W(\mathcal{E} - \mathcal{D}))$, and second, $(p_2)_*(\mathcal{O}_W(\mathcal{E} - \mathcal{D})) \cong \mathcal{O}_U$. For the first, note that the fibres of $(p_2)_*(\mathcal{O}_{\mathcal{C}\times U}(\mathcal{E} - \mathcal{D}))$ are one-dimensional, so by Grauert's theorem (see [H], p.288), $(p_2)_*(\mathcal{O}_{\mathcal{C}\times U}(\mathcal{E} - \mathcal{D}))$ is a line bundle on U. It maps pointwise nontrivially into $(p_2)_*(\mathcal{O}_W(\mathcal{E} - \mathcal{D}))$, which is also a line bundle on U. For the second, note that $\mathcal{O}_W(\mathcal{E}) \cong \mathcal{O}_W$ and $\mathcal{O}_W(\mathcal{D}) \cong \mathcal{O}_W$ since $\mathcal{D}$ and $\mathcal{E}$ are disjoint from W. However, $(p_2)_*\mathcal{O}_W \cong \mathcal{O}_U$ since $\varphi : U \to W \subset \mathcal{C} \times U$ is a section, and so $(p_2)_*(\mathcal{O}_W(\mathcal{E} - \mathcal{D})) \cong \mathcal{O}_U$. The rationality assertion is evident from the properties of $f_{w,\vec{p},\vec{q}}(z)$. □

[1]The author thanks Varley for permission to include his construction here. The only machinery it requires from algebraic geometry is Grauert's theorem. Varley also noted that $f(z, w; \vec{p}, \vec{q})$ can be defined even when $\mathcal{C}(F) = \phi$.

Unfortunately, the author does not see how to make Varley's construction yield the estimates in Theorem C.2 below. Hence, we give an alternate construction of the Universal Function using the theory of the Picard scheme. The part of the theory we need goes back to Weil and Matsusaka. A modern reference for this is Kleiman ([**31**]); see also Milne ([**43**]).

We follow Kleiman's notation. Given a separated map of locally Noetherian schemes $F : X \to S$, let $\mathrm{Pic}_{X/S}$ be the relative Picard functor, defined by

$$\mathrm{Pic}_{X/S}(T) \;=\; \mathrm{Pic}(X_T)/\mathrm{Pic}(T)$$

for any locally Noetherian scheme T/S ([**31**], Definition 2.2); here $X_T = X \times_S T$.

If $F : X \to S$ is projective and flat, with reduced, connected geometric fibres, and if $X(S) \neq \phi$, then $\mathrm{Pic}_{X/S}$ is represented by a commutative group scheme $\mathbf{Pic}_{X/S}$ which is separated and locally of finite type over S (see [**31**], Theorems 2.5 and 4.8, and Exercise 3.11).

The scheme $\mathbf{Pic}_{X/S}$ commutes with base change: for any locally Noetherian scheme S'/S, $\mathbf{Pic}_{X_{S'}/S'}$ exists and equals $\mathbf{Pic}_{X/S} \times_S S'$ ([**31**], Exercise 4.4). Points of $\mathbf{Pic}_{X/S}$ correspond in a natural bijective way to classes of invertible sheaves on the fibres of X/S ([**31**], Exercise 4.5). There is an invertible sheaf $\mathcal{P}$ on $X \times \mathbf{Pic}_{X/S}$, called a Poincaré sheaf, such that for any locally Noetherian scheme T/S, and any invertible sheaf $\mathcal{L}$ on X_T, there exists a unique S-map $h : T \to \mathbf{Pic}_{X/S}$ such that for some invertible sheaf $\mathcal{N}$ on T,

$$\mathcal{L} \;\cong\; (1 \times h)^*\mathcal{P} \otimes f_T^*(\mathcal{N})$$

(see [**31**], Exercise 4.3). In general, the Poincaré sheaf is not unique.

Let $\mathrm{Div}_{X/S}$ be the functor defined by

$$\mathrm{Div}_{X/S}(T) \;=\; \{\text{relative effective divisors on } X_T/T\}\ ;$$

see ([**31**], §3) for details. It is represented by an open subscheme $\mathbf{Div}_{X/S}$ of the Hilbert scheme $\mathbf{Hilb}_{X/S}$ ([**31**], Theorem 3.7). By ([**31**], §3.10 and Exercise 4.7), there is a coherent $\mathcal{O}_{\mathbf{Pic}_{X/S}}$-module $\mathcal{Q}$ for which $\mathbf{Div}_{X/S} \cong \mathbf{P}(\mathcal{Q})$. There is a natural map of functors $A_{X/S}(T) : \mathrm{Div}_{X/S}(T) \to \mathrm{Pic}_{X/S}(T)$, called the "Abel map", which sends a relative effective divisor D on X_T/T to the sheaf $\mathcal{O}_{X_T}(D)$, and there is a corresponding Abel map of S-schemes $\mathbf{A}_{X/S} : \mathbf{Div}_{X/S} \to \mathbf{Pic}_{X/S}$.

If $S = \mathrm{Spec}(F)$ where F is a field, and $X = \mathcal{C}$ is a smooth, projective, geometrically integral curve of genus $g > 0$ with $\mathcal{C}(F) \neq \phi$, the connected component of the identity, $\mathbf{Pic}^0_{X/S}$, is an abelian variety of dimension g ([**31**], Exercise 5.23) which is F-isomorphic to the Jacobian $\mathrm{Jac}(\mathcal{C})/F$. In this setting, we will write $\mathbf{Pic}_{\mathcal{C}/F}$ for $\mathbf{Pic}_{X/S}$ and $\mathbf{Div}_{\mathcal{C}/F}$ for $\mathbf{Div}_{X/S}$. Here $\mathbf{Pic}_{\mathcal{C}/F}$ is a disjoint union of open subschemes $\mathbf{Pic}^d_{\mathcal{C}/F}$ representing invertible sheaves $\mathcal{L}$ of degree d, and $\mathbf{Pic}^d_{\mathcal{C}/F}$ is a $\mathbf{Pic}^0_{\mathcal{C}/F}$-torsor for each d ([**31**], Exercise 6.21). Similarly $\mathbf{Div}_{\mathcal{C}/F}$ is a disjoint union of open subschemes $\mathbf{Div}^d_{\mathcal{C}/F}$ for $d \geq 1$, and for each d the Abel map takes $\mathbf{Div}^d_{\mathcal{C}/F}$ to $\mathbf{Pic}^d_{\mathcal{C}/F}$. Let $\mathcal{C}^d$ be the d-fold product of $\mathcal{C}$ with itself, and let $\mathrm{Sym}^{(d)}(\mathcal{C})$ be the d-fold symmetric product; it is smooth, since $\mathcal{C}$ is a curve ([**43**], Proposition 3.2, p.94). For each $d \geq 1$ there is a canonical surjective morphism $\alpha_d : \mathcal{C}^d \to \mathbf{Div}^d_{\mathcal{C}/F}$ given on $\overline{F}$-points by $\alpha_d(p_1, \ldots, p_d) = \sum(p_i)$. It induces an isomorphism $\mathrm{Sym}^d(\mathcal{C}) \cong \mathbf{Div}^d_{\mathcal{C}/F}$ (see [**31**], Remark 3.9).

Second Proof of Theorem C.1. We first carry out the construction over $\overline{F}$. Make a base change to $\overline{F}$, and until further notice, replace $\mathcal{C}$ by $\overline{\mathcal{C}} = \mathcal{C}_{\overline{F}}$ and Y by $\overline{Y} = Y_{\overline{F}}$. We will denote the function in the theorem constructed over $\overline{F}$ by $\overline{f}(z, w; \vec{p}, \vec{q})$.

Clearly $\overline{\mathcal{C}}(\overline{F}) \neq 0$. If $\mathcal{C}$ has genus 0, then $\overline{\mathcal{C}} \cong \mathbb{P}^1/\overline{F}$, so we can construct $\overline{f}(z, w; \vec{p}, \vec{q})$ by using the cross-ratio as in (C.1). In the argument below we will assume that $g > 0$.

Note that for each $(\vec{p}, \vec{q}) \in \overline{Y}(\overline{F})$, and each $w \in \overline{\mathcal{C}}(\overline{F})$ distinct from the p_i, q_i, there is a unique function $\overline{f}_{w,\vec{p},\vec{q}}(z) \in \overline{F}(\overline{\mathcal{C}})$ for which $\mathrm{div}_{\overline{\mathcal{C}}}(\overline{f}_{w,\vec{p},\vec{q}}) = \sum(p_i) - \sum(q_i)$ and $\overline{f}_{w,\vec{p},\vec{q}}(w) = 1$. We must show that these functions glue to give a globally defined rational function on $\overline{\mathcal{C}} \times \overline{\mathcal{C}} \times \overline{Y}$.

Put $\overline{\mathcal{Z}} = \overline{\mathcal{C}} \times \mathbf{Pic}^d_{\overline{\mathcal{C}}/\overline{F}}$. If $p_2 : \overline{\mathcal{Z}} \to \mathbf{Pic}^d_{\overline{\mathcal{C}}/\overline{F}}$ is the projection on the second factor, and $y = [D] \in \mathbf{Pic}^d_{\overline{\mathcal{C}}/\overline{F}}(\overline{F})$, write $\overline{\mathcal{Z}}_y$ for the fibre $p_2^*(y) \cong \overline{\mathcal{C}}$. Let $\overline{\mathcal{P}}$ be a Poincaré sheaf on $\overline{\mathcal{Z}}$. Then if $i_y : \overline{\mathcal{C}} \hookrightarrow \overline{\mathcal{Z}}$ is the inclusion $i_y(P) = (P, y) \in \overline{\mathcal{Z}}$, we have

$$\overline{\mathcal{P}}_y := i_y^*(\overline{\mathcal{P}}) \cong \overline{\mathcal{P}}|_{\overline{\mathcal{Z}}_y} \cong \mathcal{O}_{\overline{\mathcal{C}}}(D) .$$

Since $d \geq 2g-1$, the Riemann-Roch theorem shows that $\dim(H^0(\overline{\mathcal{C}}, \mathcal{O}_{\overline{\mathcal{C}}}(D))) = d - g + 1$ for all $D \in \mathrm{Div}^d_{\overline{\mathcal{C}}/\overline{F}}(\overline{F})$. The projection $p_2 : \overline{\mathcal{Z}} \to \mathbf{Pic}^d_{\overline{\mathcal{C}}/\overline{F}}$ is a flat, projective morphism of Noetherian schemes, and $\overline{\mathcal{P}}$ is flat over $\mathcal{O}_{\overline{\mathcal{Z}}}$, hence also flat over $\mathcal{O}_{\mathbf{Pic}^d_{\overline{\mathcal{C}}/\overline{F}}}$. By Grauert's theorem (see [H], p.288), $(p_2)_*(\overline{\mathcal{P}})$ is locally free of rank $d - g + 1$ over $\mathcal{O}_{\mathbf{Pic}^d_{\overline{\mathcal{C}}/\overline{F}}}$, and for each $y = [D] \in \mathbf{Pic}^d_{\overline{\mathcal{C}}/\overline{F}}(\overline{F})$, the natural map

$$(p_2)_*(\overline{\mathcal{P}}) \otimes k(y) \to H^0(\overline{\mathcal{Z}}_y, \overline{\mathcal{P}}_y) \cong H^0(\overline{\mathcal{C}}, \mathcal{O}_{\overline{\mathcal{C}}}(D)) \tag{C.2}$$

is an isomorphism.

Via the isomorphism $\mathrm{Sym}^{(d)}(\overline{\mathcal{C}}) \cong \mathbf{Div}^d_{\overline{\mathcal{C}}/\overline{F}}$, we can identify $\mathrm{Sym}^{(d)}(\overline{\mathcal{C}})(\overline{F})$ with the set of effective divisors $D = \sum_{i=1}^d (p_i)$ of degree d supported on $\overline{\mathcal{C}}(\overline{F})$. Let $Q : \overline{\mathcal{C}}^d \to \mathrm{Sym}^{(d)}(\overline{\mathcal{C}})$ be the quotient map, and let $P = \mathbf{A}_{\overline{\mathcal{C}}/\overline{F}} : \mathrm{Sym}^{(d)}(\overline{\mathcal{C}}) \to \mathbf{Pic}^d_{\overline{\mathcal{C}}/\overline{F}}$ be the Abel map, so that $P(D) = [D]$. Then Q is finite of degree $d!$, and the fibres of P are isomorphic to $\mathbb{P}^{d-g}$. Indeed, if $y = [D] \in \mathbf{Pic}^d_{\overline{\mathcal{C}}/\overline{F}}(\overline{F})$,

$$\begin{aligned} P^{-1}(y) &= \{D' \in \mathrm{Sym}^{(d)}(\overline{\mathcal{C}})(\overline{F}) : D' \sim D\} \\ &= \{D' \in \mathrm{Div}^d_{\overline{\mathcal{C}}/\overline{F}}(\overline{F}) : D' \geq 0, D' \sim D\} \\ &\cong \mathrm{Proj}(H^0(\overline{\mathcal{C}}, \mathcal{O}_{\overline{\mathcal{C}}}(D))) . \end{aligned}$$

Let $U \subset \mathbf{Pic}^d_{\overline{\mathcal{C}}/\overline{F}}$ be an affine subset small enough that $(p_2)_*(\overline{\mathcal{P}})$ is free over $\mathcal{O}_{\mathbf{Pic}^d_{\overline{\mathcal{C}}/\overline{F}}}(U)$. For each $y_0 \in \mathbf{Pic}^d_{\overline{\mathcal{C}}/\overline{F}}(\overline{F})$, there is such an affine containing y_0.

Put $\overline{\mathcal{Z}}|_U = p_2^{-1}(U) \cong \overline{\mathcal{C}} \times U \subset \overline{\mathcal{Z}}$, and let

$$\begin{aligned} \mathrm{Sym}^{(d)}(\overline{\mathcal{C}})|_U &= P^{-1}(U) \subset \mathrm{Sym}^{(d)}(\overline{\mathcal{C}}) , \\ \overline{\mathcal{C}}^d|_U = (P \circ Q)^{-1}(U) &= Q^{-1}(\mathrm{Sym}^{(d)}(\overline{\mathcal{C}})|_U) \subset \overline{\mathcal{C}}^d . \end{aligned}$$

Then $\mathrm{Sym}^{(d)}(\overline{\mathcal{C}})|_U \cong \mathbb{P}^{d-g} \times U$. Let $\mathcal{F}_0, \ldots, \mathcal{F}_{d-g}$ be a set of $\mathcal{O}_{\mathbf{Pic}^d_{\overline{\mathcal{C}}/\overline{F}}}(U)$-generators for $H^0(\overline{\mathcal{Z}}|_U, \overline{\mathcal{P}}) \cong H^0(U, (p_2)_*(\overline{\mathcal{P}}))$. Then for each $y = [D] \in U(\overline{F})$, the sections $i_y^*(\mathcal{F}_j) = \mathcal{F}_j|_{\overline{\mathcal{Z}}_y}$ form a basis for $H^0(\overline{\mathcal{C}}, \mathcal{O}_{\overline{\mathcal{C}}}(D))$. Recall that $\mathrm{Sym}^{(d)}(\overline{\mathcal{C}}) \cong \mathbf{Div}^d_{\overline{\mathcal{C}}/\overline{F}} \subset$

$\mathbf{Hilb}_{\overline{\mathcal{C}}/\overline{F}}$, and let $((a_0 : \cdots : a_{d-g}), y)$ vary over $\mathbb{P}^{d-g} \times U$. By the universal property of the Hilbert scheme, the flat family of divisors

$$\text{(C.3)} \qquad \widetilde{\mu}((a_0 : \cdots : a_{d-g}), y) \;=\; \operatorname{div}_{\overline{\mathcal{C}}}(i_y^*(\sum a_j \mathcal{F}_j)) \;=:\; \sum_{i=1}^{d} (p_i)$$

corresponds to a morphism $\mu : \mathbb{P}^{d-g} \times U \to \mathrm{Sym}^{(d)}(\overline{\mathcal{C}})|_U$. However, this morphism is simply a realization of the fibration of $\mathrm{Sym}^{(d)}(\overline{\mathcal{C}})|_U$ as a trivial $\mathbb{P}^{d-g}$-bundle over U, and hence is an isomorphism.

Let $\lambda : \mathrm{Sym}^{(d)}(\overline{\mathcal{C}})|_U \to \mathbb{P}^{d-g} \times U$ be the isomorphism inverse to μ. The surjective morphism

$$\varphi_U = \lambda \circ Q : \overline{\mathcal{C}}^d|_U \to \mathbb{P}^{d-g} \times U$$

provides a means of parametrizing sections of $\overline{\mathcal{P}}$ by their zeros: using the homogeneous coordinates $a_0, \ldots, a_{d-g}$ on $\mathbb{P}^{d-g}$, given $\vec{p} \in \overline{\mathcal{C}}|_U^d(\overline{F})$, write

$$\varphi_U(\vec{p}) \;=\; \big((a_0(\vec{p}) : \cdots : a_{d-g}(\vec{p})), [\vec{p}]\big) \;\in\; (\mathbb{P}^{d-g} \times U)(\overline{F}) \; .$$

Then by (C.3), tautologically

$$\text{(C.4)} \qquad \operatorname{div}_{\overline{\mathcal{C}}}(i_{[\vec{p}]}^*(\sum a_j(\vec{p}) \mathcal{F}_j)) \;=\; \mu(\varphi_U(\vec{p})) \;=\; \sum (p_i) \; .$$

For each pair (j,k) with $(j,k) \in \{0, \ldots, d-g\}$, consider the rational function on $\overline{\mathcal{C}}^d$ defined by $h_{j,k}(\vec{p}) = a_j(\vec{p})/a_k(\vec{p})$ on $\overline{\mathcal{C}}^d|_U(\overline{F})$. Its domain includes all points in U where $a_k(\vec{p}) \neq 0$. Put

$$\text{(C.5)} \qquad \mathcal{G}_{k,U}(z, \vec{p}) \;=\; \sum_{j=0}^{d-g} h_{j,k}(\vec{p}) \mathcal{F}_j(z, [\vec{p}]) \;\in\; \overline{F}(\overline{\mathcal{C}}^d) \otimes_{\overline{F}} H^0(\overline{\mathcal{Z}}|_U, \overline{\mathcal{P}}) \; .$$

Formula (C.4) shows that for each $\vec{p} \in \overline{\mathcal{C}}^d|_U(\overline{F})$ with $a_k(\vec{p}) \neq 0$, the pullback

$$G_{k,\vec{p}}(z) \;=\; i_{[\vec{p}]}^*(\mathcal{G}_{k,U}(z, \vec{p}))$$

is a section of $H^0(\overline{\mathcal{C}}, \mathcal{O}_{\overline{\mathcal{C}}}([\vec{p}]))$ satisfying

$$\text{(C.6)} \qquad \operatorname{div}_{\overline{\mathcal{C}}}(G_{k,\vec{p}}(z)) \;=\; \sum_{i=1}^{d} (p_i) \; .$$

Put $\overline{Y}|_U = \overline{Y} \cap (\overline{\mathcal{C}}^d|_U \times \overline{\mathcal{C}}^d|_U)$. For each (k, ℓ) with $0 \leq k, \ell \leq d-g$, let

$$\text{(C.7)} \qquad \overline{f}_{k,\ell,U}(z, w; \vec{p}, \vec{q}) \;=\; \frac{\mathcal{G}_{k,U}(z, \vec{p})}{\mathcal{G}_{\ell,U}(z, \vec{q})} \frac{\mathcal{G}_{\ell,U}(w, \vec{q})}{\mathcal{G}_{k,U}(w, \vec{p})} \; .$$

This a rational function on $\overline{\mathcal{C}} \times \overline{\mathcal{C}} \times \overline{Y}$, defined at least for $\vec{p}, \vec{q} \in \overline{\mathcal{C}}^d|_U(\overline{F})$ where $a_k(\vec{p}), a_\ell(\vec{q}) \neq 0$, and for $z, w \in \overline{\mathcal{C}}(\overline{F})$ distinct from the p_i, q_i.

Suppose $\vec{p} \in \overline{\mathcal{C}}^d|_U(\overline{F})$. If $\vec{q} \in \overline{\mathcal{C}}^d(\overline{F})$ satisfies $\sum(q_i) \sim \sum(p_i)$, that is, if $[\vec{q}] = [\vec{p}] \in U$ (so in particular $\vec{q} \in \overline{\mathcal{C}}^d|_U(\overline{F})$), there is an index ℓ such that $a_\ell(\vec{q}) \neq 0$. For each $w \in \overline{\mathcal{C}}(\overline{F})$ distinct from the p_i, q_i,

$$\overline{f}_{w,\vec{p},\vec{q}}(z) \;=\; i_{w,\vec{p},\vec{q}}^*(\overline{f}_{k,\ell,U}) \;=\; \frac{G_{k,\vec{p}}(z)}{G_{\ell,\vec{q}}(z)} \frac{G_{\ell,\vec{q}}(w)}{G_{k,\vec{p}}(w)} \;\in\; \overline{F}(\overline{\mathcal{C}})$$

is the unique rational function on $\overline{\mathcal{C}}$ for which $\mathrm{div}_{\overline{\mathcal{C}}}(\overline{f}_{w,\vec{p},\vec{q}}) = \sum(p_i) - \sum(q_i)$ and $\overline{f}_{w,\vec{p},\vec{q}}(w) = 1$. Hence as U and k, ℓ vary, the $\overline{f}_{k,\ell,U}$ glue to give the desired function $\overline{f}(z, w; \vec{p}, \vec{q})$.

Now consider the field of definition of $\overline{f}(z, w; \vec{p}, \vec{q})$. Recall that a field M is pseudoalgebraically closed (PAC) if every absolutely irreducible variety V/M has an M-rational point. It is well-known that for any field F, the separable closure F^{sep} is PAC (see [**27**], p.130, and [**35**], p.76). Since $\mathcal{C}(F^{\mathrm{sep}})$ is nonempty, there is a finite separable extension $\widehat{F}/F$ for which $\mathcal{C}(\widehat{F})$ is nonempty. Thus the theory of the Picard scheme applies over $\widehat{F}$. Let $\widehat{f}(z, w; \vec{p}, \vec{q})$ be the function $\overline{f}(z, w; \vec{p}, \vec{q})$ constructed above, regarding $\widehat{F}$ as the ground field.

We will first show that $\widehat{f}(z, w; \vec{p}, \vec{q})$ is $\widehat{F}$-rational. Put $\widehat{\mathcal{C}} = \mathcal{C}_{\widehat{F}}$, $\widehat{Y} = Y_{\widehat{F}}$, $\widehat{\mathcal{Z}} = \mathcal{Z}_{\widehat{F}}$. If $g = 0$, then $\widehat{\mathcal{C}} \cong \mathbb{P}^1_{\widehat{F}}$, $\widehat{Y} \cong (\mathbb{P}^1_{\widehat{F}})^d \times (\mathbb{P}^1_{\widehat{F}})^d$, and $\widehat{\mathcal{Z}} \cong \mathbb{P}^1_{\widehat{F}} \times \mathbb{P}^1_{\widehat{F}} \times (\mathbb{P}^1_{\widehat{F}})^d \times (\mathbb{P}^1_{\widehat{F}})^d$. In this case $\widehat{f}(z, w; \vec{p}, \vec{q})$ is defined using the crossratio and is $\widehat{F}$-rational by construction. If $g > 0$, then $\widehat{\mathcal{Z}} = \widehat{\mathcal{C}} \times \mathbf{Pic}^d_{\widehat{\mathcal{C}}/\widehat{F}}$; let $\widehat{\mathcal{P}}$ be a Poincaré sheaf on $\widehat{\mathcal{Z}}$. Then $\overline{\mathcal{Z}} = (\widehat{\mathcal{Z}})_{\overline{F}}$ and we can take $\overline{\mathcal{P}} = (\widehat{\mathcal{P}})_{\overline{F}}$. The varieties $\widehat{\mathcal{C}}$ and $\widehat{Y}$ are $\widehat{F}$-rational, so $\widehat{\mathcal{C}} \times \widehat{\mathcal{C}} \times \widehat{Y}$ is $\widehat{F}$-rational. For each $y_0 \in \mathbf{Pic}^d_{\widehat{\mathcal{C}}/\widehat{F}}(\widehat{F})$ there is an $\widehat{F}$-rational affine neighborhood U of y_0 such that $(p_2)_*(\widehat{\mathcal{P}})$ is free over $\mathcal{O}_{\mathbf{Pic}^d_{\widehat{\mathcal{C}}/\widehat{F}}}(U)$, so we can assume the affines U in the construction above are $\widehat{F}$-rational. The invertible sheaf $\widehat{\mathcal{P}}$ is $\widehat{F}$-rational, so the sections $\widehat{\mathcal{F}}_0, \ldots, \widehat{\mathcal{F}}_d$ can be chosen to be $\widehat{F}$-rational, and then the maps φ_U and the functions $h_{j,k}(\vec{p})$ will be $\widehat{F}$-rational. Hence $\widehat{f}(z, w; \vec{p}, \vec{q})$ is $\widehat{F}$-rational.

Recall that $\mathrm{Aut}(\overline{F}/F) \cong \mathrm{Gal}(F^{\mathrm{sep}}/F)$. Given $\sigma \in \mathrm{Aut}(\overline{F}/F)$, let $\widehat{\mathcal{C}}^\sigma$ (resp. $\widehat{Y}^\sigma$) be the conjugate variety to $\widehat{\mathcal{C}}$ (resp. $\widehat{Y}$). Similarly, put $\widehat{f}^\sigma = \sigma \circ f \circ \sigma^{-1}$; it is a function on $\widehat{\mathcal{C}}^\sigma \times \widehat{\mathcal{C}}^\sigma \times \widehat{Y}^\sigma$. It has properties analogous to those of $\widehat{f}$: for each $(w, \vec{p}, \vec{q}) \in \widehat{\mathcal{C}}^\sigma \times \widehat{Y}^\sigma$ with $w \neq p_i, q_i$ for all i, if we put $\widehat{f}^\sigma_{w,\vec{p},\vec{q}}(z) = \widehat{f}^\sigma(z, w; \vec{p}, \vec{q})$ then $\mathrm{div}_{\widehat{\mathcal{C}}^\sigma}(\widehat{f}^\sigma_{w,\vec{p},\vec{q}}) = \sum(p_i) - \sum(q_i)$ and $\widehat{f}^\sigma_{w,\vec{p},\vec{q}}(w) = 1$. Regarding $\mathcal{C}$ and $\widehat{\mathcal{C}}$ as projective varieties cut out by $\widehat{F}$-rational equations, fix an $\widehat{F}$-rational isomorphism $\gamma : \mathcal{C} \to \widehat{\mathcal{C}}$. By abuse of notation, we will denote the induced isomorphisms $\mathcal{C}^d \to \widehat{\mathcal{C}}^d$ and $\mathcal{C} \times \mathcal{C} \times Y \to \widehat{\mathcal{C}} \times \widehat{\mathcal{C}} \times \widehat{Y}$ by γ as well.

Define $f(z, w; \vec{p}, \vec{q})$ on $\mathcal{C} \times \mathcal{C} \times Y$ by $f = \widehat{f} \circ \gamma$. We claim that f is F-rational. It suffices to show that $f^\sigma = f$ for all $\sigma \in \mathrm{Aut}(\overline{F}/F)$. For this, note that by the defining properties of $\widehat{f}$, for each $(\vec{p}, \vec{q}) \in Y(\overline{F})$ and each $w \in \mathcal{C}(\overline{F})$ distinct from the p_i, q_i, the function $f_{w,\vec{p},\vec{q}} \in \overline{F}(\mathcal{C})$ given by $f_{w,\vec{p},\vec{q}}(z) = f(z, w; \vec{p}, \vec{q})$ satisfies $f_{w,\vec{p},\vec{q}}(w) = 1$ and $\mathrm{div}_{\mathcal{C}}(f_{w,\vec{p},\vec{q}}) = \sum(p_i) - \sum(q_i)$. Indeed, $f_{w,\vec{p},\vec{q}}(w) = \widehat{f}(\gamma(w), \gamma(w); \gamma(\vec{p}), \gamma(\vec{q})) = 1$ and

$$\begin{aligned} \mathrm{div}_{\mathcal{C}}(f_{w,\vec{p},\vec{q}}) &= \gamma^{-1}\big(\mathrm{div}_{\widehat{\mathcal{C}}}(\widehat{f}_{\gamma(w),\gamma(\vec{p}),\gamma(\vec{q})})\big) \\ &= \gamma^{-1}\big(\sum(\gamma(p_i)) - \sum(\gamma(q_i))\big) \;=\; \sum(p_i) - \sum(q_i)\,. \end{aligned}$$

These properties uniquely determine $f_{w,\vec{p},\vec{q}}$. However, for each σ, an analogous computation using the representation $f^\sigma = \widehat{f}^\sigma \circ \gamma^\sigma$ shows that $(f^\sigma)_{w,\vec{p},\vec{q}}$ has the same properties. Hence $f_{w,\vec{p},\vec{q}} = (f^\sigma)_{w,\vec{p},\vec{q}}$. Since $f(z, w; \vec{p}, \vec{q}) = f^\sigma(z, w; \vec{p}, \vec{q})$ for a Zariski-dense set of points $(z, w, \vec{p}, \vec{q})$, it follows that $f = f^\sigma$.

The final assertion in the theorem is that for any subfield $F \subset M \subset \overline{F}$, if $\sum(p_i) - \sum(q_i)$ is M-rational and $w \in \mathcal{C}(M)$, then $f_{w,\vec{p},\vec{q}}$ is M-rational. This is clear, since there exists a $g \in M(\mathcal{C})$ for which $\mathrm{div}_{\mathcal{C}}(g) = \sum(p_i) - \sum(q_i)$, and then the function $g(w)^{-1} \cdot g$ is M-rational and has the properties that characterize $f_{w,\vec{p},\vec{q}}$. □

We now specialize to the case $F = K_v$, where K_v is a nonarchimedean local field.

Our main result is the following. Let $\mathcal{C}_v/K_v$ be a smooth, projective, geometrically integral curve of genus $g \geq 0$. Fix a spherical metric $\|z, w\|_v$ on $\mathcal{C}_v(\mathbb{C}_v)$, and recall that for each $p \in \mathcal{C}_v(\mathbb{C}_v)$ and each $r > 0$,

$$\begin{aligned} B(p,r)^- &= \{z \in \mathcal{C}_v(\mathbb{C}_v) : \|z,p\|_v < r\} , \\ B(p,r) &= \{z \in \mathcal{C}_v(\mathbb{C}_v) : \|z,p\|_v \leq r\} . \end{aligned}$$

We will show that if $\vec{p}, \vec{q} \in \mathcal{C}_v(\mathbb{C}_v)^d$ are sufficiently near each other, and $[\vec{p}] = [\vec{q}]$, then $f(z, w; \vec{p}, \vec{q})$ is close to 1 outside fixed balls containing the p_j and q_j.

THEOREM C.2. *Let K_v be a nonarchimedean local field. Suppose $\mathcal{C}_v/K_v$ is a smooth, projective, geometrically integral curve of genus $g \geq 0$, and let $E_v \subset \mathcal{C}_v(\mathbb{C}_v)$ be compact. Then for each $d \geq \max(1, 2g)$, there are a radius $r_0 = r_0(E_v, d) > 0$ and a constant $D = D(E_v, d) > 0$ with the following property:*

Given $0 < \varepsilon < r \leq r_0$, suppose $\vec{p}, \vec{q} \in E_v^d$ are such that $\sum(p_j) \sim \sum(q_j)$ and $\|p_j, q_j\|_v \leq \varepsilon$ for each $j = 1, \ldots, d$. Put $r_j = \|p_j, q_j\|_v$. Then

(A) for all z, w in $\mathcal{C}_v(\mathbb{C}_v) \backslash \big((\bigcup_{j=1}^d B(p_j, r_j)^-) \cup (\bigcup_{j=1}^d B(q_j, r_j)^-)\big)$ we have

$$|f(z, w; \vec{p}, \vec{q})|_v = 1 ;$$

(B) for all z, w in $\mathcal{C}_v(\mathbb{C}_v) \backslash \big((\bigcup_{j=1}^d B(p_j, r)^-) \cup (\bigcup_{j=1}^d B(q_j, r)^-)\big)$ we have

$$|f(z, w; \vec{p}, \vec{q}) - 1|_v \leq \frac{D}{r^d} \cdot \varepsilon .$$

We will use Theorem C.2 in the proof of Lemma 11.10, the "First Moving Lemma" in the patching process in the nonarchimedean compact case. In the application r will be fixed, and the important factor governing $|f(z, w; \vec{p}, \vec{q}) - 1|_v$ will be $\varepsilon = \max_j(\|p_j, q_j\|_v)$.

PROOF OF THEOREM C.2 WHEN $g = 0$. Fixing an isomorphism of $\overline{\mathcal{C}}_v/\mathbb{C}_v$ with $\mathbb{P}^1/\mathbb{C}_v$, we can assume that $f(z, w; \vec{p}, \vec{q}) = \prod_{j=1}^d \chi(z, w; p_j, q_j)$. Since any two spherical metrics on $\mathcal{C}_v(\mathbb{C}_v)$ are comparable ([**51**], Theorem 1.1.1) we can assume that $\|x, y\|_v$ is the standard metric on $\mathbb{P}^1(\mathbb{C}_v)$ given for $x, y \in \mathbb{C}_v$ by

$$\text{(C.8)} \qquad \|x, y\|_v = \frac{|x - y|_v}{\max(1, |x|_v) \max(1, |y|_v)} .$$

By simple algebraic manipulations, one sees that that for $z, w, p, q \in \mathbb{C}_v$,

$$\text{(C.9)} \qquad |\chi(z, w; p, q)|_v = \left| \frac{(z - p)(w - q)}{(z - q)(w - p)} \right|_v = \frac{\|z, p\|_v \|w, q\|_v}{\|z, q\|_v \|w, p\|_v} ,$$

and that

$$\text{(C.10)} \quad \chi(z, w; p, q) - 1 = \frac{(z - w)(p - q)}{(z - q)(w - p)} , \quad |\chi(z, w; p, q) - 1|_v = \frac{\|z, w\|_v \|p, q\|_v}{\|z, q\|_v \|w, p\|_v} .$$

By continuity, these formulas extend to $z, w, p, q \in \mathbb{P}^1(\mathbb{C}_v)$. Furthermore, by a telescoping argument

$$(C.11) \quad f(z,w;\vec{p},\vec{q}) - 1 = \sum_{j=1}^{d} \Big(\chi(z,w;p_j,q_j) - 1\Big) \cdot \left(\prod_{k=j+1}^{d} \chi(z,w;p_k,q_k) \right) .$$

Take $r_0 = 1$ and $D = 1$.

If $z, w \in \mathbb{P}^1(\mathbb{C}_v) \backslash \big((\bigcup_{j=1}^d B(p_j,r_j)^-) \cup (\bigcup_{j=1}^d B(q_j,r_j)^-)\big)$ then since $\|x,y\|_v$ satisfies the ultrametric inequality and $\|p_j,q_j\| = r_j$, we have $\|z,p_j\|_v = \|z,q_j\|_v$ and $\|w,p_j\|_v = \|w,q_j\|_v$ for each j. From (C.9) it follows that

$$|f(z,w,\vec{p},\vec{q})|_v = \prod_{j=1}^{d} |\chi(z,w;p_j,q_j)|_v = 1 ,$$

which is assertion (A).

If $z, w \in \mathbb{P}^1(\mathbb{C}_v) \backslash \big((\bigcup_{j=1}^d B(p_j,r)^-) \cup (\bigcup_{j=1}^d B(q_j,r)^-)\big)$, then since $\|x,y\|_v$ satisfies the ultrametric inequality, and $\|z,q_j\|_v, \|w,p_j\|_v \geq r$ while $\|p_j,q_j\|_v = r_j \leq r$, it follows that

$$\|z,w\|_v \leq \max(\|z,q_j\|, \|q_j,p_j\|_v, \|w,p_j\|_v) = \max(\|z,q_j\|, \|w,p_j\|_v) .$$

It then follows from (C.10) that

$$(C.12) \qquad |\chi(z,w;p_j,q_j) - 1|_v = \frac{\|z,w\|_v \|p_j,q_j\|_v}{\|z,q_j\|_v \|w,p_j\|_v} \leq \frac{\|p_j,q_j\|_v}{r} .$$

Hence by (C.9), (C.11), (C.12) and the ultrametric inequality,

$$|f(z,w;\vec{p},\vec{q}) - 1|_v \leq \frac{1}{r} \cdot \max_j(\|p_j,q_j\|_v)$$

which is stronger than the inequality claimed in (B). □

Note that in the proof when $g = 0$, we did not use anything about the compact set E_v, and the bound in Theorem C.2(B) holds for all $\vec{p}, \vec{q} \in \mathcal{C}_v(\mathbb{C}_v)^d$ with each $\|p_j,q_j\|_v \leq r$. It seems likely that this remains true when $g > 0$ as well. For our application we only need the bound when $\vec{p}, \vec{q} \in E_v^d$, so we have not pursued it.

For the remainder of this appendix, we will assume that $g > 0$. In this case the proof of Theorem C.2 requires much more machinery. The idea is to first locally control the functions $G_{k,\mathcal{F},U}(z,\vec{p})$ in the factorization (C.16) below, using power series expansions, and then extend that control to all of $\mathcal{C}_v(\mathbb{C}_v)$ using various forms of the Maximum Modulus Principle and the theory of the canonical distance.

Before giving the proof, we will need several technical lemmas. We wish to apply the theory of rigid analysis, so we work with $\overline{\mathcal{C}}_v$ rather than $\mathcal{C}_v$. The first three lemmas prepare the way to use the Maximum Modulus Principle of rigid analysis on $\overline{\mathcal{C}}_v \times \overline{\mathcal{C}}_v^d$.

LEMMA C.3. *Let $p_1, \ldots, p_d \in \overline{\mathcal{C}}_v(\mathbb{C}_v)$. Suppose $0 < r < 1$ belongs to value group of $\mathbb{C}_v^\times$ and is small enough that each ball $B(p_j,r)$ is isometrically parametrizable. Then for each $\zeta \in \overline{\mathcal{C}}_v(\mathbb{C}_v) \backslash (\bigcup_{j=1}^d B(p_j,r))$, there is a function $g(z) \in \mathbb{C}_v(\overline{\mathcal{C}}_v)$ with* *poles only at ζ, such that*

$$\bigcup_{j=1}^{d} B(p_j, r) \quad = \quad \{z \in \overline{\mathcal{C}}_v(\mathbb{C}_v) : |g(z)|_v \le 1\} \ , \tag{C.13}$$

$$\bigcup_{j=1}^{d} B(p_j, r)^- \quad = \quad \{z \in \overline{\mathcal{C}}_v(\mathbb{C}_v) : |g(z)|_v < 1\} \ . \tag{C.14}$$

PROOF. If the balls $B(p_j, r)$ are pairwise disjoint, the result follows from ([**51**], Theorem 4.2.16) and its proof. In the general case, note that since $\|z, w\|_v$ satisfies the ultrametric inequality, then any two balls $B(p_i, r)$ and $B(p_j, r)$ either coincide, or are disjoint. For each $\ell = 1, \ldots, d$ we can represent $\bigcup_j B(p_j, r)$ as a disjoint union of a subset of the balls $B(p_j, r)$, in such a way that $B(p_\ell, r)$ occurs in the representation. Let $g_\ell(z)$ be the function obtained for this representation, and put $g(z) = \prod_{\ell=1}^{d} g_\ell(z)$. Then (C.13) and (C.14) hold for this $g(z)$. □

Recall that a subset $W \subset \overline{\mathcal{C}}_v(\mathbb{C}_v)$ is called an RL-domain ("Rational Lemniscate domain"; see [**51**], p.220) if there is nonconstant function $h \in \mathbb{C}_v(\overline{\mathcal{C}}_v)$ for which

$$W \ = \ \{z \in \overline{\mathcal{C}}_v(\mathbb{C}_v) : |h(z)|_v \le 1\} \ .$$

In that case

$$\partial W \ = \ \partial W(h) \ = \ \{z \in \overline{\mathcal{C}}_v(\mathbb{C}_v) : |h(z)|_v = 1\}$$

is called its boundary (with respect to h). By ([**51**], Corollary 4.2.14), a finite intersection (or union) of RL-domains is again an RL-domain. However, that corollary does not explicitly give a boundary.

LEMMA C.4. *Under the hypotheses of Lemma* C.3,
(A) $\overline{\mathcal{C}}_v(\mathbb{C}_v) \backslash (\bigcup_{j=1}^{d} B(p_j, r)^-)$ *is an* RL*-domain with boundary*

$$(\bigcup_{j=1}^{d} B(p_j, r)) \backslash (\bigcup_{i=1}^{d} B(p_j, r)^-) \ ;$$

(B) $(\bigcup_{j=1}^{d} B(p_j, r)) \backslash (\bigcup_{j=1}^{d} B(p_j, r)^-)$ *is an* RL*-domain.*

(C) Each isometrically parametrizable ball $B(p_j, r_j)$ *with* r_j *in the value group of* $\mathbb{C}_v^\times$, *is an* RL*-domain with boundary* $B(p_j, r_j) \backslash B(p_j, r_j)^-$.

PROOF. Let $g(z) \in \mathbb{C}_v(\overline{\mathcal{C}}_v)$ be the function from Lemma C.3. Then the RL-domain

$$\{z \in \overline{\mathcal{C}}_v(\mathbb{C}_v) : |1/g(z)|_v \le 1\} \ = \ \overline{\mathcal{C}}_v(\mathbb{C}_v) \backslash (\bigcup_{j=1}^{d} B(p_j, r)^-)$$

has boundary

$$\{z \in \overline{\mathcal{C}}_v(\mathbb{C}_v) : |1/g(z)|_v = 1\} \ = \ (\bigcup_{j=1}^{d} B(p_i, r)) \backslash (\bigcup_{j=1}^{d} B(p_j, r)^-)$$

which proves (A). For part (B), note that

$$(\bigcup_{j=1}^{d} B(p_j, r)) \backslash (\bigcup_{j=1}^{d} B(p_j, r)^-) \ = \ \{z \in \overline{\mathcal{C}}_v(\mathbb{C}_v) : |1/g(z)|_v \le 1, |g(z)|_v \le 1\}$$

and apply ([**51**], Corollary 4.2.14). Part (C) follows by applying Lemma C.3 to each $B(p_j, r_j)$ by itself. □

Recall that there is a faithful functor from the category of varieties over $\mathbb{C}_v$ to the category of rigid analytic spaces over $\mathbb{C}_v$ (see [**11**], p.363). If $\overline{X}_v/\mathbb{C}_v$ is a variety, we will say that a subset of $\overline{X}_v(\mathbb{C}_v)$ is a *affinoid domain* if its image in the rigid analytic space $\overline{X}_v^{an}$ associated to $\overline{X}_v$ is an admissible affinoid in the sense of rigid analysis (see [**11**], p.277, p.357): essentially, if its image under the functor above is the underlying point set of $\mathrm{Sp}(T)$ for some Tate algebra $T = \mathbb{C}_v\langle\langle z_1, \ldots, z_k\rangle\rangle/\mathcal{I}$ associated to an affine subset of $\overline{X}_v$. Here $\mathbb{C}_v\langle\langle z_1, \ldots, z_k\rangle\rangle$ is the ring of power series converging on the unit polydisc, and $\mathcal{I}$ is a finitely generated ideal.

An RL-domain on a curve is an affinoid domain in the sense above:

LEMMA C.5. *If $\overline{\mathcal{C}}_v/\mathbb{C}_v$ is a curve, and $W \subset \overline{\mathcal{C}}_v(\mathbb{C}_v)$ is an* RL*-domain, then W is an affinoid domain.*

PROOF. Let $h(z) \in \mathbb{C}_v(\overline{\mathcal{C}}_v)$ be a nonconstant function for which $W = \{z \in \overline{\mathcal{C}}_v(\mathbb{C}_v) : |h(z)|_v \leq 1\}$. The function field $\mathbb{C}_v(\overline{\mathcal{C}}_v)$ is finite over $\mathbb{C}_v(h)$. If $\mathrm{char}(\mathbb{C}_v) = 0$, then $\mathbb{C}_v(\overline{\mathcal{C}}_v)/\mathbb{C}_v(h)$ is separably algebraic. If $\mathrm{char}(\mathbb{C}_v) = p > 0$, then since $\mathbb{C}_v$ is algebraically closed, $\mathbb{C}_v(\overline{\mathcal{C}}_v)$ is separably generated. After replacing h by h^{1/p^m} for some m we can assume that $\mathbb{C}_v(\overline{\mathcal{C}}_v)/\mathbb{C}_v(h)$ is separably algebraic. By the primitive element theorem, there is a function $G(z) \in \mathbb{C}_v(\overline{\mathcal{C}}_v)$ for which $\mathbb{C}_v(\overline{\mathcal{C}}_v) = \mathbb{C}_v(h, G)$. Let $f(x, h) = x^d + a_1(h)x^{d-1} + \ldots + a_d(h)$ be the minimal polynomial of G over $\mathbb{C}_v(h)$, and let $\pi_v \in \mathbb{C}_v^\times$ satisfy $\mathrm{ord}_v(\pi_v) > 0$.

After multiplying G by a power of the product of denominators of the rational functions $a_i(h)$ and an appropriate power of π_v, we can assume that the $a_i(h)$ are polynomials in h with coefficients in $\widehat{\mathcal{O}}_v$. By the ultrametric inequality, for each $z \in W$ we must have $|G(z)|_v \leq 1$. Hence the map which sends z to $(x, y) = (G(z), h(z))$ induces an isomorphism of W with the underlying point set of $\mathrm{Sp}(\mathbb{C}_v\langle\langle x, y\rangle\rangle/(f(x, y))$. Examining the construction in ([**11**], Example 2, p.363), one sees that this map is the one realizing W as an affinoid domain under the functor above. □

COROLLARY C.6. *Let $p_1, \ldots, p_d \in \overline{\mathcal{C}}_v(\mathbb{C}_v)$; suppose that $r, r_1, \ldots, r_d$ belong to the value group of $\mathbb{C}_v^\times$ and are small enough that each ball $B(p_j, r)$ and $B(p_j, r_j)$ is isometrically parametrizable. Then*

$$W := ((\bigcup_{j=1}^{d} B(p_j, r)) \backslash (\bigcup_{j=1}^{d} B(p_j, r)^-)) \times \prod_{j=1}^{d} B(p_j, r_j)$$
$$\subset \overline{\mathcal{C}}_v(\mathbb{C}_v) \times \overline{\mathcal{C}}_v^d(\mathbb{C}_v)$$

is an affinoid domain in $\overline{\mathcal{C}}_v \times \overline{\mathcal{C}}_v^d$.

PROOF. A product of admissible affinoids is again one. ([**11**], §9.3.5). □

In order to study $f(z, w; \vec{p}, \vec{q})$, it will be useful to reformulate (C.7) using functions rather than sections of a line bundle. We begin with the following lemma, keeping the notations in the second proof of Theorem C.1.

LEMMA C.7. *Let $\mathcal{C}/F$ be a smooth, projective, geometrically integral curve of genus $g > 0$. Suppose $d \geq \max(1, 2g)$. Let $U \subset \mathbf{Pic}^d_{\overline{\mathcal{C}}/\overline{F}}$ be an affine subset over which $(p_2)_*(\overline{\mathcal{P}})$ is free. Then for any given $\vec{p}, \vec{q} \in \overline{\mathcal{C}}^d|_U(\overline{F})$ with $[\vec{p}] = [\vec{q}]$, and any finite set of points $z_1, \ldots, z_k \in \overline{\mathcal{C}}(\overline{F})$, there is a section $\mathcal{F} \in H^0(\overline{\mathcal{Z}}|_U, \overline{\mathcal{P}})$ for which the support of*

$$\mathrm{div}_{\mathcal{C}}(i^*_{[\vec{p}]}(\mathcal{F})) = \sum_{i=1}^{d}(p'_i)$$

is disjoint from the p_i, q_i, and z_i.

PROOF. Put $D = \sum(p_i)$. Our assumption on d assures there is movement in the linear system on $\overline{\mathcal{C}}$ associated to D. Consider the $\overline{F}$-vector space $\Gamma(D) = \{h(z) \in \overline{F}(\mathcal{C}) : \mathrm{div}_{\mathcal{C}}(h) \geq D\}$. Since $d \geq \max(1, 2g)$, the Riemann-Roch theorem shows that $\dim_{\overline{F}}\big(\Gamma(D)\big) = d - g + 1$, while for each $p \in \overline{\mathcal{C}}(\overline{F})$

$$\dim_{\overline{F}}\big(\Gamma(D - (p))\big) = d - g .$$

The set

$$\Gamma'(D) := \Gamma(D)\backslash\Big(\bigcup_{i=1}^{d}\Gamma(D - (p_i)) \cup \bigcup_{i=1}^{d}\Gamma(D - (q_i)) \cup \bigcup_{i=1}^{k}\Gamma(D - (z_i))\Big)$$

is nonempty because $\overline{F}$ is infinite, and each of the finitely many subspaces removed is a proper $\overline{F}$-subspace. For any function $h \in \Gamma'(D)$, the polar divisor of h is precisely $\sum(p_i)$, and the zeros of h are distinct from the p_i, q_i and z_i. Fix such an h and write

$$\mathrm{div}_{\mathcal{C}}(h) = \sum_{i=1}^{d}(p'_i) - \sum_{i=1}^{d}(p_i) .$$

Then $D' = \sum(p'_i)$ is linearly equivalent to $\sum(p_i)$ and $\sum(q_i)$, and the p'_i are distinct from the p_i, q_i, and z_i.

Now let $\mathcal{F}_0, \ldots, \mathcal{F}_{d-g}$ be generators for $H^0(\overline{\mathcal{Z}}|_U, \overline{\mathcal{P}})$ over $\mathcal{O}_{\mathbf{Pic}^d_{\overline{\mathcal{C}}/\overline{F}}}(U)$. Let $\mathcal{H}$ be the $\overline{F}$-vector space generated by the $\mathcal{F}_i$. By our assumptions, the map

$$i^*_{[\vec{p}]} : \mathcal{H} \to H^0(\overline{\mathcal{C}}, \mathcal{O}_{\overline{\mathcal{C}}}(\sum(p_i)))$$

is an isomorphism. Thus there is an $\mathcal{F} \in \mathcal{H}$ with $\mathrm{div}_{\mathcal{C}}(i^*_{[\vec{p}]}(\mathcal{F})) = \sum(p'_i)$. □

Henceforth, assume $d \geq \max(1, 2g)$. Given an affine $U \subset \overline{\mathcal{Z}}$ and a basis of sections $\mathcal{F}_0, \ldots, \mathcal{F}_{d-g} \in H^0(\overline{\mathcal{Z}}_U, \overline{\mathcal{P}})$ as above, consider the sections $\mathcal{G}_{k,U}(z, \vec{p})$ defined in (C.5). Let $0 \neq \mathcal{F} \in H^0(\overline{\mathcal{Z}}_U, \overline{\mathcal{P}})$ be an arbitrary section and put

$$G_{k,\mathcal{F},U}(z, \vec{p}) = \frac{\mathcal{G}_{k,U}(z, \vec{p})}{\mathcal{F}(z, [\vec{p}])} = \sum_{j=0}^{d-g}\frac{a_j(\vec{p})}{a_k(\vec{p})} \cdot \frac{\mathcal{F}_j(z, [\vec{p}])}{\mathcal{F}(z, [\vec{p}])} . \tag{C.15}$$

Then $G_{k,\mathcal{F},U}$ is an $\overline{F}$-rational function on $\mathcal{C} \times \mathcal{C}^d$, defined at least for $(z, \vec{p}) \in \mathcal{C}(\overline{F}) \times \mathcal{C}^d|_U(\overline{F})$ where $a_k(\vec{p}) \neq 0$ and $\mathcal{F}(z, [\vec{p}]) \neq 0$. The important point is that $\mathcal{F}$ depends on $\vec{p}$ only through $[\vec{p}]$, so for each $\vec{p}$ the polar divisor of $G_{k,\mathcal{F},U}(\vec{p})$ depends only on $[\vec{p}]$.

Fix $\vec{p}, \vec{q} \in \mathcal{C}^d|_U(\overline{F})$ with $[\vec{p}] = [\vec{q}]$, fix $z_0 \in \mathcal{C}(\overline{F})$, and fix $w \in \mathcal{C}(\overline{F})$ distinct from the p_i and q_i. Choose k with $a_k(\vec{p}) \neq 0$ and ℓ with $a_\ell(\vec{q}) \neq 0$. By Lemma C.7, there is an $\mathcal{F}$ for which the support of $\mathrm{div}_{\mathcal{C}}(i^*_{[\vec{p}]}(\mathcal{F}))$ is disjoint from z_0, w, and the p_i and

q_i. Since $\mathcal{F}(z,[\vec{p}])$ depends on $\vec{p}$ only through $[\vec{p}]$, we have $\mathcal{F}(z,[\vec{p}]) = \mathcal{F}(z,[\vec{q}])$ for all z. It follows from (C.6) that

$$\operatorname{div}_{\mathcal{C}}\left(\frac{G_{k,\mathcal{F},U}(z,\vec{p})}{G_{\ell,\mathcal{F},U}(z,\vec{q})}\right) = \sum(p_i) - \sum(q_i) .$$

Thus, for an appropriate choice of U, k, ℓ and $\mathcal{F}$, we can represent $f_{w,\vec{p},\vec{q}}(z) = f(z,w;\vec{p},\vec{q})$ in a neighborhood of z_0 by

$$f(z,w;\vec{p},\vec{q}) = \frac{G_{k,\mathcal{F},U}(z,\vec{p})}{G_{\ell,\mathcal{F},U}(z,\vec{q})} \cdot \frac{G_{\ell,\mathcal{F},U}(w,\vec{q})}{G_{k,\mathcal{F},U}(w,\vec{p})} . \tag{C.16}$$

This means that to understand $f(z,w;\vec{p},\vec{q})$, it suffices to understand the $G_{k,\mathcal{F},U}(z,\vec{p})$, which is simpler to do because only the zeros of $G_{k,\mathcal{F},U}(z,\vec{p})$ are controlled.

LEMMA C.8. *Let $E_v \subset \overline{\mathcal{C}}_v(\mathbb{C}_v)$ be compact, and let $d \geq \max(1,2g)$. Then there are a radius $R = R(E_v,d) > 0$ in the value group of $\mathbb{C}_v^\times$, and a number $B = B(E_v,d) > 0$, with the following properties:*

There are finitely many affine subsets $U_i \subset \mathbf{Pic}^d_{\overline{\mathcal{C}}_v/\mathbb{C}_v}$ such that $(p_2)_*(\overline{\mathcal{P}})$ is free over U_i, with functions $G_i(z,\vec{p}) = G_{k_i,\mathcal{F}_i,U_i}(z,\vec{p})$ as in (C.15)*, such that for each $\vec{p} = (p_1,\ldots,p_d) \in E_v^d$, the balls $B(p_j,R)$, for $j = 1,\ldots,d$, are isometrically parametrizable, and there is some i for which $(\bigcup_{j=1}^d B(p_j,R)) \times \prod_{j=1}^d B(p_j,R) \subset \overline{\mathcal{C}}_v(\mathbb{C}_v) \times \overline{\mathcal{C}}_v^d|_{U_i}(\mathbb{C}_v)$ and*

(A) $|G_i(z,\vec{q})|_v \leq 1$ for all $(z,\vec{q})$ in $(\bigcup_{j=1}^d B(p_j,R)) \times \prod_{j=1}^d B(p_j,R)$;

(B) $|G_i(z,\vec{q})|_v \geq B$ for all $(z,\vec{q})$ in

$$(\bigcup_{j=1}^d B(p_j,R)) \backslash (\bigcup_{j=1}^d B(p_j,R)^-) \times \prod_{j=1}^d B(p_j,\frac{1}{2}R) .$$

PROOF. By Theorem 3.9 there is a number $0 < R_0 \leq 1$ (depending on the spherical metric $\|z,w\|_v$) such that each ball $B(a,r)$ with $a \in \overline{\mathcal{C}}_v(\mathbb{C}_v)$ and $0 < r \leq R_0$ is isometrically parametrizable.

Fix $\vec{p} \in E_v^d$. Choose an affine set $U \subset \mathbf{Pic}^d_{\overline{\mathcal{C}}_v/\mathbb{C}_v}$ for which $\vec{p} \in \overline{\mathcal{C}}_v^d|_U(\mathbb{C}_v)$ and which is small enough that $(p_2)_*(\overline{\mathcal{P}})$ is free over U. Then for each sufficiently small $r > 0$ we will have $\prod_{j=1}^d B(p_j,r) \subset \overline{\mathcal{C}}_v^d|_U(\mathbb{C}_v)$ and all of the balls $B(p_j,r)$ will be isometrically parametrizable.

By Lemma C.7 there is a section $\mathcal{F} = \sum c_j \mathcal{F}_j$ of $(p_2)_*(\overline{\mathcal{P}})(U)$ with coefficients $c_j \in \mathbb{C}_v$ for which $\operatorname{div}_{\overline{\mathcal{C}}_v}(i^*_{[\vec{p}]}(\mathcal{F})) = \sum(p'_i)$ is coprime to $\sum(p_i)$. Thus, we can find k and $\mathcal{F}$ so that if $G_{k,\mathcal{F},U}$ is as in (C.15)), then $i^*_{[\vec{p}]}(G_{k,\mathcal{F},U})$ has polar divisor $\sum(p'_i)$ with support disjoint from $\{p_1,\ldots,p_d\}$. Since $G_{k,\mathcal{F},U}(p_j,\vec{p}) = 0$ for each $j = 1,\ldots,d$, by continuity there is an $r > 0$ such that $(\bigcup_{j=1}^d B(p_j,r)) \times \prod_{j=1}^d B(p_j,r)$ is contained in

$$\{(z,\vec{q}) \in \overline{\mathcal{C}}_v(\mathbb{C}_v) \times \overline{\mathcal{C}}_v^d|_U(\mathbb{C}_v) : |G_{k,\mathcal{F},U}(z,\vec{q})|_v \leq 1\} .$$

Without loss we can take $r \leq R_0$, so the balls $B(p_j,r)$ are isometrically parametrizable.

By compactness, there are a finite number of points $\vec{p}^{(i)}$ and radii $r^{(i)}$ such that the sets $\prod_{j=1}^d B(p_j^{(i)},r^{(i)})$ cover E_v^d. Let the U_i and $G_i = G_{k_i,\mathcal{F}_i,U_i}(z,\vec{p})$ be the corresponding affine sets and functions, and let $R = R(E_v,d)$ be the minimum of the $r^{(i)}$. After shrinking R if necessary, we can assume R belongs to the value group of $\mathbb{C}_v^\times$. Then for any $\vec{p} \in E_v^d$, there is some $\vec{p}^{(i)}$ for which

$$\prod_{j=1}^d B(p_j, R) \subset \prod_{j=1}^d B(p_j^{(i)}, r^{(i)})$$

and so (A) holds for this R.

Again fix $\vec{p} \in E_v^d$, and choose $U = U_i$ and $G(z, \vec{q}) = G_i(z, \vec{q})$ so that (A) holds. Fix $1/2 < C < 1$ in the value group of $\mathbb{C}_v^\times$. By Lemma C.4, the set

$$\begin{aligned} W &:= ((\bigcup_{j=1}^d B(p_j, R)) \backslash (\bigcup_{j=1}^d B(p_j, R)^-)) \times \prod_{j=1}^d B(p_j, CR) \\ &\subset \overline{\mathcal{C}}_v(\mathbb{C}_v) \times \overline{\mathcal{C}}_v^d|_{U_i}(\mathbb{C}_v) \end{aligned}$$

is an affinoid domain. Moreover, for each $\vec{q} = (q_1, \ldots, q_d) \in \prod_{j=1}^d B(p_j, cR)$, the function $G_{\vec{q}}(z) = G(z, \vec{q})$ has zeros only at $q_1, \ldots, q_d$, and so in particular it does not vanish on $((\bigcup_{j=1}^d B(p_j, R)) \backslash (\bigcup_{j=1}^d B(p_j, R)^-))$. It follows that $1/G(z, \vec{q})$ is a rigid analytic function on W. By the Maximum Modulus Principle of rigid analysis (see [**11**], p.237), there is a number $B(\vec{p})$ with $0 < B(\vec{p}) < 1$ such that for all $(z, \vec{q}) \in W$,

$$\left|\frac{1}{G(z, \vec{q})}\right|_v \le \frac{1}{B(\vec{p})} ;$$

equivalently, $|G(z, \vec{q})|_v \ge B(\vec{p})$.

Again by compactness, there are finitely many points $\vec{p}^{(\ell)} \in E_v^d$ such that the sets $\prod_{j=1}^d B(p_j^{(\ell)}, CR)$ cover E_v^d. Moreover, for any two points $\vec{p}, \vec{q} \in E_v^d$, the sets $\prod_{j=1}^d B(p_j, CR)$ and $\prod_{j=1}^d B(q_j, CR)$ either coincide or are disjoint. Hence, if B is the minimum of the corresponding numbers $B(\vec{p}^{(\ell)})$, then part (B) holds for all $\vec{p} \in E_v^d$, with this B. □

The following lemma uses power series to obtain local control of $|G(z, \vec{p})|_v$.

LEMMA C.9. *With the notation and hypotheses of Lemma* C.8, *write* $B = B(E_v, d)$ *and* $R = R(E_v, d)$. *Fix* $\vec{p} \in E_v^d$, *and take* $U = U_i$ *and* $G = G_i$ *so that the assertions of Lemma* C.8 *hold for* $\vec{p}$ *with respect to* U *and* G. *Put*

$$r_0 = r_0(E_v, d) := \min(1/2, B(E_v, d)) \cdot R(E_v, d) < R = R(E_v, d) .$$

Then:

(A) For each $\ell = 1, \ldots, d$, *put* $\mathcal{M}_\ell = \mathcal{M}_\ell(\vec{p}) = \max_{z \in B(p_\ell, R)} |G(z, \vec{p})|_v$. *Then for each* $\vec{q} \in \prod_{j=1}^d B(p_j, r_0)$, *and each* ℓ, *we have*

$$\max_{z \in B(q_\ell, R)} |G(z, \vec{q})|_v = \mathcal{M}_\ell .$$

(B) For each $\ell = 1, \ldots, d$, *there is a constant* $C_\ell = C_\ell(\vec{p})$ *with the following property: For each* $\vec{q} \in \prod_{j=1}^d B(p_j, r_0)$, *each* ℓ, *and each* $z \in B(q_\ell, R)$,

$$|G(z, \vec{q})|_v = C_\ell \cdot \prod_{q_j \in B(q_\ell, R)} \|z, q_j\|_v .$$

PROOF. Note that if $\vec{q} \in \prod_{j=1}^d B(p_j, r_0)$, then $B(q_\ell, R) = B(p_\ell, R)$ for each $\ell = 1, \ldots, d$.

For part (A), fix ℓ, and note that $\mathcal{M}_\ell = \max_{z \in B(p_\ell, R)} |G(z, \vec{p})|_v \geq B(E_v, d)$. Let $z_\ell \in B(p_\ell, R)$ be a point where $|G(z_\ell, \vec{p})|_v = \mathcal{M}_\ell$. Choose isometric parametrizations of the balls $B(p_\ell, R)$ and $B(p_1, R), \ldots, B(p_d, R)$ in terms of local coordinate functions $Z, P_1, \ldots, P_d$ on $D(0, R) = \{z \in \mathbb{C}_v : |z|_v \leq 1\}$, in such a way that $z_\ell = Z(0)$, and $p_j = P_j(0)$. Let $Q_1, \ldots, Q_d \in D(0, R)$ be such that $q_j = P_j(Q_j)$. For each $j = 1, \ldots, d$, since $q_j \in B(p_j, r_0)$, the definition of isometric parametrizability shows that $|Q_j|_v = \|p_j, q_j\|_v \leq r_0$.

Using these parametrizations, on $B(p_\ell, R) \times \prod_{j=1}^d B(p_j, R)$ we can expand G as a power series

$$\mathcal{G}(Z, \vec{P}) = \sum_{i,k} a_{i,k} Z^i \vec{P}^k .$$

Here $a_{0,0} = G(z_\ell, \vec{p})$, so $|a_{0,0}| = \mathcal{M}_\ell$. Moreover, since $\mathcal{G}(Z, \vec{P})$ converges on $D(0, R) \times D(0, R)^d$ and $|\mathcal{G}(Z, \vec{P})|_v \leq 1$ for all $(Z, \vec{P}) \in D(0, R) \times D(0, R)^d$, we have

$$|a_{i,k}|_v \leq \frac{1}{R^{i+|k|}}$$

for all i, k. Consequently, for each (i, k) with $|k| > 0$,

$$\begin{aligned} |a_{i,k} Z^i \vec{Q}^k|_v &\leq \frac{1}{R^{i+|k|}} \cdot R^i \cdot r_0^{|k|} = \left(\frac{r_0}{R}\right)^{|k|} \\ &\leq \frac{r_0}{R} \leq B < \mathcal{M}_\ell = |a_{0,0}|_v . \end{aligned}$$

For each $z \in B(q_\ell, R)$ we can write

$$G(z, \vec{q}) = \sum_{i=0}^\infty a_{i,0} Z^i + \sum_{i=0}^\infty \sum_{|k|>0} a_{i,k} Z^i \vec{Q}^k = G(z, \vec{p}) + \sum_{i=0}^\infty \sum_{|k|>0} a_{i,k} Z^i \vec{Q}^k ,$$

so

$$|G(z, \vec{q})|_v \leq \max(|G(z, \vec{p})|_v, \max_i(\max_{|k|>0}(|a_{i,k} Z^i \vec{Q}^k|_v))) \leq \mathcal{M}_\ell .$$

On the other hand, when $z = z_\ell$

$$\begin{aligned} |G(z_\ell, \vec{q})|_v &= |\mathcal{G}(0, \vec{Q})|_v = |a_{0,0} + \sum_{|k|>0} a_{0,k} \vec{Q}^k|_v \\ &= |a_{0,0}|_v = \mathcal{M}_\ell . \end{aligned}$$

This proves (A).

For part (B), write

$$\operatorname{div}_{\overline{\mathcal{C}}_v}(G(z, \vec{p})) = \sum_{j=1}^d (p_j) - \sum_{j=1}^d (\delta_j), \quad \operatorname{div}_{\overline{\mathcal{C}}_v}(G(z, \vec{q})) = \sum_{j=1}^d (q_j) - \sum_{j=1}^d (\Delta_j) .$$

By the definition of R in Lemma C.8, we have $|G(z, \vec{p})|_v \leq 1$ and $|G(z, \vec{q})|_v \leq 1$ on $\mathcal{D} := \bigcup_{j=1}^d B(p_j, R)$, so the δ_j and Δ_j lie outside $\mathcal{D}$. Fix a point $\zeta \in \overline{\mathcal{C}}_v(\mathbb{C}_v) \backslash \mathcal{D}$, distinct from the δ_j and Δ_j, and consider the canonical distance $[z, w]_\zeta$. By the factorization property of the canonical distance (see §3.5), there are constants C and D such that for all $z \in \overline{\mathcal{C}}_v(\mathbb{C}_v)$,

$$|G(z,\vec{p})|_v = C \cdot \frac{\prod_{j=1}^d [z,p_j]_\zeta}{\prod_{j=1}^d [z,\delta_j]_\zeta}, \qquad |G(z,\vec{q})|_v = D \cdot \frac{\prod_{j=1}^d [z,q_j]_\zeta}{\prod_{j=1}^d [z,\Delta_j]_\zeta} .$$

By Proposition 3.11(B.2) (applied with $\mathfrak{X} = \{\zeta\}$), for each isometrically parametrizable ball $B(a,r_a)$ not containing ζ, there is a constant c_a such that $[z,w]_\zeta = c_a\|z,w\|_v$ for all $z,w \in D(a,r_a)$. By Proposition 3.11(B.1), if $B(a,r_a)$ and $B(b,r_b)$ are disjoint isometrically parametrizable balls not containing ζ, then $[z,w]_\zeta$ is constant for $z \in B(a,r_a)$ and $w \in B(b,r_b)$. It follows that there are constants C_ℓ and D_ℓ such that for all $z \in B(p_\ell,R) = B(q_\ell,R)$,

$$|G(z,\vec{p})|_v = C_\ell \cdot \prod_{p_j \in B(p_\ell,R)} \|z,p_j\|_v , \quad |G(z,\vec{q})|_v = D_\ell \cdot \prod_{q_j \in B(q_\ell,R)} \|z,p_j\|_v .$$

Clearly $|G(z,\vec{p})|_v$ achieves its maximum $\mathcal{M}_\ell$ at a point $z_\ell \in B(p_\ell,R)$ if and only if $\|z_\ell,p_j\|_v = R$ for all $p_j \in B(p_\ell,R)$, and then $\mathcal{M}_\ell(\vec{p}) = C_\ell R^{m_\ell}$ where m_ℓ is the number of points p_j (or q_j) in $B(p_\ell,R)$. Similarly $|G(z_\ell,\vec{q})|_v$ achieves its maximum $\mathcal{M}_\ell = D_\ell R^{m_\ell}$ on $B(q_\ell,R) = B(p_\ell,R)$ if and only $\|z_\ell,q_j\|_v = R$ for all $q_j \in B(q_\ell,R)$. Since there are infinitely many $z \in B(p_\ell,R)$ satisfying both conditions simultaneously, we must have $C_\ell = D_\ell$. □

Observe that in the notation of Lemma C.9, for each $\vec{q} \in \prod_{j=1}^d B(p_j,r_0)$ and each $\ell = 1,\ldots,d$, we have $\mathcal{M}_\ell(\vec{q}) = \mathcal{M}_\ell(\vec{p}) < 1$ and $C_\ell(\vec{q}) = C_\ell(\vec{p})$.

LEMMA C.10. *With the notation and hypotheses of Lemmas* C.8 *and* C.9, *there is a constant $C = C(E_v,d)$ such that for each $\vec{p} \in E_v^d$, if $0 < r \le r_0$ belongs to the value group of $\mathbb{C}_v$, and if $U = U_i$ and $G = G_i$ are chosen for $\vec{p}$ as in Lemma* C.9, *then*

$$|G(z,\vec{p})|_v \ \ge \ C \cdot r^d$$

for all $z \in (\bigcup_{j=1}^d B(p_j,r))\backslash(\bigcup_{j=1}^d B(p_j,r)^-)$.

PROOF. We can cover E_v^d with finitely many sets of the form $\prod_{j=1}^d B(p_j,r_0)$. For each of these sets, Lemma C.9 gives constants $C_\ell = C_\ell(\vec{p})$ such that for all $\vec{q} \in \prod_{j=1}^d B(p_j,r_0)$, and all $z \in B(q_\ell,r)$

$$|G(z,\vec{q})|_v = C_\ell \cdot \prod_{q_j \in B(q_\ell,r)} \|z,q_j\|_v .$$

Let C be the minimum of these constants, for all the representative sets and all ℓ.

If $z \in B(p_\ell,r)\backslash(\bigcup_{j=1}^d B(p_j,r)^-)$, then $\|z,p_j\|_v = r$ for all j. There are at most d points p_j in $B(p_\ell,r)$, so $|G(z,\vec{p})|_v \ge Cr^d$. □

The lemma below uses the Maximum Modulus Principle for RL-domains to control $|\frac{G(z,\vec{p})}{G(z,\vec{q})} - 1|_v$ outside $\bigcup_{j=1}^d B(q_j,r_j)^-$.

LEMMA C.11. *With the notation and hypotheses of Lemmas* C.8 *and* C.9, *there is a constant $D = D(E_v,d)$ with the following property. Let $0 < r \le r_0$ belong to the value group of $\mathbb{C}_v^\times$. Suppose $\vec{p},\vec{q} \in E_v^d$ are such that $\max_j(\|p_j,q_j\|_v) \le r$, and*

$\sum(p_j) \sim \sum(q_j)$. *Put* $r_j = \|p_j, q_j\|_v$. *Take* $U = U_i$ *and* $G = G_i$ *as in Lemma* C.8. *Then for each* $z \in \overline{\mathcal{C}}_v(\mathbb{C}_v)\backslash(\bigcup_{j=1}^d B(p_j, r)^-)$,

$$\left|\frac{G(z,\vec{q})}{G(z,\vec{p})} - 1\right|_v \le \frac{D}{r^d} \cdot \max_j(r_j) .$$

PROOF. Suppose $G(z,\vec{p}) = G_{k,\mathcal{F},U}(z,\vec{p})$ corresponds to a section $\mathcal{F}$ and an affine set U as in (C.15). Fix $\vec{p}$, $\vec{q}$ as in the lemma. Noting that the polar divisors $s_{\mathcal{F}}([\vec{p}])$ and $s_{\mathcal{F}}([\vec{q}])$ of $G(z,\vec{p})$ and $G(z,\vec{q})$ depend only on the class $[\vec{p}] = [\vec{q}] \in \mathrm{Pic}^d_{\mathcal{C}_v/\mathbb{C}_v}$, we have

$$s_{\mathcal{F}}([\vec{p}]) = s_{\mathcal{F}}([\vec{q}]) = \sum_{j=1}^d (\Delta_j) .$$

Hence $G(z,\vec{q})/G(z,\vec{p})$ has poles only at the points $p_j \in \mathrm{supp}(\vec{p})$.

By Lemma C.3 and the Maximum Modulus Principle for RL-domains with boundary (see [**51**], Theorem 1.4.2, p. 51), it suffices to establish the bound in the Lemma for each $z_0 \in (\bigcup_{j=1}^d B(p_j,r))\backslash(\bigcup_{j=1}^d B(p_j,r)^-)$.

Fix ℓ, fix $z_0 \in B(p_\ell, r)\backslash(\bigcup_{j=1}^d B(p_j,r)^-)$, and introduce local coordinate functions $Z, \vec{P}$ on $B(p_\ell, R)$ and the $B(p_j, R)$ as in Lemma C.9 so that $z_0 = Z(0)$, $p_j = P_j(0)$, and $q_j = P_j(Q_j)$. On $B(p_\ell, R) \times \prod_j B(p_j, R)$, expand $G_{k,\mathcal{F},U}$ as a power series

$$\mathcal{G}(Z, \vec{P}) = \sum_{i,k} a_{i,k} Z^i \vec{P}^k$$

where $|a_{i,k}|_v \le 1/R^{i+|k|}$ for all i, k. Then $G(z_0,\vec{p}) = \mathcal{G}(0,\vec{0}) = a_{0,0}$, and $G(z_0,\vec{q}) = \mathcal{G}(0,\vec{Q}) = a_{0,0} + \sum_{|k|>0} a_{0,k}\vec{Q}^k$. Hence

$$\frac{G(z_0,\vec{q})}{G(z_0,\vec{p})} - 1 = \sum_{|k|>0} \frac{a_{0,k}}{a_{0,0}} \vec{Q}^k .$$

By Lemma C.10, $|a_{0,0}|_v \ge Cr^d$. Hence, term by term,

$$\begin{aligned}\left|\frac{a_{0,k}}{a_{0,0}}\vec{Q}^k\right|_v &\le \frac{1}{Cr^d} \cdot \frac{1}{R^{|k|}} \cdot (\max_j(r_j))^{|k|} \\ &\le \frac{1}{Cr^d} \max(\frac{r_j}{R}) = \frac{1}{CRr^d} \max(r_j) .\end{aligned}$$

This gives the result, with $D = 1/(C \cdot R)$. □

LEMMA C.12. *With the notation and hypotheses of Lemmas* C.8 *and* C.9, *let* $\vec{p}, \vec{q} \in E_v^d$ *be such that* $\max_j \|p_j, q_j\|_v \le r_0$, *and assume* $\sum(p_j) \sim \sum(q_j)$. *Put* $\|p_j, q_j\|_v = r_j$ *for* $j = 1, \ldots, d$. *Then*

$$\left|\frac{G(z,\vec{q})}{G(z,\vec{p})}\right|_v = 1$$

for all $z \in \overline{\mathcal{C}}_v(\mathbb{C}_v)\backslash((\bigcup_{j=1}^d B(p_j, r_j)^-) \cup (\bigcup_{j=1}^d B(q_j, r_j)^-))$.

PROOF. As in Lemma C.11, $G(z,\vec{p})$ and $G(z,\vec{q})$ have common polar divisor $\sum(\Delta_j)$. Fix $\zeta \in \overline{\mathcal{C}}_v(\mathbb{C}_v)\backslash((\bigcup_{j=1}^d B(p_j,r_j)^-) \cup (\bigcup_{j=1}^d B(q_j,r_j)^-))$, distinct from the

Δ_j. By the theory of the canonical distance there are constants $C_{\vec{p}}$ and $C_{\vec{q}}$ such that for all $z \in \overline{\mathcal{C}}_v(\mathbb{C}_v)$,

$$G(z,\vec{p}) = C_{\vec{p}} \cdot \frac{\prod_{j=1}^d [z,p_j]_\zeta}{\prod_{j=1}^d [z,\Delta_j]_\zeta}, \qquad G(z,\vec{q}) = C_{\vec{q}} \cdot \frac{\prod_{j=1}^d [z,q_j]_\zeta}{\prod_{j=1}^d [z,\Delta_j]_\zeta} .$$

Hence

$$\left|\frac{G(z,\vec{q})}{G(z,\vec{p})}\right|_v = \frac{C_{\vec{p}}}{C_{\vec{q}}} \cdot \frac{\prod_{j=1}^d [z,q_j]_\zeta}{\prod_{j=1}^d [z,p_j]_\zeta} .$$

As noted in the proof of Lemma C.9, $[z,w]_\zeta$ is constant for z and w belonging to isometrically parametrizable balls disjoint from each other and from ζ, while on a given isometrically parametrizable ball disjoint from ζ it is a constant multiple of $\|z,w\|_v$. Hence for $z \notin (\bigcup_{j=1}^d B(p_j,r_j)^-) \cup (\bigcup_{j=1}^d B(q_j,r_j)^-) \cup \{\zeta\}$, we have $[z,p_j]_\zeta = [z,q_j]_\zeta$ for each j, and it follows that

$$\left|\frac{G(z,\vec{q})}{G(z,\vec{p})}\right|_v = \frac{C_{\vec{p}}}{C_{\vec{q}}} .$$

This holds for $z = \zeta$ as well, by continuity, if we view $G(z,\vec{q})/G(z,\vec{p})$ as a rational function with divisor $\sum(q_j) - \sum(p_j)$. By the proof of Lemma C.9(B) there are points $z_\ell \notin \bigcup_{j=1}^d B(p_j,r_j)^- \cup \bigcup_{j=1}^d B(q_j,r_j)^-$ where $|G(z_\ell,\vec{p})|_v = |G(z_\ell,\vec{q})|_v$, so $C_{\vec{p}}/C_{\vec{q}} = 1$, and the result follows. □

We can now prove the main theorem.

Proof of Theorem C.2 when $g > 0$. Let r_0 be as in Lemma C.9. Given $\vec{p},\vec{q} \in E_v^d$ with $\sum(p_m) \sim \sum(q_j)$ and $\max_j(\|p_j,q_j\|_v) \le r \le r_0$, choose $U = U_i$ and $G = G_i$ as in Lemma C.8. Then

$$f(z,w;\vec{p},\vec{q}) = \frac{G(z,\vec{p})}{G(z,\vec{q})} \cdot \frac{G(w,\vec{q})}{G(w,\vec{p})}$$

so (A) follows from Lemma C.12. Furthermore,

$$f(z,w;\vec{p},\vec{q}) - 1 = \frac{G(z,\vec{p})}{G(z,\vec{q})} \cdot \left(\frac{G(w,\vec{q})}{G(w,\vec{p})} - 1\right) + \left(\frac{G(z,\vec{p})}{G(z,\vec{q})} - 1\right) .$$

Let D be the constant from Lemma C.11. Then by Lemma C.11, for all $z,w \notin ((\bigcup_{j=1}^d B(p_j,r)^-) \cup (\bigcup_{j=1}^d B(q_j,r)^-))$

$$\left|\frac{G(z,\vec{p})}{G(z,\vec{q})} - 1\right|_v \le \frac{D}{r^d} \max_j(r_j) , \qquad \left|\frac{G(w,\vec{q})}{G(w,\vec{p})} - 1\right|_v \le \frac{D}{r^d} \max_j(r_j) ,$$

while by Lemma C.12

$$\left|\frac{G(z,\vec{p})}{G(z,\vec{q})}\right|_v = 1 .$$

Combining these gives (B). □

APPENDIX D

The Local Action of the Jacobian

Let K_v be a nonarchimedean local field, and let $\mathcal{C}_v/K_v$ be a smooth, projective, geometrically integral curve of genus $g > 0$. Write $\mathrm{Jac}(\mathcal{C}_v)$ for the Jacobian of $\mathcal{C}_v$ over K_v.

In this appendix we will show that for a dense set of points $\vec{a} \in \mathcal{C}_v(\mathbb{C}_v)^g$, there is an action of a neighborhood of the origin in $\mathrm{Jac}(\mathcal{C}_v)(\mathbb{C}_v)$ on a sufficiently small neighborhood of $\vec{a}$ in $\mathcal{C}_v(\mathbb{C}_v)^g$, which makes that neighborhood into a principal homogeneous space. This action is used in §6.4 in the construction of the initial local approximating functions in the nonarchimedean compact case, and in §11.3, the patching process in the nonarchimedean compact case, in moving the roots of the partially patched function away from each other.

Fix a spherical metric $\|x, y\|_v$ on $\mathcal{C}_v(\mathbb{C}_v)$. We will be working simultaneously with balls in $\mathcal{C}_v(\mathbb{C}_v)$, $\mathcal{C}_v(\mathbb{C}_v)^g$, and $\mathrm{Jac}(\mathcal{C}_v)(\mathbb{C}_v)$, so we will write $B_{\mathcal{C}_v}(a, r)$ for the ball $\{z \in \mathcal{C}_v(\mathbb{C}_v) : \|z, a\|_{\mathcal{C}_v,v} \le r\}$ simply denoted $B(a, r)$ elsewhere. As usual, we put $D(0, r) = \{z \in \mathbb{C}_v : |z|_v \le r\}$ and $D(0, r)^- = \{z \in \mathbb{C}_v : |z|_v < r\}$. Given a point $\vec{a} = (a_1, \ldots, a_g) \in \mathcal{C}_v(\mathbb{C}_v)^g$, and radii $r_1, \ldots, r_g > 0$, we write $B_{\mathcal{C}_v^g}(\vec{a}, \vec{r}) = \prod_{i=1}^g B_{\mathcal{C}_v}(a_i, r_i)$. We also put $D(\vec{0}, \vec{r}) = \prod_{i=1}^g D(0, r_i)$, and if $r_1 = \cdots = r_g = R$ we write $D(\vec{0}, R) = \prod_{i=1}^g D(0, R)$, $D(\vec{0}, R)^- = \prod_{i=1}^g D(0, R)^-$.

If $r_1, \ldots, r_g > 0$ are small enough, then by Theorem 3.9 (proved in [**51**], Theorem 1.2.3) each ball $B_{\mathcal{C}_v}(a_i, r_i)$ can be isometrically parametrized by power series, that is, there is an analytic isomorphism $\varphi_i : D(0, r_i) \to B_{\mathcal{C}_v}(a_i, r_i)$ defined by convergent power series (which are F_u-rational provided $a_1, \ldots, a_g \in \mathcal{C}_v(F_u)$, where $K_v \subseteq F_u \subseteq \mathbb{C}_v$ and F_u is complete), such that $\|\varphi_i(x), \varphi_i(y)\|_{\mathcal{C}_v,v} = |x - y|_v$ for all $x, y \in D(0, r_i)$. It follows that

$$\Phi_{\vec{a}} := (\varphi_1, \ldots, \varphi_g) : D(\vec{0}, \vec{r}) \to B_{\mathcal{C}_v^g}(\vec{a}, \vec{r}) \tag{D.1}$$

is an analytic isomorphism. The construction in ([**51**], Theorem 1.2.3) shows that the maps φ_i can be chosen in such a way that $\varphi_i^{-1} : B(a_i, r_i) \to D(0, r_i)$ is projection on one of the coordinates, followed by a translation, and we will assume that that is the case.

The Jacobian $\mathrm{Jac}(\mathcal{C}_v)/K_v$ is an abelian variety characterized by the property that $\mathrm{Jac}(\mathcal{C}_v) \times_{K_v} \mathrm{Spec}(F_u)$ is isomorphic to $\mathbf{Pic}^0_{\mathcal{C}_u/F_u}$ over any extension F_u/K_v such that $\mathcal{C}_v(F_u) \ne \phi$. For each $\vec{a} = (a_1, \ldots, a_g) \in \mathcal{C}_v(\mathbb{C}_v)^g$, there is a morphism $\mathbf{J}_{\vec{a}} : \mathcal{C}_v^g \to \mathrm{Jac}(\mathcal{C}_v)$, defined over $K_v(a_1, \ldots, a_g)$, which takes $\vec{x} = (x_1, \ldots, x_g)$ to the linear equivalence class of the divisor $\sum(x_i) - \sum(a_i)$. It induces a birational morphism from $\mathrm{Sym}^{(g)}(\mathcal{C}_v)$ onto $\mathrm{Jac}(\mathcal{C}_v)$. This was the idea behind Weil's algebraic construction of $\mathrm{Jac}(\mathcal{C}_v)$: using the Riemann-Roch theorem, he showed that there was a birational, commutative law of composition defined on an open subset of

$\mathrm{Sym}^{(g)}(\mathcal{C}_v)$ (a "group chunk"), which could be extended to an addition law on an abelian variety. Later, he showed that every abelian variety is projective, and Matsusaka showed that $\mathrm{Jac}(\mathcal{C}_v)$ and its group law were defined over K_v. A modern account of this theory can be found in ([**43**]).

Since $\mathrm{Jac}(\mathcal{C}_v)$ is smooth and projective, each point of $\mathrm{Jac}(\mathcal{C}_v)(\mathbb{C}_v)$ has a neighborhood in the v-topology which is isometrically parametrizable by power series (Theorem 3.9). If $x \in \mathrm{Jac}(\mathcal{C}_v)(F_u)$, where $K_v \subseteq F_u \subseteq \mathbb{C}_v$ and F_u is complete, then those power series are defined over F_u. It follows that each point of $\mathrm{Jac}(\mathcal{C}_v)(F_u)$ has a neighborhood in $\mathrm{Jac}(\mathcal{C}_v)(F_u)$ analytically isomorphic to $\mathcal{O}_u^g$.

Write $J_{\mathrm{Ner}}(\mathcal{C}_v)$ for the Néron model of $\mathrm{Jac}(\mathcal{C}_v)$. The Néron model (see [**5**], [**12**]) is a smooth, separated group scheme of finite type over $\mathrm{Spec}(\mathcal{O}_v)$ whose generic fibre is isomorphic to $\mathrm{Jac}(\mathcal{C}_v)$, characterized by the property that each point of $\mathrm{Jac}(\mathcal{C}_v)(K_v)$ extends to a section of $J_{\mathrm{Ner}}(\mathcal{C}_v)/\mathrm{Spec}(\mathcal{O}_v)$. By ([**12**], Theorem 1, p.153), $J_{\mathrm{Ner}}(\mathcal{C}_v)$ is quasi-projective. Let it be embedded in $\mathbb{P}^N/\mathcal{O}_v$, for an appropriate N. Fix a corresponding system of homogeneous coordinates on $\mathbb{P}^N/K_v$. We identify $\mathrm{Jac}(\mathcal{C}_v)$ with generic fibre of $J_{\mathrm{Ner}}(\mathcal{C}_v)$, viewing it as locally cut out of $\mathbb{P}^N$ by the equations defining $J_{\mathrm{Ner}}(\mathcal{C}_v)$.

Let $\|x,y\|_{J,v}$ be the induced spherical metric on $\mathrm{Jac}(\mathcal{C}_v)(\mathbb{C}_v)$, and let O be the origin of $\mathrm{Jac}(\mathcal{C}_v)$. Then the ball $B_J(O,1)^- := \{z \in \mathrm{Jac}(\mathcal{C}_v)(\mathbb{C}_v) : \|z,O\|_{J,v} < 1\}$ is a subgroup. Since O is nonsingular on the special fibre of the Néron model, $B_J(O,1)^-$ can be isometrically parametrized by parametrized by power series converging on $D(\vec{0},1)^-$, taking $\vec{0}$ to O (Theorem 3.9). Let

$$\Psi : D(\vec{0},1)^- \to B_J(O,1)^- \tag{D.2}$$

be such an isometric parametrization. By the construction in ([**51**], Theorem 1.2.3) we can assume Ψ has been chosen in such a way that $\Psi^{-1} : B_J(O,1)^- \to D(\vec{0},1)^-$ is projection on some of the coordinates. Pulling the group action back to $D(\vec{0},1)^-$ using Ψ yields the formal group of $\mathrm{Jac}(\mathcal{C}_v)$ over $\mathcal{O}_v$.

Writing $\vec{X} = (X_1,\ldots,X_g)$ and $\vec{Y} = (Y_1,\ldots,Y_g)$, let $S(\vec{X},\vec{Y}) \in \mathcal{O}_v[[\vec{X},\vec{Y}]]^g$ and $M(\vec{X}) \in \mathcal{O}_v[[\vec{X}]]^g$ be the vectors of power series defining addition and negation in the formal group. Since $S(\vec{X},\vec{0}) = \vec{X}$, $S(\vec{X},\vec{Y}) = S(\vec{Y},\vec{X})$, and $S(\vec{X},M(\vec{X})) = \vec{0}$, modulo terms of degree ≥ 2 we have

$$S(\vec{X},\vec{Y}) \equiv \vec{X}+\vec{Y}, \qquad M(\vec{X}) \equiv -\vec{X}\ . \tag{D.3}$$

The following facts are well known:

PROPOSITION D.1. *Let p be the residue characteristic of K_v. Then*
(A) For each $0 < r < 1$ the ball

$$B_J(O,r) := \{z \in \mathrm{Jac}(\mathcal{C}_v)(\mathbb{C}_v) : \|z,O\|_{J,v} \leq r\}$$

is an open subgroup of $B_J(O,1)^-$.
(B) $B_J(O,1)^-$ is a topological pro-p-group.
(C) There is an $R > 0$ such that $B_J(O,R)$ is torsion-free.

PROOF. For the convenience of the reader we recall the proofs. It suffices to prove the assertions for $D(\vec{0},1)^-$ with the group law defined by $S(\vec{X},\vec{Y})$. Let $[2](\vec{X}) = S(\vec{X},\vec{X})$ and inductively put $[n](\vec{X}) = S(\vec{X},[n-1](\vec{X}))$ for $n = 3,4,\ldots$.

For (A), fix $0 < r < 1$ and suppose $\vec{x},\vec{y} \in D(\vec{0},r)$. It follows easily from (D.3) that $S(\vec{x},\vec{y})$ and $M(\vec{x},\vec{y})$ belong to $D(\vec{0},r)$.

For (B), note that by (D.3) we have $[p](\vec{X}) \equiv p\vec{X}$ modulo terms of degree ≥ 2. If $\vec{x} \in D(\vec{0},1)^-$, then $\vec{x} \in D(\vec{0},r)$ for some $r < 1$, and each term in the series defining $[p](\vec{x})$ has absolute value at most $R = \max(|p|_v r, r^2)$, so $[p](\vec{x}) \in D(\vec{0},R)$. Iterating this we see that

$$\lim_{k\to\infty} [p^k](\vec{x}) \;=\; \vec{0}\ .$$

For (C), note that by (A) and (B) above, $B_J(O,1)^-$ can only have p-power torsion. By the general theory of abelian varieties, $\mathrm{Jac}(\mathcal{C}_v)(\mathbb{C}_v)$ has at most $p^{2g}-1$ elements of order p, so the same is true for $B_J(O,1)^-$. If $R > 0$ is small enough, then $B_J(O,R)$ has no elements of order p, and hence is torsion free. □

Remark. When $\mathrm{char}(K_v) = 0$, it follows from the existence of the v-adic 'logarithm map' (see [**59**], Corollary 4, p.LG5.36) that there is a subgroup $B_J(0,R)$ analytically isomorphic to the additive group $\widehat{\mathcal{O}}_v^g$. However, when $\mathrm{char}(K_v) = p > 0$, no logarithm map exists, and no subgroup $B_J(0,R)$ can be isomorphic to $\widehat{\mathcal{O}}_v^g$, since $B_J(0,r)$ is torsion-free for all small r, while $\widehat{\mathcal{O}}_v^g$ is purely p-torsion. Nonetheless, by considering the form of the power series $S(\vec{X},\vec{Y})$ and $M(\vec{X})$, one sees easily that for any $0 \neq \pi \in \widehat{\mathcal{O}}_v$, there is an $R_0 > 0$ such that if $0 < R \leq R_0$ and $R \in |\mathbb{C}_v^\times|_v$, then $B_J(O,R)/B_J(O,|\pi|_v R)$ is isomorphic to $\widehat{\mathcal{O}}_v^g/\pi\widehat{\mathcal{O}}_v^g$.

1. The Local Action of the Jacobian on $\mathcal{C}_v^g$

Write $\overline{\mathcal{C}}_v = \mathcal{C}_v \times_{K_v} \mathrm{Spec}(\mathbb{C}_v)$. Then $\mathrm{Jac}(\overline{\mathcal{C}}_v) \cong \mathrm{Jac}(\mathcal{C}_v) \times_{K_v} \mathrm{Spec}(\mathbb{C}_v)$. We will identify $\mathcal{C}_v(\mathbb{C}_v)$ with $\overline{\mathcal{C}}_v(\mathbb{C}_v)$, and $\mathrm{Jac}(\mathcal{C}_v)(\mathbb{C}_v)$ with $\mathrm{Jac}(\overline{\mathcal{C}}_v)(\mathbb{C}_v)$.

Let $\mathrm{Div}_{\overline{\mathcal{C}}_v/\mathbb{C}_v}(\mathbb{C}_v)$ be the divisor group of $\overline{\mathcal{C}}_v$, and let $\sim$ denote the relation of linear equivalence for divisors. Let $\mathrm{Pic}_{\overline{\mathcal{C}}_v/\mathbb{C}_v}(\mathbb{C}_v) = \mathrm{Div}_{\overline{\mathcal{C}}_v/\mathbb{C}_v}(\mathbb{C}_v)/\sim$ be the relative Picard group (see the discussion after Theorem C.1 in Appendix C), and let $\mathbf{Pic}_{\overline{\mathcal{C}}_v/\mathbb{C}_v}$ be the associated Picard scheme. Let $\mathbf{Pic}^{\nu}_{\overline{\mathcal{C}}_v/\mathbb{C}_v}$ be its degree ν component, regarded as an algebraic variety. Then $\mathbf{Pic}^0_{\overline{\mathcal{C}}_v/\mathbb{C}_v} \cong \mathrm{Jac}(\overline{\mathcal{C}}_v)$ as a group scheme, and $\mathbf{Pic}^{\nu}_{\overline{\mathcal{C}}_v/\mathbb{C}_v}(\mathbb{C}_v)$ is a principal homogeneous space for $\mathrm{Jac}(\overline{\mathcal{C}}_v)(\mathbb{C}_v) = \mathrm{Jac}(\mathcal{C}_v)(\mathbb{C}_v)$.

On the product $\overline{\mathcal{C}}_v^g$ we have the cycle class map $[\;] : \overline{\mathcal{C}}_v^g \to \mathbf{Pic}^g_{\overline{\mathcal{C}}_v/\mathbb{C}_v}$, defined on $\mathcal{C}_v(\mathbb{C}_v)^g$ by

$$[\vec{x}] \;=\; [(x_1,\ldots,x_g)] \;=\; ((x_1)+\ldots+(x_g))/\sim\ .$$

In the notation of Appendix C, $[\vec{x}] = P\circ Q(\vec{x})$ where $Q : \overline{\mathcal{C}}_v^g \to \mathrm{Sym}^{(g)}(\overline{\mathcal{C}}_v) \cong \mathbf{Div}^g_{\overline{\mathcal{C}}_v/\mathbb{C}_v}$ is the quotient by the symmetric group S_g, and $P : \mathrm{Sym}^{(g)}(\overline{\mathcal{C}}_v) \to \mathbf{Pic}^g_{\overline{\mathcal{C}}_v/\mathbb{C}_v}$ is the Abel map $P(\sum_{i=1}^g(x_i))) = [\sum_{i=1}^g(x_i)]$. The morphism Q is flat and finite of degree $g!$, and the morphism P is a birational isomorphism.

Let $\dot{+}$ and $\dot{-}$ denote addition and subtraction for the group law on $\mathbf{Pic}_{\overline{\mathcal{C}}_v/\mathbb{C}_v}(\mathbb{C}_v)$, and by restriction, on $\mathrm{Jac}(\mathcal{C}_v)(\mathbb{C}_v)$. For each $\vec{a} \in \mathcal{C}_v(\mathbb{C}_v)^g$, let $\mathbf{J}_{\vec{a}} : \mathcal{C}_v(\mathbb{C}_v)^g \to \mathrm{Jac}(\mathcal{C}_v)(\mathbb{C}_v)$ be the map

$$\text{(D.4)} \qquad \mathbf{J}_{\vec{a}}(\vec{x}) \;=\; [\vec{x}]\dot{-}[\vec{a}]\ .$$

Now suppose F_u/K_v is a separable finite extension, and let H_w be the Galois closure of F_u over K_v. Put $d = [F_u : K_v]$, and let $\sigma_1,\ldots,\sigma_d$ be the distinct embeddings of F_u into H_w. Extend each σ_i to an automorphism of H_w. Since the addition law $\dot{+}$ in $\text{Jac}(\mathcal{C}_v)$ is defined over K_v, for each $x \in \text{Jac}(\mathcal{C}_v)(F_u)$ the trace $\text{Tr}_{F_u/K_v} : \text{Jac}(\mathcal{C}_v)(F_u) \to \text{Jac}(\mathcal{C}_v)(K_v)$ is given by

$$Tr_{F_u/K_v}(x) = \sigma_1(x)\dot{+}\sigma_2(x)\dot{+}\cdots\dot{+}\sigma_d(x) .$$

Our local action theorem is as follows:

THEOREM D.2. *Let K_v be a nonarchimedean local field, and let $\mathcal{C}_v/K_v$ be a smooth, projective, geometrically integral curve of genus $g > 0$. Then the points $\vec{a} = (a_1, \ldots, a_g) \in \mathcal{C}_v(\mathbb{C}_v)^g$ such that $\mathbf{J}_{\vec{a}} : \mathcal{C}_v(\mathbb{C}_v)^g \to \text{Jac}(\mathcal{C}_v)(\mathbb{C}_v)$ is nonsingular at $\vec{a}$ are dense in $\mathcal{C}_v(\mathbb{C}_v)^g$ for the v-topology. If F_u/K_v is a finite extension and $\mathcal{C}_v(F_u)$ is nonempty, they are dense in $\mathcal{C}_v(F_u)^g$.*

Fix such an $\vec{a}$; then $a_1, \ldots, a_g$ are distinct, and for each $0 < \eta < 1$, there is an R with $0 < R < 1$ (depending on $\vec{a}$ and η) such that the balls $B_{\mathcal{C}_v}(a_1, R), \ldots, B_{\mathcal{C}_v}(a_g, R)$ are pairwise disjoint and isometrically parametrizable, and for each $\vec{r} = (r_1, \ldots, r_g)$ satisfying

$$0 < r_1, \ldots, r_g \le R \quad \textit{and} \quad \eta \cdot \max(r_i) \le \min(r_i) , \tag{D.5}$$

(A) (Subgroup) *$\mathbf{J}_{\vec{a}} : \mathcal{C}_v(\mathbb{C}_v)^g \to \text{Jac}(\mathcal{C}_v)(\mathbb{C}_v)$ is injective on $\prod_{i=1}^g B_{\mathcal{C}_v}(a_i, r_i)$, and the image $W_{\vec{a}}(\vec{r}) := \mathbf{J}_{\vec{a}}\big(\prod_{i=1}^g B_{\mathcal{C}_v}(a_i, r_i)\big)$ is an open subgroup of $\text{Jac}(\mathcal{C}_v)(\mathbb{C}_v)$.*

(B) (Limited Distortion) *For $i = 1, \ldots g$, let $\varphi_i : D(0, R) \to B(a_i, R)$ be isometric parametrizations with $\varphi_i(0) = a_i$. Given $0 < r_1, \ldots, r_g \le R$, let $\Phi_{\vec{a}} = (\varphi_1, \ldots, \varphi_g) : \prod_{i=1}^g D(0, r_i) \to \prod_{i=1}^g B(a_i, r_i)$ be the associated map. Let $\Psi : D(\vec{0}, 1)^- \to B_J(O, 1)^-$ be the isometric parametrization inducing the formal group, and let $L_{\vec{a}} : \mathbb{C}_v^g \to \mathbb{C}_v^g$ be the linear map $(\Psi^{-1} \circ \mathbf{J}_{\vec{a}} \circ \Phi_{\vec{a}})'(\vec{0})$.*

Then $W_{\vec{a}}(\vec{r}) = \Psi(L_{\vec{a}}(D(\vec{0}, \vec{r})))$. Giving $D(\vec{0}, \vec{r})$ its structure as an additive subgroup of $\mathbb{C}_v^g$, the map $\Psi \circ L_{\vec{a}}$ induces an isomorphism of groups

$$D(\vec{0}, \vec{r})/D(\vec{0}, \eta\vec{r}) \cong W_{\vec{a}}(\vec{r})/W_{\vec{a}}(\eta\vec{r}),$$

with the property that for each $\vec{x} \in D(\vec{0}, \vec{r})$,

$$\mathbf{J}_{\vec{a}}(\Phi_{\vec{a}}(\vec{x})) \equiv \Psi(L_{\vec{a}}(\vec{x})) \pmod{W_{\vec{a}}(\eta\vec{r})} .$$

(C) (Action) *There is an action $(w, \vec{x}) \mapsto w\ddot{+}\vec{x}$ of $W_{\vec{a}}(\vec{r})$ on $\prod_{i=1}^g B_{\mathcal{C}_v}(a_i, r_i)$ which makes $\prod_{i=1}^g B_{\mathcal{C}_v}(a_i, r_i)$ into a principal homogeneous space for $W_{\vec{a}}(\vec{r})$. It is defined by $w\ddot{+}\vec{x} = \mathbf{J}_{\vec{a}}^{-1}(w\dot{+}\mathbf{J}_{\vec{a}}(\vec{x}))$ if we restrict the domain of $\mathbf{J}_{\vec{a}}$ to $\prod_{i=1}^g B_{\mathcal{C}_v}(a_i, r_i)$, and has the property that for each $w \in W_{\vec{a}}(\vec{r})$ and each $\vec{x} \in \prod_{i=1}^g B_{\mathcal{C}_v}(a_i, r_i)$,*

$$[w\ddot{+}\vec{x}] = w\dot{+}[\vec{x}] . \tag{D.6}$$

(D) (Uniformity) *For each $\vec{b} \in \prod_{i=1}^g B_{\mathcal{C}_v}(a_i, r_i)$,*

$$W_{\vec{a}}(\eta\vec{r}) \ddot{+} \vec{b} = \prod_{i=1}^g B_{\mathcal{C}_v}(b_i, \eta r_i) \quad \textit{and} \quad \mathbf{J}_{\vec{b}}\Big(\prod_{i=1}^g B_{\mathcal{C}_v}(b_i, \eta r_i)\Big) = W_{\vec{a}}(\eta\vec{r}) . \tag{D.7}$$

(E) (Rationality) *If F_u/K_v is a finite extension, and $\vec{a} \in \mathcal{C}_v(F_u)^g$, then*

$$\mathbf{J}_{\vec{a}}(\prod_{i=1}^g(B_{\mathcal{C}_v}(a_i, r_i) \cap \mathcal{C}_v(F_u))) = W_{\vec{a}}(\vec{r}) \cap \text{Jac}(\mathcal{C}_v)(F_u) , \tag{D.8}$$

$$(W_{\vec{a}}(\vec{r}) \cap \text{Jac}(\mathcal{C}_v)(F_u)) \ddot{+} \vec{a} = \prod_{i=1}^g \big(B_{\mathcal{C}_v}(a_i, r_i) \cap \mathcal{C}_v(F_u)\big) . \tag{D.9}$$

(F) (Trace) *If F_u/K_v is finite and separable, there is a constant $C = C(F_u, \vec{a}) > 0$, depending on F_u and $\vec{a}$ but not on $\vec{r}$, such that if $r = \min_i(r_i)$ then*

$$B_J(O, Cr) \cap \mathrm{Jac}(\mathcal{C}_v)(K_v) \subseteq \mathrm{Tr}_{F_u/K_v}\left(W_{\vec{a}}(\vec{r}) \cap \mathrm{Jac}(\mathcal{C}_v)(F_u)\right) . \tag{D.10}$$

For the proof of the Fekete-Szegö theorem with local rationality we will need one more property of the subgroups $W_{\vec{a}}(\vec{r})$: Lipschitz continuity of the Abel maps $\mathbf{j}_x(z) = [(z) - (x)]$ on compact sets $E_v \subset \mathcal{C}_v(\mathbb{C}_v)$.

PROPOSITION D.3. *Let K_v be a nonarchimedean local field, and let $\mathcal{C}_v/K_v$ be a smooth, projective, geometrically integral curve of genus $g = g(\mathcal{C}_v) > 0$. Let $E_v \subset \mathcal{C}_v(\mathbb{C}_v)$ be compact. Suppose that $\vec{a} \in E_v^g$ is a point such that $\mathbf{J}_{\vec{a}} : \mathcal{C}_v(\mathbb{C}_v)^g \to \mathrm{Jac}(\mathcal{C}_v)(\mathbb{C}_v)$ is nonsingular at $\vec{a}$, and let $0 < \eta < 1$, $0 < R < 1$ and $0 < r_1, \ldots, r_g \leq R$ be numbers satisfying the conditions* (D.5)*, so $W_{\vec{a}}(\vec{r}) = \mathbf{J}_{\vec{a}}\big(\prod_{i=1}^g B_{\mathcal{C}_v}(a_i, r_i)\big)$ is an open subgroup of $\mathrm{Jac}(\mathcal{C}_v)(\mathbb{C}_v)$ with the properties in Theorem* D.2.

Then there are constants $\varepsilon_0, C_0 > 0$ (depending on $\vec{a}$ and E_v) such that if $0 < \varepsilon \leq \varepsilon_0$, then for all $x, z \in E_v$ with $\|x, z\|_v \leq \varepsilon$, the divisor class $\mathbf{j}_x(z) = [(z) - (x)]$ belongs to $W_{\vec{a}}(C_0 \varepsilon \vec{r})$.

The proofs, given in §D.3, involve expanding the maps in question in terms of power series, and applying properties of power series proved below.

2. Lemmas on Power Series in Several Variables

In this section we recall some facts about power series in several variables. All the results are standard. Fix $0 < d \in \mathbb{N}$; in the application we will take $d = g$.

Given variables $X_1, \ldots, X_d$ and natural numbers $k_1, \ldots, k_d$, we write $\vec{X} = (X_1, \ldots, X_d)$, $k = (k_1, \ldots, k_d)$, and $\vec{X}^k = X_1^{k_1} \cdots X_d^{k_d}$. Put $\vec{0} = (0, \ldots, 0) \in \mathbb{C}_v^d$. If $\vec{a} = (a_1, \ldots, a_d) \in \mathbb{C}_v^d$ and $\vec{r} = (r_1, \ldots, r_d) \in \mathbb{R}^d$, with $r_1, \ldots, r_d > 0$, we write

$$D(\vec{a}, \vec{r}) = \prod_{i=1}^d D(a_i, r_i) , \qquad D(\vec{a}, \vec{r})^- = \prod_i D(a_i, r_i)^- .$$

If $r > 0$ we put $D(\vec{0}, r) = D(0, r)^d$ and $D(\vec{0}, r)^- = (D(0,1)^-)^d$. The norm $|\vec{x}|_v = \max_i(|x_i|_v)$ induces a metric $|\vec{x} - \vec{y}|_v$ on $\mathbb{C}_v^d$ and on each polydisc $D(\vec{a}, \vec{r})$, $D(\vec{a}, \vec{r})^-$.

LEMMA D.4. *Suppose $g(\vec{X}) \in \mathbb{C}_v[[\vec{X}]]$ converges on $D(\vec{0}, \vec{r})$, with $g(\vec{a}) = 0$ for each $\vec{a} \in D(\vec{0}, \vec{r}) \cap K_v^d$. Then $g(\vec{X})$ is the zero power series.*

PROOF. If $g(\vec{X}) \not\equiv 0$, after making a K_v-rational change of variables and shrinking the r_i if necessary, we can apply the Weierstrass Preparation Theorem (see [**11**], Theorem 1, p.201, and Proposition 2, p.205). This means we can factor $g(\vec{X})$ as $G(\vec{X})h(\vec{X})$ where $h(\vec{X})$ is an invertible power series converging in $D(\vec{0}, \vec{r})$ and

$$G(\vec{X}) = X_d^M + \sum_{i=1}^M c_i(X_1, \ldots, X_{d-1}) X_d^{M-i}$$

is a monic polynomial in X_d with coefficients $c_i(X_1, \ldots, X_{d-1}) \in \mathbb{C}_v[[X_1, \ldots, X_{d-1}]]$ which converge in $D(\vec{0}, (r_1, \ldots, r_{d-1}))$. Since $g(\vec{X})$ vanishes on $D(\vec{0}, \vec{r}) \cap K_v^d$, so does $G(\vec{X})$. Fixing $a_1, \ldots, a_{d-1} \in K_v$ with $|a_i|_v \leq r_i$ for each i, we see that $G(a_1, \ldots, a_{d-1}, X_d)$ is a monic polynomial in X_d with infinitely many roots.

This is impossible, so $g(\vec{X}) \equiv 0$. □

Suppose $h(\vec{X}) = \sum a_k \vec{X}^k \in \mathbb{C}_v[[\vec{X}]]$ converges on the polydisc $D(\vec{0},1)$. It follows from the Maximum Modulus Principle (see [**11**], p.201) that

$$\|h\|_{D(\vec{0},1)} = \max_k(|a_k|_v) . \tag{D.11}$$

Here $\|h\|_{D(\vec{0},1)} = \sup_{\vec{z}\in D(\vec{0},1)}(|h(\vec{z})|_v)$ is the sup norm of h, and $\max_k(|a_k|_v)$ is the so-called "Gauss norm". If $h(\vec{X})$ converges on the polydisc $D(\vec{0},\vec{r})$, an analogous result holds:

$$\|h\|_{D(\vec{0},\vec{r})} = \max_k(|a_k|_v \vec{r}^{\,k}) . \tag{D.12}$$

Uniform convergence of values implies convergence in the Gauss norm:

LEMMA D.5. *Fix a polydisc $D(\vec{0},\vec{r})$. Let $g^{(\ell)}(\vec{X}) \in \mathbb{C}_v[[\vec{X}]]$, for $\ell = 1,2,\ldots$, be power series for which there is a function $G(\vec{X})$ on $D(\vec{0},\vec{r})$ such that uniformly for $\vec{a} \in D(\vec{0},\vec{r})$, the values $g^{(\ell)}(\vec{a})$ converge to $G(\vec{a})$. Then there is a power series $g(\vec{X}) \in \mathbb{C}_v[[\vec{X}]]$ converging on $D(\vec{0},\vec{r})$ such that the coefficients of the $g^{(\ell)}(\vec{X})$ converge to the coefficients of $g(\vec{X})$, and $g(\vec{a}) = \lim_{\ell\to\infty} g^{(\ell)}(\vec{a}) = G(\vec{a})$ for each $\vec{a} \in D(\vec{0},\vec{r})$.*

PROOF. Write $g^{(\ell)}(\vec{X}) = \sum_k b_k^{(\ell)} \vec{X}^k$. By the Maximum Modulus Principle, applied to $D(\vec{0},\vec{r})$, for each $k \in \mathbb{N}^d$ the coefficients $b_k^{(\ell)}$ converge to a number b_k, and if $g(\vec{X}) = \sum_k b_k \vec{X}^k$ then in the Gauss norm for $D(\vec{0},\vec{r})$ the $g^{(\ell)}(\vec{X})$ converge to $g(\vec{X})$. But convergence in the Gauss norm implies convergence of values on $D(\vec{0},\vec{r})$. □

Recall that $\widehat{\mathcal{O}}_v$ denotes the ring of integers of $\mathbb{C}_v$.

PROPOSITION D.6. *Let $H(\vec{X}) = (h_1(\vec{X}),\ldots,h_d(\vec{X})) \in \widehat{\mathcal{O}}_v[[\vec{X}]]^d$ be such that $H(\vec{X}) \equiv \vec{X}$ modulo terms of degree ≥ 2. Then*

(A) H induces an isometry from $D(\vec{0},1)^-$ onto $D(\vec{0},1)^-$.

(B) For each polydisc $D(\vec{0},\vec{r}) \subset D(\vec{0},1)^-$ satisfying $\max_i(r_i)^2 \leq \min_i(r_i)$, H induces an isometry from $D(\vec{0},\vec{r})$ onto $D(\vec{0},\vec{r})$, and if $F_u \subseteq \mathbb{C}_v$ is a complete field such that $H(\vec{X})$ is rational over F_u, then $H(D(\vec{0},\vec{r}) \cap F_u^d) = D(\vec{0},\vec{r}) \cap F_u^d$.

PROOF. First, note that the form of $H(\vec{X})$ shows that H converges on $D(\vec{0},1)^-$ and that $H(D(\vec{0},1)^-) \subset D(\vec{0},1)^-$.

Next we claim that H preserves distances on $D(\vec{0},1)^-$. To see this, fix $\vec{p},\vec{q} \in D(\vec{0},1)^-$ with $\vec{p} \neq \vec{q}$. Write $h_i(\vec{X}) = \sum_k a_{i,k}\vec{X}^k$, and put $|k| = k_1 + \ldots + k_d$. For each i we have

$$h_i(\vec{p}) - h_i(\vec{q}) = (p_i - q_i) + \sum_{|k|\geq 2} a_{i,k}(\vec{p}^{\,k} - \vec{q}^{\,k}) .$$

Using the ultrametric property of $\mathbb{C}_v$, it is easy to see that for each k

$$|a_{i,k}(\vec{p}^{\,k} - \vec{q}^{\,k})|_v \leq (\max_j |p_j - q_j|_v) \cdot \big(\max(|\vec{p}|_v, |\vec{q}|_v)\big)^{|k|-1} . \tag{D.13}$$

Consequently, $|H(\vec{p}) - H(\vec{q})|_v \leq \max_i |p_i - q_i|_v = |\vec{p} - \vec{q}|_v$.

Now take i so that $|p_i - q_i|_v$ is maximal. Noting that $\max(|\vec{p}|_v, |\vec{q}|_v) < 1$, it follows from (D.13) that $|p_i - q_i|_v > |a_{i,k}(\vec{p}^{\,k} - \vec{q}^{\,k})|_v$ for all k with $|k| \geq 2$.

Consequently $|h_i(\vec{p}) - h_i(\vec{q})|_v = |p_i - q_i|_v \neq 0$, so $|H(\vec{p}) - H(\vec{q})|_v = |\vec{p} - \vec{q}|_v$. In particular, H is $1-1$ on $D(\vec{0},1)^-$.

Next we will show that $H(\vec{X})$ has a right inverse $G(\vec{X}) = (g_1(\vec{X}), \ldots, g_d(\vec{X}))$ belonging to $\widehat{\mathcal{O}}_v[[\vec{X}]]$, such that $G(\vec{X}) \equiv \vec{X}$ modulo terms of degree ≥ 2. To do this, we apply Newton's method to power series, and use convergence of values on $D(\vec{0},1)^-$ to deduce convergence of coefficients of the power series.

Let $J_H(\vec{X}) = (\frac{\partial h_i}{\partial z_j}(\vec{X}))$ be the Jacobian matrix of $H(\vec{X})$, computed using formal partial derivatives of series. By hypothesis, $J_H(\vec{X}) \equiv I$ modulo terms of degree ≥ 1, so $\det(J_H(\vec{X})) \in \widehat{\mathcal{O}}_v[[\vec{X}]]$ is a power series with constant term 1. Thus, the formal geometric series for $\det(J_H(\vec{X}))^{-1}$ belongs to $\widehat{\mathcal{O}}_v[[\vec{X}]]$ and converges for all $\vec{z} \in D(\vec{0},1)^-$, and $J_H(\vec{X})^{-1} = \det(J_H(\vec{X}))^{-1} \cdot \mathrm{Adj}(J_H(\vec{X}))$ has components given by power series in $\widehat{\mathcal{O}}_v[[\vec{X}]]$ which also converge for all $\vec{z} \in D(\vec{0},1)^-$.

Define a sequence of maps $G^{(\ell)}(\vec{X}) = (g_1^{(\ell)}(\vec{X}), \ldots, g_d^{(\ell)}(\vec{X}))$ with coordinate functions belonging to $\widehat{\mathcal{O}}_v[[\vec{X}]]$, by setting $G^{(0)}(\vec{X}) = \vec{X}$ and putting

$$G^{(\ell+1)}(\vec{X}) = G^{(\ell)}(\vec{X}) - J_H(G^{(\ell)}(\vec{X}))^{-1} \cdot (H(G^{(\ell)}(\vec{X})) - \vec{X}) \tag{D.14}$$

for each ℓ. Inductively one sees that $G^{(\ell)}(\vec{X}) \equiv \vec{X}$ modulo terms of degree ≥ 2, so the substitutions make sense formally and all the component functions $g_i^{(\ell)}(\vec{X})$ belong to $\widehat{\mathcal{O}}_v[[\vec{X}]]$. In particular, they converge on $D(\vec{0},1)^-$.

Next fix $\vec{q} \in D(\vec{0},1)^-$ and put $r = |\vec{q}|_v < 1$. The usual sequence of Newton iterates $\vec{p}_0, \vec{p}_1, \ldots \in D(\vec{0},r)$ converging to a solution of $H(\vec{p}) = \vec{q}$ is defined by setting $\vec{p}_0 = \vec{q}$ and putting

$$\vec{p}_{\ell+1} = \vec{p}_\ell - J_H(\vec{p}_\ell)^{-1} \cdot (H(\vec{p}_\ell) - \vec{q}) . \tag{D.15}$$

By the form of $H(\vec{X})$, clearly $|H(\vec{p}_0) - \vec{q}|_v \leq r^2$. Assume inductively that $\vec{p}_\ell \in D(\vec{0},r)$ and $|H(\vec{p}_\ell) - \vec{q}|_v \leq r^{\ell+2}$. By (D.15) we have $\vec{p}_{\ell+1} \in D(\vec{0},r)$. Expanding $H(\vec{p}_{\ell+1})$ and using (D.15) we find that

$$|H(\vec{p}_{\ell+1}) - \vec{q}|_v \leq r^{2(\ell+2)} \leq r^{(\ell+1)+2} .$$

From (D.15) we see that the $\vec{p}_\ell$ converge to a vector $\vec{p} \in D(\vec{0},r)$ such that $H(\vec{p}) = \vec{q}$.

Comparing (D.14) and (D.15) shows that that $\vec{p}_\ell = G^{(\ell)}(\vec{q})$ for each ℓ, that is, the values of the $G^{(\ell)}(\vec{q})$ converge for each $\vec{q} \in D(\vec{0},1)^-$. Moreover by the ultrametric inequality, for each $r < 1$ the convergence is uniform on $D(\vec{0},r)$, with $|G^{(\ell+1)}(\vec{q}) - G^{(\ell)}(\vec{q})|_v \leq r^{\ell+2}$ for all ℓ and all $\vec{q} \in D(\vec{0},r)$. Hence, Lemma D.5 produces a function $G(\vec{X}) = (g_1(\vec{X}), \ldots, g_d(\vec{X}))$ with components $g_i(\vec{X}) \in \widehat{\mathcal{O}}_v[[\vec{X}]]$, such that $H(G(\vec{q})) = \vec{q}$ for all $\vec{q} \in D(\vec{0},1)^-$. The fact that each $G^{(\ell)}(\vec{X}) \equiv \vec{X}$ modulo terms of degree ≥ 2 means that $G(\vec{X})$ has this property as well. From this, it follows that $G(D(\vec{0},1)^-) \subset D(\vec{0},1)^-$. Hence, H gives a surjection from $D(\vec{0},1)^-$ onto $D(\vec{0},1)^-$. To see that $G(\vec{X})$ is a left inverse to $H(\vec{X})$ as well as a right inverse, note that if $\vec{p} \in D(\vec{0},1)^-$ and $\vec{q} = H(\vec{p})$ then

$$H(\vec{p}) = \vec{q} = H(G(\vec{q})) ;$$

however, since $H(\vec{z})$ is $1-1$ on $D(\vec{0},1)^-$, necessarily $\vec{p} = G(\vec{q})$. That is, $\vec{p} = G(H(\vec{p}))$ for all $\vec{p} \in D(\vec{0},1)^-$.

The fact that H and G define inverse functions on $D(\vec{0},1)^-$ means $G(H(\vec{X}))-\vec{X}$ and $H(G(\vec{X}))-\vec{X}$ are identically equal to $\vec{0}$ on $D(\vec{0},1)^-$, and then Lemma D.4 shows their component power series are identically 0. Hence, H and G are formal inverses.

If $D(\vec{0},\vec{r}) \subset D(\vec{0},1)^-$ is a polydisc satisfying $\max_i(r_i)^2 \leq \min_i(r_i)$, then the fact that $H(\vec{X}) \in \widehat{\mathcal{O}}_v[[\vec{X}]]^d$ and $H(\vec{X}) \equiv \vec{X}$ modulo terms of degree ≥ 2 shows that H maps $D(\vec{0},\vec{r})$ into itself. However, $G(\vec{X})$ has the same form, so G maps $D(\vec{0},\vec{r})$ into itself as well. Since H and G are inverses, and H preserves distances, H induces an isometry from $D(\vec{0},\vec{r})$ onto itself.

Finally, if $H(\vec{X})$ is rational over F_u for some subfield $F_u \subset \mathbb{C}_v$, the formulas (D.14) show that the $G^{(\ell)}(\vec{X})$ are rational over F_u, and if F_u is complete then $G(\vec{X})$ is also rational over F_u. It follows that $H(D(\vec{0},\vec{r}) \cap F_u^d) \subseteq D(\vec{0},\vec{r}) \cap F_u^d$ and $G(D(\vec{0},\vec{r}) \cap F_u^d) \subseteq D(\vec{0},\vec{r}) \cap F_u^d$, so $H(D(\vec{0},\vec{r}) \cap F_u^d) = D(\vec{0},\vec{r}) \cap F_u^d$ □

Remark. Under the hypotheses of Proposition D.6, in general $H(\vec{X})$ will not converge on the closed unit polydisc $D(\vec{0},1)$, and even if it does, it will not in general be $1-1$ and $G(\vec{X})$ will not converge on $D(\vec{0},1)$. For example, in one variable, if $H(X) = X - X^2$ then $H(0) = H(1) = 0$, while the power series expansion for its inverse $G(X) = (1-\sqrt{1+4X}))/2$ only converges on $D(0,1)^-$.

For the proof of Proposition D.3, we will need the following lemma.

LEMMA D.7. *Let $R \in |\mathbb{C}_v^\times|_v$, and suppose $h : D(0,R) \times D(0,R) \to D(0,r)$ is a map defined by a power series $h(X,Y) = \sum_{j,k=0}^\infty c_{jk}X^jY^k \in \mathbb{C}_v[[X,Y]]$ which converges on $D(0,R) \times D(0,R)$, and satisfies $h(x,x) = 0$ for all $x \in D(0,R)$. Then for all $x,y \in D(0,R)$,*

$$|h(x,y)|_v \ \leq \ \frac{r}{R} \cdot |x-y|_v \ .$$

PROOF. By the Maximum Modulus Principle for power series, we have $|c_{jk}|_v \leq r/R^{j+k}$ for all j,k, and

$$\lim_{j,k\to\infty} |c_{jk}|_v \cdot R^{j+k} \ = \ 0 \tag{D.16}$$

since $h(X,Y)$ converges on $D(0,R) \times D(0,R)$. Fix $x,y \in D(0,R)$. Then

$$\begin{aligned} h(x,y) \ &= \ h(x,y) - h(x,x) \ = \ \sum_{j,k=0}^{\infty} c_{jk}x^j(y^k - x^k) \\ &= \ (y-x) \cdot \sum_{j,k=0}^{\infty} c_{jk}x^j \Big(\sum_{\ell=0}^{k-1} x^\ell y^{k-1-\ell}\Big) \ . \end{aligned} \tag{D.17}$$

By our estimate for $|c_{ij}|_v$ and the ultrametric inequality, each term in the sum on the right in (D.17) has absolute value at most r/R, and by (D.16) the sum converges and has absolute value at most r/R. Thus $|h(x,y)|_v \leq |x-y|_v \cdot r/R$. □

3. Proof of the Local Action Theorem

In this section we prove Theorem D.2 and Proposition D.3. We use the notation established before the statement of the theorem, and begin with four lemmas.

In the first lemma, we show that certain subsets of the formal group of $\mathrm{Jac}(\mathcal{C}_v)$ are subgroups. We denote a group structure by a triple consisting of the underlying set and two vectors of power series, representing addition and negation, which converge on the set.

LEMMA D.8. *Let* $\mathcal{D}(\vec{0},1)^- := (D(\vec{0},1)^-, S(\vec{X},\vec{Y}), M(\vec{X}))$ *be the formal group of* $\mathrm{Jac}(\mathcal{C}_v)$. *Let* $L : \mathbb{C}_v^g \to \mathbb{C}_v^g$ *be a nonsingular linear map, and fix* $0 < \eta < 1$. *Then there is an* $R_1 > 0$ *such that for each* $\vec{r} = (r_1, \dots, r_g)$ *satisfying* $0 < r_1, \dots, r_g \le R_1$ *and* $\eta \cdot \max_i(r_i) \le \min_i(r_i)$, *and each* $0 < \lambda \le 1$:

(A) $\mathcal{D}_L(\vec{0},\lambda\vec{r}) := (L(D(\vec{0},\lambda\vec{r})), S(\vec{X},\vec{Y}), M(\vec{X}))$ *is a subgroup of* $\mathcal{D}(\vec{0},1)^-$.

(B) If $D(\vec{0},\vec{r})$ *is given its structure as an additive subgroup of* $\mathbb{C}_v^g$, *the map of sets* $L : D(\vec{0},\vec{r}) \to L(D(\vec{0},\vec{r}))$ *induces an isomorphism of groups*

$$D(\vec{0},\vec{r})/D(\vec{0},\eta\vec{r}) \;\cong\; \mathcal{D}_L(\vec{0},\vec{r})/\mathcal{D}_L(\vec{0},\eta\vec{r}) \ .$$

PROOF. Let $(\alpha_{ij}), (\beta_{ij}) \in M_g(\mathbb{C}_v)$ be the matrices associated to L^{-1} and L, respectively, and put $A = \max_{i,j} |\alpha_{ij}|_v$, $B = \max_{i,j} |\beta_{ij}|_v$. Let

$$\widetilde{S}(\vec{X},\vec{Y}) \;=\; L^{-1}(S(L(\vec{X}),L(\vec{Y}))) \ , \qquad \widetilde{M}(\vec{X}) = L^{-1}(M(L(\vec{X}))) \ .$$

Then $\mathcal{D}_L(\vec{0},\vec{r})$ is a group if and only if $\widetilde{\mathcal{D}}(\vec{0},\vec{r}) := (D(\vec{0},\vec{r}), \widetilde{S}(\vec{X},\vec{Y}), \widetilde{M}(\vec{X}))$ is a group, and if they are groups, the map $L : D(\vec{0},\vec{r}) \to L(D(\vec{0},\vec{r}))$ induces isomorphism between them. Note that if $\vec{r}$ satisfies the conditions in the Lemma, then so does $\lambda\vec{r}$ for each $0 < \lambda \le 1$. Hence it suffices to consider the $\widetilde{\mathcal{D}}(\vec{0},\vec{r})$.

Write $\widetilde{S}(\vec{X},\vec{Y}) = (\widetilde{S}_1(\vec{X},\vec{Y}), \dots, \widetilde{S}_g(\vec{X},\vec{Y}))$, $\widetilde{M}(\vec{X}) = (\widetilde{M}_1(\vec{X}), \dots, \widetilde{M}_g(\vec{X}))$, and for each i, expand

$$\widetilde{S}_i(\vec{X},\vec{Y}) = X_i + Y_i + \sum_{|k|+|\ell|\ge 2} c_{i,k,\ell}\vec{X}^k\vec{Y}^\ell \ , \qquad \widetilde{M}_i(\vec{X}) = -X_i + \sum_{|k|\ge 2} d_{i,k}\vec{X}^k \ .$$

Since the power series defining $S(\vec{X},\vec{Y})$ and $M(\vec{X})$ have coefficients in $\widehat{\mathcal{O}}_v$, it is easy to see that for each $i = 1, \dots, g$ and all $k, \ell \in \mathbb{N}^g$ with $|k| + |\ell| \ge 2$ (resp. all $k \in \mathbb{N}^g$ with $|k| \ge 2$), we have $|c_{i,k,\ell}|_v \;\le\; AB^{|k|+|\ell|}$ and $|d_{i,k}|_v \le AB^{|k|}$. Hence for all $\vec{x},\vec{y} \in D(\vec{0},\vec{r})$

$$|c_{i,k,\ell}\vec{x}^k\vec{y}^k|_v \;\le\; A\big(B\max_i(r_i)\big)^{|k|+|\ell|} \ , \qquad |d_{i,k}\vec{x}^k|_v \;\le\; A\big(B\max_i(r_i)\big)^{|k|} \ .$$

If $\max_i(r_i) < 1/B$ then $\widetilde{S}(\vec{X},\vec{Y})$ and $\widetilde{M}(\vec{X})$ converge for $\vec{x},\vec{y} \in D(\vec{0},\vec{r})$. If also $\max_i(r_i) \le \eta/(AB^2)$ and $\eta\max_i(r_i) \le \min_i(r_i)$, then for all $|k| + |\ell| \ge 2$,

$$|c_{i,k,\ell}\vec{x}^k\vec{y}^\ell|_v \le A\big(B\max_i(r_i)\big)^{|k|+|\ell|} \le AB^2\max_i(r_i)^2 \;\le\; \eta\max_i(r_i) \;\le\; \min_i(r_i) \ .$$

Similarly for all $|k| \ge 2$, we have $|d_{i,k}\vec{x}^k|_v \le \min_i(r_i)$, and so $\widetilde{S}(\vec{X},\vec{Y})$ and $\widetilde{M}(\vec{X})$ map $D(\vec{0},\vec{r})$ into itself. Since $\widetilde{S}(\widetilde{S}(\vec{x},\widetilde{M}(\vec{y})),\vec{y}) = \vec{x}$ and $\widetilde{M}(\widetilde{M}(\vec{x})) = \vec{x}$ for all $\vec{x},\vec{y} \in D(\vec{0},\vec{r})$, they are surjective. Thus $\widetilde{\mathcal{D}}(\vec{0},\vec{r})$ is a group.

Finally, if $\max(r_i) \le \eta^2/AB^2$ then by an argument similar to the one above, for all $\vec{x},\vec{y} \in D(\vec{0},\vec{r})$ and all k,ℓ with $|k|+|\ell| \ge 2$, one has $|c_{i,k,\ell}\vec{x}^k\vec{y}^\ell|_v \le \eta\min_i(r_i)$. Likewise, for all $\vec{x} \in D(\vec{0},\vec{r})$ and all k with $|k| \ge 2$, one has $|d_{i,k}\vec{x}^k| \le \eta\min_i(r_i)$. Thus

$$\widetilde{S}(\vec{x},\vec{y}) = \vec{x} + \vec{y} + \delta_1(\vec{x},\vec{y}) \ , \qquad \widetilde{M}(\vec{x}) = -\vec{x} + \delta_2(\vec{x}) \ ,$$

where $\delta_1(\vec{x},\vec{y})$ and $\delta_2(\vec{x})$ belong to $D(\vec{0},\eta\vec{r})$. This means that if $D(\vec{0},\vec{r})$ is viewed as an additive subgroup of $\mathbb{C}_v^g$, then $\widetilde{\mathcal{D}}(\vec{0},\vec{r})/\widetilde{\mathcal{D}}(\vec{0},\eta\vec{r}) \;\cong\; D(\vec{0},\vec{r})/D(\vec{0},\eta\vec{r})$.

Put $R_1 = \frac{1}{2}\min(1/B, \eta^2/(AB^2))$. Then if $0 < r_1, \ldots, r_g \leq R_1$ and $\eta \max(r_i) \leq \min(r_i)$, both $\widetilde{\mathcal{D}}(\vec{0}, \vec{r})$ and $\widetilde{\mathcal{D}}(\vec{0}, \eta\vec{r})$ are groups, and

$$D(\vec{0}, \vec{r})/D(\vec{0}, \eta\vec{r}) \ \cong\ \widetilde{\mathcal{D}}(\vec{0}, \vec{r})/\widetilde{\mathcal{D}}(\vec{0}, \eta\vec{r}) \ \cong\ \mathcal{D}_L(\vec{0}, \vec{r})/\mathcal{D}_L(\vec{0}, \eta\vec{r}) \ .$$

This yields the result. □

Our next lemma shows that if $H(\vec{X}) : \mathbb{C}_v^g \to \mathbb{C}_v^g$ is a map defined by convergent power series, whose derivative $L = H'(\vec{0})$ is nonsingular, then for suitable polydiscs $D(\vec{0}, \vec{r})$ the image $H(D(\vec{0}, \vec{r}))$ coincides with $L(D(\vec{0}, \vec{r}))$.

LEMMA D.9. *Let $H(\vec{X}) \in \mathbb{C}_v[[\vec{X}]]^g$ be a vector of power series which converges on $D(\vec{0}, r)$ for some $r > 0$, and maps $D(\vec{0}, r)$ into $D(0, 1)^-$. Assume the derivative $L = H'(\vec{0}) : \mathbb{C}_v^g \to \mathbb{C}_v^g$ is nonsingular.*

Then for each $0 < \eta < 1$, there is an R_2 with $0 < R_2 \leq r$ such that for each $D(\vec{0}, \vec{r})$ with $0 < r_1, \ldots, r_g \leq R_2$ and $\eta \cdot \max_i(r_i) \leq \min_i(r_i)$:

(A) $H(\vec{X})$ gives an analytic isomorphism from $D(\vec{0}, \vec{r})$ onto $L(D(\vec{0}, \vec{r}))$, with the property that for each $\vec{z} \in D(\vec{0}, \vec{r})$,

$$H(\vec{z}) \ \equiv\ L(\vec{z}) \pmod{L(D(\vec{0}, \eta\vec{r}))} \ .$$

(B) For each complete field F_u such that $K_v \subseteq F_u \subseteq \mathbb{C}_v$ and $H(\vec{X})$ is rational over F_u,

$$H(D(\vec{0}, \vec{r}) \cap F_u^g) \ =\ L(D(\vec{0}, \vec{r})) \cap F_u^g \ .$$

PROOF. Let $(\alpha_{ij}) \in M_g(\mathbb{C}_v)$ be the matrix associated to L^{-1}, and put $A = \max_{i,j}(|\alpha_{ij}|_v)$. Choose $\pi \in K_v^\times$ so that $|\pi|_v < \min(r, r^2/A)$, and put $B = |\pi|_v$.

Our plan is to apply Proposition D.6 to

$$\widetilde{H}(\vec{X}) \ :=\ L^{-1}(\frac{1}{\pi}H(\pi\vec{X})) \ . \tag{D.18}$$

By construction, $\widetilde{H}'(\vec{0}) = \mathrm{id}$. Write

$$H(\vec{X}) \ =\ L(\vec{X}) + \sum_{|k|\geq 2} \vec{c}_k\vec{X}^k \ , \qquad \widetilde{H}(\vec{X}) \ =\ \vec{X} + \sum_{|k|\geq 2} \widetilde{c}_k\vec{X}^k \ ,$$

where $\vec{c}_k = (c_{1,k}, \ldots, c_{g,k})$ and $\widetilde{c}_k = (\widetilde{c}_{1,k}, \ldots, \widetilde{c}_{g,k})$. Since $H(\vec{X})$ converges on $D(\vec{0}, r)$ and maps $D(\vec{0}, r)$ into $D(\vec{0}, 1)^-$, we have $|c_{i,k}|_v \leq 1/r^{|k|}$ for all i, k. Since $B/r \leq 1$ and $B \leq r^2/A$, for each i and each k with $|k| \geq 2$ we have

$$|\widetilde{c}_{i,k}|_v \ \leq\ \frac{A}{B}\big(\frac{B}{r}\big)^{|k|} \ <\ \frac{A}{B}\big(\frac{B}{r}\big)^2 \ \leq\ 1 \ .$$

Thus $\widetilde{H}(\vec{X}) \in \widehat{\mathcal{O}}_v[[\vec{X}]]^g$ and $\widetilde{H}(\vec{X}) \equiv \vec{X}$ modulo terms of degree ≥ 2. By Proposition D.6, if $0 < s_1, \ldots, s_d < 1$ and $\max(s_i)^2 \leq \min(s_i)$, then $\widetilde{H}(D(\vec{0}, \vec{s})) = D(\vec{0}, \vec{s})$. Furthermore, if F_u is a complete field with $K_v \subset F_u \subset \mathbb{C}_v$, and if $H(\vec{X})$ is rational over F_u, then $\widetilde{H}(\vec{X})$ is rational over F_u, so $\widetilde{H}(D(\vec{0}, \vec{s}) \cap F_u^g) = D(\vec{0}, \vec{s}) \cap F_u^g$.

Given $0 < \eta < 1$, put $R_2 = \eta^2 B$. Then $R_2 < r$.

Suppose $0 < r_1, \ldots, r_g \leq R_2$, with $\eta \cdot \max_i(r_i) \leq \min_i(r_i)$. We first show that $H(D(\vec{0}, \vec{r})) = L(D(\vec{0}, \vec{r}))$. Put $\vec{s} = (1/B)\vec{r}$. Then $0 < s_i < 1$ for each i, since $\max_i(r_i) \leq R_2$ and $R_2/B = \eta^2 < 1$, and

$$\max_i(s_i)^2 \ \leq\ \frac{R_2}{B^2}\max_i(r_i) \ \leq\ \frac{\eta}{B}\big(\eta \max_i(r_i)\big) \ <\ \frac{1}{B}\min_i(r_i) \ =\ \min_i(s_i) \ .$$

Since $H(\pi\vec{X}) = \pi L(\widetilde{H}(\vec{X}))$ and $D(\vec{0},\vec{r}) = D(\vec{0},B\vec{s}) = \pi D(\vec{0},\vec{s})$, it follows that $H(D(\vec{0},\vec{r})) = L(D(\vec{0},\vec{r}))$ and $H(D(\vec{0},\vec{r}) \cap F_u^g) = L(D(\vec{0},\vec{r})) \cap F_u^g$.

Third, we show that $H(\vec{x}) \equiv L(\vec{x}) \pmod{L(D(\vec{0},\eta\vec{r}))}$ for each $\vec{x} \in D(\vec{0},\vec{r})$. Note that $H(\vec{X}) = L(\pi\widetilde{H}(\frac{1}{\pi}\vec{X}))$, and that if $\vec{x} \in D(\vec{0},\vec{r})$ then $\frac{\vec{x}}{\pi} \in D(\vec{0},1)^-$. By the inequalities $\max_i(r_i) \le R_2$ and $R_2 \le \eta^2 B < B$, and our estimate $|\widetilde{c}_{i,k}|_v \le 1$ above, for each i and each k with $|k| \ge 2$ we have

$$\begin{aligned} |\pi\widetilde{c}_{i,k}\left(\frac{\vec{x}}{\pi}\right)^k|_v &\le B\left(\frac{\max_i(r_i)}{B}\right)^{|k|} < B\left(\frac{\max_i(r_i)}{B}\right)^2 \le \frac{R_2}{B}\max_i(r_i) \\ &\le \eta^2 \max_i(r_i) \le \eta \min_i(r_i) . \end{aligned}$$

Here $\widetilde{H}(\vec{X}) \equiv \vec{X}$ modulo terms of degree ≥ 2, so $\pi\widetilde{H}(\frac{\vec{x}}{\pi}) \equiv \vec{x} \pmod{D(\vec{0},\eta\vec{r})}$. Since $H(\vec{x}) = L(\pi\widetilde{H}(\frac{\vec{x}}{\pi}))$, it follows that $H(\vec{x}) \equiv L(\vec{x}) \pmod{L(D(\vec{0},\eta\vec{r}))}$. □

Our third lemma shows that for a dense set of points $\vec{a} \in \mathcal{C}_v(\mathbb{C}_v)^g$, the map $[\,] : \overline{\mathcal{C}}_v^g \to \mathbf{Pic}^g_{\mathcal{C}_v/\mathbb{C}_v}$ is nonsingular at $\vec{a}$.

LEMMA D.10. *There is a nonempty Zariski-open subset $U \subset \overline{\mathcal{C}}_v^g$ such that for each $\vec{a} \in U(\mathbb{C}_v)$, the map $[\,] : \overline{\mathcal{C}}_v^g \to \mathbf{Pic}^g_{\mathcal{C}_v/\mathbb{C}_v}$ is nonsingular at $\vec{a}$. Moreover, for each complete field F_u with $K_v \subseteq F_u \subseteq \mathbb{C}_v$ such that $\mathcal{C}_v(F_u)$ is nonempty, $U(F_u)$ is dense in $\mathbb{C}_v(F_u)^g$ for the v-topology.*

PROOF. Fix $\vec{a} \in \mathcal{C}_v^g(\mathbb{C}_v)$. The function $[\,] : \overline{\mathcal{C}}_v^g \to \mathbf{Pic}^g_{\mathcal{C}_v/\mathbb{C}_v}$, where $[\vec{x}]$ is the linear equivalence class of the divisor $(x_1) + \cdots + (x_g)$, factors as

$$[\,] : \overline{\mathcal{C}}_v^g \xrightarrow{Q} \mathrm{Sym}^g(\overline{\mathcal{C}}_v) \xrightarrow{P} \mathbf{Pic}^g_{\mathcal{C}_v/\mathbb{C}_v}$$

where P is the quotient map, and Q is the Abel map.

First consider $Q : \overline{\mathcal{C}}_v^g \to \mathrm{Sym}^{(g)}(\overline{\mathcal{C}}_v)$. Let S_g be the symmetric group on the letters $\{1,\ldots,g\}$. For each $\pi \in S_g$, write $\pi(\vec{x}) = (x_{\pi(1)},\ldots,x_{\pi(g)})$. If $\vec{x} \in \mathcal{C}_v^g(\mathbb{C}_v)$ has distinct coordinates, then $Q(\vec{x})$ has $g!$ distinct preimages, namely the points $\pi(\vec{x})$ for $\pi \in S_g$. Since Q is a finite, flat morphism of degree $g!$, it is étale at $\vec{x}$. In particular, it is nonsingular at $\vec{x}$. Thus, Q is nonsingular in the complement of the generalized diagonal $X = \{\vec{x} = (x_1,\ldots,x_g) : x_i = x_j \text{ for some } i \ne j\}$.

The morphism $P : \mathrm{Sym}^{(g)}(\overline{\mathcal{C}}_v) \to \mathbf{Pic}^g_{\mathcal{C}_v/\mathbb{C}_v}$ is a birational isomorphism, and so it is nonsingular outside a proper Zariski-closed set $Y \subset \mathrm{Sym}^{(g)}(\overline{\mathcal{C}}_v)$. Since Q is dominant, $Q^{-1}(Y) \subset \overline{\mathcal{C}}_v^g$ is a proper, Zariski-closed subset of $\overline{\mathcal{C}}_v^g$. Thus $P \circ Q$ is nonsingular on the nonempty, Zariski-dense set $U = \overline{\mathcal{C}}_v^g \backslash (X \cup Q^{-1}(Y))$.

We now show that if F_u is a complete field with $K_v \subseteq F_u \subseteq \mathbb{C}_v$ for which $\mathcal{C}_v(F_u)$ is nonempty, then $U(F_u)$ is dense in $\mathcal{C}_v^g(F_u)$ for the v-topology. Suppose to the contrary that there was a point $\vec{a} \in \mathcal{C}_v^g(F_u)$ such that $B_{\mathcal{C}_v^g}(\vec{a},\vec{r}) \cap U(F_u)$ is empty, for some $r_1,\ldots,r_g > 0$. Then $\vec{a} \in X \cup Q^{-1}(Y)$. Let h be one of the equations cutting out $X \cup Q^{-1}(Y)$ in a neighborhood of $\vec{a}$. After shrinking the r_i, if necessary, we can assume that each ball $B(a_i,r_i)$ is isometrically parametrizable. Let $\Phi_{\vec{a}} : D(\vec{0},\vec{r}) \to B_{\mathcal{C}_v^g}(\vec{a},\vec{r})$ be a product of F_u-rational isometric parametrization maps as in (D.1). By composing h with $\Phi_{\vec{a}}$, we would obtain a power series $h \circ \Phi_{\vec{a}}(\vec{z}) \in \mathbb{C}_v[[\vec{X}]]$ converging on $D(\vec{0},\vec{r})$ and vanishing identically on $D(\vec{0},\vec{r}) \cap F_u^g$. By Lemma D.4, $h \circ \Phi_{\vec{a}} \equiv 0$, so h would vanish on all of $B_{\mathcal{C}_v^g}(\vec{a},\vec{r}) = \Phi_{\vec{a}}(D(\vec{0},\vec{r}))$. This is impossible by what has been shown above. □

Our final lemma concerns the trace $\mathrm{Tr}_{F_u/K_v} : \mathrm{Jac}(\mathcal{C}_v)(F_u) \to \mathrm{Jac}(\mathcal{C}_v)(K_v)$ when F_u/K_v is finite and separable. Let H_w be the Galois closure of K_v in $\mathbb{C}_v$. The subgroup $B_J(O,1)^- \cap \mathrm{Jac}(\mathcal{C}_v)(H_w)$ is stable under $\mathrm{Gal}(H_w/K_v)$, so we can pull back the trace Tr_{F_u/K_v} to the formal group $\mathcal{D}(\vec{0},1)^-$. Let $\vec{X}_1 = (X_{11},\ldots,X_{1g})$, $\ldots$, $\vec{X}_d = (X_{d1},\ldots,X_{dg})$ be d vectors of independent variables, and put

$$S^{(d)}(\vec{X}_1,\ldots,\vec{X}_d) \;=\; S(\vec{X}_1, S(\vec{X}_2,\ldots,S(\vec{X}_{d-1},\vec{X}_d))) \;\in\; \mathcal{O}_v[[\vec{X}_1,\ldots,\vec{X}_d]]\ .$$

Since $S(\vec{X},\vec{Y})$ is commutative and associative, for each permutation π on the letters $\{1,\ldots,d\}$ we have $S^{(d)}(\vec{X}_{\pi(1)},\ldots,\vec{X}_{\pi(d)}) = S^{(d)}(\vec{X}_1,\ldots,\vec{X}_d)$. Let $\sigma_1,\ldots,\sigma_d$ be the distinct embeddings of F_u into H_w, and extend each σ_i to an automorphism of H_w. Define $\mathrm{Tr}_{F_u/K_v,S} : D(\vec{0},1)^- \cap F_u^g \to D(\vec{0},1)^- \cap K_v^g$ by

$$\mathrm{Tr}_{F_u/K_v,S}(\vec{x}) \;=\; S^{(d)}(\sigma_1(\vec{x}),\ldots,\sigma_d(\vec{x}))\ .$$

Then $\mathrm{Tr}_{F_u/K_v}(\Psi(\vec{x})) = \Psi(\mathrm{Tr}_{F_u/K_v,S}(\vec{x}))$ for each $\vec{x} \in D(\vec{0},1)^- \cap F_u^g$.

LEMMA D.11. *Let F_u/K_v be a separable finite extension. Then there are constants R_u and C_u, depending on F_u, with $0 < R_u < 1$ and $0 < C_u \le 1$, such that for each r with $0 < r \le R_u$ we have*

$$\mathrm{Tr}_{F_u/K_v}(B_J(O,r) \cap \mathrm{Jac}(\mathcal{C}_v)(F_u)) \;\supseteq\; B_J(O,C_u r) \cap \mathrm{Jac}(\mathcal{C}_v)(K_v)\ . \tag{D.19}$$

PROOF. We keep the notation above. Since $\Psi : D(\vec{0},1)^- \to B_J(O,1)^-$ is a K_v-rational isometric parametrization, for each $0 < r < 1$ it takes $D(\vec{0},r) \cap F_u^g$ to $B_J(0,r) \cap \mathrm{Jac}(\mathcal{C}_v)(F_u)$ and $D(\vec{0},r) \cap K_v^g$ to $B_J(0,r) \cap \mathrm{Jac}(\mathcal{C}_v)(K_v)$. Thus it suffices to prove the assertion corresponding to (D.19) for $\mathrm{Tr}_{F_u/K_v,S}$.

The idea is that one can describe the image of $\mathrm{Tr}_{F_u/K_v,S}$ by restricting $\mathrm{Tr}_{F_u/K_v,S}$ to a carefully chosen K_v-rational subspace. Since F_u/K_v is separable, there is a $0 \ne \beta \in \mathcal{O}_v$ such that $\mathrm{Tr}_{F_u/K_v}(\mathcal{O}_u) = \beta\mathcal{O}_v$. Fix an $\alpha \in \mathcal{O}_u$ with $\mathrm{Tr}_{F_u/K_v}(\alpha) = \beta$; then $|\alpha|_v = 1$, $|\beta|_v \le 1$. Define

$$T(\vec{X}) \;=\; S^{(d)}(\sigma_1(\alpha)\vec{X},\cdots,\sigma_d(\alpha)\vec{X})\ .$$

Then for each $\vec{x} \in D(\vec{0},1)^- \cap K_v^g$, we have $\alpha\vec{x} \in D(\vec{0},1)^- \cap F_u^g$, and

$$\mathrm{Tr}_{F_u/K_v,S}(\alpha\vec{x}) \;=\; T(\vec{x})\ . \tag{D.20}$$

À priori $T(\vec{X}) \in \mathcal{O}_w[[\vec{X}]]$. However, for each $\sigma \in \mathrm{Gal}(H_w/K_v)$ there is a permutation $\pi = \pi_\sigma$ of $\{1,\ldots,d\}$ such that $\sigma\sigma_i = \sigma_{\pi(i)}$, so (writing σT for the power series obtained by letting σ act on the coefficients of T)

$$\begin{aligned}(\sigma T)(\vec{X}) \;&=\; S^{(d)}(\sigma(\sigma_1(\alpha))\vec{X},\cdots,\sigma(\sigma_d(\alpha))\vec{X}) \\ &=\; S^{(d)}(\sigma_{\pi(1)}(\alpha)\vec{X},\cdots,\sigma_{\pi(d)}(\alpha)\vec{X}) \;=\; T(\vec{X}).\end{aligned}$$

Thus $T(\vec{X}) \in \mathcal{O}_v[[\vec{X}]]$. Since $S(\vec{X},\vec{Y}) \equiv \vec{X} + \vec{Y}$ modulo terms of degree ≥ 2, it follows that $T(\vec{X}) \equiv \mathrm{Tr}_{F_u/K_v}(\alpha)\vec{X} = \beta\vec{X}$ modulo terms of degree ≥ 2.

Put $\widetilde{T}(\vec{X}) = \frac{1}{\beta^2}T(\beta\vec{X})$. Then $\widetilde{T}(\vec{X}) \in \mathcal{O}_v[[\vec{X}]]$ and $\widetilde{T}(\vec{X}) \equiv \vec{X}$ modulo terms of degree ≥ 2. By Proposition D.6, for each $0 < s < 1$ we have $\widetilde{T}(D(\vec{0},s) \cap K_v^g) = D(\vec{0},s) \cap K_v^g$. Suppose $0 < r < |\beta|_v$, and put $s = r/|\beta|_v$. Then $0 < s < 1$. Since $T(\beta\vec{X}) = \beta^2\widetilde{T}(\vec{X})$ and $\beta \in K_v^\times$, it follows that

$$\begin{aligned}T(D(\vec{0},r) \cap K_v^g) \;&=\; T(\beta(D(\vec{0},s) \cap K_v^g)) \;=\; \beta^2\widetilde{T}(D(\vec{0},s) \cap K_v^g) \\ &=\; \beta^2(D(\vec{0},s) \cap K_v^g) \;=\; D(\vec{0},|\beta|_v r) \cap K_v^g\ .\end{aligned} \tag{D.21}$$

Set $R_u = \frac{1}{2}|\beta|_v$ and $C_u = |\beta|_v$. By (D.20) and (D.21), if $0 < r \le R_u$, then

$$\mathrm{Tr}_{F_u/K_v,S}(D(\vec{0},r) \cap F_u^g) \ \supseteq \ T(D(\vec{0},r) \cap K_v^g) \ = \ D(\vec{0},C_u r) \cap K_v^g \ .$$

This yields (D.19). □

PROOF OF THEOREM D.2. Fix a point $\vec{a} = (a_1,\ldots,a_g) \in \mathcal{C}_v^g(\mathbb{C}_v)$ for which $[\ ] : \overline{\mathcal{C}}_v^g \to \mathbf{Pic}^g_{\overline{\mathcal{C}}_v/\mathbb{C}_v}$ is nonsingular at $\vec{a}$. By Lemma D.10, such $\vec{a}$ are dense in $\mathcal{C}_v^g(\mathbb{C}_v)$ for the v-topology, and if F_u is a complete field with $K_v \subseteq F_u \subseteq \mathbb{C}_v$ such that $\mathcal{C}_v(F_u)$ is nonempty, they are dense in $\mathcal{C}_v^g(F_u)$.

The associated map $\mathbf{J}_{\vec{a}} : \mathcal{C}_v(\mathbb{C}_v)^g \to \mathrm{Jac}(\mathcal{C}_v)$ given by $\mathbf{J}_{\vec{a}}(\vec{x}) = [\vec{x}]\dot{-}[\vec{a}]$ is nonsingular at $\vec{a}$, since the subtraction morphism $\dot{-}[\vec{a}]$ in $\mathbf{Pic}(\overline{\mathcal{C}}_v/\mathbb{C}_v)$ is an isomorphism. For a suitably small $r > 0$, let $\Phi_{\vec{a}} : D(\vec{0},r) \to B_{\mathcal{C}_v^g}(\vec{a},r)$ be the isometric parametrization from (D.1). Without loss, we can assume r is small enough that $\mathbf{J}_{\vec{a}}(\Phi_{\vec{a}}(D(\vec{0},r))) \subset B_J(O,1)^-$. Let $\Psi : D(\vec{0},1)^- \to B_J(O,1)^-$ be the isometric parametrization from (D.2), and put

$$H_{\vec{a}}(\vec{X}) \ = \ \Psi^{-1} \circ \mathbf{J}_{\vec{a}} \circ \Phi_{\vec{a}}(\vec{X}) \ .$$

Then $H_{\vec{a}}(\vec{X}) \in \mathbb{C}_v[[\vec{X}]]^g$, and $H_{\vec{a}}(\vec{X})$ converges on $D(\vec{0},r)$. If $\vec{a} \in \mathcal{C}_v(F_u)^g$, with F_u as above, then $H_{\vec{a}}(\vec{X}) \in F_u[[\vec{X}]]^g$. Put $L_{\vec{a}} = H'_{\vec{a}}(\vec{0})$, viewing it as a linear map $L_{\vec{a}} : \mathbb{C}_v^g \to \mathbb{C}_v^g$. By our discussion, $L_{\vec{a}}$ is invertible, and if $H_{\vec{a}}(\vec{X}) \in F_u[[\vec{X}]]$ with F_u as above, then $L_{\vec{a}} \in \mathrm{GL}_g(F_u)$.

Fix $0 < \eta < 1$, and let $R_1, R_2 > 0$ be the numbers given by Lemmas D.8 and D.9 applied to $H_{\vec{a}}(\vec{X})$, $L_{\vec{a}}$ and η. Put $R = \min(1/2, R_1, R_2)$. Suppose $0 < r_1, \ldots, r_g \le R$ and $\eta \max_i(r_i) \le \min(r_i)$. By Lemma D.9, $H_{\vec{a}}$ induces an analytic isomorphism from $D(\vec{0},\vec{r})$ onto $L_{\vec{a}}(D(\vec{0},\vec{r}))$. In particular, $H_{\vec{a}}$ is injective on $D(\vec{0},\vec{r})$, which means (since $\Phi_{\vec{a}}$ and Ψ are isomorphisms and $\mathbf{J}_{\vec{a}}(\vec{x}) = [\vec{x}]\dot{-}[\vec{a}]$) that the balls $B_{\mathcal{C}_v}(a_1,r_1), \ldots, B_{\mathcal{C}_v}(a_g,r_g)$ must be pairwise disjoint, and especially that the points $a_1, \ldots, a_g$ must be distinct.

By Lemma D.8, $L_{\vec{a}}(D(\vec{0},\vec{r}))$ (equipped with the addition law $S(\vec{X},\vec{Y})$ and negation $M(\vec{X})$) is a subgroup of the formal group of $\mathrm{Jac}(\mathcal{C}_v)$, denoted by $\mathcal{D}_{L_{\vec{a}}}(\vec{0},\vec{r})$. By Lemma D.9, for each $\vec{z} \in D(\vec{0},\vec{r})$,

$$H_{\vec{a}}(\vec{z}) \ \equiv \ L_{\vec{a}}(\vec{z}) \quad (\mathrm{mod}\ L_{\vec{a}}(D(\vec{0},\eta\vec{r}))) \ , \tag{D.22}$$

and the map $L_{\vec{a}} : D(\vec{0},\vec{r}) \to L_{\vec{a}}(D(\vec{0},\vec{r}))$ induces an isomorphism from the additive group $D(\vec{0},\vec{r})/D(\vec{0},\eta\vec{r})$ onto $\mathcal{D}_{L_{\vec{a}}}(\vec{0},\vec{r})/\mathcal{D}_{L_{\vec{a}}}(\vec{0},\eta\vec{r})$.

We now define an action $\ddot{+}_0$ of $\mathcal{D}_{L_{\vec{a}}}(\vec{0},\vec{r})$ on the polydisc $D(\vec{0},\vec{r})$ by setting, for $\vec{w} \in L_{\vec{a}}(D(\vec{0},\vec{r}))$ and $\vec{z} \in D(\vec{0},\vec{r})$,

$$\vec{w}\ddot{+}_0\vec{z} \ = \ H_{\vec{a}}^{-1}(S(\vec{w},H_{\vec{a}}(\vec{z}))) \ .$$

It is easy to check that for all $\vec{w}_1, \vec{w}_2 \in L_{\vec{a}}(D(\vec{0},\vec{r}))$ we have $\vec{w}_1\ddot{+}_0(\vec{w}_2\ddot{+}_0\vec{z}) = S(\vec{w}_1,\vec{w}_2)\ddot{+}_0\vec{z}$. Thus, $\mathcal{D}_{L_{\vec{a}}}(\vec{0},\vec{r})$ acts on $D(\vec{0},\vec{r})$. The action is transitive, since if $\vec{z}_1, \vec{z}_2 \in D(\vec{0},\vec{r})$ and $w = S(H_{\vec{a}}(\vec{z}_2), M(H_{\vec{a}}(\vec{z}_1)))$ then $w\ddot{+}_0\vec{z}_1 = \vec{z}_2$. It is simple, since if $w_1\ddot{+}\vec{z} = w_2\ddot{+}\vec{z}$ then $S(w_1,H_{\vec{a}}(\vec{z})) = S(w_2,H_{\vec{a}}(\vec{z}))$, and hence $w_1 = w_2$. Thus, $D(\vec{0},\vec{r})$ is a principal homogeneous space for $\mathcal{D}_{L_{\vec{a}}}(\vec{0},\vec{r})$ under $\ddot{+}_0$.

If in addition $\vec{a} \in \mathcal{C}_v(F_u)^g$ where F_u is as above, then by Lemma D.9, $H_{\vec{a}}$ maps $D(\vec{0},\vec{r}) \cap F_u^g$ isomorphically onto $L_{\vec{a}}(D(\vec{0},\vec{r})) \cap F_u^g$. Since $S(\vec{X},\vec{Y})$ and $M(\vec{X})$ are rational over K_v, $\mathcal{D}_{L_{\vec{a}}}(\vec{0},\vec{r}) \cap F_u^g$ acts simply and transitively on $D(\vec{0},\vec{r}) \cap F_u^g$.

We can now prove parts (A) and (B) of Theorem D.2. Given $\vec{a}$ and $\vec{r}$ as above, define $W_{\vec{a}}(\vec{r}) \subset \mathrm{Jac}(\mathcal{C}_v)(\mathbb{C}_v)$ by $W_{\vec{a}}(\vec{r}) = \mathbf{J}_{\vec{a}}(B_{\mathcal{C}_v^g}(\vec{a},\vec{r}))$. Since $\Psi \circ H_{\vec{a}} = \mathbf{J}_{\vec{a}} \circ \Phi_{\vec{a}}$, we also have

$$W_{\vec{a}}(\vec{r}) \;=\; \mathbf{J}_{\vec{a}}(\Phi_{\vec{a}}(D(\vec{0},\vec{r}))) \;=\; \Psi(H_{\vec{a}}(D(\vec{0},\vec{r}))) \;=\; \Psi(\mathcal{D}_{L_{\vec{a}}}(\vec{0},\vec{r}))\ .$$

Since $\mathcal{D}_{L_{\vec{a}}}(\vec{0},\vec{r}))$ is a subgroup of the formal group $(D(\vec{0},1)^-, S(\vec{X},\vec{Y}), M(\vec{X}))$, it follows that $W_{\vec{a}}(\vec{r})$ (equipped with the group operations $\dot{+}$, $\dot{-}$) is a subgroup of $\mathrm{Jac}(\mathcal{C}_v)(\mathbb{C}_v)$. Furthermore, by Lemma D.8 the map $\Psi \circ L_{\vec{a}}$ induces an isomorphism

$$D(\vec{0},\vec{r})/D(\vec{0},\eta\vec{r}) \;\cong\; W_{\vec{a}}(\vec{r})/W_{\vec{a}}(\eta\vec{r})\ . \tag{D.23}$$

For each $\vec{z} \in D(\vec{0},\vec{r})$, we have $L_{\vec{a}}(\vec{z}) \equiv H_{\vec{a}}(\vec{z}) \pmod{L_{\vec{a}}(D(\vec{0},\eta\vec{r}))}$, and $H_{\vec{a}} = \Psi^{-1} \circ \mathbf{J}_{\vec{a}} \circ \Phi_{\vec{a}}$. Hence (D.22) shows that

$$\mathbf{J}_{\vec{a}}(F_{\vec{a}}(\vec{z})) \;\equiv\; \Psi(L_{\vec{a}}(\vec{z})) \pmod{W_{\vec{a}}(\eta\vec{r})}\ . \tag{D.24}$$

To prove (C), we define the action $\ddot{+}$ of $W_{\vec{a}}(\vec{r})$ on $B_{\mathcal{C}_v^g}(\vec{a},\vec{r}) = \prod_{i=1}^g B_{\mathcal{C}_v}(a_i,r_i)$ by restricting the domain of $\mathbf{J}_{\vec{a}}$ to $B_{\mathcal{C}_v^g}(\vec{a},\vec{r})$ and setting, for $w \in W_{\vec{a}}(\vec{r})$ and $\vec{x} \in B_{\mathcal{C}_v^g}(\vec{a},\vec{r})$,

$$w \ddot{+} \vec{x} \;=\; \mathbf{J}_{\vec{a}}^{-1}(w \dot{+} \mathbf{J}_{\vec{a}}(\vec{x}))\ . \tag{D.25}$$

Applying $\mathbf{J}_{\vec{a}}$ to (D.25) shows that $[w \ddot{+} \vec{x}] \dot{-} [\vec{a}] = w \dot{+} ([\vec{x}] \dot{-} [\vec{a}])$, or equivalently that

$$[w \ddot{+} \vec{x}] = w \dot{+} [\vec{x}]\ . \tag{D.26}$$

Tracing through the definitions, one sees that if $w = \Psi(\vec{w})$ and $\vec{x} = \Phi_{\vec{a}}(\vec{z})$ with $\vec{w} \in L_{\vec{a}}(D(\vec{0},\vec{r}))$ and $\vec{z} \in D(\vec{0},\vec{r})$, then $w \ddot{+} \vec{x} = \Phi_{\vec{a}}(\vec{w} \ddot{+}_0 \vec{z})$. Thus, $\ddot{+}$ is the pushforward of the action $\ddot{+}_0$ to $\mathcal{C}_v^g$ and $\mathrm{Jac}(\mathcal{C}_v)$, using the maps $\Phi_{\vec{a}}$ and Ψ. Since $D(\vec{0},\vec{r})$ is a principal homogeneous space for $\mathcal{D}_{L_{\vec{a}}}(\vec{0},\vec{r}))$ under $\ddot{+}_0$, it follows that $\prod_{i=1}^g B_{\mathcal{C}_v}(a_i,r_i)$ is a principal homogeneous space for $W_{\vec{a}}(\vec{r})$ under $\ddot{+}$.

To prove (D), given $\vec{b} \in \prod_{i=1}^g B_{\mathcal{C}_v}(a_i,r_i)$, write $\vec{b} = \Phi_{\vec{a}}(\vec{\beta})$ with $\vec{\beta} \in D(\vec{0},\vec{r})$. Since the component functions of $\Phi_{\vec{a}}$ are isometric parametrization of the balls $B_{\mathcal{C}_v}(a_i,r_i)$, for each $\vec{z} \in D(\vec{0},\vec{r})$ we will have $\Phi_{\vec{a}}(\vec{z}) \in \prod_{i=1}^g B_{\mathcal{C}_v}(b_i,\eta r_i)$ if and only if $\vec{z} = \vec{\beta} + \vec{\delta}$ with $\vec{\delta} \in D(\vec{0},\eta\vec{r})$. Suppose $\Phi_{\vec{a}}(\vec{z}) \in \prod_{i=1}^g B_{\mathcal{C}_v}(b_i,\eta r_i)$. Since $W_{\vec{a}}(\eta\vec{r}) = \Psi(L_{\vec{a}}(D(\vec{0},\eta\vec{r})))$, by (D.23) and (D.24)

$$\mathbf{J}_{\vec{a}}(\Phi_{\vec{a}}(\vec{z})) \;\equiv\; \Psi(L_{\vec{a}}(\vec{b}) + L_{\vec{a}}(\vec{\delta})) \;\equiv\; \Psi(L_{\vec{a}}(\vec{b})) \;\equiv\; \mathbf{J}_{\vec{a}}(\vec{b}) \pmod{W_{\vec{a}}(\eta\vec{r})}\ .$$

Hence $\Phi_{\vec{a}}(\vec{z}) \in W_{\vec{a}}(\eta\vec{r}) \ddot{+} \vec{b}$. Conversely, if $\Phi_{\vec{a}}(\vec{z}) \in W_{\vec{a}}(\eta\vec{r}) \ddot{+} \vec{b}$, then

$$\Psi(L_{\vec{a}}(\vec{z})) \equiv \mathbf{J}_{\vec{a}}(\Phi_{\vec{a}}(\vec{z})) \;\equiv\; \mathbf{J}_{\vec{a}}(\vec{b}) \;\equiv\; \Psi(L_{\vec{a}}(\vec{\beta})) \pmod{W_{\vec{a}}(\eta\vec{r})}\ ,$$

which means that $\vec{z} - \vec{\beta} \in D(\vec{0},\eta\vec{r})$, and hence that $\Phi_{\vec{a}}(\vec{z}) \in \prod_{i=1}^g B_{\mathcal{C}_v}(b_i,\eta r_i)$. Thus $W_{\vec{a}}(\eta\vec{r}) \ddot{+} \vec{b} = \prod_{i=1}^g B_{\mathcal{C}_v}(b_i,\eta r_i)$. It follows that $J_{\vec{b}}(\prod_{i=1}^g B_{\mathcal{C}_v}(b_i,\eta r_i)) = W_{\vec{a}}(\eta\vec{r})$.

Next, we prove (E). Let F_u/K_v be a finite extension in $\mathbb{C}_v$. If $\vec{a} \in \mathcal{C}_v(F_u)^g$, then $\Phi_{\vec{a}}$, $L_{\vec{a}}$, $\mathbf{J}_{\vec{a}}$ and Ψ are F_u-rational. Since $\Phi_{\vec{a}}$ and Ψ are isometric parametrizations, this means that $\Phi_{\vec{a}}(D(\vec{0},\vec{r}) \cap F_u^g) = \prod_{i=1}^g \big(B_{\mathcal{C}_v}(a_i,r_i) \cap \mathcal{C}_v(F_u)\big)$ and $\Psi\big(L_{\vec{a}}(D(\vec{0},\vec{r}) \cap F_u^g)\big) = W_{\vec{a}}(\vec{r}) \cap \mathrm{Jac}(\mathcal{C}_v)(F_u)$. Since $\mathbf{J}_{\vec{a}}$ is F_u-rational, $\prod_{i=1}^g \big(B_{\mathcal{C}_v}(a_i,r_i) \cap \mathcal{C}_v(F_u)\big)$ is a principal homogeneous space for $W_{\vec{a}}(\vec{r}) \cap \mathrm{Jac}(\mathcal{C}_v)(F_u)$ under $\ddot{+}$. This implies (D.8) and (D.9).

To prove (F), assume in addition that F_u/K_v is separable. Let $(\alpha_{ij}) \in M_g(F_u)$ be the matrix of the linear map $L_{\vec{a}}^{-1}$, and let $A = \max_{ij}(|\alpha_{ij}|_v)$ be its operator norm. Write $r = \min_i(r_i)$, and put $s = (1/A)r$. Then $L_{\vec{a}}^{-1}(D(\vec{0},s)) \subseteq$

$D(\vec{0},\vec{r})$, so $D(\vec{0},s) \subseteq L_{\vec{a}}(D(\vec{0},\vec{r}))$, hence $B_J(O,s)$ is contained in $W_{\vec{a}}(\vec{r})$. Let $0 < R_u, C_u \leq 1$ be the constants from Lemma D.11. If $s \leq R_u$, then by Lemma D.11, $\mathrm{Tr}_{F_u/K_v}\big(W_{\vec{a}}(\vec{r}) \cap \mathrm{Jac}(\mathcal{C}_v)(F_u)\big)$ contains $B_J(O, C_u s) \cap \mathrm{Jac}(\mathcal{C}_v)(K_v)$. If $s > R_u$, it contains $B_J(0, C_u R_u) \cap \mathrm{Jac}(\mathcal{C}_v)(K_v)$. Put $C = C_u \min(1/A, R_u)$. Since $r \leq 1$, in either case

$$\mathrm{Tr}_{F_u/K_v}\big(W_{\vec{a}}(\vec{r}) \cap \mathrm{Jac}(\mathcal{C}_v)(F_u)\big) \;\supseteq\; B_J(O, Cr) \cap \mathrm{Jac}(\mathcal{C}_v)(K_v)\ .$$

This completes the proof. □

Finally, we establish the Lipschitz continuity of the Abel map:

PROOF OF PROPOSITION D.3. Let $W_{\vec{a}}(\vec{r}) = \mathbf{J}_{\vec{a}}\big(\prod_{i=1}^{g} B_{\mathcal{C}_v}(a_i, r_i)\big)$ (which coincides with $L_{\vec{a}}(D(\vec{0},\vec{r}))$) be an open subgroup of $\mathrm{Jac}(\mathcal{C}_v)(\mathbb{C}_v)$ with the properties in Theorem D.2, and let $E_v \subset \mathcal{C}_v(\mathbb{C}_v)$ be a nonempty compact subset.

Next consider the map $\mathbf{j}_x(y) = [(y) - (x)]$, which takes $\mathcal{C}_v(\mathbb{C}_v) \times \mathbb{C}_v(\mathbb{C}_v)$ to $\mathrm{Jac}(\mathcal{C}_v)(\mathbb{C}_v)$. We claim that for each $\tau \in E_v$, there are an isometrically parametrizable ball $B_{\mathcal{C}_v}(\tau, R_\tau)$ with $\mathbf{j}_\tau(B_{\mathcal{C}_v}(\tau, R_\tau)) \subseteq W_{\vec{a}}(\vec{r})$, such that if $0 < \varepsilon \leq R_\tau$ then for all $x, y \in B_{\mathcal{C}_v}(\tau, R_\tau)$ with $\|x,y\|_v \leq \varepsilon$, we have $\mathbf{j}_x(y) \in W_{\vec{a}}((\varepsilon/R_\tau)\cdot\vec{r})$.

To see this, let $\varphi_\tau : D(0,R_\tau) \to B_{\mathcal{C}_v}(\tau, R_\tau)$ be an isometric parametrization. Without loss we can assume that $R_\tau \in |\mathbb{C}_v^\times|_v$. Let $g_\tau(X)$ be the composite of the sequence of maps

$$D(0,R_\tau) \xrightarrow{\varphi_\tau} B_{\mathcal{C}_v}(\tau, R_\tau) \xrightarrow{\mathbf{j}_\tau} W_{\vec{a}}(\vec{r}) \xrightarrow{\Psi^{-1}} D(\vec{0},1)^- \;\subset\; \mathbb{C}_v^g\ .$$

Here φ_τ is an analytic map defined by convergent power series in $\mathbb{C}_v[[X]]$, $\mathbf{j}_\tau$ is an algebraic morphism, and Ψ^{-1} is the inverse of the isometric parametrization $\Psi : D(\vec{0},1)^- \to B_J(O,1)^-$; it is given by projection on g of the coordinates (see D.2). Thus $g_\tau(X) = (g_{\tau,1}(X), \ldots, g_{\tau,g}(X))$ is a map whose coordinate functions are defined by convergent power series in $\mathbb{C}_v[[X]]$.

Since $\mathbf{j}_x(y) = \mathbf{j}_\tau(y)\dot{-}\mathbf{j}_\tau(x)$ on $B(\tau, R_\tau)$, and since the group operations $\dot{+}$, $\dot{-}$ on $B_J(O,1)^-$ correspond to $S(\vec{X},\vec{Y})$, $M(\vec{X})$ in the formal group $D(\vec{0},1)^-$, when $\mathbf{j}_x(y)$ is pulled back to $D(0,R_\tau) \times D(0,R_\tau)$ using φ_τ, it is represented by the power series map $S(g_\tau(X), M(g_\tau(Y))$. By hypothesis, $\mathbf{j}_\tau(B(\tau,R_\tau)$ is contained in $W_{\vec{a}}(\vec{r}) = L_{\vec{a}}(D(0,\vec{r}))$, where $L_{\vec{a}} = (\Psi^{-1} \circ \mathbf{J}_{\vec{a}} \circ \Phi_{\vec{a}})'(0) : \mathbb{C}_v^g \to \mathbb{C}_v^g$ is a nonsingular linear map. Since $W_{\vec{a}}(\vec{r})$ is a group the image of $S(g_\tau(X), M(g_\tau(Y))$ is contained in $L_{\vec{a}}(D(0,\vec{r}))$.

Now define $h_\tau(X,Y) : D(0,R_\tau) \times D(0,R_\tau) \to \mathbb{C}_v^g$ by

$$h_\tau(X,Y) \;=\; L_{\vec{a}}^{-1}(S(g_\tau(X), M(g_\tau(Y))) \;=\; (h_{\tau,1}(X,Y), \ldots, h_{\tau,g}(X,Y))\ .$$

The image of $h_\tau(X,Y)$ is contained in $D(0,\vec{r}) = \prod_{j=1}^{g} D(0,r_j)$, so each $h_{\tau,j}(X,Y) \in \mathbb{C}_[[X,Y]]$ is a map defined by power series converging on $D(0,R_\tau) \times D(0,R_\tau)$, with image contained in $D(0,r_j)$. Since $\mathbf{j}_x(x) = 0$ for each x, it follows that $h_{\tau,j}(x,x) = 0$ for all $x \in D(0,R_\tau)$. It follows from Lemma D.3 that $|h_{\tau,j}(x,y)|_v \leq |x-y|_C \cdot (r_j/R_\tau)$ for all $x, y \in D(0,R_\tau)$. In particular, if $0 < \varepsilon \leq R_\tau$ and $x, y \in D(0,R_\tau)$ satisfy $|x - y|_v \leq \varepsilon$, then $h_\tau(x,y) \in D(0,(\varepsilon/R_\tau)\vec{r})$. Since φ_τ is an isometric parametrization, and since $L_{\vec{a}}(D(0,(\varepsilon/R_\tau)\vec{r}) = W_{\vec{a}}((\varepsilon/R_\tau)\vec{r})$, for all $x, y \in B(\tau, R_\tau)$ with $\|x,y\|_v \leq \varepsilon$ we have $\mathbf{j}_x(y) \in W_{\vec{a}}((\varepsilon/R_\tau)\cdot\vec{r})$.

To conclude the proof, consider the nonempty compact set $E_v \subset \mathcal{C}_v(\mathbb{C}_v)$. Cover E_v with finitely many balls $B(\tau_1, R_{\tau_1}), \ldots, B(\tau_n, R_{\tau_n})$ satisfying the conditions above. Put $\varepsilon_0 = \min_{1 \leq j \leq n}(R_{\tau_j})$, and put $C_0 = 1/\varepsilon_0$. Then for each $0 < \varepsilon \leq \varepsilon_0$, and all $x, y \in E_v$ with $\|x,y\|_v \leq \varepsilon$, we have $\mathbf{j}_x(y) \in W_{\vec{a}}(C_0\varepsilon\cdot\vec{r})$. □

Bibliography

[1] L. Ahlfors, Complex Analysis, McGraw-Hill, 1979.

[2] N.I. Akhiezer, Elements of the Theory of Elliptic Functions, AMS Translations of Mathematical Monographs 79, AMS, Providence, 1990.

[3] Y. Amice, Les nombres p-adiques, Presses Universitaires de France, Paris, 1975.

[4] E. Artin, Algebraic Numbers and Function Fields, Gordon and Breach, New York, 1967.

[5] M. Artin, Néron Models, chapter VIII in: G. Cornell, J. Silverman, eds., Arithmetic Geometry, Springer-Verlag, New York, 1986.

[6] P. Autissier, Pointes entiers et théorèmes de Bertini arithmétiques, Thèse de l'Université Paris XI (Décembre 2001).

[7] M. Baker and R. Rumely, Potential Theory and Dynamics on the Berkovich Projective Line, Mathematical Surveys and Monographs 159, AMS, Providence, 2010.

[8] M. Baker, S. Ih, and R. Rumely, A finiteness property of torsion points, Algebra and Number Theory 2 (2008), 217-248.

[9] M. L. Balinski and H. P. Young, The quota method of apportionment, American Mathematical Monthly 82 (1975), 701-730.

[10] V. Berkovich, Spectral theory and Analytic Geometry over non-Archimedean Fields, Mathematical Surveys and Monographs 33, AMS, Providence, 1990.

[11] S. Bosch, U. Güntzer, R. Remmert, Non-Archimedean Analysis, Grundlehren der Mathematischen Wissenschaften 261, Springer-Verlag, New York, 1984.

[12] S. Bosch, W. Lütkebohmert, M. Raynaud, Néron Models, Ergebnisse der Mathematik 21, Springer-Verlag, New York, 1990.

[13] R. Courant and D. Hilbert, Methods of Mathematical Physics (Volume 1), Interscience, New York, 1953.

[14] D. Cantor, On approximation by polynomials with algebraic integer coefficients, in: Proc. Symposia in Pure Math 12 (W. Leveque and E. Strauss, eds.), AMS Providence, R.I., 1969.

[15] D. Cantor, On certain algebraic integers and approximation by rational functions with integral coefficients, Pacific J. Math 67 (1976), 323-338.

[16] D. Cantor, On an extension of the definition of transfinite diameter and some applications, J. Reine Angew. Math 316 (1980), 160-207.

[17] D. Cantor and P. Roquette, On diophantine equations over the ring of all algebraic integers, J. Number Theory 18 (1984), 1-26.

[18] J. W. Cassels and A. Frölich, Algebraic Number Theory, Academic Press, London, 1967.

[19] T. Chinburg and R. Rumely, The capacity pairing, J. Reine Angew. Math 434 (1993), 1-44.

[20] P. Clark, Rational points on Atkin-Lehner Quotients of Shimura Curves, Doctoral Thesis, Harvard University, 2003.

[21] J. Cremona, Algorithms for Modular Elliptic Curves, Cambridge University Press, Cambridge, 1992.

[22] Th. Falliero, Séries d'Eisenstein Hyperboliques et Capacité d'Unions d'Intervalles, Ph.D. Thesis, University of Bordeaux I, 1998.

[23] Th. Falliero and A. Sebbar, Capacité de trois intervalles et Fonctions thêta de genre 2, J. Math. Pures. Appl 80 (2001), 409–443.

[24] Th. Falliero and A. Sebbar, Capacities and Jacobi Matrices, Proceedings of the Edinburgh Mathematical Society 46 (2003), 719-745.

[25] M. Fekete and G. Szegö, On algebraic equations with integral coefficients whose roots belong to a given point set, Math. Zeit. 63 (1955), 158-172.

[26] K.-H. Fieseler, Zariski's Main Theorem für affinoide Kurven, Math. Ann. 251 (1980), 97-110.

[27] M. D. Fried and M. Jarden, Field Arithmetic, Ergebnisse der Mathematik, 3 Folge, Band 11, Springer-Verlag, Berlin-Heidelberg-New York, 1986.

[28] F. R. Gantmacher, The Theory of Matrices, vol. II, Chelsea, 1974.

[29] A. Grothendieck and J. Dieudonné, Eléments de Geometrie Algébrique, IHES Publications Mathématiques 17, 29, 24, 28, 32, Presses Universitaires de France, Paris, 1961-1967.

[30] B. Green, M. Matignon and F. Pop, On the local Skolem property, J. Reine Angew. Math. 458 (1995), 183-199.

[31] S. Kleiman, The Picard scheme, pp. 235–321 in: Fundamental Algebraic Geometry, Mathematical Surveys and Monographs 123, AMS, Providence, 2005.

[32] M. Klimek, Pluripotential theory, London Mathematical Society Monographs (New Series) 6, Oxford Science Publications, 1991.

[33] H. Kober, Dictionary of Conformal Representations, Dover, New York, 1952.

[34] S. Lang, Algebraic Number Theory, Addison-Wesley, Reading, 1968.

[35] S. Lang, Introduction to algebraic geometry, Addison-Wesley, Reading; 4th printing, 1973.

[36] S. Lang, Elliptic Functions, Addison-Wesley, Reading, 1973.

[37] H. Matsumura, Commutative Algebra (2nd edition), Lecture Notes in Mathematics Series, Benjamin/Cummings Publishing Company, Reading, 1980.

[38] L. Moret-Bailly, Groupes de Picard et problèmes de Skolem I, Ann. Scient. École. Norm. Sup (4 ser.) 22 (1989), 161-179.

[39] L. Moret-Bailly, Groupes de Picard et problèmes de Skolem II, Ann. Scient. École. Norm. Sup (4 ser.) 22 (1989), 181-194.

[40] L. Moret-Bailly, Problémes de Skolem sur les champs algebraiques, Compositio Math. 125 (2001), 1-30.

[41] B. Mazur, Modular Curves and the Eisenstein Ideal, IHES Publ. Math 47 (1977), 33-186: Appendix by B. Mazur and M. Rapoport, "Behavior of the Néron model of the Jacobian of $X_0(N)$ at bad primes".

[42] W. McCallum, The degenerate fibre of the Fermat Curve, in: N. Koblitz, ed., Number Theory Related to Fermat's Last Theorem, Progress in Math. 26, Birkhaüser, Boston (1982), 57–70.

[43] J.S. Milne, Jacobian Varieties, chapter VII in: G. Cornell, J. Silverman, eds., Arithmetic Geometry, Springer-Verlag, New York, 1986.

[44] Z. Nehari, Conformal Mapping, Dover, New York, 1952.

[45] D. Park, The Fekete-Szegö theorem with splitting conditions on the projective line of positive characteristic p, Doctoral thesis, University of Georgia, 2006.

[46] G. Pólya, Über ganzwertige Polynome in algebraischen Zahlkörpern, J. Reine Angew. Math. 149 (1919), 97-116.

[47] A. Prestel and J. Schmid, Decidability of the rings of real algebraic and p-adic algebraic integers, J. Reine Angew. Math. 414 (1991), 141-148.

[48] R. Robinson, Conjugate algebraic numbers in real point sets, Math Zeit. 84 (1964), 415-427.

[49] R. Robinson, Intervals containing infinitely many conjugate sets of algebraic units, Ann. Math. (2) 80 (1968), 532-543.

[50] P. Roquette, B. Green and F. Pop, On Rumely's local-global principle, Jahresber. Deutch. Math.-Verein 97 (1995), 43-74.

[51] R. Rumely, Capacity Theory on Algebraic Curves, Lecture Notes in Mathematics 1378, Springer-Verlag, Berlin-Heidelberg-New York, 1989.

[52] R. Rumely, The Fekete-Szegö theorem with splitting conditions, part I, Acta Arithmetica 93 (2000), 99-116.

[53] R. Rumely, The Fekete-Szegö theorem with splitting conditions, part II, Acta Arithmetica 103 (2002), 347–410.

[54] R. Rumely, On the relation between Cantor's capacity and the sectional capacity, Duke Mathematical Journal 70 (1993), 517-574.

[55] R. Rumely, Arithmetic over the ring of all algebraic integers, J. Reine Ange. Math. 368 (1986), 127-133.

[56] R. Rumely, An intersection theory for curves with analytic contributions from nonarchimedean places, in: Canadian Mathematical Society Conference Proceedings 15(1995), AMS, Providence, 325-357.

[57] E. Saff and V. Totik, Logarithmic Potentials with External Fields, Grundlehren 316, Springer-Verlag, Berlin-Heidelberg-New York, 1991.

[58] M. Schiffer and D. Spencer, Functionals of Finite Riemann Surfaces, Princeton University Press, Princeton, 1954.

[59] J.-P. Serre, Lie Algebras and Lie Groups, 1964 Lectures given at Harvard University, W. A. Benjamin, New York, 1965.

[60] G. Shimura, Theta functions with complex multiplication, Duke Mathematical Journal 43 (1976), 673–696.

[61] J. Silverman, Advanced Topics in the Arithmetic of Elliptic Curves, Graduate Texts in Mathematics 151, Springer-Verlag, New York, 1994.

[62] N. Steinmetz, Rational Iteration, Walter de Gruyter, New York, 1993.

[63] A. Tamagawa, Unramified Skolem Problems and Unramified Bertini Theorems in Positive Characteristic, Documenta Mathematica, Extra Volume Kato (2003), 789–831.

[64] A. Thuillier, Théorie du potentiel sur les courbes en géometrie analytique non-archimédienne. Applications à la théorie d'Arakelov. Ph.D. Thesis, University of Rennes, 2005. Preprint available at http://tel.ccsd.cnrs.fr/documents/archives0/00/01/09/90/index.html.

[65] M. Tsuji, Potential Theory in Modern Function Theory, Maruzen, Tokyo, 1959 (reprinted: Chelsea, New York, 1975).

[66] L. van den Dries, Elimination theory for the ring of all algebraic integers, J. Reine Angew. Math. 388 (1988), 189-205.

[67] E. T. Whittaker and G. N. Watson, A Course of Modern Analysis (4th) edition, Cambridge University Press, 1927; reprinted 1973.

[68] H. Widom, Extremal polynomials associated with a system of curves in the complex plane, Advances in Math. 3 (1969), 127–232.

Index